Machine
Tool
Practices

Machine Tool Practices

Richard R. Kibbe

Machine Technology
Lane Community College

John E. Neely

Machine Technology
Lane Community College

Roland O. Meyer

Warren T. White

New York
Chichester
Brisbane
Toronto

John Wiley & Sons

Library of Congress Cataloging in Publication Data:

Main entry under title:
Machine tool practices.

   Previous ed. (© 1977) published under title:
Machine tools and machining practices.
   Includes index.
   1. Machine-tools. 2. Machine-shop practice.

I. Kibbe, Richard R.
TJ1185.M224   1979   621.9′02   78-18533
ISBN 0-471-04331-1

Printed in the United States of America.

10 9 8 7 6 5 4 3

preface

This text is intended for students who are seriously thinking of becoming machinists, either through apprenticeship training, vocational schools, or community college programs. The content deals with topics usually presented in a combined lecture and laboratory program of 1200 to 1600 hours extending over an appropriate number of school terms. In writing this text, we have attempted to overcome some of the limitations of conventional books in the field of machine tools and machining practices.

The structure of the book gives the instructor maximum flexibility in shop and classroom application. Although the order of topic presentation represents traditional sequencing derived nationally from numerous course outlines, the internal structure of the book allows the material to be taken up in an order best suited to the individual teacher. The traditional structure of chapters has been eliminated and replaced by lettered sections representing major divisions of the topic. Each section is prefaced by an introductory component that provides an overview of the section topic. To stimulate interest, some historical information is included, followed by a general description of the types and applications of that specific group of tools or machines.

The detailed treatments of the topics are contained in the instructional units following each section introduction. Each unit follows a standardized format. It begins with a statement of purpose that explains why the information is important to a machinist. This statement is followed by a list of specific objectives to be accomplished by the student. It should be noted that in a number of units, the stated objectives are met only if the student accomplishes specific tasks detailed in the unit worksheets. However, it is not our intent to make those units totally dependent on the specific worksheets that we have created. We realize that in many cases, individual instructors have prepared their own worksheets or projects specific to their particular teaching programs. Therefore, our unit worksheets do not appear in the text. They are found in the accompanying Instructor's Manual and may be freely reproduced by anyone who desires to use them.

It has been our experience that learning is reinforced if the student has the opportunity to test himself. Therefore, each instructional unit is followed by a self-test; answers appear in Appendix II. When the student is confident that he knows the material in the unit, he may then take the unit post-test.

The work of machining is highly visual. Thus, for the benefit of the students using this text, we have created a visually intensive format. This arrangement offers some of the advantages of other external visual media with the added advantage that the reader may study an illustration as long as necessary to extract the maximum amount of information. Whenever possible, illustrations of the tools and machines in use are seen as opposed to static views. The use of line drawings has been minimized and overlaid photographs substituted in order to add a further degree of realism. We have attempted to keep the series as up to date as possible. Archaic machining practices have been deemphasized.

The accompanying Instructor's Manual contains the unit post-tests, answers to post-tests, and suggested sources for audio-visual aids. The Manual also contains suggestions regarding the organization and management for those instructors who might be interested in using this text in an individualized instructional setting.

Richard R. Kibbe
John E. Neely
Roland O. Meyer
Warren T. White

acknowledgments

The authors wish to thank the following for their contributions to this textbook.

Alan Hancock College, Santa Maria, California
American Society for Metals, Metals Park, Ohio
American Society of Mechanical Engineers, New York, New York
California State University at Fresno, Department of Industrial Arts and Technology, Fresno, California
Community College of Denver, Denver, Colorado
DeAnza College, Engineering and Technology, Cupertino, California
Epsilon Pi Tau Fraternity, Alpha Lambda Chapter, California State University at Fresno, Fresno, California
Ferris State College, Big Rapids, Michigan
Foothill College District, Office of Technical Education, Individualized Machinist's Curriculum Project, Los Altos Hills, California
Paul C. Hays Technical School, Grove City, Ohio
Henry Kappel
Lane Community College, Mechanics Department, Eugene, Oregon
Larrimer County Voc-Tech, Fort Collins, Colorado
Linn-Benton Community College, Business and Industrial Division, Albany, Oregon
Richard L. McKee
Muskingun Area Joint Vocational School, Zanesville, Ohio
National Machine Tool Builders Association, McLean, Virginia
National Screw Machine Products Association, Cleveland, Ohio
North County Technical School, Florissant, Missouri
North County Technical School, St. Louis County, Missouri
Pasadena City College, Pasadena, California
Spartanburg Technical College, Spartanburg, South Carolina
Triton College, River Grove, Illinois
Utah State Technical College, Salt Lake City, Utah
Yuba College, Applied Arts Department, Marysville, California

Our special thanks go to the following firms and their employees with whom we corresponded and visited and who supplied us with invaluable technical information and illustrations.

Accurate Diamond Tool Corporation, Hackensack, New Jersey
Aloris Tool Company, Inc., Clifton, New Jersey
American Chain & Cable Company, Inc., Wilson Instrument Division, Bridgeport, Connecticut
American Iron & Steel Institute, Washington, D.C.
American SIP Corporation, Elmsford, New York
Ameropean Industries, Inc., Hamden, Connecticut
Ames Research Center, National Aeronautics and Space Administration (NASA), Mountain View, California
Barber-Colman Company, Rockford, Illinois
Barnes Drill Company, Rockford, Illinois
Barret Centrifugals, Worcester, Massachusetts

Bay State Abrasives, Division of Dresser Industries, Westborough, Massachusetts
Bendix Corporation, South Bend, Indiana
Bethlehem Steel Corporation, Bethlehem, Pennsylvania
Boyer-Schultz Corporation, Broadview, Illinois
Bridgeport Milling Machine Division, Textron, Inc., Bridgeport, Connecticut
Brown & Sharpe Manufacturing Company, North Kingstown, Rhode Island
Bryant Grinder Corporation, Springfield, Vermont
Buck Tool, Kalamazoo, Michigan
The Carborundum Company, Niagara Falls, New York
Cincinnati Incorporated, Cincinnati Ohio
Cincinnati Milacron, Inc., Cincinnati, Ohio
Clausing Corporation, Kalamazoo, Michigan
Cleveland Twist Drill Company, Cleveland, Ohio
Cone-Blanchard Machine Company, Windsor, Vermont
Dake Corporation, Grand Haven, Michigan
Desmond-Stephan Manufacturing Company, Urbana, Ohio
Diamond Abrasive Corporation, New York, New York
DoAll Company, Des Plaines, Illinois
Dover Publications Inc., New York, New York
The du Mont Corporation, Greenfield, Massachusetts
El-Jay Inc., Eugene, Oregon
Elm Systems, Inc., Arlington Heights, Illinois
Enco Manufacturing Company, Chicago, Illinois
Engis Corporation, Morton Grove, Illinois
Ex-Cell-O Corporation, Troy, Michigan
Exolon Company, Tonawanda, New York
Federal Products Corporation, Providence, Rhode Island
Fellows Corporation, Springfield, Vermont
Floturn, Inc., Division of Lodge & Shipley Company, Cincinnati, Ohio
Gaertner Scientific Corporation, Chicago, Illinois
General Electric Company, Detroit, Michigan, and Specialty Materials Department, Worthington, Ohio
Geometric Tool, New Haven, Connecticut
Giddings & Lewis, Fond Du Lac, Wisconsin
Gleason Works, Rochester, New York
Great Lakes Screw, Chicago, Illinois
Greenfield Tap & Die, Division of TRW, Inc., Greenfield, Massachusetts
Hammond Machinery Builders, Kalamazoo, Michigan
Hardinge Brothers, Inc., Elmira, New York
Harig Products, Inc., Elgin, Illinois
Harry M. Smith & Associates, Santa Clara, California
Heald Machine Division, Cincinnati Milacron Company, Worcester, Massachusetts
Hewlett-Packard Company, Palo Alto and Santa Clara, California
Hitachi Magna-Lock Corporation, Big Rapids, Michigan
Illinois/Eclipse, Division of Illinois Tool Works, Inc., Chicago, Illinois
Industrial Plastics Products, Inc., Forest Grove, Oregon
Industrial Press, New York, New York
Ingersoll Milling Machine Company, Rockford, Illinois
Jarvis Products Corporation, Middletown, Connecticut
K & M Tool, Inc., Eugene, Oregon

Kasto-Racine, Inc., Monroeville, Pennsylvania
Kennametal, Latrobe, Pennsylvania
Landis Tool Company, Division of Litton Industries, Waynesboro,
    Pennsylvania
Lapmaster Division, Crane Packing Company, Morton Grove, Illinois
LeBlond, Inc., Cincinnati, Ohio
Lodge & Shipley Company, Cincinnati, Ohio
Louis Levin & Son, Inc., Culver City, California
M & M Tool Manufacturing Company, Dayton, Ohio
Madison Industries, Division of Amtel, Inc., Providence, Rhode Island
Mahr Gage Company, New York, New York
Mattison Machine Works, Rockford, Illinois
Megadiamond Industries, New York, New York
Minnesota Mining and Manufacturing Company (3M), St. Paul, Minnesota
The MIT Press, Cambridge, Massachusetts
Monarch Machine Tool Company, Sidney, Ohio
Moog, Inc., Hydra Point Division, Buffalo, New York
Moore Special Tool Company, Bridgeport, Connecticut
MTI Corporation, New York, New York
National Broach & Machine Division, Lear Siegler, Inc., Detroit, Michigan
National Twist Drill & Tool Division, Lear Siegler, Inc., Rochester, Michigan
Newcomer Products, Inc., Latrobe, Pennsylvania
Norton Company, Worcester, Massachusetts
PMC Industries, Wickliffe, Ohio
Pratt & Whitney Machine Tool Division of The Colt Industries Operating
    Corporation, West Hartford, Connecticut
Precision Diamond Tool Company, Elgin, Illinois
Production Components, Inc., Ranchita, California
Ralmike's Tool-A-Rama, South Plainfield, New Jersey
Rank Precision Industries, Inc., Des Plaines, Illinois
Rockford Machine Tool Company, Rockford, Illinois
Sipco Machine Company, Marion, Massachusetts
Snap-On Tools Corporation, Kenosha, Wisconsin
Southwestern Industries, Inc., Los Angeles, California
Speedfam Corporation, Des Plaines, Illinois
Standard Gage Company, Poughkeepsie, New York
L. S. Starret Company, Athol, Massachusetts
Sunnen Products Company, St. Louis, Missouri
Superior Electric Company, Bristol, Connecticut
Surface Finishes, Inc., Addison, Illinois
Taft-Pierce Manufacturing Company, Woonsocket, Rhode Island
Taper Micrometer Corporation, Worcester, Massachusetts
Thompson Vacuum Company, Sarasota, Florida
Tinius Olsen Testing Machine Company, Inc., Willow Grove, Pennsylvania
Ultramatic Equipment Company, Addison, Illinois
Unison Corporation, Madison Heights, Michigan
Waldes Kohinoor, Inc., Long Island City, New York
Walton Company, Hartford, Connecticut
Warner & Swasey Company, Cleveland, Ohio, and King of Prussia,
    Pennsylvania
Weldon Tool Company, Cleveland, Ohio
Whitman & Barnes Division, TRW, Inc., Plymouth, Michigan
Whitnon Spindle Division, Mite Corporation, Farmington, Connecticut

Acknowledgments

J. H. Williams Division, TRW, Inc., Buffalo, New York
Wilton Tool Division, Wilton Corporation, Des Plaines, Illinois

R.R.K.
J.E.N.
R.O.M.
W.T.W.

contents

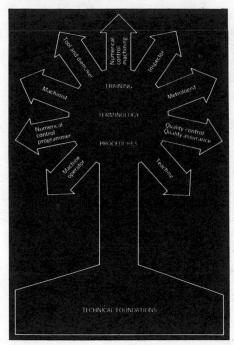

section a
introduction

In our highly industrialized and mechanized society, we are indeed surrounded by many mechanical marvels. These devices are so common that we tend to take them for granted. Technical advances have come so fast in recent years that in less than a century humankind has learned to fly, explored the deepest oceans, and begun the exploration of space.

Such technical achievements would not have been possible had human beings not learned to extract metals from the planet on which they live and then shape these metals into useful products. Furthermore, metalworking is essential to the extraction or creation of many other nonmetallic materials that are equally important to the building of technical hardware. Although metalworking has been done since ancient times, only in quite recent history has it developed into a scientifically based technology.

The most important aspect of metalworking is that of shaping metals into useful products. Many methods have been and are used today to shape metals. However, few of these methods produce the wide variety of items as do the processes of **machining.** In fact, machining is probably the most important method by which metals and other materials are transformed into the many products enjoyed by so many. Machining technology is at the very base of our way of life in an industrialized society.

Machining is basically a process of shaping materials using a variety of cutting tools. The material is shaved away in small pieces or chips uncovering the final size and shape of the workpiece. This may be done by direct contact of a cutting tool material that is harder than the workpiece material. Machining may be accomplished by other means as well. Large pieces may be removed by various sawing processes. Material may also be removed by using an electric spark, electrochemical, or ultrasonic sound processes. However, most machining is accomplished by direct contact of a cutting tool. These cutting tools are frequently motor driven. Thus, machining is done with a motor driven tool or **machine tool.** There are many types of machine tools, and they are generally grouped together in a **machine shop.**

Machine shops vary greatly in size and types of work. A machine shop may have many types of machine tools and produce many different products. Almost every mechanical marvel that is enjoyed today began as a model built in a machine shop. In this book, you will learn about many machine tools, and you will see the great potential of these tools for making the hardware of technology.

TECHNICAL FOUNDATIONS No technical study, machining included, stands entirely by itself. For example, some degree of mathematics is needed in every application of technology whether it be building a house, putting a person on the moon, or operating a machine tool. Mathematics, then, can be thought of as a technical foundation of almost universal application. Equally important are technical foundations such as communication skills that include speaking, writing, and drawing.

Through your educational experiences up to now, you have acquired many of the basic technical foundations such as speaking, writing, and fundamental mathematics. As you begin your study of machine tools and machining practices, you will study other foundations that are specific to this technical study. Many of these, including machine shop related mathematics, drafting, blueprint reading, metallurgy, and metrology, will be presented in specific related courses of study.

The first section of this text deals with two technical foundations that are significantly important and closely related to a study of machining practices:

1. General shop safety.

2. Introduction to mechanical hardware.

Shop Safety   Shop safety is probably the most important technical foundation. Safety is not only important in the machine shop, it is equally important in anything that you do. You must develop a safety attitude, and you must constantly remind yourself of the possible consequences of an accident. A useful career as a machinist can be ended or reduced in an instant if you should have an accident. Furthermore, you can be left with a permanent physical handicap.

Mechanical Hardware   Most tools and machines are made from many parts and subassemblies. These are connected together by a large variety of mechanical hardware including bolts, nuts, and screws. Some of your work as a machinist will be concerned with making many of these hardware items. In addition, you should be familiar with mechanical hardware in order to properly maintain a machine tool.

The unit on mechanical hardware will familiarize you with many of the common fasteners and other devices used is all tools and machinery. It will be most helpful to you as you study the subject of machine tool practices if you are familiar with the terminology of mechanical hardware.

unit 1 shop safety

Safety is not often thought about as you proceed through your daily tasks. Often you expose yourself to needless risk because you have experienced no harmful effects in the past. Unsafe habits become almost automatic. You may drive your automobile without wearing a seat belt. You know this to be unsafe, but you have done it before and so far no harm has resulted. None of us really likes to think about the possible consequences of an unsafe act. However, safety can and does have an important effect on anyone who makes his living in a potentially dangerous environment such as a machine shop. An accident can reduce or end your career as a machinist. You may spend several years learning the trade and more years gaining experience. Experience is a particularly valuable asset. It can only be gained through time spent on the job. This becomes economically valuable to you and to your employer. Years spent in training and gaining experience can be wasted in an instant if you should have an accident, not to mention a possible permanent physical handicap for you and hardship on your family. Safety is an attitude that should extend far beyond the machine shop and into every facet of your life. You must constantly think about safety in everything you do.

objectives After completing this unit, you will be able to:

1. Identify and use safety equipment designed for a machinist.
2. Identify hazards in the machine shop.
3. Describe safe procedures around the machine shop.

PERSONAL SAFETY

Eye Protection
Eye protection is a primary safety consideration around the machine shop. Machine tools produce metal chips, and there is always a possibility that these may be ejected from a machine at high velocity. Sometimes they can fly many feet. Furthermore, most cutting tools are made from hard materials. They can occasionally break or shatter from the stress applied to them during a cut. The result can be more flying metal particles.

Eye protection must be worn **at all times** in the machine shop. There are several types of eye protection available. Plain safety glasses are all that is required in most shops. These have shatterproof lenses that may be changed if they become scratched. The lenses have a high resistance to impact. Common types include fixed bow safety glasses (Figure 1) and the flexible bow safety glasses (Figure 2). The flexible bow may be adjusted to the most comfortable position for the wearer.

Side shield safety glasses must be worn around any grinding operation. The side shield

3

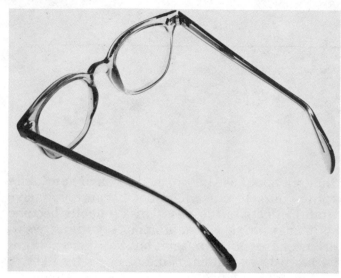

Figure 1. Common fixed bow safety glasses (Epsilon Pi Tau Fraternity, California State University at Fresno).

Figure 3. Solid side shield safety glasses (Epsilon Pi Tau Fraternity, California State University at Fresno).

protects the side of the eye from flying particles. Side shield safety glasses may be of the solid (Figure 3) or perforated type (Figure 4). The perforated side shield fits closer to the eye. Bows may wrap around the ear. This prevents the safety glasses from falling off.

If you wear prescription glasses, you may want to cover them with a safety goggle (Figure 5). The full face shield may also be used (Figure 6). Prescription glasses can be made as safety glasses. In industry, prescription safety

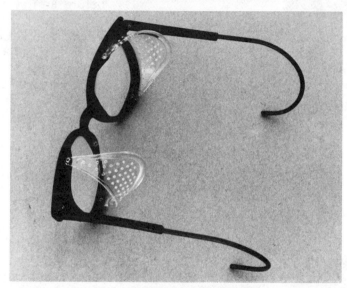

Figure 4. Perforated side shield safety glasses (Epsilon Pi Tau Fraternity, California State University at Fresno).

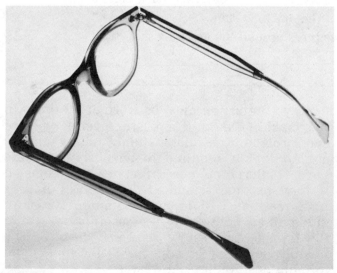

Figure 2. Flexible bow safety glasses (Epsilon Pi Tau Fraternity, California State University at Fresno).

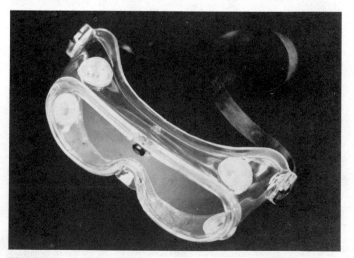

Figure 5. Safety goggles.

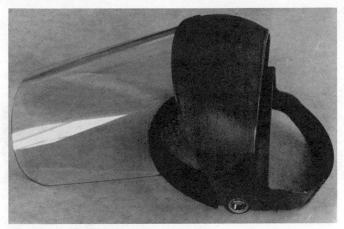

Figure 6. Safety face shield.

glasses are sometimes provided free to employees.

Foot Protection
Generally, the machine shop does not present too great a hazard to the feet. However, there is always a possibility that you could drop something on your foot. A safety shoe is available. This will have a steel toe shield designed to resist impacts. Some safety shoes also have an instep guard. Shoes must be worn at all times in the machine shop. A solid leather shoe is recommended. Tennis shoes and sandals should not be worn. You must never even enter a machine shop with bare feet. Remember that the floor is often covered with razor sharp metal chips.

Ear Protection
The instructional machine shop usually does not present a noise problem. However, an industrial machine shop may be adjacent to a fabrication or punch press facility. New safety regulations are quite strict regarding exposure to noise. Several types of sound suppressors and noise reducing ear plugs may be worn. Excess noise can cause a permanent hearing loss. Usually this occurs over a period of time, depending on the intensity of the exposure. Noise is considered an industrial hazard if it is continuously above 85 *decibels*, the units used in measuring sound waves. If it is over 115 decibels for short periods of time, ear protection must be worn (Figure 7). Ear muffs or plugs should be used wherever high intensity noise is likely. A considerate workman will not create excessive noise when it is not necessary. Table 1

Figure 7. Sound suppressors are designed to protect the ears from damage caused by loud noises.

shows the decibel level of various sounds; sudden sharp or high intensity noises are the most disturbing to your eardrums.

Grinding Dust and Hazardous Fumes
Grinding dust is produced by abrasive wheels and consists of extremely fine metal particles and abrasive wheel particles. These should not be inhaled. In the machine shop, most grinding machines have a vacuum dust collector (Figure 8). Grinding may be done with coolants that aid in dust control. A machinist may be involved in portable grinding operations. This is common in such industries as shipbuilding. You should wear an approved respirator if you are exposed to grinding dust. Change the respirator filter at regular intervals. Grinding dust can present a great danger to health. Examples include the dust of such metals as beryllium, or the presence of radioactivity in nuclear systems. In these situations, the spread of grinding dust must be carefully controlled.

Some metals such as zinc give off toxic fumes when heated above their boiling point. Some of these fumes when inhaled cause tempo-

Table 1

The Decibel Level of Various Sounds

130—Painful sounds: jet engine on ground
120—Airplane on ground: Reciprocating engine
110—Boiler factory
    —Pneumatic riveter
100—
    —Maximum street noise
    —Roaring lion
90—
    —Loud shout
80—Disesl truck
    —Piano practice
    —Average city street
70—
    —Dog barking
    —Average conversation
60—
    —Average city office
50—
    —Average city residence
40—One typewriter
    —Average country residence
30—Turning page of newspaper
    —Purring cat
20—
    —Rustle of leaves in breeze
    —Human heartbeat
10—
  0—Faintest audible sound

Figure 8. Vacuum dust collector on grinders (California State University at Fresno).

rary sickness, but other fumes can be severe or even fatal. The fumes of mercury and lead are especially dangerous, as their effect is cumulative in your body and can cause irreversible damage. Cadmium and beryllium compounds are also very poisonous. Therefore, when welding, burning, or heat treating metals, adequate ventilation is an absolute necessity. This is also true when parts are being carburized with compounds containing potassium cyanide. These *cyanogen compounds* are deadly poisonous and every precaution should be taken when using them. Kasenite, a trade name for a carburizing compound that is not toxic, is often found in school shops and in machine shops. Uranium salts are toxic and all radioactive materials are extremely dangerous.

Clothing, Hair, and Jewelry
Wear a short sleeve shirt or roll up long sleeves above the elbow. Keep your shirt tucked in and remove your necktie. It is recommended that

you wear a shop apron. If you do, keep it tied behind you. If apron strings become entangled in the machine, you may be reeled in as well. A shop coat may be worn as long as you roll up long sleeves. Do not wear fuzzy sweaters around machine tools.

If you have long hair, keep it secured properly. In industry, you may be required to wear a hair net so that your hair cannot become entangled in a moving machine. The result of this can be disastrous (Figure 9).

Remove your wristwatch and rings before operating any machine tool. These can cause serious injury if they should be caught in a moving machine part.

Figure 9. Long hair may be caught and reeled into the machine (Courtesy of John Allan, Jr.).

Hand Protection

There is really no device that will totally protect your hands from injury. Next to your eyes, you hands are the most important tools that you have. It is up to you to keep them out of danger. Use a brush to remove chips from a machine (Figure 10). Do not use your hands. Chips are not only razor sharp, they are often extremely hot. Resist the temptation to grab chips as they come from a cut. Long chips are extremely dangerous. These can often be eliminated by properly sharpening your cutting tools. Chips should *not* be removed with a rag. The metal particles become inbedded in the cloth and they may cut you. Furthermore, the rag may be caught in a moving machine. Gloves must not be worn around most machine tools, although they are acceptable when working with a band saw blade. If a glove should be caught in a moving part, it will be pulled in, along with the hand inside it.

Various cutting oils, coolants, and solvents may affect your skin. The result may be a rash or possible infection. Avoid direct contact with these products as much as possible and wash your hands as soon as possible after contact.

Lifting

Improper lifting can result in a permanent back injury that can limit or even end your career. Back injury can be avoided if you lift properly at all times. If you must lift a large or heavy object, get some help or make use of a hoist or forklift. Don't try to be a "superman" and lift something that you know is too heavy. It is not worth the risk.

Figure 10. Use a brush to clear chips (Courtesy of California Community Colleges IMC Project).

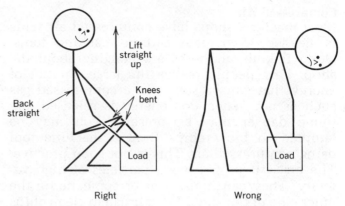

Figure 11. Proper and improper lifting.

Objects within your lifting capability can be lifted safely by the following procedure (Figure 11):

1. Keep your back straight.
2. Squat down, bending your knees.
3. Lift smoothly using the muscles in your legs to do the work. Keep your back straight. Bending over the load puts an excessive stress on your spine.
4. Position the load so that it is comfortable to carry. Watch where you are walking when carrying a load.
5. If you are placing the load back to floor level, lower it in the same manner you picked it up.

Scuffling and Horseplay

The machine shop is no place for scuffling and horseplay. This activity can result in a serious injury to you, a fellow student, or worker. Practical joking is also very hazardous. What might appear to be a comical situation to you could result in a disastrous accident to someone else. In industry, horseplay and practical joking are often grounds for dismissal of an employee.

Injuries

If you should be injured, report it immediately to your instructor.

IDENTIFYING SHOP HAZARDS

A machine shop is not so much a dangerous place as a potentially dangerous place. One of the best ways to be safe is to be able to identify shop hazards before they involve you in an accident. By being aware of potential danger, you can better make safety part of your work in the machine shop.

Compressed Air

Most machine shops have compressed air. This is needed to operate certain machine tools. Often flexible air hoses are hanging about the shop. Few people realize the large amount of energy that can be stored in a compressed gas such as air. When this energy is released, extreme danger may be present. You may be tempted to blow chips from a machine tool using compressed air. This is not good practice. The air will propel metal particles at high velocity. They can injure you or someone on the other side of the shop. Use a brush to clean chips from the machine. Do not blow compressed air on your clothing or skin. The air can be dirty and the force can implant dirt and germs into your skin. Air can be a hazard to ears as well. An eardrum can be ruptured.

Should an air hose break, or the nozzle on the end come unscrewed, the hose will whip wildly. This can result in an injury if you happen to be standing nearby. When an air hose is not in use, it is good practice to shut off the supply valve. The air trapped in the hose should be vented. When removing an air hose from its supply valve, be sure that the supply is turned off and the hose has been vented. Removing a charged air hose will result in a sudden venting of air. This can surprise you and an accident might result.

Housekeeping

Keep floor and aisles clear of stock and tools. This will insure that all exits are clear if the building should have to be evacuated. Material on the floor, especially round bars, can cause falls. Clean oils or coolants that may spill on the floor. Several preparations designed to absorb oil are available. These may be used from time to time in the shop. Keep oily rags in an approved safety can (Figure 12). This will prevent possible fire from spontaneous combustion.

Electrical

Electricity is another potential danger in a machine shop. Your exposure to electrical hazard will be minimal unless you become involved with machine maintenance. A machinist is mainly concerned with the on and off switch on a machine tool. However, if you are adjusting the machine or accomplishing maintenance, you should unplug it from the electrical service. If it is permanently wired, the circuit breaker may

Figure 12. Store oil soaked rags in an approved safety can (DeAnza College).

be switched off and tagged with an appropriate warning. In industry, this procedure often means that the operator must sign a clearance stating that electrical service has been secured. Service cannot be restored until the operator signs a restoration order. Normally you will not disconnect the electrical service for routine adjustments such as changing speeds. However, when a speed change involves a belt change, you must insure that no other person is likely to turn on the machine while your hands are in contact with belts and pulleys.

Carrying Objects

Carry long stock in the vertical position. Be careful of light fixtures and ceilings. A better way is to have someone carry each end of a long piece of material. Do not carry sharp tools in your pockets. They can injure you or someone else.

MACHINE HAZARDS

There are many machine hazards. Each section of this book will discuss the specific hazards applicable to that type of machine tool. Remember that a machine has no intelligence of its own. It cannot distinguish between cutting metal and cutting fingers. Do not think that you are strong enough to stop a machine should you become tangled in moving parts. You are not. When operating a machine, think about what you are going to do before you do it. Go over a safety checklist.

1. Do I know how to operate this machine?
2. What are the potential hazards involved?
3. Are all guards in place?
4. Are my procedures safe?
5. Am I doing something that I probably should not do?
6. Have I made all the proper adjustments and tightened all locking bolts and clamps?
7. Is the workpiece secured properly?
8. Do I have proper safety equipment?
9. Do I know where the stop switch is?
10. Do I think about safety in everything that I do?

INDUSTRIAL SAFETY AND FEDERAL LAW

In 1970, Congress passed the **Williams-Steiger Occupational Safety and Health Act.** This act took effect on April 28, 1971. The purpose and policy of the act are "to assure so far as possible every working man and woman in the Nation safe and healthful working conditions and to preserve our human resources."

The **Occupational Safety and Health Act** is commonly known as **OSHA.** Prior to its passage, industrial safety was the individual responsibility of each state. The establishment of OSHA added a degree of standardization to industrial safety throughout the nation. OSHA encourages states to assume full responsibility in administration and enforcement of federal occupational safety and health regulations.

Duties of Employers and Employees
Each employer under OSHA has the general duty to furnish employment and places of employment free from recognized hazards causing or likely to cause death or serious physical harm. The employer has the specific duty of complying with safety and health standards as defined under OSHA. Each employee has the duty to comply with safety and health standards and all rules and regulations established by OSHA.

Occupational Safety and Health Standards
Job safety and health standards consist of rules for avoiding hazards that have been proven by research and experience to be harmful to personal safety and health. These rules may apply to all employees as in the case of fire protection standards. Many standards apply only to workers engaged in specific types of work. A typical standard states that aisles and passageways shall be kept clear and in good repair, with no obstruction across or in aisles that could create a hazard.

Complaints of Violations
Any employee who believes that a violation of job safety or health standard exists may request an inspection by sending a signed written notice to OSHA. This includes anything that threatens physical harm or represents an imminent danger. A copy must also be provided to the employer. However, the name of the person complaining need not be revealed to the employer.

Enforcement of OSHA Standards
OSHA inspectors may enter a plant or school at any reasonable time and conduct an inspection. They are not permitted to give prior notice for this. They may question any employer, owner, operator, agent, or employee in regard to any safety violation. The employer and a representative of the employees have the right to accompany the inspector during the inspection.

If a violation is discovered, a written citation is issued to the employer. A reasonable time is permitted to correct the condition. The citation must be posted at or near the place of the violation. If, after a reasonable time, the condition has not been corrected, a fine may be imposed on the employer. If the employer has made an attempt to correct the unsafe condition but has exceeded the time limit, a hearing may be held to determine progress.

Penalties
Willful or repeated violations may incur penalties up to $10,000. Citations issued for serious violations incur mandatory penalties of $1,000. A serious violation where extreme danger exists may be penalized up to $1,000 for each day the violation exists.

OSHA Education and Training Programs
The Occupational Safety and Health Act provides for programs to be conducted by the Secretary

of Labor in consultation with the Department of Health, Education and Welfare. These programs provide for education and training of employers and employees in recognizing, avoiding, and preventing unsafe and unhealthful working conditions. The act also provides for training an adequate supply of qualified personnel to carry out OSHA's purpose.

self-evaluation

SELF-TEST
1. What is the primary piece of safety equipment in the machine shop?
2. What can you do if you wear prescription glasses?
3. Describe proper dress for the machine shop.
4. What can be done to control grinding dust?
5. What hazards exist from coolants, oils, and solvents?
6. Describe proper lifting procedure.
7. Describe at least two compressed air hazards.
8. Describe good housekeeping procedures.
9. How should long pieces of material be carried?
10. List at least five points from the safety checklist for a machine tool.

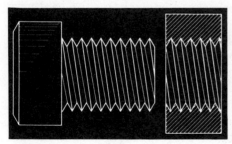

unit 2
introduction to mechanical hardware

Many precision machined products produced in the machine shop would be useless until assembled into a machine, tool, or other mechanism. This assembly requires many types of fasteners and other mechanical hardware. In this unit, you will be introduced to many of these important hardware items.

objective After completing this unit, you will be able to:
1. Identify most common mechanical hardware used to assemble tools, machines, and other mechanisms.

THREADS

The **thread** is an extremely important mechanical device. It derives its usefulness from the inclined plane, one of the six simple machines. Almost every mechanical device is assembled with threaded fasteners. A thread is a spiral or helical groove that is formed on the outside or inside diameter of a cylinder (Figure 1). These spiral grooves take several forms. Furthermore, they have specific and even spacing. One of the fundamental tasks of a machinist is to produce both external and internal threads using several machine tools and hand tools. The majority of threads appear on **threaded fasteners.** These include many types of **bolts, screws,** and **nuts.** However, threads are used for a number of other applications aside from fasteners. These include threads for adjustment purposes, measuring tool applications, and the transmission of power. A close relative to the thread, the spiral auger, is used to transport material.

THREAD FORMS

There are a number of thread forms. In later units, you will examine these in detail, and you will have the opportunity to make several of them on a machine tool. As far as the study of machined hardware is concerned, you will be most concerned with the **unified thread form** (Figure 2). The unified thread form was an outgrowth of the American National Standard form. In order to help standardize manufacturing in the United States, Canada, and Great Britain, the unified form was developed. Unified threads are a combination of the American National and the British Standard Whitworth forms. Unified threads are divided into the following series:

> **UNC—Unified National Coarse**
> **UNF—Unified National Fine**
> **UNS—Unified National Special**

IDENTIFYING THREADED FASTENERS

Unified coarse and unified fine refers to the number of threads per inch of length on standard threaded fasteners. A specific diameter of bolt or nut will have a specific number of threads per inch of length. For example, a $\frac{1}{2}$ in. diameter Unified National Coarse bolt will have

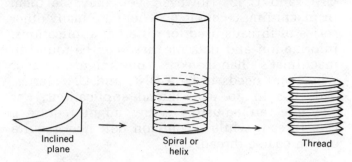

Figure 1. Thread helix.

13 threads per inch of length. This bolt will be identified by the following marking:

$$\frac{1}{2} \text{ in.} - 13 \text{ UNC}$$

One half is the **major diameter** and **13** is the **number of threads per inch of length.** A $\frac{1}{2}$ in. diameter Unified National Fine bolt will be identified by the following marking:

$$\frac{1}{2} \text{ in.} - 20 \text{ UNF}$$

One-half is the major diameter and 20 is the number of threads per inch.

The Unified National Special Threads are identified in the same manner. A $\frac{1}{2}$ in. diameter UNS bolt may have 12, 14, or 18 threads per inch. These are less common than the standard

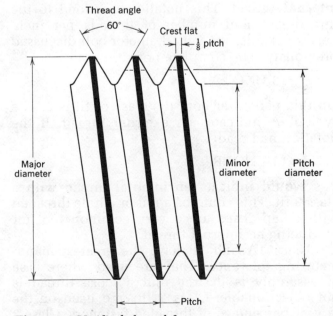

Figure 2. Unified thread form.

UNC and UNF. However, you may see them in machining technology. There are many other series of threads used for different applications. Information and data on these can be found in machinist's handbooks. You might wonder why there needs to be a UNC and UNF series. This has to do with thread applications. For example, an adjusting screw might require a fine thread, while a common bolt may require only a coarse thread.

CLASSES OF THREAD FITS

The preceding information was necessary for an understanding of **thread fit classes.** Some thread applications can tolerate loose threads, while other applications require tight threads. For example, the head of your car's engine is held down by a threaded fastener called a stud bolt or simply a stud. A stud is threaded on both ends. One end is threaded into the engine block. The other end receives a nut that bears against the cylinder head. When the head is removed, it is desirable to have the stud remain screwed into the engine block. This end requires a tighter thread fit than the end of the stud accepting the nut. If the fit on the nut end is too tight, the stud may unscrew as the nut is removed.

Unified **thread fits** are classified as **1A, 2A, 3 A,** or, **1 B, 2 B, 3 B.** The **A symbol** indicates an **external** thread. The **B symbol** indicates an **internal** thread. This notation is added to the thread size and number of threads per inch. Let us consider the $\frac{1}{2}$ in. diameter bolt discussed previously. The complete notation reads:

$$\frac{1}{2} - 13 \text{ UNC 2A}$$

On this particular bolt, the class of fit is 2. The symbol A indicates an external thread. If the notation had read:

$$\frac{1}{2} - 13 \text{ UNC 3B}$$

this would indicate an internal thread with a class 3 fit. This could be a nut or a hole threaded with a tap. Taps are a very common tool for producing an internal thread.

Class 1A and 1B have the greatest manu-facturing tolerance. They are used where ease of assembly is desired and a loose thread is not objectionable. Class 2 fits are used on the largest percentage of threaded fasteners. Class 3

fits will be tight when assembled. Each class of fit has a specific tolerance on major diameter and pitch diameter. This data may be found in machinist's handbooks and is required for the manufacture of threaded fasteners.

STANDARD SERIES OF THREADED FASTENERS

Threaded fasteners, including all common bolts and nuts, range from quite small machine screws through quite large bolts. Below a diameter of $\frac{1}{4}$ in. threaded fasteners are given a number. Common UNC and UNF series threaded fasteners are listed in the following table.

UNC and UNF Threaded Fasteners

UNC		UNF	
Size	Threads/Inch	Size	Threads/Inch
		0	80
1	64	1	72
2	56	2	64
3	48	3	56
4	40	4	48
5	40	5	44
6	32	6	40
8	32	8	36
10	24	10	32
12	24	12	28

From here, the major diameter is expressed in fractional form.

$\frac{1}{4}$ in.	20	$\frac{1}{4}$ in.	28
$\frac{5}{16}$ in.	18	$\frac{5}{16}$ in.	24
$\frac{3}{8}$ in.	16	$\frac{3}{8}$ in.	24
$\frac{7}{16}$ in.	14	$\frac{7}{16}$ in.	20
$\frac{1}{2}$ in.	13	$\frac{1}{2}$ in.	20
$\frac{9}{16}$ in.	12	$\frac{9}{16}$ in.	18
$\frac{5}{8}$ in.	11	$\frac{5}{8}$ in.	18
$\frac{3}{4}$ in.	10	$\frac{3}{4}$ in.	16
$\frac{7}{8}$ in.	9	$\frac{7}{8}$ in.	14
1 inch	8	1 inch	12

Both series continue up to about 4 inches.

All of the sizes listed in the table are very common fasteners in all types of machines, automobiles, and other mechanisms. Your contact with these common sizes will be so frequent that you will soon begin to recall them from memory.

COMMON EXTERNALLY THREADED FASTENERS

Common mechanical hardware includes threaded fasteners such as bolts, screws, nuts, and thread inserts. All of these are used in a variety of ways to hold parts and assemblies together. Complex assemblies such as an airplane, ship, or automobile may have many thousands of fasteners taking many forms.

Bolts and Screws

A general definition of a **bolt** is an externally threaded fastener that is inserted through holes in an assembly. A bolt is tightened with a **nut** (Figure 3 right). A **screw** is an externally threaded fastener that is inserted into a threaded hole and tightened or released by turning the head (Figure 3 left). From these definitions, it is apparent that a bolt can become a screw or the reverse can be true. This depends on the application of the hardware. Bolts and screws are the most common of the threaded fasteners. These fasteners are used to assemble parts quickly and they make disassembly possible.

The strength of an assembly of parts depends to a large extent on the diameter of the screws or bolts used. In the case of screws, strength depends on the amount of **thread engagement.** Thread engagement is the distance

Figure 4. Square head bolt and hex head bolt. Square nut and hex nut.

that a screw extends into a threaded hole. The minimum thread engagement should be a distance equal to the diameter of the screw used; preferably it would be one and one-half times the screw diameter. Should an assembly fail, it is better that the screw break than to have the internal thread stripped from the hole. It is generally easier to remove a broken screw than to drill and tap for a larger screw size. With a screw engagement of $1\frac{1}{2}$ times its diameter, the screw will usually break rather than strip the thread in the hole.

Machine bolts (Figure 4) are made with **hexagonal** or **square** heads. These bolts are often used in the assembly of parts that do not require a precision bolt. The body diameter of machine bolts is usually slightly larger than the nominal or standard size of the bolt. Body diameter is the diameter of the unthreaded portion of a bolt below the head. A hole that is to accept a common bolt must be slightly larger than the body diameter. When machine bolts are purchased, nuts are frequently included. Common bolts are made with a class 2A unified thread and come in both UNC and UNF series. Sizes in hexagonal head machine bolts range from a $\frac{1}{4}$ in. diameter to a 4 in. diameter. Square head machine bolts are standard to $1\frac{1}{2}$ in. diameter.

Stud bolts (Figure 5) have threads on both ends. Stud bolts are used where one end is semipermanently screwed into a threaded hole. A good example of the use of stud bolts is an automobile engine. The stud bolts are tightly

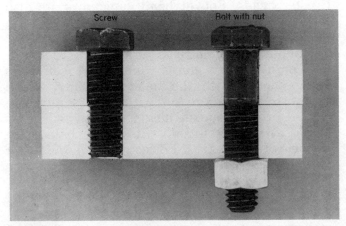

Figure 3. Screw and bolt with nut.

Figure 5. Stud bolt.

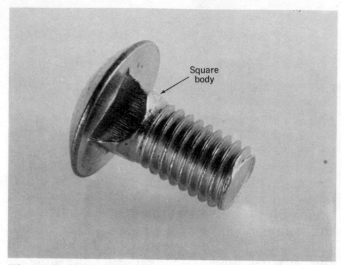

Figure 6. Carriage bolt.

held in the cylinder block and easily changed nuts hold the cylinder heads in place. The end of the stud bolt screwed into the tapped hole has a class 3A thread, while the nut end is a class 2A thread.

Carriage bolts (Figure 6) are used to fasten wood and metal parts together. Carriage bolts have round heads with a square body under the head. The square part of the carriage bolt, when pulled into the wood, keeps the bolt from turning while the nut is being tightened. Carriage bolts are manufactured with class 2A coarse threads.

Machine screws are made with either coarse or fine thread and are used for general assembly work. The heads of most machine

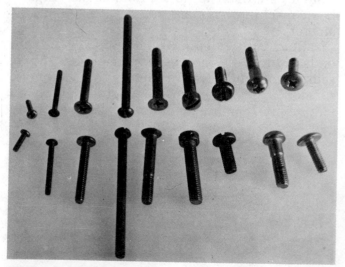

Figure 7. Machine screws.

screws are slotted to be driven by screw drivers. Machine screws are available in many sizes and lengths (Figure 7). Several head styles are also available (Figure 8). Machine screw sizes fall into two categories. Fraction sizes range from diameters of $\frac{1}{4}$ to $\frac{3}{4}$ in. Below $\frac{1}{4}$ in. diameter, screws are identified by numbers from 0 to 12. A number 0 machine screw has a diameter of .060 in. (sixty thousandths of an inch). For each number above zero add .013 in. to the diameter.

EXAMPLE
Find the diameter of a number 6 machine screw:

$$\#0 \text{ diameter} = .060 \text{ in.}$$
$$\#6 \text{ diameter} = .060 \text{ in.} + (6 \times .013 \text{ in.})$$
$$= .060 \text{ in.} + .078 \text{ in.}$$
$$= .138 \text{ in.}$$

Machine screws 2 in. long or shorter have threads extending all the way to the head. Longer machine screws have a $1\frac{3}{4}$ in. thread length.

Cap screws (Figure 9) are made with a variety of different head shapes and are used where precision bolts or screws are needed. Capscrews are manufactured with close tolerances and have a finished appearance. Capscrews are made with coarse, fine, or special threads. Capscrews with a 1 in. diameter have a class 3A thread. Those greater than a 1 in. diameter have a class 2A thread. The strength of screws depends mainly on the kind of material used to make the screw. Different screw materials are aluminum, brass, bronze, low carbon steel, medium carbon steel, alloy steel, stainless, steel, and titanium. Steel hex head cap screws come in diameters from $\frac{1}{4}$ to 3 in., and their strength is indicated by symbols on the hex head (Figure 10). Slotted head capscrews can have flat heads, round heads, or fillister heads. Socket head capscrews are also made with socket flat heads, and socket button heads.

Set screws (Figure 11) are used to lock pulleys or collars or shafts. Set screws can have square heads with the head extending above the surface or, more often, the set screws are slotted or have socket heads. **Slotted or socket head set screws** usually disappear below the surface of the part to be fastened. A pulley or collar where the set screws are below the surface is much safer for persons working around them. Socket head set screws may have hex socket heads or spline socket heads. Set screws are manufactured in number sizes from 0 to 10

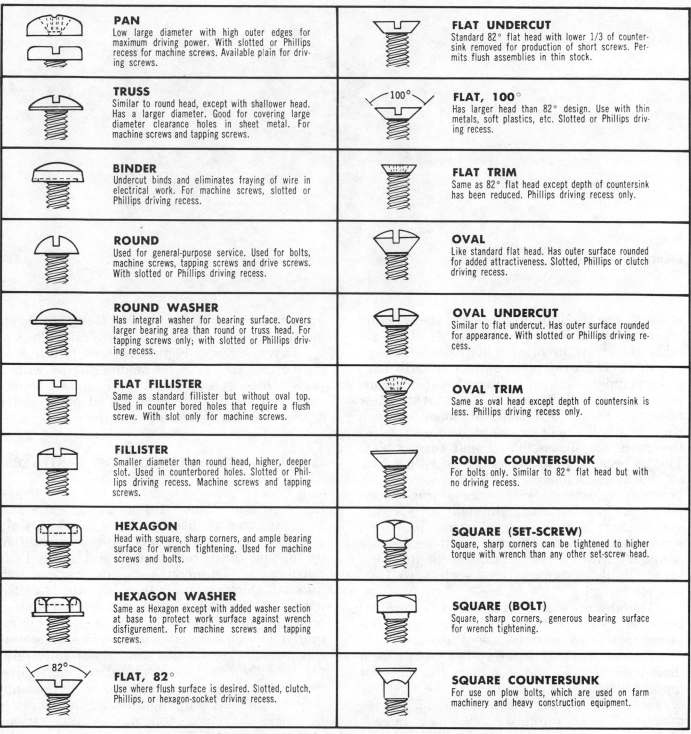

Figure 8. Machine screw head styles (Courtesy of Great Lakes Screw).

and in fractional sizes from $\frac{1}{4}$ to 2 in. Set screws are usually made from carbon or alloy steel and hardened.

Square head set screws are often used on tool holders (Figure 12) or as jackscrews in level-ing machine tools (Figure 13). Set screws have several different points (Figure 14). The flat point set screw will make the least amount of indentation on a shaft and is used where frequent adjustments are made. A flat point set

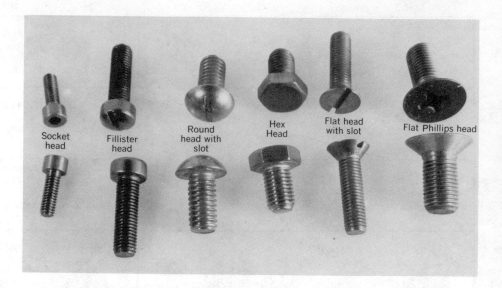

Figure 9. Cap screws.

screw is also used to provide a jam screw action when a second set screw is tightened on another set screw to prevent its release through vibration. The oval point set screw will make a slight indentation as compared with the cone point. With a half dog or full dog point set screw holding a collar to a shaft, alignment between shaft and collar will be maintained even when the parts are disassembled and reassembled. This is because the shaft is drilled with a hole of the same diameter as the dog point. Cup pointed set screws will make a ring-shaped depression in the shaft and will give a very slip-resistant connection. Square head set screws have a class 2A thread and are usually supplied with a coarse thread. Slotted and socket head set screws have a class 3A UNC or UNF thread.

Thumbscrews and wingscrews (Figure 15) are used where parts are to be fastened or adjusted rapidly without the use of tools.

Thread forming screws (Figure 16) form their own threads and eliminate the need for tapping. These screws are used in the assembly of sheet metal parts, plastics, and nonferrous material. Thread forming screws form threads by displacing material with no cutting action. These screws require an existing hole of the correct size.

Thread cutting screws (Figure 17) make threads by cutting and producing chips. Because of the cutting action these screws need less driving torque than thread forming screws, Applications are similar as those for thread

forming screws. These include fastening sheet-metal, aluminum brass, diecastings, and plastics.

Drive screws (Figure 18) are forced into the correct size hole by hammering or with a press. Drive screws make permanent connections and are often used to fasten name plates or identification plates on machine tools.

COMMON INTERNALLY THREADED FASTENERS

Nuts
Common nuts (Figure 19) are manufactured in as many sizes as there are bolts. Most nuts are either hex or square in shape. Nuts are identified by the size of the bolt they fit and not by their outside size. Common hex nuts are made in different thicknesses. A **thin hex nut** is called a **jam nut.** They are used where space is limited or where the strength of a regular nut is not required. Jam nuts are often used to lock other nuts (Figure 20). Regular hex nuts are slightly thinner than their size designation. A $\frac{1}{2}$ in. regular hex nut is $\frac{7}{16}$ in. thick. A $\frac{1}{2}$ in. heavy hex nut is $\frac{31}{64}$ in. thick. A $\frac{1}{2}$ in. high hex nut measures $\frac{11}{16}$ in. thick. Other common nuts include various stop or lock nuts. Two common types are the **elastic stop nut** and the **compression stop nut.** They are used in applications where the nut might vibrate off the bolt. Wing nut and thumb nuts are used where quick assembly or disassembly by hand is desired. Other hex nuts are slotted and castle nuts. These nuts have slots cut into them. When the slots are aligned with holes

Bolt head marking	SAE — Society of Automotive Engineers ASTM — American Society for Testing and Materials SAE — ASTM Definitions	Material	Minimum tensile strength in pounds per square inch (PSI)
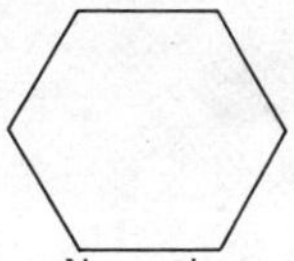 No marks	SAE grade 1 SAE grade 2 Indeterminate quality	Low carbon steel Low carbon steel	65,000 PSI
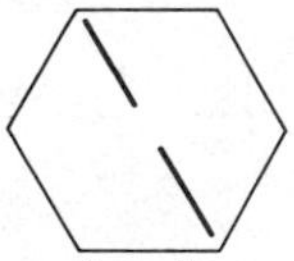 2 marks	SAE grade 3	Medium carbon steel, cold worked	110,000 PSI
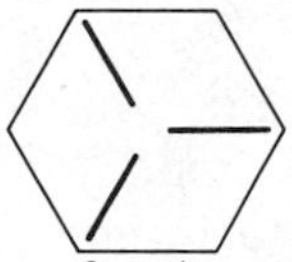 3 marks	SAE grade 5 ASTM — A 325 Common commercial quality	Medium carbon steel, quenched and tempered	120,000 PSI
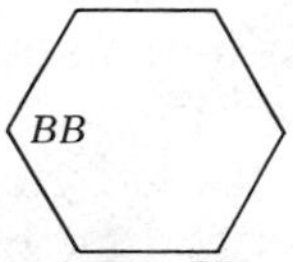 Letters BB	ASTM — A 354	Low alloy steel or medium carbon steel, quenched and tempered	105.000 PSI
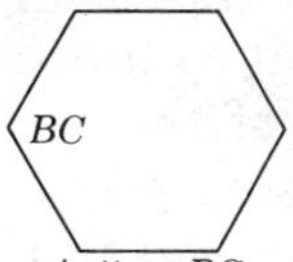 Letters BC	ASTM — A 354	Low alloy steel or medium carbon steel, quenched and tempered	125,000 PSI
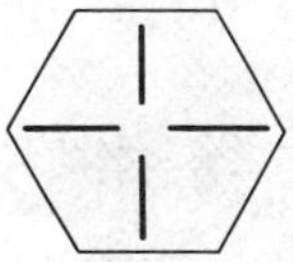 4 marks	SAE grade 6 Better commercial quality	Medium carbon steel, quenched and tempered	140,000 PSI
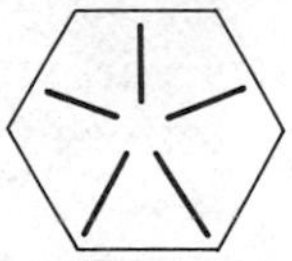 5 marks	SAE grade 7	Medium carbon alloy steel, quenched and tempered, roll threaded after heat treatment	133,000 PSI
 6 marks	SAE grade 8 ASTM — A 345 Best commercial quality	Medium carbon alloy steel, quenched and tempered	150,000 PSI

Figure 10. Grade markings for bolts.

in a bolt, a cotter pin may be used to prevent the nut from turning. Axles and spindles on vehicles have slotted nuts to prevent wheel bearing adjustments from slipping.

Cap or acorn nuts are often used where decorative nuts are needed. These nuts also protect projecting threads from accidental damage. Nuts are made from many different

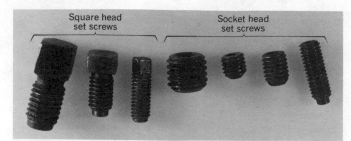

Figure 11. Socket and square head set screws.

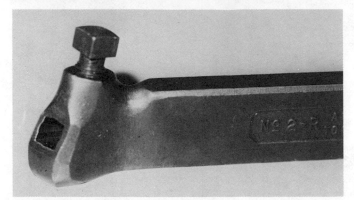

Figure 12. Square head set screws are found in tool holders (Lane Community College).

Figure 13. Square head jack screw (Lane Community College).

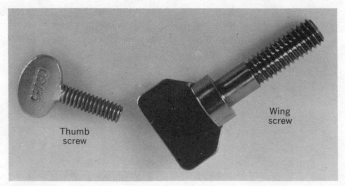

Figure 15. Thumb screw and wingscrew.

materials, depending on their application and strength requirements.

INTERNAL THREAD INSERTS

Internal thread inserts may be used where an internal thread is damaged or stripped and it is not possible to drill and tap for a larger size. A thread insert retains the original thread size. However, it is necessary to drill and tap a somewhat larger hole to accept the thread insert.

One common type of internal thread insert is the wedge type. The thread insert has both external and internal threads. This type of thread insert is screwed into a hole tapped to the same size as the thread on the outside of the insert. The four wedges are driven in using a special driver (Figure 21). This holds the insert in place. The internal thread in the insert is the same as the original hole.

A second type of internal thread insert is also used in repair applications as well as in new installations. Threaded holes are often required in products made from soft metals such as aluminum. If bolts, screws, or studs were to be screwed directly into the softer mate-

Figure 14. Set screw points.

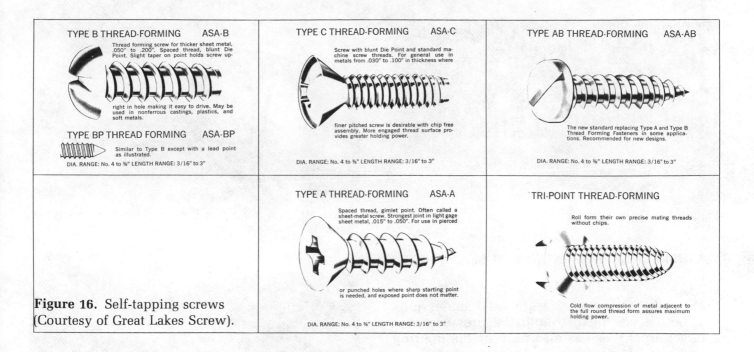

Figure 16. Self-tapping screws (Courtesy of Great Lakes Screw).

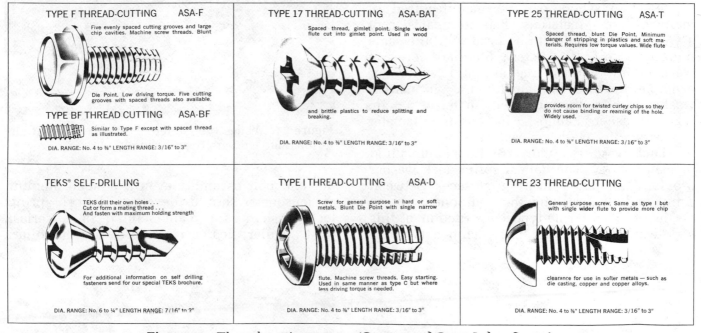

Figure 17. Thread cutting screws (Courtesy of Great Lakes Screw).

rial, excessive wear could result, especially if the bolt is taken in and out a number of times. To overcome this problem, a thread insert made from a more durable material may be used. Stainless steel inserts are frequently used in aluminum (Figure 22). This type of thread insert requires an insert tap, an insert driver, and a thread insert (Figure 23). After the hole for the thread insert is tapped, the insert driver is used to screw the insert into the hole (Figure 24). The end of the insert coil must be broken off and removed after the insert is screwed into place. The insert in the illustration is used to repair sparkplug threads in engine blocks.

Figure 18. Drive screw.

Figure 20. Jam nuts (California State University at Fresno).

WASHERS, PINS, RETAINING RINGS, AND KEYS

Washers

Flat washers (Figure 25) are used under nuts and bolt heads to distribute the pressure over a larger area. Washers also prevent the marring of a finished surface when nuts or screws are tightened. Washers can be manufactured of many different materials. The nominal size of a washer is intended to be used with the same nominal size bolt or screw. Standard series of washers are narrow, regular, and wide. For example, the outside diameter of a $\frac{1}{4}$ in. narrow washer is $\frac{1}{2}$ in. the outside diameter of a $\frac{1}{4}$ in. regular washer is almost $\frac{3}{4}$ in., and the diameter of a wide $\frac{1}{4}$ in. washer measures 1 in.

    Lock washers (Figure 26) are manufactured in many styles. The helical spring lock washer provides hardened bearing surfaces between a nut or bolt head and the components of an assembly. The spring-type construction of this lock washer will hold the tension between a

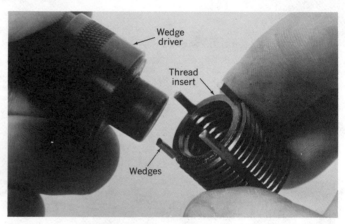

Figure 21. Wedge-type internal thread insert (California State University at Fresno).

nut and bolt assembly even if a small amount of looseness should develop. Helical spring lock washers are manufactured in 5 series: light, regular, heavy, extra duty, and hi-collar.

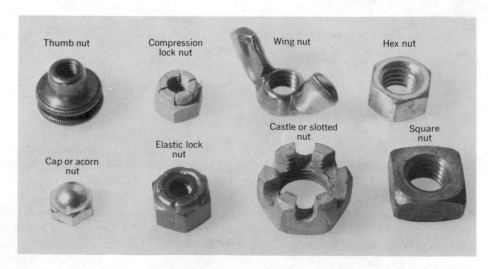

Figure 19. Common nuts.

Figure 22. Stainless steel thread insert used in an aluminum valve housing (California State University at Fresno).

The hi-collar lock washer has an outside diameter equal to the same nominal size socket head cap screw. This makes the use of these lock washers in a counterbored bolt hole possible. Counterbored holes have the end enlarged to accept the bolt head. A variety of standard tooth lock washers are produced, the external type providing the greatest amount of friction or locking effect between fastener and assembly. For use with small head screws and where a smooth appearance is desired, an internal tooth lock washer is used. When large bearing area is desired or where the assembly holes are oversized, an internal-external tooth lock washer is available. A countersunk tooth lock washer is used for a locking action with flat head screws.

Pins
Pins (Figure 27) find many applications in the assembly of parts. **Dowel pins** are heat treated

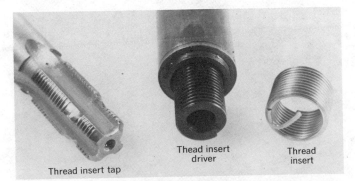

Figure 23. Thread insert tap, driver, and repair insert for sparkplug holes (California State University at Fresno).

Figure 24. Thread insert driver (California State University at Fresno).

and precision ground. Their diameter varies from the nominal dimension by only plus or minus .0001 in. ($\frac{1}{10,000}$ of an inch). Dowel pins are used where very accurate alignments must be maintained between two or more parts. Holes for dowel pins are reamed to provide a slight press fit. Reaming is a machining process during which a drilled hole is slightly enlarged to provide a smooth finish and accurate diameter. Dowel pins only locate. Clamping pressure is supplied by the screws. Dowel pins may be driven into a blind hole. A blind hole is closed at one end. When this kind of hole is used, provision must be made to let the air that is displaced by the pin escape. This can be done by drilling a small through hole or by grinding a narrow flat the full length of the pin. Always use the correct lubricant when screw and pin assemblies are made.

One disadvantage of dowel pins is that they tend to enlarge the hole in an unhardened workpiece if they are driven in and out several times. When parts are intended to be disassembled frequently, **taper pins** will give accurate alignment. Taper pins have a taper of $\frac{1}{4}$ in. per foot of length and are fitted into reamed taper holes.

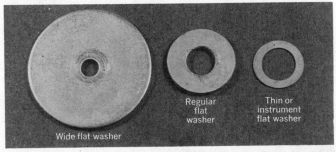

Figure 25. Wide, regular, and thin (or instrument) flat washers.

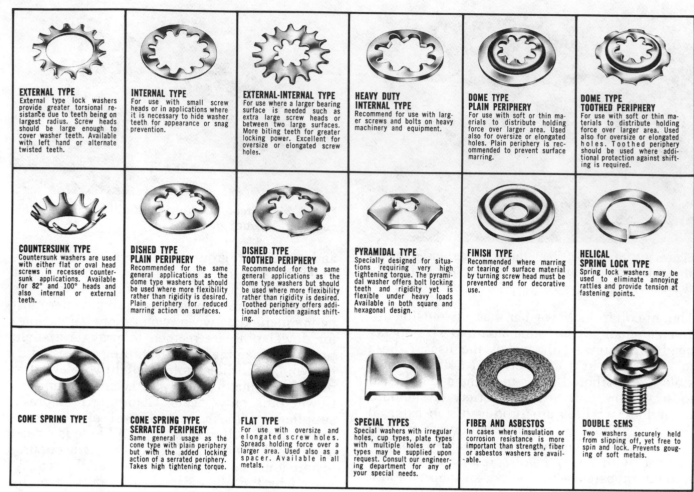

EXTERNAL TYPE
External type lock washers provide greater torsional resistance due to teeth being on largest radius. Screw heads should be large enough to cover washer teeth. Available with left hand or alternate twisted teeth.

INTERNAL TYPE
For use with small screw heads or in applications where it is necessary to hide washer teeth for appearance or snag prevention.

EXTERNAL-INTERNAL TYPE
For use where a larger bearing surface is needed such as extra large screw heads or between two large surfaces. More biting teeth for greater locking power. Excellent for oversize or elongated screw holes.

HEAVY DUTY INTERNAL TYPE
Recommend for use with larger screws and bolts on heavy machinery and equipment.

DOME TYPE PLAIN PERIPHERY
For use with soft or thin materials to distribute holding force over larger area. Used also for oversize or elongated holes. Plain periphery is recommended to prevent surface marring.

DOME TYPE TOOTHED PERIPHERY
For use with soft or thin materials to distribute holding force over larger area. Used also for oversize or elongated holes. Toothed periphery should be used where additional protection against shifting is required.

COUNTERSUNK TYPE
Countersunk washers are used with either flat or oval head screws in recessed countersunk applications. Available for 82° and 100° heads and also internal or external teeth.

DISHED TYPE PLAIN PERIPHERY
Recommended for the same general applications as the dome type washers but should be used where more flexibility rather than rigidity is desired. Plain periphery for reduced marring action on surfaces.

DISHED TYPE TOOTHED PERIPHERY
Recommended for the same general applications as the dome type washers but should be used where more flexibility rather than rigidity is desired. Toothed periphery offers additional protection against shifting.

PYRAMIDAL TYPE
Specially designed for situations requiring very high tightening torque. The pyramidal washer offers bolt locking teeth and rigidity yet is flexible under heavy loads Available in both square and hexagonal design.

FINISH TYPE
Recommended where marring or tearing of surface material by turning screw head must be prevented and for decorative use.

HELICAL SPRING LOCK TYPE
Spring lock washers may be used to eliminate annoying rattles and provide tension at fastening points.

CONE SPRING TYPE

CONE SPRING TYPE SERRATED PERIPHERY
Same general usage as the cone type with plain periphery but with the added locking action of a serrated periphery. Takes high tightening torque.

FLAT TYPE
For use with oversize and elongated screw holes. Spreads holding force over a larger area. Used also as a spacer. Available in all metals.

SPECIAL TYPES
Special washers with irregular holes, cup types, plate types with multiple holes or tab types may be supplied upon request. Consult our engineering department for any of your special needs.

FIBER AND ASBESTOS
In cases where insulation or corrosion resistance is more important than strength, fiber or asbestos washers are available.

DOUBLE SEMS
Two washers securely held from slipping off, yet free to spin and lock. Prevents gouging of soft metals.

Figure 26. Lock washers (Courtesy of Great Lakes Screw).

If a taper pin hole wears larger because of frequent disassembly, the hole can be reamed larger to receive the next larger size of taper pin. Diameters of taper pins range in size from $\frac{1}{16}$ in. to $\frac{11}{16}$ in. measured at the large end. Taper pins are identified by a number from 7/0 (small diameter) to number 10 (large diameter) as well as by their length. The large end diameter is constant for a given size pin, but the small diameter changes with the length of the pin.

A **grooved pin** is either a cylindrical or tapered pin with longitudinal grooves pressed into the pin body. This causes the pin to deform. A groove pin will hold securely in a drilled hole even after repeated removal.

Roll pins can also be used in drilled holes with no reaming required. These pins are manufactured from flat steel bands and rolled into cylindrical shape. Roll pins, because of their spring action, will stay tight in a hole even after repeated disassemblies.

Cotter pins are used to retain parts on a shaft or to lock a nut or a bolt as a safety precaution. Cotter pins make a quick assembly and disassembly possible.

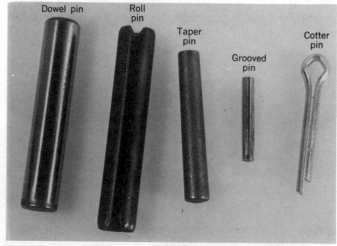

Figure 27. Pins (California State University at Fresno).

Selector Guide: Standard Truarc® Ring Series

DESIGN FEATURES

RING TYPES FOR AXIAL ASSEMBLY

Series N5000, 5100: Tapered section assures constant circularity and groove pressure. Secure against heavy thrust loads and high rotational speeds.

Series 5008, 5108: Lugs inverted to abut groove bottom. Rings form high circular shoulder, concentric with bore or shaft. Good for parts having large corner radii or chamfers.

Series 5160: Heavy-duty ring resists high thrust, impact loads. Eliminates spacer washers in bearing assemblies.

Series 5560: New miniature, high-strength ring. Forms tamper-proof shoulder on small diameter shafts subject to heavy thrust, impact loads.

Series 5590: Permanent-shoulder ring for small diameter shafts. When compressed into groove, notches deform to close gaps, reducing both I.D. and O.D.

RING TYPES FOR RADIAL ASSEMBLY

Series 5103: Forms narrow, uniformly concentric shoulder. Excellent for assemblies where clearance is limited.

Series 5133: Provides large shoulder on small diameter shafts. Installed in deep groove for added thrust capacity.

Series 5144: Reinforced to provide five times greater gripping strength, 50% higher rpm limits than conventional E-rings. Secure against rotation.

Series 5107: Two-part ring balanced to withstand high rpm's, heavy thrust loads, relative rotation between parts.

Series 5304: New high-strength ring for large bearing surface. Can be installed quickly with pliers or mallet, removed with ordinary screw driver.

Series T5304: Thinner model of 5304. Can be seated in same width grooves as E-rings, has more gripping power. Good for cast or molded grooves.

RING TYPES FOR TAKING UP END-PLAY

Series N5001, 5101: Bowed cylindrically to accommodate large tolerances, provide resilient end-play take-up.

Series N5002, 5102: Rings beveled 15° on groove-engaging edge for use in groove with similar bevel. Wedge action provides rigid end-play take-up.

Series 5131: Provides large shoulder on small diameter shafts. Bowed for resilient end-play take-up.

Series 5139: Bowed ring designed for use as shoulder against rotating parts. Prongs lock against shaft, prevent ring from being forced from groove.

SELF-LOCKING TYPE RINGS (No groove required)

Series 5115: Push-on type fastener for ungrooved shafts and studs. Has arched rim for extra strength, long prongs for wide shaft tolerances.

Series 5105, 5005: Flat rim, shorter prongs, smaller O.D. than 5115. For flat contact surface, better clearance.

Series 5555: Secure against axial displacement from either direction. No groove needed. Adjustable, reusable.

Series 5305: Dished body, three heavy prongs lock on shaft under spring tension. Withstands heavy thrust loads.

Series 5300: Free-spinning nut. Dished body flattens under torque, eliminating need for separate lock washers.

INTERNAL	**BASIC N5000** For housings and bores — Size Range .250—10.0 in. / 6.4—254.0 mm.	EXTERNAL	**BOWED 5101** For shafts and pins — Size Range .188—1.750 in. / 4.8—44.4 mm.	EXTERNAL	**REINFORCED 5115** — Size Range .094—1.0 in. / •	EXTERNAL	**TRIANGULAR NUT 5300** For threaded parts — Size Range 6-32 and 8-32 10-24 and 10-32 1/4-20 and 1/4-28
INTERNAL	**BOWED N5001** For housings and bores — Size Range .250—1.750 in. / 6.4—44.4 mm.	EXTERNAL	**BEVELED 5102** For shafts and pins — Size Range 1.0—10.0 in. / 25.4—254.0 mm.	EXTERNAL	**BOWED E-RING 5131** For shafts and pins — Size Range .110—1.375 in. / 2.8—34.9 mm.	EXTERNAL	**KLIPRING 5304 T-5304** For shafts and pins — Size Range .156—1.000 in. / 4.0—25.4 mm.
INTERNAL	**BEVELED N5002** For housings and bores — Size Range 1.0—10.0 in. / 25.4—254.0 mm.	EXTERNAL	**CRESCENT® 5103** For shafts and pins — Size Range .125—2.0 in. / 3.2—50.8 mm.	EXTERNAL	**E-RING 5133** For shafts and pins — Size Range .040—1.375 in. / 1.0—34.9 mm.	EXTERNAL	**TRIANGULAR 5305** For shafts and pins — Size Range .062—.438 in.
INTERNAL	**CIRCULAR 5005** For housings and bores — Size Range .312—2.0 in. / •	EXTERNAL	**CIRCULAR 5105** For shafts and pins — Size Range .094—1.0 in. / •	EXTERNAL	**PRONG-LOCK® 5139** For shafts and pins — Size Range .092—.438 in. / •	EXTERNAL	**GRIPRING® 5555** For shafts and pins — Size Range .079—.750 in. / 2.0—19.0 mm.
INTERNAL	**INVERTED 5008** For housings and bores — Size Range .750—4.0 in. / 19.0—101.6 mm.	EXTERNAL	**INTERLOCKING 5107** For shafts and pins — Size Range .469—3.375 in. / 11.9—85.7 mm.	EXTERNAL	**REINFORCED E-RING 5144** For shafts and pins — Size Range .094—.562 in. / 2.4—14.3 mm.	EXTERNAL	**HIGH-STRENGTH 5560** For shafts and pins — Size Range .101—.328 in. / •
EXTERNAL	**BASIC 5100** For shafts and pins — Size Range .125—10.0 in. / 3.2—254.0 mm.	EXTERNAL	**INVERTED 5108** For shafts and pins — Size Range .500—4.0 in. / 12.7—101.6 mm.	EXTERNAL	**HEAVY-DUTY 5160** For shafts and pins — Size Range .394—2.0 in. / 10.0—50.8 mm.	EXTERNAL	**PERMANENT SHOULDER 5590** For shafts and pins — Size Range .250—.750 / 6.4—19.0 mm.

Figure 28. Retaining rings (Copyright © 1966 Waldes Kohinoor, Inc. Reprinted with permission).

Figure 29. External retaining ring used on a shaft (Courtesy of Waldes Kohinoor, Inc.).

Retaining Rings

Retaining rings (Figure 28) are fasteners used in many assemblies. Retaining rings can easily be installed in machined grooves, internally in housings, or externally on shafts or pins (Figure 29). Some types of retaining rings do not require grooves but have a self-locking spring-type action. The most common application of a retaining ring is to provide a shoulder to hold and retain a bearing or other part on an otherwise smooth shaft. They may also be used in a bearing housing (Figure 30). Special pliers are used to install and remove retaining rings.

Keys

Keys (Figure 31) are used to prevent the rotation of gears or pulleys on a shaft. Keys are fitted into key seats in both the shaft and the external part. Keys should fit the key seats rather snugly. **Square keys,** where the width and the height are equal, are preferred on shaft sizes up to a $6\frac{1}{2}$ in. diameter Above a $6\frac{1}{2}$ in. diameter rectangular keys are recommended. **Woodruff keys,** which are almost in the shape of a half circle, are used where relatively light loads are transmitted. One advantage of woodruff keys is that they cannot change their axial location on a shaft because they are retained in a pocket. A key fitted into an endmilled pocket will also retain its axial position on the shaft. Most of these keys are held under tension with one or more set screws threaded through the hub of the pulley or gear. Where extremely heavy shock loads or high torques are encountered, a **taper key** is used. Taper keys have a taper of $\frac{1}{8}$ in. per foot. Where a tapered key is used, the key seat in the shaft is parallel to the shaft axis and a taper to match the key is in the hub. Where only one side of an assembly is accessible, a **gib head taper key** is used instead of a plain taper key. When a gib head taper key is driven into the key seat as far as possible, a gap remains between the gib and the hub of the pulley or gear. The key is removed for disassembly by driving a wedge into the gap to push the key out. A **feathered key** is a key that is secured in a key seat with screws. A feathered key is often a part of a sliding gear or sliding pulley.

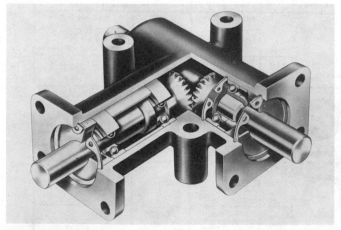

Figure 30. Internal retaining rings used to retain bearings (Courtesy of Waldes Kohinoor, Inc.).

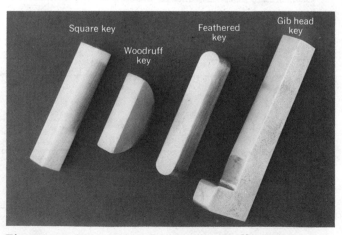

Figure 31. Keys (Lane Community College).

self-evaluation

SELF-TEST
1. What is the difference between a bolt and a screw?
2. How much thread engagement is recommended when a screw is used in an assembly?
3. When are class 3 threads used?
4. What is the difference between a machine bolt and a cap screw?
5. What is the outside diameter of a No. 8 machine screw?
6. Where are set screws used?
7. When are stud bolts used?
8. Explain the difference between thread-forming and thread-cutting screws.
9. Where are castle nuts used?
10. Where are cap nuts used?
11. Explain two reasons why flat washers are used.
12. What is the purpose of a helical spring lock washer?
13. When is an internal-external tooth lock washer used?
14. When are dowel pins used?
15. When are taper pins used?
16. When are roll pins used?
17. What are retaining rings?
18. What is the purpose of a key?
19. When is a woodruff key used?
20. When is gib head key used?

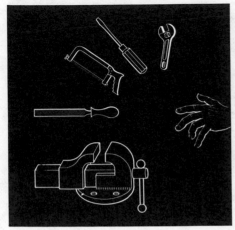

section b
hand tools

Man's first tools were crudely made of flint and wood (Figures 1a and 1b). After the discovery of metals, bronze, iron, and steel were the chief materials of hand tools. Many different types of tools were developed; saws, hammers, chisels, and files are all very ancient tools that are still in common use today, having changed little until the last few centuries. Tools used in medieval times had changed very little from the ancient patterns (Figure 2). Eighteenth century European metalworkers, however, showed an inventiveness that brought about the development of many new tools and techniques (Figure 3).

Workholding devices were not developed in early times. Craftsmen in many Middle East and Asian countries still preferred to use their feet instead of a vise to hold the workpiece. Machinists today tend to take the bench vise for granted, seldom realizing that they could hardly get along without it.

Arbor presses and hydraulic shop presses are very useful and powerful shop tools (Figure 4). If they are used incorrectly, however, they can be very hazardous to the operator, and workpieces can be ruined.

Noncutting tools such as screwdrivers, pliers, and wrenches should be properly identified. It is impossible to request a particular tool from the toolroom without knowing its correct name.

Cutting hand tools such as hacksaws, files, hand reamers, taps, and dies are very important to a machinist. In this section you will also be introduced to the pedestal grinder and its important functions in the machine shop.

The units that follow in this section will instruct you in the identification, selection, use, and safety of these important hand tools and hand operated machines.

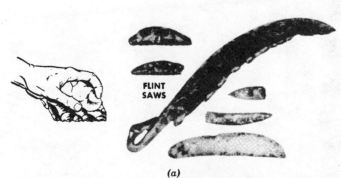

FLINT SAWS

(a)

Figure 1a. Stone saws (Courtesy of DoAll Company).

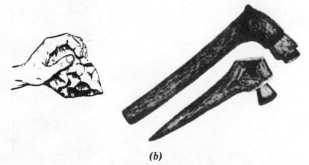

(b)

Figure 1b. Stone hand hatchet (Courtesy of DoAll Company).

Figure 2. Wooden molds for the casting of metal buttons are being made using very crude tools and equipment such as the bow drill in the foreground and the polishing machine in the background. A primitive vise with a wooden screw is shown on the far left (Dover Publications, Inc.).

Figure 3. In the eighteenth century a much more advanced level of machining is evident in this view of a small toolmaker's shop. A metal vise is shown in the left background. A primitive screw cutting lathe is being operated by hand power (Dover Publications, Inc.).

Figure 4. The small hydraulic shop press (Lane Community College).

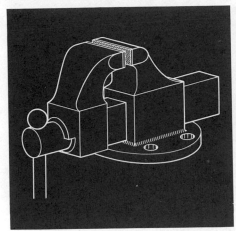

unit 1
workholding for hand operations

The bench vise is a basic but very necessary tool in the shop. With proper care and use, this workholding tool will give many years of faithful service.

objectives After completing this unit, you will be able to:
1. Identify various types of vises and their uses.
2. Explain the procedures used for the care and maintenance of vises.

TYPES OF VISES

Vises of various types are used by machinists when doing hand or bench work. They should be mounted in such a way that a long workpiece can be held in a vertical position extending alongside the bench (Figure 1). Some bench vises have a solid base (Figure 2), and others have a swivel base (Figure 3). The machinist's bench vise is measured by the width of the jaws (Figure 4).

Toolmakers often use small vises that pivot on a ball and socket for holding delicate work. Hand-held vises, called pin vises, are made for holding very small or delicate parts.

Most bench vises have hardened insert jaws that are serrated for greater gripping power (Figure 5). These criss-cross serrations are sharp and will dig into finished workpieces enough to mar them beyond repair. Soft jaws (Figure 6) made of copper, other soft metals, or wood are used to protect a finished surface on a workpiece. These soft jaws are made to slip over the vise jaws. Some vises used for sheet metal work have smooth, deep jaws (Figure 7).

USES OF VISES

Vises are used to hold work for filing, hacksawing, chiseling, and bending light metal. They are also used for holding work when assembling and disassembling parts.

Vises should be placed on the workbench at the correct working height for the individual. The top of the vise jaws should be at elbow

Figure 1. When long work is clamped in the vise vertically, it should clear the workbench (Lane Community College).

29

Figure 2. A solid base bench vise (Lane Community College).

Figure 5. View of the hardened, serrated insert jaws on the vise (Lane Community College).

Figure 3. A swivel base bench vise (Lane Community College).

Figure 6. View of the soft jaws placed on the vise (Lane Community College).

Figure 4. How to measure a vise (Lane Community College).

Figure 7. Smooth-jawed vise for working with sheet metal (Lane Community College).

height. Poor work is produced when the vise is mounted too high or too low. A variety of vise heights should be provided in the shop or skids made available to stand on.

CARE OF VISES

Like any other tool, vises have limitations. "Cheater" bars or pipes should not be used on the handle to tighten the vise. Heat from a torch should not be applied to work held in the jaws as the hardened insert jaws will then become softened. There is usually one vise in a shop that is reserved for heating and bending.

Heavy hammering should not be done on a bench vise. The force of bending or pounding should be against the fixed jaw rather than the movable jaw of the vise. Bending light, flat stock or small round stock in the jaws is permissible if a light hammer is used. The movable jaw slide bar (Figure 8) *should never be hammered upon* as it is usually made of thin cast iron and can be cracked quite easily. An anvil is often provided behind the solid jaw for the purpose of light hammering.

Bench vises should occasionally be taken apart so that the screw, nut, and thrust collars

Figure 8. Hammering on the slide bar should never be done to a vise. This may crack or distort it (Lane Community College).

may be cleaned and lubricated. The screw and nut should be cleaned in solvent. A heavy grease should be packed on the screw and thrust collars before reassembly.

self evaluation

SELF-TEST

1. What clamping position should be considered when mounting a vise on a workbench?
2. Name two types of bench vises.
3. How is the machinist's bench vise measured for size?
4. Small, delicate work may be held in a _hand_ or a _Pin_ vise.
5. Explain two characteristics of the insert jaws on vises.
6. How can a finished surface be protected?
7. In what way are vises that are used for sheet metal work different from a machinist's vise?
8. What are vises usually used for?
9. Name three things that should never be done to a vise.
10. How should a vise be lubricated?

This unit has no post-test.

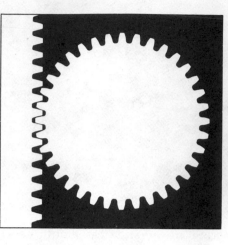

unit 2
arbor
and
shop presses

The arbor press and the small shop press are very common sights in most machine shops. It would be difficult indeed to get along without these machines. You will find these tools extremely useful when you know how to use them, but if you are not instructed in their use, they can be dangerous to you and destructive to the workpiece.

objectives   After completing this unit, you will be able to:

1. Install and remove a bronze bushing using an arbor press.
2. Press on and remove a ball bearing from a shaft on an arbor press using the correct tools.
3. Press on and remove a ball bearing from a housing using an arbor press and correct tooling.
4. Install and remove a mandrel using an arbor press.
5. Install and remove a shaft with key in a hub using the arbor press.
6. Broach a $\frac{1}{4}$ inch keyway in a $1\frac{1}{4}$ inch bore.

TYPES

The arbor press is an essential piece of equipment in the small machine shop. Without it a machinist would be forced to resort to the use of a hammer or sledge to make any forced fit, a process that could easily damage the part.

Two basic types of arbor presses are manufactured and used: the hydraulic (Figure 1) and the mechanical (Figure 2). Both types are hand-powered with a lever. The lever gives a "feel" or a sense of pressure applied, which is not possible with power-driven presses. This pressure sensitivity is needed when small delicate parts are being pressed so that a workman will know where to stop before collapsing the piece.

USES

The major uses of the arbor press are bushing installation and removal, ball and roller bearing installation and removal (Figure 3), pressing shafts into hubs (Figure 4), pressing mandrels into workpieces, broaching keyways (Figure 5), and straightening and bending (Figure 6).

PROCEDURES

Installing Bushings

A bushing is a short metal tube, machined inside and out to precision dimensions, and usually made to fit into a bore, or accurately machined hole. Many kinds of bushings are used

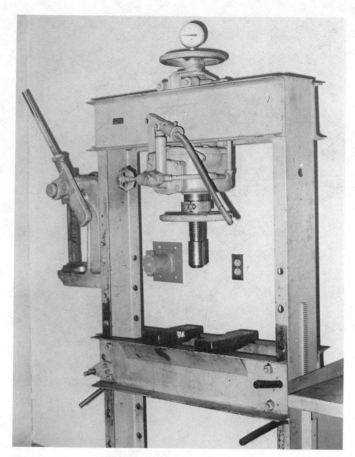

Figure 1. Fifty ton capacity hydraulic shop presses (Lane Community College).

Figure 2. Simple ratchet floor type arbor press (Courtesy of Dake Corporation).

for various purposes and are usually installed with an **interference fit** or press fit. This means that the bushing is slightly larger than the hole into which it is pressed. The amount of interference will be considered in greater detail in a later unit. There are many bushings made of many materials including bronze and hardened steel, but they all have one thing in common: they must be lubricated with high pressure lube before they are pressed into the bore. Oil is not used as it will simply wipe off and cause the bushing to seize the bore. Seizing is the condition where two unlubricated metals tend to weld together under pressure. In this case it may cause the bushing to be damaged beyond repair.

The bore should always have a strong chamfer, that is, an angled or beveled edge, since a sharp edge would cut into the bushing and damage it (Figures 7a and b). The bushing should also have a long tapered chamfer or **start** so it will not **dig in** and enter mis-

aligned. Bushings are prone to go in crooked if there is a sharp edge, especially if it is a hardened steel bushing. Care should be taken to see that the bushing is straight entering the bore, and that it continues into the bore in proper alignment. This should not be a problem if the tooling is right; that is, if the end of the press ram is square and if it is not loose and worn. The proper bolster plate should also be used under the part so that it cannot tilt out of alignment. Sometimes special tooling is used to guide the bushing (Figure 8). Only the pressure needed to force the bushing into place should be applied, especially if the bushing is longer than the bore length. Excessive pressure might distort the bushing and cause it to be undersized (Figure 9).

Ball and Roller Bearings
Ball and roller bearings pose special problems when they are installed and removed by pressing. This is because the pressure must be applied

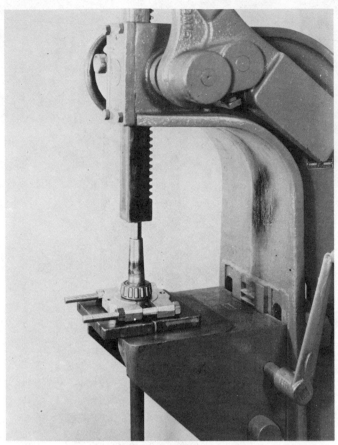

Figure 3. Roller bearing being removed from axle (Lane Community College).

Figure 5. Keyway being broached in gear hub with a mechanical arbor press (Courtesy of The duMont Corporation).

Figure 4. Shaft being pressed into hub (Lane Community College).

Figure 6. Shaft being straightened in hydraulic shop press (Lane Community College).

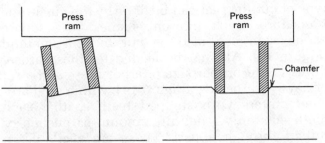

Figure 7a. Bushing being pressed where bore is not chamfered and bushing is misaligned. **Figure 7**b. Bushing being pressed into correctly chamfered hole in correct alignment.

directly against the race and not through the balls or rollers, since this could destroy the bearing. Frequently, when removing ball bearings from a shaft, the inner race is hidden by a shoulder and cannot be supported in the normal way. In this condition a special tool is used (Figure 10). On the inner and outer races, bearings may be installed by pressing on the race with a steel tube of the proper diameter. As with bushings, high pressure lubricant should be used.

Sometimes there is no other way to remove an old ball bearing except by exerting pressure through the balls. When this is done, there is a real danger that the race may be violently shattered. In this case a scatter shield must be used. A scatter shield is a heavy steel tube about 8 to 12 inches long and is set up to cover the work. The shield is placed around the bearing during

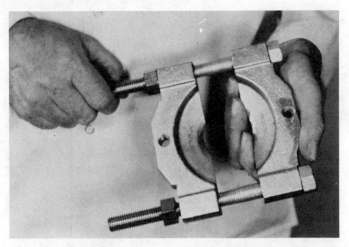

Figure 10. Special tool for supporting inner bearing race. (See Figure 3.) (Lane Community College).

pressing to keep shattered parts from injuring the operator. It is a good safety practice to always use a scatter shield when ball bearings are removed from a shaft by pressing. Safety glasses should be worn during all pressing operations.

Bores and Shafts
Holes in the hubs of gears, sprockets and other machine parts are also frequently designed for a force fit. In these instances, there is usually a keyway that needs to be aligned. A keyway is a groove in which a key is placed. This key, in turn, also fits into a slot in the hub of a gear or pulley, and secures the part against the shaft, keeping it from rotating. When pressing shafts with keys into hubs with keyways, it is sometimes helpful to chamfer the leading edge of the key so it will align itself properly (Figure 11). Seizing will occur in this operation, as with the installation of bushings, if high pressure lubricant is not used.

Mandrels
Mandrels, cylindrical pieces of steel with a slight taper, are pressed into bores in much the same way that shafts are pressed into hubs. There is one important difference, however; since the mandrel is tapered about .006 of an inch per foot, it can be installed only with the small end in first. Determine which end is the small one by measuring with a micrometer or by trying the mandrel in the bore. The small end should start into the hole, but the large end should not. Apply lubricant and press the mandrel in until definite resistance is felt (Figure 12).

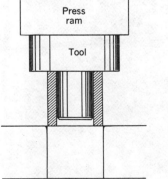

Figure 8. Special tool to keep bushing square to press ram.

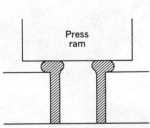

Figure 9. Effect of excessive pressure on bushing that exceeds bore length.

Figure 11. Chamfer on key helps in alignment of parts being pressed together (Lane Community College).

Keyway Broaching

Although many types of keyseating machines are in use in many machine shops, keyway broaching is often done on arbor presses. Broaching is the process of cutting out shapes on the interior of a metal part. Keyways are only one

type of cutting that can be done by the push-type procedure. Such internal shapes as a square or hexagon can also be cut by this method (Figure 13). All that is needed for these procedures is the proper size of arbor press and a set of keyway broaches (Figure 14), which are hardened cutters with stepped teeth so that each tooth cuts only a definite amount as pushed or pulled through a part. These are available in inch and metric dimensions.

Broaching keyways (multiple pass method) is done as follows:

1. Choose the bushing that fits the bore and the broach, and put it in place in the bore.
2. Insert the correct size broach into the bushing slot (Figure 15).
3. Place this assembly in the arbor press (Figure 16).
4. Lubricate.
5. Push the broach through.
6. Clean the broach.
7. Place second-pass shim in place.
8. Insert broach.
9. Lubricate.
10. Push the broach through.
11. If more than one shim is needed to obtain the correct depth, repeat the procedure (Figure 17).

Figure 12. Mandrel being lubricated and pressed into part for further machining (Lane Community College).

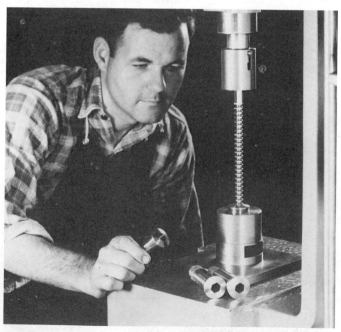

Figure 13. Hexagonal shape being push broached (Courtesy of The duMont Corporation).

Figure 14. A typical set of keyway broaches (Courtesy of The duMont Corporation).

The tools should be cleaned and returned to their box and the finished keyway should be deburred and cleaned.

Production or single-pass broaching requires no shims or second-pass cuts, and with some types no bushings need be used (Figure 18).

Two important things to remember when push broaching are alignment and lubrication. Misalignment, caused by a worn or loose ram, can cause the broach to hog (dig in) or break. Sometimes this can be avoided by facing the teeth of the broach toward the back of the press and permitting the bushing to protrude above the work to provide more support for the broach. After starting the cut, relieve the pressure to allow the broach to center itself. Repeat this procedure during each cut.

Figure 16. Broach, guide bushing placed in arbor press that is ready to lubricate and to perform first pass (Lane Community College).

At least two or three teeth should be in contact with the work. If needed, stack two or more workpieces to lengthen the cut. The cut should never exceed the length of the standard bushing used with the broach. Never use a broach on material harder than Rockwell C35, one of many grades of a hardness test you will meet later in this book. If it is suspected that a part is harder than mild steel, its hardness should be determined before any broaching is attempted.

Use a good high pressure lubricant. Also apply a sulfur base cutting oil to the teeth of the broach. Always lubricate the back of the keyway broach to reduce friction, regardless of the material to be cut. Brass is usually broached dry, but bronzes cut better with oil or soluble oil. Cast iron is broached dry, and kerosene or cutting oil is recommended for aluminum.

Bending and Straightening
Bending and straightening are frequently done on hydraulic shop presses. Mechanical arbor

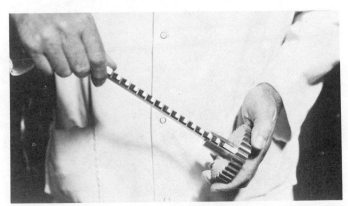

Figure 15. Broach with guide bushing inserted into gear (Lane Community College).

Figure 17. Shims in place behind broach that is ready to lubricate and make final cut on part (Lane Community College).

Figure 18. Production push broaching without bushings on shims (Courtesy of The duMont Corporation).

presses are not usually used for this purpose. There is a definite safety hazard in this type of operation as a poor setup can allow pieces under pressure to suddenly fly out of the press. Brittle materials such as cast iron or hardened steel bearing races can suddenly break under pressure and explode into fragments.

A shaft to be straightened is placed between two nonprecision vee blocks, steel blocks with a vee-shaped groove running the length of the blocks that support a round workpiece. In the vee blocks, the shaft is rotated to detect runout, or the amount of bend in the shaft. The rotation is measured on a dial indicator, which is a device capable of detecting very small mechanical movements, and read from a calibrated dial. The high point is found and marked on the shaft (Figure 19). After removing the indicator, a soft metal pad such as copper is placed between the shaft and the ram and pressure is applied (Figure

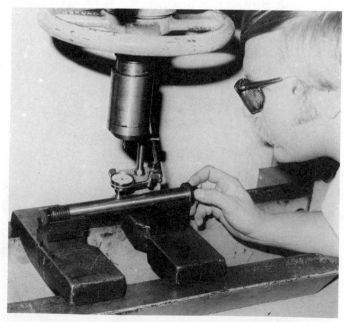

Figure 19. Part being indicated for runout prior to straightening (Lane Community College).

20). The shaft should be bent back to a straight position and then very slightly beyond that point. The pressure is then removed and the dial indicator is again put in position. The shaft is rotated as before, and the position of the mark noted, as well as the amount of runout. If improvement has been found, continue the process; but if the first mark is now opposite the high point, too much pressure has been applied. Repeat the same steps, applying less pressure.

Other straightening jobs on flat stock and other shapes are done in a similar fashion. Frequently, two or more bends will be found that may be opposite or are not in the same direction. This condition is best corrected by straightening one bend at a time and checking with a straightedge and feeler gage. Special shop press tooling is sometimes used for simple bending jobs in the shop.

Note: Be sure to stand to one side when applying pressure in all pressing operations. Be especially careful when using the press for straightening and always use safety glasses or an eye shield.

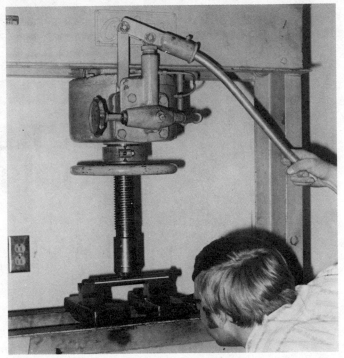

Figure 20. Pressure being applied to straighten shaft (Lane Community College).

self-evaluation

SELF-TEST

1. Is it important to know how to use the arbor press properly and how to correctly set up pressing operations? Why?

2. What kinds of arbor presses are made? What makes them different from large commercial presses?

3. List several uses of the arbor press.

4. A newly machined steel shaft with an interference fit is pressed into the bore of a steel gear. The result is a shaft ruined beyond repair; the bore of the gear is also badly damaged. What has happened? What caused this failure?

5. The ram of an arbor press is loose in its guide and the pushing end is rounded off. What kind of problems could be caused by this?

6. A $\frac{1}{2}$ inch diameter bronze bushing is $\frac{1}{8}$ inch longer than the bore. Should you apply 30 tons of pressure to make sure it has seated on the press plate? If your answer is no, how much pressure should you apply?

7. If the inner race on a ball bearing is pressed onto a shaft, why should you not support the outer race while pushing the shaft off?

8. What difference is there in the way a press fit is obtained between mandrels and ordinary shafts?

9. Prior to installing a bushing with the arbor press, what two important steps must be taken?

10. Name five ways to avoid tool breakage and other problems when using push broaches for making keyways in the arbor press.

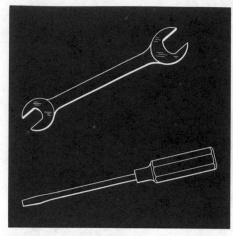

unit 3
noncutting
hand
tools

Hand tools are essential in all of the mechanical trades. This unit will help you learn the names and uses of most of the noncutting tools used by machinists.

objectives   After completing this unit, you will be able to:
1. **Identify the proper tool for a given job.**
2. **Determine the correct use of a selected tool.**

CLAMPING DEVICES

A machinist has to do many jobs that require the use of clamping devices. These clamping devices are varied in types and tasks that they can perform, but they all serve one general purpose: to hold a workpiece while machining operations are being performed.

A machinist's bench vise, introduced in the first unit of this section, holds a workpiece while such operations as filing or sawing are done.

Figure 2. Swivel base vise (Courtesy of Wilton Corporation).

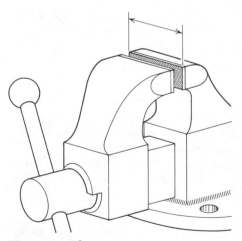

Figure 1. The way to measure a vise.

They are made of cast iron with one jaw attached to the base and the other jaw adjustable by a handle or lever. The size of a vise is determined by the width of these jaws (Figure 1). Some vises have fixed bases while others have swivel bases (Figure 2). The inner jaw faces, made of hardened steel, are usually serrated and can often damage finished workpieces or ones made of soft metals such as aluminum. To prevent this, soft jaws

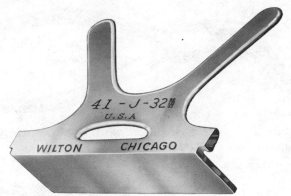

Figure 3. Jaw caps for use on a vise (Courtesy of Wilton Corporation).

are made to slip over the regular vise jaws (Figure 3).

C-clamps (Figure 4) are used to hold work-pieces on machines such as drill presses, and are also used for clamping parts together. The size of the clamp is determined by the largest opening of its jaws. Parallel clamps (Figure 5) are used to hold small parts. Since they do not

Figure 5. Single size parallel clamps (DeAnza College).

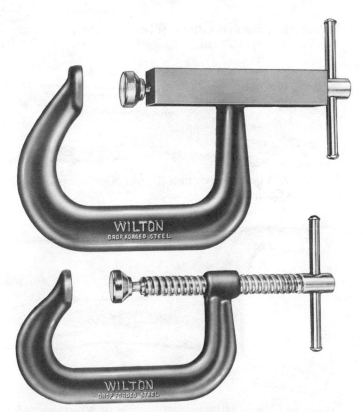

Figure 4. Two types of C-clamps (Courtesy of Wilton Corporation).

have as much holding power as C-clamps, this usually limits the use of parallel clamps to delicate work. Precision measuring setups are usually held in place with parallel clamps.

PLIERS

Pliers come in several shapes and with several types of jaw action. Simple combination or slip joint pliers (Figure 6) will do most jobs for which you need pliers. The slip joint allows the jaws to expand to grasp a larger size work. They are measured by overall length and are made in 5, 6, 8 and 10-inch sizes.

Interlocking joint pliers (Figure 7), or water pump pliers, were made to tighten packing gland nuts on water pumps on cars and trucks, but are useful for a variety of jobs. Pliers should never be used as a substitute for a wrench, as the nut or bolt head will be permanently deformed by the serrations in the plier jaws and the wrench will no longer fit properly. Round nose pliers (Figure 8) are used to make loops in wire and to shape light metal. Needle nose pliers are used for holding small delicate workpieces in tight spots. They are available in both straight (Figure 9) and bent nose (Figure 10) types. Linemen's pliers (Figure 11) can be used for wire cutting and bending. Some types have wire stripping grooves and insulated handles. Diagonal cutters (Figure 12) are only used for wire cutting.

The lever-jawed locking wrench has an unusually high gripping power. The screw in the handle adjusts the lever action to the work

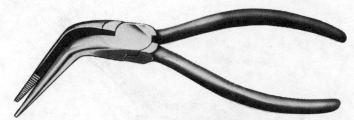

Figure 10. Needlenose pliers, bent (Photo courtesy of Snap-on Tools Corporation).

Figure 11. Side cutting pliers (Photo courtesy of Snap-on Tools Corporation).

Figure 12. Diagonal cutters (Photo courtesy of Snap-on Tools Corporation).

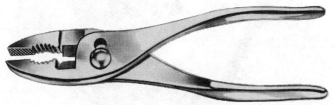

Figure 6. Slip joint or combination pliers (Photo courtesy of Snap-on Tools Corporation).

Figure 7. Interlocking joint or water pump pliers (Photo courtesy of Snap-on Tools Corporation).

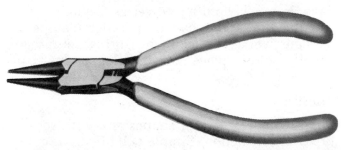

Figure 8. Round nose or wire looper pliers (Photo courtesy of Snap-on Tools Corporation).

Figure 9. Needlenose pliers, straight (Photo courtesy of Snap-on Tools Corporation).

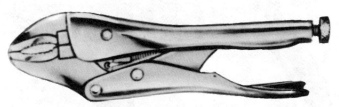

Figure 13. Vise grip wrench (Photo courtesy of Snap-on Tools Corporation).

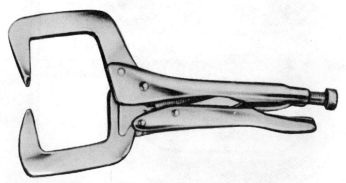

Figure 14. Vise grip C-clamp (Photo courtesy of Snap-on Tools Corporation).

size (Figure 13). They are made with special jaws for various uses such as the C-clamp type used in welding (Figure 14).

HAMMERS

Hammers are classified as either hard or soft. Hard hammers have steel heads such as black-

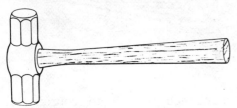

Figure 15. Mall.

Figure 16. Ball peen hammer.

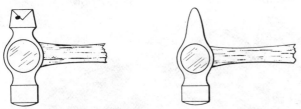

Figure 17. Straight peen hammer.   **Figure 18.** Cross peen hammer.

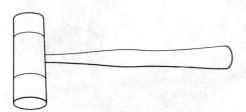

Figure 19. Plastic hammer.

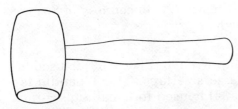

Figure 20. Lead hammer.

smith types or mauls made for heavy hammering (Figure 15). The ball peen hammer (Figure 16) is the one most frequently used by machinists. It has a rounded surface on one end of the head, which is used for upsetting or riveting metal and a hardened striking surface on the other. Two hammers should never be struck together on the face, as pieces could break off. Hammers are specified according to the weight of the head. Ball peen hammers range from two ounces to three pounds. Those under 10 ounces are used for layout work. Two other shop hammers are the straight peen (Figure 17) and the cross peen (Figure 18).

Soft hammers are made of plastic (Figure 19), brass, copper, lead (Figure 20), or rawhide and are used to properly position workpieces, that have finishes that would be damaged by a hard hammer. The movable jaw on most machine tool vises tends to move slightly upward when tightened against the workpiece. Thus, the workpiece is moved upward and out of position. The machinist must then use a soft, heavy mallet to reposition it.

WRENCHES

A large variety of wrenches is made for different uses such as turning cap screws, bolts, and nuts. The adjustable wrench (Figure 21) is a general purpose tool and will not suit every job, especially those requiring work in close quarters. The wrench should be rotated toward the movable jaw and should fit the nut or bolt tightly. The size of the wrench is determined by its overall length in inches.

Open end wrenches (Figure 22) are best suited to square-headed bolts, and usually fit two sizes, one on each end. The ends on this type of wrench are also angled so they can be used in close quarters. Box wrenches (Figure 23) are also

Figure 21. Adjustable wrench showing the correct direction of pull. Moveable jaw should always face the direction of rotation (Photo courtesy of Snap-on Tools Corporation).

double ended and offset to clear the user's hand. The box completely surrounds the nut or bolt and usually has 12 points so that the wrench can be reset after rotating only a partial turn. Mostly used on hex-headed bolts, these wrenches have the advantage of precise fit. Combination and open end wrenches are made with a box at one end and an open end at the other (Figure 24).

Socket wrenches are similar to box wrenches in that they also surround the bolt or nut and usually are made with 12 points contacting the six-sided nut. Sockets are made to be detached from various types of drive handles (Figure 25).

Pipe wrenches, as the name implies, are used for holding and turning pipe. These wrenches have sharp serrated teeth and will damage any finished part on which they are used (Figure 26). Strap wrenches (Figure 27) are used for extremely large parts or to avoid marring the surface of tubular parts.

Spanner wrenches come in several basic types including face and hook. Face types are sometimes called pin spanners (Figure 28). Spanners are made in fixed sizes or adjustable types (Figures 29, 30, and 31).

Socket head wrenches (Figure 32) are six-sided bars having a 90-degree bend near one end. They are used with socket head cap screws and socket set screws.

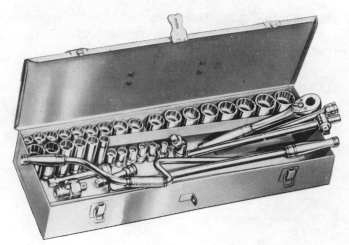

Figure 25. Socket wrench set (Photo courtesy of Snap-on Tools Corporation).

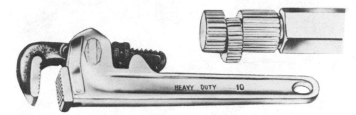

Figure 26. Pipe wrenches, external and internal (Photo courtesy of Snap-on Tools Corporation).

Figure 22. Open end wrench (Photo courtesy of Snap-on Tools Corporation).

Figure 23. Box wrench (Photo courtesy of Snap-on Tools Corporation).

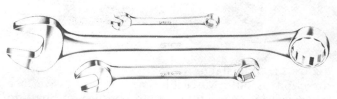

Figure 24. Combination wrench (Photo courtesy of Snap-on Tools Corporation).

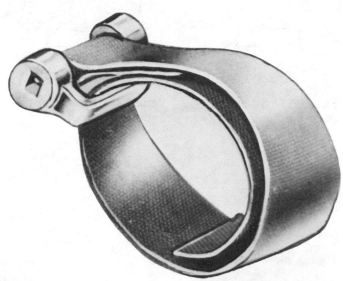

Figure 27. Strap wrench (Photo courtesy of Snap-on Tools Corporation).

The hand tap wrench (Figure 33) is used for medium and large size taps. The T-handle tap wrench (Figure 34) is used for small taps $\frac{1}{4}$-inch and under, as its more sensitive "feel" results in less tap breakage.

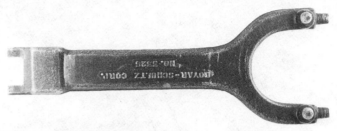

Figure 28. Fixed face spanner.

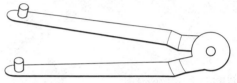

Figure 29. Adjustable face spanner.

Figure 30. Hook spanner.

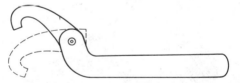

Figure 31. Adjustable hook spanner.

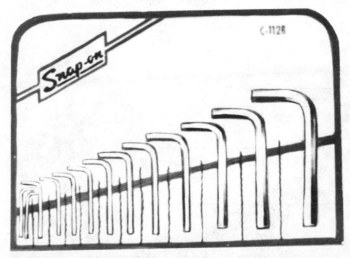

Figure 32. Socket head wrench set (Photo courtesy of Snap-on Tools Corporation).

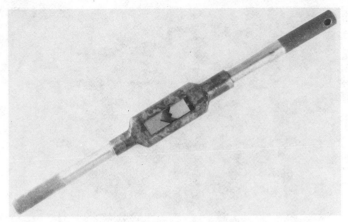

Figure 33. Hand tap wrench.

Here are safety hints for using wrenches:

1. Make sure the wrench you select fits properly. If it is a loose fit, it may round off the corners of the nut or bolt head.
2. Pull on a wrench instead of pushing to avoid injury.
3. Never use a wrench on moving machinery.
4. Do not hammer on a wrench or extend the handle for additional leverage. Use a larger wrench.

Chuck keys (Figure 35) are used to open and close chucks on drill presses and electric hand drills. Their size is determined by the size of the chuck. They should never be left in a chuck,

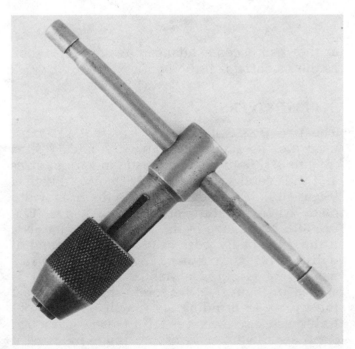

Figure 34. T-handle tap wrench.

Figure 35. Chuck key and drill chuck in use (Lane Community College).

as they can become a dangerous missile as soon as the machine is turned on.

SCREWDRIVERS

The two types of screwdrivers that are most used are the standard (Figure 36) and Phillips (Figure 37). Both types are made in various sizes and in several styles, straight, shank, and offset (Figure 38). It is important to use the right width blade when installing or removing screws (Figure 39). The shape of the tip is important also. If the tip is badly worn or incorrectly ground, it will tend to jump out of the slot. The correct method of grinding a standard screwdriver is shown in Figure 40. Be careful not to overheat the tip when grinding, as it will become soft. Never use a screwdriver for a chisel or pry bar. Keep a screwdriver in proper shape by using it only on the screws for which it was meant.

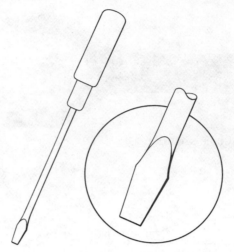

Figure 36. Screwdriver, standard.

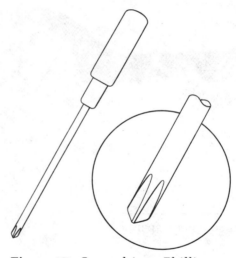

Figure 37. Screwdriver, Phillips.

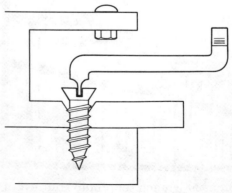

Figure 38. Offset screwdriver in use.

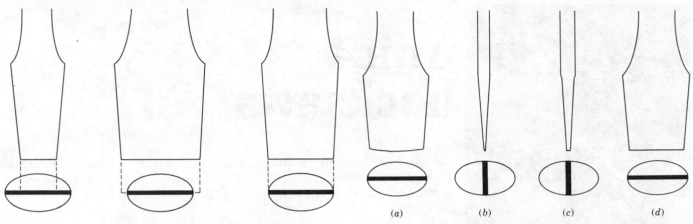

(a)   (b)   (c)   (a)   (b)   (c)   (d)

Figure 39. The proper width of a screwdriver blade: (a) Too narrow, (b) too wide, and (c) correct width.

Figure 40. The proper shape of the end of screwdriver blade. Blades (a) and (b) are badly worn; blades (c) and (d) are ground correctly.

self-evaluation

SELF-TEST

1. A four-inch machinist bench vise has jaws four inches on each side.

   True _____ False _____

2. What is the purpose of false jaws or jaw caps?

3. Parallel clamps are used for heavy duty clamping work, and C-clamps are used for holding precision setups. True _____ False _____

4. In order to remove a nut or bolt, slip joint or water pump pliers make a good substitute for a wrench when a wrench is not handy. True _____

   False _____

5. What advantage does the lever jawed wrench offer over other similar tools such as pliers?

6. Would you use a three-pound ball peen hammer for layout work? If not, what size do you think is right?

7. Some objects should never be struck with a hard hammer, a finished machine surface or the end of a shaft, for instance. What could you use to avoid damage?

8. A machine has a capscrew that needs to be tightened and released quite often. Which wrench would be best to use in this case, the adjustable or box-type wrench? Why?

9. Why should pipe wrenches never be used on bolts, nuts, or shafts?

10. What are two important things to remember about standard screwdrivers that will help you avoid problems in their use?

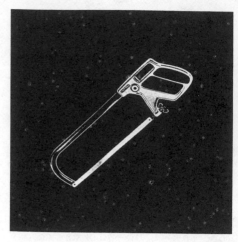

unit 4
hacksaws

Hacksaws are one of the more frequently used hand tools. The hand hacksaw is a relatively simple tool to use, but the facts and rules contained in this unit will help you improve your use of the hacksaw.

objective After completing this unit, you will be able to:
Identify, select, and use hand hacksaws.

The hacksaw consists of three parts: the frame, the handle, and the saw blade (Figure 1). Frames are either the solid or adjustable type. The solid frame can only be used with one length of saw blade. The adjustable frame can be used with hacksaw blades from 8 to 12 inches in length. The blade can be mounted to cut in line with the frame or at a right angle to the frame (Figures 2$\bar{a}$ and 2b).

Most hacksaw blades are made from high speed steel, and in standard lengths are 8, 10, and 12 inches. Blade length is the distance between the centers of the holes at each end. Hand hacksaw blades are generally $\frac{1}{2}$ inch wide and .025 inch thick. The kerf or cut produced by the hacksaw is wider than the .025 inch thickness of the blade because of the set of the teeth (Figure 3).

The set refers to the bending of teeth outward from the blade itself. Two kinds of set are found on hand hacksaw blades. First, with the straight or alternate set (Figure 4) one tooth is bent to the right and the next tooth to the left for the length of the blade. The second kind of set is the wavy set in which a number of teeth are gradually bent to the right and then to the left (Figure 5). A wavy set is found on most fine tooth hacksaw blades.

The spacing of the teeth on a hand hacksaw blade is called the pitch and is expressed in teeth per inch of length (Figure 6). Standard pitches are 14, 18, 24, and 32 teeth per inch,

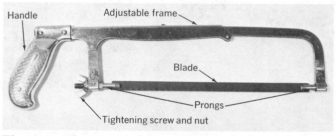

Figure 1. The parts of a hacksaw.

Figure 2a. Straight sawing with a hacksaw (Lane Community College).

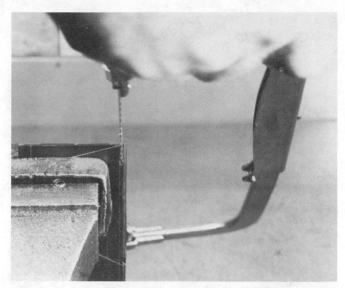

Figure 2*b*. Sawing with the blade set at 90° to the frame (Lane Community College).

with the 18 pitch blade used as a general purpose blade.

The hardness and size or thickness of a workpiece determine to a great extent which pitch blade to use. As a rule, you should use a coarse tooth blade on soft materials, to have sufficient clearance for the chips, and a fine tooth blade on harder materials. But you also should have at least three teeth cutting at any

Figure 3. The kerf is wider than the blade because of the set of the teeth (Lane Community College).

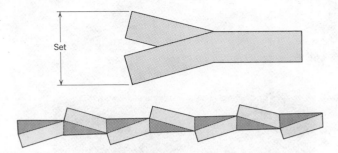

Figure 4. The straight (alternate) set.

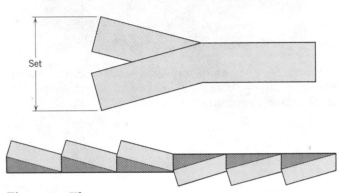

Figure 5. The wavy set.

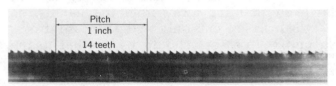

Figure 6. The pitch of the blade is expressed as the number of teeth per inch (Lane Community College).

time, which may require a fine tooth blade on soft materials with thin cross sections.

Hand hacksaw blades fall into two categories: soft-backed or flexible blades and all-hard blades. On the flexible blades only the teeth are hardened, the back being tough and flexible. The flexible blade is less likely to break when used in places that are difficult to get at such as in cutting off bolts on machinery. The all-hard blade is, as the name implies, hard and very brittle and should be used only where the workpiece can be rigidly supported, as in

Figure 7. A new blade must be started on the opposite side of the work, not in the same kerf as the old blade (Lane Community College).

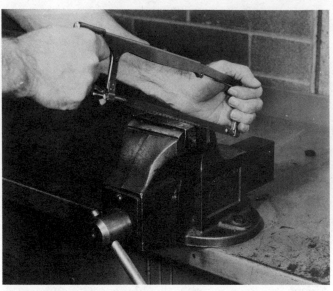

Figure 8. The workpiece is being sawed close to the vise to avoid vibration and chatter (Lane Community College).

a vise. On an all-hard blade even a slight twisting motion may break the blade. All-hard blades, in the hands of a skilled person, will cut true straight lines and give long service.

The blades are mounted in the frame with the teeth pointing away from the handle, so that the hacksaw cuts only on the forward stroke. No cutting pressure should be applied to the blade on the return stroke as this tends to dull the teeth. The sawing speed with a hacksaw should be from 40 to 60 strokes per minute. To get the maximum performance from a blade, make long, slow and steady strokes using the full length of the blade. Sufficient pressure should be maintained on the forward stroke to keep the teeth cutting. Teeth on a saw blade will dull rapidly if too little or too much pressure is put on the saw. The teeth will dull also if too fast a cutting stroke is used; a speed in excess of 60 strokes a minute will dull the blade because friction will overheat the teeth.

The saw blade may break if it is too loose in the frame or if the workpiece slips in the vise while sawing. Too much pressure may also cause the blade to break. A badly worn blade where the set has been worn down, will cut a too narrow kerf, which will cause binding and perhaps breakage of the blade. When this hap-

pens and a new blade is used to finish the cut, turn the workpiece over and start with the new blade from the opposite side and make a cut to meet the first one (Figure 7). The set on the new blade is wider than the old kerf. Forcing the new blade into an old cut will immediately ruin a new blade by wearing the set down.

A cut on a workpiece should be started with only light cutting pressure, with the thumb or fingers on one hand acting as a guide for the blade. Sometimes it helps to start a blade when a small vee-notch is filed in the workpiece. When a workpiece is supported in a vise, make sure that the cutting is done close to the vise jaws for a rigid setup free of chatter (Figure 8). Work should be positioned in a vise so that the saw cut is vertical. This makes it easier for the saw to follow a straight line. At the end of a saw cut, just before the pieces are completely parted, reduce the cutting pressure or you may be caught off balance when the pieces come apart and cut your hands on the sharp edges of the workpiece. To saw thin material, sandwich it between two pieces of wood for a straight cut. Avoid bending the saw blades because they are likely to break, and when they do, they usually shatter in all directions and could injure you or others nearby.

self-evaluation

SELF-TEST
1. What is the kerf?
2. What is the set on a saw blade?
3. What is the pitch of a hacksaw blade?
4. What determines the selection of a saw blade for a job?
5. Hand hacksaw blades fall into two basic categories. What are they?
6. What speed should be used in hand hacksawing?
7. Give four causes that make saw blades dull.
8. Give two reasons why hack saw blades break.
9. A new hacksaw blade should not be used in a cut started with a blade that has been used. Why?
10. What dangers exist when a hacksaw blade breaks while it is being used?

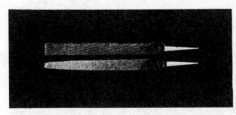

unit 5
files and off-hand grinding

Files are often used to put the finishing touches on a machined workpiece, either to remove burrs or sharp edges or as a final fitting operation. Intricate parts or shapes are often entirely produced by skilled craftsmen using files. In this unit you are introduced to the types and uses of files in metalworking.

Although really a machine tool, the pedestal grinder is used for many hand grinding operations, especially in sharpening and shaping tool bits. In this unit you will study the setup, use, and safety aspects of this important machine.

objective After completing this unit, you will be able to:

1. Identify eight common files and some of their uses.
2. Describe setup, use, and safety on the pedestal grinder.

TYPES OF FILES

Files are tools that anyone in metalwork will use. Often, through lack of knowledge, these tools are misused. Files are made in many different lengths ranging from 4 to 18 inches (Figure 1). Files are manufactured in many different **shapes** and are used for many specific purposes. Figure 2 shows the parts of a file. When a file is measured, the length is taken from the heel to the point, with the tang excluded. Most files are made from high carbon steel and are heat-treated to the correct hardness range. They are manufactured in four different cuts, single, double, curved tooth, and rasp. The single cut, double cut, and curved tooth are commonly encountered in machine shops (Figure 3). Rasps are usually used with wood. Curved tooth files will give excellent results with soft materials such as aluminum, brass, plastic, or lead.

Files also vary in their **coarseness**: rough, coarse, bastard, second cut, smooth, and dead smooth. The files most often used are the bastard, second cut, and smooth grades. Different sizes of files within the same coarseness designation will have varying sizes of teeth (Figure 4): the longer the file, the coarser the teeth. For maximum metal removal a double cut file is used. If the emphasis is on a smooth finish, a single cut file is recommended.

The face of most files is slightly convex (Figure 5) because they are made thicker in the middle than on the ends. Through this curva-

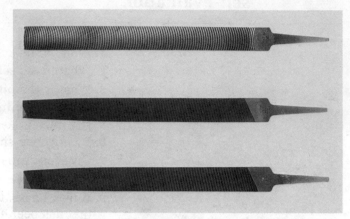

Figure 3. Three files that are frequently found in machine shops are the curved tooth, double cut, and single cut files. (DeAnza College).

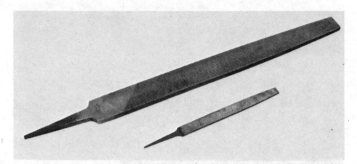

Figure 4. These two files are both bastard cut, but since they are of different lengths, they have different coarsenesses. (Lane Community College).

Figure 5. The edge of this file shows the convex face (Lane Community College).

Figure 1. Files are made in several different lengths (Lane Community College).

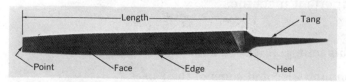

Figure 2. The parts of a file.

ture only some teeth are cutting at any one time, which makes them penetrate better. If the face were flat, it would be difficult to obtain an even surface because of the tendency to rock a file while filing. Some of this curvature is also offset by the pressure applied to make the file cut. New files do not cut as well as slightly used ones, since on new files some teeth are longer than most others and leave scratches on a work-piece.

Files are either blunt or tapered (Figure 6). A blunt file has the same cross-sectional area from heel to point, where a tapered file narrows toward the point.

Files fall into five basic categories: mill and saw files, machinists' files, Swiss pattern files, curved tooth files, and rasps. Machinists' and mill and saw files are classified as American pattern files. Mill files (Figure 7) were originally designed to sharpen large saws in lumber mills, but now they are used for draw filing, filing on a lathe (Figure 8), or filing a finish on a work-piece. Mill files are single cut and work well on brass and bronze. Mill files are slightly thinner than an equal-sized flat file, a machinists' file (Figure 9) that is usually double cut. Double cut files are used when fast cutting is needed. The finish produced is relatively rough.

Pillar files (Figure 10) have a narrower but

Figure 9. The flat file is usually a double cut file (DeAnza College).

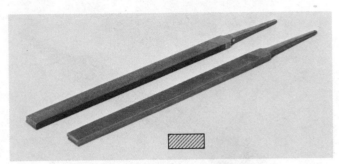

Figure 10. Two pillar files (Lane Community College).

Figure 11. The square file (Lane Community College).

Figure 6. Blunt and tapered file shapes (DeAnza College).

Figure 7. A mill file (DeAnza College).

Figure 8. The lathe file has a longer angle on the teeth to clear the chips when filing on the lathe (DeAnza College).

Figure 12. Warding file (Lane Community College).

thicker cross section than flat files. Pillar files are parallel in width and taper slightly in thickness. They also have one or two safe edges that allow filing into a corner without damaging the shoulder. Square files (Figure 11) usually are double cut and are used to file in keyways, slots, or holes.

If a very thin file is needed with a rectangular cross section, a warding file (Figure 12) is used. This file is often used by locksmiths when filing notches into locks and keys. Another file that will fit into narrow slots is a knife file (Figure 13). The included angle between the two faces of this file is approximately 10 degrees.

Three-square files (Figure 14), also called three-cornered files, are triangular in shape with the faces at 60-degree angles to each other. These files are used for filing internal angles between 60 and 90 degrees as well as to make sharp corners in square holes. Half-round files

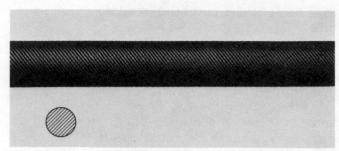

Figure 16. Round files are used to file a small radius or to enlarge a hole (DeAnza College).

(Figure 15) are available to file large internal curves. Half-round files, because of their tapered construction, can be used to file many different radii. Round files (Figure 16) are used to file small radii or to enlarge holes. These files are available in many diameter sizes.

Swiss pattern files (Figure 17) are manufactured to much closer tolerances than American pattern files, but are made in the same shapes. Swiss pattern files are more slender as they taper to finer points and their teeth extend to the extreme edges. Swiss pattern files range in length from 3 to 10 inches and their coarseness is indicated by numbers from 00 (coarse) to 6 (fine). Swiss pattern files are made with tangs to be used with file handles or as

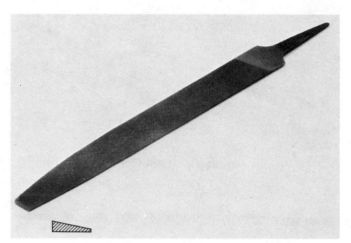

Figure 13. Knife file (Lane Community College).

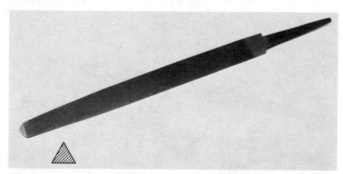

Figure 14. Three-square files are used for filing angles between 60° and 90° (Lane Community College).

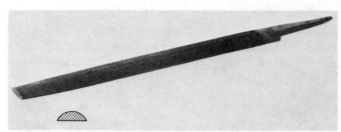

Figure 15. Half-round files are used for internal curves (Lane Community College).

Figure 17. A set of Swiss pattern files. Since these small files are very delicate and can be broken quite easily, great care must be exercised in their use (DeAnza College).

Figure 18. Die sinker's rifflers.

needle files with round or square handles that are part of the files. Another type of Swiss pattern files are die sinkers' rifflers (Figure 18). These files are double-ended with cutting surfaces on either end. Swiss pattern files are used primarily by tool and die makers, mold makers, and other craftsmen engaged in precision filing on delicate instruments.

Curved tooth files (Figure 19) cut very freely and remove material rapidly. The teeth on curved tooth files are all of equal height and the gullets or valleys between teeth are deep and provide sufficient room for the filings to curl and drop free. Curved tooth files are manufactured in three grades of cut: standard, fine, and smooth, and in length from 8 to 14 inches. These files are made as rigid tang types for use with a file handle, or as rigid or flexible blade types used with special handles. Curved tooth file shapes are flat, half-round, pillar, and square.

The bastard cut file (Figure 20) has a safe edge that is smooth. Flat filing may be done up to the shoulders of the workpiece without fear of damage. Files of other cuts and coarseness are also available with safe edges on one or both sides.

Thread files (Figure 21) are used to clean up and reshape damaged threads. They are square in cross section and have eight different thread pitches on each file. The thread file of the correct pitch is most effectively used when held or stroked against the thread while it is

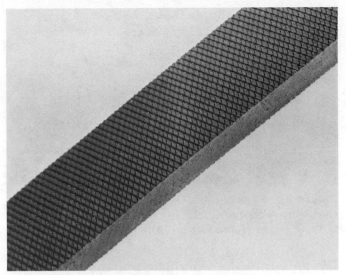

Figure 20. A file with a safe edge will not cut into shoulders or corners when filing is being done (De-Anza College).

rotating in a lathe. A thread can be repaired, however, even when it cannot be turned in a lathe.

CARE AND USE OF FILES

Files do an efficient job of cutting only while they are sharp. Files and their teeth are very hard and brittle. Do not use a file as a hammer or as a pry bar. When a file breaks, particles will fly quite a distance at high speed and may cause an injury. Files should be stored so that they are not in contact with any other file. The same applies to files on a workbench. Do not let files lie on top of each other because one file will break teeth on the other file (Figure 22). Teeth on files will also break if too much pressure is put on them while filing. On the other hand, if not enough pressure is applied while filing, the

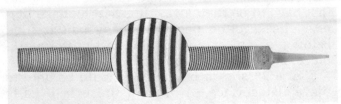

Figure 19. Curved tooth files are used on soft metals (DeAnza College).

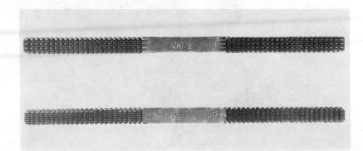

Figure 21. Thread files (Lane Community College).

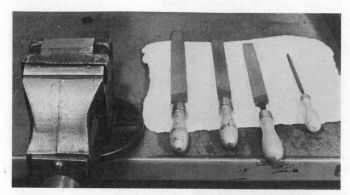

Figure 22. Files should be kept neatly arranged so that they will not strike each other and damage the cutting edges (Lane Community College).

file only rubs the workpiece and dulls the teeth. A dull file can be identified by its shiny, smooth teeth and by the way it slides over the work without cutting. Dulling of teeth is also caused by the filing of hard materials or when filing too fast. A good filing speed is 40 to 50 strokes per minute, but remember that the harder the material is, the slower the strokes should be; the softer the materials, the coarser the file should be.

Too much pressure or a new file may cause **pinning,** that is, filings wedged in the teeth; the result is deep scratches on the work surface. If the pins cannot be removed with a file card, try a piece of brass, copper, or mild steel and push it through the teeth (Figure 23). Do not use a scriber or other hard object for this operation. A file will not pin as much if some blackboard chalk is applied to the face (Figure 24). Never

Figure 23. Using a file card to clean a file (DeAnza College).

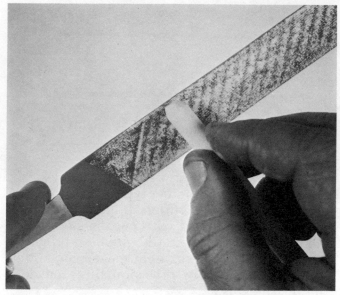

Figure 24. Using chalk on the file to help reduce pinning (DeAnza College).

use a file without a file handle or the pointed tang may cause a serious hand or wrist injury (Figures 25a and 25b).

Many filing operations are performed with the workpiece held in a vise. Clamp the workpiece securely, but remember to protect it from the serrated vise jaws with some soft piece of material such as copper, brass, wood, or paper (Figure 26). The workpiece should extend out of the vise so that the file clears the vise jaws by $\frac{1}{8}$ to $\frac{1}{4}$ inch. Since a file cuts only on the forward stroke, no pressure should be applied on the return stroke. Letting the file drag over the workpiece on the return stroke helps release the small chips so that they can fall from the file. Use a stroke as long as possible; this will make the file wear out evenly instead of just in the middle. To file a flat surface, change the direction of the strokes frequently to produce a cross-hatch pattern (Figure 27). By using a straightedge steel rule to test for flatness, we can easily determine where the high spots are that have to be filed away. It is best to make flatness checks often because, if any part is filed below a given layout line, the rest of the workpiece may have to be brought down just as far.

Figure 28 shows how a file should be held to file a flat surface. A smooth finish is usually obtained by draw filing (Figure 29), where a single cut file is held with both hands and drawn back and forth on a workpiece. The file should

Figure 25*a*. A file should never be used without a file handle (DeAnza College).

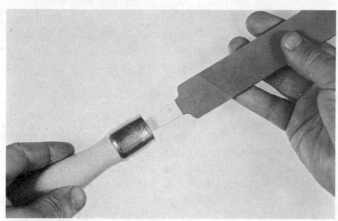

Figure 25*b*. This style of handle is designed to screw on rather than be driven on the tang (DeAnza College).

Figure 26. Workpiece in vise with protective jaws. The work extends only $\frac{1}{4}$ inch from the vise jaws for better rigidity (Lane Community College).

Figure 27. The cross-hatch pattern shows that this piece has been filed from two directions thus producing a flatter surface (Lane Community College).

Figure 28. Proper filing position (Lane Community College).

not be pushed over the ends of the workpiece as this would leave rounded edges. To get a smooth finish it sometimes helps to hold the file as shown in Figure 30, making only short strokes. The pressure is applied by a few fingers and does not extend over the ends of the workpiece. When a round file or half-round file is used, the forward stroke should also include a clockwise rotation for deeper cuts and a smoother finish. A tendency of people who are filing is to run their hands or fingers over a newly filed surface. This deposits a thin coat of skin oil

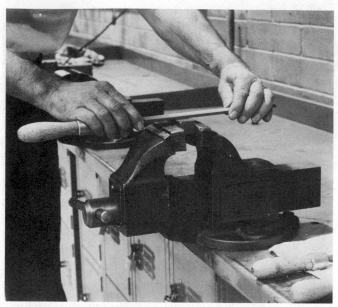

Figure 29. Draw filing (Lane Community College).

Figure 30. Use this procedure to correct high spots on curvatures on the workpiece. Apply pressure with short strokes only where cutting is needed (Lane Community College).

on the surface. When filing is resumed, the file will not cut for several strokes, but will only slip over the surface causing the file to dull more quickly.

OFF-HAND GRINDING ON PEDESTAL GRINDERS

The **pedestal grinder** is really a machine tool. However, since the workpiece is hand held, it is more logical to discuss this machine in conjunction with cutting hand tools. Furthermore, you must be familiar with the pedestal grinder, as you will be using it very early in your study of machine tool practices.

The pedestal grinder gets its name from the floor stand or pedestal that supports the motor and abrasive wheels. The pedestal grinder is a common machine tool that you will use almost daily in the machine shop. This grinding machine is used for general purpose, **off-hand grinding** where the workpiece is hand held and applied to the rapidly rotating abrasive wheel. One of the primary functions of the pedestal grinder is the shaping and sharpening of tool bits and drills in machine shop work. Pedestal grinders are often modified for use with rotary wire brushes or buffing wheels.

Setup of the Pedestal Grinder

The pedestal grinder in your shop stands ready for use most of the time. If it becomes necessary to replace a worn wheel, the side of the guard must be removed and the tool rest moved out of the way. A piece of wood may be used to prevent the wheel from rotating so that the spindle nut can be turned and removed (Figure 31). Remember that the **left side** of the spindle has **left-handed** threads, while those on the **right side** are **right-handed.**

A new wheel should be **ring tested** to determine if there are any cracks or imperfections (Figure 32). Gently tap the wheel near its rim with a screwdriver handle or a piece of wood and listen for a clear ringing sound like a bell. A clear ring indicates a sound wheel that is safe to use, but if a dull thud is heard, the wheel may be cracked and **should not be used.** The flanges and the spindle should be clean before mounting the wheel. Be sure that the center hole in the wheel is the correct size for the grinder spindle. If a bushing must be used, be sure that it is the correct size and properly installed (Figure 33). Place a clean, undamaged blotter on each side between wheel and flanges. The spindle nut should be tightened just enough to hold the wheel firmly. **Excessive tightening will break the wheel.**

After the guard and cover plate have been replaced, the tool rest should be brought up to the wheel so that between $\frac{1}{16}$ and $\frac{1}{8}$ **inch clearance**

Figure 31. Using a piece of wood to hold the wheel while removing the spindle nut (Lane Community College).

Figure 32. The ring test is made before mounting the wheel (Lane Community College).

Figure 33. The wheel is mounted with the proper bushing in place (Lane Community College).

exists between the rest and wheel (Figure 34.) If there is excessive space between the tool rest and the wheel, a small workpiece, such as a tool bit that is being ground, may flip up and catch between the wheel and tool rest. Your finger may be caught between workpiece and grinding wheel resulting in a serious injury. The **clearance** between the tool rest and wheel should **never exceed $\frac{1}{8}$ inch.** The **spark guard,** located on the upper side of the wheel guard (Figure 35), should be adjusted to **within $\frac{1}{16}$ inch** of the wheel. This protects the operator if the wheel should shatter.

Using the Pedestal Grinder

Stand aside out of line with the rotation of the grinding wheel and turn on the grinder. **Let the wheel run idle for one full minute.** A new wheel does not always run exactly true and therefore must be dressed (Figure 36). A Desmond **dresser** may be used to true the face of the wheel. Pedestal grinder wheels often become grooved, out of round, glazed, or misshapen, and must be frequently dressed to obtain proper grinding results.

Bring the workpiece into contact with the wheel gently without bumping. Grind only **on the face** of the wheel. The workpiece will heat from friction during the grinding operation. It may become too hot to hold in just a few seconds. To prevent this, cool the workpiece in the water pot attached to the grinder. Be especially careful when grinding drills and tool bits so that they do not become overheated. Excessive heat may permanently affect tool steel metallurgical properties.

Safety on the Pedestal Grinder

Always wear appropriate **eye protection** when dressing wheels or grinding on the pedestal grinder. Be sure that grinding wheels are rated at the proper speed for the grinder that you are using. The safety shields, wheel guards, and spark guard must be kept in place at all times while grinding. The tool rest must be adjusted and the setting corrected as the diameter of the wheel decreases from use. Grinding wheels and rotary wire brushes **may catch loose clothing or long hair** (Figures 37a and 37b). Long hair should be contained in an industrial type hair net. Wire wheels **often throw out** small pieces of wire at high velocities.

Nonferrous metals such as aluminum and brass should never be ground on the aluminum oxide wheels found on most pedestal grinders. These metals fill the voids or spaces between the abrasive particles in the grinding wheel so that more pressure is needed to accomplish the desired grinding. This additional **pressure** sometimes causes the **wheel to break or shatter.** Pieces of grinding wheel may be thrown out of the machine at extreme velocities. Always use **silicon carbide** abrasive wheels **for grinding nonferrous metals.** Excessive pressure should **never** be

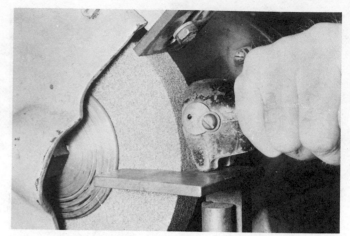

Figure 36. The wheel is being dressed (Lane Community College).

used in any grinding operation. If this seems to be necessary, the improper grit or grade of abrasive is being used, or the wheel is glazed and needs to be dressed. Always use the **correct** abrasive **grit** and **grade** for the particular grinding that you are doing.

Figure 34. The tool rest is adjusted (Lane Community College).

Figure 35. The spark guard is adjusted (Lane Community College).

Figure 37a. A wire wheel on this grinder caught the loose hair of the operator and instantly pulled it from the operator's head.

Figure 37b. The force and speed of this action was such that the operator's head was jerked suddenly into the guard. Note that the cast aluminum guard was shattered as a result of the impact.

self-evaluation

SELF-TEST 1. How is a file identified?

2. What are the four different cuts found on files?

3. Name four coarseness designations for files.

4. Which of the two kinds of files, single cut or double cut, is designed to remove more material?

5. Why are the faces of most files slightly convex?

6. What difference is there between a blunt and a tapered file?

7. What difference exists between a mill file and an equal-sized flat file?

8. What is a warding file?

9. An American pattern file differs in what way from a Swiss pattern file?

10. What are the coarseness designations for needle files?

11. Why should files be stored so they do not touch each other?

12. What happens if too much pressure is applied when filing?

13. What causes a file to get dull?

14. Why should a handle be used on a file?

15. Why should workpieces be measured often?

16. What happens when a surface being filed is touched with the hand or fingers?
17. How does the hardness of a workpiece affect the selection of a file?
18. How can rounded edges be avoided when a workpiece is drawfiled?
19. Should pressure be applied to a file on the return stroke?
20. Why is a round file rotated while it is being used?

unit 6
hand reamers

Holes produced by drilling are seldom accurate in size and often have rough surfaces. A reamer is used to finish a hole to an exact dimension with a smooth finish. This unit will describe some commonly used hand reamers and how they are used.

objectives   After completing this unit, you will be able to:
1. Identify at least five types of hand reamers.
2. Hand ream a hole to a specified size.

Hand reamers are often used to finish a previously drilled hole to an exact dimension and a smooth surface. When parts of machine tools are aligned and fastened with cap screws or bolts, the final operation is often the hand reaming of a hole in which a dowel pin is placed to maintain the alignment. Hand reamers are designed to remove only a small amount of material from a hole—usually from .001 to .005 in. These tools are made from high carbon or high speed steel.

FEATURES OF HAND REAMERS

Figure 1 shows the major features of the most common design of hand reamer. Another design is available with a pilot ahead of the starting taper (see *Machinery's Handbook* for details). The square on the end of the shank permits the

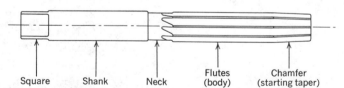

Square   Shank   Neck   Flutes (body)   Chamfer (starting taper)

Figure 1. Major features of the hand reamer.

clamping of a tap wrench or T-handle wrench to provide the driving torque for reaming. The diameter of this square is between .004 and .008 in. smaller than the reamer size, and the shank of the reamer is between .001 and .006 in. smaller, to guide the reamer and to permit it to pass through a reamed hole without marring it. It is very important that these tools *not* be put into a drill chuck, because a burred shank can ruin a reamed hole as the shank is passed through it.

Hand reamers have a long starting taper that is usually as long as the diameter of the reamer, but may be as long as one third of the fluted body. This starting taper is usually very slight and may not be apparent at a casual glance. Hand reamers do their cutting on this tapered portion. The gentle taper and length of the taper help to start the reamer straight and keep it aligned in the hole.

Details of the cutting end of the hand reamer are shown in Figure 2. The full diameter or actual size of the hand reamer is measured where the starting taper ends and the margin of the land appears. The diameter of the reamer should only be measured at this junction, as the hand reamer is generally back tapered or reduced in outside diameter by about .0005 to .001 inch per inch of length toward the shank. This back tapering is done to reduce tool contact with the workpiece. When hand reamers become dull, they are resharpened at the starting taper, using a tool and cutter grinder.

The function of the hand reamer is like that of a scraper, rather than an aggressive cutting tool like most drills and machine reamers. For this reason hand reamers typically have zero or negative radial rake on the cutting face, rather than the positive radial rake characteristic of most machine reamers. (See Section G Unit 7, "Reaming in the Drill Press"). The right-hand cut with a left-hand spiral or helix is considered standard for hand reamers. The left-hand spiral produces a negative axial rake for the tool, which contributes to a smooth cutting action.

Figure 3. Straight flute hand reamer (Courtesy of Whitman and Barnes Division, TRW, Inc.).

Figure 4. Helical flute hand reamer (Courtesy of Whitman and Barnes Division, TRW, Inc.).

Figure 5. Straight flute expansion hand reamer (Courtesy of Whitman and Barnes Division, TRW, Inc.).

Most reamers, hand or machine types, are also incrementally cut, which means that the flutes or body channels are not precisely of uniform spacing. The difference is very small, only a degree or two, but it tends to reduce chatter by reducing harmonic effects between cutting edges. Harmonic chatter is especially a problem with adjustable hand reamers, which often leave a tooth pattern in the work.

Hand reamers are made with straight flutes (Figure 3) or with helical flutes (Figure 4). Most hand reamers are manufactured with a right-hand cut, which means they will cut when rotated in a clockwise direction. Helical or spiral fluted reamers are available with a right-hand helix or a left-hand helix. Helical flute reamers are especially useful when reaming a hole having keyways or grooves cut into it, as the helical flutes tend to bridge the gaps and reduce binding or chattering.

Hand reamers for cylindrical holes are made as solid (Figures 3 and 4) or as expansion types (Figure 5). Expansion reamers are designed for use where it is necessary to enlarge a hole slightly for proper fit such as in maintenance applications. These reamers have an adjusting screw that allows limited expansion to an exact size. The maximum expansion of these reamers is approximately .006 in. for diameters up to $\frac{1}{2}$ in., .010 in. for diameters between $\frac{1}{2}$ and 1 in., and .012 in. for diameters for between 1 and $1\frac{1}{2}$ in. These tools are frequently broken by attempts to expand them beyond these limits.

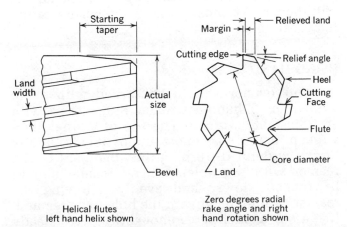

Figure 2. Functional details of the hand reamer (Courtesy of Bendix Industrial Tools Division).

Helical flute expansion reamers are especially adapted for the reaming of bushings or holes having a keyway or straight grooves because of their bridging and shearing cutting action. Expansion reamers have a slightly undersized pilot on the end that guides the reamer and helps to keep it in alignment.

The adjustable hand reamer (Figure 6) is different from the expansion reamer in that it has inserted blades. These cutting blades fit into tapered slots in the body of the reamer and are held in place by two locking nuts. The blades have a taper corresponding to the taper of the slots, which keeps them parallel at any setting. Adjustments in reamer size are made by loosening one nut while tightening the other. Adjustable hand reamers are available in diameters from $\frac{1}{4}$ to 3 in. The adjustment range varies from $\frac{1}{32}$ in. on the smaller diameter reamers to $\frac{5}{16}$ in. on the larger size reamers. Only a small amount of material should be removed at one time, as too large a cut will usually cause chatter.

Taper pin reamers (Figures 7 and 8) are used for reaming holes for standard taper pins used in the assembly of machine tools and other parts. Taper pin reamers have a taper of $\frac{1}{4}$ in. per foot of length, and are manufactured in 18 different sizes numbered from 8/0 to 0 and on up to size 10. The smallest size, number 8/0, has a large end diameter of .0514 in. and the largest reamer, a number 10, has a large end diameter of .7216 in. The sizes of these reamers are designed to allow the small end of each reamer to enter a hole reamed by the next smaller size reamer. As with other hand reamers, the helical flute reamer will cut with more shearing action and less chattering, especially on interrupted cuts.

Morse taper socket reamers are designed to produce holes for American Standard Morse

Figure 7. Straight flute taper pin hand reamer (Courtesy of Whitman and Barnes Division, TRW, Inc.).

Figure 8. Spiral flute taper pin hand reamer (Courtesy of Whitman and Barnes Division, TRW, Inc.).

Figure 9. Morse taper socket roughing reamer (Courtesy of Whitman and Barnes Division, TRW, Inc.).

Figure 10. Morse taper socket finishing reamer (Courtesy of Whitman and Barnes Division, TRW, Inc.).

taper shank tools. These reamers are available as roughing reamers (Figure 9) and as finishing reamers (Figure 10). The roughing reamer has notches ground at intervals along the cutting edges. These notches act as chip-breakers and make the tool more efficient at the expense of fine finish. The finishing reamer is used to impart the final size and finish to the socket. Morse taper socket reamers are made in sizes from number 0, with a large end diameter of .356 in. to number 5, with a large end diameter of 1.8005 in. There are two larger Morse tapers, but they are typically sized by boring rather than reaming.

USING HAND REAMERS

A hand reamer should be turned with a tap wrench or T-handle wrench rather than with an adjustable wrench. The use of a single end wrench makes it almost impossible to apply torque without disturbing the alignment of the reamer with the hole. A hand reamer should be rotated slowly and evenly, allowing the reamer to align itself with the hole to be reamed. Use a tap wrench large enough to give a steady torque and to prevent vibration and chatter. Use a steady and large feed; feeds up to one-

Figure 6. Adjustable hand reamers. The lower reamer is equipped with a pilot and tapered guide bushing for reaming in alignment with a second hole (Lane Community College).

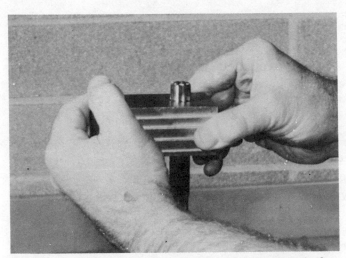

Figure 11. Hand reaming a small workpiece with the reamer held in a vise (Lane Community College).

quarter of the reamer diameter per revolution can be used. Small and lightweight workpieces can be reamed by fastening the reamer vertically in a bench vise and rotating the work over the reamer by hand (Figure 11).

In all hand reaming with solid, expansion, or adjustable reamers, never rotate the reamer backwards to remove it from the hole, as this will dull it rapidly. If possible, pass the reamer through the hole and remove it from the far side without stopping the forward rotation. If this is not possible, it should be withdrawn while maintaining the forward rotation.

The preferred stock allowance for hand reaming is between .001 and .005 in. Reaming more material than this would make it very difficult to force the reamer through the workpiece. Reaming too little, on the other hand, results in excessive tool wear because it forces the reamer to work in the zone of material work-hardened during the drilling operation. This stock allowance does not apply to taper reamers, for which a hole has to be drilled at least as large as the small diameter of the reamer. The hole size for a taper pin is determined by the taper pin number and its length. This data can be found in machinist's handbooks.

Since cylindrical hand reaming is restricted to small stock allowances, it is most important that you be able to drill a hole of predictable size and of a surface finish that will assure a finished cleanup cut by the reamer. It is a good idea to drill a test hole in a piece of scrap of similar composition and carefully measure it

both for size and for an enlarged or bell-mouth entrance. You may find it necessary to drill a slightly smaller hole before drilling the correct reaming size to assure a more accurate hole size. Carefully spot drill the location before drilling the hole in your actual workpiece. The hole should then be lightly chamfered with a countersinking tool to remove burrs and to promote better reamer alignment.

The use of a cutting oil also improves the cutting action and the surface finish when reaming most metals. Exceptions are cast iron and brass, which should be reamed dry.

When a hand reamer is started it should be checked for squareness on two sides of the reamer, 90 degrees apart. Another way to assure alignment of the reamer with the drilled hole is to use the drill press as a reaming fixture. Put a piece of cylindrical stock with a 60 degree center in the drill chuck (Figure 12) and use it to guide and follow the squared end of the reamer as you turn the tool with the tap wrench. Be sure to plan ahead so that you can drill, countersink, and ream the hole without moving the table or head of the drill press between operations.

On deep holes, or especially on holes reamed with taper reamers, it becomes necessary to remove the chips frequently from the reamer flutes to prevent clogging. Remove these chips with a brush to avoid cutting your hands.

Figure 12. Using the drill press as a reaming fixture (Lane Community College).

Reamers should be stored so they do not contact one another to avoid burrs on the tools that can damage a hole being reamed. They should be kept in their original shipping tubes or set up in a tool stand. Always check reamers for burrs or for pickup of previous material before you use them. Otherwise the reamed hole can be oversized or marred with a rough finish.

self-evaluation

SELF-TEST
1. How is a hand reamer identified?
2. What is the purpose of a starting taper on a reamer?
3. What is the advantage of a spiral flute reamer over a straight flute reamer?
4. How does the shank diameter of a hand reamer compare with the diameter measured over the margins?
5. When are expansion reamers used?
6. What is the difference between an expansion and an adjustable reamer?
7. What is the purpose of coolant used while reaming?
8. Why should reamers not be rotated backwards?
9. How much reaming allowance is left for hand reaming?

unit 7
identification and uses of taps

Most internal threads produced today are made with taps. These taps are available in a variety of styles, each one designed to perform a specific type of tapping operation efficiently. This unit will help you identify and select taps for threading operations.

objectives After completing this unit, you will be able to:
1. Identify common taps.
2. Select taps for specific applications.

IDENTIFYING COMMON TAP FEATURES

Taps are used to cut internal threads into holes. This process is called tapping. Tap features are illustrated in Figures 1 and 2. The active cutting part of the tap is the chamfer, which is produced by grinding away the tooth form at an angle, with relief back of the cutting edge, so that the cutting action is distributed progressively over a number of teeth. The fluted portion of the tap provides a space for chips to accumulate and for the passage of cutting fluids. Two, three, and four flute taps are common.

The major diameter (Figure 2) is the outside diameter of the tool as measured over the thread crests at the first full thread behind the chamfer. This is the largest diameter of the cutting portion of the tap, as most taps are back tapered or reduced slightly in thread diameter toward the shank. This back taper reduces the amount of tool contact with the thread during the tapping process, hence making the tap easier to turn.

The pitch diameter (Figure 2) is the diameter of an imaginary cylinder where the width of the spaces and the width of the threads are equal. The pitch of the thread is the distance between a point on one thread and the same point on the next thread.

Taps are made from either high carbon steel or high speed steel and have a hardness of about Rockwell C63. High speed steel taps are far more common in manufacturing plants than carbon steel taps. High speed steel taps typically are ground after heat treatment to ensure accurate thread geometry.

Another identifying characteristic of taps is the amount of chamfer at the cutting end of a tap (Figure 3). A set consists of three taps, taper, plug, and bottoming taps, which are identical except for the number of chamfered threads. The taper tap is useful in starting a tapped thread square with the part. The most

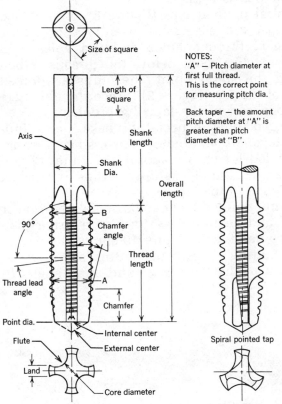

Figure 1. General tap terms (Courtesy of Bendix Industrial Tools Division).

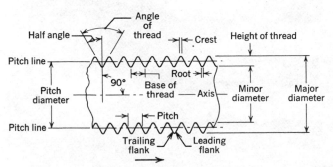

Figure 2. Detailed tap terms (Courtesy of Bendix Industrial Tools Division).

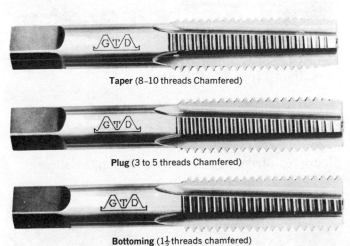

Taper (8–10 threads Chamfered)

Plug (3 to 5 threads Chamfered)

Bottoming (1½ threads chamfered)

Figure 3. Chamfer designations for cutting taps (Photo courtesy of TRW Greenfield Tap & Die Division).

commonly used tap, both in hand and machine tapping, is a plug tap. Bottoming taps are used to produce threads that extend almost to the bottom of a blind hole. A blind hole is one that is not drilled clear through a part.

Serial taps (Figure 4) are also made in sets of three taps for any given size of tap. Each of these taps has one, two, or three rings cut on the shank near the square. The number 1 tap has smaller major and pitch diameters and is used for rough cutting the thread. The number 2 tap cuts the thread slightly deeper, and the number 3 tap finishes it to size. Serial taps are used when tough metals are to be tapped by hand. Another tap used for tough metal such as stainless steel is the interrupted thread tap (Figure 5). This tap has alternate teeth removed to reduce tapping friction.

Figure 6 shows the identifying markings of a tap, where $\frac{5}{8}$ in. is the nominal size, 11 is the number of threads per inch, and NC refers to the standardized National Coarse thread series. G is the symbol used for ground taps.

H3 identifies the tolerance range of the tap. HS means that the tap material is high speed steel. Left-handed taps will also be identified by an LH or left-hand marking on the shank.

OTHER KINDS AND USES OF TAPS

Figure 7 illustrates the most commonly used kind of tap, the hand tap. The hand tap is manufactured to produce threads in both machine screw sizes and fractional sizes. Hand taps were originally intended for hand tapping of threads, but are now commonly used in machine production jobs as well.

Spiral pointed taps (Figure 8), often called gun taps, are especially useful for machine tapping of through holes or blind holes with sufficient chip room below the threads. When turning the spiral point, the chips are forced ahead of the tap (Figure 9). Since the chips are pushed ahead of the tap, the problems caused by clogged flutes, especially breakage and dulling of taps, are eliminated. Also, since they are not needed for chip disposal, the flutes of gun taps can be made shallower, thus increasing the strength of the tap.

Spiral pointed taps can be operated at higher speeds and require less torque to drive than ordinary hand taps. Figure 10 shows the design of the cutting edges. The cutting edges (A) at the point of the tap are ground at an angle (B)

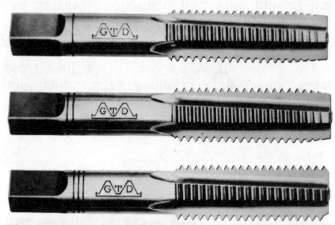

Figure 4. Set of serial taps (Photo courtesy of TRW Greenfield Tap & Die Division).

Figure 5. Interrupted thread tap (DeAnza College).

Figure 6. Identifying marking on a tap.

Figure 7. Hand tap.

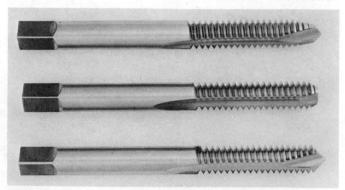

Figure 8. Set of spiral pointed (or gun) taps (DeAnza College).

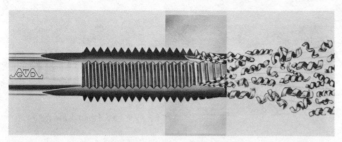

Figure 9. Cutting action of spiral pointed taps (Photo courtesy of TRW Greenfield Tap & Die Division).

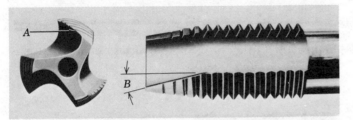

Figure 10. Detail of spiral pointed tap (Photo courtesy of TRW Greenfield Tap & Die Division).

to the axis. Fluteless spiral pointed taps (Figure 11) are recommended for production tapping of through holes in sections no thicker than the tap diameter. This type of tap is very strong and rigid, which reduces tap breakage caused by misalignment. Fluteless spiral point taps give excellent results when tapping soft and stringy materials or sheet metal.

Spiral fluted taps are made with helical instead of straight flutes (Figure 12), which

Figure 11. Fluteless spiral pointed tap for thin materials (DeAnza College).

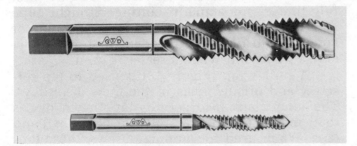

Figure 12. Spiral fluted taps—regular spiral (Photo courtesy of TRW Greenfield Tap & Die Division).

draw the chips out of the hole. This kind of tap is also used when tapping a hole that has a keyway or spline as the helical lands of the tap will bridge the interruptions. Spiral fluted taps are recommended for tapping deep blind holes in ductile materials such as aluminum, magnesium, brass, copper, and die-cast metals. Fast spiral fluted taps (Figure 13) are similar to regular spiral fluted taps, but the faster spiral flutes increase the chip lifting action and permit the spanning of comparably wider spaces.

Thread forming taps (Figure 14) are fluteless and do not cut threads in the same manner as conventional taps. They are forming tools and

Figure 13. Spiral fluted tap—fast spiral. The action of the tap lifts the chips out of hole to prevent binding (Photo courtesy of TRW Greenfield Tap & Die Division).

Figure 14. Fluteless thread forming tap (DeAnza College).

their action can be compared with external thread rolling. On ductile materials such as aluminum, brass, copper, die castings, lead, and leaded steels these taps give excellent results. Thread forming taps are held and driven just as are conventional taps, but because they do not cut the threads no chips are produced. Problems of chip congestion and removal often associated with the tapping of blind holes are eliminated. Figure 15 shows how the forming tap displaces metal. The crests of the thread that are at the minor diameter may not be flat but will be slightly concave because of the flow of the displaced metal. Threads produced in this manner have improved surface finish and increased strength because of the cold working of the metal. The size of the hole to be tapped must be closely controlled, since too large a hole will result in a poor thread form and too small a hole will result in the breaking of the tap.

A tapered pipe tap (Figure 16) is used to tap holes with a $\frac{3}{4}$ in. per foot taper for pipes with a matching thread and to produce a leakproof fit. The nominal size of a pipe tap is that of the pipe fitting and not the actual size of the tap.

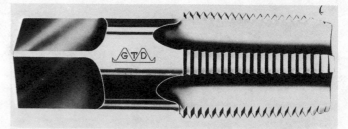

Figure 17. Straight pipe tap (Photo courtesy of TRW Greenfield Tap & Die Division).

Figure 18. Pulley tap (Photo courtesy of TRW Greenfield Tap & Die Division).

Figure 19. Nut tap (Photo courtesy of TRW Greenfield Tap & Die Division).

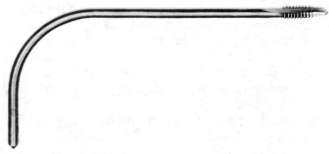

Figure 20. Bent shank tapper tap (Photo courtesy of TRW Greenfield Tap & Die Division).

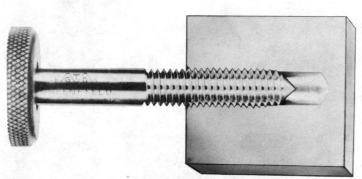

Figure 15. The thread forming action of a fluteless thread forming tap (Photo courtesy of TRW Greenfield Tap & Die Division).

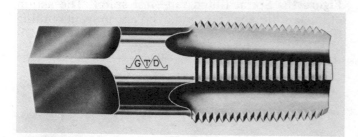

Figure 16. Taper pipe tap (Photo courtesy of TRW Greenfield Tap & Die Division).

When tapping taper pipe threads, every tooth of the tap engaged with the work is cutting until the rotation is stopped. This takes much more torque than does the tapping of a straight thread in which only the chamfered end and the first full thread are actually cutting. Straight pipe taps (Figure 17) are used for tapping holes or couplings to fit taper threaded pipe and to secure a tight joint when a sealer is used.

A pulley tap (Figure 18) is used to tap set-screw and oilcup holes in the hubs of pulleys. The long shank also permits tapping in places that might be inaccessible for regular hand taps. When used for tapping pulleys, these taps are inserted through holes in the rims, which are slightly larger than the shanks of the taps.

These holes serve to guide the taps and assure proper alignment with the holes to be tapped.

Nut taps (Figure 19) differ from pulley taps in that their shank diameters are smaller than the root diameter of the thread. The smaller shank diameter makes the tapping of deep holes possible. Nut taps are used when small quantities of nuts are made or when nuts have to be made from tough materials such as some stainless steels or similar alloys. Bent shank tapper taps (Figure 20) are designed for the mass production of nuts in automatic tapping machines. Tapping in these machines is continuous; nuts are automatically fed to the tap by a hopper and after the nut has been tapped it passes on over the shank and is automatically ejected. Automatic nut tapping and tapping of irregularly shaped small parts is also done in machines, as the one shown in Figure 21. In this type of machine the tap reverses to clear the workpiece.

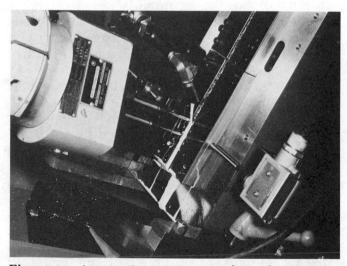

Figure 21. Automatic tapping machine for tapping nuts and irregularly shaped small parts.

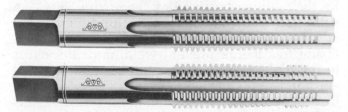

Figure 22. Set of Acme thread taps. The upper tap is used for roughing, the lower tap for finishing (Photo courtesy of TRW Greenfield Tap & Die Division).

Figure 23. A tandem Acme tap designed to rough and finish cut the thread in one pass (Lane Community College).

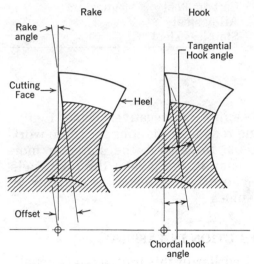

Figure 24. Rake and hook angles on cutting taps (Courtesy of Bendix Industrial Tools Division).

Figure 22 shows Acme taps for roughing and finishing. Acme threads are used to provide accurate movement such as lead screws on machine tools and for applying pressure in various mechanisms. On some Acme taps the roughing and finishing operation is performed with one tap (Figure 23). The length of this tap usually requires a through hole.

RAKE AND HOOK ANGLES ON CUTTING EDGES

When selecting a tap for the most efficient cutting, the cutting face geometry will be an important factor. It should vary depending on the material to be tapped. Cutting face geometry is expressed in terms of rake and hook (Figure 24). The rake of a tap is the angle between a line through the flat cutting face and a radial line from the center of the tool to the tooth tip. The rake can be negative, neutral, or positive. Hook angle, on the other hand, relates to the concavity of the cutting face. It is defined by the intersection of the radial line with the tangent line through the tooth tip (tangential hook) or by the average angle of the tooth face from crest to root (chordal). Unlike the rake angle,

Table 1
Recommended Tap Rake Angles

0-5 Degrees	8-12 Degrees	16-20 Degrees
Bakelite	Bronze	Aluminum and alloys
Plastics	Hard rubber	Zinc die castings
Cast iron	Cast steel	Copper
Brass	Carbon steel	Magnesium
Hard rubber	Alloy steel	
	Stainless steel	

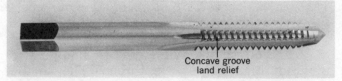

Figure 26. Tap with concave groove land relief (De-Anza College).

hook angle cannot be negative. Table 1 gives the rake angle recommendations for some work-piece materials. In general, the softer or more ductile the material, the greater the rake angle. Harder and more brittle materials call for reduced rake angles.

REDUCING FRICTION IN TAPPING

As discussed earlier in this unit, a tap is usually back tapered along the thread to relieve the friction between the tool and the workpiece. There is another form of relief often applied to taps with the same results. When the fully threaded portion of the tap is cylindrical (other than back taper), it is called a concentric thread (Figure 25). If the pitch diameter of the fully threaded portion of the tap is brought uniformly closer to the axis of the tap as measured from face to back (heel), it has eccentric relief. This means less tool contact with the workpiece and less friction. A third form of friction relief combines the concentric thread and the eccentric thread relief, and is termed con-eccentric. The concentric margin gives substantial guidance and the relief following the margin reduces friction. Relief is also provided behind the chamfer of the tap to provide radial clearance for the cutting edge. Relief may also be provided in the form of a channel that runs lengthwise down the center of the land (Figure 26), termed a concave groove land relief.

Other steps may also be taken to reduce friction and to increase tap life. Surface treatment of taps is often an answer if poor thread forming or tap breakage is caused by chips adhering to the flutes or welding to the cutting faces. These treatments generally improve the wear life of taps by increasing their abrasion resistance.

Three different kinds of surface treatments are used by tap manufacturers. Liquid nitride produces a very hard shallow surface on high speed steel tools when these tools are immersed in cyanide salts at closely controlled temperatures. Oxide finishes are usually applied in steam tempering furnaces and can be identified by their bluish-black color. The oxide acts as a solid lubricant. It also holds liquid lubricant at the cutting edges during a tapping operation. Oxide treatments prevent welding of the chips to the tool and reduce friction between the tool and the work. Chrome plating is a very effective treatment for taps used on nonferrous metals and some soft steels. The chromium deposit is very shallow and often referred to as flash chrome plating.

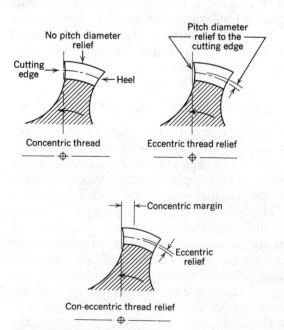

Figure 25. Pitch diameter relief forms on taps (Courtesy of Bendix Industrial Tools Division).

self-evaluation

SELF-TEST
1. What difference exists between a set of taps and serial taps?
2. Where is a spiral pointed tap used?
3. When is a fluteless spiral pointed tap used?
4. When is a spiral fluted tap used?
5. How are thread forming taps different from conventional taps?
6. How are taper pipe taps identified?
7. What is the difference between a pulley tap and a nut tap?
8. Why are finishing and roughing Acme taps used?
9. Why are the rake angles varied on taps for different materials?
10. Name at least three methods used by tap manufacturers to reduce friction between the tap and the workpiece material.

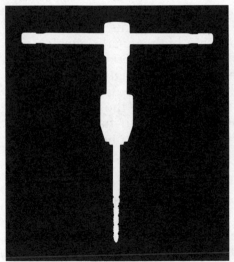

unit 8
tapping
procedures

Today's mass production of consumer goods depends to a large extent on the efficient and secure assembly of parts, for which we often rely on threaded fasteners. It takes skill to produce usable tapped holes, so a craftsman in the metal trades must have an understanding of the factors that affect the tapping of a hole, such as the work material and its cutting speed, the proper coolant, and the size and condition of the hole. A good machinist can analyze a tapping operation, determine whether or not it is satisfactory, and usually find a solution if it is not. In this unit, you will learn about common tapping procedures.

objectives After completing this unit, you will be able to:
1. Select the correct tap drill for a specific percentage of thread.
2. Determine the cutting speed for a given work material-tool combination.
3. Select the correct coolant for tapping.
4. Tap holes by hand or with a drill press.
5. Identify and correct common tapping problems.

Taps are used to cut internal threads in holes. The actual cutting process is called tapping and can be performed by hand or with a machine. A tap wrench (Figure 1) or a T-handle tap wrench (Figure 2) attached to the tap is used to provide driving torque while hand tapping. To obtain a greater accuracy in hand tapping, a hand tapper (Figure 3) is used. This fixture acts as a guide for the tap to insure that it stays in alignment and cuts concentric threads.

Holes can also be tapped in a drill press that has a spindle reverse switch, which is often foot operated for convenience. Drill presses without reversing switches can be used for tapping with a tapping attachment (Figure 4). Some of these tapping attachments have an internal friction clutch where downward pressure on the tap turns the tap forward and feeds it into the work. Releasing downward pressure will automatically reverse the tap and back it out of the workpiece. Some tapping attachments have lead screws that provide tap feed rates equal to the lead of the tap. Most of these attachments also have an adjustment to limit the torque to match the size of tap, which eliminates most tap breakage.

THREAD PERCENTAGE AND HOLE STRENGTH

The strength of a tapped hole depends largely on the workpiece material, the percentage of full thread used, and the length of the thread. The workpiece material is usually selected by the designer, but the machinist can often control the percentage of thread produced and the depth of the thread. The percentage of thread produced is dependent on the diameter of the drilled hole. Tap drill charts generally give tap drill sizes to produce 75 percent thread.

An example will illustrate the relationships between the percentage of thread, torque required to drive the tap, and resulting thread strength. An increase in thread depth from 60 to 72 percent in AISI 1020 steel required twice the torque to drive the tap, but it increased the strength of the thread by only 5 percent. The practical limit seems to be 75 percent of full

Figure 2. T-handle tap wrench.

thread, since greater percentage of thread does not increase the strength of the threaded hole in most materials.

In some difficult-to-machine materials such as titanium alloys, high tensile steels, and some stainless steels, 50 to 60 percent thread depth will give sufficient strength to the tapped hole. Threaded assemblies are usually designed so that the bolt breaks before the threaded hole strips. Common practice is to have a bolt engage a tapped hole by 1 to $1\frac{1}{2}$ times its diameter.

DRILLING THE RIGHT HOLE SIZE

The condition of the drilled hole affects the quality of the thread produced, as an out-of-round hole leads to an out-of-round thread. Bell-mouthed holes will produce bell-mouthed threads. When an exact hole size is needed, the hole should be reamed before tapping. This is especially important for large diameter taps and when fine pitch threads are used. The size of the hole to be drilled is usually obtained from tap drill charts, which usually show a 75 percent thread depth. If a thread depth other than 75 percent is wanted, use the following formula to determine the proper hole size:

$$\text{Outside diameter of thread} - \frac{.01299 \times \text{percentage of thread}}{\text{Number of threads per inch}}$$

$$= \text{Hole size}$$

Figure 1. Tap wrench.

For example, calculate the hole size for a 1 in.–12 thread fastener with a 70 percent thread depth:

$$1 - \frac{.01299 \times 70}{12} = .924 \text{ in.}$$

SPEEDS FOR TAPPING

The quality of the thread produced also depends on the speed at which a tap is operated. The selection of the best speed for tapping is limited, unlike the varying speeds and feeds possible with other cutting tools, because the feed per revolution is fixed by the lead of the thread. Excessive speed develops high temperatures that cause rapid wear of the tap's cutting edge. Dull taps produce rough or torn and off-size threads. High cutting speeds prevent adequate lubrication at the cutting edges and often create a problem of chip disposal.

When selecting the best speed for tapping, you should consider not only the material being tapped, but also the size of the hole, the kind of tap holder being used, and the lubricant being used. Table 1 gives some guidelines in selecting a speed and a lubricant for some materials when using high speed steel taps.

These cutting speeds in feet per minute have to be translated into RPM to be useful. For example, calculate the RPM when tapping a $\frac{3}{8}$–24 UNF hole in free machining steel. The cutting speed chart gives a cutting speed between 60

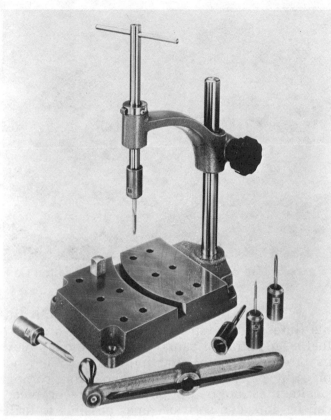

Figure 3. Hand tapper (Courtesy of Ralmike's Tool-A-Rama).

Figure 4. Drill press tapping attachment (Lane Community College).

Table 1
Recommended Cutting Speeds and Lubricants for Machine Tapping

Material	Speeds in Feet per Minute	Lubricant
Aluminum	90–100	Kerosene and light base oil
Brass	90–100	Soluble oil or light base oil
Cast iron	70– 80	Dry or soluble oil
Magnesium	20– 50	Light base oil diluted with kerosene
Phosphor bronze	30– 60	Mineral oil or light base oil
Plastics	50– 70	Dry or air jet
Steels		
Low carbon	40– 60	Sulphur base oil
High carbon	25– 35	Sulphur base oil
Free machining	60– 80	Soluble oil
Molybdenum	10– 35	Sulphur base oil
Stainless	10– 35	Sulphur base oil

and 80 feet per minute. Use the lower figure; you can increase the speed once you see how the material taps. The formula for calculating RPM is:

$$\frac{\text{Cutting Speed} \times 4}{\text{Diameter}} \quad \text{or} \quad \frac{60 \times 4}{\frac{3}{8}} = 640 \text{ RPM}$$

Lubrication is one of the most important factors in a tapping operation. Cutting fluids used when tapping serve as coolants, but are more important as lubricants. It is important to select the correct lubricant because the use of a wrong lubricant may give results that are worse than if no lubricant was used. For lubricants to be effective, they should be applied in sufficient quantity to the actual cutting area in the hole.

SOLVING TAP PROBLEMS

In Table 2, common tapping problems are presented with some possible solutions.

Occasionally it becomes necessary to remove a broken tap from a hole. If a part of the broken tap extends out of the workpiece, removal is relatively easy with a pair of pliers. If the tap breaks flush with or below the surface of the workpiece, a tap extractor can be used (Figure 5). Before trying to remove a broken tap, the chips in the flutes should be removed. A jet of compressed air or cutting fluid can be used for this. Always stand aside when cleaning out holes with compressed air as chips and particles tend to fly out at high velocity.

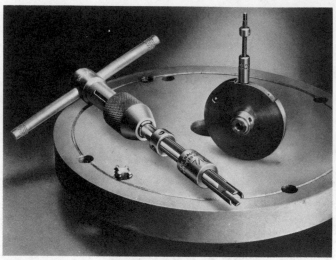

Figure 5. Tap extractor (Courtesy of the Walton Co.).

When the chips are packed so tightly in the flutes or the tap is jammed in the work so that a tap extractor cannot be used, the tap may be broken up with a pin punch and removed piece by piece. If the tap is made from carbon steel and cannot be pin punched, the workpiece can be annealed so it becomes possible to drill the tap out.

On high speed steel taps it may be necessary to use an electrical discharge machine (EDM) to remove the broken tap. These machines erode away material from extremely hard workpieces while they are immersed in a fluid. The shape of the hole conforms precisely to that of the electrode.

Table 2
Common Tapping Problems and Possible Solutions

Causes of Tap Breakage	Solutions
Tap hitting bottom of hole or bottoming on packed chips	Drill hole deeper. Eject chips with air pressure. (*Caution:* Stand aside when you do this and always wear safety glasses.) Use spiral fluted taps to pull chips out of hole. Use a thread forming tap.
Chips are packing in flutes	Use tap style with more flute space. Tap to a lesser depth or use a smaller percentage of threads. Select a tap that will eject chips forward (spiral point) or backward (spiral fluted).

Table 2 (continued)

Causes of Tap Breakage	Solutions
Hard materials or hard spots	Anneal the workpiece. Reduce cutting speed. Use longer chamfers on tap. Use taps with more flutes.
Inadequate lubricant	Use the correct lubricant and apply a sufficient amount of it under pressure at the cutting zone.
Tapping too fast	Reduce cutting speed.
Excessive wear: Abrasive materials	Improve lubrication. Use surface treated taps. Check the alignment of tap and hole to be tapped.
Chips clogging flutes: Insufficient lubrication	Use better lubricant and apply it with pressure at the cutting zone.
Excessive speed	Reduce cutting speed.
Wrong-style tap	Use a more free cutting tap such as spiral pointed tap, spiral fluted tap, interrupted thread tap, or surface treated taps.
Torn or rough threads: Dull tap	Resharpen.
Chip congestion	Use tap with more chip room. Use lesser percentage of thread. Drill deeper hole. Use a tap that will eject chips.
Inadequate lubrication and chips clogging flutes	Correct as previously suggested.
Hole improperly prepared	Torn areas on the surface of the drilled, bored, or cast hole will be shown in the minor diameter of the tapped thread.
Undersize threads: Pitch diameter of tap too small	Use tap with a larger pitch diameter.
Excessive speed	Reduce tapping speed.
Thin wall material	Use a tap that cuts as freely as possible. Improve lubrication. Hold the workpiece so that it cannot expand while it is being tapped. Use an oversize tap.
Dull tap	Resharpen.
Oversize or bellmouth threads: Loose spindle or worn holder	Replace or repair spindle or holder.
Misalignment	Align spindle, fixture, and work.
Tap oversize	Use smaller pitch diameter tap.
Dull tap	Resharpen.
Chips packed in flutes	Use tap with deeper flutes, spiral flutes, or spiral points.
Buildup on cutting edges of tap	Use correct lubricant and tapping speed.

TAPPING PROCEDURE, HAND TAPPING

1. Determine the size of the thread to be tapped and select the tap.
2. Select the proper tap drill with the aid of a tap drill chart. A taper tap should be selected for hand tapping; or if a drill press or tapping machine is to be used for alignment, use a plug tap.
3. Fasten the workpiece securely in a drill press vise. Calculate the correct RPM for the drill used:

$$RPM = \frac{CS \times 4}{D}$$

   Drill the hole using the recommended coolant. Check the hole size.
4. Countersink the hole entrance to a diameter slightly larger than the major diameter of the threads (Figure 6). This allows the tap to be

started more easily, and it protects the start of the threads from damage.

5. Mount the workpiece in a bench vise so that the hole is in a vertical position.
6. Tighten the tap in the tap wrench.
7. Cup your hand over the center of the wrench (Figure 7) and place the tap in the hole in a vertical position. Start the tap by turning two or three turns in a clockwise direction for a right-hand thread. At the same time keep a steady pressure downward on the tap. When the tap is started, it may be turned as shown in Figure 8.
8. After the tap is started for several turns, remove the tap wrench without disturbing the tap. Place the blade of a square against the solid shank of the tap to check for squareness (Figure 9). Check from two positions 90 degrees apart. If the tap is not square with the

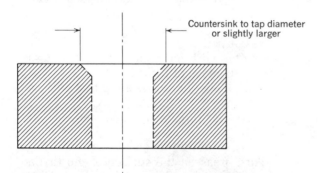

Figure 6. Preparing the workpiece.

Figure 7. Starting the tap (Lane Community College).

Figure 8. Tapping a thread by hand (Lane Community College).

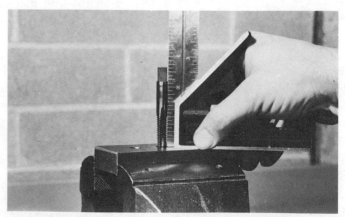

Figure 9. Checking the tap for squareness (Lane Community College).

work, it will ruin the thread and possibly break in the hole if you continue tapping. Back the tap out of the hole and restart.

9. Use the correct cutting oil on the tap when cutting threads.

10. Turn the tap clockwise one-quarter to one-half turn and then turn it back a three-quarter turn to break the chip. This is done with a steady motion to avoid breaking the tap.

11. When tapping a blind hole, use the taps in the order of starting, plug, and then bottoming. Remove the chips from the hole before using the bottoming tap and be careful not to hit the bottom of the hole with the tap.

12. Figure 10 shows a 60 degree point center chucked in a drill press to align a tap squarely with the previously drilled hole. Only very slight follow-up pressure should be applied to the tap. Too much downward pressure will cut a loose, oversize thread.

Figure 10. Using the drill press as a tapping fixture (Lane Community College).

self-evaluation

SELF-TEST
1. What kind of tools are used to drive taps when hand tapping?
2. What is a hand tapper?
3. What is a tapping attachment?
4. Which three factors affect the strength of a tapped hole?
5. How deep should the usable threads be in a tapped hole?
6. When should tap drill holes be reamed?
7. What causes taps to break while tapping?
8. What causes rough and torn threads?
9. What causes oversize threads in a hole?
10. Give three methods of removing broken taps from holes.

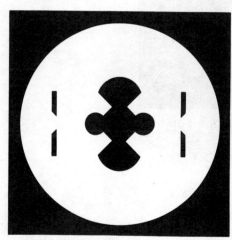

unit 9
thread cutting dies and their uses

A die is used to cut external threads on the surface of a bolt or rod. Many machine parts and mechanical assemblies are held together with threaded fasteners, most of which are mass produced. Occasionally, however, a machinist has to make a bolt or extend the threads on a bolt for which he uses a die. In this unit you will be introduced to some dies and their hand threading uses.

objectives After completing this unit, you will be able to:

1. Identify dies used for hand threading.
2. Select and prepare a rod for threading.
3. Cut threads with a die.

Dies are used to cut external threads on round materials. Some dies are made from carbon steel, but most are made from high speed steel. Dies are identified by the markings on the face as to the size of thread, number of threads per inch, and form of thread, such as NC, UNF, or other standard designations (Figure 1).

COMMON TYPES OF HAND THREADING DIES

The die shown in Figure 1 is an example of a round split adjustable die, also called a button die. These dies are made in all standardized thread sizes up to $1\frac{1}{2}$ in. thread diameters and $\frac{1}{2}$ in. pipe threads. The outside diameters of these dies vary from $\frac{5}{8}$ to 3 in.

Adjustments on these dies are made by turning a fine pitch screw that forces the sides of the die apart or allows them to spring together. The range of adjustment of round split adjustable dies is very small, allowing only for a loose or tight fit on a threaded part. Adjustments made to obtain threads several thousandths of an inch oversize will result in poor die performance because the heel of the cutting edge will drag on the threads. Excessive expansion may cause the die to break in two.

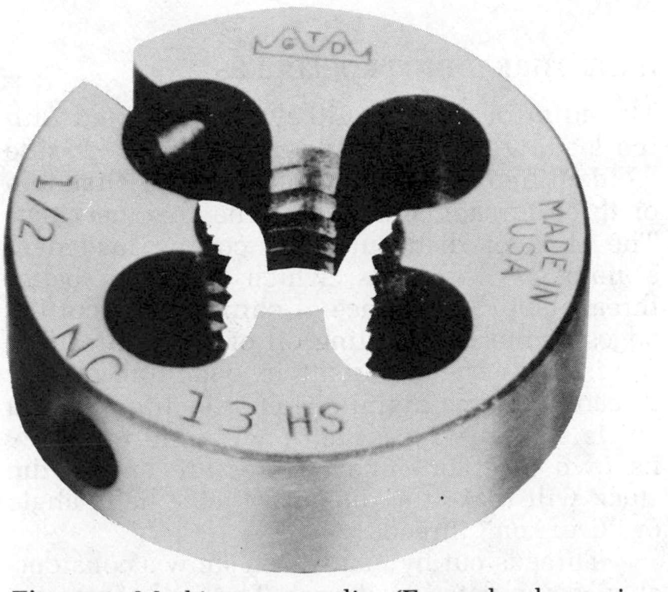

Some round split adjustable dies do not have the built-in adjusting screw. Adjustments are then made with the three screws in the die stock (Figure 2). Two of these screws on opposite sides of the die stock hold the die in the die stock and also provide closing pressure. The third screw engages the split in the die and provides opening pressure. These dies are used in a die stock for hand threading or in a machine holder for machine threading.

Another type of threading die is the two piece die, whose halves (Figure 3) are called blanks. These blanks are assembled in a collet consisting of a cap and the guide (Figure 4). The normal position of the blanks in the collet is indicated by witness marks (Figure 5). The adjusting screws allow for precise control of the cut thread size. The blanks are inserted in the cap with the tapered threads toward the guide. Each of the two die halves is stamped with a serial number. Make sure the halves you select have the same numbers. The guide used in the collet serves as an aid in starting and holding the

Figure 2. Diestock for round split adjustable dies (Photo courtesy of TRW Greenfield Tap & Die Division).

Figure 3. Die halves for two piece die (Photo courtesy of TRW Greenfield Tap & Die Division).

Figure 1. Markings on a die. (Example shown is a round split adjustable die).

Cap   Guide   Collet

Figure 4. Components of a split adjustable die collet (Photo courtesy of TRW Greenfield Tap & Die Division).

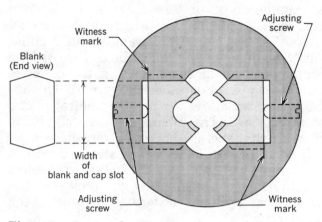

Figure 5. Setting the die position to the witness marks on the die and collet assembly (Courtesy of TRW Greenfield Tap & Die Division).

Figure 6. Diestock for adjustable die and collet assembly (Photo courtesy of TRW Greenfield Tap & Die Division).

dies square with the work being threaded. Each thread size uses a guide of the same nominal or indicated size. Collets are held securely in die stocks (Figure 6) by a knurled setscrew that seats in a dimple in the cap.

Hexagon rethreading dies (Figure 7) are used to recut slightly damaged or rusty threads. Occasionally it may be necessary to cut a new thread with these dies, but it is difficult to start

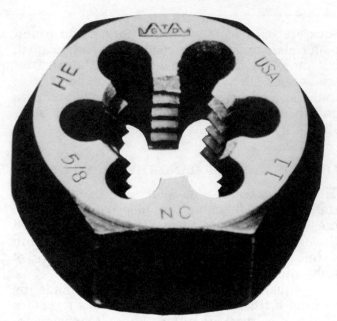

Figure 7. Hexagon rethreading die (Photo courtesy of TRW Greenfield Tap & Die Division).

the die square with the workpiece and to maintain this alignment. Rethreading dies are driven with a wrench large enough to fit the die. Solid square dies (Figure 8) have the same uses and limitations as hexagon rethreading dies.

All of the previously discussed die types are also available in bolt and pipe sizes. Spring dies, die head thread chasers, and other production methods of threading will be covered in later volumes.

HAND THREADING PROCEDURES

Threading of a rod should always be started with the leading or throat side of the die. This side is identified by the chamfer on the first two or three threads and also by the size markings. The chamfer distributes the cutting load over a number of threads, which produces better threads and less chance of chipping the cutting edges of the die. Cutting oil or other threading fluids are very important in obtaining quality threads and maintaining long die life. Once a cut is started with a die, it will tend to follow its own lead, but uneven pressure on the die stock will make the die cut variable helix angle or "drunken" threads.

Threads cut by hand often show a considerable accumulated lead error. The lead of a screw thread is the distance a nut will move on the

Figure 8. Solid square die (Photo courtesy of TRW Greenfield Tap & Die Division).

Figure 9. Threading a rod with a hand die in a lathe (Lane Community College).

screw if it is turned one full revolution. This problem is caused by the dies being relatively thin when compared to the diameter of thread that they cut. Only a few threads in the die can act as a guide on the already cut threads. This error usually does not cause problems when standard or thin nuts are used on the threaded part. However, when an item with a long internal thread is assembled with a threaded rod, it usually gets tight and then locks, not because the thread depth is insufficient, but because there is a lead error. This lead error can be as much as one-fourth of a thread in one inch of length.

The outside diameter of the material to be threaded should not be over the nominal size of the thread and preferably a few thousandths of an inch (.002 to .005 in.) undersize. After a few full threads are cut, the die should be removed so that the thread can be tested with a nut or thread ring gage. A thread ring gage set usually consists of two gages, a go and a no go gage. As the names imply, a go gage should screw on the thread, while the no go gage will not go more than $1\frac{1}{2}$ turns on a thread of the correct size. Do not assume that the die will cut the correct size thread; always check by gaging or assembling. Adjustable dies should be spread open for the first cut and set progressively smaller for each pass after checking the thread size.

It is very important that a die is started squarely on the rod to be threaded. A lathe can be used as a fixture for cutting threads with a

die (Figure 9). The rod is fastened in a lathe chuck for rotation, while the die is held square because it is supported by the face of the tailstock spindle. The carriage or the compound rest prevents the die stock from turning while the chuck is rotated *by hand*. As the die advances, the tailstock spindle is also advanced to stay in contact with the die. Do not force the die with the tailstock spindle, or a loose thread may result. A die may be used to finish to size a long thread that has been rough threaded on the lathe.

Figure 10. Chamfer workpiece before using die (Lane Community College).

Occasionally, a die is used to extend the thread on a bolt. Make certain that the bolt is not hardened or the die will be ruined. To cut full, usable threads close to a shoulder, first cut the thread normally until the die touches the shoulder, then reverse the die and use the unchamfered side to finish the last few threads.

It is always good practice to chamfer the end of a workpiece before starting a die (Fig-ure 10). The chamfer on the end of a rod can be made by grinding on a pedestal grinder, by filing or with a lathe. This will help in starting the cut and it will also leave a finished thread end. While cutting threads with a hand die, the die rotation should be reversed after each full turn forward to break the chips into short pieces that will fall out of the die. Chips jammed in the clearance holes will tear the thread.

THREADING PROCEDURE, THREADING DIES

1. Select the workpiece to be threaded and measure its diameter. Then chamfer the end. This may be done on a grinder or with a file. The chamfer should be at least as deep as the thread to be cut.

2. Select the correct die and mount it in a die stock.

3. Mount the workpiece in a bench vise. Short workpieces are mounted vertically and the long pieces usually are held horizontally.

4. To start the thread, place the die over the workpiece. Holding the die stock with one hand (Figure 11), apply downward pressure and turn the die.

5. When the cut has started, apply cutting oil to the workpiece and die and start turning the die stock with both hands (Figure 12). After each complete revolution forward, reverse the die one-half turn to break the chips.

6. Check to see that the thread is started square, using a machinist's square. Corrections can be made by applying slight downward pressure on the high side while turning.

7. When several turns of the thread have been completed, you should check the fit of the thread with a nut, thread ring gage, thread micrometer, or the mating part. If the thread fit is incorrect, adjust the die with the adjustment screws and take another cut with the adjusted die. Continue making adjustments until the proper fit is achieved.

8. Continue threading to the required thread length. To cut threads close to a shoulder, invert the die after the normal threading operation and cut the last two or three threads with the side of the die that has no chamfer.

Figure 11. Start the die with one hand (Lane Community College).

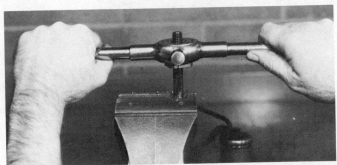

Figure 12. Use both hands to turn the threading die (Lane Community College).

self-evaluation

SELF-TEST

1. What is a die?
2. What tool is used to drive a die?
3. How much adjustment is possible with a round split adjustable die?
4. What is the purpose of the guide in a two-piece adjustable die collet?
5. What are important points to watch when assembling two-piece dies in a collet?
6. Where are hexagon rethreading dies used?
7. Why do dies have a chamfer on the cutting end?
8. Why are cutting fluids used?
9. What diameter should a rod be before being threaded?
10. Why should a rod be chamfered before being threaded?

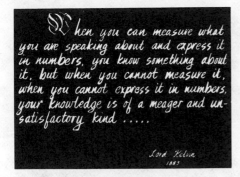

section c
dimensional
measurement

All of us, no matter what we may be doing, are totally immersed in a world of measurement. Measurement can be generally defined as: the assignment of a value to time, length, and mass. We cannot escape measurement. Our daily lives are greatly influenced by the clock, a device that measures time. Mass or weight is measured in almost every product we buy, and the measure of length is incorporated in every creation of man, ranging from the minute components of a watch to many thousands of miles of superhighways extending across a continent.

Measurement, in the modern age, has been developed to an exacting science known as metrology. As the hardware of technology has become ever more complex, a machinist is ever more concerned with that branch of the science called dimensional metrology. Furthermore, mass production of goods has made necessary very complex systems of metrology to check and control the critical dimensions that control standardization and interchangeability of parts. Components of an automobile, for example, may be manufactured at locations far removed from each other and then brought to a central assembly point, with the assurance that all parts will fit as intended by the designer. In addition, the development and maintenance of a vast system of carefully controlled measurement have permitted manufacturing to locate its factories in close proximity to raw materials and the availability of abundant labor. Because of the standardization of measurement, industry has been able to diversify its products. Thus, a manufacturer can do what he does best, and manufacturing effort can be directed toward the quality of a product and its production at a competitive price. As a result, metrology not only affects the technical aspects of production but also the economic aspects. Metrology is a common thread woven through the entire fabric of manufacturing from the drafting room to the shipping dock.

HISTORICAL DEVELOPMENT OF DIMENSIONAL MEASUREMENT

As man began to build, he soon found that some type of dimensional measurement was essential. It was logical that he use parts of his body as measurement standards. The early Egyptians established the cubit as the length of the arm from elbow to finger tips (Figure 1). Subdivisions of the cubit included the span (Figure 2), the palm (Figure 3), and the digit (Figure 4). The disadvantages of these standards soon became apparent. No two people were exactly alike.

87

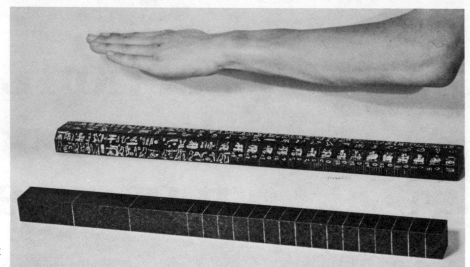

Figure 1. Royal Egyptian Cubit (Courtesy of the DoAll Company).

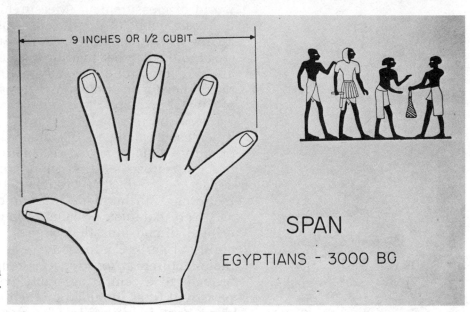

Figure 2. Span (Courtesy of Glenn Herreman, Hewlett Packard Company).

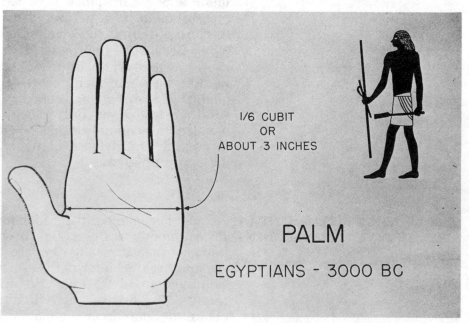

Figure 3. Palm (Courtesy of Glenn Herreman. Hewlett Packard Company).

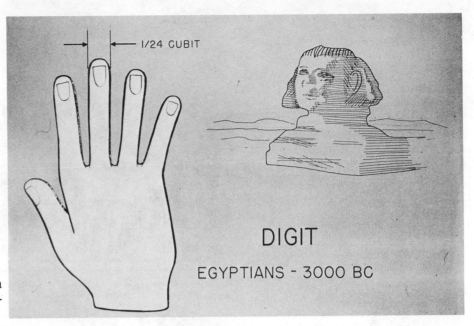

Figure 4. Digit (Courtesy of Glenn Herreman, Hewlett Packard Company).

This brought the development of a more permanent standard. The Royal Egyptian Cubit (Figure 1) was constructed from granite. Copies were made for distribution with the requirement that they be periodically returned and checked against the royal master. This is the process of calibration and is still carried on today in all areas of measurement.

The inch was taken to be a thumb breadth during the Roman era (Figure 5). The rod was also defined in terms of parts of the body (Figure 6). Other objects were used as standards. Three barley corns defined the inch in early England (Figure 7).

As industry began to serve more people and machines replaced men in the production of goods, the need for extensive standardization of measurement became readily apparent. Today, total standardization of measurement throughout the world still has not been fully realized. However, measurement in the industrialized world

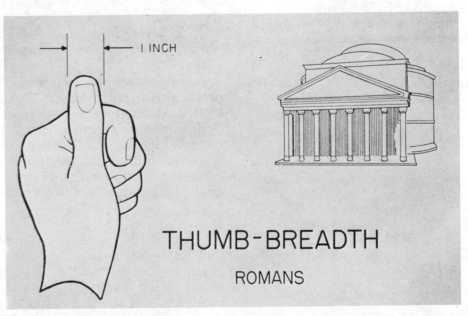

Figure 5. Thumb breadth (Courtesy of Glenn Herreman, Hewlett Packard Company).

89

Figure 6. Rod (Courtesy of Glenn Herreman, Hewlett Packard Company).

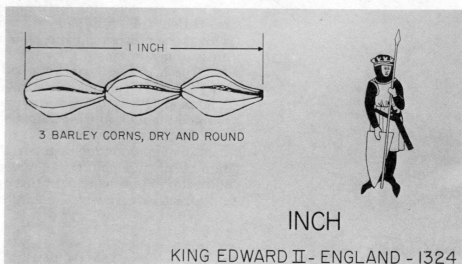

Figure 7. Inch (Courtesy of Glenn Herreman, Hewlett Packard Company).

does conform to the inch or metric system. Metric measurement is predominant but, because of the vast amount of industry in the United States, inch measurement is still the fundamental system. Today, manufacturing is an international effort. This is due to many factors, including the location of natural resources and world economic conditions. In the future you may see a complete transition to metric measurement. However, for the present, you are in an environment that uses two systems.

MEASUREMENT NEEDS OF THE MACHINIST A machinist is mainly concerned with the measurement of **length**; that is, the distance along a line between two points (Figure 8). It is length that defines the **size** of most objects. **Width** and **depth** are simply other names for length. A machinist measures length in the basic units of linear measure such as **inches**, **millimeters** and, in

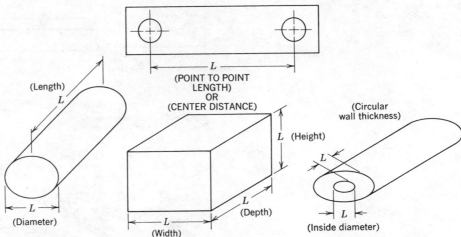

Figure 8. The measurement of length may appear under several different names.

advanced metrology, wavelengths of light. In addition, he sometimes needs to measure the relationship of one surface to another, which is commonly called **angularity** (Figure 9). **Squareness**, which is closely related to angularity, is the measure of deviation from true perpendicularity. A machinist will measure angularity in the basic units of angular measure, **degrees**, **minutes**, and **seconds of arc**.

In addition to the measure of length and angularity, a machinist also needs to measure such things as surface finish, concentricity, straightness, and flatness. He also occasionally comes in contact with measurements that involve circularity, sphericity, and alignment (Figure 10). However, many of these more specialized measurement techniques are in the realm of the inspector or laboratory metrologist and appear infrequently in general machine shop work.

GENERAL PRINCIPLES OF METROLOGY A machinist has available to him a large number of measuring tools designed for use in many different applications. However, not every tool is equally suited for a specific measurement. As with all the tools of a machinist, he must be able to select the proper measuring tool for the specific application at hand. The successful outcome of his work may indeed depend upon his choice of measuring tools. In this regard, a machinist must be familiar with several important terms and principles of metrology.

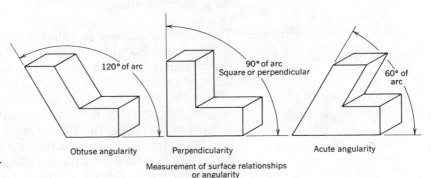

Figure 9. Measurement of surface relationships or angularity.

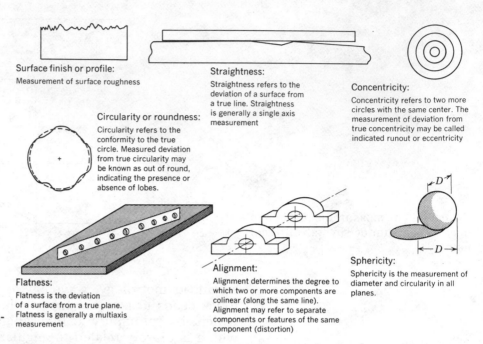

Surface finish or profile:
Measurement of surface roughness

Straightness:
Straightness refers to the
deviation of a surface from
a true line. Straightness
is generally a single axis
measurement

Concentricity:
Concentricity refers to two more
circles with the same center. The
measurement of deviation from
true concentricity may be called
indicated runout or eccentricity

Circularity or roundness:
Circularity refers to the
conformity to the true
circle. Measured deviation
from true circularity may
be known as out of round,
indicating the presence or
absence of lobes.

Flatness:
Flatness is the deviation
of a surface from a true plane.
Flatness is generally a multiaxis
measurement

Alignment:
Alignment determines the degree to
which two or more components are
colinear (along the same line).
Alignment may refer to separate
components or features of the same
component (distortion)

Sphericity:
Sphericity is the measurement of
diameter and circularity in all
planes.

Figure 10. Other measurements encountered by the machinist.

Accuracy

Accuracy in metrology has a twofold meaning. First, accuracy can refer to whether or not a specific measurement is actually within its stated size. For example, a certain drill has its size stamped on its shank. A doubtful machinist decides to verify the drill size using a properly adjusted micrometer. The size is found to be as stated. Therefore, the size stamped on the drill is accurate. Second, accuracy refers to the act of measurement itself with regard to whether or not the specific measurement taken is within the capability of the measuring tool selected. A machinist obtains a drill with the size marked on the shank and decides to verify it using his steel rule. He selects the edge of his rule with the finest graduations and lays the drill over the marks. Sighting along the drill, he discovers that it is really three graduations, on his rule, smaller than the size stamped on the shank. He then reasons that the size marked on the drill must be in error. In this example, the act of measurement is not accurate because the inappropriate measuring tool was selected and the improper procedure was used. **User accuracy** is also an important consideration. If, when the machinist measured the drill with his micrometer, as described in the first example, he did not bother to confirm the accuracy of the instrument prior to making the measurement, an inaccuracy which can be attributed to the user, may have resulted.

Precision

The term **precision** is relative to the specific measurement being made, with regard to the degree of exactness required. For example, the distance from the earth to the moon, measured to within one mile, would indeed be a precise measurement. Likewise, a clearance of five-thousandths of an inch between a certain bearing and journal might be precise for that specific application. However, five-thousandths of an inch clearance between ball and race on a ball bearing would not be considered precise, as this clearance would be only a very few millionths of an inch. There are many degrees of precision dependent on application and design requirements. For a machinist, any measurement made to a degree finer than one sixty-fourth of an inch or one half millimeter, can be considered a **precision measurement** and is made with the appropriate precision measuring instrument.

Reliability **Reliability** in measurement refers to the ability to obtain the desired result to the degree of precision required. Reliability is most important in the selection of the proper measuring tool. A certain tool may be reliable for a certain measurement, but totally unreliable in another application. For example, if it were desired to measure the distance to the next town, the odometer on an automobile speedometer would yield quite a reliable result, providing that a degree of precision of less than one-tenth mile is not required. On the other hand, to measure the length of a city lot with an odometer is much less likely to yield a reliable result. This is explained by examining another important principle of metrology, that of the discrimination of a measuring instrument.

Discrimination **Discrimination** refers to the degree to which a measuring instrument divides the basic unit of length that it is using for measurement. The automobile odometer divides the mile into 10 parts; therefore, it discriminates to the nearest tenth of a mile. A micrometer, one of the most common measuring instruments of a machinist, subdivides an inch into 1000 or, in some cases, 10,000 parts. Therefore, the micrometer discriminates to .001 or .0001 of an inch. If a measuring instrument is used beyond its discrimination, a loss of reliability will result. Consider the example cited previously regarding the measurement of a city lot. Most lots are less than one-tenth of a mile in length; therefore, the discrimination of the auto odometer for this measurement is not sufficient for reliability.

The 10 : 1 Ratio for Discrimination In general, a measuring instrument should **discriminate 10 times finer** than the smallest unit that it will be asked to measure. The odometer, which discriminates to a tenth of a mile, is most reliable for measuring whole miles. To measure the length of a city lot in feet requires an instrument that discriminates at least to one-tenth of a foot. Since most surveyor's measuring tapes used for this application discriminate to tenths and in some cases to hundredths of a foot, they are an appropriate tool for the measurement.

The Position of a Linear Measuring Instrument with Regard to the Axis of Measurement A large portion of the measurements made by a machinist is linear in nature. These measurements attempt to determine the shortest distance between two points. In order to obtain an accurate and reliable linear measurement, **the measuring instrument must be exactly in line with the axis of that measurement.** If this condition is not met, reliability will be in question. The alignment of the measuring instrument with the axis of measurement applies to all linear measurements (Figure 11). The figure illustrates the alignment of the instrument with the axis of measurement using a simple graduated measuring device. Only under the reliable condition can the measurement approach accuracy. Misalignment of the instrument, as illustrated in the unreliable condition, will result in inaccurate measurements.

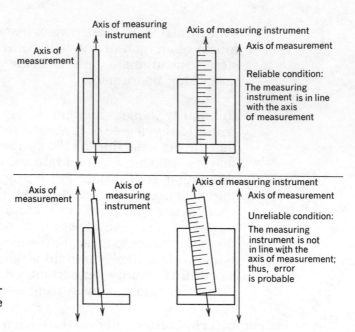

Figure 11. The axis of a linear measuring instrument must be in line with the axis of measurement.

Responsibility of the Machinist in Measurement

The following units in this section discuss most of the common measuring tools available to a machinist. The capabilities, discrimination, and reliability as well as procedures for use are examined. It is, of course, the responsibility of a machinist to select the proper measuring instrument for the job at hand. When faced with a need to measure, a machinist should ask the following questions:

1. **What degree of accuracy and precision must this measurement meet?**
2. **What degree of measuring tool discrimination does this required accuracy and precision demand?**
3. **What is the most reliable tool for this application?**

CALIBRATION. Accurate and reliable measurement places a considerable amount of responsibility on a machinist. He is responsible for the conformity of his measuring tools to the appropriate standards. This is the process known as **calibration**. In a large industrial facility, all measuring tools are periodically cycled through the metrology laboratory where they are calibrated against appropriate standards. Any adjustments are then made to bring the tools into conformity to the standards. Only through this process can standardized measure within an individual plant or within an entire industrial nation be maintained. Most of the common measuring instruments provide a factory standard. Even though calibration cannot be carried out under laboratory conditions, the instruments should at least be checked periodically against available standards.

VARIABLES IN MEASURING. A machinist should be further aware that any measurement is **relative to the conditions under which**

it is taken. A common expression that is often used around the machine shop is: "the measurement is right on." There is, of course, little probability of obtaining a measurement that is truly exact. Each measurement has a certain degree of deviation from the theoretical exact size. This degree of error is dependent on many variables including the measuring tool selected, the procedure used, the temperature of the part, the temperature of the room, the cleanliness of the room air, and the cleanliness of the part at the time of measurement. The deviation of a measurement from exact size is taken into consideration by the designer. Every measurement has a tolerance, meaning that the measurement is acceptable within a specific range. Tolerance can be quite small depending on design requirements. When this condition exists, reliable measurement becomes more difficult because it is more heavily influenced by the many variables present. Therefore, before a machinist makes any measurement, he should stop for a moment and consider the possible variables involved. He should then consider what might be done to control as many of these variables as possible.

If you understand these basic principles of dimensional metrology and you assume the proper responsibility in the selection, calibration, and application of measuring instruments, you will experience little difficulty in performing the many measurements to be encountered in the science of machine tools and machining practices.

TOOLS FOR DIMENSIONAL MEASUREMENT There are many hundreds of measuring instruments available to a machinist. In this modern age there is a measuring instrument that can be applied to almost any conceivable measurement. Many instruments are simply variations and combinations of a few common precision measuring tools. As you begin your study of machine tool practices, you will be initially concerned with the use, care, and applications of the common measuring instruments found in the machine shop. These will be discussed in detail within this section.

In addition to these, there is a large variety of instruments that are designed for many specialized uses. Some of these are rarely seen in the school or general purpose machine shop. Others are intended for use in the tool room or metrology laboratory where they are used in the calibration process. Your contact with these instruments will depend on the particular path that you take while learning the trade.

Many measuring instruments have undergone modernization in recent years. Even though the function of these tools is basically the same, many have been redesigned and equipped with mechanical or electronic digital displays. These features make the instruments easier to read and improve accuracy. As a machinist, you must be skilled in the use of all the common measuring instruments. In addition, you should be familiar with the many important instruments used in production machining, inspection, and calibration. In the following pages many of these tools will be briefly described so that you may become familiar with the wide selection of measuring instruments available to the machinist.

Fixed Gages and FIXED GAGES. In production machining, where large numbers of
Air Gages duplicate parts are produced, it may only be necessary to determine
if the part is within acceptable tolerance. Many types of fixed gages
are used. The adjustable limit snap gage (Figure 12) is used to check
outside diameter. One anvil is set to the minimum limit of the tolerance to be measured. The other anvil is set to the maximum limit of
the tolerance. If both anvils slip over the part, an undersized
condition is indicated. If neither anvil slips over the part, an oversized condition is indicated. The gage is set initially to a known
standard such as gage blocks.

Threaded products are often checked with fixed gages. The
thread plug gage (Figure 13) is used to check internal threads. The
thread ring gage (Figure 14) is used to check external threads. These
are frequently called **go** and **no-go** or **not-go** gages. One end of the
plug gage is at the low limit of the tolerance, while the other end is at
the high limit of the tolerance. The thread gage functions in the same
manner. Thread gages appear in many different forms (Figure 15).

Fixed gages are also used to check internal and external tapers
(Figure 16). Hole plug gages are used for internal holes (Figure 17). A
ring gage is used for external diameters (Figure 18).

AIR GAGES. **Air gages** are also known as pneumatic comparators.
Two types of air gages are used in comparison measuring applications. In the pressure-type air gage (Figure 19), filtered air flows
through a reference and measuring channel. A sensitive differential
pressure meter is connected across the channels (Figure 20). The
gage head is adjusted to a master setting gage. Air gage heads may be
ring, snap (Figure 21), or plug types (Figure 22). Air flowing through
the reference and measuring channel is adjusted until the differential pressure meter reads zero with the setting master in place. A
difference in workpiece size above or below the master size will
cause more or less air to escape from the gage head. This, in turn,
will change the pressure in the reference channel. The pressure
change will be indicated on the differential pressure meter.

Figure 12. Adjustable limit snap
gage (Courtesy of Rank Precision
Industries, Inc.).

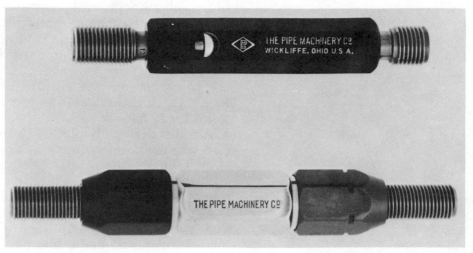

Figure 13. Thread plug gage (Courtesy of PMC Industries).

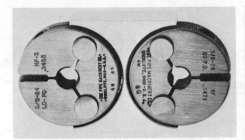

Figure 14. Thread ring gage (Courtesy of PMC Industries).

Figure 15. Fixed thread gages appear in many different forms (Courtesy of PMC Industries).

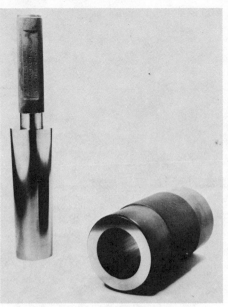

Figure 16. Taper plug and taper ring gage (Courtesy of PMC Industries).

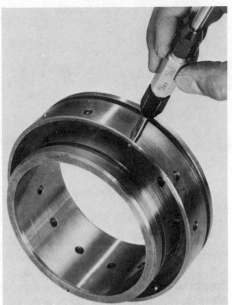

Figure 17. Using the cylindrical plug gage (Courtesy of PMC Industries).

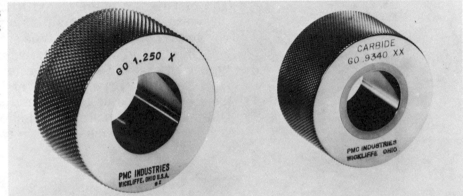

Figure 18. Cylindrical ring gages (Courtesy of PMC Industries).

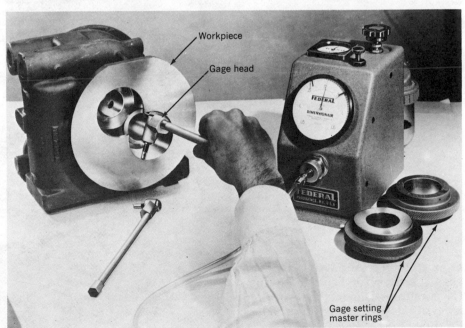

Workpiece

Gage head

Gage setting master rings

Figure 19. Pressure-type air gage. (Courtesy of Federal Products Corporation—Dimensionair is a registered trademark of Federal Products Corp.).

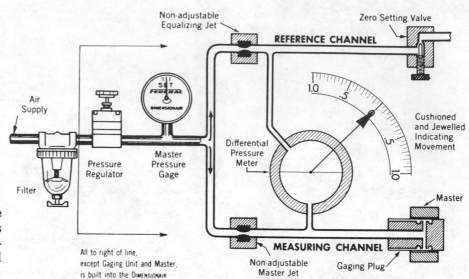

Figure 20. Pressure-type air gage system (Courtesy of Federal Products Corporation—Dimensionair is a registered trademark of Federal Products Corp.).

The meter scale is graduated in suitable linear units. Thus, workpiece size above or below the master can be directly determined.

In the column or flow-type air gage (Figure 23) air flow from the gage head is indicated on a flow meter or rotameter (Figure 24). This type of air gage is also set to master gages. In the case of the plug gage shown, if the workpiece is oversized, more air will flow from the gage head. An undersized condition will permit less air to flow. Differences in flow are indicated on a suitably graduated flowmeter scale. Workpiece size deviation can be read directly.

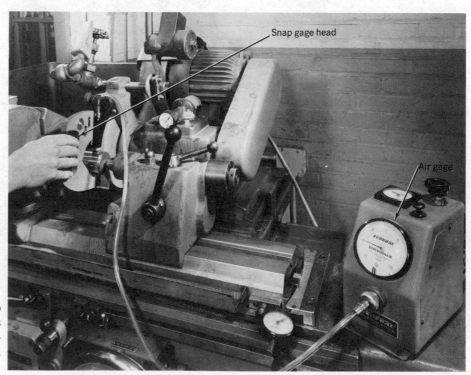

Figure 21. Pressure-type air snap gage (Courtesy of Federal Products Corporation—Dimensionair is a registered trademark of Federal Products Corp.).

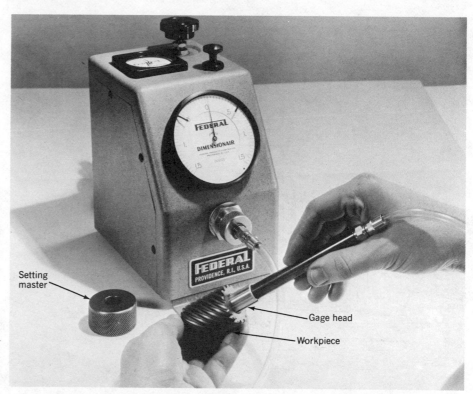

Figure 22. Air plug gage (Courtesy of Federal Products Corporation—Dimensionair is a registered trademark of Federal Products Corp.).

Setting master

Gage head

Workpiece

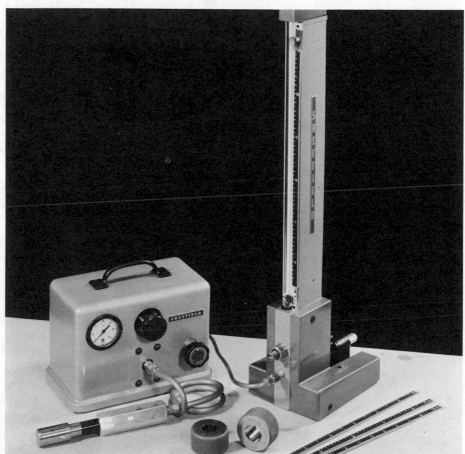

Figure 23. Column or flow-type air gage (Courtesy of the Automation and Measurement Division—Bendix Corporation).

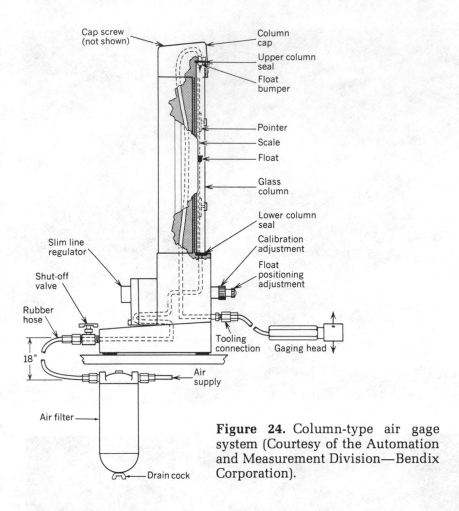

Figure 24. Column-type air gage system (Courtesy of the Automation and Measurement Division—Bendix Corporation).

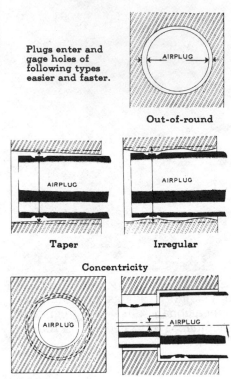

Plugs enter and gage holes of following types easier and faster.

Out-of-round

Taper   Irregular

Concentricity

Figure 25. Hole geometry detectable by air gaging (Courtesy of Federal Products Corporation).

Air gages have several advantages. The gage head does not touch the workpiece. Consequently there is no wear on the gage head and no damage to the finish of the workpiece. Variations in workpiece geometry that would be difficult to measure by mechanical means can be detected by air gaging (Figure 25).

Mechanical Dial Measuring Instruments

Measuring instruments that show a measurement on a dial have become very popular in recent years. Several of the common dial instruments are outgrowths from vernier instruments of the same type. Dial instruments have an advantage over their vernier counterparts in that they are easier to read. Dial measuring equipment is frequently found in the inspection department where many types of measurements must be made quickly and accurately.

DIAL THICKNESS GAGE. The dial thickness gage (Figure 26) is used to measure the thickness of paper, leather, sheet metal, and rubber. Discrimination is .0005 in.

DIAL INDICATING SNAP GAGES. **Dial indicating snap gages** (Figure 27) are used for determining whether workpieces are within acceptable tolerances. They are first set to a gage block standard. Part size deviation is noted on the dial indicator.

Figure 26. Dial thickness gage (Courtesy of Rank Precision Industries, Inc.).

Figure 27. Using the indicating snap gage (Courtesy of Federal Products Corporation).

DIAL BORE GAGE. The dial bore gage (Figure 28) uses a three point measuring contact. This more accurately measures the true shape of a bore (Figure 29). The dial bore gage is useful for checking engine block cylinders for size, taper, bell mouth, ovality, barrel shape, and hour glass shape (Figure 30). Dial bore gages are set to a master ring and then compared to a bore diameter. Discrimination ranges from .001 to .0001 in.

Figure 28. Dial bore gage (Courtesy of Rank Precision Industries, Inc.).

Figure 29. Using the dial bore gage (Courtesy of the L. S. Starrett Company).

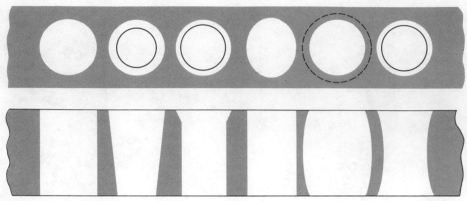

Figure 30. Hole geometry detectable with the dial bore gage.

DIAL INDICATING EXPANSION PLUG BORE GAGE. The indicating expansion plug gage (Figure 31) is used to measure the inside diameter of a hole or bore. This type gage is built to check a single dimension. It can detect ovality, bell-mouth, barrel shape, and taper. The expanding plug is retracted and the instrument inserted into the hole to be measured (Figure 32).

Figure 31. Dial indicating expansion plug bore gage (Courtesy of Federal Products Corporation).

Figure 32. Using the expansion plug bore gage (Courtesy of Federal Products Corporation).

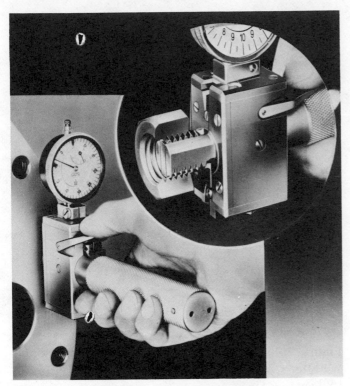

Figure 34. Dial indicating thread snap gage (Courtesy of the Mahr Gage Company).

Figure 33. Using the indicating thread plug gage (Courtesy of the Mahr Gage Company).

DIAL INDICATING THREAD PLUG GAGE. The indicating thread plug gage (Figure 33) is used to measure internal threads. This type of gage need not be screwed into the thread. The measuring anvils retract, so that the gage may be inserted into a threaded hole.

DIAL INDICATING SCREW THREAD SNAP GAGE. The dial indicating screw thread snap gage (Figure 34) is used to measure an external thread. The instrument may be fitted with suitable anvils for measuring the major, minor, or pitch diameter of screw threads. Discrimination is .0005 or .00005 in. depending on the dial indicator used.

DIAL INDICATING INPROCESS GRINDING GAGE. Gages can be built into machining processes. The indicating inprocess grinding gage is used to measure the workpiece while it is still running in the machine tool (Figure 35). The instrument swings down over the part to be measured (Figure 36). The machine can remain running. These instruments are used in such applications as cylindrical grinding. Discrimination can be .0005 or .00005 in., depending on the dial indicator used.

MECHANICAL DIAL INDICATING TRAVEL INDICATORS. Mechanical dial indicators can be used to indicate the travel of machine tool components. This is very valuable to the machinist in controlling machine movement that in turn controls the dimensions of the parts produced. Mechanical dial travel indicators are used in many applications such as indicating table and saddle travel on a milling

Figure 35. Dial indicator-inprocess grinding gage, in retracted position (Courtesy of Federal Products Corporation).

Figure 36. Inprocess grinding gage measuring the workpiece (Courtesy of Federal Products Corporation).

machine (Figure 37). They may also be used to indicate vertical travel of quills and spindles. Mechanical dial travel indicators are also useful for indicating travel in metric dimensions. Discrimination is .001 in., .005 in., and .01 mm.

Figure 37. Mechanical dial travel indicators installed on a milling machine (Courtesy of Southwestern Industries, Inc., Trav-A-Dial®).

Figure 38. Indicating bench micrometer or "supermicrometer" (Courtesy of Colt Industries, Pratt and Whitney Cutting Tool and Gage Division).

Inspection and Calibration Through Mechanical Measurement

All measuring instruments must be periodically checked against accepted standards if the control that permits interchangeability of parts is to be maintained. Without control of measurement there could be no diverse mass production of parts that will later fit together to form the many products that we now enjoy.

Parts produced on a machine tool must be inspected to determine if their size meets design requirements. Parts produced out of tolerance can greatly increase the dollar cost of production. These must be kept to a minimum, and this is the purpose of part inspection and calibration of measuring instruments.

INDICATING BENCH MICROMETER. The indicating bench micrometer, commonly called a "supermicrometer" (Figure 38), is used to inspect tools, parts, and gages. This instrument has a discrimination of .00002 in. (20 millionths).

SURFACE FINISH VISUAL COMPARATOR. Surface finish may be approximated by visual inspection using the surface roughness gage (Figure 39). Samples of finishes produced by various machining operations

Figure 39. Visual surface roughness comparator gage (Courtesy of the DoAll Company).

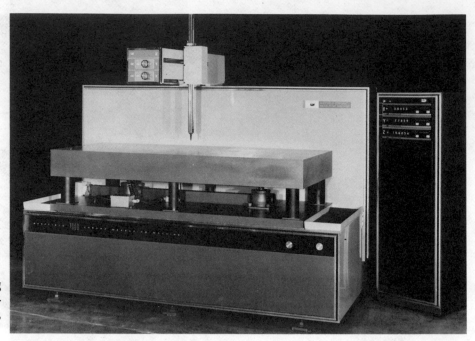

Figure 40. Coordinate measuring machine (Courtesy of the Automation and Measurement Division, Bendix Corporation).

are indicated on the gage. These can be visually compared to a machined surface to determine the approximate degree of surface finish.

COORDINATE MEASURING MACHINES. The coordinate measuring machine (Figure 40) is an extremely accurate instrument that can measure the workpiece in three dimensions. Coordinate measuring machines are very useful for determining the location of a part feature relative to a reference plane, line, or point.

Measurement with Electronics REMOTE GAGING. Electronic technology has come into wide use in measurement. Electronic equipment can be designed with greater sensitivity than mechanical equipment. Thus, higher discrimination

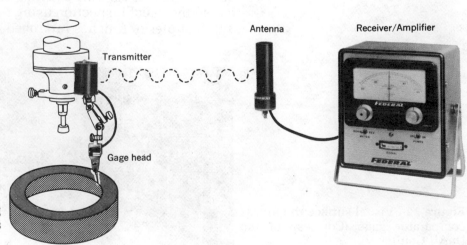

Figure 41. Remote electronic gaging system (Courtesy of Federal Products Corporation).

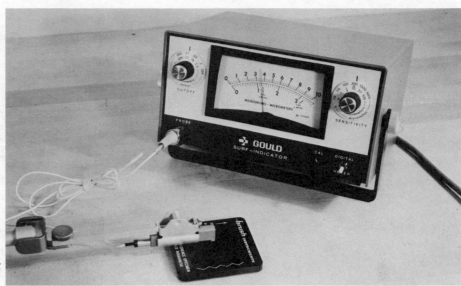

Figure 42. Surface finish indicator being calibrated (DeAnza College).

can be achieved. Electronics can be applied in remote gaging applications (Figure 41). In this application, there is no direct connection to the gage head. The head is free to move with the mechine tool since there are no attached wires. This facilitates use of the gaging instruments.

SURFACE FINISH INDICATORS. Surface finish is critical on many parts such as bearings, gears, and hydraulic cylinders. Surface finish is a measure of surface roughness or profile. The measurement is in microinches. A microinch is one millionth of an inch. A surface finish indicator (Figure 42) consists of a diamond stylus connected to a suitably graduated dial (meter). The stylus records surface deviations and these are indicated on the dial.

ELECTRONIC DIGITAL TRAVEL INDICATORS. Electronic digital travel indicators use a sensor attached to the machine tool. These systems will discriminate to .0001 in. and can be switched to read in metric dimensions. The travel of the machine component is indicated on a digital display (Figure 43). They are very useful for the accurate positioning of machine tables on such tools as milling machines and jig borers (Figure 44). A sensor on the machine tool detects movement of the machine components. The amount of travel is displayed on an electronic digital display. Discrimination may be as close as .0001 in.

ELECTRONIC COMPARATORS. **Electronic comparators** (Figure 45) take advantage of the sensitivity of electronic equipment. They are used to make comparison measurements of parts and other measuring tools. For example, gage blocks may be calibrated using a suitable electronic comparator.

Measurement with Light TOOLMAKER'S MICROSCOPE. The toolmaker's microscope (Figure 46) is used to inspect parts, cutting tools, and measuring tools. The microscope has a stage that can be precisely rotated and that can be

Figure 43. Electronic digital travel indicator display (Courtesy of Elm Systems, Inc.).

Figure 44. Electronic digital travel indicator system installed on a jib borer (Courtesy of Elm Systems, Inc.).

moved in two perpendicular axes. The instrument may be equipped with an electronic accessory measuring system that discriminates to .0001 in. Thus, stage movement can be recorded permitting measurements of a workpiece to be made.

OPTICAL COMPARATORS. The optical comparator (Figure 47) is used in the inspection of parts, cutting tools, and other measuring instruments. Optical comparators project a greatly magnified shadow of the object on a screen. The surface of the workpiece may also be illuminated. Shape patterns or graduated patterns can be placed on the screen and used to make measurements on the workpiece projection.

OPTICAL FLATS. Optical flats are used in the inspection of other measuring instruments and for the measurement of flatness. They can be used, for example, to reveal the surface geometry of a gage block or measuring faces of a micrometer (Figure 48). Optical flats take advantage of the principles of light interferometry to make extremely small measurements in millionths of an inch.

AUTOCOLLIMATORS. The autocollimator in Figure 49 is being used to check the flatness of a surface plate. The mirror on the left is moved

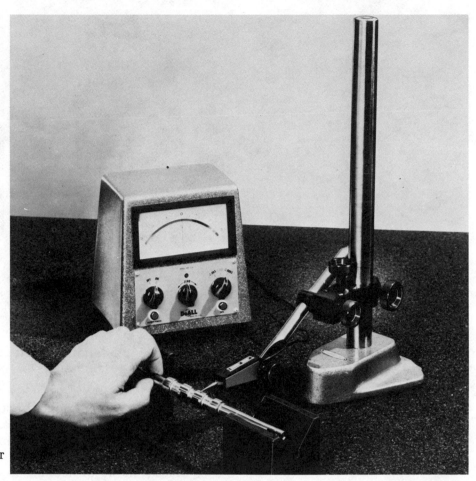

Figure 45. Electronic comparator (Courtesy of the DoAll Company).

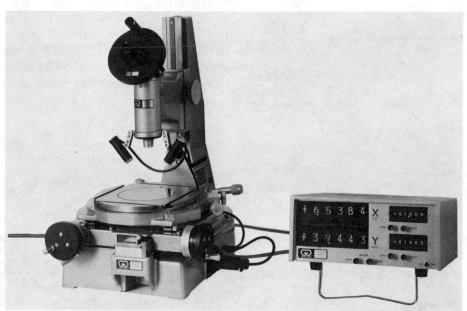

Figure 46. Toolmaker's microscope (Courtesy of Gaertner Scientific Corporation).

Figure 47. Optical comparator (Courtesy of Rank Precision Industries, Inc.).

Figure 48. Using optical flats to check micrometer measuring faces (Courtesy of the DoAll Company).

along the straightedge in small increments. Deviations from flatness are shown by angular changes of the mirror. This change is recorded by the instrument.

ALIGNMENT TELESCOPE. A machinist may accomplish alignment tasks by optical means. Optical alignment may be used on such applications as ship propeller shaft bearings. Portable machine tools such as boring bars may be positioned by optical alignment. The dual micrometer alignment telescope (Figure 50) is a very useful alignment instrument. The micrometers permit the deivation of the workpiece from the line of sight to be determined.

LASER INTERFEROMETER. The term laser is an acronym for "light activated stimulation of radiation." A laser light beam is a coherent beam. This means that each ray of light follows the same path. Thus, it does not disperse over long distances. This property makes the laser beam very useful in many measurement and alignment applications. For example, the laser beam may be used to determine how straight a machine tool table travels (Figure 51). Other uses include the checking of machine tool measuring systems (Figure 52).

Figure 49. Autocollimator checking a surface plate for flatness (DeAnza College).

Figure 50. Dual micrometer alignment telescope (Courtesy of the DoAll Company).

Figure 51. Laser interferometer being used for straightness determination (Courtesy of the Hewlett-Packard Company).

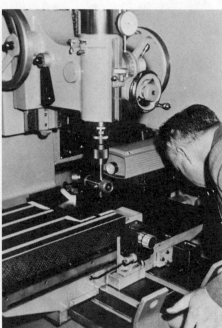

Figure 52. Laser interferometer checking the measuring system on a jig boring machine (DeAnza College).

INCH / METRIC

unit 1 systems of measurement

Throughout history there have been many systems of measurement. Prior to the era of national and international industrial operations, an individual craftsman was often responsible for the manufacture of a complete product. Since he made all the necessary parts and did the required assembly, he needed to conform only to his particular system of measurement. However, as machines replaced men and diversified mass production was established on a national and international basis, the need for standardization of measurement became readily apparent. Total standardization of measurement throughout the world still has not been fully realized. Most measurement in the modern world does, however, conform to either the English (inch-pound-second) or the metric (meter-kilogram-second) system. Metric measurement is now predominant in most of the industrialized nations of the world. The inch system is still used to a great extent in the United States. However, because of the interdependence of the world's industrial community, even the United States is turning more and more toward the use of metric measurement.

Today's machinist must now begin to think in terms of metric measurement. During his career he will come in contact with more and more metric specifications. However, for the present and the near future he will also be using inch measurement. Until a full transition to metrics takes place, you, as a machinist, may have to convert from one system to the other. Since you are primarily concerned with length measurement, this unit will review the basic length standards of both systems, examine mathematical and other methods of converting from system to system, look at techniques by which a machine tool can be converted to work in metrics.

objectives After completing this unit, you will be able to:

1. Identify common methods of measurement conversion.
2. Convert inch dimensions to metric equivalents and convert metric dimensions to inch equivalents.

THE ENGLISH SYSTEM OF MEASUREMENT

The English system of measurement uses the units of inches, pounds, and seconds to represent the measurement of time, length, and mass. Since we are primarily concerned with the measurement of length in the machine shop, we will simply refer to the English system as the **inch** system. Most of us are thoroughly familiar with inch measurement.

Subdivisions and Multiples of the Inch
The following table shows the common subdivisions and multiples of the inch that are used by the machinist.

Common Subdivisions

.000001	millionth
.00001	hundred thousandth
.0001	ten thousandth
.001	thousandth
.01	hundredth
.1	tenth
1.00	*Unit inch*

Common Multiples

12.00	1 foot
36.00	1 yard

Other common subdivisions of the inch are:

$\frac{1}{128}$	.007810 (decimal equivalent)
$\frac{1}{64}$	.015625
$\frac{1}{32}$	.031250
$\frac{1}{20}$	.050000
$\frac{1}{16}$	.062500
$\frac{1}{8}$	.125000
$\frac{1}{4}$	.250000
$\frac{1}{2}$	.500000

Multiples of Feet
3 feet = 1 yard
5280 feet = 1 mile

Multiples of Yards
1760 yards = 1 mile

THE METRIC SYSTEM AND THE INTERNATIONAL SYSTEM OF UNITS—SI

The basic unit of length in the metric system is the **meter**. Originally the length of the meter was defined by a natural standard, specifically a portion of the earth's circumference. Later, more convenient metal standards were constructed. In 1886, the metric system was legalized in the United States, but its use was not

made mandatory. Since 1893 the yard has been defined in terms of the metric meter by the ratio

$$1 \text{ yard} = \frac{3600}{3937} \text{ meter}$$

Although the metric system has been in use for many years in many different countries, it still lacked complete standardization among its users. Therefore, an attempt was made to modernize and standardize the metric system. From this effort has come the **Systeme International d'Unites**, known as **SI** or the **International Metric System.**

The basic unit of length in SI is the meter or metre (in the common international spelling). The SI meter is defined by a physical standard that can be reproduced anywhere with unvarying accuracy.

1 meter = 1,650,763.73 wavelengths in a vacuum of the orange-red light spectrum of the Krypton-86 atom

Probably the primary advantage of the metric system is that of convenience in computation. All subdivisions and multiples use 10 as a divisor or multiplier. This can be seen in the following table.

.000001	(one-millionth meter or micrometer)
.001	(one-thousandth meter or millimeter)
.01	(one-hundredth meter or centimeter)
.1	(one-tenth meter or decimeter)
1.00	*Unit meter*
10	(ten meters or one dekameter)
100	(100 meters or one hectometer)
1000	(1000 meters or one kilometer)
1,000,000	(one million meters or one megameter)

METRIC SYSTEM EXAMPLES

1. One meter (m) = _____ millimeters (mm).
Since a mm is $\frac{1}{1000}$ part of an m, there are 1000 mm in a meter.

2. 50 mm = _____ centimeters (cm).
Since 1 cm = 10 mm, $\frac{50}{10}$ = 5 cm in 50 mm.

3. Four kilometers (km) = _____ m.
Since 1 km = 1000 m then 4 km = 4000 m.

4. 582 mm = _____ cm.
Since 10 mm = 1 cm, $\frac{582}{10}$ = 58.2 cm.

CONVERSION BETWEEN SYSTEMS

Much of the difficulty with working in a two-system environment is experienced in converting from one system to the other. This can be of particular concern to the machinist as he must exercise due caution in making conversions. Arithmetic errors can be easily made. Therefore the use of a calculator is recommended.

Conversion Factors and Mathematical Conversion
Since the historical evolution of the inch and metric systems is quite different, there are no obvious relationships between length units of the two systems. You simply have to memorize the basic conversion factors. We know from the preceding discussion that the yard has been defined in terms of the meter. Knowing this relationship, you can derive mathematically any length unit in either system. However, the conversion factor

$$1 \text{ yard} = \frac{3600}{3937} \text{ meter}$$

is a less common factor for the machinist. A more common factor can be determined by the following:

$$1 \text{ yard} = \frac{3600}{3937} \text{ meter}$$

Therefore

$$1 \text{ yard} = .91440 \text{ meter}$$

$$\left(\frac{3600}{3937} \text{ expressed in decimal form} \right)$$

Then

$$1 \text{ inch} = \frac{1}{36} \text{ of .91440 meter}$$

So

$$\frac{.91440}{36} = .025400$$

We know that

$$1 \text{ m} = 1000 \text{ mm}$$

Therefore

$$1 \text{ inch} = .025400 \times 1000$$

Or

1 inch = 25.4000 mm

The conversion factor 1 inch = 25.4 mm is very common and should be memorized. From the example shown it should be clear that in order to find inches knowing millimeters, you must divide inches by 25.4.

$$1000 \text{ mm} = \underline{\hspace{1cm}} \text{ inches}$$

$$\frac{1000}{25.4} = 39.37 \text{ inches}$$

In order to simplify the arithmetic, any conversion can always take the form of a multiplication problem.

EXAMPLE

Instead of $\frac{1000}{25.4}$, multiply by the reciprocal of

25.4, which is $\frac{1}{25.4}$ or .03937.

Therefore, $1000 \times .03937 = 39.37$ inches

EXAMPLES OF CONVERSIONS [INCH TO METRIC]

1. 17 in. = \underline{\hspace{1cm}} cm.
   Knowing inches, to find centimeters multiply inches by 2.54: 2.54×17 in. = 43.18 cm.
2. .807 in. = \underline{\hspace{1cm}} mm.
   Knowing inches, to find millimeters multiply inches by 25.4: $25.4 \times .807$ in. = 20.49 mm

EXAMPLES OF CONVERSIONS [METRIC TO INCH]

1. .05 mm = \underline{\hspace{1cm}} in.
   Knowing millimeters, to find inches multiply millimeters by .03937: $.05 \times .03937$ = .00196 inches.
2. 1.63 m = \underline{\hspace{1cm}} in.
   Knowing meters, to find inches, multiply meters by 39.37: 1.63×39.37 m = 64.173 in.

CONVERSION FACTORS TO MEMORIZE

1 in. = 25.4 mm or 2.54 cm

1 mm = .03937 in.

Other Methods of Conversion
The **conversion chart** (Figure 1) is a popular device for making conversions between systems. Conversion charts are readily available from many manufacturers. However, most conversion charts give equivalents for whole millimeters or standard fractional inches. If you must find an equivalent for a factor that does not appear on the chart, you must interpolate. In this instance, knowing the common conversion factors and determining the equivalent mathematically is more efficient.

Several electronic calculators designed to convert directly from system to system are available. Of course, any calculator can and should

MILLIMETERS TO INCHES
(Basis: 1 inch = 25.4 millimeters)

Millimeters	Inches	Millimeters	Inches	Millimeters	Inches	Millimeters	Inches
1	0.039370	26	1.023622	51	2.007874	76	2.992126
2	.078740	27	1.062992	52	2.047244	77	3.031496
3	.118110	28	1.102362	53	2.086614	78	3.070866
4	.157480	29	1.141732	54	2.125984	79	3.110236
5	.196850	30	1.181102	55	2.165354	80	3.149606
6	.236220	31	1.220472	56	2.204724	81	3.188976
7	.275591	32	1.259843	57	2.244094	82	3.228346
8	.314961	33	1.299213	58	2.283465	83	3.267717
9	.354331	34	1.338583	59	2.322835	84	3.307087
10	.393701	35	1.377953	60	2.362205	85	3.346457
11	.433071	36	1.417323	61	2.401575	86	3.385827
12	.472441	37	1.456693	62	2.440945	87	3.425197
13	.511811	38	1.496063	63	2.480315	88	3.464567
14	.551181	39	1.535433	64	2.519685	89	3.503937
15	.590551	40	1.574803	65	2.559055	90	3.543307
16	.629921	41	1.614173	66	2.598425	91	3.582677
17	.669291	42	1.653543	67	2.637795	92	3.622047
18	.708661	43	1.692913	68	2.677165	93	3.661417
19	.748031	44	1.732283	69	2.716535	94	3.700787
20	.787402	45	1.771654	70	2.755906	95	3.740157
21	.826772	46	1.811024	71	2.795276	96	3.779528
22	.866142	47	1.850394	72	2.834646	97	3.818898
23	.905512	48	1.889764	73	2.874016	98	3.858268
24	.944882	49	1.929134	74	2.913386	99	3.897638
25	.984252	50	1.968504	75	2.952756	100	3.937008

Note: The above table is approximate: 1/25.4 0.039370078740

Figure 1. Metric conversion chart (Courtesy of the MTI Corporation).

be used to do a conversion problem. The direct converting calculator does not require that any conversion constant be remembered. These constants are permanently programmed into the calculator memory.

Converting Machine Tools

With the increase in metric measurement in industry, which predominantly uses the inch system, several devices have been developed that permit a machine tool to function in either system. These conversion devices eliminate the need to convert all dimensions prior to beginning a job.

Conversion equipment includes **conversion dials** (Figure 2) that can be attached to lathe cross slide screws as well as milling machine saddle and table screws. The dials are equipped with gear ratios that permit a direct metric reading to appear on the dial.

Metric mechanical and electronic travel indicators can also be used. The mechanical dial travel indicator (Figure 3) uses a roller that contacts a moving part of a machine tool. Travel of the machine component is indicated on the dial. This type of travel indicator discriminates to .01 millimeter. Whole millimeters are counted on the one millimeter counting wheel. Mechanical dial travel indicators are used in many ap-

Figure 2. Inch/metric conversion dials for machine tools (Courtesy of the Sipco Machine Co.).

Figure 3. Metric mechanical dial travel indicator (Courtesy of Southwestern Industries, Inc., Trav-A-Dial®).

Figure 4. Metric mechanical dial travel indicators reading milling machine saddle and table movement (Courtesy of Southwestern Industries, Inc., Trav-A-Dial®).

plications such as reading the travel of a milling machine saddle and table (Figure 4).

The electronic travel indicator (Figure 5) uses a sensor that is attached to the machine tool. Machine tool component travel is indicated on an electronic digital display. The equipment can be switched to read travel in inch or metric dimensions.

Metric conversion devices can be fitted to existing machine tools for a moderate expense. Many new machine tools, especially those built abroad, have dual system capability built into them.

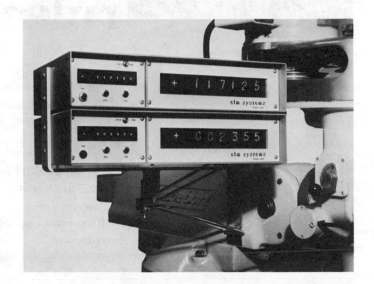

Figure 5. Inch/metric electronic travel indicator with digital display (Courtesy of Elm Systems, Inc.).

self-evaluation

SELF-TEST Perform the following conversions:

1. 35 mm =

2. 125 in. =

3. 6.273 in. =

4. Express the tolerance ± .050 in metric terms to the nearest mm.

5. To find cm knowing mm, _____ (multiply/divide) by 10.

6. Express the tolerance $\pm$ .02 mm in terms of inches to the nearest $\frac{1}{10\,000}$ inch.
7. What is meant by SI?
8. Describe methods by which conversions between metric and inch measurement systems may be accomplished.
9. How is the yard presently defined?
10. Can an inch machine tool be converted to work in metric units?

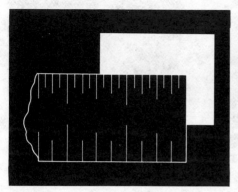

unit 2
using
steel
rules

One of the most practical and common measuring tools available in the machining and inspection of parts is the steel rule. It is a tool that the machinist uses daily in different ways. It is important that anyone engaged in machining be able to select and use steel rules.

objectives After completing this unit, you will be able to:

1. **Identify various kinds of rules and their applications.**

2. **Measure and record the dimensions of 10 objects that you have obtained from your instructor, using a fractional rule, decimal rule, and a metric rule.**

SCALES AND RULES

The terms **scale** and **rule** are often used interchangeably and often incorrectly. A rule is a linear measuring instrument whose graduations represent **real units** of lengths and their subdivisions. In contrast a **scale** is graduated into **imaginary** units that are either smaller or larger than the real units they represent. This is done for convenience where proportional measurements are needed. For example, an architect uses a scale that has graduations representing feet and inches. However, the actual length of the graduations on the architect's scale are quite different from full-sized dimensions.

DISCRIMINATION OF STEEL RULES

The general concept of **discrimination** was discussed in the introduction to this section. Discrimination refers to the extent to which a unit of length has been divided. If the smallest graduation on a specific rule is $\frac{1}{32}$ in., then the rule has a discrimination of, or discriminates to $\frac{1}{32}$ in. Likewise, if the smallest graduation of the rule is $\frac{1}{64}$ in., then this rule discriminates to $\frac{1}{64}$ in.

The maximum discrimination of a steel rule is generally $\frac{1}{64}$ in. or in the case of the decimal inch rule, $\frac{1}{100}$ in. The metric rule has a discrimination of $\frac{1}{2}$ mm (millimeter). Remembering that a measuring tool should never be used beyond its discrimination, the steel rule will not be reliable in trying to ascertain a measurement increment smaller than $\frac{1}{64}$ or $\frac{1}{100}$ in. If a specific measurement falls between the markings on the rule, only this can be said of this reading: it is more or less than the amount of the nearest mark. No further data as to how much more or less can be reliably determined. It is not recommended practice to attempt to read between the graduations on a steel rule with the intent of obtaining reliable readings.

RELIABILITY AND EXPECTATION OF ACCURACY IN STEEL RULES

For reliability, great care must be taken if the steel rule is to be used at its maximum discrimination. Remember that the markings on the rule occupy a certain width. A good quality steel rule has engraved graduations. This means that the markings are actually fine cut in the metal from which the rule is made. Of all types of graduations, engraved ones occupy the least width along the rule. Other rules, graduated by other processes, may have markings that occupy greater width. These rules are not necessarily any less accurate, but they may require more care in reading. Generally, the reliability of the rule will diminish as its maximum discrimination is approached. The smaller graduations are more difficult to see without the aid of a magnifier. Of particular importance is the point from which the measurement is taken. This is the **reference point** and must be carefully aligned at the point where the length being measured begins.

From a practical standpoint, the steel rule finds widest application for measurements no smaller than $\frac{1}{32}$ in. on a fractional rule or $\frac{1}{50}$ in. on a decimal rule. This does not mean that the rule cannot measure to its maximum discrimination, because under the proper conditions it certainly can. However, at or very near maximum discrimination, the time consumed to insure reliable measurement is really not justified. You will be more productive if you make use of a type of measuring instrument with considerably finer discrimination for measurements below the nearest $\frac{1}{32}$ or $\frac{1}{50}$ in. It is good practice to take more than one reading when using a steel rule. After determining the desired measurement, apply the rule once again to see if the same result is obtained. By this procedure, the reliability factor is increased.

TYPES OF RULES

Rules may be selected in many different shapes and sizes, depending on the need. The common **rigid steel rule** is six inches long, $\frac{3}{4}$ inch wide and $\frac{3}{64}$ in. thick. It is engraved with number 4 standard rule graduations. A number 4 graduation consists of $\frac{1}{8}$ and $\frac{1}{16}$ in. on one side (Figure 1) and $\frac{1}{32}$ and $\frac{1}{64}$ in. divisions on the reverse side (Figure 2). Other common graduations are summarized in the following table.

Graduation number	Front Side	Back Side
Number 3	32nds 64ths	10ths 50ths
Number 16	50ths 100ths	32nds 64ths

The number 16 graduated rule is often found in the aircraft industry where dimensions are specified in decimal fraction notations, based on 10 or a multiple of 10 divisions of an inch rather than 32 or 64 divisions as found on common rules. Many rigid rules are one inch wide.

Another common rule is the **flexible type** (Figure 3). This rule is six inches long, $\frac{1}{2}$ inch wide, and $\frac{1}{64}$ inch thick. Flexible rules are made

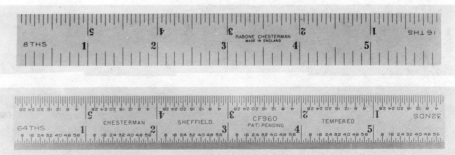

Figure 1. Six inch rigid steel rule (front side).

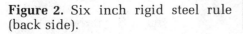

Figure 2. Six inch rigid steel rule (back side).

Figure 3. Flexible steel rule (metric).

Figure 4. Narrow rule (decimal inch).

Figure 5. Standard hook rule.

from hardened and tempered spring steel. One advantage of a flexible rule is that it will bend, permitting measurements to be made in a space shorter than the length of the rule. Most flexible rules are six or 12 inches long.

The **narrow rule** (Figure 4) is very convenient when measuring in small openings, slots, or holes. Most narrow rules have only one set of graduations on each side. These can be number 10, which is 32nds and 64ths, or number 11, which is 64ths and 100ths.

The **standard hook rule** (Figure 5) makes it possible to reach through an opening; the rule is hooked on the far side in order to measure a

thickness or the depth of a slot (Figure 6). When a workpiece has a chamfered edge, a hook rule will be advantageous over a common rule. Providing that the hook is not loose or excessively worn, it will provide an easy to locate reference point.

The **short rule set** (Figure 7) consists of a set of rules with a holder. Short rule sets have a range of $\frac{1}{4}$ to 1 in. They can be used to measure shoulders in holes or steps in slots, where space is extremely limited. The holder will attach to the rules at any angle, making these very versatile tools.

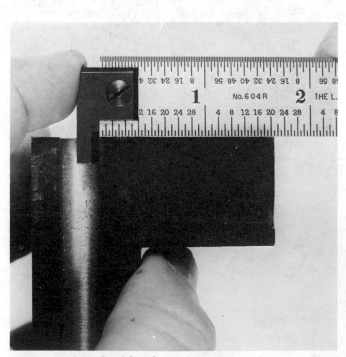

Figure 6. Standard hook rule in use.

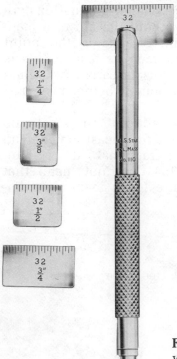

Figure 7. Short rule set with holder (Courtesy of the L. S. Starrett Co).

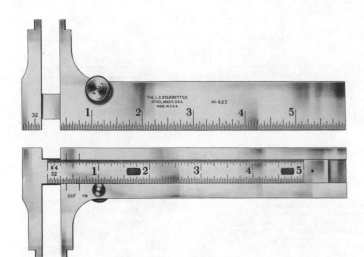

Figure 8. Slide caliper rule (Courtesy of the L. S Starrett Co.).

The **slide caliper rule** (Figure 8) is a versatile tool used to measure round bars, tubing, and other objects where it is difficult to measure at the ends and difficult to estimate the diameter with a rigid steel rule. The small slide

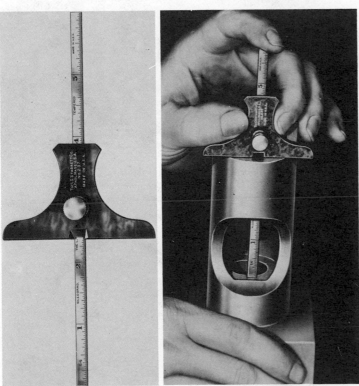

Figure 9. (*left*) Rule depth gage (Courtesy of the L. S. Starrett Co.).

Figure 10. (*right*) Rule depth gage in use (Courtesy of the L. S. Starrett Co.).

caliper rule can also be used to measure internal dimensions from $\frac{1}{4}$ in. up to the capacity of the tool.

The **rule depth gage** (Figure 9) consists of a slotted steel head in which a narrow rule slides. For depth measurements the head is held securely against the surface with the rule extended into the cavity or hole to be measured (Figure 10). The locking nut is tightened and the rule depth gage can then be removed and the dimension determined.

CARE OF RULES

Rules are precision tools, and only those that are properly cared for will provide the kind of service they are designed to give. A rule should not be used as a screwdriver. Rules should be kept separate from hammers, wrenches, files, and other hand tools to protect them from possible damage. An occasional wiping of a rule with a lightly oiled shop towel will keep it clean and free from rust.

APPLYING STEEL RULES

When using a steel rule in close proximity to a machine tool, **always keep safety in mind.** Stop the machine before attempting to make any measurements of the workpiece. Attempting to measure with the machine running may result in the rule being caught by a moving part. This may damage the rule, but worse, may result in serious injury to the operator.

One of the problems associated with the use of rules is that of **parallax error.** Parallax error is the error that results when the observer making the measurement is not in line with the workpiece and the rule. You may see the graduation either too far left or too far right of its real position (Figure 11). Parallax error occurs when the rule, is read from a point other than one directly above the point of measurement. The **point of measurement** is the point at which the measurement is read. It may or may not be the true reading of the size depending on what location was used as the reference point on the rule. Parallax can be controlled by always observing the point of measurement from **directly above.** Furthermore, the graduations on a rule should be placed as close as possible to the surface being measured. In this regard, a thin rule is preferred over a thick rule.

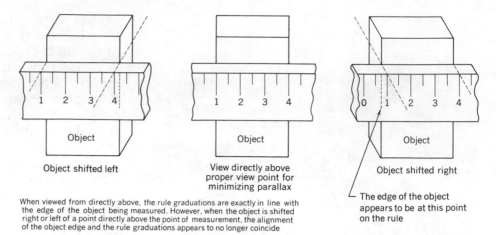

Figure 11. Parallax error.

When viewed from directly above, the rule graduations are exactly in line with the edge of the object being measured. However, when the object is shifted right or left of a point directly above the point of measurement, the alignment of the object edge and the rule graduations appears to no longer coincide

Object shifted left

View directly above proper view point for minimizing parallax

Object shifted right

The edge of the object appears to be at this point on the rule

As a rule is used it becomes worn, usually on the ends. The outside inch markings on a worn rule are less than one inch from the end. This has to be considered when measurements are made. A reliable way to measure (Figure 12) is to **use the one inch mark on the rule as the reference point.** In the figure, the measured point is at $2\frac{1}{32}$. Subtracting one inch results in a size of $1\frac{1}{32}$ for the part.

Round bars and tubing should be measured with the rule applied on the end of the tube or bar (Figure 13). Select a reference point and set it carefully at a point on the circumference of the round part to be measured. Using the reference point as a pivot, move the rule back and forth slightly to find the largest distance across the diameter. When the largest distance is determined, read the measurement at that point.

Rules are also used for transfer measurements with calipers. The caliper is set to the part, and the reading is obtained by use of the rule. Both inside and outside calipers can be used in this manner (Figures 14 and 15).

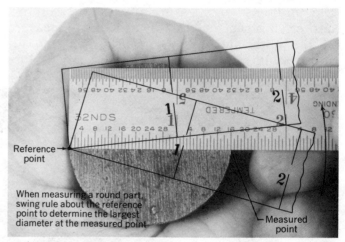

Reference point

When measuring a round part, swing rule about the reference point to determine the largest diameter at the measured point

Measured point

Figure 13. Measuring round objects.

Figure 12. Using the one inch mark as the reference point.

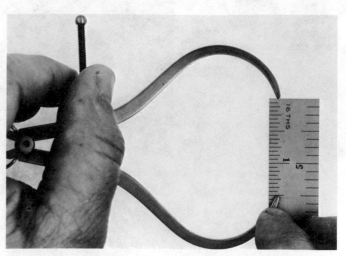

Figure 14. Using a rule to set an outside caliper.

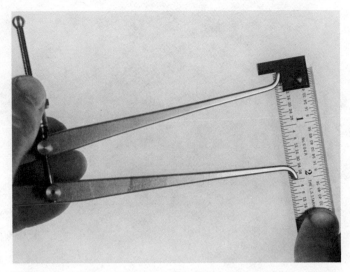

Figure 15. Using a rule to set an inside caliper.

READING FRACTIONAL INCH RULES

Most dimensions are expressed in inches and fractions of inches. These dimensions are mea-sured with fractional inch rules. The typical ma-chinist's rule is broken down into $1, \frac{1}{2}, \frac{1}{4}, \frac{1}{8}, \frac{1}{16}, \frac{1}{32}$, and $\frac{1}{64}$ in. graduations. In order to facil-itate reading, the $1, \frac{1}{2}, \frac{1}{4}, \frac{1}{8}$, and $\frac{1}{16}$ in. graduations appear on one side of the rule (Figure 16). The reverse side of the rule has one edge graduated in $\frac{1}{32}$ in. increments and the other edge grad-uated in $\frac{1}{64}$ in. increments. On the $\frac{1}{32}$ in. side, every fourth mark is numbered and on the $\frac{1}{64}$ in. side, every eighth mark is numbered (Figure 17). This eliminates the need to count gradua-tions from the nearest whole inch mark. On these rules, the length of the graduation line varies with the one inch line being the longest, the $\frac{1}{2}$ in. line being next in length, the $\frac{1}{4}, \frac{1}{8}$, and $\frac{1}{16}$ in. lines each being consecutively shorter. The difference in line lengths is an important aid in reading a rule. The smallest graduation on any edge of a rule is marked by small num-bers on the end. Note that the words 8THS and 16THS appear at the ends of the rule. The num-bers 32NDS and 64THS appear on the reverse side of the rule, thus indicating thirty-seconds and sixty-fourths of an inch.

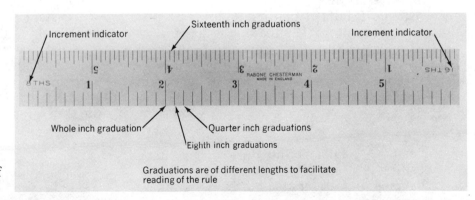

Figure 16. Front side graduations of the typical machinist's rule.

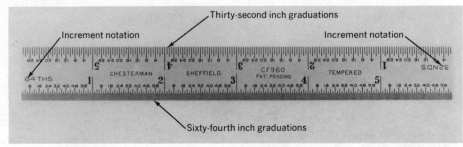

Figure 17. Back side graduations of the typical machinist's rule.

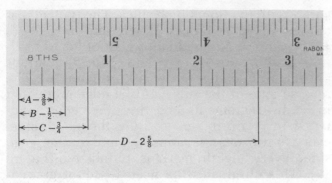

Figure 18. Examples of readings on the $\frac{1}{8}$ in. discrimination edge.

EXAMPLES OF FRACTIONAL INCH READINGS:
FIGURE 18

Distance A falls on the third $\frac{1}{8}$ in. graduation. This reading would be $\frac{3}{8}$ in.

Distance B falls on the longest graduation between the end of the rule and the first full inch mark. The reading is $\frac{1}{2}$ in.

Distance C falls on the sixth $\frac{1}{8}$ in. graduation making it $\frac{6}{8}$ or $\frac{3}{4}$ in.

Distance D falls at the fifth $\frac{1}{8}$ in. mark beyond the 2 in. graduation. The reading is $2\frac{5}{8}$ in.

FIGURE 19

Distance A falls at the thirteenth $\frac{1}{16}$ in. mark making the reading $\frac{13}{16}$ in.

Distance B falls at the first $\frac{1}{16}$ in. mark past the 1 in. graduation. The reading is $1\frac{1}{16}$ in.

Distance C falls at the seventh $\frac{1}{16}$ in. mark past the 1 in. graduation. The reading is $1\frac{7}{16}$ in.

Distance D falls at the third $\frac{1}{16}$ in. mark past the 2 in. graduation. The reading is $2\frac{3}{16}$ in.

FIGURE 20

Distance A falls at the third $\frac{1}{32}$ in. mark. The reading is $\frac{3}{32}$ in.

Distance B falls at the ninth $\frac{1}{32}$ in. mark. The reading is $\frac{9}{32}$ in.

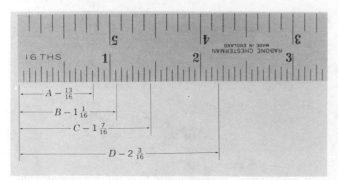

Figure 19. Examples of readings on the $\frac{1}{16}$ in. discrimination edge.

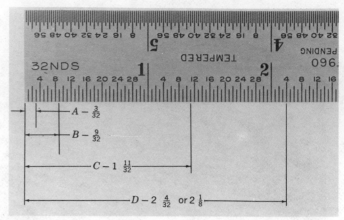

Figure 20. Examples of readings on the $\frac{1}{32}$ in. discrimination edge.

Distance C falls at the eleventh $\frac{1}{32}$ in. mark past the 1 in. graduation. The reading is $1\frac{11}{32}$ in.

Distance D falls at the fourth $\frac{1}{32}$ in. mark past the 2 in. graduation. The reading is $2\frac{4}{32}$ in., which reduced to lowest terms becomes $2\frac{1}{8}$ in.

FIGURE 21

Distance A falls at the ninth $\frac{1}{64}$ in. mark making the reading $\frac{9}{64}$ in.

Distance B falls at the fifty seventh $\frac{1}{64}$ in. mark making the reading $\frac{57}{64}$ in.

Distance C falls at the thirty-third $\frac{1}{64}$ in. mark past the 1 in. graduation. The reading is $1\frac{33}{64}$ in.

Distance D falls at the first $\frac{1}{64}$ in. mark past the 2 in. graduation, making the reading $2\frac{1}{64}$ in.

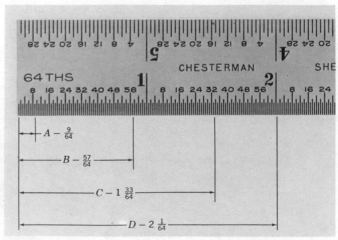

Figure 21. Examples of readings on the $\frac{1}{64}$ in. discrimination edge.

self-evaluation

SELF-TEST Read and record the dimensions indicated by the letters *A* to *H* in Figures 22.

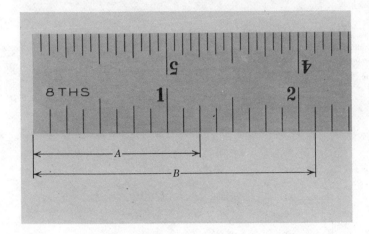

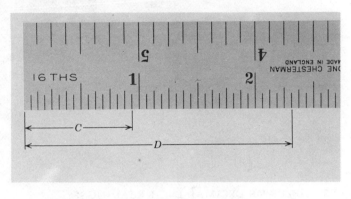

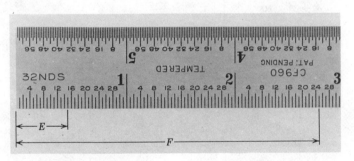

Figure 22. *E* and *F*.

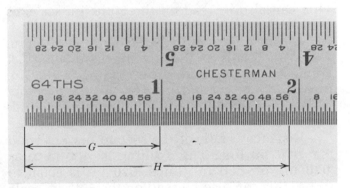

Figure 22. *G* and *H*.

READING DECIMAL INCH RULES

Many dimensions in the auto, aircraft, and missile industries are specified in **decimal notations,** which refers to the division of the inch into 10 parts or a multiple of 10 parts, such as 50 or 100 parts. In this case, a **decimal rule** would be used. Decimal inch dimensions are specified and read as thousandths of an inch. Decimal rules, however, do not discriminate to the individual thousandth because the width of an engraved or etched division on the rule is approximately .003 in. (three thousandths of an

inch). Decimal rules are commonly graduated in increments of $\frac{1}{10}$ in., $\frac{1}{50}$ in., or $\frac{1}{100}$ in.

A typical decimal rule may have $\frac{1}{50}$ in. divisions on the top edge and $\frac{1}{100}$ in. divisions on the bottom edge (Figure 23). The inch is divided into 10 equal parts, making each numbered division $\frac{1}{10}$ in. or .100 in. (100 thousandths of an inch). On the top scale each $\frac{1}{10}$ increment is further subdivided into five equal parts, which makes the value of each of these divisions .020 in. (20 thousandths of an inch).

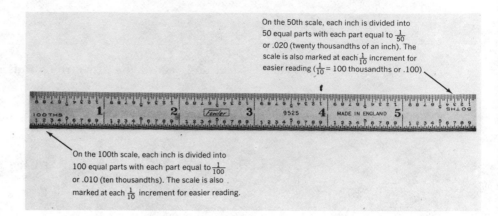

Figure 23. Six inch decimal rule.

EXAMPLES OF DECIMAL INCH READINGS:
FIGURE 24

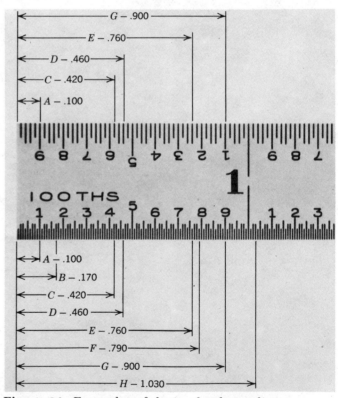

Figure 24. Examples of decimal rule readings.

Distance *A* falls on the first marked graduation. The reading is $\frac{1}{10}$ or .100 thousandths in. This can also be read on the 50ths in. scale, as seen in the figure.

Distance *B* can only be read on the 100th in. scale, as it falls at the seventh graduation beyond the .10 in. mark. The reading is .100 in. plus .070 in., or .170 in. This distance cannot be read on the 50th in. scale because discrimination of the 50th in. scale is not sufficient.

Distance *C* falls at the second mark beyond the .400 in. line. This reading is .400 in. plus .020 in., or .420 in. Since .020 in. is equal to $\frac{1}{50}$ in., this can also be read on the 50th in. scale as shown in the figure.

Distance *D* falls at the sixth increment beyond the .400 in. line. The reading is .400 in. plus .060 in., or .460 in. This can also be read on the 50th in. scale, as seen in the figure.

Distance *E* falls at the sixth division beyond the .700 in. mark. The reading is .700 in. plus .060 in., or .760 in. This can also be read on the 50th in. scale.

Distance *F* falls at the ninth mark beyond the .700 in. line. The reading is .700 in. plus .090 in., or .790 in. This cannot be read on the 50th in. scale.

Distance *H* falls three marks past the first full inch mark. The reading is 1.00 in. plus .030 in., or 1.030 in. This cannot be read on the 50th in. scale.

self-evaluation

SELF-TEST Read and record the dimensions indicated by the letters *A* to *E* in Figures 25.

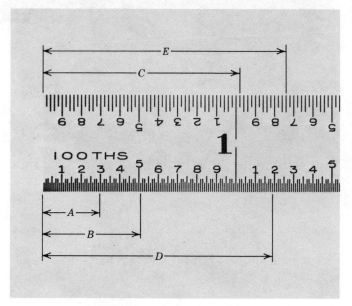

Figure 25. Decimal inch rule.

READING METRIC RULES

Many products are made in metric dimensions requiring a machinist to use a **metric rule.** The typical metric rule has millimeter (mm) and half millimeter graduations (Figure 26).

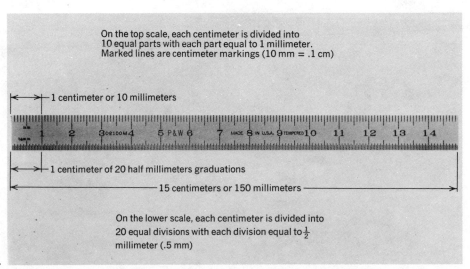

Figure 26. 150 millimeter metric rule.

EXAMPLES OF READING METRIC RULES:
FIGURE 27

Distance *A* falls at the 53rd graduation on the mm scale. The reading is 53 mm.

Distance *B* falls at the 22nd graduation on the mm scale. The reading is 22 mm.

Distance *C* falls at the sixth graduation on the mm scale. The reading is 6 mm.

Distance *D* falls at the 18th half mm mark. The reading is 8 mm plus an additional half mm, giving a total of 8.5 mm.

Distance *E* falls one half mm beyond the 3 centimeter (cm) graduation. Since 3 cm is equal to 30 mm, the reading is 30.5 mm.

Distance *F* falls one half mm beyond the 51 mm graduation. The reading is 51.5 mm.

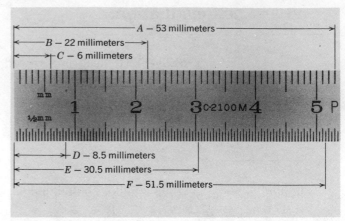

Figure 27. Examples of metric rule readings.

In machine design, all dimensions are specified in mm. Hence, 1.5 meters (m) would be 1500 mm.

self-evaluation

SELF-TEST Read and record the dimensions indicated by the letters *A* to *F* in Figure 28.

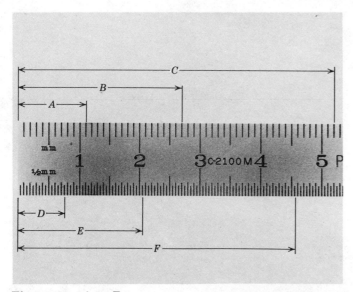

Figure 28. *A* to *F*.

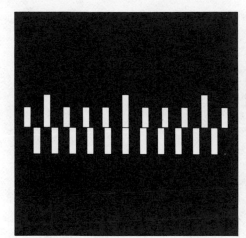

unit 3
using vernier calipers and vernier depth gages

The inspection and measurement of machined parts requires various kinds of measuring tools. Often the discrimination of a rule is sufficient, but, in many cases the discrimination of a rule with a vernier scale is required. This unit explains the types, use and applications of common vernier instruments.

objectives After completing this unit, you will be able to:

1. Measure and record the dimensions of 10 objects to an accuracy of plus or minus .001 in. with a vernier caliper.
2. Measure and record the dimensions of 10 objects to an accuracy of plus or minus .02 mm using a metric vernier caliper.
3. Measure and record the dimensions of 10 objects using a vernier depth gage.

PRINCIPLE OF THE VERNIER

The principle of the **vernier** may be used to increase the discrimination of all graduated scale measuring tools used by a machinist. A vernier system consists of a **main scale** and a **vernier scale.** The vernier scale is placed adjacent to the main scale so that graduations on both scales can be observed together. The spacing of the vernier scale graduations is shorter than the spacing of the main scale graduations. For example, consider a main scale divided as shown (Figure 1a). It is desired to further subdivide each main scale division into 10 parts with the use of a vernier. The spacing of each vernier scale division is made $\frac{1}{10}$ of a main scale division shorter than the spacing of a main scale division. This may sound confusing but,

think of it as 10 vernier scale divisions corresponding to nine main scale divisions (Figure 1a). The vernier now permits the main scale to discriminate to $\frac{1}{10}$ of its major divisions. Therefore, $\frac{1}{10}$ is known as the **least count** of the vernier.

The vernier functions in the following manner. Assume that the zero line on the vernier scale is placed as shown (Figure 1b). The reading on the main scale is two, plus a fraction of a division. It is desired to know the amount of the fraction over two, to the nearest tenth or least count of the vernier. As you inspect the alignment of the vernier scale and the main scale lines, you will note that they move closer together until one line on the vernier scale **coincides** with a line on the main scale. This is the

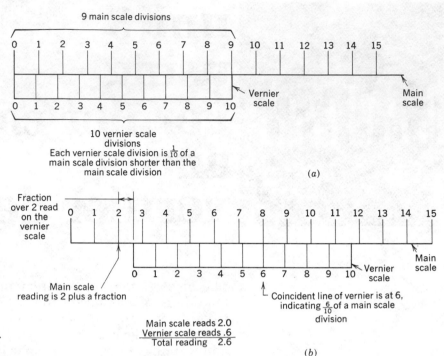

Figure 1a, 1b. Principle of the vernier.

coincident line of the vernier and indicates the fraction in tenths that must be added to the main scale reading. The vernier is coincident at the sixth line. Since the least count of the vernier is $\frac{1}{10}$, the zero vernier line is six-tenths past two on the main scale. Therefore, the main scale reading is 2.6 (Figure 1b).

DISCRIMINATION AND APPLICATIONS OF VERNIER INSTRUMENTS

Vernier instruments used for linear measure in the inch system discriminate to .001 in. ($\frac{1}{1000}$). Metric verniers generally discriminate to .02 ($\frac{1}{50}$) of a millimeter.

The most common vernier instruments include several styles of **calipers**. The common vernier caliper is used for outside and inside linear measurement. Another style of vernier caliper has the capability depth measurement in addition to outside and inside capacity. The vernier also appears on a variety of depth gages.

Beyond its most common applications, the vernier also appears on a height gage, which is an extremely important layout tool for a machinist. The vernier is also used on the gear tooth caliper, a special vernier caliper used in gear measurement. As the principle of the vernier can be used to subdivide a unit of angular measure as well as linear measure, it appears

on various types of protractors used for angular measurement.

RELIABILITY AND EXPECTATION OF ACCURACY IN VERNIER INSTRUMENTS

Reliability in vernier calipers and depth gages is highly dependent on proper use of the tool. The simple fact that the caliper or depth gage has increased discrimination over a rule does not necessarily provide increased reliability. The improved degree of discrimination in vernier instruments requires more than the mere visual alignment of a rule graduation against the edge of the object to be measured. The zero reference point of a vernier caliper is the positively placed contact of the solid saw with the part to be measured. On the depth gage, the base is the zero reference point. Positive contact of the zero reference is an important consideration in vernier reliability.

The vernier scale must be read carefully if a reliable measurement is to be determined. On many vernier instruments the vernier scale should be read with the aid of a magnifier. Without this aid, the coincident line of the vernier is difficult to determine. Therefore, the reliability of the vernier readings can be in question. The typical vernier caliper has very narrow jaws and thus must be carefully aligned with the axis of measurement. On the plain slide vernier caliper, no provision is made for the "feel" of the measuring pressure. Some calipers and the depth gage are equipped with a screw thread fine adjustment that gives them a slight advantage in determining the pressure applied during the measurement.

Generally, the overall reliability of vernier instruments for measurement at maximum discrimination of .001 is fairly low. The vernier should never be used in an attempt to discriminate below .001. The instrument does not have that capability. Vernier instruments are a popular tool on the inspection bench, and they can serve very well for measurement in the range of plus or minus .005 of an inch. With proper use and an understanding of the limitations of a vernier instrument, this tool can be a valuable addition to the many measuring tools available to you.

VERNIER CALIPERS

With a rule, measurements can be made to the nearest $\frac{1}{64}$ or $\frac{1}{100}$ in., but often this is not sufficiently accurate. A measuring tool based on a rule but with much greater discrimination is the **vernier caliper** (Figure 2). Vernier calipers have a discrimination of .001 in. The **beam** or **bar** is engraved with the **main scale.** This is also called the **true scale,** as each inch marking

is exactly one inch apart. The beam and the solid jaw are square, or at 90° to each other.

The movable jaw contains the **vernier scale.** This scale is located on the sliding jaw of a vernier caliper or it is part of the base on the vernier depth gage. The function of the vernier scale is to subdivide the minor divisions on the beam scale into the smallest increments that the vernier instrument is capable of measuring. For example, a 25 division vernier subdivides the minor divisions of the beam scale into 25 parts. Since the minor divisions are equal to .025 thousandths of an inch, the vernier divides them into increments of .001 of an inch. This is the finest discrimination of the instrument.

Most of the longer vernier calipers have a fine adjustment clamp for precise adjustments of the movable jaw. Inside measurements are made over the **nibs** on the jaw and are read on the top scale of the vernier caliper (Figure 2). The top scale is a duplicate of the lower scale, with the exception that it is offset to compensate for the size of the nibs.

The "Mauser pattern or Cross Horn" vernier caliper is very common (Figure 3). This is a versatile tool because of its capacity to make outside, inside, and depth measurements. Many different measuring applications are made with this particular design of vernier caliper (Figure 4).

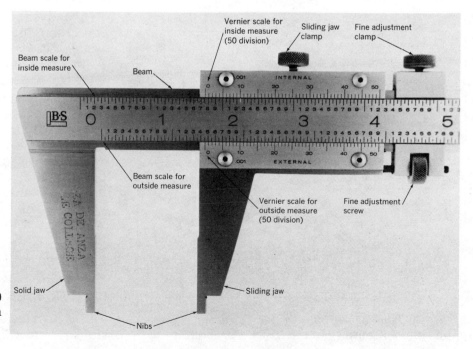

Figure 2. Typical inside-outside, 50 division vernier caliper (DeAnza College).

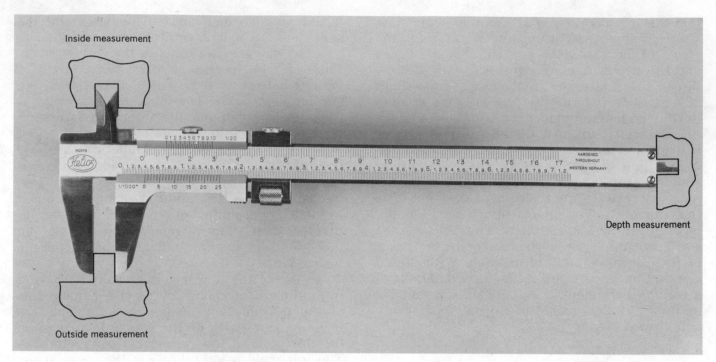

Inside measurement

Outside measurement

Depth measurement

Figure 3. Typical "Cross Horn" or "Mauser Pattern" vernier caliper.

VERNIER CALIPER PROCEDURES

To test a vernier caliper for accuracy, clean the contact surfaces of the two jaws. Bring the movable jaw with normal gaging pressure into contact with the solid jaw. Hold the caliper against a light source and examine the alignment of the solid and movable jaws. If wear exists, a line of light will be visible between the

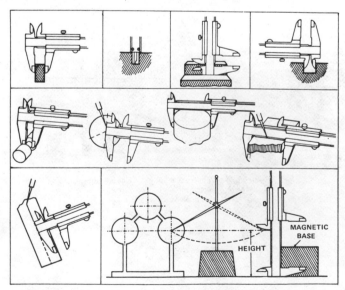

Figure 4. This design of vernier caliper has many applications (Courtesy of the M.T.I. Corporation).

jaw faces. A gap as small as .0001 ($\frac{1}{10,000}$) of an inch can be seen against a light. If the contact between the jaws is satisfactory, check the vernier scale alignment. The vernier scale zero mark should be in alignment with the zero on the main scale. Realignment of the vernier scale to adjust it to zero can be accomplished on some vernier calipers.

A vernier caliper is a delicate precision tool and should be treated as such. It is very important that the correct amount of pressure or feel is developed while taking a measurement. The measuring jaws should contact the workpiece firmly. However, excessive pressure will spring the jaws and give inaccurate readings. When measuring an object, use the solid jaw as the reference point. Then move the sliding jaw until contact is made. When measuring with the vernier caliper make certain that the beam of the caliper is in line with the surfaces being measured. Whenever possible, read the vernier caliper while it is still in contact with the workpiece. Moving the instrument may change the reading. Any measurement should be taken at least twice to assure reliability.

READING INCH VERNIER CALIPERS

Vernier scales are engraved with 25 or 50 divisions (Figures 5 and 6). On a 25 division vernier

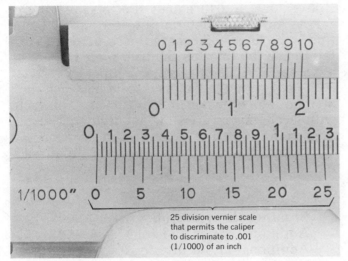

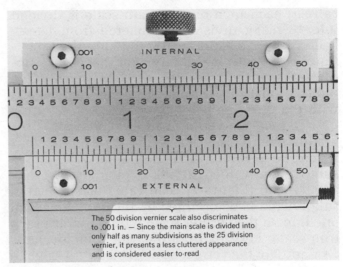

25 division vernier scale that permits the caliper to discriminate to .001 (1/1000) of an inch

1/1000"

The 50 division vernier scale also discriminates to .001 in. — Since the main scale is divided into only half as many subdivisions as the 25 division vernier, it presents a less cluttered appearance and is considered easier to read

Figure 5. Lower scale is a 25 division vernier.

Figure 6. Fifty division vernier caliper.

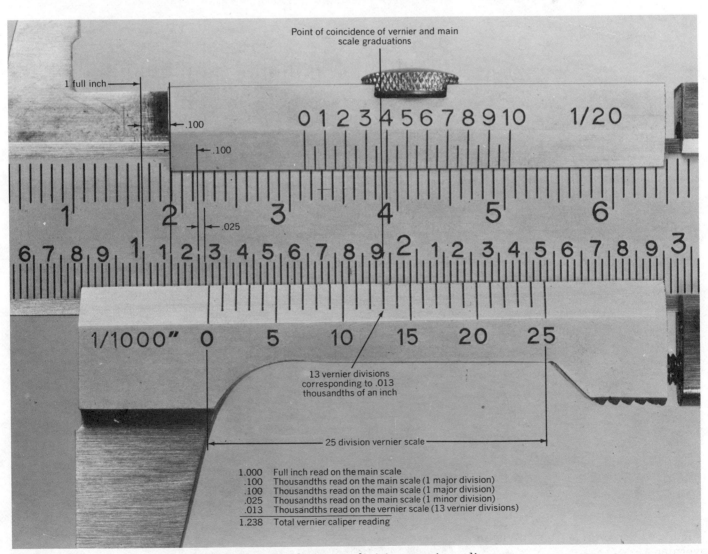

Point of coincidence of vernier and main scale graduations

1 full inch

.100

.100

.025

1/20

13 vernier divisions corresponding to .013 thousandths of an inch

25 division vernier scale

1/1000"

1.000	Full inch read on the main scale
.100	Thousandths read on the main scale (1 major division)
.100	Thousandths read on the main scale (1 major division)
.025	Thousandths read on the main scale (1 minor division)
.013	Thousandths read on the vernier scale (13 vernier divisions)
1.238	Total vernier caliper reading

Figure 7. Reading a 25 division vernier caliper.

caliper, each inch on the main scale is divided into 10 major divisions numbered from 1 to 9. Each major division is .100 (one hundred thousandths). Each major division has four subdivisions with a spacing of .025 (twenty-five thousandths). The vernier scale has 25 divisions with the zero line being the index.

To read the vernier caliper, count all of the graduations to the left of the index line. This would be 1 whole inch plus $\frac{2}{10}$ or .200, plus 1 subdivision valued at .025, plus part of one subdivision (Figure 7). The value of this partial subdivision is determined by the coincidence of one line on the vernier scale with one line of the true scale. For this example, the coincidence is on line 13 of the vernier scale. This is the value in thousandths of an inch that has to be added to the value read on the beam. Therefore, 1 + .100 +

.100 + .025 + .013 equals the total reading of 1.238. An aid in determining the coincidental line is the **lines adjacent to the coincidental line fall inside the lines on the true scale** (Figure 8).

The 50 division vernier caliper is read as follows (Figure 9). The true scale has each inch divided into 10 major divisions of .100 in. each, with each major division subdivided in half, thus being .050 in. The vernier scale has 50 divisions. The 50 division vernier caliper reading shown is read as follows:

Beam whole inch reading	1.000
Additional major divisions	.400
Additional minor divisions	.050
Vernier scale reading	.009
Total caliper reading	1.459

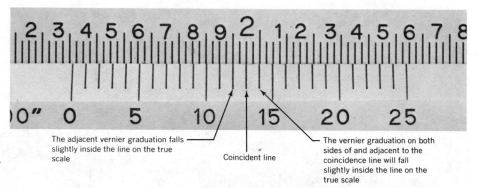

Figure 8. Determining the coincident line on a vernier.

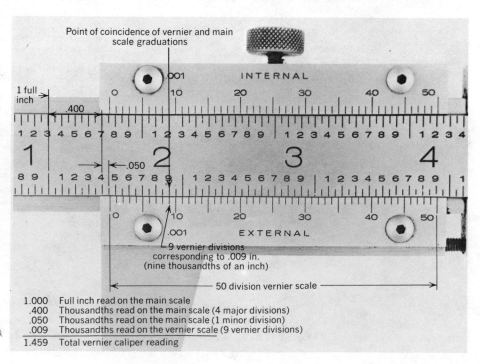

Figure 9. Reading a 50 division vernier.

DIAL CALIPER

An outgrowth from the vernier caliper is the dial caliper (Figure 10). However, this instrument does not employ the principle of the vernier. The beam scale on the dial caliper is graduated only into .10 in. increments. The caliper dial is either graduated into 100 or 200 divisions. The dial hand is operated by a pinion gear that engages a rack on the caliper beam. On the 100 division dial, the hand makes one complete revolution for each .10 in. movement of the sliding jaw along the beam. Therefore, each dial graduation represents $\frac{1}{100}$ of .10 in., or .001 in. maximum discrimination. On the 200 division dial the hands makes only one half a revolution for each .10 in. of movement along the beam. Discrimination is also .001 in.

Since the dial caliper is direct reading, the need to determine the coincident line of a vernier scale is eliminated. This greatly facilitates reading of the instruments and, for this reason, the dial caliper has all but replaced its vernier counterpart in many applications. When using

Figure 10. Dial caliper (Harry Smith & Associates).

the dial caliper, remember what you have learned about the expectation of accuracy in caliper instruments.

self-evaluation

SELF-TEST—READING INCH VERNIER CALIPERS

Determine the dimensions in the vernier caliper illustrations (Figures 11a to d):

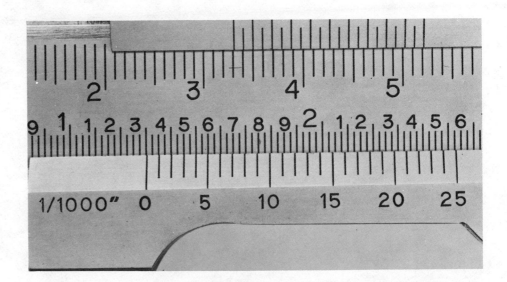

Figure 11a.

136

Figure 11*b*.

Figure 11*c*.

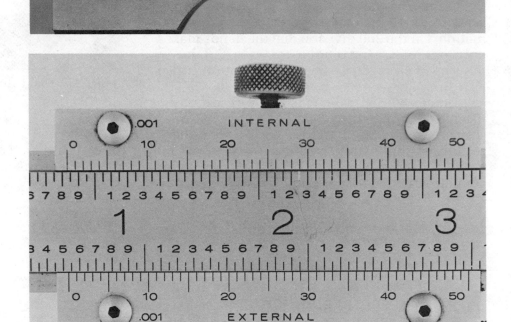

Figure 11*d*.

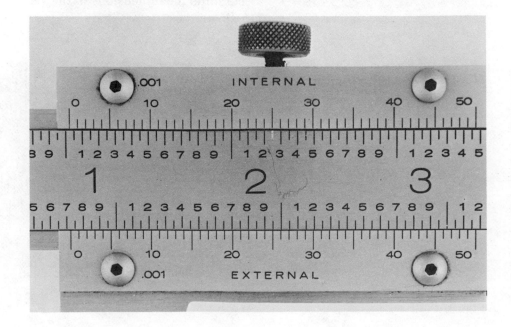

READING METRIC VERNIER CALIPERS

The applications for a metric vernier caliper are exactly the same as those described for an inch system vernier caliper. The discrimination of metric vernier caliper models varies from .02 millimeter, .05 mm, or .1 mm. The most commonly used type discriminates to .02 mm. The main scale on a metric vernier caliper is divided into millimeters with every tenth millimeter mark numbered. The 10 millimeter line is numbered 1, the 20 millimeter line is numbered 2 and so on, up to the capacity of the tool (Figure 12). The vernier scale on the sliding jaw is divided into 50 equal spaces with every 5th space numbered. Each numbered division on the vernier represents one tenth of a milli-

meter. The five smaller divisions between the numbered lines represent two hundredths (.02 mm) of a millimeter.

To determine the caliper reading, read, on the main scale, whole millimeters to the left of the zero or the index line on the sliding jaw. The example (Figure 11) shows 27 mm plus part of an additional millimeter. The vernier scale coincides with the main scale at the 18th vernier division. Since each vernier scale spacing is equal to .02 mm, the reading on the vernier scale is equal to 18 times .02, or .36 mm. Therefore, .36 mm must be added to the amount showing on the main scale to obtain the final reading. The result is equal to 27 mm + .36 mm or 27.36 mm. (Figure 11).

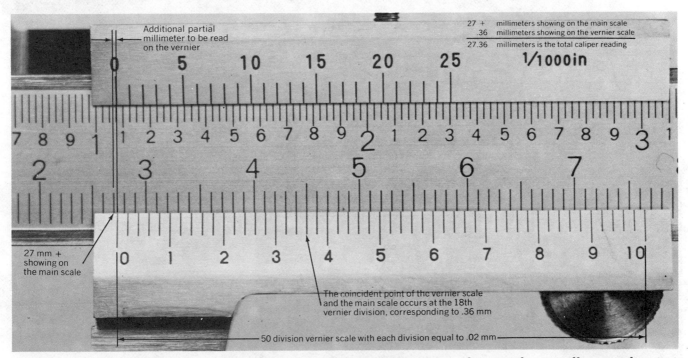

Figure 12. Reading a metric vernier caliper with .02 millimeter discrimination.

self-evaluation

SELF-TEST—READING METRIC VERNIER CALIPERS

Determine the metric vernier caliper dimensions illustrated in Figures 13a to d:

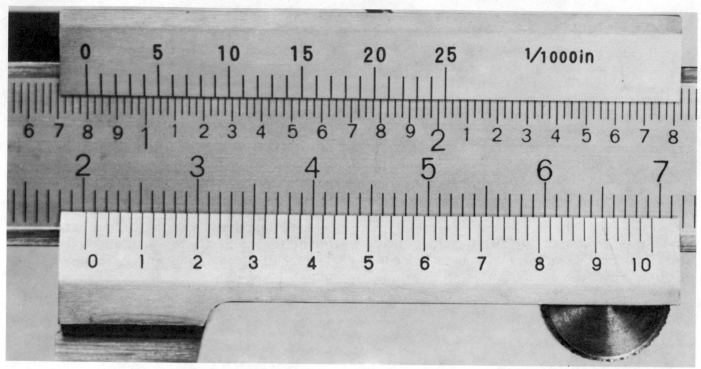

Figure 13a.

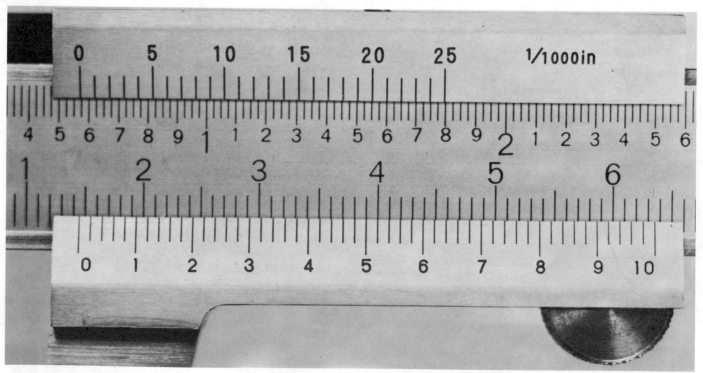

Figure 13b.

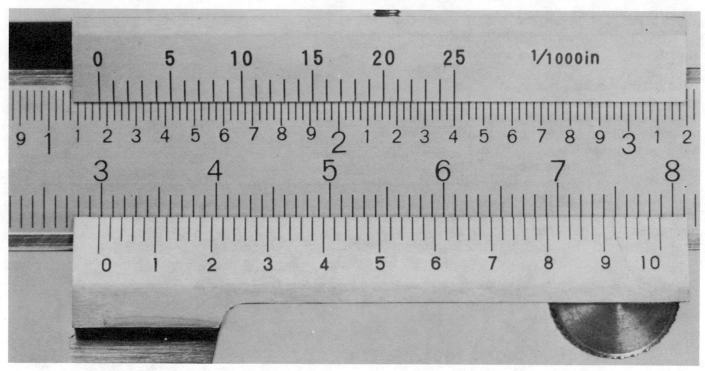

Figure 13c.

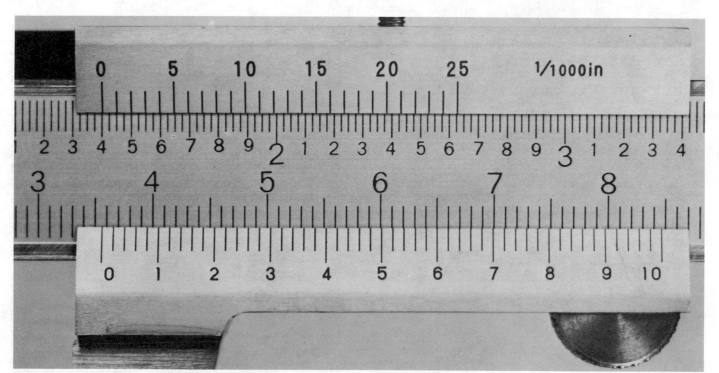

Figure 13d.

READING VERNIER DEPTH GAGES

These measuring tools are designed to measure the depth of holes, recesses, steps, and slots. Basic parts of a vernier depth gage include the base or anvil with the vernier scale and the fine adjustment screw (Figure 14). Also shown is the graduated beam or bar that contains the true scale. To make accurate measurements the reference surface needs to be flat and free from nicks and burrs. The base should be held firmly against the reference surface while the beam is brought in contact with the surface being measured. The measuring pressure should approximately equal the pressure exerted when making a light dot on a piece of paper with a pencil. On a vernier depth gage, dimensions are read in the same manner as on a vernier caliper.

Figure 15. Dial depth gage (Harry Smith & Associates).

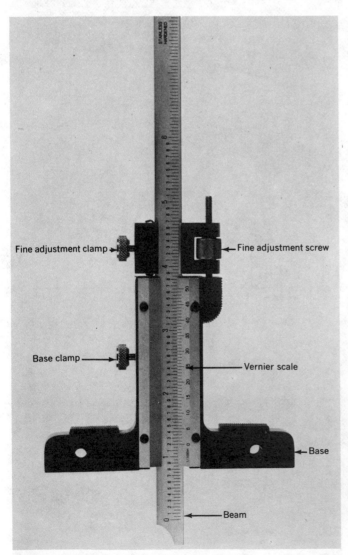

Figure 14. Vernier depth gage with .001 in. discrimination.

Figure 16. Dial indicator depth gage (Courtesy of the L. S. Starrett Company).

DIAL DEPTH GAGES

As with vernier calipers, vernier depth gages have their dial counterparts (Figure 15). The dial depth gage functions in the same manner as the dial caliper. Readings are direct without the need to use a vernier scale. The dial depth gage has the capacity to measure over several inches of range, depending on the length of the beam. Discrimination is .001 in.

Another type of dial depth gage uses a dial indicator (Figure 16). However, the capacity and discrimination of this instrument is dependent on the range and discrimination of the dial indicator used. The tool is primarily used in comparison measuring applications.

self-evaluation

SELF-TEST—READING VERNIER DEPTH GAGES

Determine the depth measurements illustrated in Figures 17a to d:

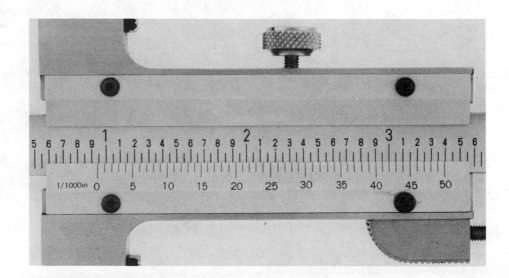

Figure 17a.

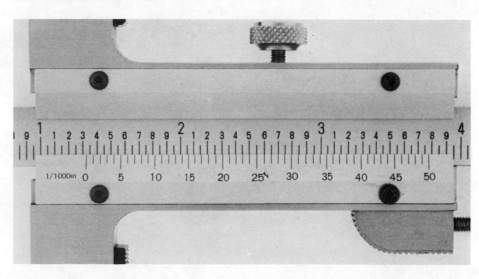

Figure 17b.

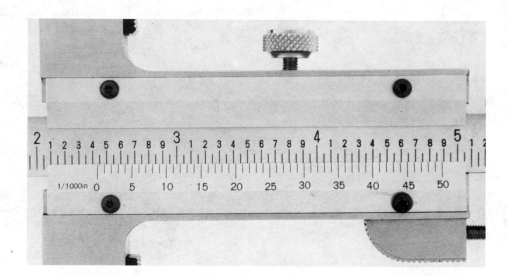

Figure 17c.

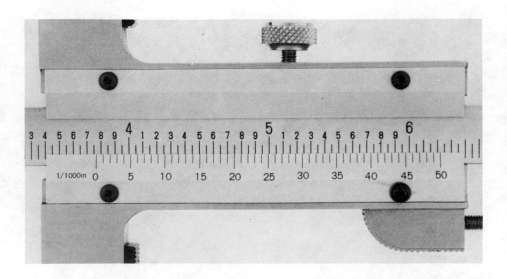

Figure 17d.

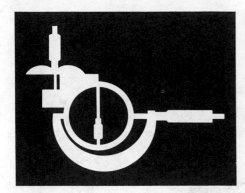

unit 4
using micrometer instruments

Micrometer measuring instruments are the most commonly used precision measuring tools found in industry. Correct use of them is essential to anyone engaged in the making or inspecting of machined parts.

objectives After completing this unit and with the use of appropriate measuring kits, you will be able to:

1. Measure and record the dimensions of 10 objects, using outside micrometers, to an accuracy of plus or minus .001 of an inch.

2. Measure and record the diameters of five holes in test objects to an accuracy of plus or minus .001 in., using an inside micrometer.

3. Measure and record five depth measurements on a test object using a depth micrometer to an accuracy of plus or minus .001 in.

4. Measure and record the dimensions of 10 objects, using a metric micrometer, to an accuracy of plus or minus .01 mm.

5. Measure and record the dimensions of five objects, using a vernier micrometer, to an accuracy of plus or minus .0001 in. (assuming proper measuring conditions).

TYPES OF MICROMETER INSTRUMENTS

The common types of micrometer instruments, **outside, inside,** and **depth,** are discussed in detail within this unit. The micrometer appears in many other forms in addition to these common types.

Blade Micrometer

The blade micrometer (Figure 1), so-called because of its thin spindle and anvil, is used to measure narrow slots and grooves (Figure 2) where the standard micrometer spindle and anvil could not be accommodated because of their diameter.

Combination Metric/Inch or Inch/Metric Micrometer

The combination micrometer (Figure 3) is designed for dual system use in metric and inch measurement. The tool has a digital reading scale for one system while the other system is read from the sleeve and thimble.

Point Micrometer and Comparator Micrometer

The point micrometer (Figure 4) is used in applications where limited space is available or where it might be desired to take a measurement at an exact location. Several point angles are available. The 60 degree comparator micrometer (Figure 5) is usually called a screw thread com-

Figure 1. Blade micrometer (DeAnza College).

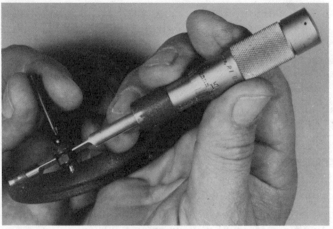

Figure 2. Blade micrometer measuring a groove (DeAnza College).

Figure 3. Combination inch/metric micrometer.

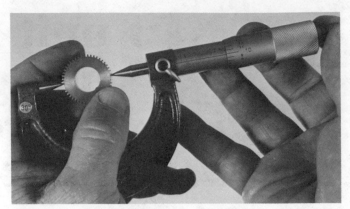

Figure 4. Thirty degree point comparator micrometer.

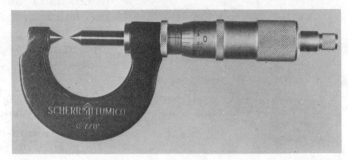

Figure 5. Screw thread comparison micrometer (DeAnza College).

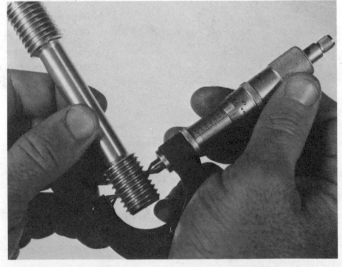

Figure 6. Screw thread comparison micrometer measuring a screw thread (DeAnza College).

parator micrometer. It is most often used to compare screw threads to some known standard like a thread plug gage (Figure 6).

Disc Micrometer

The disc micrometer (Figure 7) finds application in measuring thin materials such as paper where a measuring face with a large area is needed. It is also useful for such measurements as the one shown in the figure where the distance from the slot to the edge is to be determined.

Direct Reading Micrometer

The direct reading micrometer, which may also be known as a high precision micrometer, reads directly to .0001 ($\frac{1}{10,000}$) of an inch (Figure 8).

Hub Micrometer

The frame of the hub micrometer (Figure 9) is designed such that the instrument may be put through a hole or bore in order to measure the hub thickness of a gear or sprocket (Figure 10).

Indicating Micrometer

The indicating micrometer (Figure 11) is useful in inspection applications where a determination of acceptable tolerance is to be made. The instrument has an indicating mechanism built into the frame that permits a dial reading discriminating to .0001 of an inch. When an object is measured, the size deviation above or below the micrometer setting will be indicated on

Figure 9. Hub micrometer (DeAnza College).

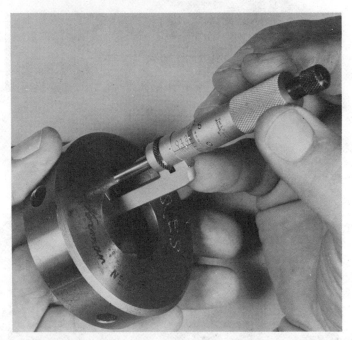

Figure 10. Hub micrometer measuring through a bore (DeAnza College).

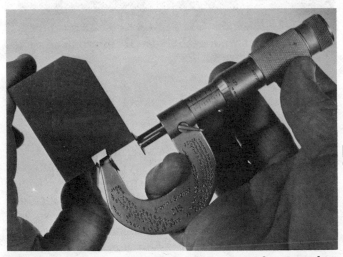

Figure 7. Disc micrometer measuring slot to edge distance (DeAnza College).

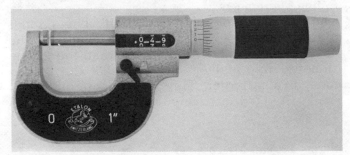

Figure 8. Direct reading digit micrometer (Harry Smith & Associates).

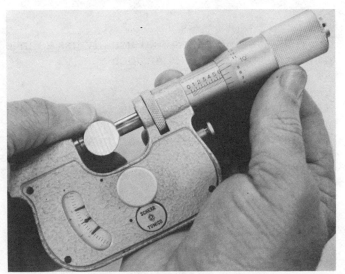

Figure 11. Indicating micrometer (Harry Smith & Associates).

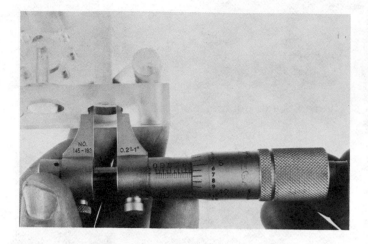

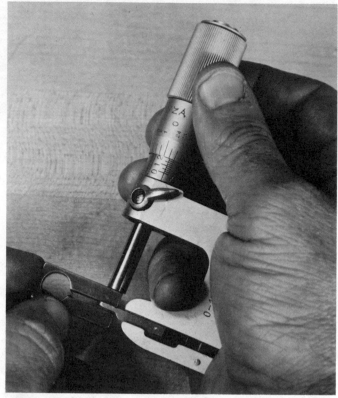

Figure 14. Interchangeable anvil micrometer with flat anvil (DeAnza College).

Figure 13. Internal micrometer (Harry Smith & Associates).

the dial. The indicating dial usually has a range of plus or minus .001 of an inch.

Inside Micrometer Caliper
The inside micrometer caliper (Figure 12) has jaws that resemble those on a vernier caliper. This instrument is designed for inside measurement. Thus, the versatility of the caliper and the reliability of the micrometer are combined.

Internal Micrometer
The internal micrometer (Figure 13) uses a three point measuring contact system to determine the size of a bore or hole. The instrument is direct reading and is more likely to yield a reliable reading because its three point measuring contacts make the instrument self-centering

as compared to a tool making use of only two contacts.

Interchangeable Anvil Type Micrometer
The interchangeable anvil type micrometer is often called a multi-anvil micrometer. It can be used in a variety of applications. A straight anvil is used to measure into a slot (Figure 14). A cylindrical anvil may be used for measuring into a hole (Figure 15). Various shaped anvils may be clamped into position to meet special measuring requirements.

Spline Micrometer
The spline micrometer (Figure 16) has a small diameter spindle and anvil. The length of the anvil is also considerably longer than that of the standard micrometer. The frame of the instrument is also larger. This type of micrometer is well suited to measuring the minor diameter of a spline.

Screw Thread Micrometer
The screw thread micrometer (Figure 17) is specifically designed to measure the pitch di-

ameter of a screw thread. The anvil and spindle tips are shaped to match the form of the thread to be measured.

V-Anvil Micrometer

The V-anvil micrometer (Figure 18) is used to measure the diameter of an object with odd-numbered symmetrical or evenly spaced features. They are designed for specific numbers of these features. The type shown is for three-sided objects like the three fluted end mill being measured (Figure 19). This design is also very useful in checking out-of-round conditions in centerless grinding that cannot be determined

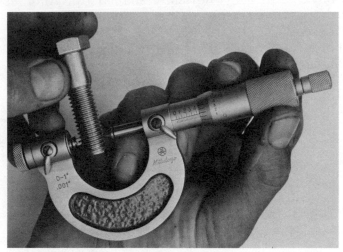

Figure 17. Screw thread micrometer (Yuba College).

Figure 15. Interchangeable anvil micrometer with pin anvil (DeAnza College).

Figure 18. V-anvil micrometer (Yuba College).

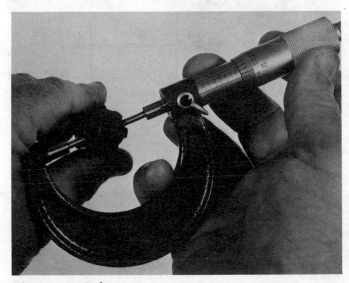

Figure 16. Spline micrometer.

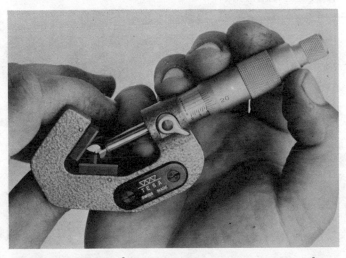

Figure 19. V-anvil micrometer measuring a three fluted end mill (Yuba College).

with a conventional outside micrometer caliper. The next most common type of V-anvil micrometer is for five fluted tools.

Tubing Micrometer

One type of tubing micrometer has a vertical anvil with a cylindrically shaped tip. Another design is like the ordinary micrometer caliper except that the anvil is a half sphere instead of a flat surface. This instrument is designed to measure the wall thickness of tubing (Figure 20). The tubing micrometer can also be applied in other applications such as determining the distance of a hole from an edge (Figure 21).

A standard outside micrometer may also

be used to determine the wall thickness of tube or pipe (Figure 22). In this application, a ball adaptor is placed on the anvil. The diameter of the ball must be subtracted from the micrometer reading in order to determine the actual reading.

Caliper-Type Outside Micrometer

The caliper type outside micrometer is used where measurements to be taken are inaccessible to a regular micrometer (Figure 23).

Taper Micrometer

The taper micrometer can measure inside tapers (Figure 24) or outside tapers (Figure 25).

Groove Micrometer

The groove micrometer (Figure 26) is well-suited to measuring grooves and slots, especially in inaccessible places.

DISCRIMINATION OF MICROMETER INSTRUMENTS

The standard micrometer will discriminate to .001 ($\frac{1}{1000}$) of an inch. In its vernier form, the

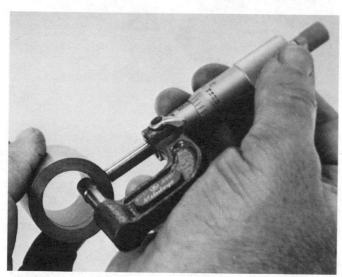

Figure 20. Tubing micrometer measuring a tubewall (DeAnza College).

Figure 22. Ball attachment for tubing measurement (Harry Smith & Associates).

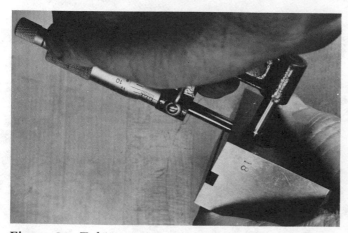

Figure 21. Tubing micrometer measuring hole to edge distance (DeAnza College).

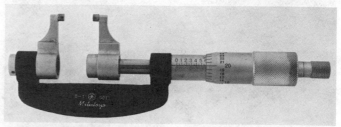

Figure 23. Caliper-type outside micrometer (Harry Smith & Associates).

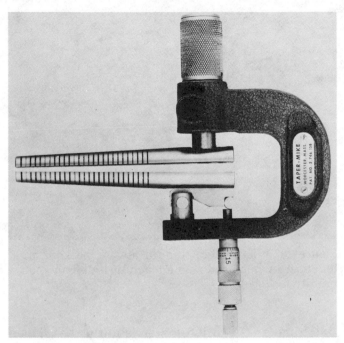

Figure 24. Inside taper micrometer (Courtesy of the Taper Micrometer Corp.).

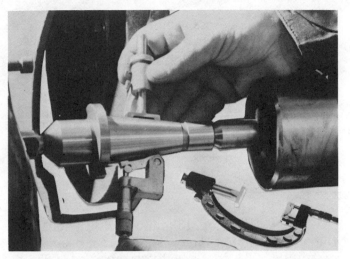

Figure 25. Outside taper micrometer in use (Courtesy of the Taper Micrometer Corp.).

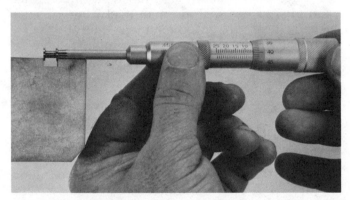

Figure 26. Groove micrometer (Harry Smith & Associates).

discrimination is increased to .0001 $(\frac{1}{10,000})$ of an inch. The common metric micrometer discriminates to .01 $(\frac{1}{100})$ of a millimeter. The same rules apply to micrometers as apply to all measuring instruments. The tool should not be used beyond its discrimination. A standard micrometer with .001 discrimination should not be used in an attempt to ascertain measurements beyond that point. In order to measure to a discrimination of .0001 with the vernier micrometer, certain special conditions must be met. These will be discussed in more detail within this unit.

RELIABILITY AND EXPECTATION OF ACCURACY IN MICROMETER INSTRUMENTS

The micrometer has increased reliability over the vernier. One reason for this is readability of the instruments. The .001 graduations that dictate the maximum discrimination of the micrometer are placed on the circumference of the thimble. The distance between the marks is therefore increased, making them easier to see.

The micrometer will yield very reliable results to .001 discrimination if the instrument is properly cared for, properly calibrated, and correct procedure for use is followed. Care and procedure will be discussed in detail within

this unit. **Calibration** is the process by which any measuring instrument is compared to a known standard. If the tool deviates from the standard, it may then be adjusted to conformity. This is an additional advantage of the micrometer over the vernier. The micrometer must be periodically calibrated if reliable results are to be obtained.

Can a micrometer measure reliably to within .001? The answer is "no" for the standard micrometer, as this violates the 10 to 1 rule for discrimination. The answer is "yes" for the vernier micrometer, but only under controlled conditions. What then, is an acceptable expectation of accuracy that will yield maximum reliability? This is dependent to some degree on the tolerance specified and can be summarized in the following table.

Tolerance Specified	Acceptability of the Standard Micrometer	Acceptability of the Vernier Micrometer
−.000 −.001 or +.001 +.000	No	Yes (under controlled conditions)
±.001	Yes	Yes (vernier will not be required)

Figure 28. Micrometers should always be kept on a tool board when used near a machine tool (Lane Community College).

For a specified tolerance within .001 in., the vernier micrometer should be used. Plus or minus .001 in. is a total range of .002 in. or within the capability of the standard micrometers.

The micrometer is indeed a marvelous example of precision manufacturing. These rugged tools are produced in quantity with each one conforming to equally high standards. Micrometer instruments, in all their many forms, constitute one of the fundamental measuring instruments for the machinist.

CARE OF OUTSIDE MICROMETERS

You should be familiar with the names of the major parts of the typical outside micrometer (Figure 27). The micrometer uses the movement of a precisely threaded rod turning in a nut for precision measurements. The accuracy of micrometer measurements is dependent on the quality of its construction, the care the tool receives, and the skill of the user. Consider some of the important factors in the care of the

micrometer. A micrometer should be wiped clean of dust and oil before and after it is used. A micrometer should not be opened or closed by holding it by the thimble and spinning the frame around the axis of the spindle. Make sure that the micrometer is not dropped. Even a fall of a short distance can spring the frame. This will cause misalignment between the anvil and spindle faces and destroy the accuracy of this precision tool. A micrometer should be kept away from chips on a machine tool. The instrument should be placed on a clean tool board (Figure 28) or on a clean shop towel (Figure 29) close to where it is needed.

Always remember that the machinist is responsible for any measurements that he may make. To excuse an inaccurate measurement on the grounds that a micrometer was not properly adjusted or cared for would be less than professional. When a micrometer is stored after use, make sure that the spindle face does not touch the anvil. Perspiration, moisture from the air, or even oils promote corrosion between the measuring faces with a corresponding reduction in accuracy.

Prior to using a micrometer, clean the measuring faces. The measuring faces of many newer micrometers are made from an extremely hard metal called tungsten carbide. These instruments are often known as carbide-tipped micrometers. If you examine the measuring faces of a carbide-tipped micrometer, you will see where the carbide has been attached to the face of the anvil and spindle. Carbide-tipped

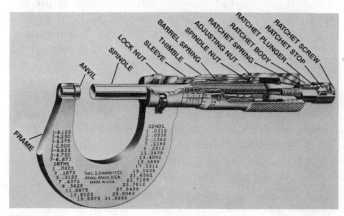

Figure 27. Parts of the outside micrometer (Courtesy of the L. S. Starrett Co.).

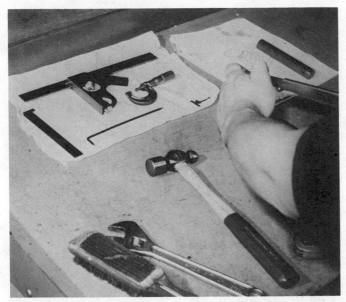

Figure 29. Micrometers should be kept on a clean shop towel when used on the bench (Lane Community College).

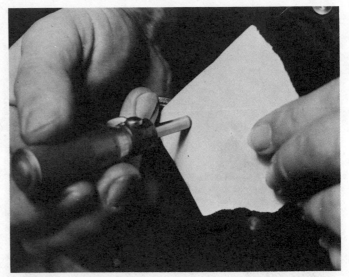

Figure 30. Cleaning the measuring faces (Lane Community College).

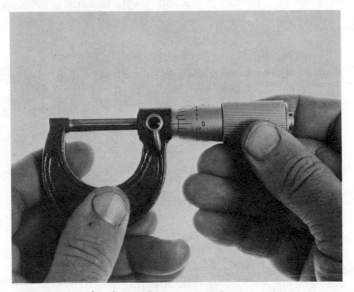

Figure 31. Checking the zero reading.

micrometers have very durable and long-wearing measuring faces. Screw the spindle down lightly against a piece of paper held between it and the anvil (Figure 30). Slide the paper out from between the measuring faces and blow away any fuzz that clings to the spindle or anvil. At this time, you should test the zero reading of the micrometer by bringing the spindle slowly into contact with the anvil (Figure 31). Use the ratchet stop or friction thimble to perform this operation. The ratchet stop or friction thimble found on most micrometers is designed to to equalize the gaging force. When the spindle and anvil contact the workpiece, the ratchet stop or friction thimble will slip as a predetermined amount of torque is applied to the micrometer thimble. If the micrometer does not have a ratchet device, use your thumb and index finger to provide a slip clutch effect on the thimble. Never use more pressure when checking the zero reading than when making actual measurements on the workpiece. If there is a small error, it may be corrected by adjusting the index line to the zero point (Figure 32). The manufacturer's instructions provided with the micrometer should be followed when making this adjustment. Also, follow the manufacturer's instructions for correcting a loose thimble to spindle connection or incorrect friction thimble or ratchet stop action. One drop of instrument oil

applied to the micrometer thread at monthly intervals will help it to provide many years of reliable service. A machinist is often judged by his associates on the way he handles and cares for his tools. Therefore, if he cares for his tools properly, he will more likely be held in higher professional regard than if he is careless.

READING INCH MICROMETERS

Dimensions requiring the use of micrometers will generally be expressed in decimal form to

Figure 32. Adjusting the index line to zero.

exposed on the sleeve. The edge of the thimble exposes 3 major divisions. This represents .300 in. (three hundred thousandths). However, there are also 2 minor divisions showing on the sleeve. The value of these is .025, for a total of .050 in. (fifty thousandths). The reading on the thimble is 9, which indicates .009 in. (nine thousandths). The final micrometer reading is determined by adding the total of the sleeve and thimble readings. In the example shown (Figure 34), the sleeve shows a total of .350 in.. Adding this to the thimble, the final reading becomes .350 in. + .009 in., or .359 in..

USING THE MICROMETER

A micrometer should be gripped by the **frame** (Figure 35) leaving the thumb and forefinger free to operate the thimble. When possible, take micrometer readings while the instrument is in contact with the workpiece (Figure 36). Use only enough pressure on the **spindle** and **anvil** to yield a reliable result. This is what the machinist refers to as "**feel.**" The proper feel of a micrometer will come only from experience. Obviously, excessive pressure will not only result in an inaccurate measurement, it will also distort the frame of the micrometer and possibly damage it permanently. You should also remember that too light a pressure on the part by the measuring faces can yield an unreliable result.

The micrometer should be held in both hands whenever possible. This is especially true when measuring cylindrical workpieces (Figure 37). Holding the instrument in one hand does not permit sufficient control for reliable readings. Furthermore, cylindrical workpieces should be checked at least twice with **measurements made 90 degrees apart.** This is to check for an out-of-round condition (Figure 38). When critical dimensions are measured, that is, any dimension where a very small amount of tolerance is acceptable, make at least two consecutive measurements. Both readings should indicate identical results. If two identical readings cannot be determined, then the actual size of the part cannot be stated reliably. **All critical measurements should be made at a temperature of 68° Fahrenheit (20° Celsius or Centigrade).** A workpiece warmer than this temperature will be larger because of heat expansion.

Outside micrometers usually have a measuring range of one inch. They are identified

three decimal places. In the case of inch instrument, this would be the thousandths place. You should think in terms of thousandths whenever reading decimal fractions. For example, the decimal .156 of an inch would be read as one hundred and fifty-six thousandths of an inch. Likewise, .062 would be read as sixty-two thousandths.

On the **sleeve** of the micrometer is a graduated scale with 10 numbered divisions, each one being $\frac{1}{10}$ of one inch or .100 (100 thousandths) apart. Each of these major divisions is further subdivided into four equal parts, which makes the distance between these graduations $\frac{1}{4}$ of .100 or .025 (25 thousandths) (Figure 33). The **spindle screw** of a micrometer has 40 threads per inch. When the spindle is turned one complete revolution, it has moved $\frac{1}{40}$ of one inch, or expressed as a decimal, .025 (25 thousandths).

When we examine the **thimble,** we find 25 evenly spaced divisions around its circumference (Figure 33). Because each complete revolution of the thimble causes it to move a distance of .025 in., each thimble graduation must be equal to $\frac{1}{25}$ of .025, or .001 in. (one thousandth). On most micrometers, each thimble graduation is numbered to facilitate reading the instrument. On older micrometers only every fifth line may be numbered.

When reading the micrometer (Figure 34), first determine the value indicated by the lines

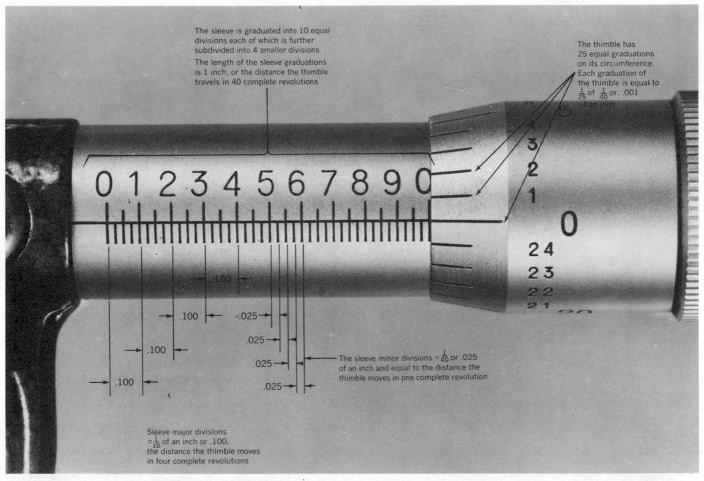

The sleeve is graduated into 10 equal divisions each of which is further subdivided into 4 smaller divisions

The length of the sleeve graduations is 1 inch, or the distance the thimble travels in 40 complete revolutions

The thimble has 25 equal graduations on its circumference. Each graduation of the thimble is equal to $\frac{1}{25}$ of $\frac{1}{40}$ or .001 of an inch

.100

.100

.100

.100

.025

.025

.025

.025

The sleeve minor divisions = $\frac{1}{40}$ or .025 of an inch and equal to the distance the thimble moves in one complete revolution

Sleeve major divisions = $\frac{1}{10}$ of an inch or .100, the distance the thimble moves in four complete revolutions

Figure 33. Graduations on the inch micrometer.

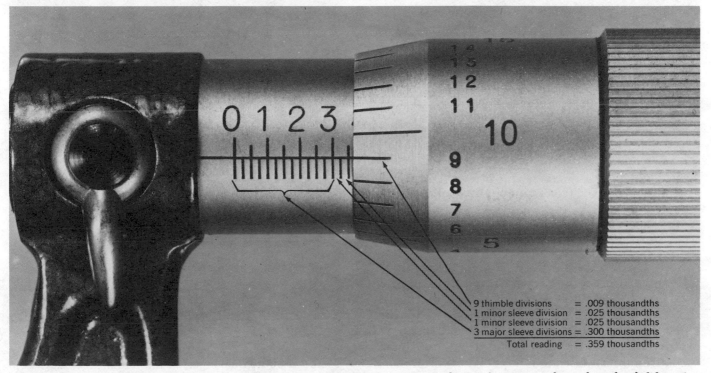

9 thimble divisions	= .009 thousandths
1 minor sleeve division	= .025 thousandths
1 minor sleeve division	= .025 thousandths
3 major sleeve divisions	= .300 thousandths
Total reading	= .359 thousandths

Figure 34. Inch micrometer reading of .359 or three hundred fifty-nine thousandths.

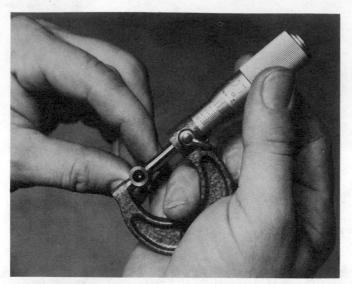

Figure 35. Proper way to hold a micrometer (Lane Community College).

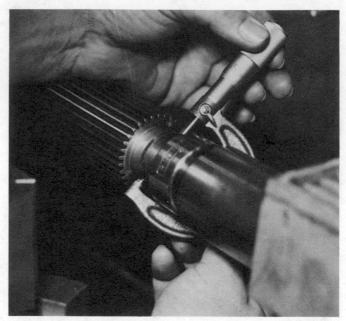

Figure 37. Hold a micrometer in both hands when measuring a round part (Lane Community College).

Figure 36. Read a micrometer while still in contact with the workpiece (Lane Community College).

by size as to the largest dimensions they measure. A two inch mircometer will measure from one to two inches. A three inch micrometer will measure from two to three inches. The capacity of the tool is increased by increasing the size of the frame. Typical outside micrometers range in capacity from 0 to 168 inches. It requires a great deal more skill to get consistent measurements with large capacity micrometers.

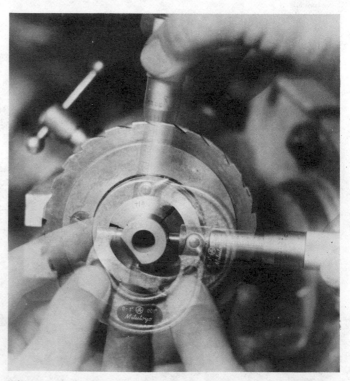

Figure 38. When measuring round parts, take two readings 90 degrees apart (Lane Community College).

self-evaluation

SELF-TEST

1. Why should a micrometer be kept clean and protected? *Any Dirt and oil could make it inaccurate*
2. Why should a micrometer be stored with the spindle out of contact with the anvil? *To Prevent corrosion and reduction of accuracy.*
3. Why are the measuring faces of the micrometer cleaned before measuring? *Because any Dirt or oils will make it in accurate.*
4. How precise is the standard micrometer? *.010 under controlled conditions*
5. What affects the accuracy of a micrometer? *Temperature*
6. What is the difference between the sleeve and thimble? *the sleeve is .100 and the thimble is .001*
7. Why should a micrometer be read while it is still in contact with the object to be measured? *Because there is always a chance of moving the measurement*
8. How often should an object be measured to verify its actual size? *Twice.*
9. What effect has an increase in temperature on the size of a part? *It can Increase or Decrease.*
10. What is the purpose of the friction thimble or ratchet stop on the micrometer? *To test the zero reading of a micrometer*
11. Read and record the five outside micrometer readings in Figure 39a to e.

A .689   E. .72
B. .787
C. .237
D. .994

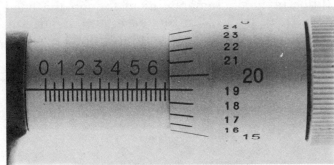

Figure 39a.

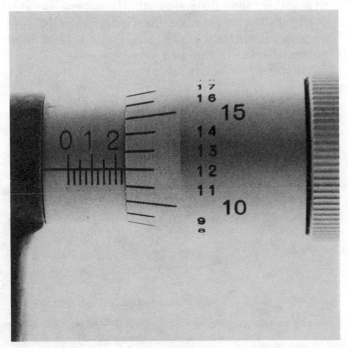

Figure 39c.

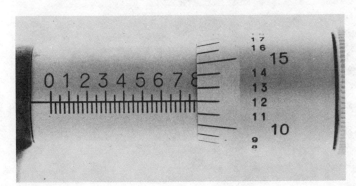

Figure 39b.

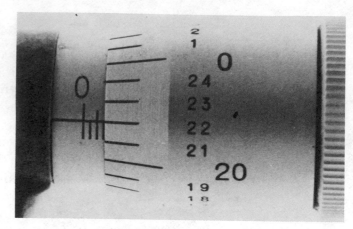

USING INSIDE MICROMETERS

Inside micrometers are equipped with the same graduations as outside micrometers. Inside micrometers discriminate to .001 in. and have a measuring capacity ranging from 1.5 to 20 in. or more. A typical **tubular type** inside micrometer set (Figure 40) consists of the **micrometer head** with detachable **hardened anvils** and several **tubular measuring rods** with **hardened contact tips.** The lengths of these rods differ in increments of .5 in. to match the measuring capacity of the micrometer head, which in this case is .5 in. A handle is provided to hold the instrument into places where holding the instrument directly would be difficult. Another common type of inside micrometer comes equipped with relatively small diameter solid rods that differ in in. increments, even though the head movement is .5 in. In this case, a .5 in. spacing collar is provided. This can be slipped over the base of the rod before it is inserted into the measuring head.

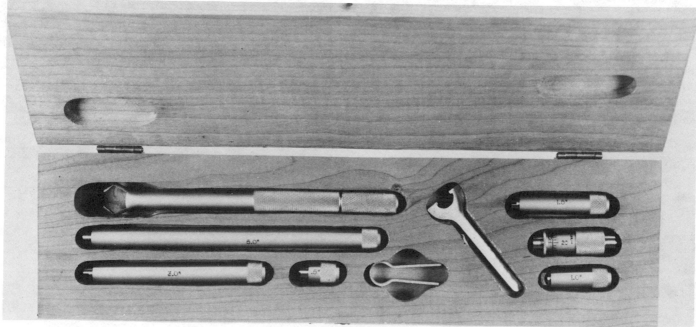

Figure 40. Tubular type inside micrometer set (Harry Smith & Associates).

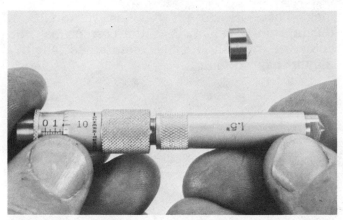

Figure 41. Attaching 1.5 in. extension rod to inside micrometer head (Harry Smith & Associates).

Inside micrometer heads have a range of .250, .500, 1.000, or 2.000 in. depending on the total capacity of the set. For example, an inside micrometer set with a head range of .500 in. will be able to measure from 1.500 to 12.500 in.

The measuring range of the inside micrometer is changed by attaching the extension rods. Extension rods may be solid or tubular. Tubular rods are lighter in weight and are often found in large range inside micrometer sets. Tubular rods are also more rigid. It is very important that all parts be **extremely clean** when changing extension rods (Figure 41). Even small dust

particles can affect the accuracy of the instrument.

When making internal measurements, set one end of the inside micrometer against one side of the hole to be measured (Figure 42). An inside micrometer should not be held in the hands for extended periods, as the resultant heat may affect the accuracy of the instrument. A handle is usually provided, which eliminates the need to hold the instrument and also facilitates insertion of the micrometer into a bore or hole (Figure 43). One end of the micrometer will become the center of the arcing movement used when finding the centerline of the hole to be measured. The micrometer should then be adjusted to the size of the hole. When the correct hole size is reached, there should be a very light drag between the measuring tip and the work when the tip is moved through the centerline of the hole. The size of the hole is determined by adding the reading of the micrometer head, the length of the extension rod, and the length of the spacing collar, if one was used. Read the micrometer **while it is still in place if possible.** If the instrument must be removed to be read, the correct range can be determined by checking with a rule (Figure 44). A skilled craftsman will usually use an **accurate outside micrometer to verify** a reading taken with an inside micrometer. In this case, the inside micrometer becomes an easily adjustable transfer

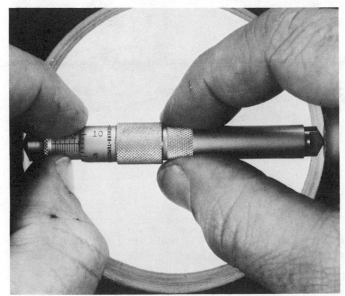

Figure 42. Placing the inside micrometer in the bore to be measured (Harry Smith & Associates).

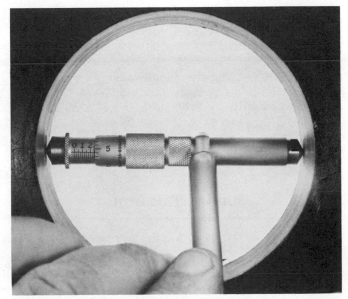

Figure 43. The inside micrometer head used with a handle (Harry Smith & Associates).

measuring tool (Figure 45). Take at least two readings 90 degrees apart to obtain the size of a hole or bore. The readings should be identical. Inside micrometers do not have a spindle lock. Therefore, to prevent the spindle from turning while establishing the correct feel, the adjusting nut should be maintained slightly tighter than normal.

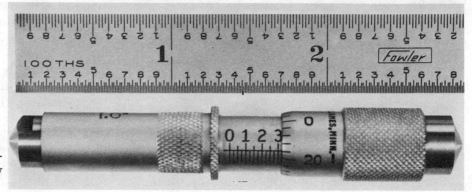

Figure 44. Confirming inside micrometer range using a rule (Harry Smith & Associates).

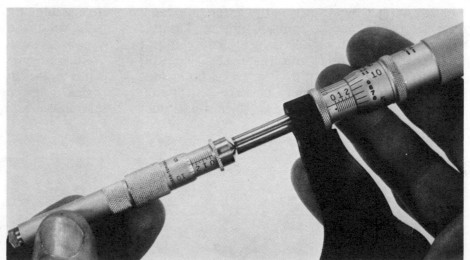

Figure 45. Checking the inside micrometer with an outside micrometer (Harry Smith & Associates).

self-evaluation

SELF-TEST Read and record the five inside micrometer readings (Figure 46a to e). Micrometer head is 1.500 in. when zeroed.

Obtain an inside micrometer set from your instructor and practice using the instrument on objects around your laboratory. Measure examples such as lathe spindle holes, bushings, bores of roller bearings, hydraulic cylinders, and tubing.

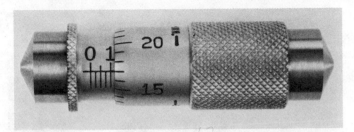

Figure 46a

Figure 46b

Figure 46c (.5 in. extension)

Figure 46d (1.0 in. extension)

Figure 46e (1.0 in. extension)

USING DEPTH MICROMETERS

A **depth micrometer** is a tool that is used to measure precisely depths of holes, grooves, shoulders, and recesses. As other micrometer instruments, it will discriminate to .001 in. Depth micrometers usually come as a set with interchangeable rods to accommodate different depth measurements (Figure 47). The basic parts of the depth micrometer are the **base, sleeve, thimble, extension rod, thimble cap,** and frequently a **ratchet stop.** The bases of a depth micrometer can be of various widths. Generally the wider bases are more stable, but in many instances, space limitations dictate the use of narrower bases. Some depth micrometers are made with only a half base for measurements in confined spaces.

The extension rods are installed or removed by holding the thimble and unscrewing the thimble cap. Make sure that the seat between the thimble cap and rod adjusting nuts is clean before reassembling the micrometer. Do not overtighten when replacing the thimble cap. Furthermore, **do not attempt to adjust the rod**

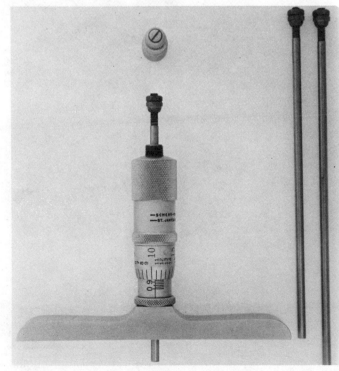

Figure 47. Depth micrometer set (DeAnza College).

length by turning the adjusting nuts. These rods are factory adjusted and matched as a set. **The measuring rods from a specific depth micrometer set should always be kept with that set.** Since these rods are factory adjusted and matched to a specific instrument, **transposing measuring rods** from set to set **will** usually **result** in **incorrect measurements.**

When making depth measurements, it is very important that the micrometer base has a smooth and flat surface on which to rest. Furthermore, sufficient pressure must be applied to keep the base in contact with the reference surface. When a depth micrometer is used without a ratchet, a slip clutch effect can be produced by letting the thimble slip while turning it between the thumb and index finger (Figure 48).

READING INCH DEPTH MICROMETERS

When a comparison is made between the sleeve of an outside micrometer and the sleeve of a depth micrometer, note that the graduations are numbered in the opposite direction (Figure 49). When reading a depth micrometer, the distance to be measured is the value covered by the

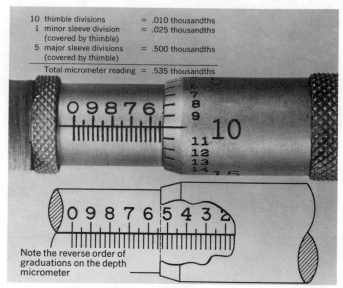

10 thimble divisions	= .010 thousandths
1 minor sleeve division (covered by thimble)	= .025 thousandths
5 major sleeve divisions (covered by thimble)	= .500 thousandths
Total micrometer reading	= .535 thousandths

Note the reverse order of graduations on the depth micrometer

Figure 49. Sleeve graduations on the depth micrometer are numbered in the opposite direction as compared to the outside micrometer.

thimble. Consider the reading shown (Figure 49). The thimble edge is between the number 5 and 6. This indicates a value of at least .500 in. on the sleeve major divisions. The thimble

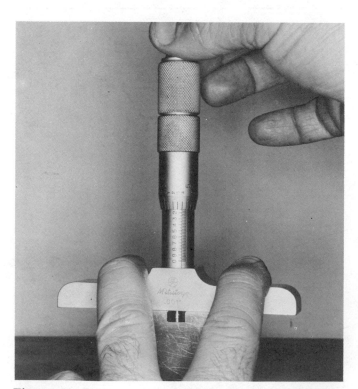

Figure 48. Proper way to hold the depth micrometer (DeAnza College).

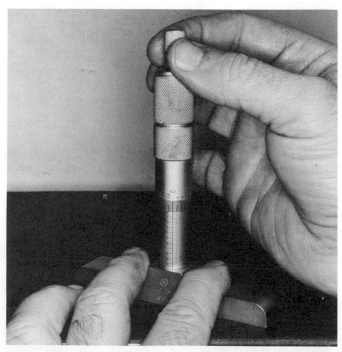

Figure 50. Checking a depth micrometer for zero adjustment using the surface plate as a reference surface (DeAnza College).

also covers the first minor division on the sleeve. This has a value of .025 in. The value on the thimble circumference indicates .010 in. Adding these three values results in a total of .535 in., or the amount of extension of the rod from the base.

A depth micrometer **should be tested for accuracy** before it is used. When the 0 to 1 in. rod is used, retract the measuring rod into the base. Clean the base and contact surface of the rod. Hold the micrometer base firmly against a flat highly finished surface, such as a surface plate, and advance the rod until it contacts the reference surface (Figure 50). If the micrometer is properly adjusted, it should read zero. When **testing for accuracy** with the **one inch extension rod**, set the base of the micrometer on a **one inch gage block** and measure to the reference surface (Figure 51). Other extension rods can be tested in a like manner.

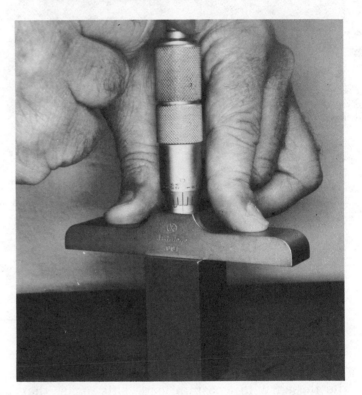

Figure 51. Checking the depth micrometer calibration at the 1.000 in. position in the 0-1 in. rod and a 1 in. square or **Hoke type** gage block (DeAnza College).

self-evaluation

SELF-TEST Read and record the five depth micrometer readings in Figures 52 *a* to *e*.

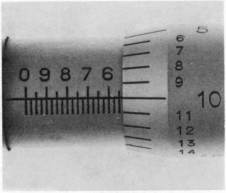

Figure 52*a*

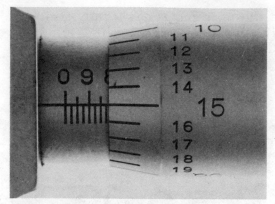

Figure 52*b*

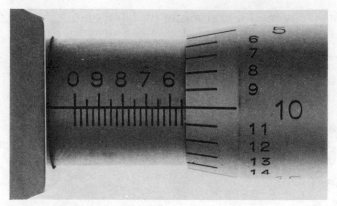

Figure 52*d*

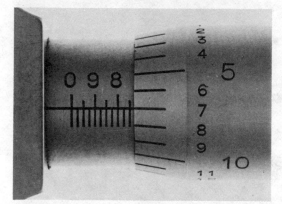

Figure 52*c*

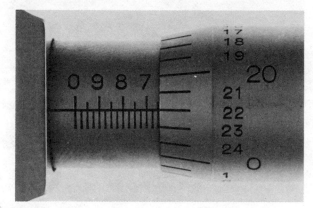

Figure 52*e*

READING METRIC MICROMETERS

The **metric micrometer** (Figure 53) has a spindle thread with a .5 mm lead. This means that the spindle will move .5 mm when the thimble is turned one complete revolution. Two revolutions of the thimble will advance the spindle one millimeter. In precision machining, metric dimensions are usually expressed in terms of .01 ($\frac{1}{100}$) of a millimeter. On the metric micrometer the thimble is graduated into 50 equal divisions with every fifth division numbered (Figure 54). If one revolution of the thimble is .5 mm, then each division on the thimble is equal to .5 mm divided by 50 or .01 mm. The sleeve of the metric micrometer is divided into 25 main divisions above the index line with

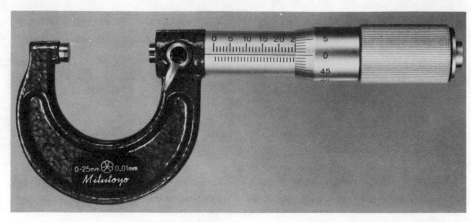

Figure 53. Metric micrometer (Lane Community College).

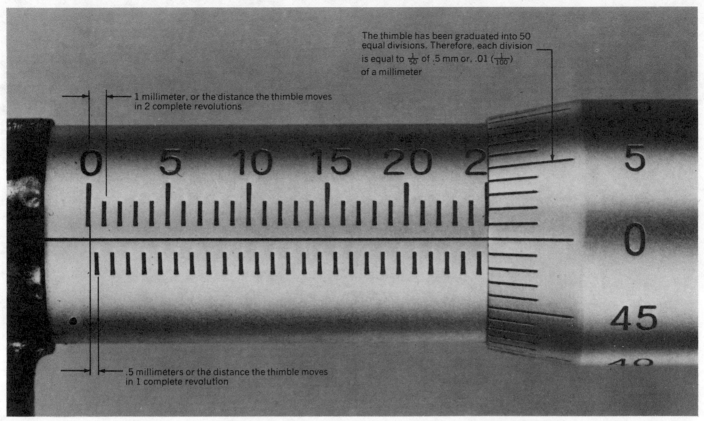

The thimble has been graduated into 50 equal divisions. Therefore, each division is equal to $\frac{1}{50}$ of .5 mm or, .01 ($\frac{1}{100}$) of a millimeter

1 millimeter, or the distance the thimble moves in 2 complete revolutions

.5 millimeters or the distance the thimble moves in 1 complete revolution

Figure 54. Graduations on the metric micrometer.

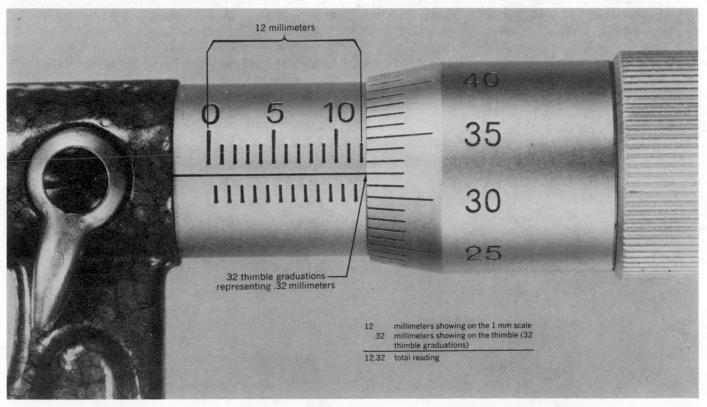

12 millimeters

32 thimble graduations representing .32 millimeters

12	millimeters showing on the 1 mm scale
.32	millimeters showing on the thimble (32 thimble graduations)
12.32	total reading

Figure 55. Metric micrometer reading of 12.32 millimeters.

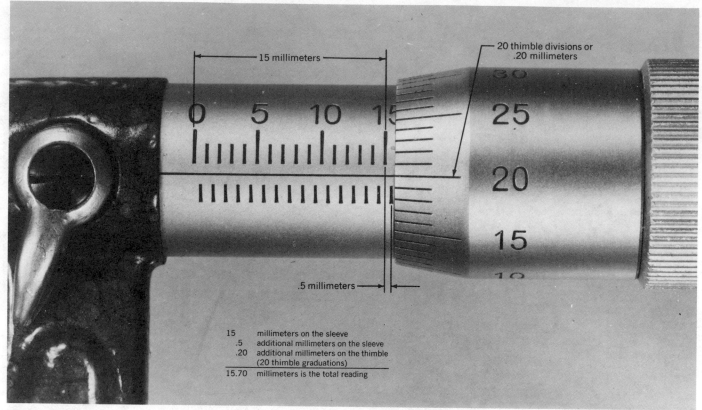

Figure 56. Metric micrometer reading of 15.70 millimeters.

every fifth division numbered. These are whole millimeter graduations. Below the index line are graduations which fall halfway between the divisions above the line. The lower graduations represent half or .5 mm values. The thimble edge (Figure 55) leaves the 12 mm line exposed with no .5 mm line showing. The thimble reading is 32, which is .32 mm. Adding the two figures results in a total of 12.32 mm.

The 15 mm mark (Figure 56) is exposed on the sleeve plus a .5 mm graduation below the index line. The thimble reads 20 or .20 mm. Adding these three values, 15.00 + .50 + .20, results in a total of 15.70 mm.

Any metric micrometer should receive the same care discussed in the section on outside micrometers.

self-evaluation

SELF-TEST Read and record the five metric micrometer readings in Figures 57 *a* to *e*.

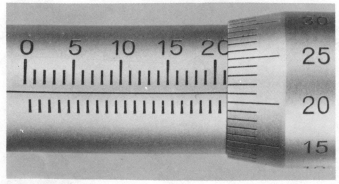

Figure 57*a*

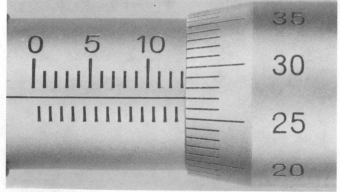

Figure 57*b*

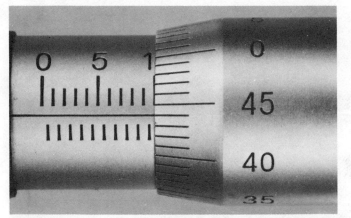

Figure 57c

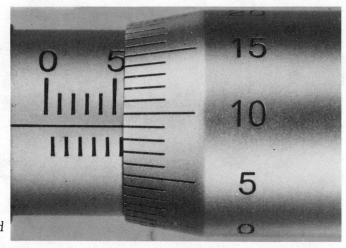

Figure 57d

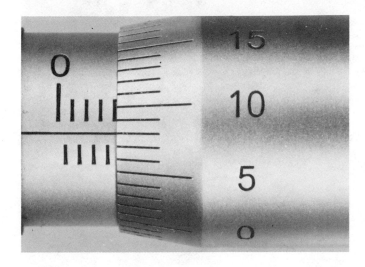

Figure 57e

READING VERNIER MICROMETERS

When measurements must be made to a discrimination greater than .001 in., a standard micrometer is not sufficient. With a **vernier micrometer,** readings can be made to a **ten-thousandth part of an inch** (.0001 in.). This kind of micrometer is commonly known as a "tenth mike." A vernier scale is part of the sleeve graduations. The vernier scale consists of 10 lines parallel to the index line and located above it (Figure 58).

If the 10 spaces on the vernier scale were compared to the spacing of the thimble graduations, the 10 vernier spacings would correspond to 9 spacings on the thimble. Therefore, the vernier scale spacing must be smaller than the thimble spacing. That is, in fact, precisely the

case. Since 10 vernier spacings compare to 9 thimble spacings, the vernier spacing is $\frac{1}{10}$ smaller than the thimble space. We know that the thimble graduations correspond to .001 in. (one thousandth). Each vernier spacing must then be equal to $\frac{1}{10}$ of .001 in., or .0001 in. (one ten-thousandth). Thus, according to the principle of the vernier, each thousandth of the thimble is subdivided into 10 parts. This permits the vernier micrometer to discriminate to .0001 in.

To read a vernier micrometer, first read to the nearest thousandth as on a standard micrometer. Then, find the line on the vernier scale that coincides with a graduation on the thimble. The value of this coincident vernier scale line is the value in ten thousandths, which must

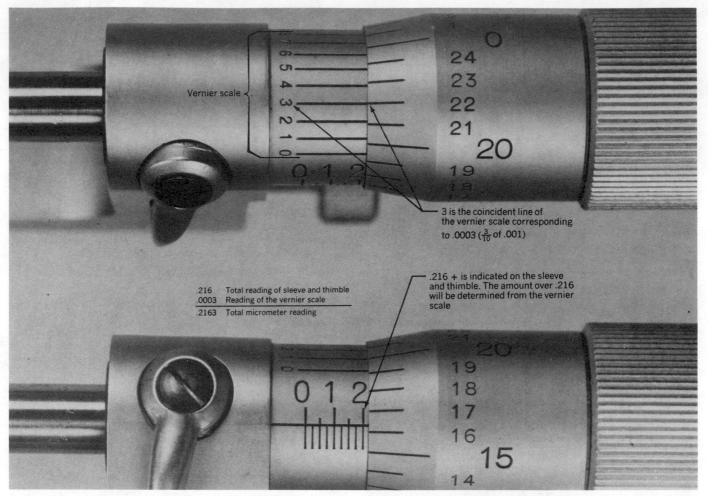

Figure 58. Inch vernier micrometer reading of .2163 in.

be added to the thousandths reading thus making up the total reading. **Remember to add the value of the vernier scale line and not the number of the matching thimble line.**

In the lower view (Figure 58), a micrometer reading of slightly more than .216 in. is indicated. In the top view, on the vernier scale, the line numbered 3 is in alignment with the line on the thimble. This indicated that .0003 (three ten-thousandths) must be added to the .216 in. for a total reading of .2163. This number is read "two hundred sixteen thousandths and three tenths."

You must exercise cautious judgment when attempting to measure to a tenth of a thousandth using a vernier micrometer. There are many conditions that can influence the reliability of such measurements. The 10 to 1 rule discussed in the section introduction states that for maxi-

mum reliability, a measuring instrument must be able to discriminate 10 times finer than the smallest measurement that it will be asked to make. A vernier micrometer meets this requirement for measurement to the nearest thousandth. However, the instrument does not have the capability to discriminate to a one-hundred thousandth, which it should have if it is to be applied in tenth of a thousandth measurement. This does not mean that vernier micrometer should not be used for tenth measurement. The modern micrometer is manufactured with this potential in mind. It does mean that tenth measure should be carried out under **controlled conditions** if truly reliable results are to be obtained. The finish of the workpiece must be extremely smooth. Contact pressure of the measuring faces must be very consistent. The workpiece and instrument must be temperature stabilized. Heat transferred to the micrometer by handling can cause it to deviate considerably. Furthermore, the micrometer must be carefully calibrated against a known standard. Only under these conditions can true reliability be realized.

self-evaluation

SELF-TEST Read and record the five vernier micrometer readings in Figures 59 *a* to *e*.

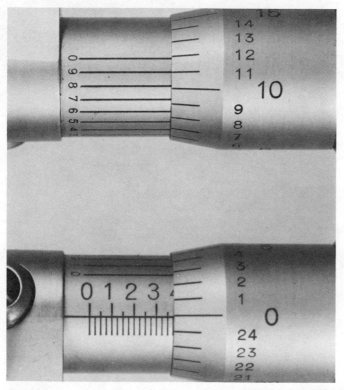

Figure 59*a*

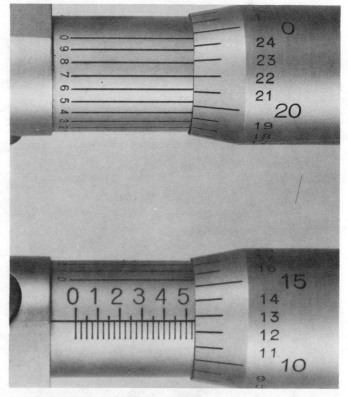

Figure 59*b*

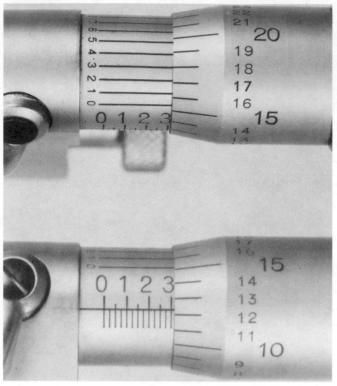

Figure 59c

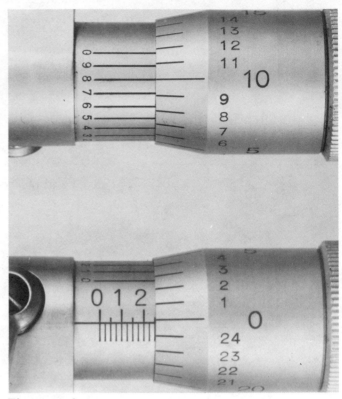

Figure 59d

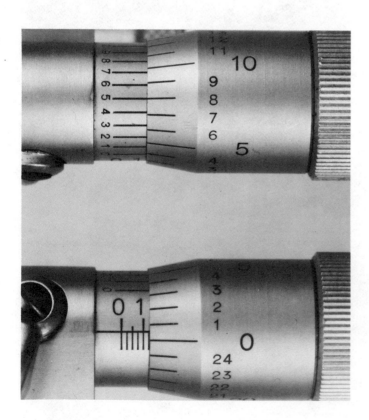

Figure 59e

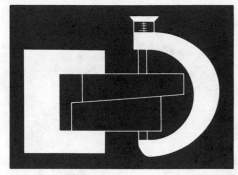

unit 5
using comparison measuring instruments

As a machinist, you will use a large number of measuring instruments that have no capacity within themselves to show a measurement. These tools will be used in comparison measurement applications where they are compared to a known standard or used in conjunction with an instrument that has the capability to show a measurement. In this unit you are introduced to the principles of comparison measurement, the common tools of comparison measurement, and their applications.

objectives **After completing this unit, you will be able to:**

1. **Define comparison measurement.**
2. **Identify common comparison measuring tools.**
3. **Given a measuring situation, select the proper comparison tool for the measuring requirement.**

MEASUREMENT BY COMPARISON

All of us, at some time, were probably involved in constructing something in which we used no measuring instruments of any kind. For example, suppose that you had to build some wooden shelves. You have the required lumber available with all boards longer than the shelf spaces. You hold a board to the shelf space and mark the required length for cutting. By this procedure, you have **compared** the length of the board **(the unknown length)** to the shelf space **(the known length or standard)**. After cutting the first board to the marked length, it is then used to determine the lengths of the remaining shelves. The board, in itself, has no capacity to show a measurement. However, in this case, it became a measuring instrument.

A great deal of comparison measurement often involves the following steps.

1. A device that has no capacity to show measurement is used to establish and represent an unknown distance.
2. This representation of the unknown is then **transferred** to an instrument that has the capability to show a measurement.

This is commonly known as **transfer measurement**. In the example of cutting shelf boards, the shelf space was transferred to the first board and then, the length of the first board was transferred to the remaining boards.

Transfer of measurements may involve some reduction in reliability. This factor must be kept in mind when using comparison tools requiring that a transfer be made. Remember that an instrument with the capability to show measurement directly is always best. **Direct reading instruments should be used whenever possible** in any situation. Measurements re-

169

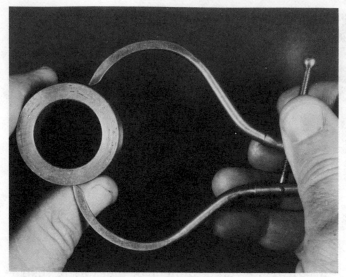

Figure 1. Set one leg of the caliper against the work-piece.

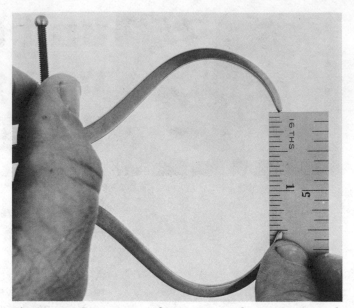

Figure 2. Comparing the spring caliper to a steel rule.

quiring a transfer must be accomplished with proper caution if reliability is to be maintained.

COMMON COMPARISON MEASURING TOOLS AND THEIR APPLICATIONS

Spring Calipers

The spring caliper is a very common comparison measuring tool for rough measurements of inside and outside dimensions. To use a spring caliper, set one jaw on the workpiece (Figure 1). Use this point as a pivot and swing the other caliper leg back and forth over the largest point on the diameter. At the same time, adjust the leg spacing. When the correct feel is obtained, remove the caliper and compare it to a steel rule to determine the reading (Figure 2). The inside spring caliper can be used in a similar manner (Figure 3).

The use of spring calipers is limited. It has been stated by some in the machining business that they can use a spring caliper to measure to .002 or even .001 of an inch. This is of questionable reliability. In modern machining technology, there is little room for crude measurement practice. The use of the spring caliper is fading, and it has been replaced by measuring instruments of much higher reliability. The spring caliper should be used only for the roughest of measurements.

Telescoping Gage

The telescoping gage is also a very common comparison measuring instrument. Telescoping gages are widely used in the machine shop, and they can accomplish a variety of measuring requirements. The telescoping gage is sometimes called a snap gage. This, however, is incorrect. A snap gage is another type of comparison measuring tool and was discussed in the introduction to this section.

Telescoping gages generally come in a set of six gages (Figure 4). The range of the set is usually $\frac{5}{16}$ to 6 inches (8 to 150 mm). The gage consists of two telescoping plungers with a handle and locking screw. The gage is inserted into a bore or slot, and the plungers are permitted to extend, thus conforming to the size of the feature. The gage is then removed and transferred to a micrometer where the reading is determined. The telescoping gage can be a reliable and versa-

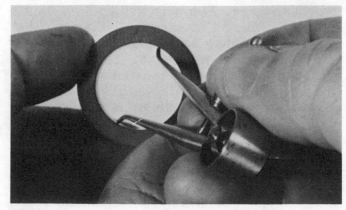

Figure 3. Using an inside spring caliper.

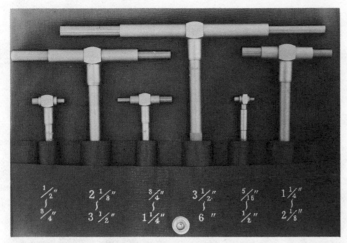

Figure 4. Set of telescoping gages (Harry Smith & Associates).

Figure 5. Inserting the telescoping gage into the bore (DeAnza College).

tile tool if proper procedure is used in its application.

PROCEDURE FOR USING THE TELESCOPING GAGE

1. Select the proper gage for the desired measurement range.
2. Insert the gage into the bore to be measured and release the handle lock screw (Figure 5). Rock the gage sideways to insure that you are

measuring at the full diameter (Figure 6). This is especially important in large diameter bores.
3. Lightly tighten the locking screw.
4. Use a downward or upward motion and roll the gage through the bore. The plungers will be pushed in, thus conforming to the bore diameter (Figure 7). Tighten the locking screw firmly and roll the gage back through the bore. Feel for a light drag.

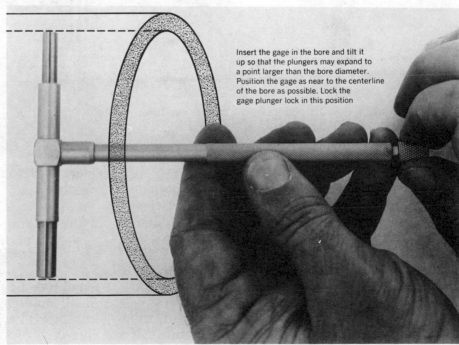

Insert the gage in the bore and tilt it up so that the plungers may expand to a point larger than the bore diameter. Position the gage as near to the centerline of the bore as possible. Lock the gage plunger lock in this position

Figure 6. Release the lock and let the plungers expand larger than the bore.

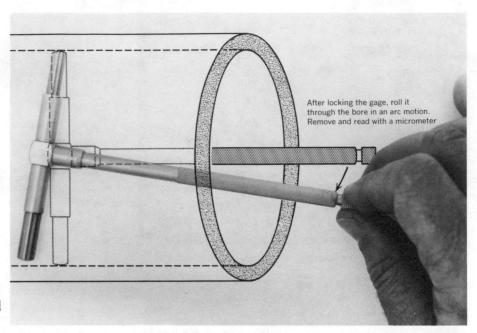

After locking the gage, roll it through the bore in an arc motion. Remove and read with a micrometer

Figure 7. Tighten the lock and roll the gage through the bore.

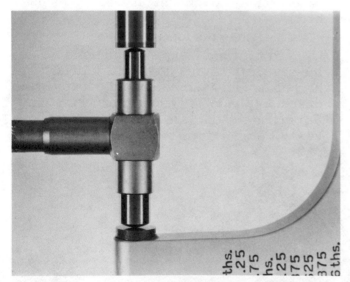

Figure 8. Checking the telescoping gage with an outside micrometer (Harry Smith & Associates).

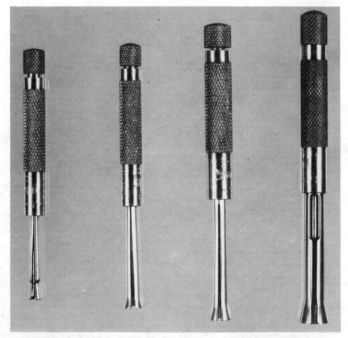

Figure 9. Set of small hole gages (Harry Smith & Associates).

5. Remove the gage and measure with an outside micrometer (Figure 8). Place the gage between micrometer spindle and anvil. Try to determine the same feel on the gage with the micrometer as you felt while the gage was in the bore. Excessive pressure with the micrometer will depress the gage plungers and cause an incorrect reading.

6. Take at least two readings or more with the telescoping gage in order to verify reliability. If the readings do not agree, repeat procedure steps 2 to 6.

Small Hole Gages

Small hole gages, like telescoping gages, come in sets with a range of $\frac{1}{8}$ to $\frac{1}{2}$ inch (4 to 12 mm). One type of small hole gage consists of a split

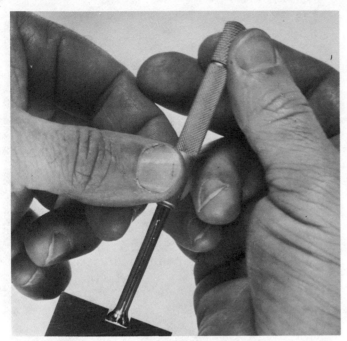

Figure 10. Insert the small hole gage in the slot to be measured (Harry Smith & Associates).

Figure 11. Withdraw the gage and measure with an outside micrometer (Harry Smith & Associates).

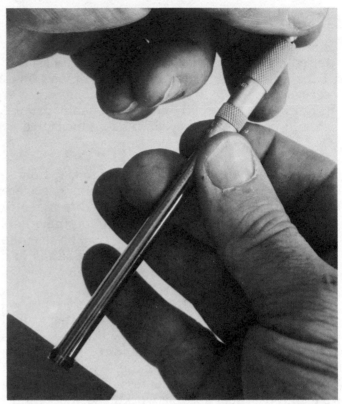

Figure 12. Using the twin ball small hole gage (Harry Smith & Associates).

gage has a flattened end so that a shallow hole or slot may be measured. After the gage has been expanded in the feature to be measured, it should be moved back and forth to determine the proper feel. The gage is then removed and measured with an outside micrometer (Figure 11).

A second type of small hole gage consists of two small balls that can be moved out to contact the surface to be measured. This type of gage is available in a set ranging from $\frac{1}{16}$ to $\frac{1}{2}$ in. (1.5 to 12 mm). Once again, the proper feel must be obtained when using this type of small hole gage (Figure 12). After the gage is set, it is removed and measured with a micrometer.

Adjustable Parallels

For the purpose of measuring slots, grooves, and keyways, the adjustable parallel may be used. Adjustable parallels are available in sets ranging from about $\frac{3}{8}$ to $1\frac{1}{2}$ in. (10 to 38 mm). They are precision ground for accuracy. The typical adjustable parallel consists of two parts that slide together on an angle. Adjusting screws are provided so that clearance in the slide may be adjusted or the parallel locked after setting for a measurement. As the halves of the parallel slide,

ball that is connected to a handle (Figure 9). A tapered rod is drawn between the split ball halves causing them to extend and contact the surface to be measured (Figure 10). The split ball small hole

the width increases or decreases depending on direction. The parallel is placed in the groove or slot to be measured and expanded until the parallel edges conform to the width to be measured. The parallel is then locked with a small screwdriver and measured with a micrometer (Figure 13). If possible, an adjustable parallel should be left in place while being measured.

Radius Gages

The typical radius gage set ranges in size from $\frac{1}{32}$ to $\frac{1}{2}$ inch (.8 to 12 mm). Larger radius gages are also available. The gage can be used to measure the radii of grooves and external or internal fillets (rounded corners). Radius gages may be separate (Figure 14) or the full set may be contained in a convenient holder (Figure 15).

Thickness Gages

The thickness gage (Figure 16) is often called a feeler gage. It is probably best known for its various automotive applications. However, a machinist may use a thickness gage for such measurements as the thickness of a shim, setting a grinding wheel above a workpiece or determining the height difference of two parts. The thickness gage is not a true comparison measuring

Figure 14. Using an individual radius gage.

Figure 15. Radius gage set.

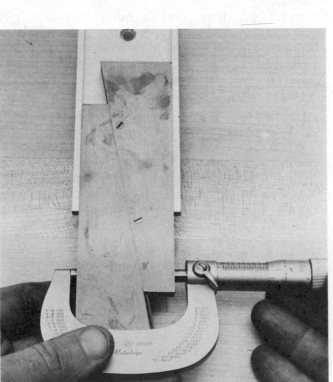

Figure 13. Using adjustable parallels (DeAnza College).

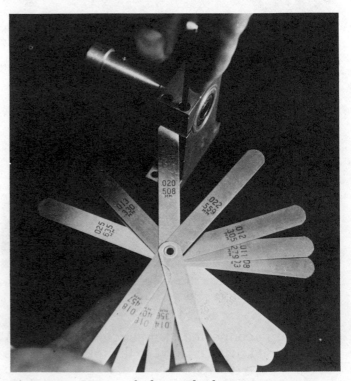

Figure 16. Using a feeler or thickness gage.

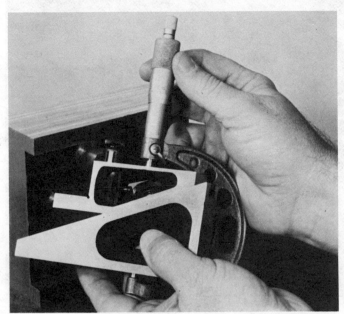

Figure 17. Setting the planer gage with an outside micrometer (DeAnza College).

instrument, as each leaf is marked as to size. However, it is good practice to check a thickness gage with a micrometer, especially when a number of leaves are stacked together.

Planer Gage

The planer gage functions much like an adjustable parallel. Planer gages were originally used to set tool heights on shapers and planers. They can also be used as a comparison measuring tool.

The planer gage may be equipped with a scriber and used in layout. The gage may be set with a micrometer (Figure 17) or in combination with a dial test indicator and gage blocks. In this application, the planer gage is set by using a test indicator set to zero on a gage block (Figure 18). This dimension is then transferred to the planer gage (Figure 19). After the gage has been set, the scriber is attached and the instrument used in a layout application (Figure 20).

Squares

The square is an important and useful tool for the machinist. A square is a comparative measuring instrument in that it compares its own degree of perpendicularity with an unknown degree of perpendicularity on the workpiece. You will use several common types of squares.

MACHINIST'S COMBINATION SQUARE. The combination square (Figure 21) is part of the combination set (Figure 22). The combination set consists of a graduated rule, square head, bevel protractor, and center head. The square head slides on the graduated rule and can be locked at any position (Figure 23). This feature makes the tool useful for layout as the square head can be set according to the rule graduations. The combination square head also has a 45 degree angle along with a spirit level and layout scriber. The combination set is one of the most versatile tools of the machinist.

SOLID BEAM SQUARE. On the solid beam square, the beam and blade are fixed. Solid beam squares range in size from 2 to 72 in. (Figure 24).

Figure 18. Setting the dial test indicator to a gage block (DeAnza College).

Figure 19. Transferring the measurement to the planer gage (DeAnza College).

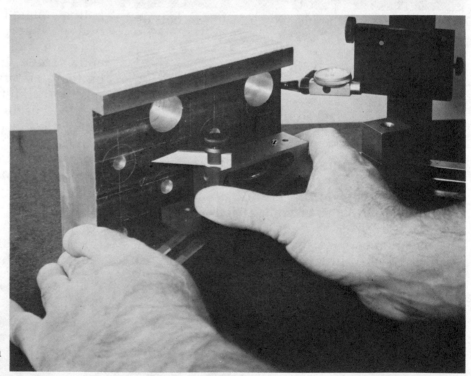

Figure 20. Using the planer gage in layout (DeAnza College).

PRECISION BEVELED EDGE SQUARE. The precision beveled edge square is an extremely accurate square used in the toolroom and in inspection applications. The beveled edge permits a single line of contact with the part to be checked. Precision squares range in size from 2 to 14 in. (Figure 25).

The squares discussed up to this point do not have any capacity to directly indicate the amount of deviation from perpendicularity. The only determination that can be made is that

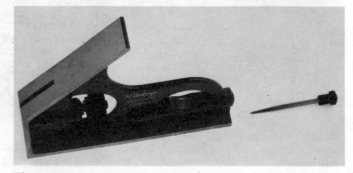

Figure 21. Combination square head with scriber.

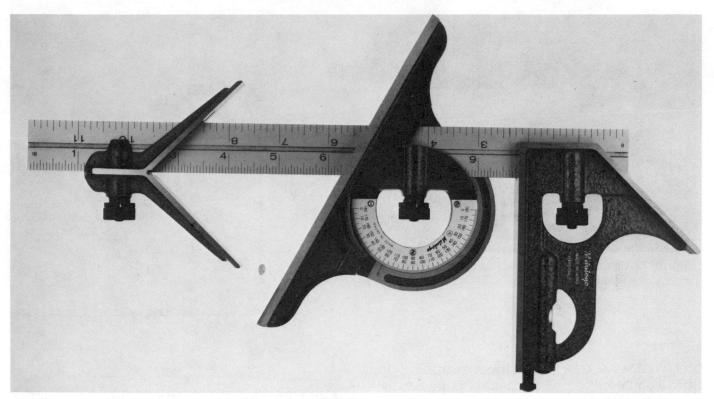

Figure 22. Machinist's combination set.

the workpiece is: as perpendicular or not as perpendicular as the square. The actual amount of deviation from perpendicularity on the workpiece must be determined by other measurements. With the following group of squares, the deviation from perpendicularity can be measured directly. In this respect, the following instruments are not true comparison tools, since they have capacity to show a measurement directly.

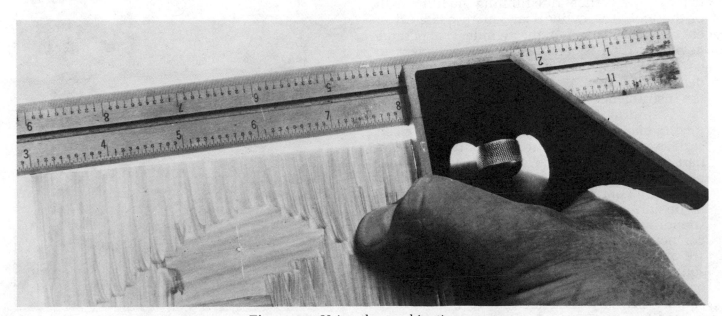

Figure 23. Using the combination square.

Figure 24. Solid beam square (California State University at Fresno).

Figure 25. Precision beveled edge square (Courtesy of the Brown and Sharpe Manufacturing Company).

CYLINDRICAL SQUARE. The direct reading cylindrical square (Figure 26) consists of an accurate cylinder with one end square to the axis of the cylinder. The other end is made slightly out of square with the cylindrical axis. When the non-square end is placed on a clean surface plate, the instrument is actually tilted slightly. As the square is rotated (Figure 27), one point on the circumference of the cylinder will eventually come into true perpendicularity with the surface plate. On a cylindrical square, this point is marked by a vertical line running the full length of the tool. The cylindrical square has a set of curved lines marked on the cylinder that permits deviation from squareness of the workpiece to be determined. Each curved line represents a deviation of .0002 of an inch over the length of the instruments.

Cylindrical squares are applied in the following manner. The square is placed on a clean surface plate and brought into contact with the part to be checked. The square is then rotated until contact is made over the entire length of the instrument. The deviation from squareness is determined by reading the amount corresponding to the line on the square that is in contact with the workpiece. Cylindrical squares are often used to check the accuracy of another square (Figure 26). When the instrument is used on its square end, it may be applied as a plain square. Cylindrical squares range in size from 4 to 12 in.

Figure 26. Cylindrical square (Yuba College).

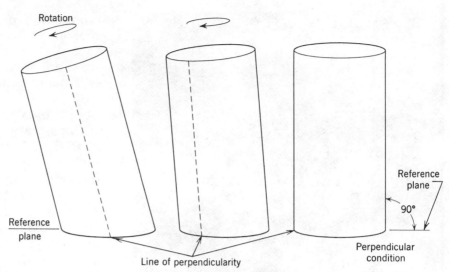

Figure 27. Principle of the cylindrical square.

Principle of the spherical square

DIEMAKER'S SQUARE. The diemaker's square (Figure 28) is used in such applications as checking clearance on a die (Figure 29). The instrument can be used with a straight or offset blade. A diemaker's square can be used to measure a deviation of 10 degrees on either side of the perpendicular.

MICROMETER SQUARE. The micrometer square (Figure 30) is another type of adjustable square. The blade is tilted by means of a micrometer adjustment to determine the deviation of the part being checked.

The square is one of the few tools used in measurement that is essentially self-checking. If you have a workpiece with accurately parallel sides, as measured with a micrometer, one end can be observed under the beam of the square and the error observed. Now the part can be rotated under the beam 180° and rechecked. If the error is identical but reversed, the square is accurate. If there is a difference, except for simple reversal, the square should be considered inaccurate. It should then be checked against a standard, such as a cylindrical square.

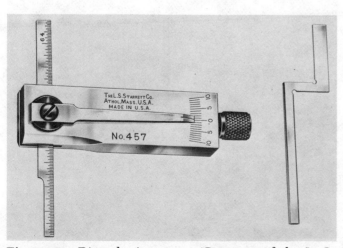

Figure 28. Diemaker's square (Courtesy of the L. S. Starrett Company).

Figure 29. Using the diemaker's square to check die clearance (Courtesy of the L. S. Starrett Company).

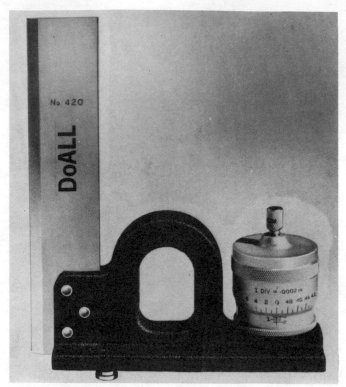

Figure 30. Micrometer square (Courtesy of the DoAll Company).

Indicators

The many types of indicators are some of the most valuable and useful tools for the machinist. There are two general types of indicators in general use. These are dial indicators and dial test indicators. Both types generally take the form of a spring loaded spindle that, when depressed, actuates the hand of an indicating dial. At the initial examination of a dial or test indicator, you will note that the dial face is usually graduated in thousandths of an inch or subdivisions of thousandths. This might lead you to the conclusion that the indicator spindle movement corresponds directly to the amount shown on the indicator face. However, this conclusion is to be arrived at only with the most cautious judgment. **Dial test indicators should not be used to make direct linear measurements.** Reasons for this will be developed in the information to follow. Dial indicators can be used to make linear measurements, but only if they are specifically designed to do so and under proper conditions.

As a machinist, you will use dial and test indicators almost daily in the machine shop. Indicators are very essential to the accurate

completion of your job. However, do not ask that an indicator do what it was not designed to do. Dial and test indicators when properly used are an invaluable member of the machinist's many tools.

DIAL INDICATORS. Dial indicators have discriminations that typically range from .00005 to .001 of an inch. In metric dial indicators, the discriminations typically range from .002 to .01 mm. Indicator ranges or the total reading capacity of the instrument may commonly range from .003 to 2.000 in., or .2 to 50 mm for metric instruments. On the **balanced** indicator (Figure 31), the face numbering goes both clockwise and counterclockwise from zero. This is convenient for comparator applications where readings above and below zero need to be indicated. The indicator shown has a lever actuated stem. This permits the stem to be retracted away from the workpiece if desired.

The continuous reading indicator (Figure 32) is numbered from zero in one direction. This indicator has a discrimination of .0005 and a total range of one inch. The small center hand counts revolutions of the large hand. Note that the center dial counts each .100 in. of spindle travel. This indicator is also equipped with **tolerance hands**

Figure 31. Balanced dial indicator (Harry Smith & Associates).

Figure 32. Dial indicator with one inch travel (Harry Smith & Associates).

Figure 33. Dial indicator with .025 in. range and .0001 in. discrimination (Harry Smith & Associates).

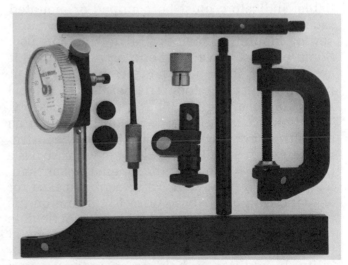

Figure 34. Back plunger indicator with mounting accessories (Harry Smith & Associates).

that can be set to mark a desired limit. Many dial indicators are designed for high discrimination and short range (Figure 33). This indicator has a .0001 discrimination and a range of .025 in.

The **back plunger** indicator (Figure 34) has the spindle in the back or at right angles to the face. This type of indicator usually has a range of about .200 in. with .001 in. discrimination. It is a very popular model for use on a machine tool. The indicator usually comes with a number of mounting accessories (Figure 34).

Indicators are equipped with a **rotating face or bezel.** This feature permits the instrument to be set to zero at any desired place. Many indicators also have a **bezel lock.** Dial indicators may have removable spindle tips, thus permitting use of different shaped tips as required by the specific application (Figure 35).

CARE AND USE OF INDICATORS. Dial indicators are precision instruments and should be treated accordingly. They **must not be dropped** and

should **not be exposed to severe shocks.** Dropping an indicator may bend the spindle and render the instrument useless. Shocks, such as hammering on a workpiece while an indicator is still in contact, may damage the delicate

Figure 35. Dial indicator tips with holder (Courtesy of Rank Precision Industries, Inc.).

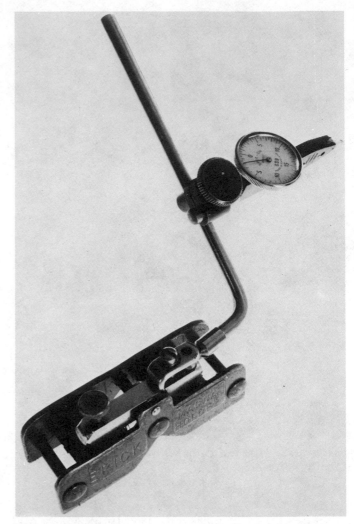

Figure 36. Permanent magnetic indicator base.

operating mechanism. The spindle should be kept free from dirt and grit. This can cause binding that results in damage and false readings. It is important to **check** indicators **for free travel** before using. When an indicator is not in use, it should be stored carefully with a protective device around the spindle.

One of the problems encountered by indicator users is **indicator mounting.** All indicators must be **mounted solidly** if they are to be reliable. Indicators must be clamped or mounted securely when used on a machine tool. A number of mounting devices are in common use. Some of these have magnetic bases that permit an indicator to be attached at any convenient place on a machine tool. The permanent magnet indicator base (Figure 36) is a useful accessory. This type of indicator base is equipped with an adjusting screw that can be used to set the instrument to zero. Another useful magnetic base has a provision for turning off the magnet by mechanical means (Figure 37). This feature makes for easy locating of the base prior to turning on the magnet. A number of bases making use of flexible link indicator holding arms are also in general use. Often they are not adequately rigid for reliability. In addition to holding an indicator on a magnetic base, it may be clamped to a machine setup by the use of any suitable clamps.

DIAL TEST INDICATORS. Dial test indicators frequently have a discrimination of .0005 in. and a range of about .030 in. The test indicator is frequently quite small (Figure 38) so that it can be used to indicate in locations inaccessible to other indicators. The spindle or tip of the test indicator can be swiveled to any desired position. Test indicators are usually equipped with a **movement reversing lever.** This means that the indicator can be actuated by pressure from either side of the tip. The instrument need not be turned around. Test indicators, like dial indicators have a rotating bezel for zero setting. Dial faces are generally of the balanced design. The same care given to dial indicators should be extended to test indicators.

POTENTIAL FOR ERROR IN USING DIAL INDICATORS. Indicators must be used with appropriate caution if reliable results are to be obtained. The spindle of a dial indicator usually consists of a gear rack that engages a pinion and other gears that drive the indicating hand. In any mechanical device, there is always some clearance be-

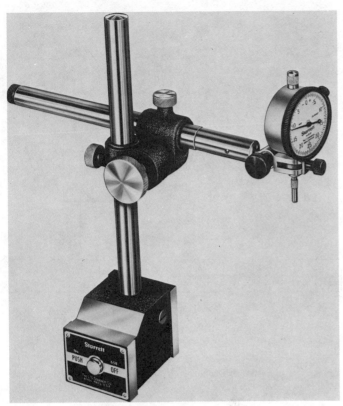

Figure 37. Magnetic base indicator holder with on/off magnet (Courtesy of the L. S. Starrett Company).

Figure 38. Dial test indicator (Harry Smith & Associates).

tween the moving parts. There are also minute errors in the machining of the indicator parts. Because of these, small errors may creep into an indicator reading. This is especially true in long travel indicators. For example, if a one-inch travel indicator with .001 discrimination had plus or minus one percent error at full travel, the following condition could exist if the instrument were to be used for a direct measurement.

You wish to determine if a certain part is within the tolerance of .750 ± .003 in. The one-inch travel indicator has the capacity for this, but remember, it is only accurate to plus or minus one percent of full travel. Therefore, .01 × 1.000 in. is equal to ±.010 in. or the total possible error. To calculate the error per thousandth of indicator travel, divide .010 in. by 1000. This is equal to .00001 in., which is the average error per thousandth of indicator travel. This means that at a travel amount of .750 in., the indicator error could be as much as .00001 in. × 750, or ± .0075 in. In a direct measurement of the part, the indicator could read anywhere from .7425 to .7575. As you can see, this is well

outside the part tolerance and would hardly be reliable (Figure 39).

The indicator should be used as a comparison measuring instrument by the following procedure (Figure 39). The indicator is set to zero on a .750 in. gage block. The part to be measured is then placed under the indicator spindle. In this case, the error caused by a large amount of indicator travel is greatly reduced, because the travel is never greater than the greatest deviation of a part from the basic size. The total part tolerance is .006 in. (± .003 in.). Therefore, 6 × .00001 in. error per thousandth is equal to only ± .00006 in. This is well within the part tolerance and, in fact, cannot even be read on a .001 in. discrimination indicator.

Of course, you will not know what the error amounts to on any specific indicator. This can only be determined by a calibration procedure. Furthermore, you would probably not use a long travel indicator in this particular application. A moderate to short travel indicator would be more appropriate. Keep in mind that any indicator may contain some **travel error** and that by using a fraction of that travel, this error can be reduced considerably.

In the introduction to this section you learned that the axis of a linear measurement instrument must be in line with the axis of measurement. If a dial indicator is misaligned with

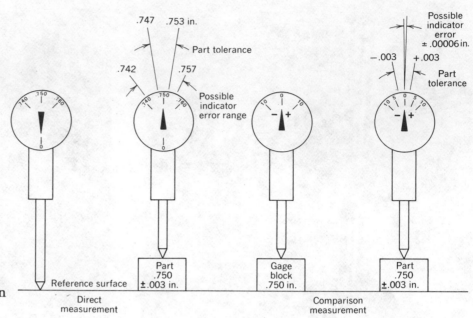

Figure 39. Potential for errors in indicator travel.

the axis of measurement, the following condition will exist.

Line *AC* represents the axis of measurement, while line *AB* represents the axis of the dial indicator (Figure 40). If the distance from *A* to *C* is .100 in., then the distance from *A* to *B* is obviously larger, since it is the hypotenuse of triangle *ABC*. The angle of misalignment, angle *A*, is equal to 20 degrees. The distance from *A* to *B* can then be calculated by the following:

$$AB = \frac{.100}{\cos A}$$

$$AB = \frac{.100}{.9396}$$

$$AB = .1064 \text{ in.}$$

This shows that a movement along the axis of measurement results in a much larger movement along the instrument axis. This **error** is known as **cosine error** and must be kept in mind when using dial indicators. Cosine error is in-

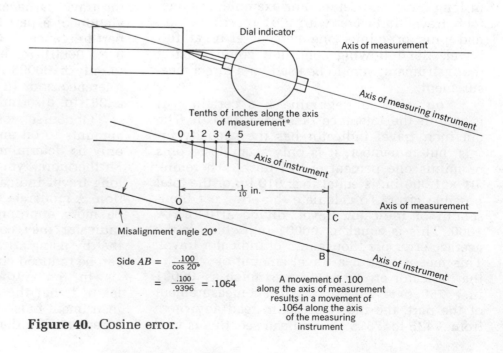

Figure 40. Cosine error.

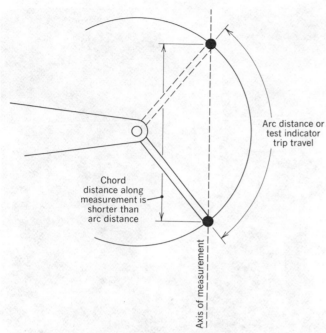

Figure 41. Potential for error in dial test indicator tip movement.

Figure 42. Setting the dial test indicator to zero on the reference surface (DeAnza College).

Figure 43. Using the dial test indicator and vernier height gage to measure the workpiece (DeAnza College).

creased as the angle of misalignment is increased.

When using dial test indicators, watch for **arc versus chord length errors** (Figure 41). The tip of the test indicator moves through an arc. This distance may be considerably greater than the chord distance of the measurement axis. Dial test indicators should **not** be used to make direct measurements. They should only be applied in comparison applications.

USING DIAL TEST INDICATORS IN COMPARISON MEASUREMENTS. The dial test indicator is very useful in making comparison measurements in conjunction with the height gage and height transfer micrometer. Comparison measurement using a vernier height gage is accomplished by the following procedure.

1. Set the height gage to zero and adjust the test indicator until it also reads zero when in contact with the surface plate (Figure 42). It is very important to use a test indicator for this procedure. Using a scriber tip on a height gage is an inferior way to attempt measurements and can lead to substantial error.

2. Raise the indicator and adjust the height gage vernier until the indicator reads zero on the workpiece (Figure 43). Read the dimension from the height scale.

Comparison measurement can also be accomplished using a precision height gage (Figure 44). The precision height gage shown consists of a series of rings that are moved by the micrometer spindle. The ring spacing is a very accurate one inch and the micrometer head has a one inch travel. Other designs use projecting gage blocks at inch intervals, or other types of measuring steps. Discrimination of the typical height micrometer is .0001 in. (Figure 45). Precision height gages come in various height capacities and sometimes have riser blocks as accessories. In Figure 47, a planer

Figure 44. Precision height gage (DeAnza College).

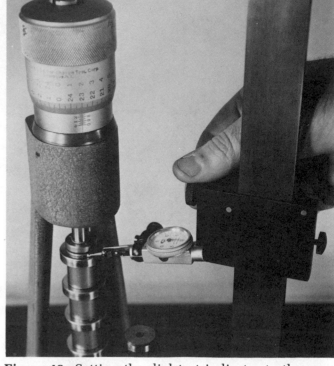

Figure 46. Setting the dial test indicator to the precision height gage (DeAnza College).

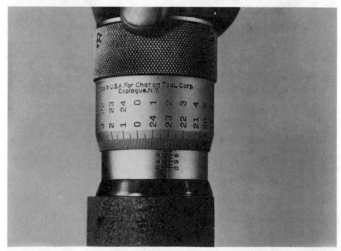

Figure 45. Reading the precision height gage (DeAnza College).

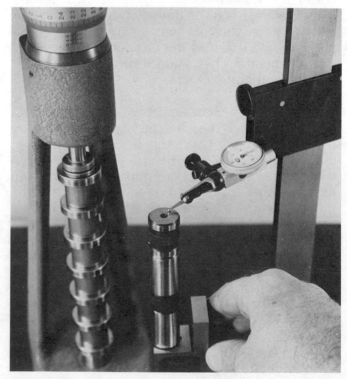

gage is being set using the test indicator and height micrometer. The following procedure is used.

1. The height transfer micrometer is adjusted to the desired height setting. The test indicator is

Figure 47. Transferring the measurement to the planer gage (DeAnza College).

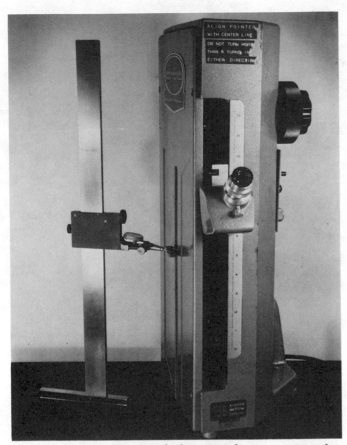

Figure 48. Setting the dial test indicator using the optical height gage (DeAnza College).

zeroed on the appropriate ring (Figure 46). A height transfer gage, vernier height gage, or other suitable means can be used to hold the indicator.

2. The indicator is then moved over to the planer gage. The planer gage is then adjusted until the test indicator reads zero (Figure 47).

Test indicators can also be used to accomplish comparison measurement in conjunction with an optical height gage (Figure 48).

Comparators

Comparators are exactly what their name implies. They are instruments that are used to compare the size or shape of the workpiece to a known standard. Types include dial indicator comparators, optical, electrical, and electronic comparators. Comparators are used where parts must be checked to determine acceptable tolerance. They may also be used to check the geometry of such things as threads, gears, and formed machine tool cutters. The electronic comparator may be found in the inspection area, toolroom, or gage laboratory and used in routine inspection and the calibration of measuring tools and gages.

DIAL INDICATOR COMPARATORS. The dial indicator comparator is no more than a dial indicator

Figure 49. Dial indicator comparator (DeAnza College).

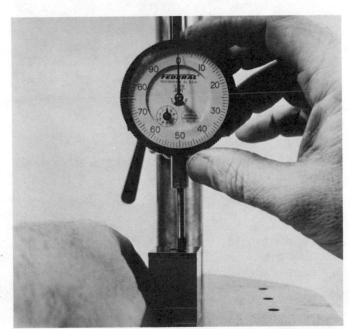

Figure 50. Setting the dial indicator to zero using gage blocks (DeAnza College).

Figure 51. Using the dial indicator comparator (DeAnza College).

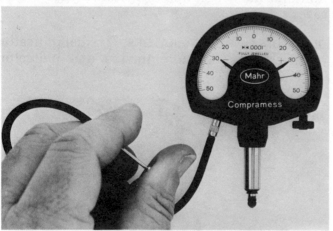

Figure 52. Dial comparator indicator with cable lift (Harry Smith & Associates).

Figure 53. Optical comparator checking a screw thread (Courtesy of Rank Precision Industries, Inc.).

attached to a rigid stand (Figure 49). These are dial indicator instruments such as the ones previously discussed. However, in their application as comparator instruments, as many errors as possible have been eliminated by the fixed design of the instrument's components. The indicator is set to zero at the desired dimension by use of gage blocks (Figure 50). When using a dial indicator comparator, keep in mind the

potential for error in indicator travel and instrument alignment along the axis of measurement. Once the indicator has been set to zero, parts can be checked for acceptable tolerance (Figure 51). A particularly useful comparator indicator for this is one equipped with tolerance hands (Figure 52). The tolerance hands can be set to establish an upper and lower limit for part size. On this type of comparator indicator, the spindle can be lifted clear of the workpiece by using the cable mechanism. This permits the indicator to always travel downward as it

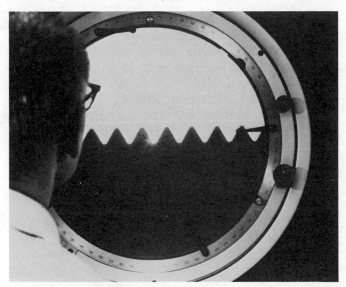

Figure 54. Shadow of the screw thread as seen on the optical comparator screen (DeAnza College).

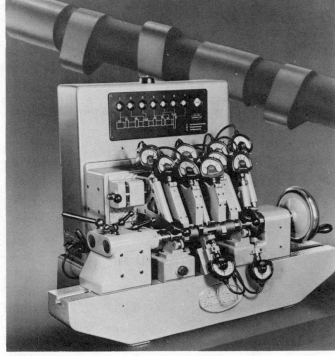

Figure 55. Electromechanical comparator inspecting a camshaft (Courtesy of the Mahr Gage Company).

Figure 56. *(left)* Electronic comparator with maximum discrimination of .00001 in. (Courtesy of the DoAll Company).

comes into contact with the work. This is an additional compensation for any mechanical error in the indicator mechanism.

OPTICAL COMPARATORS. The optical comparator (Figure 53) projects onto a screen, a greatly magnified profile of the object being measured. Various templates or patterns in addition to graduated scales can be placed on the screen and compared to the projected shadow of the part. The optical comparator is particularly useful for inspecting the geometry of screw threads, gears, and formed cutting tools.

For example, to check a screw thread, the part is mounted on the comparator stage and dusted to provide a clearly defined shadow. The stage of the comparator is adjusted for the helix angle of the thread and the magnified shadow of the part is viewed on the comparator screen (Figure 54). The surface of the workpiece may also be inspected by use of a surface illuminator.

ELECTROMECHANICAL AND ELECTRONIC COMPARATORS. Electromechanical and electronic comparators convert dimensional change into changes of electric current or voltage. These changes are read on a suitably graduated scale (meter). Economical mass production of high precision parts requires that fast and reliable measurements be made so that over- and under-sized parts can be sorted from those within tolerance. Electromechanical comparators can be used in this application (Figure 55). The comparator shown is used to check a camshaft.

The electronic comparator (Figure 56) is a very sensitive instrument. It is used in a variety of comparison measuring applications in in-spection and calibration. The comparator is set to a gage block by first adjusting the coarse adjustment. This mechanically moves the measuring probe. Final adjustment to zero is accomplished electronically. This is one of the unique advantages of such instruments. The electronic comparator shown has three scales. The first scale reads $\pm$ .003 in. at full range, with a discrimination of .0001 in. The second scale reads $\pm$ .001 in. at full range, with a discrimination of .00005 in. The third scale reads $\pm$ .0003 at full range, with a discrimination of .00001 in.

self-evaluation

SELF-TEST
1. Define comparison measurement.
2. What can be said of most comparison measuring instruments?
3. Define cosine error.
4. How can cosine error be reduced?
   Match the following measuring situations with the list of comparison measuring tools. Answers may be used more than once.
5. A milled slot two inches wide with a tolarance of $\pm$ .002 in.
6. A height transfer measurement.
7. The shape of a form lathe cutter.
8. Checking a combination square to determine its accuracy.
9. The diameter of a $1\frac{1}{2}$ in. hole.
10. Measuring a shim under a piece of machinery.

   a. Spring caliper
   b. Telescope gage
   c. Adjustable parallel
   d. Radius gage
   e. Thickness gage
   f. Planer gage
   g. Combination square
   h. Solid beam square
   i. Beveled edge square
   j. Cylindrical square
   k. Diemaker's square
   l. Micrometer square
   m. Dial indicator
   n. Dial test indicator
   o. Dial indicator comparator
   p. Optical comparator
   q. Electronic comparator

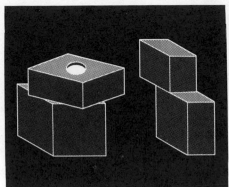

unit 6
using
gage
blocks

In the introduction to this section, we discussed the need for standardization of measurement. Today's widespread manufacturing can function only if machinists everywhere are able to check and adjust their measuring instruments to the same standards. Gage blocks are a means that permit a comparison between the working measurement instruments of manufacturing and recognized international standards of measurement. They are one of the most important measuring tools you will encounter. The practical use of gage blocks in the metrology laboratory, toolroom, and machine shop include the calibration of precision measuring instruments, establishing precise angles, and often measurements involved in the positioning of machine tool components and cutting tools.

objectives **After completing this unit, you will be able to:**

1. **Describe the care required to maintain gage block accuracy.**
2. **Wring gage blocks together correctly.**
3. **Disassemble gage block combinations and properly prepare the blocks for storage.**
4. **Calculate combinations of gage block stacks with and without wear blocks.**
5. **Describe gage blocks applications.**

GAGE BLOCK TYPES AND GRADES

Gage blocks are commonly available individually or in sets. A common gage block set will contain 81 to 88 blocks ranging in thickness from .050 to 4.000 in. The total measuring range of the set is over 25 in. (Figure 1). Also available are 121 block sets that permit measurement from .010 to 18 in. Sets with 4, 6, 9, 12 and 34 blocks are also used depending on measuring requirements. Sets of extra long blocks are available permitting measurements to 84 in. Metric gage block sets contain blocks ranging from .5 to 100 millimeters. **Angular gage blocks** can measure from 0 to 30 degrees. Gage blocks for linear measurement are either rectangular or square.

The three grades of gage blocks are **grade AA (laboratory)**, **grade A+ (inspection)**, and **grade B (shop)**. Grades AA and A+ are manufactured grades, while grade B blocks are usually out-of-

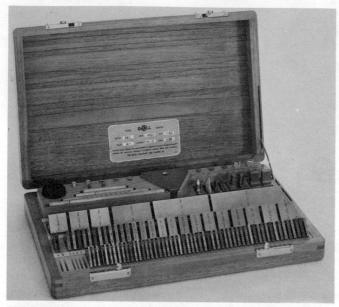

Figure 1. Gage block set with accessories (Courtesy of the DoAll Company).

tolerance AA or A+ sets that are no longer used by the inspection or metrology laboratory (Table 1).

THE VALUE OF GAGE BLOCKS

As you know, a truly exact size cannot be obtained. However, it can be quite closely approached. Gage blocks are one of the physical standards that can closely approach exact dimensions. This makes them useful as measuring instruments with which to check other measuring tools. From the table on gage block toler-

ances, you can see that the length tolerance on a grade AA block is ± .000002 in. This is only four millionths of an inch total tolerance. Such a small amount is hard to visualize. Consider that the thickness of a page of this book is about .003 in. Compare this amount with total gage block tolerance and you will note that the page is 750 times thicker than the tolerance. This should indicate that a gage block would be very useful for checking a measuring instrument with .001 or even .0001 in. discrimination.

As a further demonstration of gage block value, consider the following example. It is desired to establish a distance of 20 inches as accurately as possible. Using a typical gage block set, imagine a hypothetical situation where each block has been made to the plus tolerance of .000002 in. over the actual size. This situation would not exist in an actual gage block set, as the tolerance of each block is most likely bilateral. If it required 30 blocks to make up a 20 in. stack, the cumulative tolerance would amount to .000060 in. (sixty millionths). As you can see, the 20 in. length is still extremely close to actual size. In a real situation, because of the bilateral tolerance of the gage blocks, the 20 in. stack will actually be much closer to 20.000000 than 20.000060 in. Because the gage block is so close to actual size, cumulative tolerance has little effect even over a long distance.

PREPARING GAGE BLOCKS FOR USE

Gage blocks are, at the same time, **rugged** and **delicate.** During their manufacture, they are

Table 1

Gage Block Tolerances

Block Size	Grade AA			
	Length	Flatness	Parallel-ism	Surface Finish in Microinches (Micro-millionth)
.100 to 2.00 in.	± .000002 in.	± .000002 in.	± .000002 in.	0 − .4
Over 2.00	± .000002 in.	± .000002 in.	± .000004 in.	0 − .4
	Grade A+			
.100 to 2.00 in.	+ .000004 in. − .000002 in.	± .000002 in.	± .000002 in.	0 − .4
Over 2.00 in.	+ .000004 in. − .000002 in.	± .000005 in.	± .000005 in.	0 − .4

put through many heating and cooling cycles that stabilize their dimensions. In order for a gage block to function, its **surface** must be **extremely smooth** and **flat.**

Gage blocks are almost always used in combination with each other. This is known as the gage block stack. The secret of gage block use lies in the ability to place two or more blocks together in such a way that most of the air between them is displaced. The space or interface between wrung gage blocks is known as the wringing interval. This is the process of **wringing.** Once this is accomplished, atmospheric pressure will hold the stack together. Properly wrung gage block stacks are essential if cumulative error is to be avoided. Two gage blocks simply placed against each other will have an air layer between them. The thickness of the air layer will greatly affect the accuracy of the stack.

Before gage blocks can be wrung, they must be properly prepared. Burrs, foreign material, lint, grit, and even dust from the air can prevent proper wringing and permanently damage a gage block. The main cause of gage block wear is the wringing of poorly cleaned blocks. Preparation of gage blocks should conform to the following procedure.

1. Remove the desired blocks from the box and place them on a lintfree tissue. It is recommended that the blocks be handled with an **insulated forceps.** This will minimize heat transfer from the hands that can temporarily affect the size of the block.

2. The gage block must be cleaned thoroughly before wringing. This can be done with an appropriate cleaning solvent or commercial gage block cleaner (Figure 2). Use the solvent sparingly, especially if an aerosol is applied. The evaporation of a volatile solvent can cool the block and cause it to temporarily shrink out of tolerance.

3. Dry the block immediately with a lintfree tissue.

4. Any burrs on a gage block can prevent a proper wring and possibly damage the highly polished surface. Deburring is accomplished with a special deburring stone or dressing plate (Figure 3). The block should be lightly moved over the stone using a single back and forth motion. After deburring, the block must be cleaned again.

WRINGING GAGE BLOCKS

Gage blocks should be wrung immediately after cleaning. If more than a few seconds elapses, dust from the air will settle on the wringing surface. The block may require dusting with a camel's hair brush. To wring rectangular gage blocks, the following procedure should be used.

1. Place the freshly cleaned and deburred mating

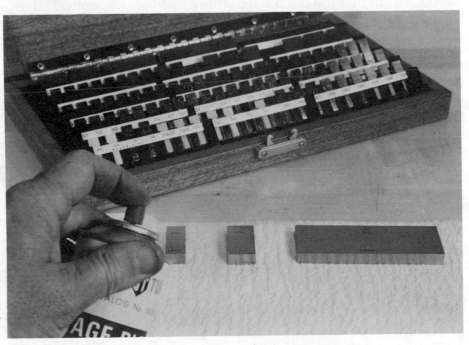

Figure 2. Applying gage block cleaner (DeAnza College).

Figure 3. Using the conditioning stone (California State University at Fresno).

surfaces together and overlap them about $\frac{1}{8}$ inch (Figure 4).

2. Slide the blocks together while lightly pressing together. During the sliding process, you should feel an increasing resistance. This resistance should then level off.
3. Position the blocks so that they are in line (Figure 5).
4. Make sure that the blocks are wrung by holding one block and releasing the other. Hold

your hand under the stack in case the block should fall (Figure 6).

Square gage blocks require the same cleaning and deburring as rectangular blocks. Square gage blocks are wrung by a slightly different technique. Since they are square, they should be placed together at a 45 degree angle. The upper block is then slid over the lower block while at the same time twisting the blocks and applying a light pressure.

During the wringing process, heat from the hands may cause the block stack to expand often well out of tolerance. The stack should be placed on a heat sink in order to normalize the temperature. Generally, gage blocks should be handled as little as possible to minimize heat problems.

If, during the wring, the blocks tend to slide freely, slip them apart immediately and recheck cleaning and deburring. If the blocks fail to wring after proper preparation has been followed, they may be warped or have a surface imperfection. A gage block may be inspected for these conditions by the use of an optical flat.

CHECKING GAGE BLOCKS WITH OPTICAL FLATS

An **optical flat** is an extremely flat piece of quartz (Figure 7). Like gage blocks, there are various grades of flats. First grade or reference optical flats are flat within .000001 in. (one millionth). Round optical flats range from 1 to

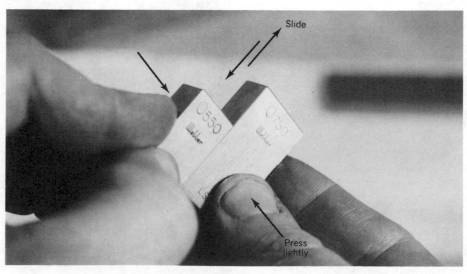

Figure 4. Overlapping gage blocks prior to wringing (DeAnza College).

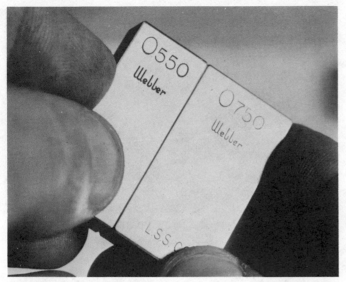

Figure 5. Wrung gage blocks in line (DeAnza College).

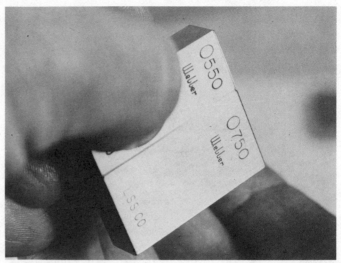

Figure 6. Making sure of a proper wring (DeAnza College).

10 in. in diameter. Square flats range from 1×1 in. to 4×4 in.

The optical flat uses the principles of **light interferometry** to make measurements and reveal surface geometry that could not be detected by other means. The working surface of the flat is placed on the gage block (Figure 8). The block and flat are then placed under a single color or **monochromatic** light source. Since it is not possible to produce a truly flat surface, some surface deviation will exist even on the most perfect of gage blocks. Therefore, a portion of the block will be in direct contact with the optical flat. Other portions will not be in contact.

In the areas of no contact, a small space exists between the optical flat and the gage block. Monochromatic light passing through the optical flat is reflected by both the lower surface of the flat and the surface of the gage block. Under certain conditions, dependent on the distance existing between block and flat, light reflected from the block surface will cancel light reflected from the lower surface of the optical flat. This cancellation effect, called **interference**, is directly related to the distance between block and flat. If this distance is the same as or proportional to the wavelength of the light used, interference can occur (Figure 9). The result of interference produces a dark band or **interference fringe** (Figure 10). Interference can occur only at $\frac{1}{2}$ wavelength or a multiple of $\frac{1}{2}$ wavelength (Figure 11). Since the wavelength of

Figure 7. Round and square optical flats (Courtesy of the DoAll Company).

Figure 8. Inspecting a gage block under monochromatic light (DeAnza College).

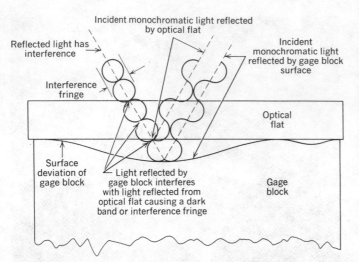

Figure 9. The optical flat in light interference.

Figure 10. Interference fringe patterns (DeAnza College).

the monochromatic light is known, by measuring the spacing of the fringe patterns the actual amount of surface deviation can be determined.

PREPARING GAGE BLOCKS FOR STORAGE

Gage block stacks should not be left wrung for extended periods of time. The surface finish can be damaged in as little as a few hours time, especially if the blocks were not exceedingly clean at the time of wringing. After use, the stack should be unwrung and the blocks cleaned once again. Blocks should be handled with tissue (Figure 12). Spray each block with a suitable gage block preservative and replace them in the box. The entire set should then be lightly sprayed with gage block preservative (Figure 13).

CALCULATING GAGE BLOCK COMBINATIONS

In making a gage block stack, a minimum number of blocks should be used, each surface or wringing interval between blocks can increase the opportunity for error. Poor wringing can make this error relatively large. In order to check your wringing ability, assemble a combination of blocks totaling 4.000 in. Compare this to the 4.000 in. block under a sensitive comparator. Make the comparison immediately after wringing to observe the effect of heat on the stack length. Place the wrung stack on a special heat sink or on the surface plate for about 15 minutes and then check the length again. This will provide some reasonable estimate of the wringing

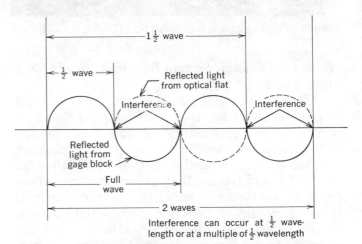

Figure 11. Points of light interference.

Figure 12. Handle gage blocks only with tissue (DeAnza College).

Figure 13. Applying gage block preservative (DeAnza College).

interval. Two millionths of an inch per interval is considered good wringing.

Table 2 gives the specifications of a typical set of 83 gage blocks. Note that they are in four series.

In the following example, it is desired to construct a gage block stack to a dimension of 3.5752 in. Wear blocks will be used on each end of the stack.

$$
\begin{array}{r}
3.5762 \\
\underline{.100} \\
3.4762
\end{array}
$$
First, eliminate the two .050 in. wear blocks

$$
\begin{array}{r}
3.4762 \\
\underline{.1002} \\
3.3760
\end{array}
$$
Then eliminate the last figure right, by subtracting the .1002 block

$$
\begin{array}{r}
3.3760 \\
\underline{.126} \\
3.250
\end{array}
$$
Once again, eliminate the last figure right, by subtracting the .126 block

$$
\begin{array}{r}
3.250 \\
\underline{.250} \\
3.000
\end{array}
$$
Eliminate the last figure right, using the .350 block

$$
\begin{array}{r}
3.000 \\
\underline{3.000} \\
0.000
\end{array}
$$
Eliminate the 3.000 in. with the 3.000 in. block

Therefore, the blocks required to construct this stack are:

Quantity	Size
2	.050 in. wear blocks
1	.1002 block
1	.126 in. block
1	.350 in. block
1	3.000 in. block

As a second example, we shall construct a gage block stack of 4.2125 without wear blocks.

$$
\begin{array}{r}
4.2125 \\
\underline{.1005} \\
4.1120
\end{array}
$$

Table 2
Typical 83 Piece Gage Block Set

First: .0001 Series—9 blocks								
.1001	.1002	.1003	.1004	.1005	.1006	.1007	.1008	.1009

Second: .001 Series—49 Blocks								
.101	.102	.103	.104	.105	.106	.107	.108	.109
.110	.111	.112	.113	.114	.115	.116	.117	.118
.119	.120	.121	.122	.123	.124	.125	.126	.127
.128	.129	.130	.131	.132	.133	.134	.135	.136
.137	.138	.139	.140	.141	.142	.143	.144	.145
.146	.147	.148	.149					

Third: .050 Series—19 Blocks									
.050	.100	.150	.200	.150	.300	.350	.400	.450	.500	.550
.600	.650	.700	.750	.800	.850	.900	.950			

Fourth: 1.000 Series—4 Blocks			
1.000	2.000	3.000	4.000

Two .050 wear blocks

.112
4.0000

4.0000
0.0000

Blocks for this stack are:

Quantity	Size
1	.1005
1	.112
1	4.000

USING WEAR BLOCKS

When gage blocks are used in applications where direct contact is made, it is advisable to use **wear blocks.** For example, if you were using a gage block stack to calibrate a large number of micrometers, wear blocks would be recommended to reduce the wear on the gage blocks. Wear blocks are usually included in typical gage block sets. They are made from a particu-

larly hard material known as tungsten carbide. A wear block is placed on one or both ends of a gage block stack to protect it from possible damage by direct contact. Wear blocks are usually .050 or .100 in. thick.

GAGE BLOCK APPLICATIONS

Gage blocks are used in setting sine bars for establishing precise angles. The use of the sine bar is discussed in the unit on angular measure. Gage blocks are used to set other measuring instruments such as a snap gage (Figure 14). The proper blocks are selected for the desired dimension and the stack assembled (Figure 15). Since this is a direct contact application, wear blocks should be used. The stack is then used to set the gage (Figure 16).

Gage block measurement is facilitated by various accessories (Figure 17). Accessories include scribers, bases, gage pins, and screw sets for holding the stack together. In any application where screws are employed to secure gage block stacks, a torque screwdriver must

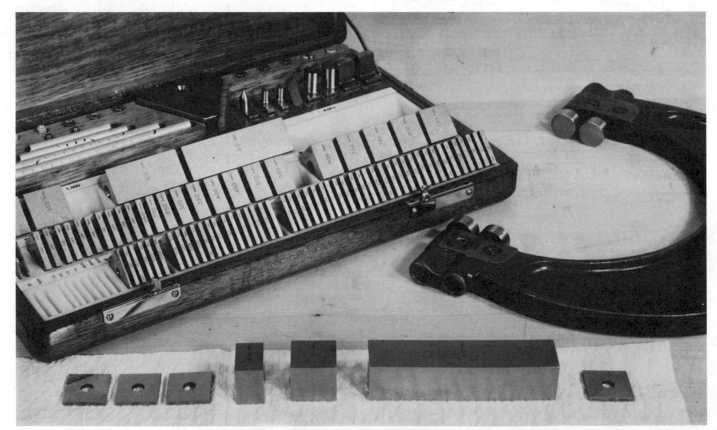

Figure 14. Gage and wear blocks for setting a snap gage (DeAnza College).

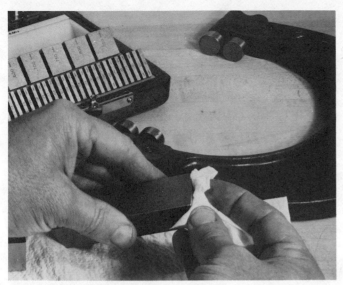

Figure 15. Cleaning blocks prior to wringing (DeAnza College).

Figure 16. Setting a snap gage using gage blocks (DeAnza College).

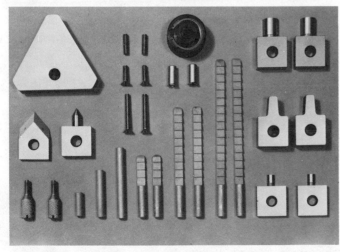

Figure 17. Gage block accessories (DeAnza College).

Figure 18. Precision height gage assembled from gage blocks (Courtesy of the DoAll Company).

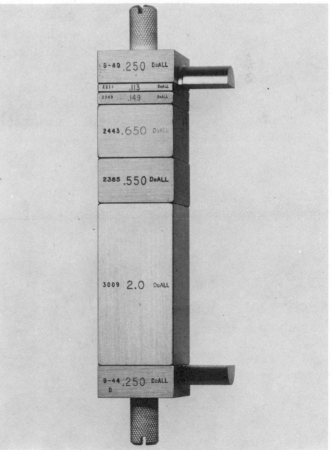

Figure 19. Gage block stack with accessory gage pins (Courtesy of the DoAll Company).

be used. This will apply the correct amount of pressure on the gage block stack. Gage block and accessories can be assembled into precision height gages for layout (Figure 18). With gage pins (Figure 19), gage blocks may be used for direct gaging or for checking other measuring instruments. Machine tool applications include the use of gage blocks as auxiliary measuring systems on milling machines, setting cutter heights, and spacing straddle milling cutters.

self-evaluation

SELF-TEST
1. What is a wringing interval?
2. Why are wear blocks frequently used in combination with gage blocks?
3. As related to gage blocks usage, what is meant by the term "normalize"?
4. What length tolerances are allowed for the following grades of gage blocks (under 2.000 in. sizes)? AA, A+, B.
5. What is a conditioning stone and how is it used?
6. What does the term microinch regarding surface finish of a gage block mean?
7. Describe the handling precautions necessary for the preservation of gage block accuracy.
8. What gage blocks are necessary in order to assemble a stack equal to 3.0213, without using wear blocks?
9. List gage blocks necessary for a stack equal to 1.9643 with wear blocks.
10. Describe at least two gage block applications.

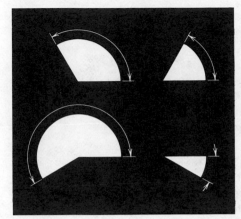

unit 7
using angular measuring instruments

Angular measurement is as important as linear measurement. The same principles of metrology apply to angular measure as apply to linear measure. Angular measuring instruments have various degrees of discrimination. They must not be used beyond their discrimination. Angular measuring instruments require the same care and handling as any of your precision tools

objectives **After completing this unit, you will be able to:**

1. **Identify common angular measuring tools.**
2. **Read and record angular measurements using a vernier protractor.**
3. **Calculate sine bar elevations and measure angles using a sine bar and adjustable parallels.**
4. **Calculate sine bar elevations and establish angles using a sine bar and gage blocks.**

As a machinist, you will find the need to measure **acute angles, right angles,** and **obtuse angles** (Figure 1). Acute angles are less than 90 degrees. Obtuse angles are more than 90 degrees but less than 180 degrees. Ninety degree or right angles are generally measured with squares. However, the amount of angular deviation from perpendicularity may have to be determined. This would require that an angular measuring instrument be used. Straight angles, or those containing 180 degrees, generally fall into the category of straightness or flatness and are measured by other types of instruments.

UNITS OF ANGULAR MEASURE

In the inch system, the unit of angular measure is the **degree.**

Full circle = 360 degrees
1 degree = 60 minutes of arc (1° = 60′)
1 minute = 60 seconds of arc (1′ = 60″)

In the metric system, the unit of angular measure is the **radian.** A radian is the length of an arc on the circle circumference that is equal in length to the radius of the circle (Figure 2). Since the circumference of a circle is

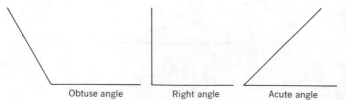

Figure 1. Acute, right, and obtuse angles.

equal to 2 pi r (radius), there are 2 pi radians in a circle. Converting radians to degrees gives the equivalent:

$$1 \text{ radian} = \frac{360}{2 \text{ pi } r}$$

Assuming a radius of 1 unit:

$$1 \text{ radian} = \frac{360}{2 \text{ pi}}$$
$$= 57° \ 17' \ 44'' \text{ (approximately)}$$

It is unlikely that you will come in contact with much radian measure. All of the common comparison measuring tools you will use read in degrees and fractions of degrees. Metric angles expressed in radian measure can be converted to degrees by the equivalent shown.

REVIEWING ANGLE ARITHMETIC

You may find it necessary to perform angle arithmetic. Use your calculator, if you have one available.

Adding Angles

Angles are added just like any other quantity. One degree contains 60 minutes. One minute contains 60 seconds. Any minute total of 60 or larger must be converted to degrees. Any second total of 60 or larger must be converted to minutes.

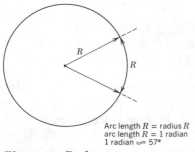

Arc length R = radius R
arc length R = 1 radian
1 radian ⇌ 57°

Figure 2. Radian measure.

$$35° + 27° = 62°$$
$$3° \ 15' + 7° \ 49' = 10° \ 64'$$

Since $64' = 1° \ 4'$, the final result is $11° \ 4'$
$$265° \ 15' \ 52'' + 10° \ 55' \ 17'' = 275° \ 70' \ 69'$$
Since $69'' = 1° \ 9''$ and $70' = 1° \ 10'$, the final result is $276° \ 11' \ 9''$

Subtracting Angles
When subtracting angles where borrowing is necessary, degrees must be converted to minutes and minutes must be converted to seconds.

EXAMPLE $15° - 8° = 7°$

EXAMPLE $15° \ 3' - 6° \ 8'$ becomes
$14° \ 63' - 6° \ 8' = 8° \ 55'$

EXAMPLE $39° \ 18' \ 13'' - 17° \ 27' \ 52''$ becomes
$38° \ 77' \ 73'' - 17° \ 27' \ 52''$
$= 21° \ 50' \ 51''$

ANGULAR MEASURING INSTRUMENTS
Plate Protractors
Plate protractors have a discrimination of one degree and are useful in such applications as layout and checking the point angle of a drill (Figure 3)

Figure 3. Plate protractor measuring a drill point angle.

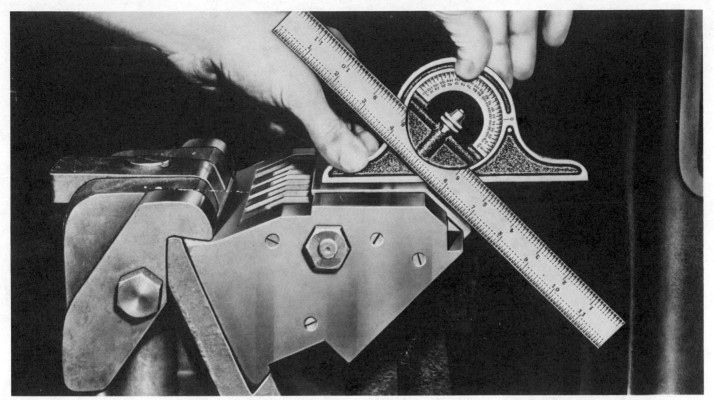

Figure 4. Using the combination set bevel protractor (Courtesy of the L. S. Starrett Company).

Bevel Protractors

The bevel protractor is part of the machinist's combination set. This protractor can be moved along the rule and locked in any position. The protractor has a flat base permitting it to rest squarely on the workpiece (Figure 4). The combination set protractor has a discrimination of one degree.

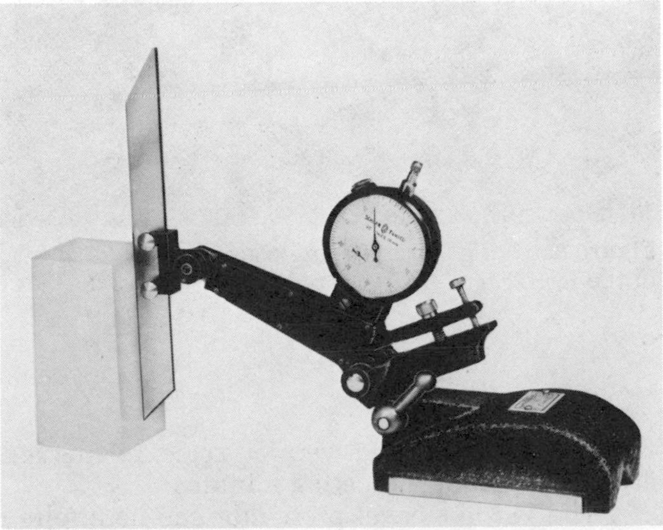

Dial Indicating Sinometer Angle Gage

The indicating sinometer (Figure 5) permits fast and accurate measurement of all angles. Discrimination is 30 seconds of arc per dial division.

Universal Bevel Vernier Protractor

The universal bevel vernier protractor (Figure 6) is equipped with a vernier that permits discrimination to $\frac{1}{12}$ of a degree or 5 minutes of arc.

The instrument can measure an obtuse angle (Figure 7). The acute attachment facilitates the measurement of angles less than 90° (Figure 8). When used in conjunction with a vernier height gage, angle measurements can be made that would be difficult by other means (Figure 9).

Vernier protractors are read like any other instrument employing the vernier. The main scale is divided into whole degrees. These are marked in four quarters each 0 to 90 degrees. The vernier divides each degree into 12 parts each equal to 5 minutes of arc.

Figure 5. Dial indicating sinometer angle gage (Courtesy of Rank Precision Industries, Inc.).

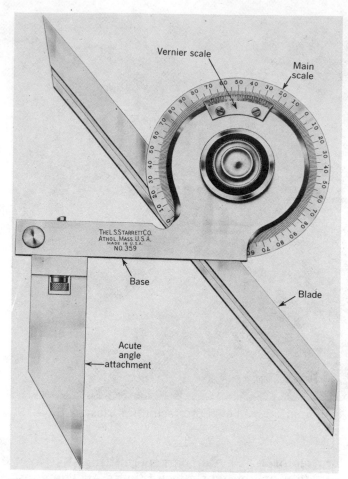

Figure 6. Parts of the universal bevel vernier protractor (Courtesy of the L. S. Starrett Company).

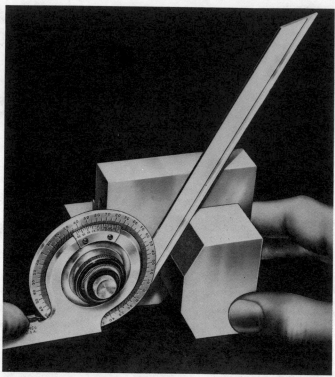

Figure 7. Measuring an obtuse angle with the vernier protractor (Courtesy of the L. S. Starrett Company).

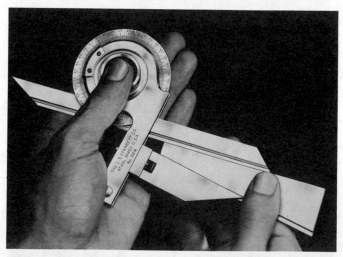

Figure 8. Using the acute angle attachment (Courtesy of the L. S. Starrett Company).

To read the protractor, determine the nearest full degree mark between zero on the main scale and zero on the vernier scale. **Always read the vernier in the same direction as you read the main scale.** Determine the number of the vernier coincident line. Since each vernier line is equal to 5 minutes, multiply the number of the coincident line by 5. Add this to the main scale reading.

EXAMPLE READING (FIGURE 10)

The protractor shown has a magnifier so that the vernier may be seen more easily.

Main scale 56°
Vernier coincident at line 6
6 × 5 minutes = 30 minutes
Total reading is 56° 30'

For convenience, the vernier scale is marked at 0, 30, and 60, indicating minutes.

The vernier bevel protractor can be applied in a variety of angular measuring applications (Figure 11).

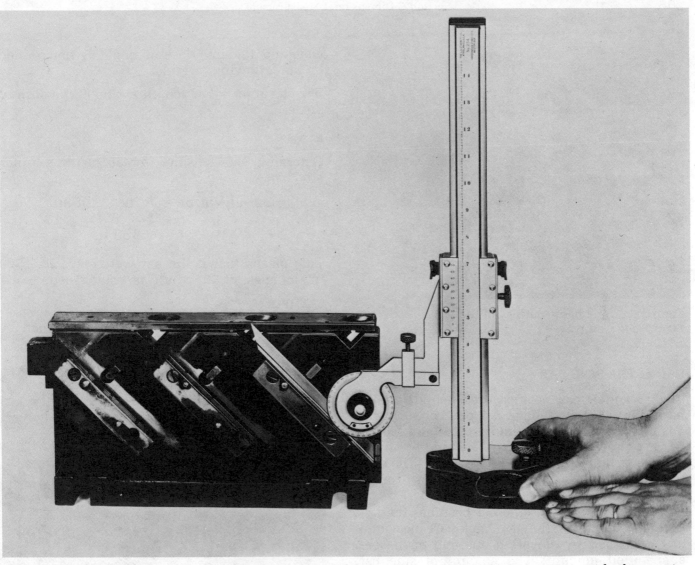

Figure 9. Using the vernier protractor in conjunction with the vernier height gage (Courtesy of the L. S. Starrett Company).

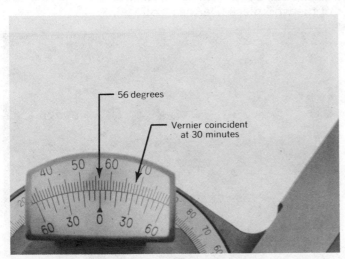

Figure 10. Vernier protractor reading of 56 degrees and 30 minutes.

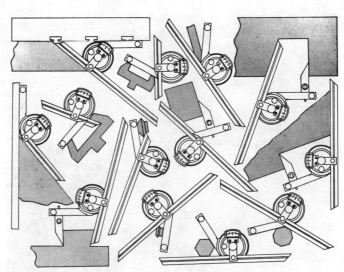

Figure 11. Applications of the vernier bevel protractor (Courtesy of the MTI Corporation).

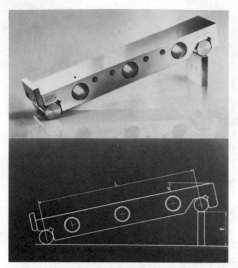

Figure 12. Sine bar (Courtesy of the Mahr Gage Company).

USING THE SINE BAR

Precise angles can be measured using the **sine bar** (Figure 12). A sine bar is a precision bar that has been hardened and then ground and lapped to very precise dimensions. The sine bar has a precise cylinder attached to each end. The center spacing of the cylinders is either 5 or 10 inches and is precisely established.

When in use, the sine bar becomes the hypotenuse of a right triangle. Angles are measured or established by elevating one end of the bar a specified amount. The amount of sine bar elevation for any desired angle is determined by the following formula.

Bar elevation = bar length × sine of the desired angle.

EXAMPLE

Determine the elevation for 30° using a 5-inch sine bar.

$$\text{Bar elevation} = 5 \text{ in.} \times \sin 30°$$
$$= 5 \times .5$$
$$= 2.500 \text{ in.}$$

This means that if the bar were elevated 2.500 in., an angle of 30° would be established.

EXAMPLE

Determine the elevation for 42° using a 5-inch sine bar.

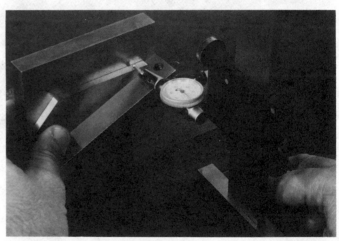

Figure 14. Setting the test indicator to zero at the end of the workpiece (DeAnza College).

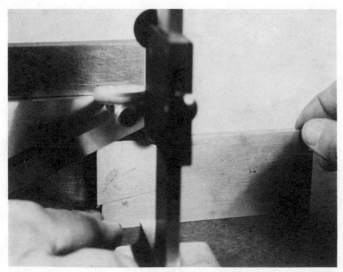

Figure 13. Placing the adjustable parallel under the sine bar (DeAnza College).

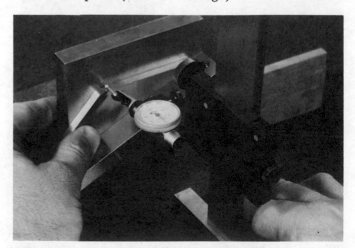

Figure 15. Checking the zero reading at the opposite end of the workpiece (DeAnza College).

Figure 16. Measuring the adjustable parallel with an outside micrometer.

Bar elevation = 5 in. × sin 42°
            = 5 × .6691
            = 3.3456 in.

Determining Workpiece Angle Using the Sine Bar and Measuring Workpiece Angle Using the Sine Bar and Adjustable Parallel

An angle may be measured using the sine bar and adjustable parallel. The adjustable parallel is used to elevate the sine bar (Figure 13). The workpiece is placed on the sine bar and a dial test indicator is set to zero on one end of the part (Figure 14). The parallel is adjusted until the dial indicator reads zero at each end of the workpiece (Figure 15). The parallel is then removed and measured with a micrometer (Figure 16). To determine the angle of the workpiece, simply transpose the sine bar elevation formula and solve for the angle.

Bar elevation = bar length × sine of the angle
                                            desired

Sine of the angle desired = elevation/bar length

Sin of angle   = 1.9935 (micrometer
                                        reading/5)

Sin   = .3987
Angle   = 23° 29′ 48″

Figure 17. Wringing the gage block stack for the sine bar elevation (DeAnza College).

Establishing Angles Using the Sine Bar and Gage Blocks

Extremely precise angles can be measured or established by using gage blocks to elevate the

Figure 18. Placing the gage block stack under the sine bar (DeAnza College).

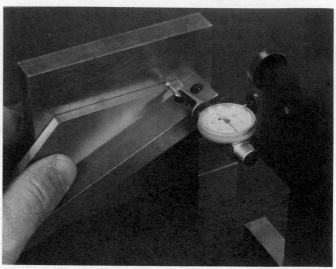

Figure 19. Checking the workpiece using the dial test indicator (DeAnza College).

sine bar. Bar elevation is calculated in the same manner. The required gage blocks are properly prepared and the stack is wrung (Figure 17). The gage block stack totalling 1.9940 in. is placed under the bar (Figure 18). This will establish an angle of 23° 30′ 11″ using a 5 in. sine bar. The angle of the workpiece is checked using a dial test indicator (Figure 19).

Sine Bar Constant Tables
The elevations for angles up to about 55 degrees can be obtained directly from a **table of sine bar constants.** Such tables can be found in machinist's handbooks. The sine bar constant table eliminates the need to perform a trigonometric calculation. The sine bar table may only discriminate to minutes of arc. If discrimination to seconds of arc is required, it is better to calculate the amount of sine bar elevation required.

self-evaluation

SELF-TEST   1. Name two angular measuring instruments with one degree of discrimination.

2. What is the discrimination of the universal bevel protractor?

3. Describe the use of the sine bar.

4.–8. Read and record the following vernier protractor readings (Figures 20 to 24).

4.

5.

6.

7.

8.

9. Calculate the required sine bar elevation for an angle of 37°. (Assume a 5 in. sine bar.)

10. A 10 in. sine bar is elevated 2.750 in. Calculate the angle established to the nearest minute.

Figure 20

Figure 21

Figure 22

Figure 23

Figure 24

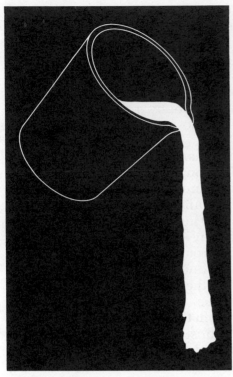

section d
materials

If there were a single thing to which we could attribute the progress of man, it would have to be his ability to make and use tools. The discovery and use of metals would follow close behind, for without metal man still would be fashioning tools of bone and stone.

Nearly everything we need for our present civilization depends upon metals. Vast amounts of iron and steel are used for automobiles, ships, bridges, buildings, machines, and a host of other products (Figure 1). Almost everything that uses electricity depends on copper and many other metals. Some metals that were impossible to smelt or extract from ores a few years ago are now being used in large quantities. These are usually called **space age** metals. There are also hundreds of combinations of metals, called alloys.

We have come a long way since the first iron was smelted, as some believe, by the Hittites about 3500 years ago (Figure 2). Their new iron tools, however, were not much better than those of the softer metals, copper and bronze, that were already in use at the time.

This was because iron, as iron wire or wrought iron bars, would bend and would not hold an edge. The steel making process, which

Figure 1. Large scale production of metal products is performed in modern steel mills (Courtesy of Bethlehem Steel Corporation).

Figure 2. One of the first methods of making iron was by heating the ore over a fire. This produced a small white hot lump of iron known as sponge, which was hammered into shape on an anvil.

uses iron to make a strong and hard material by heat treatment, was still a long way from being discovered.

In this section you will learn how metallic ores are smelted into metals, and how these metals are then formed into the many different products needed in our society. Many metallic ores exist in nature as oxides, in which metals are chemically combined with oxygen. Most iron is removed from the ore by a process called oxidation reduction. Metallic ores are also found as carbonates and silicates.

Modern metallurgy stems from man's ancient desire to fully understand the behavior of metals. Long ago, the art of the metal worker was enshrouded in mystery and folklore. Crude methods of making and heat treating small amounts of steel were discovered by trial and error only to be lost and discovered again by others (Figure 3). We have come a long way, indeed, from those early open forges that produced the soft wrought iron in amounts of 20 or 30 pounds a day to our modern production marvels that produce more than 100 million tons yearly in the United States.

The modern story of iron and steel begins with the raw materials, iron ore, coal, and limestone. From these ingredients pig iron is produced. Pig iron is the source of almost all our ferrous metals. The steel mill refines it in furnaces, after which it is cast into ingot molds to solidify. The ingot is then formed in various ways into the many steel products that are so familiar to us.

In this section, you will investigate the many materials you will work with in the machine shop. You should learn the characteristics of many metals, and become familiar with hardening and tempering, processes that strengthen and harden by repeated heating and cooling. For instance, cast iron is brittle while soft iron is easily bent, and the difference is due to how they are made. Some steels can be made as hard as needed by heat treatment. You will learn several hardness tests by which you can measure the durability and resistance of a metal.

Figure 3. Just prior to the Industrial Revolution, iron working had become a highly skilled craft (Courtesy of Dover Publications, Inc.).

All metals are classified for industrial use by their specific working qualities; you will have to be able to select and identify materials using tables in your handbook or by testing processes in the shop. Some systems are numbers to classify metals, others use color codes, which consist of a brand painted on the end of a piece of the material. Spark testing is one popular shop method for identifying metals that you will meet in this book.

It is important to be able to recognize and identify these materials of the trade in order to do a job according to its specifications. This section has been developed to help you achieve this goal.

SAFETY IN MATERIAL HANDLING Safety must be observed when handling material just as it is when using hand and machine tools.

Lifting and Hoisting Machinists were once expected to lift pieces of steel weighing a hundred pounds or more into awkward positions. This is a dangerous practice, however, that results in too many injuries. Lifting hoists and cranes are used to lift all but the smaller parts. Steel weighs about 487 pounds per cubic foot; water weighs 62.5 pounds per cubic foot, so it is evident that steel is a very heavy material for its size. You can easily be misled into thinking that a small piece of steel does not weigh much. Follow these two rules in all lifting that you do: don't lift more than you can easily handle, and bend your knees and keep your back straight. If a material is too heavy or awkward for you to position it on a machine such as a lathe, use a hoist. Once the workpiece has been hoisted to the required level, it can hang in that position until the clamps or chuck jaws on the machine have been secured.

When lifting heavy metal parts with a mechanical or electric hoist, always stand in a safe position, no matter how secure the slings and hooks seem to be. They don't often break, but it can and does happen, and if a careless foot is under the edge a painful or crippling experience is sure to follow. Slings should not have less than a 30-degree angle with the load (Figure 4). When hoisting long bars or shafts, a spreader bar (Figure 5) should be used so the slings cannot slide together and unbalance the load. When operating a crane, be careful that someone else is not standing in the way of the load or hook. If you are using a block, chain hoist, or electric winch,

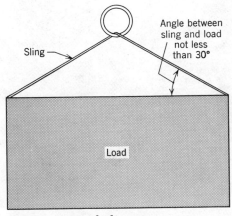

Figure 4. Load sling.

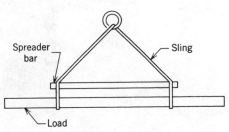

Figure 5. Sling for lifting long bars.

be sure that the lift capacity rating of the equipment and its support structure is proper for the load.

Carrying Objects Carry long stock in the vertical position. Be careful of light fixtures and ceilings. A better way is to have someone carry each end of a long piece of material. Do not carry sharp tools in your pockets. They can injure you or someone else.

Hot Metal Safety Oxy-acetylene torches are often used for cutting shapes, circles, and plates in machine shops. Safety when burning them requires proper clothing, gloves, and eye protection. It is also very important that any metal that has been heated by burning or welding be plainly marked, especially if it is left unattended. The common practice is to write the word *HOT* with soapstone on such items. Wherever arc welding is performed in a shop, the arc flash should be shielded from the other workers. *Never* look toward the arc because if the arc light enters your eye even from the side, the eye can be burned.

When handling and pouring molten metals such as babbitt, aluminum, or bronze, wear a face shield and gloves. Do not pour molten metals where there is a concrete floor unless it is covered with sand.

When heat treating, always wear a face shield and heavy gloves (Figure 6). There is a definite hazard to the face and eyes when cooling tool steel by oil quenching, that is, submerging it in oil. The oil, hot from the steel, tends to fly upward, so you should stand to one side of the oil tank.

Certain metals, when finely divided as a powder or even as coarse as machining chips, can ignite with a spark or just by the heat of machining. Magnesium and zirconium are two such metals. The fire, once started, is difficult to extinguish, and if water or a water-based fire extinguisher is used, the fire will only increase in intensity. Chloride-based power fire extinguishers are commercially available. These are effective for such fires as they prevent water absorption and form an air-excluding crust over the burning metal. Sand is also used to smother fires in magnesium.

Figure 6. Face shield and gloves are worn for protection while heat treating and grinding.

unit 1
selection and identification of steels

When the village smithy plied his trade, there were only wrought iron and carbon steel for making tools, implements, and horseshoes, so the task of separating metals was relatively simple. As industry began to need more alloy steels and special metals, they were gradually developed, so today there are many hundreds of these metals in use. Without some means of reference or identification, work in the machine shop would be chaotic. Therefore, this unit introduces you to several systems used for marking steels and some ways to choose between them.

objective After completing this unit, you will be able to:

Identify different types of metals by various means of shop testing.

STEEL IDENTIFICATION SYSTEMS

Color coding is used as one means of identifying a particular type of steel. Its main disadvantage is that there is no universal color coding system. Each manufacturer has his own system. The two identification systems most used in the United States are numerical: Society of Automotive Engineers (SAE) and American Iron and Steel Institute (AISI). See Table 1.

The first two numbers denote the alloy. Carbon, for instance, is denoted by the number 10. The third and fourth digits, represented by x, always denote the percentage of carbon in hundredths. For carbon steel it could be anywhere from .08 to 1.70 percent. For alloys the second digit designates the approximate percentage of the major alloying element.

The AISI numerical system is basically the same as the SAE system with certain capital letter prefixes. These prefixes designate the process used to make the steel. The lowercase letters from a to i as a suffix denote special conditions in the steel.

AISI prefixes:

B—Acid Bessemer, carbon steel
C—Basic open hearth carbon steel
CB—Either acid Bessemer or basic open hearth carbon steel at the option of the manufacturer
D—Acid open hearth carbon steel
E—Electric furnace alloy steel

STAINLESS STEEL

It is the element chromium (Cr) that makes stainless steels stainless. Steel must contain a minimum of about 11 percent chromium in order to gain resistance to atmospheric corrosion. Higher percentages of chromium make steel even more resistant to corrosion and high temperatures. Nickel is added to improve ductility, corrosion resistance, and other properties.

Excluding the precipitation hardening types that harden over a period of time after solution heat treatment, there are three basic types of stainless steels: the martensitic and ferritic

Table 1
SAE-AISI Numerical Designation of Alloy Steels
(x Represents Percent of Carbon in hundredths)

Carbon steels:	
Plain carbon	10xx
Free-cutting, resulfurized	11xx
Manganese steels	13xx
Nickel steels	
.50% nickel	20xx
1.50% nickel	21xx
3.50% nickel	23xx
5.00% nickel	25xx
Nickel-chromium steels	
1.25% nickel, .65% chromium	31xx
1.75% nickel, 1.00% chromium	32xx
3.50% nickel, 1.57% chromium	33xx
3.00% nickel, .80% chromium	34xx
Corrosion and heat-resisting steels	303xx
Molybdenum steels	
Chromium	41xx
Chromium-nickel	43xx
Nickel	46xx and 48xx
Chromium steels	
Low-chromium	50xx
Medium-chromium	511xx
High-chromium	521xx
Chromium-vanadium steels	6xxx
Tungsten steels	7xxx and 7xxxx
Triple alloy steels	8xxx
Silicon-manganese steels	9xxx
Leaded steels	xxLxx

types of the 400 series, and the austenitic types of the 300 series.

The martensitic, hardenable type has carbon content up to 1 percent or more, so it can be hardened by heating to a high temperature, and then quenching (cooling) in oil or air. The cutlery grades of stainless are to be found in this group. The ferritic type contains little or no carbon. It is essentially soft iron that has 11 percent or more chromium content. It is the least expensive of the stainless steels and is used for such things as building trim, pots, and pans. Both ferritic and martensitic types are magnetic.

Austenitic stainless steel contains chromium and nickel, little or no carbon, and cannot be hardened by quenching, but it readily work hardens while retaining much of its ductility. For this reason it can be work hardened until it is almost as hard as a hardened martensitic steel. Austenitic stainless steel is somewhat magnetic in its work hardened condition, but nonmagnetic when annealed or soft.

Table 2 illustrates the method of classifying the stainless steels. Only a very few of the basic types are given here. You should consult a manufacturer's catalog for further information.

Table 2
Classification of Stainless Steels

Alloy Content	Metallurgical Structure	Ability to be Heat Treated
Chromium types	Martensitic	Hardenable (Types 410, 416, 420) Nonhardenable (Types 405, 14 SF)
	Ferritic	Nonhardenable (Types 430, 442, 446)
Chromium-nickel types	Austenitic	Nonhardenable (except by cold work) (Types 301, 302, 304, 316) Strengthened by aging (Types 314, 17-14 CuMo, 22-4-9)
	Semi-austenitic	Precipitation hardening (PH 15-7 Mo, 17-7 PH)
	Martensitic	Precipitation hardening (17-4 PH, 15-5 PH)

Source. Armco Steel Corporation, Middletown, Ohio, *Armco Stainless Steels*, 1966. The following are registered trademarks of Armco Steel Corporation: 17-4 PH, 15-5 PH, 17-7 PH, and PH 15-7 Mo.

TOOL STEELS

Special carbon and alloy steels called tool steels have their own classification. There are six major tool steels for which one or more letter symbols have been assigned:

1. Water hardening tool steels
   W—high carbon steels
2. Shock resisting tool steels
   S—Medium carbon, low alloy
3. Cold work tool steels
   O—Oil hardening types
   A—Medium alloy air hardening types
   D—High carbon, high-chromium types
4. Hot work tool steels
   H—H1 to H 19, chromium base types
      H20 to H 39, tungsten base types
      H40 to H59, molybdenum base types
5. High speed tool steels
   T—Tungsten base types
   M—Molybdenum base types
6. Special purpose tool steels
   L—Low alloy types
   F—Carbon tungsten types
   P—Mold steels P1 to P19, low carbon types
      P20 to P39, other types

Several metals can be classified under each group, so that an individual type of tool steel will also have a suffix number that follows the letter symbol of its alloy group. The carbon content is given only in those cases where it is considered an identifying element of that steel.

EXAMPLES

Water hardening: straight carbon tool steel	W1
Silicon–manganese–molybdenum: punch steel	S5
Manganese–chrome–tungsten: oil hardening tool steel	O1
Chromium (5.0 percent): air hardening die steel	A2
High speed (6-5-4-2): tool steel	M2

SHOP TESTS FOR IDENTIFYING STEELS

One of the disadvantages of steel identification systems is that the marking is often lost. The end of a shaft is usually marked. If the marking is obliterated or cut off and the piece is separated from its proper storage rack, it is very difficult to ascertain its carbon content and alloy group. This shows the necessity of returning stock material to its proper rack. It is also good practice to leave the identifying mark on one end of the stock material and always cut off the other end.

Unfortunately, there are always some short ends and otherwise useful pieces in most shops that have become unidentified. Also, when repairing or replacing parts for old or nonstandard machinery, there is usually no record available for material selection. There are many shop methods a machinist may use to identify the basic type of steel in an unknown sample. By process of elimination, the machinist can then determine which of the several steels of that type in the shop is most comparable to his sample. The following are several methods of shop testing that you can use.

Observation

Some metals can be identified by direct observation of their finishes. Heat scale or black mill scale is found on all hot rolled (HR) steels. These can be either low carbon (.05 to .30 percent), medium carbon (.30 to .60 percent), high carbon (.60 to 1.75 percent), or alloy steels. Other surface coatings that might be detected are the sherardized, plated, case hardened, or nitrided surfaces. *Sherardizing* is a process in which zinc vapor is inoculated into the surface of iron or steel.

Cold finish (CF) steel usually has a metallic luster. Ground and polished (G and P) steel has a bright, shiny finish with closer dimensional tolerances than CF. Also cold drawn ebonized, or black, finishes are sometimes found on alloy and resulfurized shafting.

Chromium nickel stainless steel, which is austenitic and nonmagnetic, usually has a white appearance. Straight 12 to 13 percent chromium is ferritic and magnetic with a bluish-white color. Manganese steel is blue when polished, but copper colored when oxidized. White cast iron fractures will appear silvery or white. Gray cast iron fractures appear dark gray and will smear a finger with a gray graphite smudge when touched.

Magnet Test

All ferrous metals such as iron and steel are magnetic; that is, they are attracted to a magnet. Nickel, which is nonferrous (metals other than iron or steel), is also magnetic. United States "nickel" coins contain about 25 percent nickel and 75 percent copper, so they do not respond to the magnet test, but Canadian "nickel" coins are attracted to a magnet. Ferritic and martensitic (400 series) stainless steels are also attracted to a magnet and so cannot be separated from other steels by this method. Austenitic (300 series) stainless steel is not magnetic unless it is work hardened.

Hardness Test

Wrought iron is very soft since it contains almost no carbon or any other alloying element. Generally speaking, the more carbon (up to 2 percent) and other elements that steel contains, the harder, stronger and less ductile it becomes, even if in an annealed state. Thus, the hardness of a sample can help us to separate low carbon steel from an alloy steel or a high carbon steel. Of course, the best way to check for hardness is with a hardness tester. The Rockwell, Brinell, and other types of hardness testing will be studied in another unit. Not all machine shops have hardness testers available, in which case the following shop methods can prove useful.

Scratch Test

Geologists and "rock hounds" scratch rocks against items of known hardness for identification purposes. The same method can be used to check metals for relative hardness. Simply scratch one sample with another and the softer sample will be marked. Be sure all scale or other surface impurities have been removed before scratch testing. A variation of this method is to strike two similar edges of two samples together. The one receiving the deepest indentation is the softer of the two.

File Tests

Files can be used to establish the relative hardness between two samples, as in the scratch test, or they can determine an approximate hardness of a piece on a scale of many steels. Table 3 gives the Rockwell and Brinell hardness numbers for this file test when using new files. This method, however, can only be as accurate as the skill that the user has acquired through practice.

Care must be taken not to damage the file, since filing on hard materials may ruin the file. Testing should be done on the tip or near the edge.

Spark Testing

Spark testing is a useful way to test for carbon content in many steels. The metal tested, when held against a grinding wheel, will display a particular spark pattern depending on its content. Spark testing provides a convenient means of distinguishing between tool steel (of medium or high carbon) and low carbon steel. High carbon steel (Figure 1) shows many more bursts than low carbon steel (Figure 2).

Almost all tool steel contains some alloying elements besides the carbon, which affects the

Table 3
File Test and Hardness Table

File Reaction	Rockwell B	Rockwell C	Brinell	Type Steel
File bites easily into metal	65		100	Mild steel
File bites into metal with pressure		16	212	Medium carbon steel
File does not bite into metal except with difficulty		31	294	High alloy steel High carbon steel
Metal can only be filed with extreme pressure		42	390	Tool steel
File will mark metal but metal is nearly as hard as the file and filing is impractical		50	481	Hardened tool steel
Metal is as hard as the file		64	739	Case hardened parts

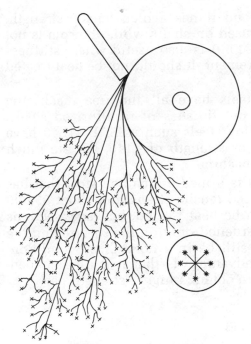

Figure 1. High carbon steel. Short, very white, or light yellow carrier lines with considerable forking, having many starlike bursts. Many of the sparks follow around the wheel.

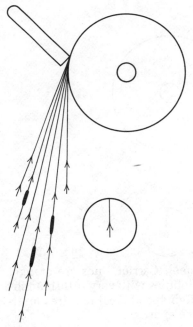

Figure 2. Low carbon steel. Straight carrier lines having a yellowish color with very small amount of branching and very few carbon bursts.

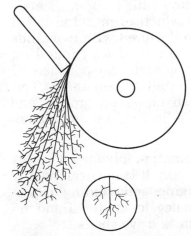

Figure 3. Cast iron. Short carrier lines with many bursts, which are red near the grinder and orange-yellow further out. Considerable pressure is required on cast iron to produce sparks.

carbon burst. Chromium, molybdenum, silicon, aluminum, and tungsten suppress the carbon burst. For this reason spark testing is not very useful in determining the content of an unknown sample of steel. It is useful, however, as a comparison test. Comparing the spark of a known sample to that of an unknown sample can be an effective method of identification for the trained observer. Cast iron may be distinguished from steel by the characteristic spark stream (Figure 3). High speed steel can also be readily identified by spark testing (Figure 4).

When spark testing always wear safety glasses or a face shield. Adjust the wheel guard so the spark will fly outward and downward, and away from you. A coarse grit wheel that has been freshly dressed to remove contaminants should be used.

Machinability
Machinability can be used in a simple comparison test to determine a specific type of steel. For example, two unknown samples identical in appearance and size can be test cut in a machine tool, using the same speed and feed for both. The ease of cutting should be compared, and chips observed for heating color and curl.

Several properties should be considered when selecting a piece of steel for a job: strength, machinability, hardness, weldability, fatigue resistance, and corrosion resistance.

Manufacturer's catalogs and handy reference books are available for selection of standard structural shapes, bars and other steel products. Others are available for the stainless steels, tool steels and finished carbon steel

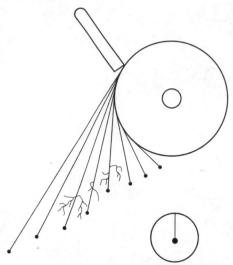

Figure 4. High speed steel. Carrier lines are orange, ending in pear-shaped globules with very little branching or carbon sparks. High speed steel requires moderate pressure to produce sparks.

and alloy shafting. Many of these steels are known by a trade name.

A machinist is often called upon to select a shaft material from which to machine finish a part. Shafting is manufactured with two kinds of surface finish: cold finished (CF) found on low carbon steel, and ground and polished (G and P) found mostly on alloy steel shafts. Tolerances are kept much closer on ground and polished shafts. The following are some common alloy steels:

1. SAE 4140 is a chromium-molybdenum alloy with .40 percent carbon. It lends itself readily to heat treating, forging, and welding. It provides a high resistance to torsional and reversing stresses such as drive shafts.
2. SAE 1140 is a resulfurized, drawn, free machining bar stock. This material has good resistance to bending stresses because of its fibrous qualities and it has a high tensile strength. It is best used on shafts while the rpm is not low. SAE 1140 is also useful where stiffness is a requirement. It should not be heat treated or welded.
3. Leaded steels have all the free machining qualities and finishes of resulfurized steels. Leaded alloy steels such as SAE 41L40 have the superior strength of 4140, but are much easier to machine.
4. SAE 1040 is a medium carbon steel that has a normalized tensile strength of about 85,000 psi. It can be heat treated, but large sections will be hardened only on the surface and the core will still be in a normalized condition. Its main advantage is that it is a less expensive way to obtain a higher strength part.

COSTS OF STEEL

Steel prices, as prices of most other products, change constantly, so costs can be shown only in example. Steel is usually priced by its weight. A cubic foot of mild steel weighs 489.60 pounds, so a square foot one inch thick weighs 40.80 pounds. From this, you can easily compute the weights for flat materials such as plate. For hexagonal and rounds, it would be much easier to consult a table in a catalog or handbook. Given a price per pound, you should then be able to figure the cost of a desired steel product.

EXAMPLE

A 1 by 6 inch mild steel bar is 48 inches long. If current steel prices are 30 cents per pound, how much does the bar cost?

$$\frac{6 \times 48}{144} = 2 \text{ sq. ft., 1 inch thick}$$

$$2 \times 40.80 = 81.6 \text{ pounds}$$

$$81.6 \times .30 = 24.48 \text{ (cost of steel in dollars)}$$

self-evaluation

SELF-TEST 1. By what universal coding system is carbon and alloy steel designated?

2. What are three basic types of stainless steels and what is the number series assigned to them? What are their basic differences?

3. If your shop stocked the following steel shafting, how would you determine the content of an unmarked piece of each, using shop tests as given in this unit?
   (a) AISI C1020 CF
   (b) AISI B1140 (G and P)
   (c) AISI C4140 (G and P)
   (d) AISI 8620 HR
   (e) AISI B1140 (Ebony)
   (f) AISI C1040

4. A small part has obviously been made by a casting process. How can you determine whether it is a ferrous or a nonferrous metal, or if it is steel, or white or gray cast iron?

5. What is the meaning of the symbols O1 and W1 when applied to tool steels?

6. A $2\frac{7}{16}$ inch diameter steel shaft weighs 1.322 pounds per linear inch, as taken from a table of weights of steel bars. A 40 inch length is needed for a job. At 30 cents per pound, what would the shaft cost?

7. When checking the hardness of a piece of steel with the file test, the file slides over the surface without cutting.
   a. Is the steel piece readily machinable?
   b. What type of steel is it most likely to be?

8. Steel that is nonmagnetic is called.

9. What nonferrous metal is magnetic?

10. List at least four properties of steel that should be kept in mind when you select the material for a job.

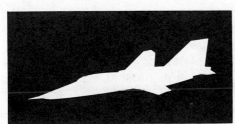

unit 2
selection and identification of nonferrous metals

Metals are designated as either ferrous or nonferrous. Iron and steel are ferrous metals, and any metal other than iron or steel is called nonferrous. Nonferrous metals such as gold, silver, copper, and tin were in use hundreds of years before the smelting of iron, and yet, some nonferrous metals have appeared relatively recently in common industrial use. For example, aluminum was first commercially extracted from ore in 1886 by the Hall-Heroult process, and titanium is a space age metal being produced in commercial quantities only after World War II.

In general, nonferrous metals are more costly than ferrous metals. It isn't always easy to distinguish a nonferrous metal from a ferrous metal, nor to separate one from another. This unit should help you to identify, select, and properly use many of these metals.

OBJECTIVES   **After completing this unit, you will be able to:**

1. Identify and classify nonferrous metals by a numerical system.

2. List the general appearance and use of various nonferrous metals.

ALUMINUM

Aluminum is white or white-gray in color and can have any surface finish from dull and anodized to shiny and polished. Aluminum weighs 168.5 pounds per cubic foot as compared to 487 pounds per cubic foot for steel, and has a melting point of 1220°F (660°C) when pure. It is readily machinable and can be manufactured into almost any shape or form (Figure 1).

Magnesium is also a much lighter metal than steel, as it weighs 108.6 pounds per cubic foot, and looks much like aluminum. In order to distinguish between the two metals, it is sometimes necessary to make a chemical test. A zinc chloride solution in water, or muriatic acids such as are used in soldering fluxes, will blacken magnesium immediately, but will not change aluminum.

Table 1
Aluminum and Aluminum Alloys

Code Number	Major Alloying Element
1xxx	None
2xxx	Copper
3xxx	Manganese
4xxx	Silicon
5xxx	Magnesium
6xxx	Magnesium and silicon
7xxx	Zinc
8xxx	Other elements
9xxx	Unused (not yet assigned)

There are several numerical systems used to identify aluminums, such as federal specifications, military specifications, the American Society for Testing Materials (ASTM), and SAE specifications. The system most used by manufacturers, however, is one which was adopted by the Aluminum Association in 1954.

From Table 1, you can see that the first digit of a number in the aluminum alloy series indicates the alloy type. The second digit, represented by an x in the table, indicates any modifications that were made to the original alloy. The last two digits indicate the numbers of similar aluminum alloys of an older marking system, except in the 1100 series, where the last two digits indicate the amount of pure aluminum above 99 percent contained in the metal.

EXAMPLES

An aluminum alloy numbered 5056 is in aluminum-magnesium alloy, where the first 5 represents the alloy magnesium, the 0 represents

Figure 1. Structural aluminum shapes used for building trim provides a pleasing appearance.

modifications to the alloy, and 56 are numbers of a similar aluminum of an older marking system. An aluminum numbered 1120 contains no major alloy, and has .20 percent pure aluminum above 99 percent.

Aluminum and its alloys are produced as castings or as wrought (cold worked) shapes such as sheets, bars, and tubing. Aluminum alloys are harder than pure aluminum and will scratch the softer (1100 series) aluminums. Pure aluminum and some of its alloys cannot be heat treated so their tempering is done by other methods. The temper designations are made by a letter that follows the four digit alloy series number:

- —F   as fabricated. No special control over strain hardening or temper designation is noted.
- —O   Annealed, recrystallized wrought products only. Softest temper.
- —H   Strain hardened, wrought products only. Strength is increased by work hardening.

This letter—H is always followed by two or more digits. The first digit, 1, 2, or 3, denotes the final degree of strain hardening,

- —H1   Strain hardened only
- —H2   Strain hardened and partially annealed
- —H3   Strain hardened and stabilized

and the second digit denotes higher strength tempers obtained by heat treatment.

2	$\frac{1}{4}$ hard
4	$\frac{1}{2}$ hard
6	$\frac{3}{4}$ hard
8	full hard

EXAMPLE

5056-H18 is an aluminum-magnesium alloy, strain hardened to a full hard temper.

Some aluminum alloys can be hardened to a great extent by a process called solution heat treatment and precipitation or aging. This process involves heating the aluminum and its alloying elements until it is a solid solution. The aluminum is then quenched in water and allowed to age or is artificially aged by heating slightly. The aging produces an internal strain that hardens and strengthens the aluminum. Some other nonferrous metals are also hardened by this process. For these aluminum alloys the letter —T follows the four digit series num-

ber. Numbers 2 to 10 follow this letter to indicate the sequence of treatment.

- —T2   Annealed (cast products only)
- —T3   Solution heat treated and cold worked
- —T4   Solution heat treated, but naturally aged
- —T6   Solution heat treated and artificially aged
- —T8   Solution heat treated, cold worked, and artificially aged
- —T9   Solution heat treated, artificially aged, and cold worked
- —T10 Artificially aged and then cold worked

EXAMPLE

2024-T6   Aluminum-copper alloy, solution heat treated and artificially aged.

OTHER NONFERROUS METALS

Babbitt Metals

A babbitt metal is a soft, antifriction alloy metal often used for bearings and usually tin or lead based. Tin babbitts usually contain from 65 to 90 percent tin with antimony, lead, and a small percentage of copper added. There are the higher grade and generally more expensive of the two types. Lead babbitts contain up to 75 percent lead, with antimony, tin, and some arsenic making up the balance.

Cadium base babbitts resist higher temperatures than other tin and lead base types. These alloys contain from 1 to 15 percent nickel or a small percentage of copper and up to 2 percent silver (Figure 2).

Figure 2. Babbitted pillow block bearings.

Beryllium Copper

Beryllium copper is an alloy of copper that can be hardened by heat treating for making tools and other products. Machining of this metal should be done after solution heat treatment and aging, and not when it is in the annealed state. Machining beryllium copper can be very hazardous if safety precautions are not taken. Machining dust should be removed by a heavy coolant flow or by a vacuum system, and a respirator type of face mask should be worn.

Brass

Zinc and copper are alloyed to make brass. Brass colors range from white-yellow to red-yellow. Brasses are easily machined.

Bronze

Many combinations of copper and other metals are called bronze, but copper and tin are the original elements combined to make bronze. Bronzes are usually harder than brasses, but are easily machined with sharp tools. Some bronze alloys are used for brazing rods.

Copper

Copper is a soft, heavy metal that has a reddish color. It has high electrical conductivity when pure, but loses this property to a great extent when alloyed. It must be strain hardened when used for electric wire. Copper is very ductile and can be easily drawn into wire or tubular products. It is so soft that it is difficult to machine and it has a tendency to adhere to tools. Copper can be work hardened or hardened by solution heat treatment when alloyed with beryllium.

Die Cast Metals

Finished castings are produced with various metal alloys by the process of die casting. Die casting is a method of forming or shaping a metal by forcing it into a mold. After the metal has solidified, the mold opens and the casting is ejected. Carburetors, door handles, and many small precision parts are manufactured with this process (Figure 3). Die cast alloys, often called "pot metals," are classified in six groups:

1. Tin base alloys
2. Lead base alloys
3. Zinc base alloys
4. Aluminum base alloys

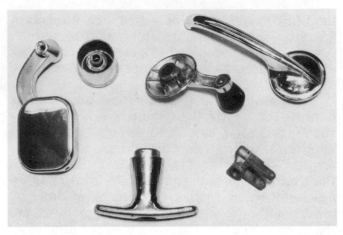

Figure 3. Die cast parts.

5. Copper, bronze, or brass alloys
6. Magnesium base alloys

The specific content of the alloying elements in each of the many die cast alloys may be found in handbooks or other references on die casting.

Lead

Lead is a heavy metal that is silvery when newly cut and gray when oxidized. It has a low tensile strength, low ductility (cannot be easily drawn into wire), and high malleability (can be easily compressed into a thin sheet). Lead has a high corrosion resistance and is alloyed with antimony and tin for various uses. Lead is alloyed with steels, brasses, and bronzes to improve machinability.

Magnesium

When pure, magnesium is a soft, silver-white metal that closely resembles aluminum. It is very light in weight at 108.6 pounds per cubic foot. Magnesium differs from aluminum in that it will readily burn, and as it burns, it gives off a brilliant white light. Cast and wrought magnesium alloys are designated by SAE numbers, which may be found in metals reference handbooks or catalogs. Many aircraft parts are made from magnesium alloys.

Molybdenum

As a pure metal, molybdenum is used for high temperature applications and when machined, it chips like gray cast iron. It is used as an alloying element in steel to promote deep hardening.

Monel Metal

Monel is an alloy of nickel and copper. It is a tough, but machinable, ductile, and corrosion resistant alloy. Its tensile strength (resistance of a metal to a force tending to tear it apart) is 70,000 to 85,000 pounds per square inch. Monel metal is used to make marine equipment such as pumps, steam valves, and turbine blades. On a spark test, monel shoots orange colored, straight sparks about 10 inches long, similar to those of nickel.

Nickel

Nickel is noted for its resistance to corrosion and oxidation. It is a whitish metal used for electroplating and as an alloying element in steel and other metals. Electroplating is the coating or covering of another material with a thin layer of metal, using electricity to deposit the layer. When spark testing nickel, it throws short orange carrier lines with no sparks or sprigs (Figure 4).

Nickel Base Alloys

Chromel and nichrome are two nickel-chromium-iron alloys used as resistance elements in electric heaters and toasters. Other nickel alloys such as inconel are used for parts that are exposed to high temperatures for extended periods. Nickel silver contains nickel and copper in similar proportions to monel, but also contain 17 percent zinc.

Precious Metals

Gold has limited industrial use and in the past has been used mostly for jewelry and coinage. Gold coinage is usually hardened by alloying with about 10 percent copper. Silver is alloyed with 8 to 10 percent copper for coinage and jewelry. Sterling silver is 92.5 percent silver in English coinage and has been 90 percent silver for American coinage.

Unlike gold, silver has many commercial uses as an alloying element for mirrors, photographic compounds, and electrical equipment. It has a very high electrical conductivity. Silver is used in silver solders that are stronger and have a higher melting point than lead-tin solders.

Platinum, palladium, and iridium, as well as other rare metals, are even more expensive than gold. These metals are used commercially because of their special properties such as extremely high resistance to corrosion, high melting points, and high hardness.

Tantalum

Tantalum is a difficult metal to machine as it is quite soft and the chip clings to the tool. It is used for high temperature operations above 2000°F (1093°C). Tantalum carbides are combined with tungsten carbides for cutting tools that have extreme wear resistance.

Tin

Since tin has a good corrosion resistance, it is used to plate steel, especially for the food processing industry (Figure 5). Tin is used as an alloying element for solder and pewter. A popu-

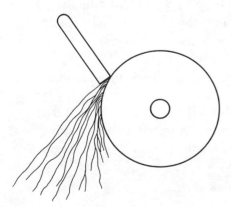

Figure 4. Spark test for nickel.

Figure 5. The most familiar tin plate product is the steel based tin can (Courtesy of American Iron & Steel Institute).

lar solder is an alloy of 50 percent tin and 50 percent lead.

Titanium
The strength and light weight of this silver-gray metal make it very useful in the aerospace industries for jet engine components, heat shrouds, and rocket parts. Pure titanium has a tensile strength of 50,000 to 90,000 pounds per square inch, similar to that of steel, and by alloying the tensile strength can be increased considerably. Titanium weighs about half as much as steel, and like stainless steel, is a relatively difficult metal to machine. Machining can be accomplished with rigid setups, sharp tools, slower surface speed, and use of proper coolants. Titanium throws a brilliant white spark with a single burst on the end of each carrier, when spark tested.

Tungsten
Typically, this metal has been used for incandescent light filaments. It has the highest known melting point (6200°F or 3427°C) of any metal, but is not resistant to oxidation at high temperatures. Tungsten is used for rocket engine nozzles and welding electrodes and as an alloying element with other metals. Machining pure tungsten is very difficult with single point tools, and grinding is preferred for finishing operations. Tungsten carbide compounds are used to make extremely hard and heat resistant lathe tools and milling cutters.

Zinc
The familiar galvanized steel is actually steel plated with zinc, used mainly for its high corrosion resistance. Zinc alloys are widely used as die casting metals.

Zirconium
Zirconium is similar to titanium in both appearance and physical properties. It was once used as an explosive primer and as a flashlight powder for photography. Machining zirconium, like titanium, requires rigid setups and slow surface speeds. Zirconium has an extremely high resistance to corrosion from acids and sea water. Zirconium alloys are used in nuclear reactors, flash bulbs, and surgical implants such as screws, pegs, and skull plates. When spark tested, it produces a spark that is similar to that of titanium.

self-evaluation

SELF-TEST

1. What advantage do aluminum and its alloys have over steel alloys? What disadvantages?
2. Describe the meaning of the letter "H" when it follows the four digit number that designates an aluminum alloy? The meaning of the letter "T"?
3. Name two ways in which magnesium differs from aluminum.
4. What is the major use of copper? How can copper be hardened?
5. What is the basic difference between brass and bronze?
6. Name two uses for nickel.
7. Lead, tin, and zinc all have one useful property in common. What is it?
8. Molybdenum and tungsten are both used in _____ steels.
9. Babbit metals, used for bearings, are made in what major basic types?
10. What type of metal can be injected under pressure into a permanent mold?

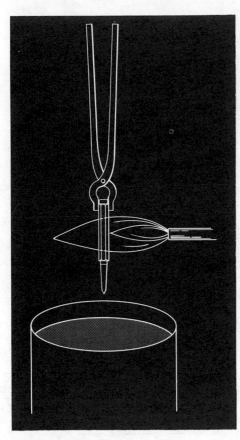

unit 3
hardening, case hardening, and tempering

Probably the most important property of carbon steels is their ability to be hardened through the process of heat treatments. Steels must be made hard if they are to be used as tools that have the ability to cut many materials including other soft steels. Various degrees of hardness are also desirable depending on the application of the tool steels. Heat treating carbon steels involves some very critical furnace operations. The proper steps must be carried out precisely, or a failure of the hardened steel will almost surely result. In come cases only surface hardening of a steel is required. This is accomplished through the process of surface and case hardening.

It is often desirable to slightly reduce the hardness of a steel tool in order to enhance other properties. For example, a chisel must have a hard cutting edge. It must also have a somewhat less hard but tough shank that will withstand hammering. The property of toughness is acquired through the process of tempering.

In this unit you will study the important processes and procedures for hardening, case hardening, and tempering of carbon steels.

objectives After completing this unit, you will be able to:

1. Correctly harden a piece of tool steel and evaluate your work.
2. Correctly temper the hardened piece of tool steel and evaluate your work.
3. Describe the proper heat treating procedures for more tool steels.
4. Correctly harden a SAE 4140 vee block or equivalent.
5. Correctly draw temper the SAE 4140 block to a prescribed hardness.

HARDENING METALS

Most metals (except copper used for electric wire) are not used commercially in their pure states because they are too soft and ductile and have low tensile strength. When they are alloyed with other elements, such as other metals, they become harder and stronger as well as more useful. A small amount (1 percent) of carbon greatly affects pure iron when alloyed with it. The alloy metal becomes a familiar tool steel used for cutting tools, files and punches. Iron with 2 to $4\frac{1}{2}$ percent carbon content yields cast iron.

Iron, steel, and other metals are composed of tiny grain structures that can be seen under

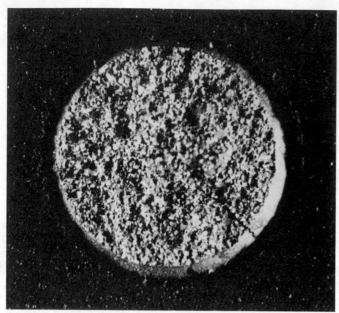

Figure 1. Single fracture of steel.

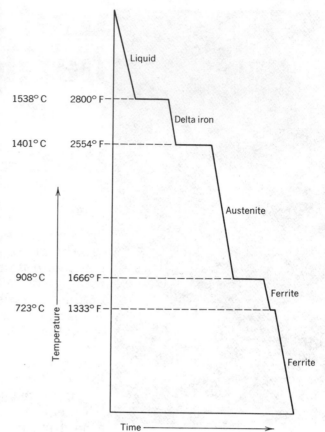

Figure 2. Cooling curve of iron.

a microscope when the specimen is polished and etched. This grain structure, which determines the strength and hardness of a steel, can be seen with the naked eye as small crystals in the rough broken section of a piece (Figure 1). These crystals or grain structures grow from a nucleus as the molten metal solidifies until the grain boundaries are formed. Grain structures differ according to the allotropic form of the iron or steel. An allotropic element is one which is able to exist in two or more forms with various properties without a change of chemical composition. Carbon exists in three *allotropic* forms: amorphous (charcoal, soot, coal), graphite, and diamond. Iron also exists in three allotropic forms (Figure 2): ferrite (at room temperature), austenite (above 1670°F, 911°C) and delta (between 2550°F and 2800°F or between 1498°C and 1371°C). The points where one phase changes to another are called critical points by heat treaters and transformation points by metallurgists. The critical points of water are the boiling point (212°F or 100°C) and the freezing point (32°F or 0°C). Figure 3 shows a critical temperature diagram for a carbon steel. The lower critical point is always about 1330°F (721°C), but the upper critical point changes as the carbon content changes (Figure 4).

HEAT TREATING STEEL

As steel is heated above the critical temperature of 1330°F, (721°C) the carbon that was in the

form of layers of iron carbide or pearlite (Figure 5) begins to dissolve in the iron and form a solid solution called austenite (Figure 6). When this solution of iron and carbon is suddenly cooled or quenched, a new microstructure is formed. This is called martensite (Figure 7). Martensite is very hard and brittle, having a much

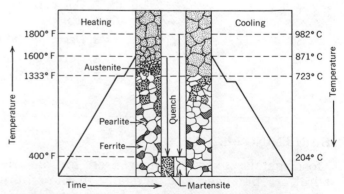

Figure 3. Critical temperature diagram of .83 percent steel showing grain structures in heating and cooling cycles. Center section shows quenching from different temperatures and the resultant grain structure.

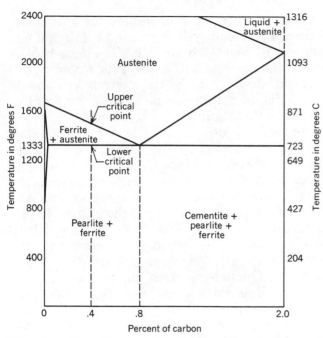

Figure 4. Simplified phase diagram for carbon steels.

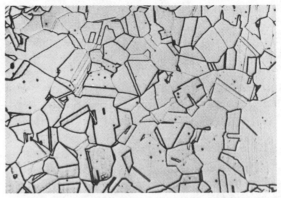

Figure 6. A microstructure of annealed 304 stainless steel that is austenitic at ordinary temperatures (250 ×). (By permission, from *Metals Handbook*, Volume 7, Copyright © American Society for Metals, 1972).

higher tensile strength than the steel with a pearlite microstructure. It is quite unstable, however, and must be tempered to relieve internal stresses in order to have the ductility and toughness needed to be useful. AISI-C1095, commonly known as water hardening (W1) steel, will begin to show hardness when quenched from a temperature just over 1330°F (721°C) but will not harden at all if quenched from a temperature lower than 1330°F (721°C). The

Figure 7. 1095 steel, water quenched from 1500°F (816°C) (1000 ×). The needlelike structure shows a pattern of fine untempered martensite. (By permission, from *Metals Handbook*, Volume 7, Copyright © American Society for Metals, 1972).

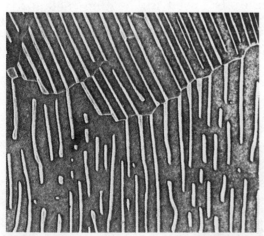

Figure 5. A replica electron micrograph. Structure consists of lamellar pearlite (11,000 ×). (By permission, from *Metals Handbook*, Volume 7, Copyright © American Society for Metals, 1972).

steel will become as hard as it can get when heated to 1450°F (788°C) and quenched in water. This quenching temperature changes as the carbon content of the steel changes. It should be 50°F (10°C) above the upper critical limit.

Low carbon steels such as AISI 1020 will not, for all practical purposes, harden when they are heated and quenched. Oil and air hardening steels harden over a longer period of time and, consequently, are deeper hardening than water hardening types, which must be cooled to 200°F (93°C) within one or two seconds. As you can see, it is quite important to know the carbon content and alloying element so that the correct temperature and quenching medium can be used. Fine grained tool steels are much stronger than coarse grained tool steels (Figure 8). If a piece of tool steel is heated above the correct temperature for its specific carbon content, a phenomenon called grain growth will occur, and a coarse, weak grain structure develops. The grain growth will remain when the part is quenched and, if used for a tool such as a punch or chisel, the end may simply drop off when the first hammer blow is struck. Tempering will not remove the coarse grain structure. If the part has been overheated, simply cooling back to the quenching temperature will not help, as the coarse grain persists well down into the hardening range (Figure 3). The part should be cooled slowly and then reheated to the correct quenching temperature. Steels containing .83 percent carbon can get as hard as any carbon steel (Rc67) containing more carbon.

AISI-C1095, or water hardening tool steel (W1), can be quenched in oil, depending on the size of the part. For example, for a piece of AISI-C1095 drill rod, if the section is thin or the diameter is small, oil should be used as a quenching medium. Oil is not as severe as water because it conducts heat less rapidly than water, and thus avoids quench cracking. Larger sections or parts would not be fully transformed into martensite if they are oil quenched, but would instead contain some softer transformation structures. Water quench should be used, but remember that W1 is shallow hardening, and will only harden about $\frac{1}{8}$ inch deep.

When using a furnace to heat for quenching, the temperature control should be set for 1450°F (788°C). If the part is small, a preheat is not necessary; but if it is thick, it should be brought up to heat slowly. If the part is left in a furnace without a controlled atmosphere for any length of time, the metal will form an oxide scale and carbon will leave the surface. This decarburization of the surface will cause it to soften, while the metal directly under the surface will remain hard. This oxidation can be avoided by painting the part with a solution of boric acid and water before heating, or wrapping it in stainless steel foil. Place the part in the furnace with tongs, and wear gloves and face shields for protection. When the part has become the same color as the furnace bricks, remove it by grasping one end with the tongs and *immediately* plunge it into the quenching bath. If the part is long like a chisel or punch, it should be inserted into the quench vertically (straight up-and-down), not at a slant. Quenching at an angle can cause unequal cooling rates and bending of the part. Also, agitate the part in an up-and-down or a figure-8 motion to remove any gases or bubbles that might cause uneven quenching.

Oil is used to quench oil hardening steels (O1). When hardening various tool steels, a manufacturer's catalog should be consulted for the correct temperature, time periods and quenching media.

FURNACES

Electric, gas, or oil fueled furnaces are used for heat treating steels (Figure 9). They use various types of controls for temperature adjustment. These controls make use of the principle of the thermocouple (Figure 10). Temperatures generally range to 2500°F (1371°C). High temperature salt baths are also used for heating metals for hardening or annealing. One of the disadvantages of most electric furnaces is that they allow the atmosphere to enter the furnace, and the oxygen causes oxides to form on the heated metal. This causes scale and decarburization of the surface of the metal. A decarburized surface will not harden. One way to control this loss of surface carbon is to keep a slightly carbonizing atmosphere or an inert gas in the furnace.

One of the most important factors when heating steel is the rate at which heat is applied. When steel is first heated, it expands. If cold steel is placed in a hot furnace, the surface expands more rapidly than the still cool core. The surface will then have a tendency to pull away from the center, thus inducing internal stress. This can cause cracking and distortion in the part. Most furnaces can be adjusted for the proper rate of

Figure 8. Fractured ends of $\frac{3}{8}$ inch diameter 1095 water quenched tool steel ranging from fine grain, quenched from 1475°F (802°C), to coarse grain, quenched from 1800°F (982°C).

Figure 9. Electric heat treating furnace. Part is being placed in furnace by heat treater wearing correct attire and using tongs (Lane Community College).

Figure 10. Thermocuple (Lane Community College).

heat input (Figure 11) when bringing the part up to the soaking temperature. (Figure 12). **Soaking** means holding the part for a given length of time at a specified temperature. Another factor is the time of soaking required for a certain size piece of steel. An old rule of thumb allows the steel to soak in the furnace for 1 hour for each inch of thickness, but there are considerable variations to this rule, since some steels require much more soaking time than others. The correct soaking period for any specific tool steel may be found in tool steel reference books.

QUENCHING MEDIA

In general six media are used to quench metals. They are listed here in their order of severity or speed of quenching.

1. Water and salt; that is, sodium chloride or sodium hydroxide. It is also called brine.
2. Tap water.
3. Fused or liquid salts.
4. Soluble oil and water.
5. Oil.
6. Air.

Liquid quenching media goes through three stages. The vapor-blanket stage occurs first because the metal is so hot it vaporizes the media. This envelops the metal with vapor, which insulates it from the cold liquid bath. This causes the cooling rate to be relatively slow during this stage. The *vapor transport cooling* stage begins when the vapor blanket collapses, allowing the liquid medium to contact the surface of the

Figure 11. Input controls on furnace (Lane Community College).

Figure 12. Temperature control (Lane Community College).

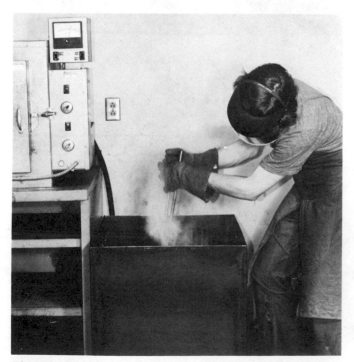

Figure 13. Heat treater is agitating part during quench (Lane Community College).

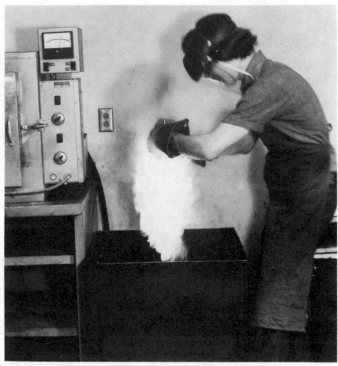

Figure 14. Beginning of quench. At this stage heat treater could be burned by hot oil if he is not adequately protected with gloves and face shield. (Lane Community College).

metal. The cooling rate is much higher during this stage. The *liquid cooling* stage begins when the metal surface reaches the boiling point of the quenching medium. There is no more boiling at this stage, so heat must be removed by conduction and convection. This is the slowest stage of cooling.

It is important in liquid quenching baths that either the quenching medium or the steel being quenched should be agitated (Figure 13). The vapor that forms around the part being quenched acts as an insulator and slows down the cooling rate. This can result in incomplete or spotty hardening of the part. Agitating the part breaks up the vapor barrier. An up and down motion works best for long, slender parts held vertically in the quench. A figure eight motion is sometimes used for heavier parts.

Gloves and face protection must be used in this operation for safety (Figure 14). Hot oil could splash up and burn the heat treater's face if he is not wearing a face shield.

Molten salt or lead is often used for isothermal quenching. This is the method of quench-

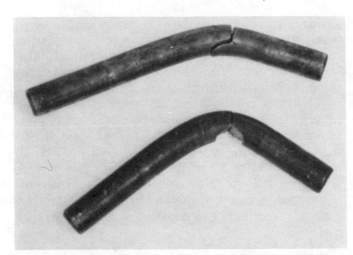

Figure 15. Austempered parts compared to the same kind of part hardened and tempered by the conventional method (Lane Community College).

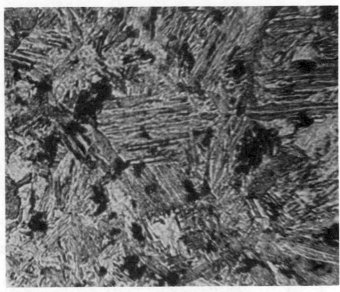

Figure 16. Lower bainite microstructure.

ing used for austempering. Austempered parts (Figure 15) are superior in strength and quality to those produced by the two-stage process of quenching and tempering. The final austempered part is essentially a fine, lower bainite microstructure (Figure 16). As a rule, only parts that are thin in cross section are austempered.

Another form of isothermal quenching is called martempering; the part is quenched in a lead or salt bath at about 400° F (204° C) until the outer and inner parts of the material are brought to the same uniform temperature. The part is next quenched below 200° F (93° C) to transform all of the austenite to martensite. Tempering is then carried out in the conventional manner.

Steels are often classified by the type of quenching medium that is used to meet the requirements of the critical cooling rate. For example, water quenched steels, which are the plain carbon steels, must have a rapid quench. Oil quench steels are alloy steels, and they must be hardened in oil. The air cooled steels are alloy steels that will harden when allowed to cool from the austenitizing temperature in still air. Air is the slowest quenching medium; however, its cooling rate may be increased by movement (by use of fans, for example).

Step or multiple quenching is sometimes used when the part consists of both thick and thin sections. A severe quench will harden the thin section before the thick section has had a chance to cool. The resulting uneven contraction often results in cracking. With this method the

part is quenched for a few seconds in a rapid quenching medium, such as water followed by a slower quench in oil. The surface is first hardened uniformly in the water quench, and time is provided by the slower quench to relieve stresses.

CASE HARDENING

Low carbon steels (.08 to .30 percent carbon) do not harden to any great extent even when combined with other alloying elements. Therefore, when a soft, tough core and an extremely hard outside surface is needed, one of several case hardening techniques is used. It should be noted that **surface** hardening is not necessarily **case hardening.** Flame hardening and induction hardening on the surfaces of gears, lathe ways, and many products depends on the carbon that is already contained in the ferrous metal. On the other hand, case hardening causes carbon from an outside source to penetrate the surface of the steel, or carburizes it. Because it raises the carbon content, it also raises the hardenability of the steel.

Carburizing for case hardening can be done by either of two methods. If only a shallow hardened case is needed, roll carburizing may be used. This consists of heating the part to 1650°F. (899°C) and rolling it in a carburizing compound, reheating and quenching in water. In roll carburizing use only a nontoxic compound

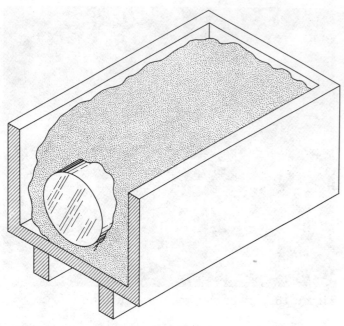

Figure 17. Pack carburizing. The workpiece to be pack carburized should be completely covered with carburizing sand. The metal box should have a close fitting lid.

such as kasenite unless special ventilation systems are used. Roll carburizing produces a maximum case of about .003 inch, but pack carburizing (Figure 17) can produce a case of $\frac{1}{16}$ inch in eight hours at 1700°F (925°C). The part is packed in carburizing compound in a metal box and placed in a furnace long enough to harden the case to the required depth. The part is then removed and quenched in water. After case hardening, tempering is not usually necessary since the core is still soft and tough. Therefore, unlike a hardened piece that is softened by tempering, the surface of a case hardened piece remains hard, usually Rc60 (as hard as a file) or above.

Some alloys and special heat treatments can be used to case harden steel parts. Liquid carburizing is an industrial method in which the parts are bathed in cyanide carbonate and chloride salts and held at a temperature between 1500 and 1700°F. Cyanide salts are extremely poisonous and adequate worker protection is essential.

Gas carburizing is a method in which the carbon is supplied from a carburizing gas atmosphere in a special furnace where the part is heated. The same principle is used when a welder accidentally case hardens a piece of low carbon steel with an oxy-acetylene cutting torch that has a carbonizing flame adjustment. Of course, this torch hardening process makes the cut surface difficult to machine.

Nitriding is a method of case hardening in which the part is heated in a special container into which ammonia gas is released. Since the temperature used is only 950°F (510°C) to 1000°F (538°C) and the part is not quenched, warpage is kept to a minimum. The iron nitrides thus formed are even harder than the iron carbides formed by conventional carburizing methods.

TEMPERING

Tempering, or drawing, is the process of reheating a steel part that has been previously hardened to transform some of the hard martensite into softer structures. The higher the tempering temperature used, the more martensite is transformed, and the softer and tougher (less brittle) the piece becomes. Therefore, tempering temperatures are specified according to the strength and ducility desired. Mechanical properties charts, which may be found in steel manufacturer's handbooks and catalogs, gives this data for each type of alloy steel. Figure 18 is an example of a mechanical properties chart for water quenched 1095.

A part can be tempered in a furnace or oven by bringing it to the proper temperature and holding it there for a length of time, then cooling it in air or water.

Some tool steels should be cooled rapidly after tempering to avoid temper brittleness. Small parts are often tempered in liquid baths such as oil, salt, or metals. Specially prepared oils that do not ignite easily can be heated to the tempering temperature. Lead and various salts are used for tempering since they have a low melting temperature.

When there are no facilities to harden and temper a tool by controlled temperatures, tempering by color is done. The oxide color used as a guide in such tempering will form correctly on steel only if it polished to the bare metal and is free from any oil or fingerprints. An oxy-acetylene torch, steel hot plate, or an electric hot plate can be used. If the part is quite small, a steel plate is heated from the underside, while the part is placed on top. Larger parts such as chisels and punches can be heated on an electric plate

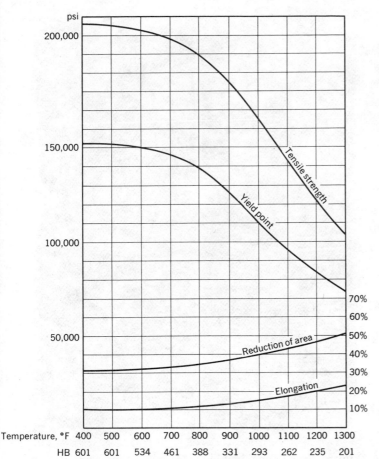

Figure 18. Mechanical properties chart, SAE 1095 steel, water quenched. The bottom two lines refer to tempering temperatures and the resultant hardness in Brinell reading. HB 601 is file hard (Courtesy of Bethlehem Steel Corporation, Bethlehem, Pa.).

Figure 19. Tempering a punch on a hot plate.

correct temperature, and the part must be dropped in water to stop further heating of the critical areas. There is no possible soaking time when this method is used.

When grinding carbon steel tools, if the edge is heated enough to produce a color, you have in effect retempered the edge. If the temperature reached was above that of the original temper, the tool has become softer than it was before you began sharpening it. Table 1 gives the hardnesses of various tools as related to their oxide colors and the temperature at which they form.

Tempering should be done as soon as possible after hardening. A part should not be allowed to cool completely, since untempered, it contains very high internal stresses and tends to split or crack. Tempering will relieve the internal stresses. A hardened part left overnight without tempering may develop cracks by itself. Furnace tempering is one of the best methods of controlling the final condition of the martensite to produce a tempered martensite of the correct hardness and toughness that the part requires.

(Figure 19) until the needed color shows, then cooled in water. With this system the tempering process must cease when the part has come to the

Table 1
Temper Color Chart

Degrees C°	F°	Oxide Color	Suggested Uses for Carbon Tool Steels	
220	425	Light straw	Steel cutting tools, files, and paper cutters	Harder
240	462	Dark straw	Punches and dies	
258	490	Gold	Shear blades, hammer faces, center punches, and cold chisels	
260	500	Purple	Axes, wood cutting tools, and striking faces of tools	
282	540	Violet	Spring and screwdrivers	
304	580	Pale blue	Springs	
327	620	Steel gray	Cannot be used for cutting tools	Softer

Tempering should be accomplished immediately after hardening. The part while still warm should be put into the furnace immediately. If it is left at room temperature for even a few minutes, it may develop a quench crack (Figure 20).

A soaking time should also be used when tempering, since it is in hardening procedures and the length of time is related to the type of tool steel used. A cold furnace should be brought up to the correct temperature for tempering. The residual heat in the bricks of a previously heated furnace may overheat the part, even though the furnace has been cooled down.

Double tempering is used for some alloy steels such as high speed steels that have incomplete transformation of the austenite when they are tempered for the first time. The second time they are tempered, the austenite transforms completely into the martensite structure.

PROBLEMS IN HEAT TREATING

Overheating of steels should always be avoided, and you have seen that if the furnace is set too high with a particular type of steel, a coarse grain can develop. The result is often a poor quality tool, quench cracking, or failure of the tool in use. Extreme overheating causes burning of the steel and damage to the grain boundaries, which cannot be repaired by heat treatment (Figure 21); the part must be scrapped. The shape of the part itself can be a contributing factor to quench failure and quench cracking. If there is a hole, sharp shoulder, or small extension from a larger cross section, a crack can develop in these areas (Figure 22). A part of the tool being held by tongs may be cooled to the extent that it may not harden. The tongs should therefore be heated prior to grasping the part for quenching (Figure 23). As mentioned before, decarburization is a problem in furnaces that do not have controlled atmosphere (Figure 24). This can be avoided in other ways, such as wrapping the part in stainless steel foil or covering it with cast iron chips.

A proper selection of tool steels is necessary to avoid failures in a particular application. If there is shock load on the tool being used, shock resisting tool steel must be selected. If there is to be heat applied in the use of the tool, a hot work type of tool steel is selected. See Table 2. If distortion must be kept to a minimum, an air hardening steel should be used.

Quench cracks have several characteristics that are easily recognized.

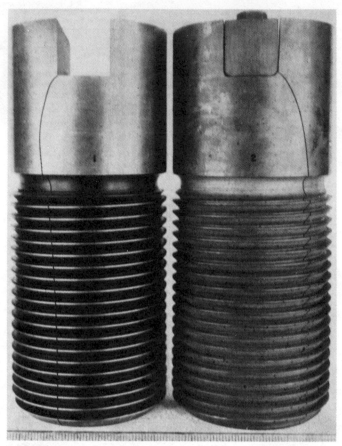

Figure 20. These two breech plugs were made of Type L6 tool steel. Plug #1 cracked in the quench through a sharp corner and was therefore not tempered. Plug #2 was redesigned to incorporate a radius in the corners of the slot, and a soft steel plug was inserted in the slot to protect it from the quenching oil. Plug #2 was oil quenched and checked for hardness (Rockwell C 62) and, after tempering at 900°F (482°C), it was found to be cracked. The fact that the as-quenched hardness was measured proves that there was a delay between the quench and the temper that was responsible for the cracking. The proper practice would be to temper immediately at a low temper, check hardness, and retemper to desired hardness (Photograph courtesy of Bethlehem Steel Corporation).

Figure 21. This tool has been overheated and the typical "chicken wire" surface markings are evident. The tool must be discarded (Lane Community College).

Figure 22. Drawing die made of Type W1 tool steel shows characteristic cracking when water quenching is done without packing the bolt holes (Photograph courtesy of Bethlehem Steel Corporation).

1. In general the fractures run from the surface toward the center in a relatively straight line. The crack tends to spread open.
2. Since quench cracking occurs at relatively low temperatures, the crack will not show any decarburization.
3. The fracture surfaces will exhibit a fine crystalline structure when tempered after quenching. The fractured surfaces may be blackened by tempering scale.

Some of the most common causes for quench cracks are:

1. Overheating during the austenitizing cycle causing the normally fine grained steel to become coarse.
2. Improper selection of the quenching medium; for example, the use of water or brine instead of oil for an oil hardening steel.
3. The improper selection of steel.
4. Time delays between quenching and tempering.
5. Improper design. Sharp changes of section such as holes and keyways (Figure 25).
6. Improper angle of the work into the quenching bath with respect to the shape of the part, causing nonuniform cooling.

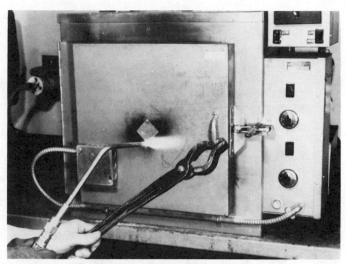

Figure 23. Heating the tongs prior to quenching a part (Lane Community College).

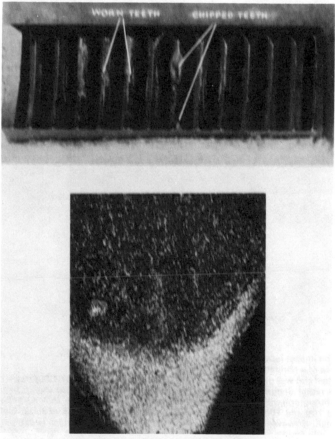

Figure 24. (Top) This thread chaser made of 18-4-1 high speed steel failed in service because of heavy decarburization on the teeth. (Below) Structure of one tooth at 150× magnification shows decarburized structure on the point of the tooth and normal structure below the decarburized zone (Photograph courtesy of Bethlehem Steel Corporation).

Table 2

Mass effect data for SAE 4140 steel. (*Source. Modern Steels and Their Properties*, Seventh Edition Handbook 2757, Bethlehem Steel Corporation, 1972.)

SINGLE HEAT RESULTS

	C	Mn	P	S	Si	Ni	Cr	Mo	
Grade	.38/.43	.75/1.00	—	—	.20/.35	—	.80/1.10	.15/.25	Grain Size
Ladle	.40	.83	.012	.009	.26	.11	.94	.21	7-8

MASS EFFECT

Size Round in.	Tensile Strength psi	Yield Point psi	Elongation % 2 in.	Reduction of Area, %	Hardness HB
Annealed (Heated to 1500 F, furnace-cooled 20 F per hour to 1230 F, cooled in air.)					
1	95,000	60,500	25.7	56.9	197
Normalized (Heated to 1600 F, cooled in air.)					
½	148,500	98,500	17.8	48.2	302
1	148,000	95,000	17.7	46.8	302
2	140,750	91,750	16.5	48.1	285
4	117,500	69,500	22.2	57.4	241
Oil-quenched from 1550 F, tempered at 1000 F.					
½	171,500	161,000	15.4	55.7	341
1	156,000	143,250	15.5	56.9	311
2	139,750	116,750	17.5	59.8	285
4	127,750	99,250	19.2	60.4	277
Oil-quenched from 1550 F, tempered at 1100 F.					
½	157,500	148,750	18.1	59.4	321
1	140,250	135,000	19.5	62.3	285
2	127,500	102,750	21.7	65.0	262
4	116,750	87,000	21.5	62.1	235
Oil-quenched from 1550 F, tempered at 1200 F.					
½	136,500	128,750	19.9	62.3	277
1	132,750	122,500	21.0	65.0	269
2	121,500	98,250	23.2	65.8	241
4	112,500	83,500	23.2	64.9	229

As-quenched Hardness (oil)

Size Round	Surface	½ Radius	Center
½	HRC 57	HRC 56	HRC 55
1	HRC 55	HRC 55	HRC 50
2	HRC 49	HRC 43	HRC 38
4	HRC 36	HRC 34.5	HRC 34

7. Failure to specify the correct size material to allow for cleaning up the outside decarburized surface of the bar before the final part is made.

It is sometimes desirable to stress relieve the part before hardening it. This is particularly appropriate for parts and tools that have been highly stressed by heavy machining or by prior heat treatment. If they are left unrelieved, the residual stresses from such operations may add to the thermal stress produced in the heating cycle and cause the part to crack even before it has reached the quenching temperature.

There is a definite relationship between grinding and heat treating. Development of surface temperatures ranging from 2000° F (1093° C) to 3000° F (1649° C) are generated during grinding. This can cause two undesirable effects on hardened tool steels: development of high internal stresses causing surface cracks to be formed, and changes in the hardness and metallurgical structure of the surface area.

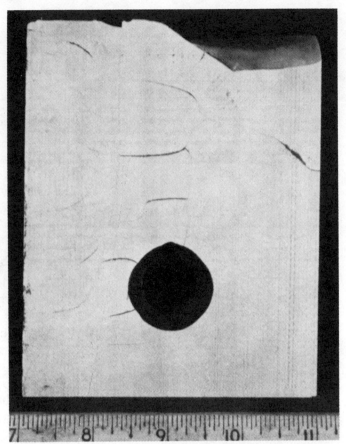

Figure 26. Severe grinding cracks in a shear blade made of Type A4 tool steel developed because the part was not tempered after quenching. Hardness was Rockwell C 64, and the cracks were exaggerated by magnetic particle test. Note the geometric scorch pattern on the surface and the fracture that developed from enlargement of the grinding cracks (Photograph courtesy of Bethlehem Steel Corporation).

Figure 25. (Top) Letter stamp made of Type S5 tool steel, which cracked in hardening through the stamped O. The other two form tools, made of Type T1 high speed steel, cracked in heat treatment through deeply stamped + marks. Stress raisers such as these deep stamp marks should be avoided. Although characters with straight lines are most likely to crack, even those with rounded lines are susceptible (Photograph courtesy of Bethlehem Steel Corporation).

One of the most common effects of grinding on hardened and tempered tool steels is that of reducing the hardness of the surface by gradual tempering where the hardness is lowest at the extreme surface but increases with distance below the surface. The depth of this tempering varies with the amount or depth of cut, the use of coolants, and the type of grinding wheel. If high

temperatures are produced locally by the grinding wheel and the surface is immediately quenched by the coolant, a martensite having a Rockwell hardness of C65 to 70 can be formed. This gradiant hardness, being much greater than that beneath the surface of the tempered part, sometimes causes very high stresses that contribute greatly to grinding cracks. Sometimes grinding cracks are visible in oblique or angling light, but they can be easily detected when present by the use of magnetic particles of fluorescent particle testing.

When a part is hardened but not tempered before it is ground, it is extremely liable to stress cracking (Figure 26). Faulty grinding procedures can also cause grinding cracks. Improper grinding operations can cause tools that have been properly hardened to fail. Sufficient stock should be allowed for a part to be heat treated so that grinding will remove any decarburized surface on all sides to a depth of .010 to .015 in.

self-evaluation

SELF-TEST

1. If you heated AISI C1080 steel to 1200° F (649° C) and quenched it in water, what would be the result?

2. If you heated AISI C1020 steel to 1500° F (815° C) and quenched it in water, what would happen?

3. List as many problems encountered with water hardening steels as you can think of.

4. Name some advantages in using air and oil hardening tool steels.

5. What is the correct temperature for quenching AISI C1095 tool steel? For any carbon steel?

6. Why is steel tempered after it is hardened?

7. What factors should you consider when you choose the tempering temperature for a tool?

8. The approximate temperature for tempering a center punch should be _____. The oxide color would be _____.

9. If a cold chisel became blue when the edge was ground on an abrasive wheel, to approximately what temperature was it raised? How would this affect the tool?

10. How soon after hardening should you temper a part?

11. What is the advantage of using low carbon steel for parts that are to be case hardened?

12. How can a deep case be made?

13. Are parts that are surface hardened always case hardened?

14. Name three methods by which carbon may be introduced into the surface of heated steel?

15. What method of case hardening uses ammonia gas?

16. Name three kinds of furnaces used for heat treating steels.

17. What can happen to a carbon steel when it is heated to high temperatures in the presence of air (oxygen)?

18. Why is it absolutely necessary to allow a soaking period for a length of time (that varies for different kinds of steels) before quenching the piece of steel?

19. Why should the part or the quenching medium be agitated when you are hardening steel?

20. Which method of tempering gives the heat treater the most control of the final product, by color or by furnace?

21. Describe two characteristics of quench cracking that would enable you to recognize them.

22. Name four or more causes of quench cracks.

23. In what ways can decarburization of a part be avoided when it is heated in a furnace?

24. Describe two types of surface failures of hardened steel when it is being ground.

25. When distortion must be kept to a minimum, which type of tool steel should be used?

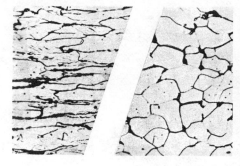

unit 4
annealing, normalizing, and stress relieving

Since the machinability of steel is so greatly affected by heat treatments, the processes of annealing, normalizing, and stress relieving are important to a machinist. You will learn about these processes in this unit.

objectives After completing this unit, you will be able to:
1. Test various steels with annealing, normalizing, and stress relieving heat treatments in order to determine their effect on machinability.
2. Explain the principles of and differences between the various kinds of annealing processes.

ANNEALING

The heat treatment for iron and steel that is generally called annealing can be divided into several different processes: full anneal, normalizing, spheroidize anneal, stress relief (anneal), and process anneal.

Full Anneal

The full anneal is used to completely soften hardened steel, usually for easier machining of tool steels that have more than .8 percent carbon content. Lower carbon steels are full annealed for other purposes. For instance, when welding has been done on a medium to high carbon steel that must be machined, a full anneal is needed. Full annealing is done by heating the part in a furnace to 50°F (10°C) above the upper critical temperature (Figure 1), and then cooling very slowly in the furnace or in an insulating material. The microstructure becomes coarse pearlite and ferrite in the process. It is necessary to heat above the critical temperature for grains containing iron carbides (pearlite) in order to recrystallize them and to reform new soft whole grains from the old hard distorted ones.

Normalizing

Normalizing is somewhat similar to annealing, but it is done for several different purposes. Medium carbon steels are often normalized to give them better machining qualities. Medium (.3 to .6 percent) carbon steel may be "gummy" when machined after a full anneal, but can be made sufficiently soft for machining by normalizing. The finer, but harder, microstructure produced by normalizing gives the piece a better surface finish. The piece is heated to 100°F (38°C) above the upper critical line, and cooled in still air. When the carbon content is above or below .8 percent, higher temperatures are required (Figure 1).

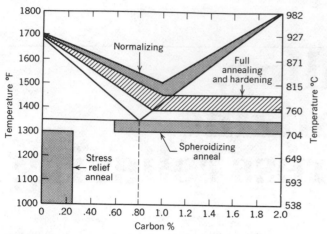

Figure 1. Temperature ranges used for heat treating carbon steels.

Forgings and castings that have unusually large and mixed grain structures are corrected by using a normalizing heat treatment. Stresses are removed, but the metal is not as soft as with full annealing. The resultant microstructure is a uniform fine grained pearlite and ferrite, including other microstructures depending on the alloy and carbon content. Normalizing also is used to prepare steel for other forms of heat treatment such as hardening and tempering.

Spheroidizing
Spheroidizing is used to improve the machinability of high carbon steels (.8 to 1.7 percent). It produces a spherical or globular carbide grain in the steel rather than the normal flat grains of pearlite (Figure 2). Low carbon steels (.08 to .3 percent) can be spheroidized, but their machinability gets poorer since they become gummy and soft, causing tool edge build up and poor finish. The spheroidization temperature is close to 1300°F (704°C).

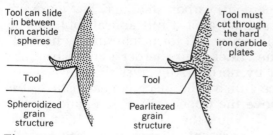

Figure 2. Comparison of cutting action between spheroidized and normal carbon steels.

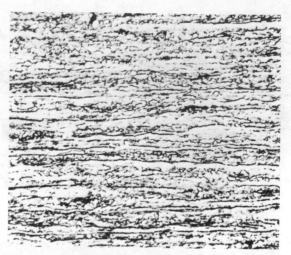

Figure 3. Microstructure of flattened grains of 0.10 percent carbon steel, cold rolled (1000 ×). (By permission, from *Metals Handbook*, Volume 7, Copyright © American Society for Metals, 1972).

Stress Relief Anneal
Stress relief annealing is a process of reheating low carbon steels to 950°F (510°C). Stresses in the ferrite (mostly pure iron) grains caused by cold working steel such as rolling, pressing, forming, or drawing are relieved by this process. The distorted grains reform or recrystallize into new softer ones (Figures 3 and 4).

The pearlite grains and some other forms of iron carbide remain unaffected by this treatment, unless done at the spheroidizing tempera-

Figure 4. The same 0.10 percent carbon steel as in Figure 3, but annealed at 1025°F (552°C) (1000 ×). Ferrite grains are mostly reformed to their original state, but the pearlite grains are still distorted. (By permission, from *Metals Handbook* Volume 7, Copyright © American Society for Metals, 1972).

ture and held long enough to effect spheroidization. In the stress relieving of weldments, often one of the aims is to spheroidize the hard carbide and martensite grains to reduce brittleness in the junction of the weld.

Process Anneal
Process annealing is essentially the same as stress relief annealing. It is done at the same temperatures and with low and medium carbon steels. In the wire and sheet steel industry, the term is used for the annealing processes used in cold rolling or drawing processes and those used to remove the final residual stresses when necessary. Sometimes referred to as bright annealing, it is often carried out in a closed container with inert gas to prevent oxidation of the surface.

self-evaluation

SELF-TEST
1. When might normalizing be necessary?
2. At what approximate temperature should you normalize .4 percent carbon steel?
3. What is the spherodizing temperature of .8 percent carbon steel?
4. What is the essential difference between the full anneal and stress relieving?
5. When should you use stress relieving?
6. What kind of carbon steels would need to be spheroidized to give them free machining qualities?
7. Explain process annealing.
8. How should the piece be cooled for a normalizing heat treatment?
9. How should the piece be cooled for the full anneal?
10. What happens to machinability in low carbon steels that are spheroidized?

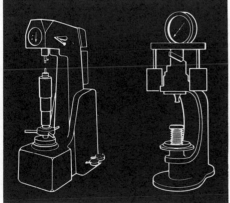

unit 5
rockwell
and brinell
hardness testers

The Rockwell Hardness Tester and the Brinell Hardness Tester are the most commonly used types of hardness testers for industrial and metallurgical purposes. Heat treaters, inspectors, and many others in industry often use these machines. This unit will direct you into a proper understanding and use of both Rockwell and Brinell hardness testers.

objectives After completing this unit, you will be able to:

1. Make a Rockwell test on three specimens using the correct penetrator, major load, and scale.
2. Make a Rockwell superficial test on two specimens using the correct penetrator, major load, and scale.
3. Make a Brinell test on three specimens, read the impression with a Brinell microscope, and determine the hardness number from a table.

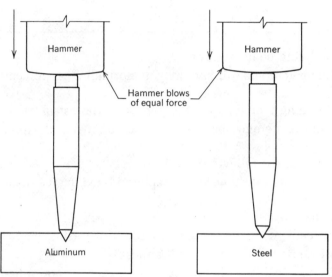

Figure 1. Indentations made by a punch in aluminum and alloy steel.

The hardness of a metal is its ability to resist being permanently deformed. There are three ways that hardness is measured: resistance to penetration, elastic hardness, and resistance to abrasion. In this unit you will study the hardness of metals by their resistance to penetration.

Hardness varies considerably between different materials. This variation can be illustrated by making an indentation in a soft metal such as aluminum and in a hard metal such as alloy tool steel. The indentation could be made with an ordinary center punch and a hammer, giving a light blow of equal force on each of the two specimens (Figure 1). Just by visual observation you can tell which specimen is hardest in this case. Of course, this is not a reliable method of hardness testing, but it does show one of the principles of both the Rockwell and Brinell hardness testers: measuring penetration of the specimen by an indenter or penetrator, such as a steel ball or diamond point.

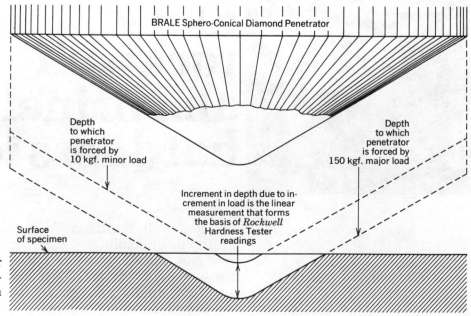

Figure 2. Schematic showing minor and major loads being applied (Courtesy of Wilson Instrument Division of Acco).

USING THE ROCKWELL HARDNESS TESTER

The Rockwell Hardness Test is made by applying two loads to a specimen and measuring the depth of penetration in the specimen between the first, or minor, load and the major load. The depth of penetration is indicated on the dial when the major load is removed (Figure 2). The amount of penetration decreases as the hardness of the specimen increases. Generally, the harder the material, the greater its tensile strength, or the ability to resist deformation and rupture when a load is applied. Table 1 compares hardness by Brinell and Rockwell testers to tensile strength.

There are two basic types of penetrators used on the Rockwell tester (Figure 3). One is a sphero-conical diamond called a *Brale* that is used only for hard materials; that is, for mate-

Table 1
Hardness and Tensile Strength Comparison Table

Hardness Conversion Table

Brinell		Rockwell		Tensile Strength, 1000 psi Approximately	Brinell		Rockwell		Tensile Strength, 1000 psi Approximately
Indentation diameter, mm	No.*	B	C		Indentation diameter, mm	No.*	B	C	
2.25	745		65.3		3.75	262	(103.0)	26.6	127
2.30	712		–		3.80	255	(102.0)	25.4	123
2.35	682		61.7		3.85	248	(101.0)	24.2	120
2.40	653		60.0		3.90	241	100.0	22.8	116
2.45	627		58.7		3.95	235	99.0	21.7	114
2.50	601		57.3		4.00	229	98.2	20.5	111
2.55	578		56.0		4.05	223	97.3	(18.8)	–
2.60	555		54.7	298	4.10	217	96.4	(17.5)	105
2.65	534		53.5	288	4.15	212	95.5	(16.0)	102
2.70	514		52.1	274	4.20	207	94.6	(15.2)	100
2.75	495		51.6	269	4.25	201	93.8	(13.8)	98
2.80	477		50.3	258	4.30	197	92.8	(12.7)	95
2.85	461		48.8	244	4.35	192	91.9	(11.5)	93
2.90	444		47.2	231	4.40	187	90.7	(10.0)	90
2.95	429		45.7	219	4.45	183	90.0	(9.0)	89
3.00	415		44.5	212	4.50	179	89.0	(8.0)	87
3.05	401		43.1	202	4.55	174	87.8	(6.4)	85
3.10	388		41.8	193	4.60	170	86.8	(5.4)	83
3.15	375		40.4	184	4.65	167	86.0	(4.4)	81
3.20	363		39.1	177	4.70	163	85.0	(3.3)	79
3.25	352	(110.0)	37.9	171	4.80	156	82.9	(0.9)	76
3.30	341	(109.0)	36.6	164	4.90	149	80.8		73
3.35	331	(108.5)	35.5	159	5.00	143	78.7		71
3.40	321	(108.0)	34.3	154	5.10	137	76.4		67
3.45	311	(107.5)	33.1	149	5.20	131	74.0		65
3.50	302	(107.0)	32.1	146	5.30	126	72.0		63
3.55	293	(106.0)	30.9	141	5.40	121	69.8		60
3.60	285	(105.5)	29.9	138	5.50	116	67.6		58
3.65	277	(104.5)	28.8	134	5.60	111	65.7		56
3.70	269	(104.0)	27.6	130					

Note 1. This is a condensation of Table 2, Report J417b, SAE 1971 Handbook. Values in ( ) are beyond normal range, and are presented for information only.

*Values above 500 are for tungsten carbide ball; below 500 for standard ball.

Note 2. The following is a formula to approximate tensile strength when the Brinell hardness is known:

$$\text{Tensile strength} = BHN \times 500$$

Source. Bethlehem Steel Corporation, *Modern Steels and Their Properties,* Seventh Edition, Handbook 2757, 1972.

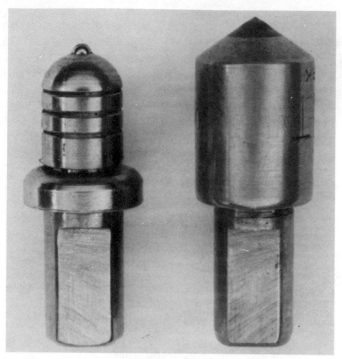

Figure 3. Brale and ball. These two penetrators are the basic types used on the Rockwell Hardness Tester. (*Note.* Brale is a registered trademark of American Chain & Cable Company, Inc., for sphero-conical diamond penetrators.)

rials over B-100, such as hardened steel, nitrided steel, and hard cast irons. When the "C" Brale diamond penetrator is used, the recorded readings should be prefixed by the letter "C." The major load used is 150 kgf (kilograms of force). The C scale is *not* used to test extremely hard materials such as cemented carbides or shallow case hardened steels and thin steel. An A Brale penetrator is used in these cases and the A scale used with 60 kgf major load.

The second type penetrator is a $\frac{1}{16}$ inch diameter ball that is used for testing material in the range of B-100 to B-0, including such relatively soft materials as brass, bronze, and soft steel. If the ball penetrator is used on materials harder than B-100, there is a danger of flattening the ball. Ball penetrators for use on very soft bearing metals are available in sizes of $\frac{1}{2}$, $\frac{1}{4}$, and $\frac{1}{8}$ inch (Table 2).

Figure 4 points out the parts used in the testing operation on the Rockwell Hardness Tester. You should learn the names of these parts before continuing with this unit.

Setting up the Rockwell Hardness Tester and making the test.

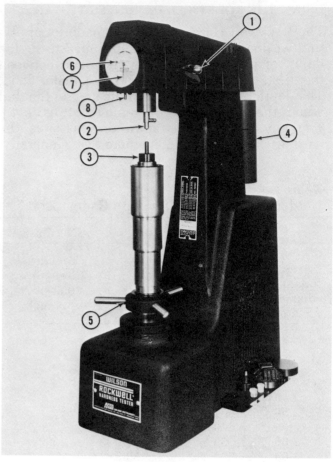

Figure 4. Rockwell Hardness Tester listing the names of parts used in the testing operations (Courtesy of Wilson Instrument Division of Acco).

1. Crank handle
2. Penetrator
3. Anvil
4. Weights
5. Capstan handwheel
6. Small pointer
7. Large pointer
8. Lever for setting the bezel

(*Note.* Rockwell is a registered trademark of American Chain & Cable Company, Inc., for hardness testers and test blocks.)

1. Using Table 2, select the proper weight (Figure 5) and penetrator. Make sure the crank handle is pulled completely forward.
2. Place the proper anvil (Figure 6) on the elevating screw, taking care not to bump the penetrator with the anvil. Make sure that the specimen to be tested is free from dirt, scale, or heavy oil on the underside.

Table 2
Penetrator and Load Selection

Scale Symbol	Penetrator	Major Load, Kgf.	Dial Figures	Typical Applications of Scales
B	$\frac{1}{16}$ in. ball	100	Red	Copper alloys, soft steels, aluminum alloys, malleable iron, etc.
C	Brale	150	Black	Steel, hard cast irons, pearlitic malleable iron, titanium, deep case hardened steel, and other materials harder than B-100.
A	Brale	60	Black	Cemented carbides, thin steel, and shallow case hardened steel.
D	Brale	100	Black	Thin steel and medium case hardened steel and pearlitic malleable iron.
E	$\frac{1}{8}$ in. ball	100	Red	Cast iron and aluminum and magnesium alloys and bearing metals.
F	$\frac{1}{16}$ in. ball	60	Red	Annealed copper alloys and thin soft sheet metals.
G	$\frac{1}{16}$ in. ball	150	Red	Phosphor bronze, beryllium copper, and malleable irons. Upper limit G-92 to avoid possible flattening of ball.
H	$\frac{1}{8}$ in. ball	60	Red	Aluminum, zinc, and lead.
K	$\frac{1}{8}$ in. ball	150	Red	
L	$\frac{1}{4}$ in. ball	60	Red	Bearing metals and other very soft or thin materials. Use the smallest ball and heaviest load that does not give anvil effect.
M	$\frac{1}{4}$ in. ball	100	Red	
P	$\frac{1}{4}$ in. ball	150	Red	
R	$\frac{1}{2}$ in. ball	60	Red	
S	$\frac{1}{2}$ in. ball	100	Red	
V	$\frac{1}{2}$ in. ball	150	Red	

Source. *Wilson Instruction Manual*, "Rockwell Hardness Tester Models OUR-a and OUS-a," American Chain & Cable Company, Inc., 1973.

3. Place the specimen to be tested on the anvil (Figure 7). Then by turning the handwheel, gently raise the specimen until it comes in contact with the penetrator (Figure 8). Continue turning the handwheel slowly until the small pointer on the dial gage is nearly vertical (near the dot). Now watch the long pointer on the gage and continue raising the work until it is approximately vertical. It should not vary from the vertical position by more than 5 divisions on the dial. Set the dial to zero on the pointer by moving the bezel until the line marked **zero set** is in line with the pointer (Figure 9). You have now applied the **minor load.** This is the actual starting point for all conditions of testing.

4. Apply the major load by tripping the crank handle clockwise (Figure 10).

5. Wait two seconds after the pointer has stopped moving, then remove the major load by pulling the crank handle forward or counterclockwise.

6. Read the hardness number in Rockwell units on the dial (Figure 11). The black numbers are for the A and C scales and the red numbers are for the B scale. The specimen should be tested in several places and an average of the test results taken, since many materials vary in hardness even on the same surface.

Superficial Testing
After testing sheet metal, examine *the underside of the sheet.* If the impression of the penetrator can be seen, then the reading is in error and the superficial test should be used. If the impression can still be seen after the superficial

Figure 5. Selecting and installing the correct weight (Lane Community College).

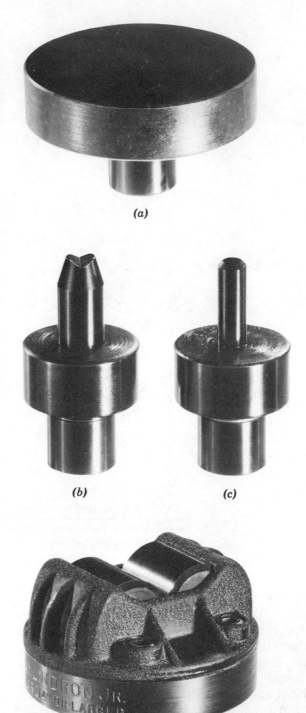

test, then a lighter load should be used. A load of 30 kgf is recommended for superficial testing. Superficial testing is also used for case hardened and nitrided steel having a very thin case.

A Brale marked N is needed for superficial testing, as A and C Brales are not suitable. Recorded readings should be prefixed by the major load and the letter **N,** when using the Brale for superficial testing; for example, 30N78. When using the $\frac{1}{16}$ inch ball penetrator, the same as that used for the B, F, and G hardness scales, the readings should always be prefixed by the major load and the letter **T**; for example, 30T85. The $\frac{1}{16}$ inch ball penetrator, however, should not be used on material harder than 30T82. Other superficial scales, such as W, X, and Y should also be prefixed with the major load when recording hardness. See Table 3 for superficial test penetrator selection.

Figure 6. Basic anvils used with Rockwell Hardness testers. (a) Plane, (b) Shallow V, (c) Spot, (d) Cylindron Jr. (Courtesy of Wilson Instrument Division of Acco).

Figure 7. Placing the test block in the machine (Courtesy of Wilson Instrument Division of Acco).

Figure 8. Specimen being brought into contact with the penetrator. This establishes the minor load (Courtesy of Wilson Instrument Division of Acco).

Figure 9. Setting the bezel (Courtesy of Wilson Instrument Division of Acco).

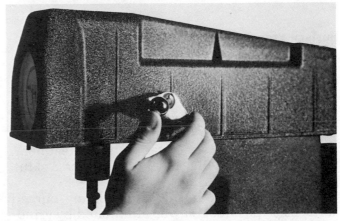

Figure 10. Applying the major load by tripping the crank handle clockwise (Courtesy of Wilson Instrument Division of Acco).

A spot anvil, as shown in Figure 6, is used when the tester is being checked on a Rockwell test block. The spot anvil should not be used for checking cylindrical surfaces. The diamond spot anvil (Figure 12) is similar to the spot anvil, but it has a diamond set into the spot. The diamond is ground and polished to a flat surface. This anvil is used **only** with the superficial tester, and then **only** in conjunction with the steel ball penetrator for testing soft metal.

Figure 11. Dial face with reading in Rockwell units after completion of the test. The reading is RC 55 (Lane Community College).

Table 3
Superficial Tester Load and Penetrator Selection

Scale Symbol	Penetrator	Load in Kilograms
15N	Brale	15 kgf
30N	Brale	30 kgf
45N	Brale	45 kgf
15T	$\frac{1}{16}$ in. ball	15 kgf
30T	$\frac{1}{16}$ in. ball	30 kgf
45T	$\frac{1}{16}$ in. ball	45 kgf
15W	$\frac{1}{8}$ in. ball	15 kgf
30W	$\frac{1}{8}$ in. ball	30 kgf
45W	$\frac{1}{8}$ in. ball	45 kgf
15X	$\frac{1}{4}$ in. ball	15 kgf
30X	$\frac{1}{4}$ in. ball	30 kgf
45X	$\frac{1}{4}$ in. ball	45 kgf
15Y	$\frac{1}{2}$ in. ball	15 kgf
30Y	$\frac{1}{2}$ in. ball	30 kgf
45Y	$\frac{1}{2}$ in. ball	45 kgf

Source. Wilson Instruction Manual, "Rockwell Hardness Tester Models OUR-a and OUS-a," American Chain & Cable Company, Inc., 1973.

Surface Preparation and Proper Use

When testing hardness, surface condition is important for accuracy. A rough or ridged surface caused from coarse grinding will not produce as reliable results as a smoother surface. Any rough scale caused from hardening must be removed before testing. Likewise, if the workpiece has been decarburized by heat treatment, the test area should have this softer "skin" ground off.

Error can also result from testing curved surfaces. This effect may be eliminated by grinding a small flat spot on the specimen. Cylindrical workpieces must always be supported in a V-type centering anvil, and the surface to be tested should not deviate from the horizontal by more than five degrees. Tubing is often so thin that it will deform when tested. It should be supported on the inside by a mandrel or gooseneck anvil to avoid this problem.

Several devices are made available for the Rockwell Hardness Tester to support overhanging or large specimens. One type, called a jack rest (Figure 13), is used for supporting long, heavy parts such as shafts. It consists of a separate elevating screw and anvil support similar to that on the tester. Without adequate support overhanging work can damage the penetrator rod and cause inaccurate readings.

Figure 12. Diamond spot anvil (Courtesy of Wilson Instrument Division of Acco).

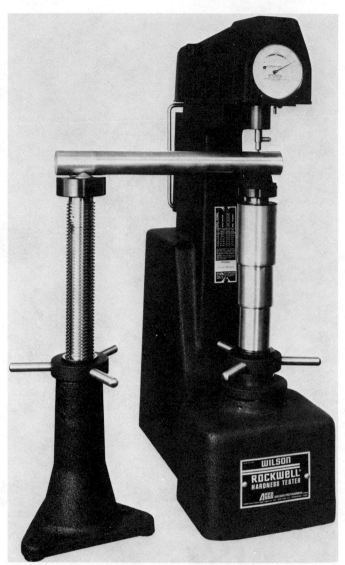

Figure 13. Correct method of testing long, heavy work requires the use of a jack rest (Courtesy of Wilson Instrument Division of Acco).

No test should be made near an edge of a specimen. Keep the penetrator at least $\frac{1}{8}$ inch away from the edge. The test block, as shown in Figure 7, should be used every day to check the calibration of the tester, if it is in constant use.

USING THE BRINELL HARDNESS TESTER

The Brinell hardness test is made by forcing a steel ball, usually 10 millimeters in diameter, into the test specimen by using a known load weight and measuring the diameter of the result-

ing impression. The Brinell hardness value is the load divided by the area of the impression, expressed as follows:

$$BHN = \frac{P}{\frac{\pi D}{2}\left(D - \sqrt{D^2 - d^2}\right)}$$

$BHN =$ Brinell Hardness Number in kilograms per square millimeter
$D =$ Diameter of the steel ball in millimeters
$P =$ Applied load in kilograms
$d =$ Diameter of the impression in millimeters

Figure 14. The Olsen Brinell microscope provides a fast, accurate means for measuring the diameter of the impression for determining the Brinell hardness number (Courtesy of the Tinius Olsen Testing Machine Co., Inc.).

A small microscope is used to measure the diameter of the impressions (Figure 14). Various loads are used for testing different materials: 500 kilograms for soft materials such as copper and aluminum, and 3000 kilograms for steels and cast irons. For convenience, Table 1 gives the Brinell hardness number and corresponding diameters of impression for a 10 millimeter ball and a load of 3000 kilograms. The related Rockwell hardness numbers and tensile strengths are also shown. Just as for the Rockwell tests, the impression of the steel ball must not show on the underside of the specimen. Tests should not be made too near the edge of a specimen.

Figure 15 shows an air operated Brinell Hardness Tester. The testing sequence is as follows.

Figure 16. Select load. Operator adjusts the air regulator as shown until the desired Brinell load in kilograms is indicated (Courtesy of the Tinius Olsen Testing Machine Co., Inc.).

Figure 15. Air-O-Brinell air-operated metal hardness tester (Courtesy of the Tinius Olsen Testing Machine Co., Inc.).

Figure 17. Apply load. Operator pulls out plunger-type control to apply load to specimen smoothly (Courtesy of the Tinius Olsen Testing Machine Co., Inc.).

Figure 18. Release load. As soon as the plunger is depressed, the Brinell ball retracts in readiness for the next test (Courtesy of the Tinius Olsen Testing Machine Co., Inc.).

1. The desired load in kilograms is selected on the dial by adjusting the air regulator (Figure 16).

2. The specimen is placed on the anvil. Make sure the specimen is clean and free from burrs. It should be smooth enough so that an accurate measurement can be taken of the impression.

3. The specimen is raised to within $\frac{5}{8}$ inch of the Brinell ball by turning the handwheel.

4. The load is then applied by pulling out the plunger control (Figure 17). Maintain the load for 30 seconds for nonferrous metals and 15 seconds for steel. Release load (Figure 18).

5. Remove the specimen from the tester and measure the diameter of the impression.

6. Determine the Brinell Hardness Number (BHN) by calculation or by using the table. Soft copper should have a BHN of about 40, soft steel from 150 to 200, and hardened tools from 500 to 600. Fully hardened high carbon steel would have a BHN of 750. A Brinell test ball of tungsten carbide should be used for materials above 600 BHN.

Brinell Hardness testers work best for testing softer metals and medium hard tool steels.

self-evaluation

SELF-TEST
1. What one specific category of the property of hardness do the Rockwell and Brinell Hardness testers use and measure? How is it measured?
2. State the relationship that exists between hardness and tensile strength.
3. Explain which scale, major load, and penetrator should be used to test a block of tungsten carbide on the Rockwell tester.
4. What is the reason that the steel ball cannot be used on the Rockwell tester to test the harder steels?
5. When testing with the Rockwell superficial tester, is the Brale used the same one that is used on the A, C, and D scales? Explain.
6. The $\frac{1}{16}$ inch ball penetrator used for the Rockwell superficial tester is a different one than that used for the B, F, and G scales. True/False.
7. What is the diamond spot anvil used for?
8. How does roughness on the specimen to be tested affect the test results?
9. How does decarburization affect the test results?
10. What does a curved surface do to the test results?
11. On the Brinell tester what load should be used for testing steel?
12. What size ball penetrator is generally used on a Brinell tester?

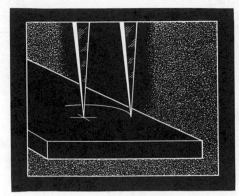

section e
layout

Layout is the process of placing reference marks on the workpiece. These marks may indicate the shape and size of a part or its features. Layout marks often indicate where machining will take place. A machinist may use layout marks as a guide for machining while checking his work by actual measurement. He may also cut to a layout mark. One of your first jobs after you have obtained material from stock will be to measure and lay out where the material will be cut. This kind of layout may be a simple pencil or chalk mark and is one of the basic tasks of semiprecision layout.

Precision layout can be a complex and involved operation making use of sophisticated tools. In the aircraft and shipbuilding industries, reference points, lines, and planes may be layed out using optical and laser instruments. In the machine shop, you will be primarily concerned with layout for stock cutoff, filing and offhand grinding, drilling, milling, and occasionally in connection with lathe work.

LAYOUT CLASSIFICATIONS The process of layout can be generally classified as **semiprecision** and **precision.** Semiprecision layout is usually done by scale measurement to a tolerance of $\pm \frac{1}{64}$ in. Precision layout is done with tools that discriminate to .001 in. or finer, to a tolerance of $\pm$ .001 in. if possible.

TOOLS OF LAYOUT
Surface Plates The surface plate is an essential tool for many layout applications. A surface plate provides an accurate reference plane from which measurements for both layout and inspection may be made. In many machine shops, where a large amount of layout work is accomplished, a large area surface plate, perhaps 4 by 8 feet, may be used. These are often known as layout tables.

Any surface plate or layout table is a precision tool and should be treated as such. They should be kept covered when not in use and kept clean when being used. No surface plate should be hammered upon, since this will impair the accuracy of the reference surface. As you study machine tool practices, measurement and layout, the surface plate will play an important part in many of your tasks.

CAST IRON AND SEMI-STEEL SURFACE PLATES. Cast iron and semisteel surface plates (Figure 1) are made from good quality castings that have been allowed to age, thus relieving internal stresses. Aging of the casting reduces distortion after its working surface has been finished to the desired degree of flatness. The cast iron or steel plate will also have several ribs on the underside to provide structural

255

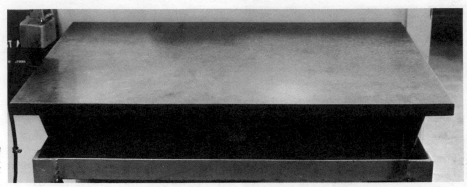

Figure 1. Cast iron surface plate (California State University at Fresno).

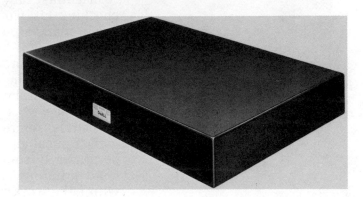

Figure 2. Granite surface plate (Courtesy of the DoAll Company).

rigidity. Cast plates vary in size from small bench models, a few square inches in area, to larger sizes that may be four by eight feet or larger. The large cast plates are usually a foot or more in thickness with appropriate ribs on the underside to provide for sufficient rigidity. The large iron plate is generally mounted on a heavy stand or legs with provision for leveling. The plate is leveled periodically to insure that its working surface remains flat.

GRANITE SURFACE PLATES. The cast iron and semisteel surface plate has all but given way to the granite plate (Figure 2). Granite is a superior material to metal because it is harder, denser, impervious to water and, if chipped, the surrounding flat surface is not affected. Furthermore, granite, because it is a natural material, has aged in the earth for a great deal of time. Therefore, it has little internal stress. Granite surface plates possess a greater temperature stability than their metal counterparts.

Granite plates range in size from about 12 by 18 inches to 4 by 12 feet. A large granite plate may be from 10 to 20 inches thick and weigh as much as 5 to 10 tons. Some granite plates are finished on two sides, thus permitting them to be turned over and their use extended.

GRADES OF GRANITE SURFACE PLATES. The granite surface plate is available in three grades. Surface plate grade specifications are an indication of the plus and minus deviation of the working surface from an average plane.

Figure 3. Applying layout dye to the workpiece (Courtesy of the L. S. Starrett Company).

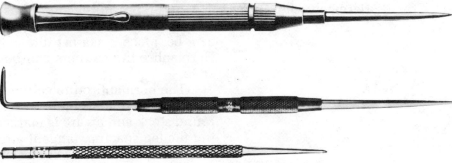

Figure 4. Pocket scriber (Courtesy of Rank Precision Industries, Inc.).

Figure 5. Engineer's scriber (Courtesy of Rank Precision Industries, Inc.).

Figure 6. Machinist's scriber (Courtesy of Rank Precision Industries, Inc.).

Grade	Type	Tolerance
AA	Laboratory grade	± 25 millionths inch
A	Inspection grade	± 50 millionths inch
B	Shop grade	± 100 millionths inch

The tolerances are proportional to the size of the plate. As the size increases, the tolerance widens.

Layout Dyes To make layout marks visible on the surface of the workpiece, a **layout dye** is used. Layout dyes are available in several colors. Among these are red, blue, and white. The blue dyes are very common. Depending on the surface color of the workpiece material, different dye colors may make layout marks more visible. Layout dye should be applied sparingly in an even coat (Figure 3).

Scribers and Dividers Several types of **scribers** are in common use. The pocket scriber (Figure 4) has a removable tip that can be stored in the handle. This

Figure 7. Rule scribe made from a high speed toolbit.

permits the scriber to be carried safely in the pocket. The engineer's scriber (Figure 5) has one straight and one hooked end. The hook permits easier access to the line to be scribed. The machinist's scriber (Figure 6) has only one end with a fixed point. **Scribers must be kept sharp.** If they become dull, they must be reground or stoned to restore their points. Scriber materials include hardened steel and tungsten carbide.

When scribing against a rule, hold the rule firmly. Tilt the scriber so that the tip marks as close to the rule as possible. This will insure accuracy. An excellent scriber can be made by grinding a shallow angle on a piece of tool steel (Figure 7). This type of scriber is particularly well suited to scribing along a rule. The flat side permits the scriber to mark very close to the rule, thus obtaining maximum accuracy.

Several types of dividers are in common use. The **spring divider** is very common (Figure 8). Spring dividers range in size from 3 to 12 in. The spacing of the divider legs is set by turning the adjusting screw. Dividers are usually set to rules. Engraved rules are best as the divider tips can be set in the engraved rule graduations (Figure 9). Like scribers, divider tips must be kept sharp and at nearly the same length.

Hermaphrodite Caliper   The **hermaphrodite caliper** has one leg similar to a regular divider. The tip is adjustable for length. The other leg has a hooked end that can be placed against the edge of the workpiece (Figure 10). Hermaphrodite calipers can be used to scribe a line parallel to an edge.

The hermaphrodite caliper can also be used to lay out the center of round stock (Figure 11). The hooked leg is placed against the round stock and an arc is marked on the end of the piece. By adjusting the leg spacing, tangent arcs can be layed out. By marking four arcs at 90 degrees, the center of the stock can be established.

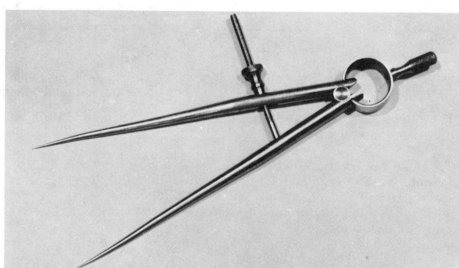

Figure 8. Spring dividers (DeAnza College).

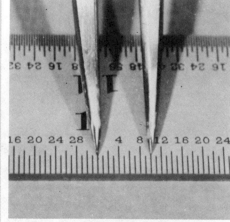

Figure 9. Setting divider points to an engraved rule (DeAnza College).

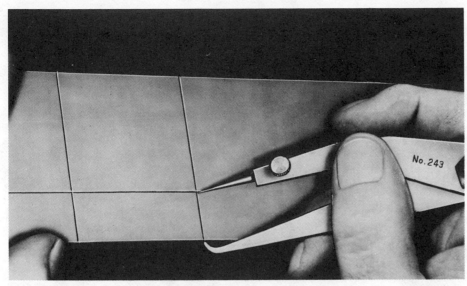

Figure 11. Scribing the centerline of round stock with the hermaphrodite caliper.

Figure 10. Scribing a line parallel to an edge using a hermaphrodite caliper (Courtesy of the L. S. Starrett Co.).

Trammel Points **Trammel points** are used for scribing circles and arcs when the distance involved exceeds the capacity of the divider. Trammel points are either attached to a bar and set to circle dimensions or they may be clamped directly to a rule where they can be set directly by rule graduations (Figure 12).

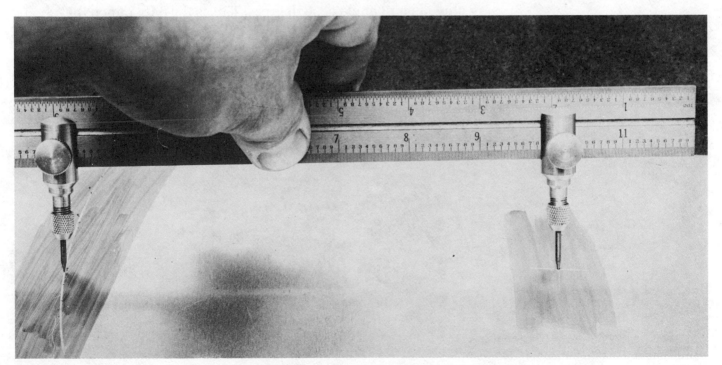

Figure 12. Trammel point attached to a rule.

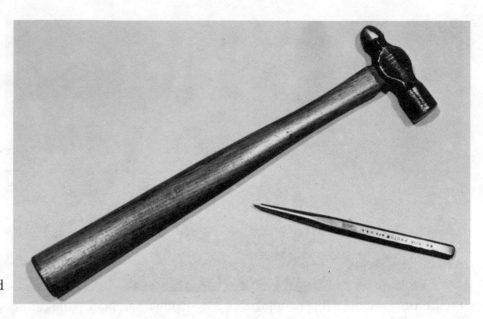

Figure 13. Layout hammer and layout prick punch.

Layout Hammers and Punches **Layout hammers** are usually light weight machinist's ball peen hammers (Figure 13). A heavy hammer should not be used in layout as it tends to create punch marks that are unnecessarily large.

The **toolmaker's hammer** is also used (Figure 14). This hammer is equipped with magnifier that can be used to help locate a layout punch on a scribe mark (Figure 15).

There is an important difference between a layout punch and a center punch. The **layout** or **prick punch** (Figure 14) **has an included point angle of 30 degrees.** This is the only punch that should be used in layout. The slim point facilitates the locating of the punch on a scribe line. A prick punch mark is only used to preserve the location of a layout mark while doing minimum damage to the workpiece. On some workpieces, depending on the material used and the part application, layout punchmarks are not acceptable

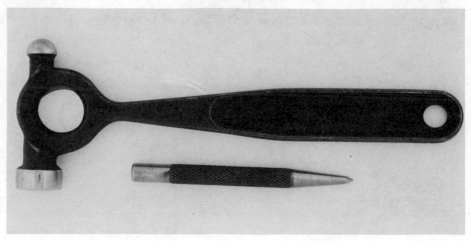

Figure 14. Toolmaker's hammer (California State University at Fresno).

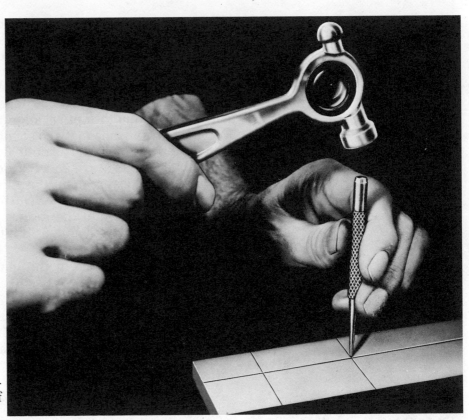

Figure 15. Using a toolmaker's hammer and layout punch (Courtesy of the L. S. Starrett Co.).

as they create a defect in the material. A punch mark may affect surface finish or metallurgical properties. Before using a layout punch, you must make sure that it is acceptable. In all cases, layout punch marks should be of minimum depth.

The **center punch** (Figure 16) **has an included point angle of 90 degrees** and is used to mark the workpiece prior to such machining operations as drilling. A center punch should not be used in place of a layout punch. Likewise, a layout punch should not be used in place of a center punch.

The **automatic center punch** (Figure 17) requires no hammer. Although called a center punch, its tip is suitably shaped for layout applications (Figure 18). Spring pressure behind the tip provides the required force. The automatic center punch may be adjusted for vari-

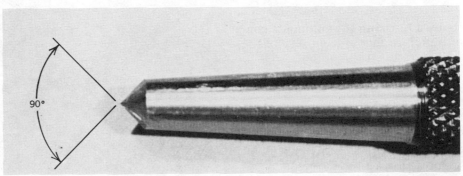

Figure 16. Center punch (DeAnza College).

able punching force by changing the spring tension. This is accomplished by an adjustment on the handle.

The **optical center punch** (Figure 19) consists of a locator, optical alignment magnifier, and punch. This type of layout punch is extremely useful in locating punch marks precisely on a scribed line or line intersection. The locator is placed over the approximate location and the optical alignment magnifier is inserted (Figure 20). The locator is magnetized so that it will remain in position when used on ferrous metals. The optical alignment magnifier has crossed lines etched on its lower end. By looking through the magnifier, you can move the locator about until the cross lines are matched to the scribe lines on the workpiece. The magnifier is then removed and the punch is inserted into the locator (Figure 21). The punch is then tapped with a layout hammer (Figure 22).

Centerhead　The **centerhead** is part of the **machinist's combination set.** Centerheads are used to lay out centerlines on round workpieces (Figure 23). When the centerhead is clamped to the combination set rule, the edge of the rule is in line with a circle center.

The other parts of the combination set are useful in layout. These include the **rule, square head, and bevel protractor.**

Figure 18. Automatic center punch (Courtesy of the L. S. Starrett Co.).

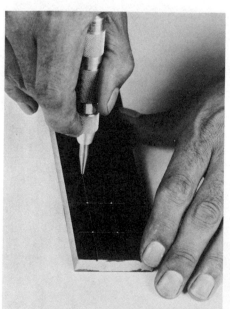

Figure 17. Using the automatic center punch in layout (Courtesy of the L. S. Starrett Co.).

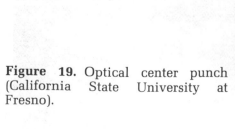

Figure 19. Optical center punch (California State University at Fresno).

Figure 20. Locating the punch holder with the optical alignment magnifier (California State University at Fresno).

Figure 21. Inserting the punch into the punch holder (California State University at Fresno).

Surface Gage The **surface gage** consists of a base, rocker, spindle adjusting screw, and scriber (Figure 24). The spindle of the surface gage pivots on the base and can be moved with the adjusting screw. The scriber can be moved along the spindle and locked at any desired position. The scriber can also swivel in its clamp. A surface gage may be used as a height transfer tool. The scribe is set to a rule dimension (Figure 25) and then transferred to the workpiece.

Figure 22. Tapping the punch with a layout hammer (California State University at Fresno).

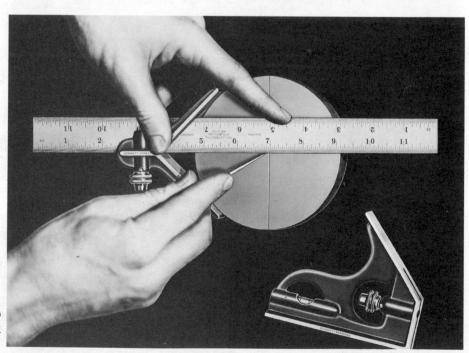

Figure 23. Using the centerhead to lay out a centerline on round stock (Courtesy of the L. S. Starrett Co.).

The hooked end of the surface gage scriber may be used to mark the centerline of a workpiece. The following procedure should be followed when doing this layout operation. The surface gage is first set as nearly as possible to a height equal to one half of the part height. The workpiece should be scribed for a short distance at this position. The part should be turned over and scribed again Figure 26. If a deviation exists, there will be two scribe lines on the workpiece. The surface gage scriber should then be adjusted so that

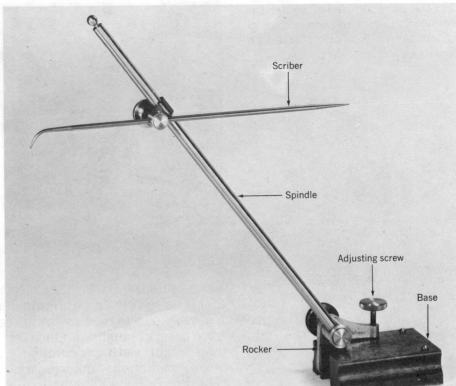

Figure 24. Parts of the surface gage (Courtesy of Rank Precision Industries, Inc.).

Figure 25. Setting a surface gage to a rule (DeAnza College).

it splits the difference between the two marks (Figure 27). This insures that the scribed line is in the center of the workpiece.

Height Gages Height gages are some of the most important instruments for precision layout. The most common layout height gage is the vernier type. Use of this instrument will be discussed in the unit on precision layout. As a machinist, you may use several other types of height gages for layout applications.

Figure 26. Finding the centerline of the workpiece using the surface gage (Courtesy of California Community Colleges IMC Project).

MECHANICAL DIAL AND ELECTRONIC DIGITAL HEIGHT GAGES. Mechanical dial (Figure 28) and electronic digital (Figure 29) height gages eliminate the need to read a vernier scale. Often these height gages do not have beam graduations. Once set to zero on the reference surface, the total height reading is cumulative on the digital display. This makes beam graduations unnecessary. The electronic digital height gage will discriminate to .0001 in.

GAGE BLOCK HEIGHT GAGES. Gage block height gages may be assembled from wrung stacks of gage blocks and accessories (Figure 30).

These height gages are extremely precise as they make use of the inherent accuracy of the gage blocks from which they are assembled.

THE PLANER GAGE AS A HEIGHT GAGE. The planer gage may be equipped with a scriber and used as a height gage (Figure 31). Dimensions are set by comparison to a precision height gage or height transfer micrometer. The planer gage can also be set with an outside micrometer.

Figure 27. Adjusting the position of the scribe line to center by inverting the workpiece and checking the existing differences in scribe marks (Courtesy of the California Community Colleges IMC Project).

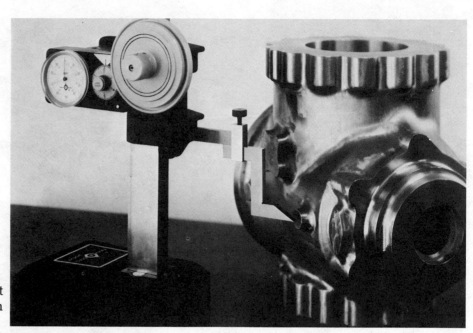

Figure 28. Mechanical dial height gage (Courtesy of Southwestern Industries, Inc.—Trav-A-Dial).

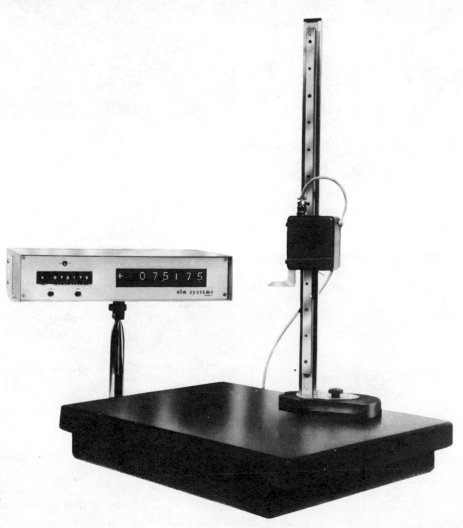

Figure 29. Electronic digital height gage (Courtesy of Elm Systems, Inc.).

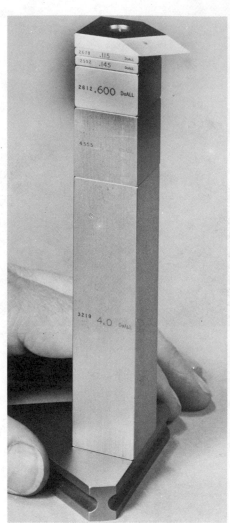

Figure 30. Height gage assembled from gage blocks (Courtesy of the DoAll Company).

Figure 31. Using the planer gage as a height gage in layout (DeAnza College).

Layout Machines   The layout machine (Figure 32) consists of a vertical column with a horizontal crossarm that can move up and down, in and out. The vertical column also moves horizontally across the layout table. From a single setup, the layout machine can accomplish layout on all sides, bottom, top, and inside of the workpiece. The instrument is equipped with an electronic digital display discriminating to .0001 in.

LAYOUT ACCESSORIES   Layout accessories are tools that will aid you in accomplishing layout tasks. They are not specifically layout tools as they are used for many other purposes. The layout plate or surface plate used for layout is the most common accessory as it provides the reference surface from which to work. Other common accessories include vee-blocks and angle plates that hold the workpiece during layout operations (Figure 33).

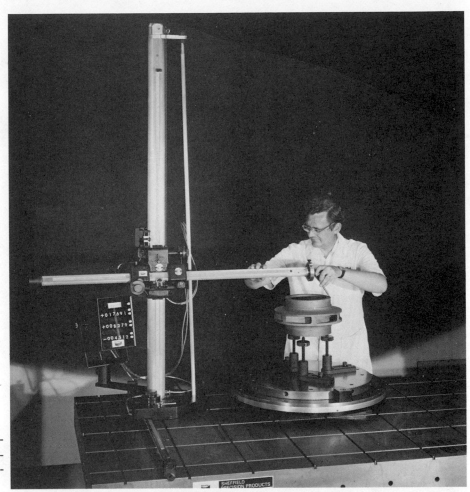

Figure 32. Layout machine (Courtesy of the Automation and Measurement Division—Bendix Corporation).

Figure 33. Universal right angle plate and vee-block used as layout accessories (California State University at Fresno).

269

unit 1
basic
semi-precision
layout practice

Before you can cut material for a certain job, you must perform a layout operation. Layout for stock cutoff may involve a simple chalk, pencil, or scribe mark on the material. No matter how simple the layout job may be, you should strive to do it neatly and accurately. In any layout, semiprecision or precision, accuracy is the watchword. Up to this point, you have been introduced to a large number of measuring and layout tools. It is now up to you to put these tools to work in the most productive manner possible. In this unit, you proceed through a typical semi-precision layout task that will familiarize you with basic layout practice.

objectives After completing this unit, you will be able to:
1. **Prepare the workpiece for layout.**
2. **Measure for and scribe layout lines on the workpiece outlining the various features.**
3. **Locate and establish hole centers using a layout prink punch and center punch.**
4. **Layout a workpiece to a tolerance of $\pm \frac{1}{64}$ in.**

PREPARING THE WORKPIECE FOR LAYOUT

After the material has been cut, all sharp edges should be removed by grinding or filing before placing the stock on the layout table. Place a paper towel under the workpiece to prevent layout dye from spilling on the layout table (Figure 1). Apply a **thin** even coat of layout dye to the workpiece. You will need a drawing of the part in order to do the required layout (Figure 2).

Study the drawing and determine the best way to proceed. The order of steps depends on the layout task. Before some features can be layed out, certain reference lines may have to be established. Measurements for other layout are made from these lines.

Figure 1. Applying layout dye with workpiece on a paper towel (Lane Community College).

271

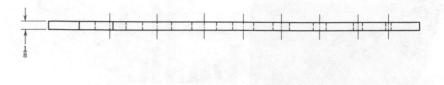

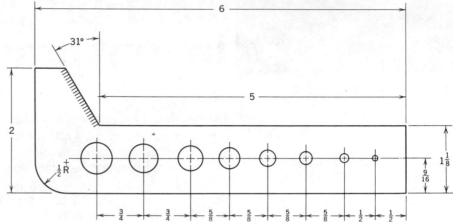

Figure 2. The drill and hole gage (Lane Community College).

LAYOUT OF THE DRILL AND HOLE GAGE

If possible, obtain a piece of material the same size as indicated on the drawing. Follow through each step as described in the reading. Refer to the layout drawings to determine where layout is to be done. The pictures will help you in selecting and using the required tools.

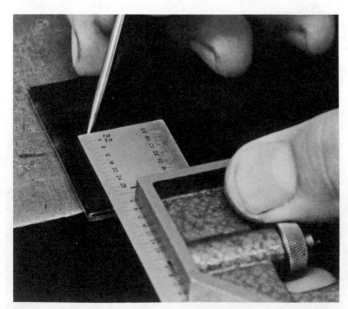

Figure 3. Measuring and marking the width of the gage using the combination square and rule (Lane Community College).

The first operation is to establish the width of the gage. Measure a distance of $1\frac{1}{8}$ in. from one edge of the material. Use the combination square and rule. Set the square at the required dimension and scribe a mark at each end of the stock (Figure 3 and Drawing A).

Remove the square and place the rule carefully on the scribe marks. Hold the rule **firmly** and scribe the line the full length of the material (Figure 4 and Drawing A). Be sure to use a sharp scribe and hold it so that the tip is against the rule. If the scribe is dull, regrind or stone it to restore its point. Scribe a clean visible line. Lay out the 5 in. length from the end of the piece to the angle vertex. Use the combination square and rule (Figure 5 and Drawing B).

Use a plate protractor to lay out the angle. The bevel protractor from the combination set is also a suitable tool for this application. Be sure that the protractor is set to the correct angle. The edge of the protractor blade must be set exactly at the 5 in. mark (Figure 6 and Drawing B). The layout of the 31 degree angle establishes its complement of 59 degrees on the drill gage. This is the correct angle for grinding drills used in general purpose drilling.

The corner radius is $\frac{1}{2}$ in. Establish this dimension using the square and rule. Two measurements will be required. Measure from the side and from the end to establish the center of the circle (Figure 7 and Drawing C). Prick

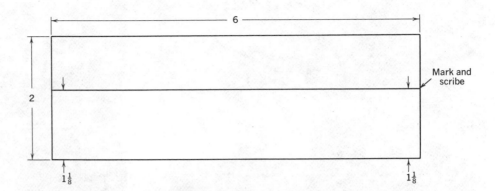

Drawing A. Width line.

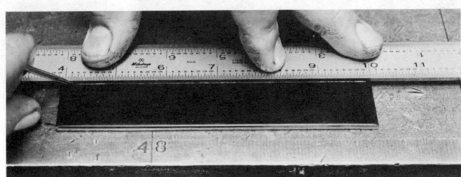

Figure 4. Scribing the width line (Lane Community College).

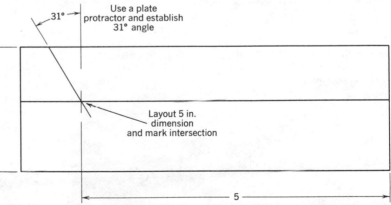

Drawing B. Angle line.

Figure 5. (*left*) Measuring the 5 in. dimension from the end to the angle vertex (Lane Community College).

punch the intersection of the two lines with the 30 degree included point angle layout punch. Tilt the punch so that it can be positioned exactly on the scribe marks (Figure 8). A magnifier will be useful here. Move the punch to its upright position and tap it lightly with the layout hammer (Figure 9 and Drawing C).

Set the dividers to a dimension of $\frac{1}{2}$ in.

Figure 6. Scribing the angle line (Lane Community College).

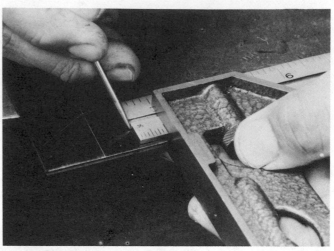

Figure 7. Establishing the center point of the corner radius (Lane Community College).

Drawing C. Corner radius.

Scribe
corner radius

½

½

Layout corner radius
center and prick punch

Figure 9. Punching the center point of the corner radius (Lane Community College).

Figure 8. Setting the layout punch on the center point of the corner radius (Lane Community College).

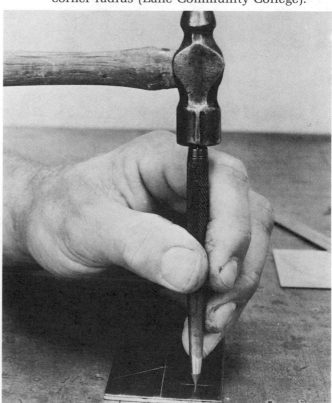

Figure 10. Setting the dividers to the rule engravings (Lane Community College).

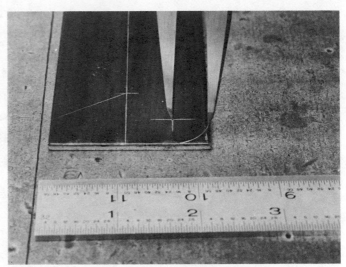

Figure 11. Scribing the corner radius (Lane Community College).

Figure 12. Center punching the hole centers (Lane Community College).

Figure 13. Completed layout for the drill and hole gage (Lane Community College).

using the rule. Adjust the divider spacing until you feel the tips drop into the rule engravings (Figure 10). Place one divider tip into the layout punch mark and scribe the corner radius (Figure 11 and Drawing C).

The centerline of the holes is $\frac{9}{16}$ in. from the edge. Use the square and rule to measure this distance. Mark at each end and scribe the line full length (Drawing D). Measure and lay out the center of each hole (Drawing D). Use

the layout punch and mark each hole center. After prick punching each hole center, set the dividers to each indicated radius and scribe all hole diameters (Drawing D).

The last step is to center punch each hole center prior to drilling. Use a 90 degree included point angle center punch (Figure 12). Layout of the drill and hole gage is now complete (Figure 13).

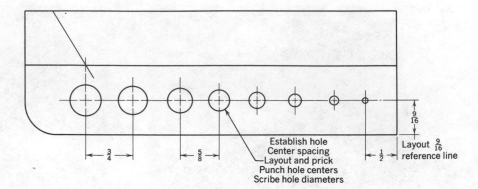

Drawing D. Hole locations.

self-evaluation

SELF-TEST 1. How should the workpiece be prepared prior to layout?
2. What is the reason for placing the workpiece on a paper towel?
3. Describe the technique of using the layout punch.
4. Describe the use of the combination square and rule in layout.
5. Describe the technique of setting a divider to size using a rule.

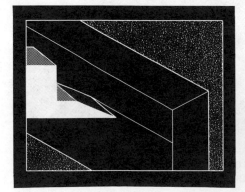

unit 2
basic precision layout practice

Precision layout is generally more reliable and accurate than layout by semiprecision practice. On any job requiring maximum accuracy and reliability, precision layout practice should be used.

The fundamental precision layout tool is the height gage. The vernier height gage is the most common type found in the machine shop. This instrument will discriminate to .001 in. With this ability, a much higher degree of accuracy and reliability is added to a layout task. Whenever possible, you should apply the height gage in all precision layout requirements.

objectives After completing this unit, you will be able to:

1. Identify the major parts of the vernier height gage.
2. Describe applications of the vernier height gage in layout.
3. Read a vernier height gage in both metric and inch dimensions.
4. Accomplish layout using the vernier height gage.

THE VERNIER HEIGHT GAGE

The vernier height gage is the most common and important tool used for precision layout. Major parts of the height gage include the base, beam, vernier slide, and scriber (Figure 1). The size of height gages is measured by the maximum height gaging ability of the instrument. Height gages range from 10 to 72 in.

Height gage scribers are made from tool steel or tungsten carbide. **Carbide scribers are subject to chipping and must be treated gently.** They do, however, retain their sharpness and scribe very clean narrow lines. Height scribers may be sharpened if they become dull. It is important that any **sharpening be done on the slanted surface so that the scriber dimensions will not be changed.**

The height gage scriber is attached to the vernier slide and can be moved up and down the beam. Scribers are either **straight** (Figure 2) or **offset** (Figure 3). The offset scriber permits direct readings with the height gage. The gage reads zero when the scriber rests down on the reference surface. With the straight scriber, the workpiece will have to be raised accordingly, if direct readings are to be obtained. This type of height gage scriber is less convenient.

READING THE VERNIER HEIGHT GAGE

On an inch height gage, the beam is graduated in inches with each inch divided into 10 parts. The tenth inch graduations are further divided into two or four parts depending on the divisions of the vernier. On the 25 division vernier, used on many older height gages, the $\frac{1}{10}$ in. divisions on the beam are graduated into four parts. The vernier permits discrimination to .001 in. Many newer height gages are making use of the 50 division vernier, which permits easier reading. On a height gage with a 50 division vernier, the $\frac{1}{10}$ in. graduations on the beam will be divided into two parts. Discrimination of this height gage is also .001 in. The metric vernier

height gage has the beam graduated in millimeters. The vernier contains 50 divisions permitting the instrument to discriminate to $\frac{1}{50}$ mm.

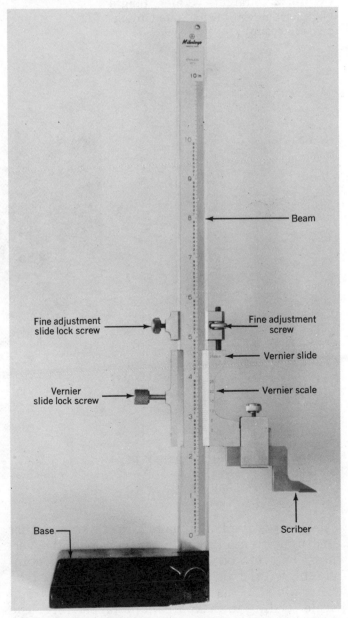

Figure 1. Parts of the vernier height gage.

Figure 2. Straight vernier height gage scriber (California State University at Fresno).

Figure 3. Offset vernier height gage scriber.

The vernier height gage is read like any other instrument employing the principle of the vernier. The line on the vernier scale that is coincident with a beam scale graduation must be determined. This value is added to the beam scale reading to make up the total reading. The inch vernier height gage with a 50 division vernier is read as follows (Figure 4, right-hand scale):

Beam reading	5.3
Vernier is coincident at 12 or .012 in.	.012
Total reading	5.312 in.

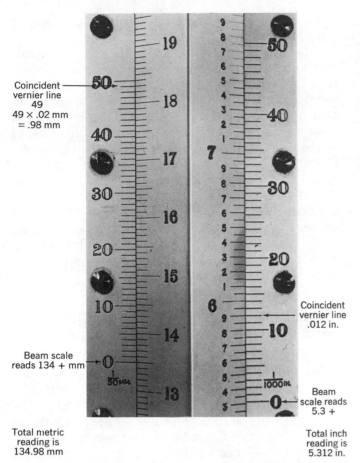

Coincident
vernier line
49
49 × .02 mm
= .98 mm

Beam scale
reads 134 + mm

Coincident
vernier line
.012 in.

Beam
scale reads
5.3 +

Total metric
reading is
134.98 mm

Total inch
reading is
5.312 in.

Figure 4. Reading the 50 division inch/metric vernier height gage (DeAnza College).

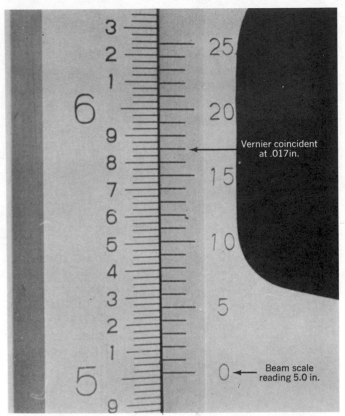

Vernier coincident
at .017in.

Beam scale
reading 5.0 in.

Figure 5. Reading the 25 division inch vernier height gage.

On the 50 division inch vernier height gage, the beam scale is graduated in $\frac{1}{10}$ in. graduations. Each $\frac{1}{10}$ in. increment is further divided into two parts. If the zero on the vernier is past the .050 in. mark on the beam, .050 in. must be added to the reading.

The metric vernier scale also has 50 divisions, each equal to $\frac{1}{50}$ or .02 millimeter (Figure 4, left-hand scale):

Beam scale reading	134 mm
Vernier coincident at line 49	
49 × .02 =	.98 mm
Total reading	134.98 mm

On the 25 division inch vernier height gage the $\frac{1}{10}$ in. beam graduations are divided into four parts, each equal to .025 in. Depending on the location of the vernier zero mark, .025, .050, or .075 in. may have to be added to the beam reading. The inch vernier height gage with a 25 division vernier is read as follows (Figure 5):

Figure 6. Checking the zero reference (DeAnza College).

Beam	5.0
Vernier coincident at 17 or .017	.017
Total reading	5.017 in.

CHECKING THE ZERO REFERENCE ON THE VERNIER HEIGHT GAGE

The height gage scriber must be checked against the reference surface before attempting to make any height measurements of layouts. Clean the surface of the layout table and the base of the gage. Slide the scriber down until it just rests on the reference surface. Check the alignment of the zero mark on the vernier scale with the zero mark on the beam scale. The two marks should coincide exactly (Figure 6). Hold the height gage base firmly against the reference surface. Be sure that you do not tilt the base of the height gage by sliding the vernier slide past the zero point on the beam scale. If the zero marks on the vernier and beam do not coincide after the scriber has contacted the reference

surface, an adjustment of the vernier scale is required.

Some height gages do not have an adjustable vernier scale. A misalignment in the vernier and the beam zero marks may indicate a loose vernier slide, an incorrect scriber dimension, or a beam that is out of perpendicular with the base. Loose vernier slides may be adjusted and scriber dimensions can be corrected. However, if the beam is out of perpendicular with the base, the instrument is unreliable because of cosine error. A determination of such a condition can be made by an appropriate calibration process. All height gages, particularly those with nonadjustable verniers, must be treated with the same respect as any precision instrument that you will use.

APPLICATIONS OF THE VERNIER HEIGHT GAGE IN LAYOUT

The primary function of the vernier height gage in layout is to measure and scribe lines of known height on the workpiece (Figure 7). Perpendicular lines may be scribed on the workpiece by the following procedure. The work is first clamped to a right angle plate if necessary and the required lines are scribed in one direction. The height gage should be set at an angle to the work and the corner of the scriber pulled across while keeping the height gage base firmly on the refer-

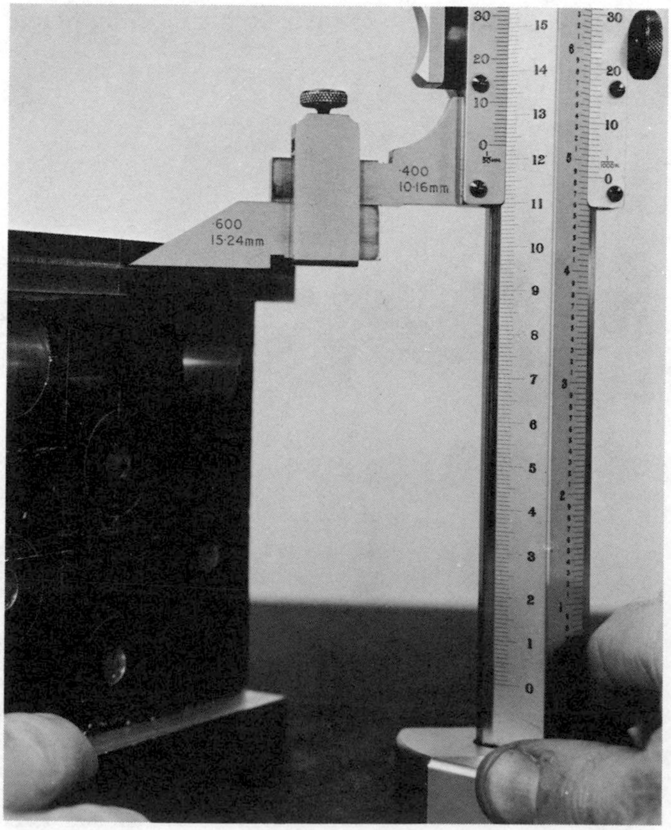

Figure 7. Scribing height lines with the vernier height gage (DeAnza College).

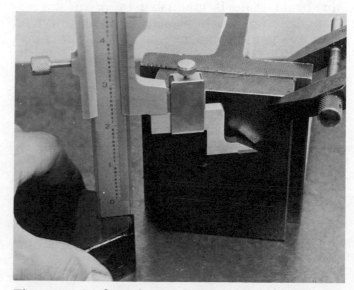

Figure 8. Scribing layout lines with the workpiece clamped to a right angle plate.

Figure 9. After turning the workpiece 90 degrees, it can be checked with a square.

Figure 10. Checking the work using a dial test indicator.

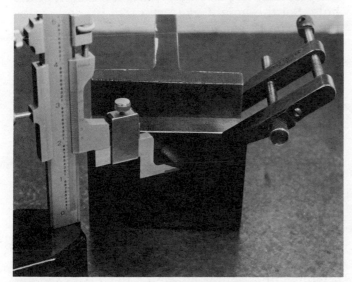

Figure 11. Scribing perpendicular lines.

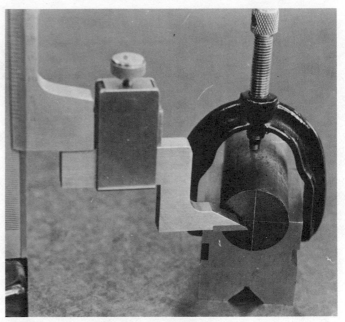

Figure 12. Scribing centerlines on round stock clamped in a vee-block.

ence surface (Figure 8). Only enough pressure should be applied with the scriber to remove the layout dye and not actually remove material from the workpiece.

After scribing the required lines in one direction, turn the workpiece by 90 degrees. Setup is quite critical if the scribe marks are to be truly perpendicular. A square (Figure 9) or a dial test indicator may be used (Figure 10) to establish the work at right angles. In both cases the edges of the workpiece must be machined smooth and square. After the clamp has been tightened, the perpendicular lines may be scribed at the required height (Figure 11).

The height gage may be used to lay out center lines on round stock (Figure 12). The stock is clamped in a vee-block and the correct dimension to center is determined. This can be done with the dial test indicator attached to the height gage. However, it must not be done with the height gage scriber.

Parallel bars (Figure 13) are a valuable and useful layout accessory. These bars are made from hardened steel or granite, and they have extremely accurate dimensional accuracy. Parallel bars are available in many sizes and lengths. In layout with the height gage they can be used to support the workpiece (Figure 14).

Angles may be laid out by placing the workpiece on the sine bar (Figure 15).

Figure 13. Hardened steel parallel bars (California State University at Fresno).

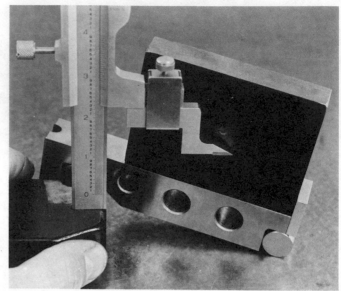

Figure 15. Laying out angle lines using the sine bar.

Figure 14. Using parallel bars in layout.

BASIC PRECISION LAYOUT PRACTICE

The workpiece should be prepared as in semi-precision layout. Sharp edges must be removed and a thin coat of layout dye applied. You will need a drawing of the part to be layed out (Figure 16). The order of steps will depend on the layout task.

Position One Layouts

In position one (Figure 17) the clamp frame is on edge. In any position, all layouts can be defined as heights above the reference surface. Refer to

the drawing on position one layouts and determine all of the layout that can be accomplished there.

Start by scribing the $\frac{3}{4}$ in. height that defines the width of the clamp frame. Set the height gage to .750 in. (Figure 18). Attach the scriber (Figure 19). Be sure that the scriber is sharp and properly installed for the height gage that you are using. Hold the workpiece and height gage firmly and pull the scriber across the work in a smooth motion (Figure 20). The height of the clamp screw hole can be layed out at this time. Refer to the part drawing and determine the height of the hole. Set the height gage at 1.625 in. and scribe the line on the end of the workpiece (Figure 21). The line may be projected around on the side of the part. This will facilitate setup in the drill press. Other layouts that can be accomplished at position one include the height equivalent of the inside corner hole centerlines.
The starting points of the corner angles on both ends may also be layed out. Refer to the drawing on position one layouts.

Position Two Layouts

In position two, the workpiece in on its side (Figure 22). Check the work with a micrometer to determine its exact thickness. Set the height gage to $\frac{1}{2}$ this amount and scribe the centerline of the clamp screw hole. This layout will also establish the center point of the clamp screw

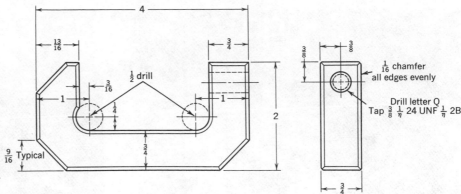

Figure 16. Clamp frame (Lane Community College).

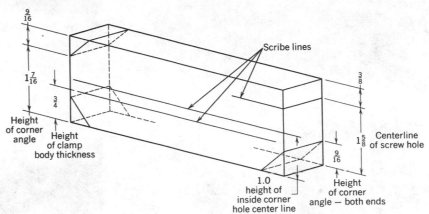

Figure 17. Clamp frame—position one layouts.

hole. A height gage setting of .375 in. will probably be adequate providing that the stock is .750 in. thick. However, if the thickness varies above or below .750 in. the height gage can be set to $\frac{1}{2}$ of whatever the thickness is. This will insure that the hole is in the center of the workpiece.

Position Three Layouts

In position three, the workpiece in on end clamped to an angle plate (Figure 23). The work must be established perpendicular using a square or dial test indicator. Set the height gage to .750 in. and scribe the height equivalent of the frame end thickness (Figure 24). Other layouts that can be done at position three include the height equivalent of the inside corner hole centerlines. This layout will also locate the center points of the inside corner holes (Figure 25). The height equivalent of the end thickness as well as the ending points of the corner angles can be scribed at position three.

HEIGHT GAGE LAYOUT BY COORDINATE MEASURE

Many layouts can be accomplished by calculating the coordinate position of the part features. Coordinate position simply means that each feature is located a certain distance from adjacent perpendicular reference lines. These are frequently known as the X and Y coordinates. You should begin to think of coordinates in terms of X and Y as this terminology will be important, especially in the area of numerical control machining. The X coordinate on a two dimensional drawing is horizontal. The Y coordinate is perpendicular to X and in the same plane. On a drawing, Y is the vertical coordinate. The X and Y coordinate lines can be and often are the edges of the workpiece, providing that the edges have been machined true and square to each other.

Coordinate lengths can be calculated by the application of appropriate trigonometric formulas. They may also be determined from

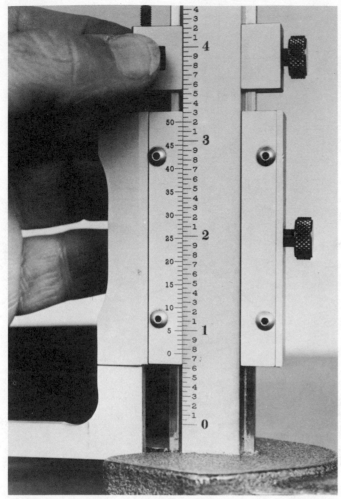

Figure 18. Setting the height gage to a dimension of .750 in. (Lane Community College).

Figure 19. Attaching the scriber (Lane Community College).

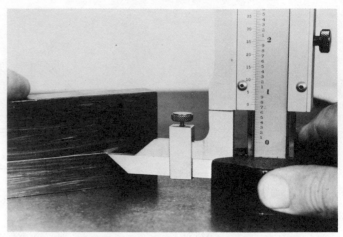

Figure 20. Scribing the height equivalent of the frame thickness (Lane Community College).

Figure 21. Scribing the height equivalent of the clamp screw hole (Lane Community College).

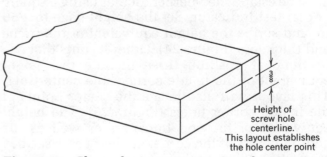

Height of screw hole centerline. This layout establishes the hole center point

Figure 22. Clamp frame—position two layouts.

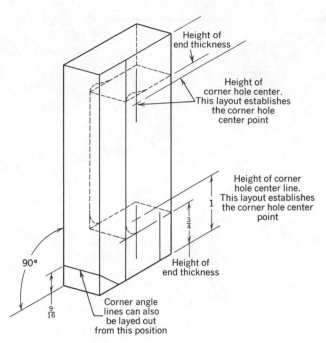

Figure 23. Clamp frame—position three layouts.

Figure 24. Scribing the height equivalent of the end thickness (Lane Community College).

tables of coordinate measure. Such tables appear in most handbooks for machinists.

Calculating Coordinate Measurements
The drawing (Figure 26) shows a five hole equally spaced pattern centered on the workpiece. Since hole one is on the centerline, its coordinate position measured from the reference edges can be easily determined (Figure 27). The X coordinate (horizontal) is 2 in. The Y coordinate (vertical) is two inches plus the radius of the hole circle. This would be $3\frac{1}{4}$ in.

The coordinate position of hole two can be calculated by the following. Since there are five equally spaced holes the central angle is $\frac{360}{5}$ or 72 degrees. Right triangle ABC (Figure 28) is formed by constructing a perpendicular line from point B to point C. Angle A equals 18 degrees $(90 - 72 = 18)$. To find the X coordinate, apply the following formula:

$$X_c = \text{circle radius} \times \cos 18°$$
$$= 1.250 \times .951$$
$$= 1.188 \text{ in.}$$

The X coordinate length from the reference edge is found by

$$2.0 - 1.188 = .812 \qquad \text{(Figure 26)}$$

Figure 25. Completed layout of the clamp frame (Lane Community College).

The Y coordinate is found by the following formula:

$$Y_c = \text{circle radius} \times \sin 18°$$
$$= 1.250 \times .309$$
$$= .386$$

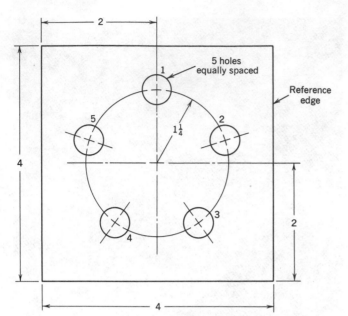

Figure 26. Equally spaced five hole circle.

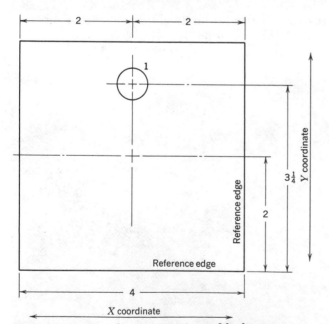

Figure 27. Coordinate position of hole one.

The *Y* coordinate length from the reference edge is found by

2.0 + .386 = 2.386 in.

The coordinate position of hole three is calculated in a similar manner. Right triangle *AEF* is formed by constructing a perpendicular line from point *F* to point *E* (Figure 28). Angle

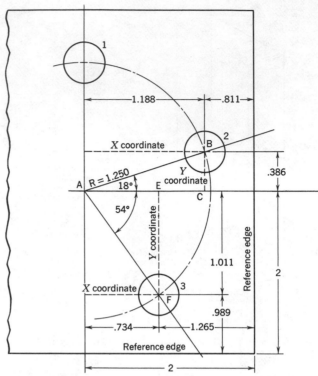

Figure 28. Coordinate positions of holes two and three.

FAE equals 54 degrees (72 − 18 = 54). To find the *X* coordinate, apply the following formula:

$$X_c = \text{circle radius} \times \cos 54°$$
$$= 1.250 \times .587$$
$$= .734$$

The *X* coordinate length from the reference edge is found by

2.0 − .734 = 1.265   (Figure 28)

To find the *Y* coordinate, apply the following formula:

$$Y_c = \text{radius} \times \sin 54°$$
$$= 1.250 \times .809$$
$$= 1.011 \text{ in.}$$

The *Y* coordinate length from the reference edge is found by

2.0 − 1.011 = .989 in.   (Figure 28)

The coordinate positions of holes four and five are the same distance from the centerlines as holes two and three. Their positions from the reference edges can be calculated easily.

Since this layout involves scribing perpendicular lines, the workpiece must be turned 90

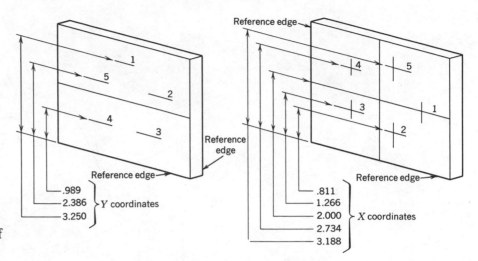

Figure 29. Height equivalents of coordinate positions for all holes.

degrees. If the edges of the work are used as reference, they must be machined square. Either coordinate may be layed out first. The workpiece is then turned 90 degrees to the adjacent reference edge. This permits the layout of the perpendicular lines (Figure 29).

LAYING OUT ANGLES

Angles may be layed out using the height gage by calculating the appropriate dimensions using trigonometry. In the example (Figure 30) the layout of height A will establish angle B at 36 degrees. Height A is calculated by the following formula:

$$\text{Height } A = 1.25 \times \tan B$$
$$= 1.25 \times .726$$
$$= .908 \text{ in.}$$

After scribing a height of .908 in., the workpiece is turned 90 degrees and the starting point of the angle established at point B. Scribing from point A to point B will establish the desired angle.

The sine bar can also be used in angular layout. In the example (Figure 31), the sine bar is elevated for the 25 degree angle. Sine bar elevation is calculated by the formula

Bar elevation = bar length × sine of require angle. If we assume a 5 in. sine bar,

$$\text{Elevation} = 5 \times \sin 25°$$
$$= 5 \times .422$$
$$= 2.113 \text{ in.}$$

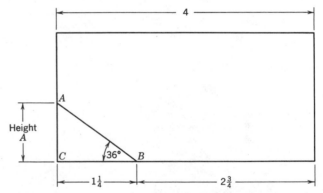

Figure 30. Laying out a 36 degree angle.

A gage block stack is assembled and placed under the sine bar. Now that the bar has been elevated, the vertical distance CD from the corner to the scribe line AB must be determined. To find distance DC, a perpendicular line must be constructed from point C to point D. Angle A is 65 degrees ($90 - 25 = 65$). Length CD is found by the following formula:

$$CD = .500 \times \sin B$$
$$= .500 \times .906$$
$$= .453 \text{ in.}$$

The height of the corner must be determined and the length of CD subtracted from this dimension. This will result in the correct height gage setting for scribing line AB. The corner height should be determined using the height gage and dial test indicator. The corner height must not be determined using the height gage scriber.

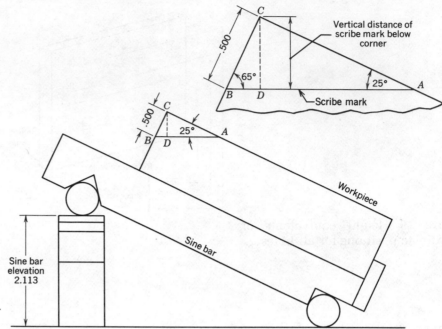

Figure 31. Laying out an angle using the sine bar.

self-evaluation

SELF-TEST Read and record the following 50 division inch/metric height gage readings.
1. (Figures 32–34)

   Read and record the following 25 division inch height gage readings:
   (Figure 35)
   (Figure 36)

2. Describe the procedure for checking the zero reference.

3. How can the zero reference be adjusted?

4. How are perpendicular lines scribed with a height gage?

5. What is the measuring range of a typical height gage?

6. When laying out angles, what tool is used in conjunction with the height gage?

Figure 32.

Figure 33.

Figure 34.

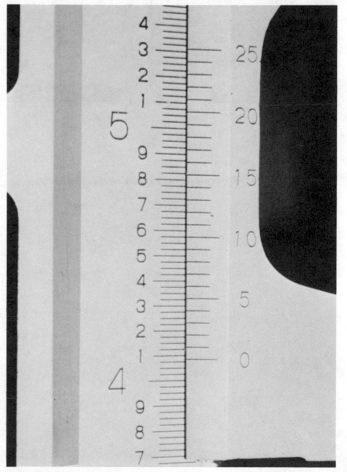

Figure 35.

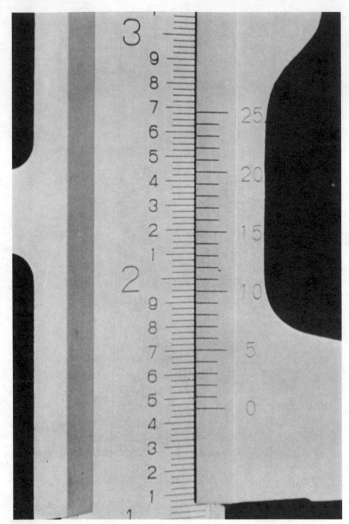

Figure 36.

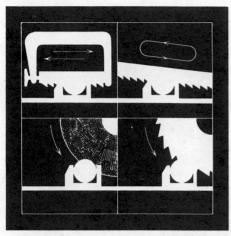

section f
sawing machines

Sawing machines constitute some of the most important machine tools found in the machine shop. These machines can generally be divided into two classifications. The first class is **cutoff machines.** Common types of cutoff machines include reciprocating saws, horizontal endless band saws, universal tilt frame band saws, abrasive saws, and cold saws. Of the cutoff machines, the **horizontal endless band saw** is the most important type. The second class is the **vertical band machine** that can be used as a band saw or with other band tools. The vertical band machine is most commonly used as a **band saw.**

The first machine tool that you will probably encounter is a cutoff machine. In the machine shop, cutoff machines are generally found near the stock supply area. The primary function of the cutoff machine is to reduce mill lengths of bar stock material into lengths suitable for holding in other machine tools. In a large production machine shop where stock is being supplied to many machine tools, the cutoff machine will be constantly busy cutting many materials. The cutoff machine that you will most likely see in the machine shop will be some type of saw.

TYPES OF CUTOFF MACHINES
Reciprocating Saws

The **reciprocating** saw is often called a power hacksaw. Early saws were hand operated by a reciprocating or back-and-forth motion. It was logical that this principle be applied to power saws (Figure 1). The reciprocating saw is still used in the machine shop. However, they are giving way to the endless band machine. The reciprocating saw is built much like the metal cutting hand hacksaw. Basically, the machine consists of a frame that holds a blade. Reciprocating hacksaw blades are wider and thicker than those used in the hand hacksaw. The reciprocating motion is provided by hydraulics or a crankshaft mechanism.

Reciprocating saws are either the hinge type (Figure 2) or the column type (Figure 3). The saw frame on the hinge type pivots around a single point at the rear of the machine. On the column type, both ends of the frame rise vertically. Column-type reciprocating saws can accommodate larger sizes of material. The size of a reciprocating saw is determined by the largest piece of square material that can be cut. Sizes range from about 5 by 5 in. to 24 by 24 in. Large capacity reciprocating saws are often of the column design.

291

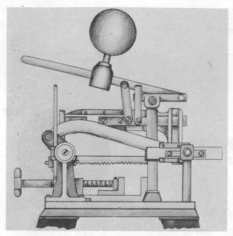

Figure 1. Early reciprocating cutoff saw (Courtesy of California Community Colleges Individualized Machinist's Curriculum Project).

Figure 2. (*right*) Hinge-type reciprocating cutoff saw (Courtesy of Kasto-Racine Inc.).

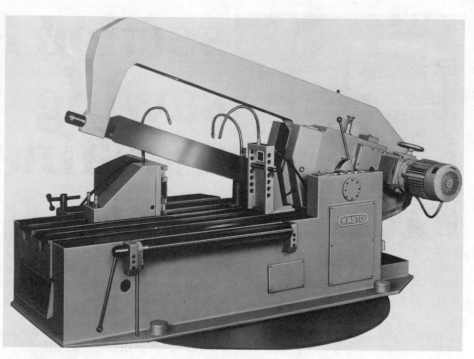

Horizontal Endless Band Cutoff Machine One disadvantage of the reciprocating saw is that it only cuts in one direction of the stroke. The endless band machine uses a steel band blade with the teeth on one edge. The endless band machine has a high cutting efficiency because the band is cutting at all times with no wasted motion. Endless band saws are the mainstay of production stock cutoff in the machine shop (Figure 4).

A modern band saw may be equipped with a variable speed drive. This permits the most efficient cutting speed to be selected for the material being cut. The feed rate through the material may also be varied. The size of the horizontal band machine is determined by the largest piece of square material that the machine can cut. Large capacity horizontal band saws (Figure 5) are designed to handle large dimension workpieces that can weigh as much as 10 tons. With a wide variety of band types available, plus many special workholding devices, the endless band saw is an extremely valuable and versatile machine tool.

Universal Tilt Frame Cutoff The universal tilt frame band saw is much like its horizontal counterpart. This machine has the band blade vertical, and the frame can be tilted from side to side (Figure 6). The tilt frame machine is particularly useful for making angle cuts on large structural shapes such as I-beams or pipe.

Abrasive Cutoff Machine The abrasive cutoff machine (Figure 7) uses a thin circular abrasive wheel for cutting. Abrasive saws are very fast cutting. They can be used to cut a number of nonmetallic materials such as glass, brick, and stone. The major advantages of the abrasive cutoff machine are speed and the ability to cut nonmetals. Each particle of

Figure 3. Column-type reciprocating cutoff saw (Courtesy of Kasto-Racine Inc.).

Figure 5. Large capacity horizontal band saw (Courtesy of the DoAll Company).

Figure 6. Tilt frame band saw (Courtesy of the DoAll Company).

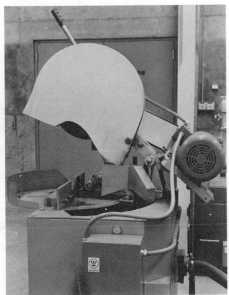

Figure 7. Typical abrasive cutoff machine (California State University at Fresno).

Figure 4. Horizontal endless band cutoff saw (Courtesy of the DoAll Company).

abrasive acts as a small tooth and actually cuts a small bit of material. Abrasive saws are operated at very high speeds. Blade speed can be as high as 10,000 to 15,000 surface feet per minute.

Cold Saw Cutoff Machines The cold saw (Figure 8) uses a circular metal saw with teeth. These machine tools can produce extremely accurate cuts and are useful where length tolerance of the cut material must be held as close as

Figure 8. Precision cold saw cutoff machine (Courtesy of Ameropean Industries, Inc.).

possible. A cold saw blade that is .040 to .080 in. thick can saw material to a tolerance of plus or minus .002 in. Large cold saws are used to cut structural shapes such as angle and flat bar. These are also fast cutting machines.

CUTOFF MACHINE SAFETY
Reciprocating Saws Be sure that all guards around moving parts are in place before starting the machine. The saw blade must be properly installed with the teeth pointed in the right direction. Check for correct blade tension. Be sure that the width of the workpiece is less than the distance of the saw stroke. The frame will be broken if it hits the workpiece during the stroke. This will damage the machine. Be sure that the speed of the stroke and the rate of feed is correct for the material being cut.

When operating a saw with coolant, see that the coolant does not run on the floor during the cutting operation. This can cause an extremely dangerous slippery area around the machine tool.

Horizontal Band Saws Recent, new regulations require that the blade of the horizontal band machine be fully guarded except at the point of cut (Figure 9).

Make sure that blade tensions are correct on reciprocating and band saws. Check endless band tensions, especially after installing a new band. New bands may stretch and loosen during their run-in period. Band teeth are sharp. When installing a new band, it should be handled with gloves. This is one of the few places that gloves may be worn around the machine shop. They must not be worn when operating any machine tool.

Endless band blades are often stored in double coils. Be careful when unwinding them, as they are under tension. The coils may spring apart and could cause an injury.

Figure 9. The horizontal band blade is guarded except in the immediate area of the cut.

Figure 11. Result of cutting stock that is too short.

Figure 10. Support both ends of the vise when cutting short material (California State University at Fresno).

Make sure that the band is tracking properly on the wheels and in the blade guides. If a band should break, it could be ejected from the machine and cause an injury.

Make sure that the material being cut is properly secured in the workholding device. If this is a vise, be sure that it is tight. If you are cutting off short pieces of material, the vise jaw must be supported at both ends (Figure 10). It is not good practice to attempt to cut pieces of material that are quite short. The stock cannot be secured properly and may be pulled from the vise by the pressure of the cut (Figure

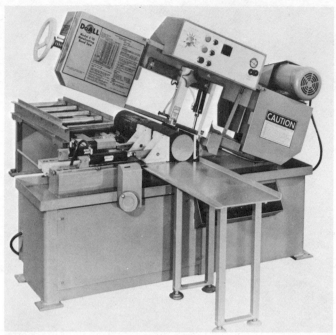

Figure 12. The material is brought into the saw on the rollcase (opposite side) and, when pieces are cut off, they are supported by the stand (this side of the saw). The stand prevents the part from falling to the floor (Courtesy of DoAll Company).

Figure 13. Inspecting the abrasive wheel for chips and cracks (California State University at Fresno).

11). This can cause damage to the machine as well as possible injury to the operator. Stock should extend at least halfway through the vise at all times.

Many cutoff machines have a rollcase that supports long bars of material while they are being cut. The stock should be brought to the saw on a rollcase (Figure 12) or a simple rollstand. The pieces being cut off can sometimes be several feet long and should be similarly supported. Sharp burrs left from the cutting should be removed immediately with a file. You can acquire a nasty cut by sliding your hand over one of these burrs.

Be careful around a rollcase, since bars of stock can roll, pinching fingers and hands. Also, be careful that heavy pieces of stock do not fall off the stock table or saw and injure feet or toes. Get help when lifting heavy bars of material. This will save your back and possibly your career.

Abrasive Saws　On an abrasive saw, inspect the cutting edge of the blade for cracks and chips (Figure 13). Replace the blade if it is damaged. Always operate an abrasive saw blade at the proper RPM (Figure 14). Overspeeding the blade can cause it to fly apart. If an abrasive saw blade should fail at high speed, pieces of the blade can be thrown out of the machine at extreme velocities. A very serious injury indeed can result if you happen to be in the path of these bulletlike projectiles.

Figure 14. Abrasive wheel must be operated at the correct RPM (California State University at Fresno).

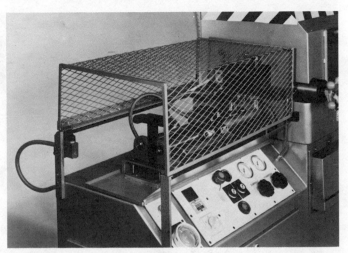

Figure 15. Wire mesh guard on the cold saw material feeding mechanism (Courtesy of Ameropean Industries, Inc.).

Cold Saws

In terms of cold saws, safety is generally the same as with all cutoff machines. Guards must be in place around the saw and the feeding mechanism (Figure 15). Before starting the saw, check to see that speeds and feeds are correct and that the workpiece is properly secured.

Safety extends to the machine as well as the operator. Never abuse any machine tool. They cost a great deal of money and in many cases are purchased with your tax dollars.

VERTICAL BAND MACHINES The vertical band machine (Figure 16) is often called the handiest machine tool in the machine shop. Perhaps the reason for this is the wide variety of work that can be accomplished on this versatile machine tool. The vertical band machine or vertical band saw is similar in general construction to its horizontal counterpart. Basically, it consists of an endless band blade or other band tool that runs on a driven and idler wheel. The band tool runs vertically at the point of the cut where it passes through a worktable on which the workpiece rests. The workpiece is pushed into the blade and the direction of the cut is guided by hand or mechanical means.

ADVANTAGES OF BAND MACHINES Shaping of material with the use of a saw blade or other band tool is often called **band machining**. The reason for this is that the band machine can perform other machining tasks aside from simple sawing. These include band friction sawing, band filing, and band polishing.

In any machining operation, a piece of stock material is cut by various processes to form the final shape and size of the part desired.

Machine Tool Practices

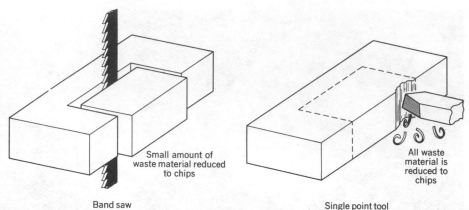

Figure 17. Sawing can uncover the workpiece shape in a minimum number of cuts.

Figure 16. Leighton A. Wilkie bandsaw of 1933 was the last basic machine tool to be developed (Courtesy of the DoAll Company).

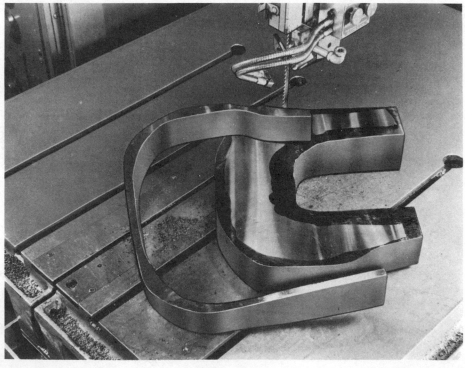

Figure 18. Curved or contour band sawing can produce part shapes that would be difficult to machine by other methods (Courtesy of the DoAll Company).

In most machining operations, all of the unwanted material must be reduced to chips in order to uncover the final shape and size of the workpiece. With a band saw, only a small portion of the unwanted material must be reduced to chips in order to uncover the final workpiece shape and size (Figure 17). A piece of stock material can often be shaped to final size by one or two saw cuts. A further advantage is gained in that the band saw cuts a very narrow kerf. A minimum amount of material is wasted. Other machining operations may require that a large amount of material be wasted as chips in order to uncover the final size and shape of the workpiece.

A second important advantage in band sawing machines is **contouring ability.** Contour band sawing is the ability of the saw to

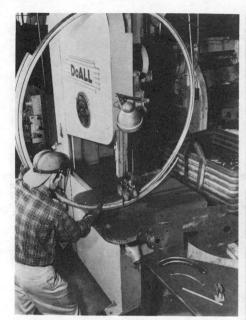

Figure 19. Splitting a large diameter ring on the vertical band machine (Courtesy of the DoAll Company).

Figure 20. Workpieces larger than the machine tool can be cut (Courtesy of the DoAll Company).

cut intricate curved shapes that would be nearly impossible to machine by other methods (Figure 18). The sawing of intricate shapes can be accomplished by a combination of hand and power feeds. On vertical band machines so equipped, the workpiece is steered by manual operation of the handwheel. The hydraulic table feed varies according to the saw pressure on the workpiece. This greatly facilitates contour sawing operations.

Band sawing and band machining have several other advantages. There is no limit to the length, angle, or direction of the cut (Figure 19). Workpieces larger than the band machine can be cut (Figure 20). Since the band tool is fed continuously past the work, cutting efficiency is high. A band tool, whether it be a saw blade, band file, or grinding band, has a large number of cutting points passing the work. In most other machining operations, only one or a fairly low number of cutting points pass the work. With the band tool, wear is distributed over these many cutting points. Tool life is prolonged.

TYPES OF BAND MACHINES
General Purpose Band Machine
with Fixed Worktable

The general purpose band machine is found in most machine shops, (Figure 21). This machine tool has a nonpower-fed worktable that can be tilted in order to make angle cuts. The table may be tilted 10 degrees left (Figure 22). Tilt on this side is limited by the saw frame. The table may be tilted 45 degrees right (Figure 23). On large machines, table tilt left may be limited to 5 degrees.

The workpiece may be pushed into the blade by hand. Mechanical (Figure 24) or mechanical-hydraulic feeding mechanisms are

Figure 22. Vertical band machine worktable can be tilted 10 degrees left (Courtesy of the California Community Colleges—Individualized Machinist's Curriculum Project).

Figure 21. General purpose vertical band machine (Courtesy of the DoAll Company).

Figure 23. Vertical band machine worktable tilted 45 degrees right (Courtesy of the California Community Colleges—Individualized Machinist's Curriculum Project).

also used. A band machine may be equipped with a hydraulic tracing attachment. This accessory uses a stylus contacting a template or pattern. The tracing accessory guides the workpiece during the cut (Figure 25).

Band Machines with Power-Fed Worktables Heavier construction is used on these machine tools. The worktable is moved hydraulically. The operator is relieved of the need to push the workpiece into the cutting blade. The direction of the cut can be guided by a steering mechanism (Figure 26). A roller chain wraps around the workpiece and passes over a sprocket at the back of the worktable. The sprocket is connected to a steering wheel at the front of the worktable. The operator can then guide the workpiece and keep the saw cutting along the proper lines. The workpiece rests on

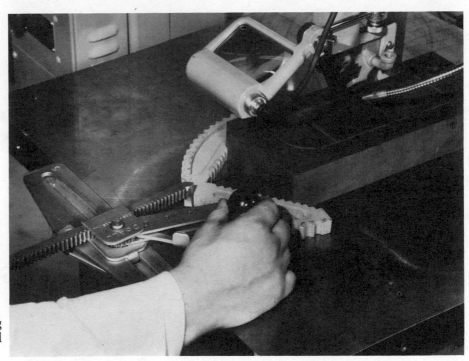

Figure 24. Mechanical work feeding mechanism (Courtesy of the DoAll Company).

Figure 25. Hydraulic tracing accessory (Courtesy of the DoAll Company).

roller bearing stands. These permit the workpiece to turn freely while it is being steered.

Figure 26. Heavy duty vertical band machine with power-fed worktable (Courtesy of the DoAll Company).

Figure 27. High tool velocity vertical band machine (Courtesy of the DoAll Company).

High Tool Velocity Band Machines On the high tool velocity band machine (Figure 27), band speeds can range as high as 10 to 15,000 feet per minute (FPM). These machine tools are used in many band machining applications. They are frequently found cutting nonmetal products. These include applications such as trimming plastic laminates (Figure 28) and cutting fiber materials (Figure 29).

Figure 28. Trimming plastic laminates on the high tool velocity band machine (Courtesy of the DoAll Company).

Figure 29. Cutting fiber material on the high tool velocity band machine (Courtesy of the DoAll Company).

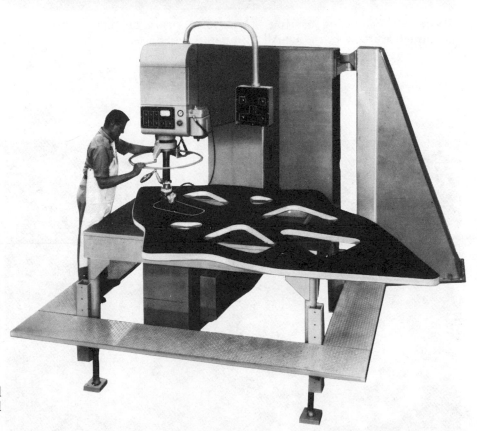

Figure 30. Large capacity vertical band machine (Courtesy of the DoAll Company).

Large Capacity Band Machines This type of band machine is used on large workpieces. The entire saw is attached to a swinging column. The workpiece remains stationary and the saw is moved about to accomplish the desired cuts (Figure 30).

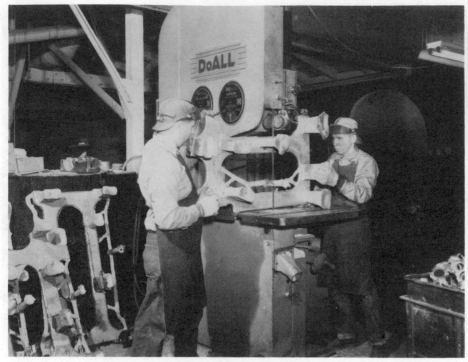

Figure 32. Production trimming of castings on the vertical band machine (Courtesy of the DoAll Company).

Figure 31. Trimming casting sprues and risers on the vertical band machine (Courtesy of the DoAll Company).

Figure 34. Trimming stainless steel wire brushes by friction sawing (Courtesy of the DoAll Company).

Figure 33. Ripping on the vertical band machine (Courtesy of the DoAll Company).

APPLICATIONS OF THE VERTICAL BAND MACHINE
Conventional and Contour Sawing

Vertical band machines are used in many conventional sawing applications. They are found in the foundry trimming sprues and risers from castings. The band machine can accommodate a large casting and make widely spaced cuts (Figure 31). Production trimming of castings is easily accomplished with the high tool velocity band machine (Figure 32). Band saws are also useful in ripping operations (Figure 32). In the machine shop, the vertical band machine is used in general purpose, straight line, and contour cutting mainly in sheet and plate stock.

Friction Sawing

Friction sawing can be used to cut materials that would be impossible or very difficult to cut by other means. In friction sawing, the workpiece is heated by friction created between it and the cutting blade. The blade melts its way through the work. Friction sawing can be used to cut hard materials such as files. Tough materials such as stainless steel wire brushes can be trimmed by friction sawing (Figure 34).

Band Filing and Band Polishing

The band file consists of file segments attached to a spring steel band (D) (Figure 35). As the band file passes through the work an interlock closes and keeps the file segment tight (B). The interlock then releases, permitting the file segment to roll around the band wheel. A space is provided for chip clearance between the band and file segment (C). The band file has a locking slot so that the ends can be joined to form a continuous loop (A). Special guides are required for both file and polishing bands. Band files can be used in both internal (Figure 36) and external (Figure 37) filing applications. They are also used in applications such as filing large gear teeth to shape and size (Figure 38).

In band polishing, a continuous abrasive strip is used (Figure 39). The grit of the abrasive can be varied depending on the surface finish desired.

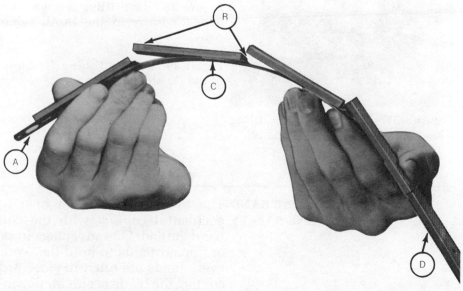

Figure 35. Band file (Courtesy of the DoAll Company).

Figure 36. Internal band filing (Courtesy of the DoAll Company).

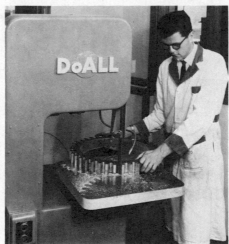

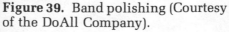

Figure 38. Band filing a large spur gear (Courtesy of the DoAll Company).

Figure 39. Band polishing (Courtesy of the DoAll Company).

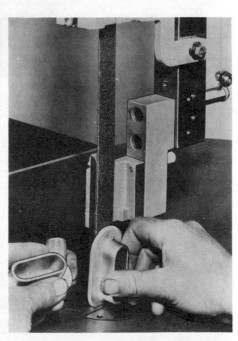

Figure 37. External band filing (Courtesy of the DoAll Company).

**VERTICAL BAND
MACHINE SAFETY** The **primary danger** in operating the vertical band machine is **accidental contact with the cutting blade.** Workpieces are often hand guided. One advantage in sawing machines is that the pressure of the cut tends to hold the workpiece against the saw table. However, hands are often in close proximity to the blade. If you should contact the blade accidentally, an injury is almost sure to occur. You

will not have time even to think about withdrawing your fingers before they are cut. Keep this in mind at all times when operating a band saw.

Always use a **pusher** against the workpiece whenever possible. This will keep your fingers away from the blade. Be careful as you are about to complete a cut. As the blade clears through the work, the pressure that you are applying is suddenly released and your hand or finger could be carried into the blade. As you approach the end of the cut, **reduce the feeding pressure** as the blade cuts through.

The vertical band machine is generally not used to cut round stock. This can be extremely hazardous and should be done on the horizontal band machine, where round stock can be secured in a vice. Hand-held round stock will turn if it is cut on the vertical band machine. This can cause an injury and may damage the blade as well. If round stock must be cut on the vertical band saw, it must be clamped securely in a vise, vee block, or other suitable workholding fixture.

Be sure to select the proper blade for the sawing requirements. Install it properly and apply the correct blade tension. Band tension should be rechecked after a few cuts. New blades will tend to stretch to some degree during their break-in period. Band tension may have to be readjusted.

The entire blade must be guarded except at the point of the cut. This is effectively accomplished by enclosing the wheels and blade behind guards that are easily opened for adjustments to the machine. Wheel and blade guard must be closed at all times during machine operation. The guidepost guard moves up and down with the guidepost (Figure 40). The operator is protected from an exposed blade at this point. For maximum safety, set the guidepost $\frac{1}{8}$ to $\frac{1}{4}$ in. above the workpiece.

Band machines may have one or two idler wheels. On machines with two idler wheels, a short blade running over only one wheel may be used. Under this condition an additional blade guard at the left side of the wheel is required (Figure 41). This guard is removed when operating over two idler wheels as the blade is then behind the wheel guard.

Roller blade guides are used in friction and high speed sawing. A roller guide shield is used to provide protection for the operator (Figure 42). Depending on the material being cut, the entire cutting area may be enclosed. This would apply to the cutting of hard, brittle materials such as granite and glass. Diamond blades are frequently used in cutting these materials. The clear shield protects the operator while permitting him to view the operation. Cutting fluids are also prevented from spilling on the floor. In any sawing operation making use of cutting fluids, see that they do not spill on the floor around the machine. This creates an extremely dangerous situation, not only for you but for others in the shop as well.

Gloves should not be worn around any machine tool. An exception to this is for friction or high speed sawing or when handling band blades. Gloves will protect hands from the sharp saw teeth. If you wear gloves during friction or high speed sawing, be extra careful that they do not become entangled in the blade or other moving parts.

Figure 41. Left side blade guard when using a short blade over one idler wheel (California State University at Fresno).

Figure 40. Guidepost guard (California State University at Fresno).

Figure 42. Roller blade guide shield (Courtesy of the California Community Colleges—Individualized Machinist's Curriculum Project).

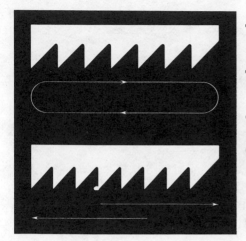

unit 1
using reciprocating and horizontal band cutoff machines

The reciprocating saw and the horizontal endless band saw are the most common cutoff machines that you will encounter. Their primary function is to cut long lengths of material into lengths suitable for other machining operations on other machine tools. The cutoff machine is often the first step in machining a part to its final shape and size. In this unit, you are introduced to saw blades and the applications and operation of these important sawing machines.

objectives After completing this unit, you will be able to:

1. Use saw blade terminology.
2. Describe the conditions that define blade selection.
3. Identify the major parts of the reciprocating and horizontal band cutoff machine.
4. Properly install blades on reciprocating and horizontal band machines.
5. Properly use reciprocating and horizontal band machines in cutoff applications.

CUTTING SPEEDS

An understanding of cutting speeds is one of the most important aspects of machining that you will encounter. Many years of machining experience have shown that certain tool materials are most effective if passed through certain workpiece materials at certain speeds. If a tool material passes through the work too quickly, the heat generated by friction can rapidly dull the tool or cause it to fail completely. Too slow a passage of the tool through a material can result in premature dulling and low productivity.

A cutting speed refers to the amount of workpiece material that passes by a cutting tool in a given amount of time. Cutting speeds are measured in feet per minute. This is abbreviated FPM. In some machining operations, the tool can pass the work. Sawing is an example. The work may pass the tool as in the lathe. In both cases, FPM is the same. The shape of the workpiece does not affect the FPM. The circumference of a round part passing a cutting tool is still in FPM. In later units, FPM is discussed in terms of revolutions per minute of a round workpiece.

In sawing, FPM is simply the speed of each saw tooth as it passes through a given length of material in one minute. If one tooth of a band saw passes through one foot of material in one minute, the cutting speed is one foot per minute. This is true of reciprocating saws as well. However, remember that this saw only cuts in one direction of the stroke. Cutting speeds are a critical factor in tool life. Productivity will be low if the sawing machine is stopped most of the time because a dull or damaged blade must be replaced frequently. The additional cost of replacement cutting tools must also be considered. In any machining operation, always keep cutting speeds in mind.

On a sawing machine, blade FPM is a function of RPM (revolutions per minute) of the saw drive. That is, the setting of a specific RPM on the saw drive will produce a specific FPM of the blade. Feet per minute is also related to the material being cut. Generally, hard, tough materials have low cutting speeds. Soft material have higher cutting speeds. In sawing, cutting speeds are affected by the material, size, and cross section of the workpiece.

SAW BLADES

The blade is the cutting tool of the sawing machine. In any sawing operation, at least three teeth on the saw blade must be in contact with the work at all times. This means that thin material requires a blade with more teeth per inch, while thick material can be cut with a blade having fewer teeth per inch. You should be familiar with the terminology of saw blades and saw cuts.

BLADE MATERIALS. Saw blades for reciprocating and band saws are made from carbon steels and

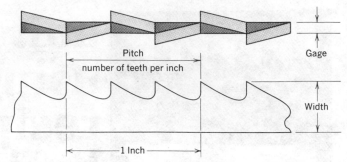

Figure 2. Gage, pitch, and width.

high speed alloy steels. Blades may also have tungsten carbide tipped teeth.

BLADE KERF. The kerf of a saw cut is the width of the cut as produced by the blade (Figure 1).

BLADE WIDTH. The width of a saw blade is the distance from the tip of the tooth to the back of the blade (Figure 2).

BLADE GAGE. Blade gage is the thickness of the blade (Figure 2). Reciprocating saw blades on large machines can be as thick as .250 in. Common band saw blades are .025 to .035 in. thick.

BLADE PITCH. The pitch of a saw blade is the number of teeth per inch (Figure 2). An eight pitch blade has eight teeth per inch (a tooth spacing of $\frac{1}{8}$ in.).

SAW TEETH

You should be familiar with saw tooth terminology (Figure 3).

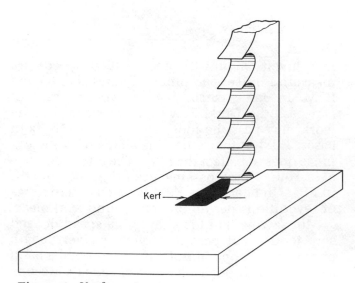

Figure 1. Kerf.

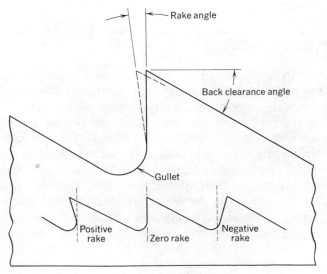

Figure 3. Saw tooth terminology.

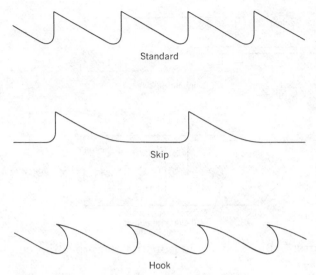

Figure 4. Tooth forms.

TOOTH FORMS. Tooth form is the shape of the saw tooth. Saw tooth forms are either standard, skip, or hook (Figure 4). Standard form gives accurate cuts with a smooth finish. Skip tooth gives additional chip clearance. Hook form provides faster cutting because of the positive rake angle.

SET. The teeth of a saw blade must be offset on each side to provide clearance for the back of the blade. This offset is called set (Figure 5). Set is equal on both sides of the blade. The set dimension is the total distance from the tip of a tooth on one side to the tip of a tooth on the other side.

SET PATTERNS. Set forms include raker, straight, and wave (Figure 5). Raker and wave are the most common. Raker set is used in general sawing. Wave set is useful where the cross-sectional shape of the workpiece varies.

SELECTING A BLADE FOR RECIPROCATING AND BAND SAWS

Blade selection will depend upon the material, thickness, and cross-sectional shape of the workpiece. Some band cutoff machines have a job selector (Figure 6). This will aid you greatly in selecting the proper blade for your sawing requirement. On a machine without a job selector, analyze the job and then select a suitable blade. For example, if you must cut thin tube, a fine pitch blade will be needed so that three teeth are in contact with the work. A particularly soft material may require a zero rake angle tooth form. Sawing through a workpiece with changing cross section may require a blade with wavy set to provide maximum accuracy.

USING CUTTING FLUIDS

Cutting fluids are an extremely important aid to sawing. The heat produced by the cutting action can become so great that the metallurgical structure of the blade teeth can be affected. Cutting fluids will dissapate much of this heat and greatly prolong the life of the blade. Besides their function as a coolant, they also lubricate

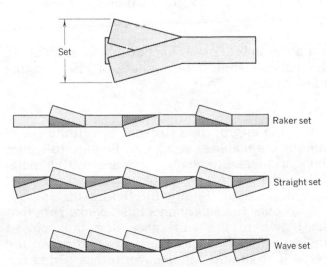

Figure 5. Set and set patterns.

Figure 6. Job selector on a horizontal band cutoff machine (Courtesy of the DoAll Company).

Figure 7. Reciprocating cutoff saw (Courtesy of Kasto-Racine, Inc.).

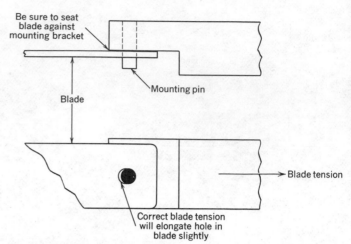

Figure 8. Blade mounting on the reciprocating cut-off saw.

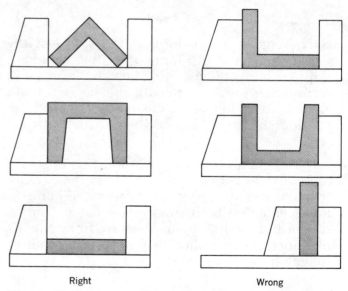

Right   Wrong

Figure 9. Cutting workpieces with sharp corners.

the blade. Sawing with cutting fluids will produce a smoother finish on the workpiece. One of the most important functions of a cutting fluid is to transport chips out of the cut. This allows the blade to work more efficiently. Common cutting fluids are oils, oils dissolved in water or soluble oils, and synthetic chemical cutting fluids

OPERATING THE RECIPROCATING CUTOFF MACHINE

The reciprocating cutoff machine (Figure 7) is often known as the power hacksaw. This machine is an outgrowth of the hand hacksaw. Basically, the machine consists of a frame supported blade that is operated in a back-and-forth motion. Power hacksaws may be hydraulically driven or driven by a crankshaft mechanism.

Installing the Blade on the Power Hacksaw

Obtain a blade of the correct length and make sure that the teeth are pointed in the direction of the cut. This will be on the back stroke. Make sure that the blade is seated against the mounting plates (Figure 8). Apply the correct tension. The blade may be tightened until a definite ring is heard when the blade is tapped. Do not overtighten the blade, as this may cause the

pin holes to break out. If a new blade is installed, the tension should be rechecked after making a few cuts.

Making the Cut

Select the appropriate strokes per minute speed rate for the material being cut. Be sure to secure the workpiece properly. If you are cutting material with a sharp corner, begin the cut on a flat side if possible (Figure 9). Before making the cut, go over the safety checklist. Make sure that the length of the workpiece does not exceed the capacity of the stroke. This can break the frame if it should hit the workpiece. Bring the saw gently down until the blade has a chance to start cutting. Apply the proper feed. On reciprocating saws, feed is regulated with a sliding weight or feeding mechanism. If chips produced in the cut are blue, too much feed is being used. The blade will be damaged rapidly. Very fine powder like chips indicate too little pressure. This will dull the blade. If the blade is replaced after starting a cut, turn the workpiece over and begin a new cut (Figure 10). Do not attempt to saw through the old cut. This will damage the new blade. After a new blade has been used for a short time, recheck the tension and adjust if necessary.

OPERATING THE HORIZONTAL BAND CUTOFF MACHINE

The horizontal band cutoff machine (Figure 11) is the most common stock cutoff machine found in the machine shop. This machine tool uses an endless steel band blade with teeth on one edge. Since the blade passes through the work continuously, there is no wasted motion. Cutting efficiency is greatly increased over the reciprocating saw.

The kerf from the band blade is quite narrow as compared to the reciprocating hacksaw or abrasive saw (Figure 12). This is an added advantage in that minimum amounts of material are wasted in the sawing operation.

The size of the horizontal band saw is determined by the largest piece of square material that can be cut. Speeds on the horizontal band machine may be set by manual belt change (Figure 13), or a variable speed drive may be used. The variable speed drive permits an in-

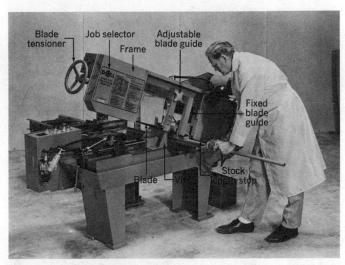

Figure 11. Horizontal endless band cutoff machine (Courtesy of the DoAll Company).

Figure 12. Comparison of kerf widths from band, reciprocating, and abrasive cutoff machines (Courtesy of the DoAll Company).

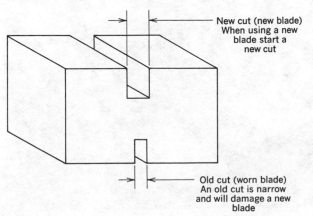

Figure 10. If the blade is changed, begin a new cut on the other side of the workpiece.

finite selection of band speeds within the capacity of the machine. Cutting speeds can be set precisely. Many horizontal band machines are of the hinge design. The saw head, containing the drive and idler wheels, hinges around a point at the rear of the machine.

Figure 13. Changing speeds by shifting belts on a horizontal band saw (Courtesy of the DoAll Company).

The saw head may be raised and locked in the up position while stock is being placed into or removed from the machine (Figure 14). Feeds are accomplished by gravity of the saw head. The feed rate can be regulated by adjusting the spring tension on the saw head (Figure 15).

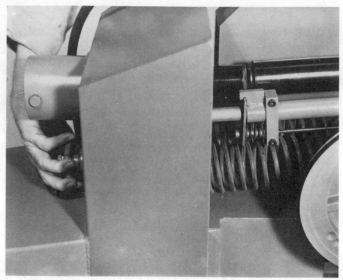

Figure 15. Adjusting the head tension on the horizontal band saw (Courtesy of the DoAll Company).

Some sawing machines use a hydraulic cylinder to regulate the feed rate. The head is held in the up position by the cylinder. A control valve permits oil to flow into the reservoir as the saw head descends. This permits the feed rate to be regulated.

Cutting fluid is pumped from a reservoir and flows on the blade at the forward guide. Additional fluid is permitted to flow on the blade at the point of the cut (Figure 16). The saw shown does not have the now required full blade guards installed. Cutting fluid flow

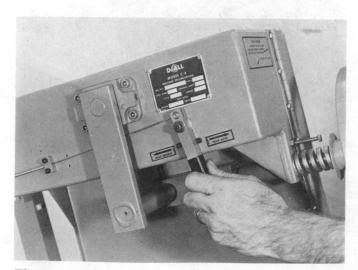

Figure 14. Horizontal band saw head release lever (Courtesy of the DoAll Company).

Figure 16. Coolant system on the horizontal band saw (Courtesy of the DoAll Company).

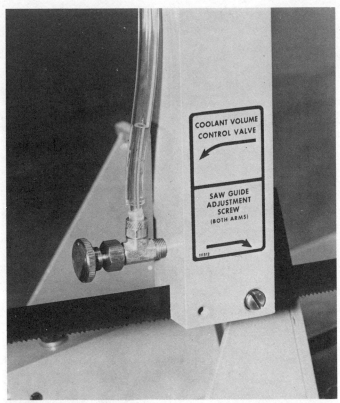

Figure 17. Coolant volume control valve on the horizontal band saw (Courtesy of the DoAll Company).

Figure 19. Setting the stock length gage (Courtesy of the DoAll Company).

Figure 18. Rotary chip brushes on the band saw blade (Courtesy of the DoAll Company).

is controlled by a control valve (Figure 17). Chips can be cleared from the blade by a rotary brush that operates as the blade runs (Figure 18).

Common accessories used on many horizontal band saws include workpiece length measuring equipment (Figure 19) and roller stock tables. The stock table shown has a hinged

section that permits the operator to reach the rear of the machine (Figure 20).

Workholding on the Horizontal Band Saw
The vise is the most common workholding fixture. Rapid adjusting vises are very popular (Figure 21). These vises have large capacity and are quickly adjusted to the workpiece. After the vise jaws have contacted the workpiece, the vise is locked by operating the lock handle. The vise may be swiveled for miter or angle cuts (Figure 22). On some horizontal band cutoff machines, the entire saw frame swivels for making angle cuts (Figure 23).

The horizontal band saw is often used to cut several pieces of material at once. Stock may be held or nested in a special vise or nesting fixture (Figure 24).

Installing Blades on the Horizontal Band Machine
Blades for the horizontal band saw may be ordered prewelded in the proper length for the

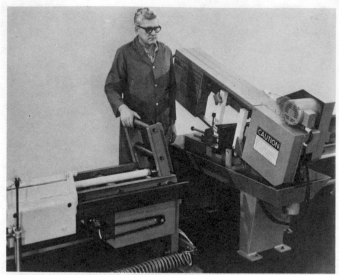

Figure 20. Roller stock table with access gate (Courtesy of the DoAll Company).

Figure 21. Quick setting vise on the horizontal band saw (Courtesy of the DoAll Company).

Figure 22. Band saw vise swiveled for angle cutting (Courtesy of the DoAll Company).

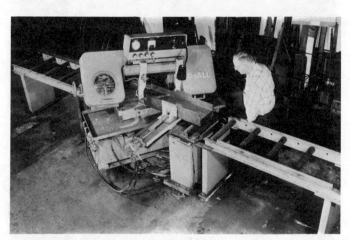

Figure 23. Horizontal cutoff saw frame swiveled for angle cuts (Courtesy of the DoAll Company).

Figure 24. "Nesting fixture" for sawing multiple workpieces (Courtesy of the DoAll Company).

machine. Band blade may also be obtained in rolls. The required length is then cut and welded at the sawing machine location.

To install the band, shut off power to the machine and open the wheel guards. Release the tension by turning the tension wheel. Place the blade around the drive and idler wheels. Be sure that the teeth are pointed in the direction of the cut. This will be toward the rear of the machine. See that the blade is tracking properly on the idler and drive wheels (Figure 25). The blade will have to be twisted slightly to fit the guides. Guides should be adjusted so that they have .001 to .002 in. clearance with the

Figure 25. Make sure that the blade is tracking properly on the band wheel (California State University at Fresno).

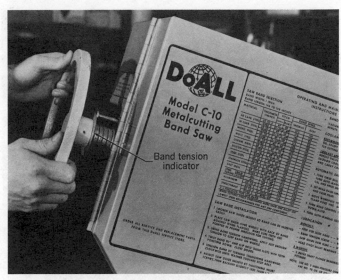

Figure 27. Band tension indicator attached to the machine (Courtesy of the DoAll Company).

blade. Adjust the blade tension using the tension gage (Figure 26) or the manual tension indicator built into the saw (Figure 27). The blade is tightened until the flange on the tension wheel contacts the tension indicator stop.

Figure 28. Blade guide should be set as close to the workpiece as possible (California State University at Fresno).

Making the Cut

Set the proper speed according to the blade type and material to be cut. If the workpiece has sharp corners, it should be positioned in the same manner as in the reciprocating saw (Figure 9). The blade guides must be adjusted so that they are as close to the work as possible (Figure 28). This will insure maximum blade support and maximum accuracy of the cut. Sufficient feed should be used to produce a good chip. Excessive feed can cause blade failure. Too little feed can dull the blade prematurely. Go over the safety checklist for horizon-

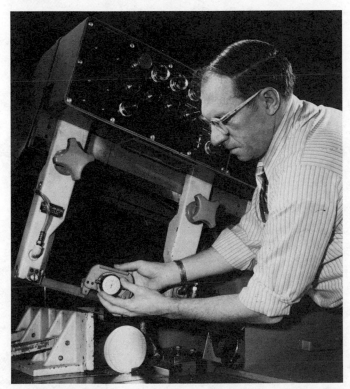

Figure 26. Dial band tension indicator gage (Courtesy of the DoAll Company).

Figure 29. Automatic shutoff mechanism (California State University at Fresno).

Figure 30. Stock length stop in position (California State University at Fresno).

Figure 31. Stock length stop must be removed before beginning the cut (California State University at Fresno).

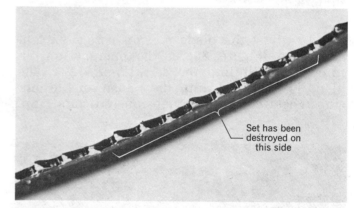

Figure 32. Band saw blade with the set worn on one side (Courtesy of California Community Colleges IMC Project).

(Figure 29). When the cut is completed, the machine will shut off automatically.

SAWING PROBLEMS ON THE HORIZONTAL BAND MACHINE

You may use the stock stop to gage the length of duplicate pieces of material (Figure 30). It is important to swing the stop clear of the work after the vise has been tightened and before the cut is begun (Figure 31). A cutoff workpiece can bind between the stop and blade. This will destroy the blade set (Figure 32). A blade with a tooth set worn on one side will drift in the direction of the side that has a set still remain-

tal band saws. Release the head and lower it by hand until the blade starts to cut. Most saws are equipped with an automatic shutoff switch

Figure 33. Using a band with the set worn on one side will cause the cut to drift toward the side of the blade with the set remaining (Courtesy of California Community Colleges IMC Project).

Figure 34. The blade may drift far enough to damage the vise (Courtesy of California Community Colleges IMC Project).

ing (Figure 33). This is the principal cause of band breakage. As the saw progresses through the cut, the side draft of the blade will place the machine under great stress. The cut may drift so far as to permit the blade to cut into the vise (Figure 34).

Very thin and quite parallel workpieces can be cut by using a sharp blade in a rigid and accurate sawing machine (Figure 35). Chip removal is important to accurate cutting. If chips are not cleared from the blade prior to it entering the guides, the blade will be scored. This will make it brittle and subject to breakage.

Figure 35. Thin and parallel cuts may be made with a sharp blade and rigid sawing machine.

self-evaluation

SELF-TEST

1. Name the most common saw blade set patterns.
2. Describe the conditions that define blade selection.
3. On a reciprocating saw, what is the direction of the cut?
4. What is set and why is it necessary?
5. What are common tooth forms?
6. What can happen if the stock stop is left in place during the cut?
7. What type of cutoff saw will most likely be found in the machine shop?
8. Of what value are cutting fluids?
9. What can result if chips are not properly removed from the cut?
10. If a blade is replaced after a cut has been started, what must be done with the workpiece?

unit 2
abrasive
and
cold saws

The abrasive saw is seldom used for stock cutoff in the machine shop. Small metal chips and abrasive wheel particles produced by the abrasive saw can damage other machine tools. However, the abrasive saw may be used in or around the machine shop or grinding room for the purpose of cutting hardened materials such as drills and endmills. Abrasive saws are very common in fabrication and welding shops. The abrasive saw has an advantage over other cutoff machines in that it can cut a number of nonmetallic materials such as slate, stone, brick, and glass.

The cold saw is seen in the machine shop. This type of cutoff machine uses a circular blade with teeth. Cold saws are useful in stock cutoff where length tolerances must be held as close as possible. You may see and possibly use a cold saw in stock cutoff applications.

objectives **After completing this unit, you will be able to:**
1. **Identify abrasive and cold saws.**
2. **Identify abrasive wheel materials.**
3. **Identify abrasive wheel bonds.**
4. **Describe the operation of abrasive cold sawing machines.**

ABRASIVE SAWING MACHINES

The **abrasive saw** (Figure 1) consists of a high speed motor driven abrasive wheel mounted on a swing arm. Abrasive saws are very fast cutting. The reason for this is that each particle of the abrasive wheel acts to cut a small bit of material. Since the abrasive wheel has a large number of abrasive particles, many "teeth" are involved in the cutting action. Abrasive wheels may operate at cutting speeds of 15 to 20 thousand feet per minute. A large number of abrasive particles delivered past the work at a high foot per minute rate makes for very fast cutting. One disadvantage of the abrasive saw is that common blades cut with a wide kerf. Considerable material may be wasted in the cut. However, some abrasive wheel materials permit a very thin wheel to be used. Abrasive sawing produces a large amount of heat in the workpiece. Cutting fluids act as coolants and are often used in abrasive sawing.

Figure 1. Typical abrasive cutoff machine (California State University at Fresno).

Abrasives

Abrasive saw wheels are made from aluminum oxide and silicon carbide. Aluminum oxide abrasives are used for most metals, including steel. Silicon carbide abrasive is used for non-metallic materials such as stone. Diamond abrasives are used for extremely hard materials, such as glass.

Abrasive Saw Wheel Bonds

The **bond** of an abrasive saw wheel is the material that holds the abrasive particles together. The bond must be strong enough to withstand the large force placed on the wheel at high revolutions per minute. However, the bond must also be able to break down as the wheel cuts. This is necessary to expose new abrasive particles to the cut.

Shellac bond abrasive wheels are suited to cutting hard steels. **Resinoid bond** wheels are suitable for cutting structural shapes and bar stock. **Rubber bond** wheels produce clean cuts when used with cutting fluids. Rubber bond wheels can be made as thin as .006 in.

Abrasive Saw Wheel Speeds

Abrasive saw wheel speeds vary as to the diameter of the wheel. A 12 in. diameter wheel can be operated at about 14,000 surface feet per minute. Twenty-inch wheels should be run at about 12,000 surface feet per minute. Surface feet is the speed of the wheel as measured at the circumference. Recommended wheel speeds will be marked on each wheel. **Abrasive saw wheels must not be operated at speeds faster than recommended.**

Selecting Abrasive Saws

The following factors affect the selection of an abrasive saw.

1. Material to be cut.
2. Size of material.
3. Cut to be made dry or with cutting fluid.
4. Degree of finish desired and acceptable burr (sharp edge left by the saw).

Quality of Abrasive Saw Cuts

Excessive heat generated in the cut may discolor the material or possibly affect metallurgical properties. This problem may be solved by selection of a different wheel. The use of a cutting fluid will produce a smooth cut with less of a burr. In some wet abrasive saw applications, the workpiece and part of the blade may be totally submerged in the cutting fluid. Fluids include a soda solution or plain water.

Abrasive Saw Feed Rates

An extremely light feed will heat the wheel excessively. Heavy feeding may result in excessive wheel wear or the wheel may break causing an extreme hazard. Feed rates depend on many factors, including the material to be cut, wheel type, speeds, and the condition of the abrasive sawing machine.

OPERATING THE ABRASIVE SAWING MACHINE

Inspect the abrasive wheel for chips and cracks. Be sure that it is rated at the proper RPM for the machine. Workholding on the abrasive saw may be a simple vise or a quick setting produc-

Figure 2. Quick setting chain vise (California State University at Fresno).

Figure 3. Chain vise operating mechanism (California State University at Fresno).

tion fixture such as an air operated chain vise. Here the chain passes over the workpiece and a link hooks into the chain catch (Figure 2). The chain is tightened by a winder operated by an air cylinder (Figure 3).

Go over the safety checklist for abrasive saws. Start the machine and bring the abrasive wheel down until the cut begins. Use a feed rate that is suitable for the material being cut and the wheel being used (Figure 4). The abrasive saw may be swiveled to the side for miter cuts (Figure 5).

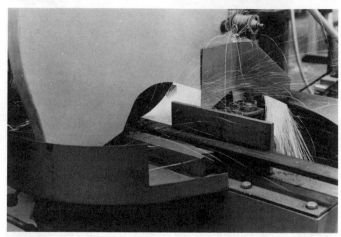

Figure 4. Making a cut with the abrasive saw (California State University at Fresno).

Figure 5. The abrasive saw may be swiveled for angle cutting (California State University at Fresno).

COLD SAWS

The **cold saw** (Figure 6) uses a circular metal blade with teeth. Cold saws are very useful in precision cutoff applications where length tolerance must be held as close as possible. A cold saw blade on a precision cutoff machine may be 7 to 8 in. in diameter with a thickness of .040 to .080 in. The blade cuts a narrow kerf. A minimum amount of material is wasted in the cut. This is important when cutting expensive materials. Length tolerance and parallelism of the cut can be held to plus or minus .002 in. To obtain this degree of accuracy, hold the stock on both sides of the blade. A precision cold saw blade may be operated at speeds of 25 to 1200 RPM, depending on the size, shape, and material of the workpiece. Cold saws can be used to make straight and angle cuts in material with different cross sections (Figure 6). Cold saws with blade diameters of about 24 inches are also used for stock cutoff in machine and fabrication shops.

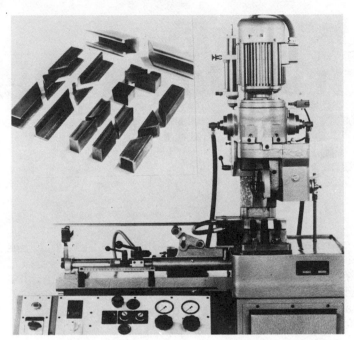

Figure 6. The precision cold saw can make straight and angle cuts in materials with different cross sections (Courtesy of Ameropean Industries, Inc.).

self-evaluation

SELF-TEST 1. Are you likely to find an abrasive saw in a machine shop?
2. What are the advantages of the abrasive saw?
3. Name two types of abrasives.
4. Name two types of abrasive wheel bonds.
5. What is one advantage of a precision cold saw?
This unit has no post-test.

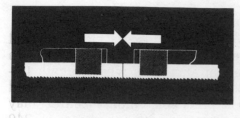

unit 3
preparing to use the vertical band machine

A machine tool can perform at maximum efficiency only if it has been properly maintained, adjusted, and set up. Before the vertical band machine can be used for a sawing or other band machining operation, several important preparations must be made. These include welding saw blades into bands and making several adjustments on the machine tool.

objectives After completing this unit, you will be able to:
1. Weld band saw blades.
2. Prepare the vertical band machine for operation.

Welding Band Saw Blades

Band saw blade is frequently supplied in rolls. The required length is measured and cut and the ends are welded together to form an endless band. Most band machines are equipped with a band welding attachment. These are frequently attached to the machine tool. They may also be separate pieces of equipment (Figure 1).

The **band welder** is a resistance-type butt welder. They are often called **flash** welders because of the bright flash and shower of sparks created during the welding operation. The metal in the blade material has a certain resistance to the flow of an electric current. This resistance causes the blade metal to heat as the electric current flows during the welding operation. The blade metal is heated to a temperature that permits the ends to be forged together under pressure. When the forging temperature is reached, the ends of the blade are pushed together by mechanical pressure. They fuse, forming a resistance weld. The band weld is then annealed or

softened and dressed to the correct thickness by grinding.

Welding band saw blades is a fairly simple operation and you should master it as soon as possible. Blade welding is frequently done in the machine shop. New blades are always being prepared. Sawing operations, where totally enclosed workpiece features must be cut, require that the blade be inserted through a starting hole in the workpiece and then welded into a band. After the enclosed cut is made, the blade is broken apart and removed.

Preparing the Blade for Welding.
The first step is to cut the required length of blade stock for the band machine that you are using. Blade stock can be cut with snips or with the band shear (Figure 2). Many band machines have a blade shear near the welder. The required length of blade will usually be marked on the saw frame. Blade length, B_L, for two wheel sawing machines can be calculated by the formula

324

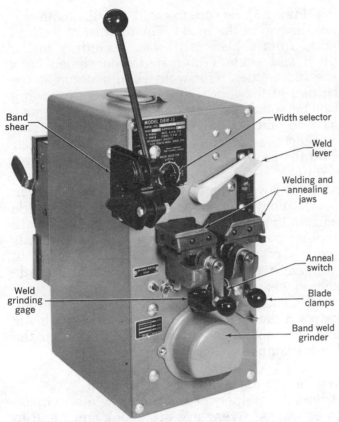

Figure 1. Band blade welder (Courtesy of the DoAll Company).

Figure 2. Blade shear (California State University at Fresno).

$$B_L = \pi D + 2L$$

where D is the diameter of the band wheel and L is the distance between band wheel centers. Set the tension adjustment on the idler wheel about midrange so that the blade will fit after welding. Most machine shops will have a permanent reference mark, probably on the floor, that can be used for gaging blade length.

After cutting the required length of stock, the ends of the blade must be ground so that they are square when positioned in the welder. Place the ends of the blade together so that the teeth are opposed (Figure 3). Grind the blade ends in this position. The grinding wheel on the blade welder may be used for this operation. Blade ends may also be ground on the pedestal grinder (Figure 4). Grinding the blade ends with the teeth opposed will insure that the ends of the blade are square when the blade is positioned in the welder. Any small error in grinding will be canceled when the teeth are placed in their normal position.

Proper grinding of the blade ends permits correct tooth spacing to be maintained. After the

Figure 3. Placing the blade ends together with the teeth opposed (California State University at Fresno).

blade has been welded, the tooth spacing across the weld should be the **same** as any other place on the band. Tooth set should be aligned as well. A certain amount of blade material is consumed

Figure 4. End grinding the blade on the pedestal grinder (California State University at Fresno).

in the welding process. Therefore, the blade must be ground correctly if tooth spacing is to be maintained. The amount consumed by the welding process may vary with different blade welders. You will have to determine this by experimentation. For example, if 1/4 in. of blade length is consumed in welding, this would amount to about one tooth on a four pitch blade. Therefore, one tooth should be ground from the blade. This represents the amount lost in weld-

ing (Figure 5). Be sure to grind only the tooth and not the end of the blade. The number of teeth to grind from a blade will vary according to the pitch and amount of material consumed by a specific welder. The weld should occur at the bottom of the tooth **gullet.** Exact tooth spacing can be somewhat difficult to obtain. You may have to practice end grinding and welding several pieces of scrap blades until you are familiar with the proper welding and tooth grinding procedure.

The jaws of the blade welder should be clean before attempting any welding. Postion the blade ends in the welder jaws (Figure 6). The saw teeth should point toward the back. This prevents scoring of the jaws when welding blades of different widths. A uniform amount of blade should extend from each jaw. The blade ends must contact squarely in the center of the gap between the welder jaws. Be sure that the blade ends are not offset or overlapped. Tighten the blade clamps.

Welding the Blade into an Endless Band

Adjust the welder for the proper width of blade to be welded. **Wear eye protection** and stand to one side of the welder during the welding operation. Depress the weld lever. A flash with a shower of sparks will occur (Figure 7). In this brief operation, the movable jaw of the welder

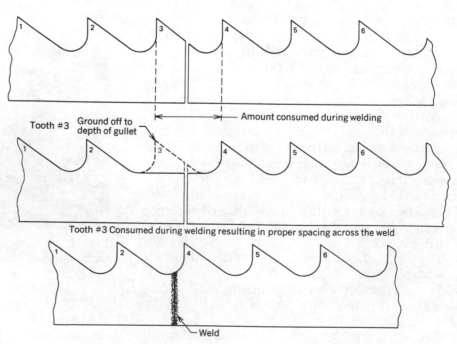

Figure 5. The amount of blade lost in welding

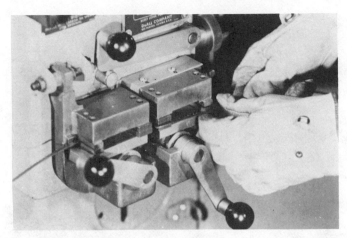

Figure 6. Placing the blade in the welder (Courtesy of the California Community Colleges — Individualized Machinist's Curriculum Project).

Figure 7. Welding the blade into a band (Courtesy of the DoAll Company).

Figure 8. Weld flash should be evenly distributed after welding (California State University at Fresno.)

tion, the weld must be **annealed** or **softened.** This improves strength qualities of the weld. Place the band in the annealing jaws with the teeth pointed out (Figure 9). This will concentrate annealing heat away from the saw teeth. A small amount of compression should be placed on the movable welder jaw prior to clamping the band. This permits the jaw to move as the annealing heat expands the band.

It is most important not to overheat the weld during the annealing process. Overheating can destroy an otherwise good weld, causing it to become brittle. The correct annealing temperature is determined by the color of the weld zone

moved toward the stationary jaw. The blade ends were heated to forging temperature by a flow of electric current, and the molten ends of the blades were pushed together, forming a solid joint.

The blade clamps should be loosened before releasing the weld lever. This prevents scoring of the welder jaws by the now welded band. A correctly welded band will have the weld **flash** evenly distributed across the weld zone (Figure 8). Tooth spacing across the weld should be the same as the rest of the band.

Annealing the Weld
The metal in the weld zone is hard and brittle immediately after welding. For the band to func-

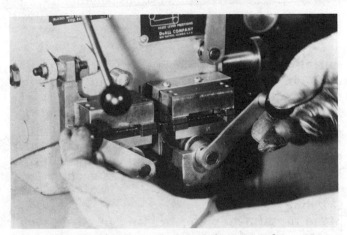

Figure 9. Positioning the band for annealing (Courtesy of the California Community Colleges — Individualized Machinist's Curriculum Project).

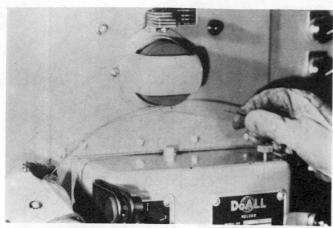

Figure 10. Grinding the band weld (Courtesy of the California Community Colleges — Individualized Machinist's Curriculum Project).

Figure 12. Band weld thickness gage (Courtesy of the California Community Colleges — Individualized Machinist's Curriculum Project).

Figure 11. The saw teeth must not be ground while grinding the band weld (California State University at Fresno).

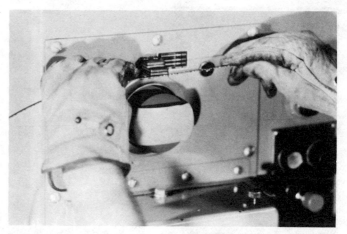

Figure 13. Gaging the weld thickness in the grinding gage (Courtesy of the California Community Colleges — Individualized Machinist's Curriculum Project).

during annealing. This should be a dull red color. Depress the anneal switch and watch the band heat. When the dull red color appears, release the anneal switch immediately and let the band begin to cool. As the weld cools, depress the anneal switch briefly several times to slow the cooling rate. Too rapid cooling can result in a band weld that is not properly annealed.

Grinding the Weld

Some machinists prefer to grind the band weld prior to annealing. This permits the annealing color to be seen more easily. More often, the weld is ground after the annealing process. However,

it is good practice to anneal the blade weld further after grinding. This will eliminate any hardness induced during the grinding operation. The grinding wheel on the band welder is designed for this operation. The top and bottom of the grinding wheel are exposed so that both sides of the weld can be ground (Figure 10). **Be careful not to grind the teeth** when grinding a band weld. This will destroy the tooth set. Grind the band weld evenly on both sides (Figure 11). The weld should be ground to the same thickness as the rest of the band. If the weld area is ground thinner, the band will be weakened at that point. As you grind, check the band thickness in the gage (Figure 12) to determine proper thickness (Figure 13).

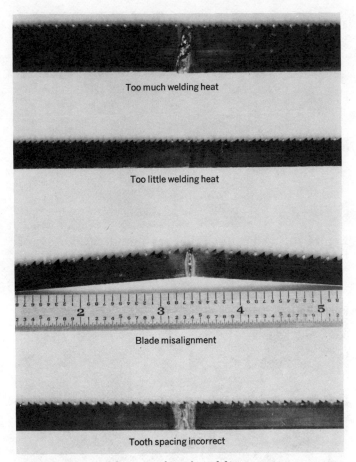

Figure 14. Problems in band welding.

Figure 15. Band guides must fully support the band but must not extend over the saw teeth (Courtesy of the California Community Colleges — Individualized Machinist's Curriculum Project).

PROBLEMS IN BAND WELDING

Several problems may be encountered in band welding (Figure 14). These include misaligned pitch, blade misalignment, insufficient welding heat, or too much welding heat. You should learn to recognize and avoid these problems. The best way to do this is to obtain some scrap blades and practice the welding and grinding operations.

INSTALLING AND ADJUSTING BAND GUIDES ON THE VERTICAL BAND MACHINE

Band guides must be properly installed if the band machine is to cut accurately and if damage to the band is to be prevented. Be sure to use the **correct width guides** for the band (Figure 15). The band must be fully supported except for the teeth. Using wide band guides with a narrow band will destroy tooth set as soon as the machine is started.

Band guides are set with a **guide setting gage.** Install the right-hand band guide and

tighten the lock screw just enough to hold the guide insert in place. Place the setting gage in the left guide slot and adjust the position of the right guide insert so that it is in contact with both the vertical and diagonal edges of the gage (Figure 16). Check the **backup bearing** at this time. Clear any chips that might prevent it from turning freely. If the backup bearing cannot turn freely, it will be scored by the band and damaged permanently.

Install the right-hand guide insert and make the adjustment for band thickness using the same setting gage (Figure 17). The thickness of the band will be marked on the tool. Be sure that this is the same as the band that will be used. The lower band guide is adjusted in a like manner. Use the setting gage on the same side as it was used when adjusting the top guides (Figure 18).

Roller band guides are used in high speed sawing applications where band velocities exceed 2000 FPM. They are also used in friction sawing operations. The roller guide should be adjusted so that it has .001 to .002 in. clearance with the band (Figure 19).

ADJUSTING THE COOLANT NOZZLE

A band machine may be equipped with flood or mist coolant. Mist coolant is liquid coolant mixed with air. Certain sawing operations may require only small amounts of coolant. With the mist system, liquid coolant is conserved and is less likely to spill on the floor. When cutting with flood coolant, be sure that the runoff returns

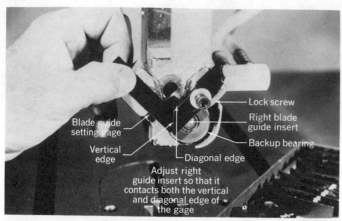

Figure 16. Using the saw guide setting gage (Courtesy of the California Community Colleges — Individualized Machinist's Curriculum Project).

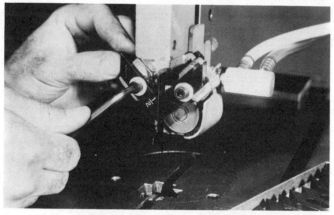

Figure 17. Adjusting the band guides for band thickness (Courtesy of the California Community Colleges — Individualized Machinist's Curriculum Project).

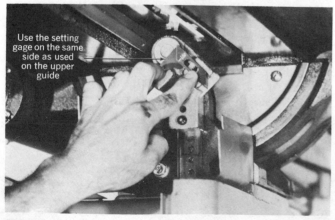

Figure 18. When adjusting the lower guide, use the setting gage on the same side as on the upper guide (Courtesy of the California Community Colleges — Individualized Machinist's Curriculum Project).

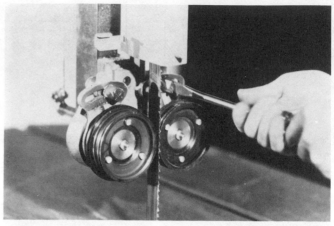

Figure 19. Adjusting roller band guides (Courtesy of the California Community Colleges — Individualized Machinist's Curriculum Project).

Figure 20. Coolant may be introduced directly ahead of the band (Courtesy of the DoAll Company).

Figure 21. Mist and flood coolant nozzles (Courtesy of the California Community Colleges — Individualized Machinist's Curriculum Project).

to the reservoir and does not spill on the floor. Flood coolant may be introduced directly ahead of the band (Figure 20).

Flood or mist coolant may be introduced through a nozzle in the upper guidepost assembly (Figure 21). Air and liquid are supplied to the inlet side of the nozzle by two hoses (Figure 22). The coolant nozzle must be installed (Figure 23) and preset (Figure 24) prior to installing the band. For mist coolant set the nozzle end $\frac{1}{2}$ in. from the face of the band guide. The setting for flood is $\frac{3}{8}$ in.

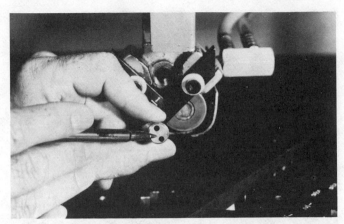

Figure 22. Inlet side of the coolant nozzle (Courtesy of the California Community Colleges — Individualized Machinist's Curriculum Project).

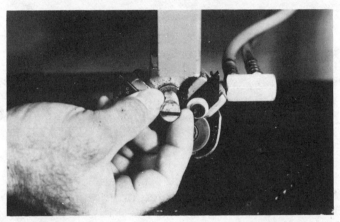

Figure 23. Installing the coolant nozzle (Courtesy of the California Community Colleges — Individualized Machinist's Curriculum Project).

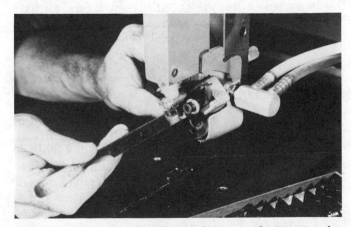

Figure 24. Presetting the coolant nozzle position before installing the band guides (Courtesy of the California Community Colleges — Individualized Machinist's Curriculum Project).

INSTALLING THE BAND ON THE VERTICAL BAND MACHINE

Open the upper and lower wheel covers and remove the filler plate for the worktable. It is **safer to handle the band with gloves to protect your hands** from the sharp saw teeth. The hand tension crank is attached to the upper idler wheel (Figure 25). Turn the crank to lower the wheel to a point where the band can be placed around the drive and idler wheels. Be sure to install the band so the teeth point in the direction of the cut. This is always in a **down direction toward the worktable.** If the saw teeth seem to be pointed in the wrong direction, the band may have to be turned inside out. This can be done easily. Place the band around the drive and idler wheels and turn the tension crank so that tension is placed on the band. Be sure that the band slips into the upper and lower guides properly. Replace the filler plate in the worktable.

Adjusting Band Tension
Proper **band tension** is important to accurate cutting. A high tensile strength band should be used whenever possible. Tensile strength refers to the strength of the band to withstand stretch. The correct band tension is indicated on the **band tension dial** (Figure 26). Adjust the tension for the width of band that you are using. After a new band has been run for a short time, recheck the tension. New bands tend to stretch during their initial running period.

Adjusting Band Tracking
Band tracking refers to the position of the band as it runs on the idler wheel tires. On the vertical band machine, the idler wheel can be tilted to adjust tracking position. The band tracking position should be set so that the back of the band just

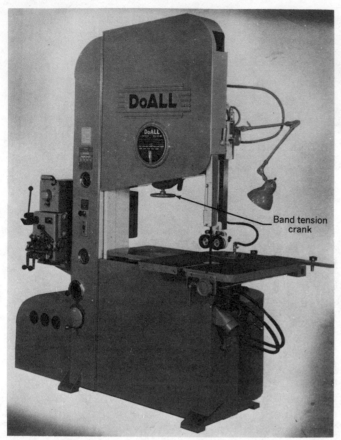

Figure 25. Band tension crank (Courtesy of the DoAll Company).

Figure 26. Band tension dial (Courtesy of the DoAll Company).

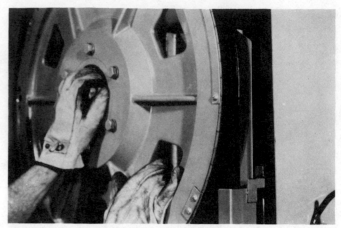

Figure 27. Adjusting the band tracking position by tilting the idler wheel (Courtesy of the California Community Colleges — Individualized Machinist's Curriculum Project).

touches the backup bearing in the guide assembly. Generally, you will not have to adjust band tracking very often. After you have installed a blade, check the tracking position. If it is incorrect, consult your instructor for help in adjusting the tracking position.

The tracking adjustment is made with the motor off and the speed range transmission in neutral. This permits the band to be rolled by hand. Two knobs are located on the idler wheel hub. The outer knob (Figure 27) tilts the wheel. The inner knob is the tilt lock. Loosen the lock knob and adjust the tilt of the idler wheel while rolling the band by hand. When the correct tracking position is reached, lock the inner knob. If the band machine has three idler wheels, adjust band tracking on the top wheel first. Then adjust tracking position on the back wheel.

OTHER ADJUSTMENTS ON THE VERTICAL BAND MACHINE

The hub of the variable speed pulley should be lubricated weekly (Figure 28). While the drive mechanism guard is open, check the oil level in the speed range transmission. The band machine may be equipped with a chip brush on the band wheel (Figure 29). This should be adjusted frequently. Chips that are transported through the band guides can score the band and make it brittle. The hydraulic oil level should be checked daily on band machines with hydraulic table feeds (Figure 30).

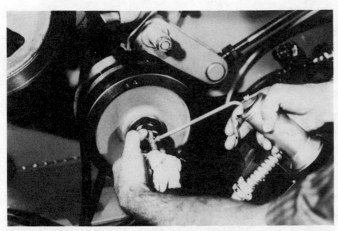

Figure 28. Lubricating the variable speed pulley hub (Courtesy of the California Community Colleges — Individualized Machinist's Curriculum Project).

Figure 29. Band wheel chip brush (Courtesy of the California Community Colleges — Individualized Machinist's Curriculum Project).

Figure 30. Checking the hydraulic oil level on the vertical band machine (Courtesy of the California Community Colleges — Individualized Machinist's Curriculum Project).

self-evaluation

SELF-TEST
1. Describe the blade end grinding procedure.
2. Describe the band welding procedure.
3. Describe the weld grinding procedure.
4. What is the purpose of the band blade guide?
5. Why is it important to use a band guide of the correct width?
6. What tool can be used to adjust band guides?
7. What is the function of annealing the band weld?
8. Describe the annealing process.
9. What is band tracking?
10. How is band tracking adjusted?

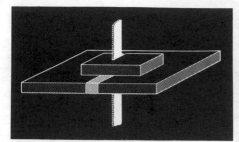

unit 4
using the vertical band machine

After a machine tool has been properly adjusted and set up, it can be used to accomplish a machining task. In the preceding unit, you had an opportunity to prepare the vertical band machine for use. In this unit, you will be able to operate this versatile machine tool.

objectives After completing this unit, you will be able to:
1. Use the vertical band machine job selector.
2. Operate the band machine controls.
3. Perform typical sawing operations on the vertical band machine.

SELECTING A BLADE
FOR THE VERTICAL BAND MACHINE

Blade materials include standard carbon steel where the saw teeth are fully hardened but the back of the blade remains soft. The standard carbon steel blade is available in the greatest combination of width, set, pitch, and gage.

The carbon alloy steel blade has hardened teeth and also a hardened back. The harder back permits sufficient flexibility of the blade but, because of increased tensile strength, a higher band tension may be used. Because of this, cutting accuracy is greatly improved. The carbon alloy blade material is well suited to contour sawing.

High speed steel and bimetallic high speed steel blade materials are used in high production and severe sawing applications where blades must have long wearing characteristics. The high speed steel blade can withstand much more heat than the carbon or carbon alloy materials. On the bimetallic blade, the cutting edge is made from one type of high speed steel, while the back is made from another type of high speed steel that has been selected for high flexibility and high tensile strength. High speed and bimetallic high speed blades can cut longer, faster, and more accurately.

Band blade selection will depend on the sawing task. You should review saw blade terminology discussed in the cutoff machine unit. The first consideration is blade pitch. The pitch of the blade should be such that at least two teeth are in contact with the workpiece. This generally means that fine pitch blades with more teeth per inch will be used in thin materials. Thick material requires coarse pitch blades so that chips will be more effectively cleared from the kerf.

Remember that there are three tooth **sets** that can be used (Figure 1). **Raker** and **wave** set are the most common in the metalworking industries. **Straight** set may be used for cutting thin materials. Wave set is best for accurate cuts through materials with variable cross sections. Raker set may be used for general purpose sawing.

You also have a choice of **tooth forms** (Figure 2). **Precision** or **regular** tooth form is best for accurate cuts where a good finish may be required. **Hook** form is fast cutting but leaves a rougher finish. **Skip** tooth is useful on deep cuts where additional chip clearance is required.

Several special bands are also used. **Straight, scalloped,** and **wavy edges** are used for cutting

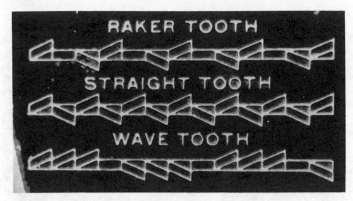

Figure 1 Blade set patterns (California State University at Fresno).

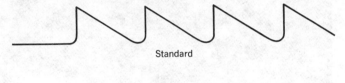

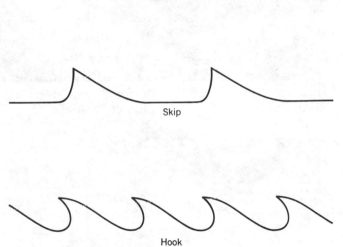

Figure 2. Saw tooth forms.

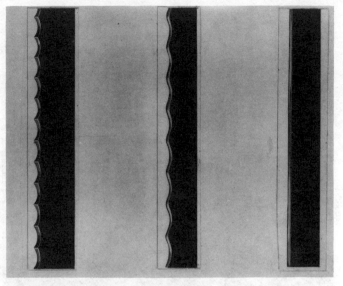

Figure 3. Straight, scalloped, and wavy edge bands (Courtesy of the DoAll Company).

Figure 4. *(left)* Continuous edge diamond band (Courtesy of the DoAll Company).

Figure 5. *(right)* Segmented diamond edge band (Courtesy of the DoAll Company.)

nonmetallic substances where saw teeth would tear the material (Figure 3). **Continuous** (Figure 4) and **segmented** (Figure 5) **diamond edged band** is used for cutting very hard nonmetallic materials.

USING THE JOB SELECTOR ON THE VERTICAL BAND MACHINE

Most vertical band machines are equipped with a **job selector.** This device will be of great aid to you in accomplishing a sawing task. Job selectors are usually attached to the machine tool. They are frequently arranged by material. The material

to be cut is located on the rim of the selector. The selector disk is then turned until the sawing data for the material can be read (Figure 6).

The job selector yields much valuable information. Sawing velocity in feet per minute is the most important. The band must be operated at the correct cutting speed for the material. If it is not, the band may be damaged or productivity will be low. Saw velocity is read at the top of the column and is dependent on material thickness. The job selector also indicates recommended pitch, set, feed, and temper. The job selector will provide information on sawing of nonmetallic materials (Figure 7). Information on band filing can also be determined from the job selector.

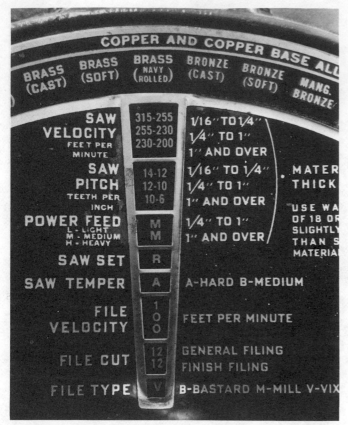

Figure 6. Job selector on the vertical band machine (California State University at Fresno).

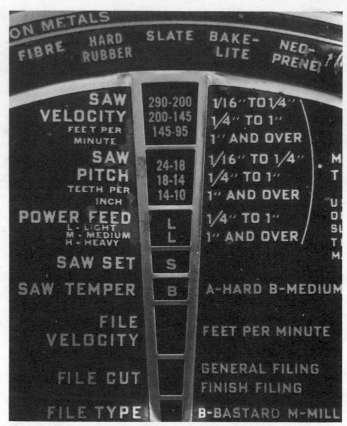

Figure 7. The job selector set for a nonferrous material (California State University at Fresno).

SETTING SAW VELOCITY
ON THE VERTICAL BAND MACHINE

Most vertical band machines are equipped with a **variable speed drive** that permits a wide selection of band velocities. This is one of the factors that make the band machine such a versatile machine tool. Saw velocities can be selected that permit successful cutting of many materials.

The typical variable speed drive uses a split flange pulley to vary the speed of the drive wheel (Figure 8). As the flanges of the pulley are spread apart by adjusting the speed control, the belt runs deeper in the pulley groove. This is the same as running the drive belt on a smaller diameter pulley. Slower speeds are obtained. As the flanges of the pulley flanges are adjusted for less spread, the belt runs toward the outside. This is equivalent to running the belt on a larger diameter pulley. Faster speeds are obtained.

Setting Band Velocity

Band velocity is indicated on the **band velocity indicator** (Figure 9). Remember that band veloc-

Figure 8. Vertical band machine variable speed drive (Courtesy of the DoAll Company).

ity is measured in **feet per minute.** The inner scale indicates band velocity in the low speed range. The outer scale indicates velocity in the high speed range. Band velocity is regulated by adjusting the speed control (Figure 10). Adjust this control only while the motor is running, as

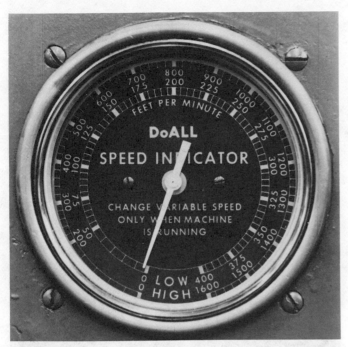

Figure 9. Band velocity indicator (California State University at Fresno).

Figure 10. Speed range and band velocity controls (California State University at Fresno).

this adjustment moves the flanges of the variable speed pulley.

Setting Speed Ranges

Most band machines with a variable speed drive have a **high** and **low speed range.** High or low speed range is selected by operating the **speed range shift lever** (Figure 10). This setting must be made while the band is stopped or is running at the lowest speed in the range. If the machine is set in high range and it is desired to go to low range, turn the band velocity control wheel until the band has slowed to the lowest speed possible. The speed range shift may now be changed to low speed. If the machine is in low speed and it is desired to shift to high range, slow the band to the lowest speed before shifting speed ranges. A speed range shift made while the band is running at a fast speed may damage the speed range transmission gears. Some band machines are equipped with an interlock to prevent speed range shifts except at low band velocity (Figure 11).

STRAIGHT CUTTING ON THE VERTICAL BAND MACHINE

Adjust the upper guidepost so that it is as close to the workpiece as possible (Figure 12). This will maximize safety by properly supporting and guarding the band. Accuracy of the cut will also be aided. The guidepost is adjusted by loosening the clamping knob and moving the post up or down according to the workpiece thickness.

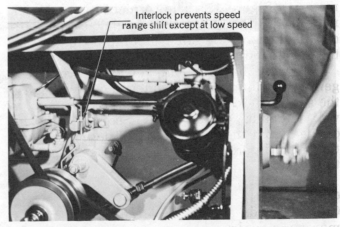

Figure 11. An interlock prevents shifting speed ranges except at low speed (Courtesy of the California Community Colleges—Individualized Machinist's Curriculum Project).

Figure 12. Adjusting the upper guide post (Courtesy of the DoAll Company).

Be sure to use a band of the **proper pitch** for the thickness of the material to be cut. If the band pitch is too fine, the teeth will clog (Figure 13). This can result in stripping and breakage of the saw teeth due to overloading (Figure 14). Cutting productivity will also be reduced. Slow cutting will result from using a fine pitch band on thick material (Figure 15). The correct pitch for thick material (Figure 16) results in much more efficient cutting in the same amount of time and at the same feeding pressure.

As you begin a cut, feed the workpiece **gently** into the band. A sudden shock will cause the saw teeth to chip or fracture (Figure 17). This will reduce band life quickly. See that chips are cleared from the band guides. These can score the band (Figure 18), making it brittle and subject to breakage.

Cutting Fluids
Cutting fluids are an important aid to sawing many materials. They **cool** and **lubricate** the band and **remove chips** from the kerf. Many band machines are equipped with a mist coolant system. Liquid cutting fluids are mixed with air to

Figure 13. Using too fine a pitch blade results in clogged teeth (Courtesy of the California Community Colleges — Individualized Machinist's Curriculum Project).

Figure 14. Stripped and broken teeth resulting from overloading the saw (Courtesy of the DoAll Company).

form a mist. With mist, the advantages of the coolant are realized without the need to collect and return large amounts of liquids to a reservoir. If your band machine uses mist coolant, set the liquid flow first and then add air to create a mist (Figure 19). Do not use more coolant than is

Figure 15. Saw cut with a fine pitch blade in thick material (Courtesy of the California Community Colleges — Individualized Machinist's Curriculum Project).

Figure 16. Saw cut with correct pitch band for thick material (Courtesy of the California Community Colleges — Individualized Machinist's Project).

Figure 17. Chipped and fractured teeth resulting from shock and vibration (Courtesy of the DoAll Company).

necessary. Overuse of air may cause a mist fog around the machine. This is both unpleasant and hazardous, as coolant mist should not be inhaled.

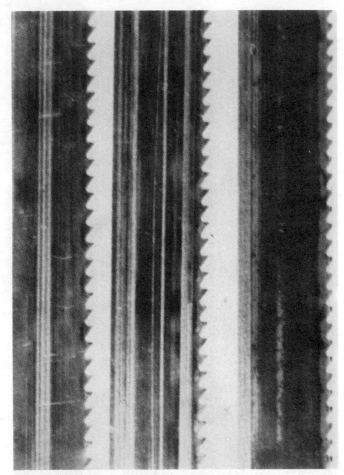

Figure 18. Scored bands can become brittle and loose flexibility (Courtesy of the DoAll Company).

CONTOUR CUTTING
ON THE VERTICAL BAND MACHINE

Contour cutting is the ability of the band machine to cut around corners and produce intricate shapes. The ability of the saw to cut a specific radius depends on its **width**. The job selector will provide information on the minimum radius that can be cut with a blade of a given width (Figure 20). As you can see, a narrow band can cut a smaller radius than a wide band.

The set of the saw needs to be adequate for the corresponding band width. It is a good idea to make a test contour cut in a piece of scrap material. This will permit you to determine if the **saw set** is adequate to cut the desired radius. If the saw set is not adequate, you may not be able to keep the saw on the layout line as you complete the radius cut (Figure 21).

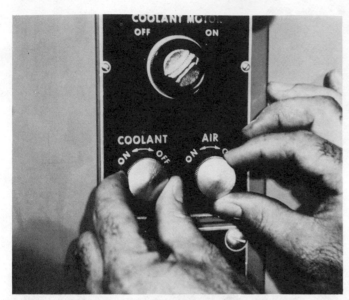

Figure 19. Adjusting the coolant and air mix on the vertical band machine (Courtesy of the California Community Colleges — Individualized Machinist's Curriculum Project).

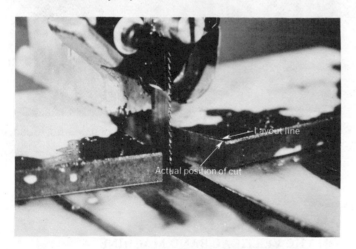

If you are cutting a totally enclosed feature, be sure to insert the saw blade through the starting hole in the workpiece before welding it into a band. Also, be sure that the teeth are pointed in the right direction.

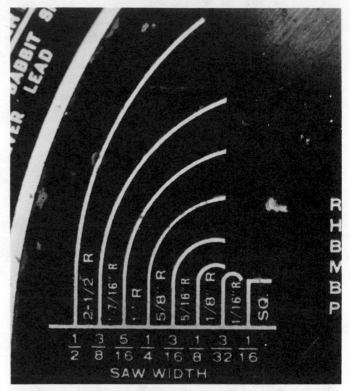

Figure 20. Minimum radius per saw width chart on the job selector (California State University at Fresno).

Figure 21. Band set must be adequate for the band width if the layed out radius is to be cut (Courtesy of the California Community Colleges —Individualized Machinist's Curriculum Project).

self-evaluation

SELF-TEST
1. Name three saw blade sets and describe the applications of each.
2. When might scalloped or wavy edged bands be used?
3. What information is found on the job selector?
4. In what units of measure are band velocities measured?
5. Explain the operating principle of the variable speed pulley.
6. Explain the selection of band speed ranges.
7. What machine safety precaution must be observed when selecting speed ranges?
8. Describe the upper guidepost adjustment.
9. What does band pitch have to do with sawing efficiency?
10. What does band set have to do with contouring?

section g
drilling machines

Primitive man drilled small holes in bone and wood with a bow string wrapped around an arrow tipped with a specially shaped flint. A block of wood was used to put pressure on the top of the arrow (Figure 1). Though the techniques are vastly different, the principle used then, pressing a rotating tool into the work, is essentially the same as that used today in modern drilling machines (Figure 2).

Unlike the stone age artisan, a modern machinist must be able to use several types of complex and powerful drilling machines. Safety is the first thing you should learn about drilling machines. Chips are produced in great quantities and must be safely handled. The operator must also be protected from these chips as they fly from the machine as well as from the rotating parts of the machine. It is most important for you to become familiar with the operation and major parts of drilling machines as soon as possible. Because of the great power exerted by these machines, workholding devices must be used to secure the work and to keep the operator safe.

There are three major types of drilling machines found in machine shops. The sensitive drill (Figure 3) is used for light drilling on small parts. The upright drill press (Figure 4) is used for heavy duty drilling. The radial drill press (Figure 5) is used for drilling large, heavy workpieces that are difficult to move.

There are a number of special purpose drilling machines, ranging from the microscopic drilling machine (Figure 6), which can drill a hole smaller than a human hair, to deep hole drilling machines and turret head drills, which are often tape controlled for automatic operation.

The gang drilling machine (Figure 7) is used when several successive operations must be performed on a workpiece. It has several drilling heads mounted over a single table, with each spindle tooled for drilling, reaming, or counterboring. Hand tooling, such as jigs and fixtures, is typically used on gang drilling machines. Tool guidance is provided by the jigs while the workpiece is clamped in the fixture. When one operation is completed, the workpiece is advanced to the next spindle. When a part requires the drilling of many holes, especially ones quite close together, a multispindle drilling machine (Figure 8) is used. This machine has a number of spindles connected to the main spindle through universal joints.

Figure 1. Primitive man with bow drill (Courtesy of Cincinnati Milacron, Inc.).

341

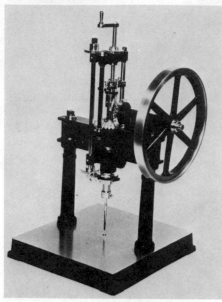

Figure 2. Power feed drill press invented by James Nasmyth of England in 1840. This was the first mechanical powered drilling machine designed for drilling accurate holes (Courtesy of DoAll Company).

Figure 3. The sensitive drill press (Courtesy of Wilton Corporation).

Figure 4. Upright drill press (Courtesy of Giddings & Lewis, Inc.).

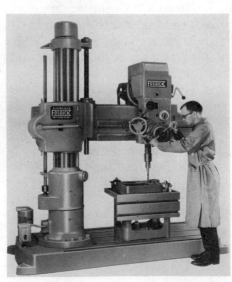

Figure 5. Radial drill press (Courtesy of LeBlond Inc., Cincinnati, Ohio).

Figure 6. Microdrill press for drilling very small holes in miniature and microminiature parts (Courtesy of Louis Levin & Sons, Inc.).

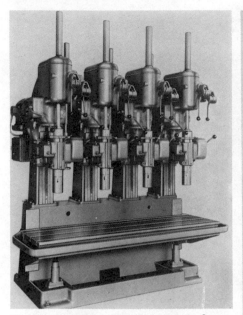

Figure 7. Gang drilling machine (Courtesy of Giddings & Lewis, Inc.).

Figure 9. Turret drilling machine with multispindle heads (Courtesy of Jarvis Products Corporation).

Figure 8. Multispindle drilling machine (Courtesy of Jarvis Products Corporation).

Turret drilling machines (Figure 9) have several drill tools on a turret, which allows a needed tool to be rotated into position for operating on a workpiece. Many of these machines are tape controlled so that the table position, spindle speed, and turret position are programmed to operate automatically. The operator simply clamps the workpiece in position and starts the machine.

Deep hole drilling machines (Figure 10) are used to drill precision holes many times longer than the bore diameter. These machines are usually horizontal with lathe-type ways and drives. The workpiece is held in clamps and is fed into a rotating drill through which coolant is pumped at high pressure. Some machines use gun drills; however, the deep drilling machine has the advantage of a very high metal removal rate.

In the following units, you will learn about tooling, how to use basic drilling machines, and how to select drills and reamers. Drills typically cut rough holes, and reamers are used to finish the holes. Reaming in the drill press is one way of producing a precision hole with a good finish. Countersinking and counterboring are also important tooling operations that will be introduced to you.

Since drills are used on a variety of materials and tend to take relatively heavy cuts, they often get dull or chipped. Sharpening can be done either by hand on a pedestal grinder or in a special

Figure 10. Deep-hole drilling machine (Courtesy of Giddings & Lewis, Inc.).

machine. You will learn these sharpening methods in this section. Resharpening is kept to a minimum when the proper speeds, feeds, and coolants are used.

We have come a long way from the bow string and arrow drill. New ways of making holes in difficult materials are still being developed and used.

DRILLING MACHINE SAFETY There is a tendency among students to dismiss the dangers involved in drilling since most drilling done in a school shop is performed in small sensitive drill presses with small diameter drills. This tendency, however, only increases the danger to the operator and has turned otherwise harmless situations into serious injuries. Clamps should be used to hold down workpieces, since hazards are always present even with small diameter drills.

One example of a safety hazard on even a small diameter drill is the "grabbing" of the drill when it breaks through the hole. If the operator is holding the workpiece with his hands and the piece suddenly begins to spin, his hand will likely be injured, especially if the workpiece is thin.

Even if the operator were strong enough to hold the workpiece, the drill can break, continue turning, and with its jagged edge become an immediate hazard to the operator's hand. The sharp chips that turn with the drill can also cut the hand that holds the workpiece.

Poor work habits produce many injuries. Chips flying into unprotected eyes, heavy tooling, or parts dropping from the drill press onto toes, slipping on oily floors, and getting hair or clothing caught in a rotating drill are all hazards that can be avoided by safe work habits.

The following are safety rules to be observed around all types of drill presses:
1. Tools to be used while drilling should never be left lying on the drill press table, but should be placed on an adjacent worktable (Figure 11).

Figure 11. Drilling operation on upright drill press with tools properly located on adjacent work table (Lane Community College).

2. Get help when lifting heavy vises or workpieces.
3. Workpieces should always be secured with bolts and strap clamps, C-clamps, or fixtures. A drill press vise should be used when drilling small parts (Figure 12). If a clamp should come loose and a "merry-go-round" results, don't try to stop it from turning with your hands. Turn off the machine quickly; if the drill breaks or comes out, the workpiece may fly off the table.
4. Never clean the taper in the spindle when the drill is running, since this practice could result in broken fingers or worse injuries.
5. Always remove the chuck key immediately after using it. A key left in the chuck will be thrown out at high velocity when the machine is turned on. It is a good practice *to never let the chuck key leave your hand when you are using it.* It should not be left in the chuck even for a moment. Some keys are spring loaded so they will automatically be ejected from the chuck when released. Unfortunately, very few of these keys are in use in the industry.

Figure 12. Properly clamped drill press vise holding work for hard steel drilling (DeAnza College).

6. Never stop the drill press spindle with your hand after you have turned off the machine. Sharp chips often collect around the chuck or spindle. Do not reach around, near, or behind a revolving drill.

7. When removing taper shank drills with a drift, use a piece of wood under the drills so they will not drop on your toes. This will also protect the drill points.

8. Interrupt the feed occasionally when drilling to break up the chip so it will not be a hazard and will be easier to handle.

9. Use a brush instead of your hands to clean chips off the machine. Never use an air jet for removing chips as this will cause the chips to fly at a high velocity and cuts or eye injuries may result. Do not clean up chips or wipe up oil while the machine is running.

10. Keep the floor clean. Immediately wipe up any oil that spills, or the floor will be slippery and unsafe.

11. Remove burrs from a drilled workpiece as soon as possible, since any sharp edges or burrs can cause severe cuts.

12. When you are finished with a drill or other cutting tool, wipe it clean with a shop towel and store it properly.

13. Oily shop towels should be placed in a closed metal container to prevent a cluttered work area and avoid a fire hazard.

14. When moving the head or table on sensitive drill presses, make sure a safety clamp is set just below the table or head on the column; this will prevent the table from suddenly dropping if the column clamp is prematurely released.

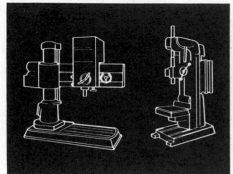

unit 1
the
drill
press

Before operating any machine a machinist must know the names and the functions of all its parts. In this unit, therefore, you should familiarize yourself with the operating mechanisms of several types of drilling machines.

objectives After completing this unit, you will be able to:
1. Identify three basic drill press types and explain their differences and primary uses.
2. Identify the major parts of the sensitive drill press.
3. Identify the major parts of the radial arm drill press.

Drilling holes is one of the most basic of machining operations and one that is very frequently done by machinists. Metal cutting requires considerable pressure of feed on the cutting edge. A drill press provides the necessary feed pressure either by hand or power drive. The primary use of the drill press is to drill holes, but it can be used for other operations such as countersinking, counterboring, spot facing, reaming and tapping, which are processes to modify the drilled hole.

There are three basic types of drill presses used for general drilling operations: the sensitive drill press, the upright drilling machine, and the radial arm drill press. The sensitive drill press (Figure 1), as the name implies, allows the operator to "feel" the cutting action of the drill as he hand feeds it into the work. These machines are either bench or floor mounted. Since these drill presses are used for light duty applications only, they usually have a maximum drill size of $\frac{1}{2}$ in. diameter. Machine capacity is measured by the diameter of work that can be drilled (Figure 2).

The sensitive drill press has four major parts, not including the motor: the head, column, table, and base. Figure 3 labels the parts of the drill press that you should remember. The spindle rotates within the quill, which does not rotate but carries the spindle up and down. The spindle shaft is driven by a stepped-vee pulley and belt (Figure 4) or by a variable speed drive (Figure 5). The *motor must be running* and the spindle turning when changing speeds with a variable speed drive.

The upright drill press is very similar to the sensitive drill press, but it is made for much heavier work (Figure 6). The drive is more powerful and many types are *gear driven*, so they are capable of drilling holes to two inches or more in diameter. The *motor must be stopped* when changing speeds on a gear drive drill press. If it doesn't shift into the selected gear, turn the spindle by hand until it meshes. Since power feeds are needed to drill these large size holes, these machines are equipped with power feed mechanisms that can be adjusted by the operator. The operator may either feed manually with a

Figure 1. A sensitive drill press. These machines are used for light duty application (Courtesy of Wilton Corporation).

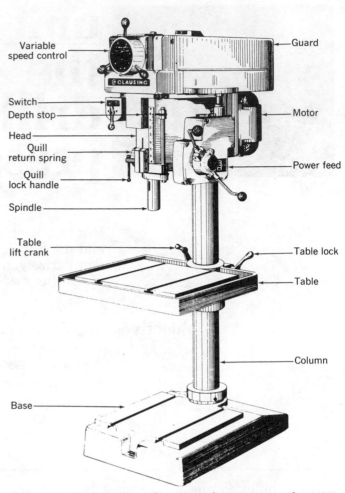

Figure 3. Drill press showing the names of major parts (Courtesy of Clausing Corporation).

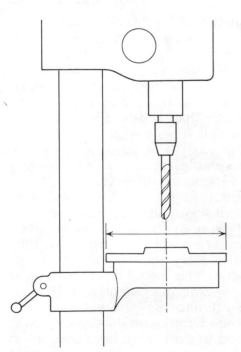

Figure 2. Drill presses are measured by the largest diameter of a circular piece that can be drilled in the center.

Figure 4. View of a vee-belt drive. Spindle speeds are highest when the belt is in the top steps and lowest at the bottom steps (Courtesy of Clausing Corporation).

Figure 5. View of a variable speed drive. Variable speed selector should only be moved when the motor is running. The exact speed choice is possible for the drill size and material with this drive (Courtesy of Clausing Corporation).

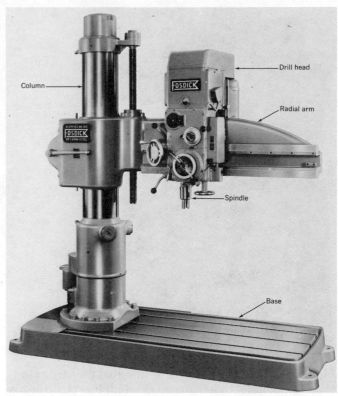

Figure 7. Radial arm drill press with names of major parts (Courtesy of LeBlond Inc., Cincinnati, Ohio).

Figure 6. Upright drill press (Courtesy of Wilton Corporation).

lever or hand wheel or he may engage the power feed. A mechanism is provided to raise and lower the table.

As Figures 7, 8, and 9 show, the radial arm drill press is the most versatile drilling machine. Its size is determined by the diameter of the column and the length of the arm measured from the center of the spindle to the outer edge of the column. It is useful for operations on large castings that are too heavy to be repositioned by the operator for drilling each hole. The work is clamped to the table or base, and the drill can then be positioned where it is needed by swinging the arm and moving the head along the arm. The arm and head can be raised or lowered on the column and then locked in place. The radial arm drill press is used for drilling small to very large holes and for boring, reaming, counterboring, and countersinking. Like the upright machine, the radial arm drill press has a power feed mechanism and a hand feed lever.

Figure 8. Heavy workpiece is mounted on a trunnion-type worktable that can be rotated for positioning (Courtesy of LeBlond Inc., Cincinnati, Ohio).

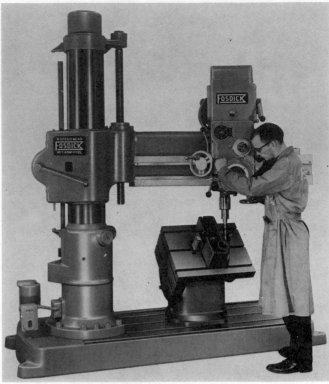

Figure 9. Small holes are usually drilled by hand feeding on a sensitive radial drill. A workpiece is clamped on the tilting table so a hole may be drilled at an angle (Courtesy of LeBlond Inc., Cincinnati, Ohio).

self-evaluation

SELF-TEST 1. List three basic types of drill presses and briefly explain their differences. Describe how the primary uses differ in each of these three drill press types.

2. Sensitive drill press. Match the correct letter with the name of that part shown on Figure 10.

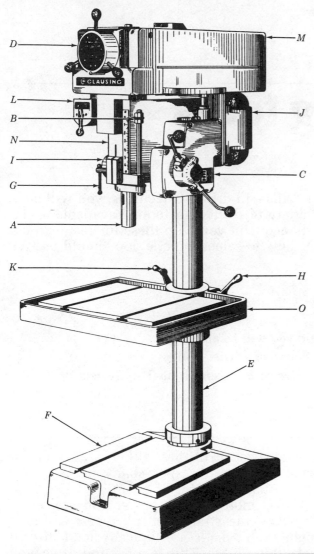

Spindle
Quill lock handle
Column
Switch
Depth stop
Head
Table
Table lock
Base
Power feed
Motor
Variable speed control
Table lift crank
Quill return spring
Guard

Figure 10. (Courtesy of Clausing Corporation).

3. Radial drill press. Match the correct letter with the name of that part

shown on Figure 11.
Column
Radial arm
Spindle
Base
Drill head

Figure 11. (Courtesy of LeBlond, Inc., Cincinnati, Ohio).

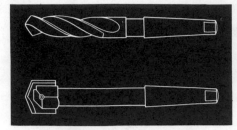

unit 2
drilling tools

Before you learn to use drills and drilling machines, you will have to know of the great variety of drills and tooling available to the machinist. This unit will acquaint you with these interesting tools as well as to show you how to select the one you should use for a given operation.

objectives After completing this unit, you will be able to:
1. Identify the various features of a twist drill.
2. Identify the series and size of 10 given decimal equivalent drill sizes.

The drill is an end cutting rotary-type tool having one or more cutting lips and one or more flutes for the removal of chips and the passage of coolant. Drilling is the most efficient method of making a hole in metals softer than Rockwell 30. Harder metals can be successfully drilled, however, by using special drills and techniques.

In the past, all drills were made of carbon steel and would lose their hardness if they became too hot from drilling. Today, however, most drills are made of high speed steel. High speed steel drills can operate at several hundred degrees Fahrenheit without breaking down and, when cooled, will be as hard as before. Carbide tipped drills are used for special applications such as drilling abrasive materials and very hard steels. Other special drills are made from cast heat-resistant alloys.

Twist Drills
The twist drill is by far the most common type of drill used today. These are made with two or more flutes and cutting lips and in many varieties of design. Figure 1 illustrates several of the most commonly used types of twist drills. The names of parts and features of a twist drill are shown in Figure 2.

The twist drill has either a straight or tapered shank. The taper shank has a Morse taper, a standard taper of about $\frac{5}{8}$ in. per foot, which has more driving power and greater rigidity than the straight shank types. Ordinary straight shank drills are typically held in drill chucks (Figure 3a). This is a friction drive, and slipping of the drill shank is a common problem. Straight shank drills with tang drives have a positive drive and are less expensive than tapered shank drills. These are held in special drill chucks with a Morse taper (Figure 3b).

Jobbers drills have two flutes, a straight shank design, and a relatively short length-to-diameter ratio that helps to maintain rigidity. These drills are used for drilling in steel, cast iron, and nonferrous metals. Center drills and spotting drills are used for starting holes in workpieces. Oil hole drills are made so that coolant can be pumped through the drill to the cutting lips. This not only cools the cutting edges, but also forces out the chips along the flutes. Core drills have from three to six flutes making heavy stock removal possible. They are generally used for roughing holes to a larger diameter or for drilling out cores in castings. Left handed drills are mostly used on multi-spindle drilling machines where some spindles

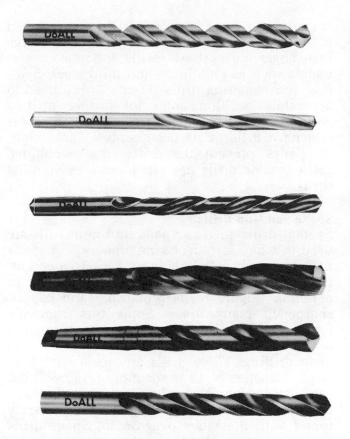

Figure 3a. Drill chuck such as this one are used to hold straight shank drills.

Figure 1. Various types of twist drills used in drilling machines (Courtesy of DoAll Company) :

   (a) High helix drill
   (b) Low helix drill
   (c) Left-hand drill
   (d) Three flute drill
   (e) Taper shank twist drill
   (f) Standard helix jobber drill
   (g) Center or spotting drill

Figure 3b. Tang drive drill chuck will fit into a Morse taper spindle (Courtesy of Illinois/Eclipse, a Division of Illinois Tool Works, Inc., Chicago, Illinois 60639).

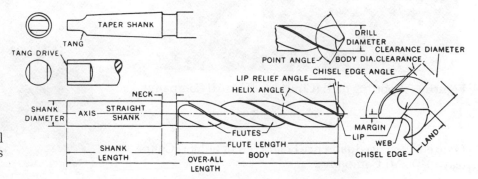

Figure 2. Features of a twist drill (Courtesy of Bendix Industrial Tools Division).

are rotated in reverse of normal drill press rotation. The step drill generally has a flat or an angular cutting edge, and can produce a hole with several diameters in one pass with either flat or countersunk shoulders.

Straight fluted drills are used for drilling brass and other soft materials because the zero rake angle eliminates the tendency for the drill to "grab" on breakthrough. For the same reason they are used on thin materials. Low helix drills, sometimes called slow spiral drills (Figure 4), are more rigid than standard helix drills and can stand more torque. Like straight fluted drills, they are less likely to "grab" when emerging from a hole, because of the small rake angle. For this reason the low helix and straight flute drills are used primarily for drilling in brass, bronze, and some other nonferrous metals. Because of the low helix angle, the flutes do not remove chips very well from deep holes, but the large chip space allows maximum drilling efficiency in shallow holes. High helix drills,

sometimes called fast spiral drills, are designed to remove chips from deep holes. The large rake angle makes these drills suitable for soft metals such as aluminum and mild steel. Spotting and centering drills (Figure 5) are used to accurately position holes for further drilling with regular drills. Centering drills are short and have little or no dead center. These characteristics prevent the drills from wobbling. Lathe center drills are often used as spotting drills.

Spade and Gun Drills
Special drills such as spade and gun drills are used in many manufacturing processes. A spade drill is simply a flat blade with sharpened cutting lips. The spade bit, which is clamped in a holder (Figure 6), is replaceable and can be sharpened many times. Some types provide for coolant flow to the cutting edge through a hole in the holder or shank for the purpose of deep drilling. These drills are made with very large diameters of 12 in. or more (Figure 7) but can also be found as microdrills, smaller than a hair. Twist drills by comparison are rarely found with diameters over $3\frac{1}{2}$ in. Spade drills are usually ground with a flat top rake and with chipbreaker grooves on the end. A chisel edge

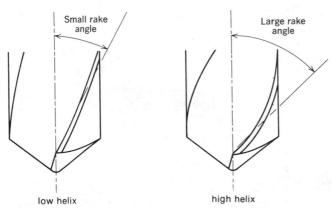

Figure 4. Low and high rake angles on drills.

Figure 5. Spotting drill (Courtesy of DoAll Company).

Figure 6. (right) Spade drill clamped in holder (Courtesy of DoAll Company).

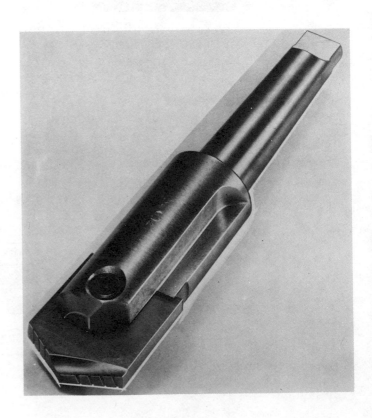

Figure 7. Large hole being drilled with spade drill. This 8 in. diameter hole, $18\frac{7}{8}$ in. deep, was spade drilled in solid SAE #4145 steel rolling mill drive coupling housing with a Brinell hardness of 200–240. The machine that did the job is a 6 ft. 19 in. Chipmaster radial with a 25 hp motor (Courtesy of Giddings & Lewis, Inc.).

Figure 9. Carbide tipped twist drill (Courtesy of DoAll Company).

Figure 8. Spade drill blades showing various grinds (Courtesy of DoAll Company).

and thinned web are ground in the dead center (Figure 8).

Some spade drills are made of solid tungsten carbide, usually only in a small diameter. Twist drills with carbide inserts (Figure 9) require a rigid drilling setup. Gun drills (Figure 10) are also carbide tipped and have a single vee-shaped flute in a steel tube through which coolant is pumped under pressure. These drills are used in horizontal machines that feed the drill with a positive guide. Extremely deep precision holes are produced with gun drills.

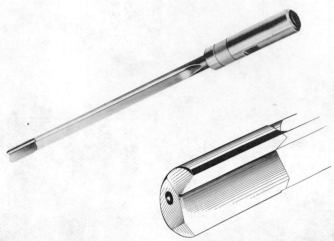

Figure 10. Single flute gun drill with inset of carbide cutting tip (Courtesy of DoAll Company).

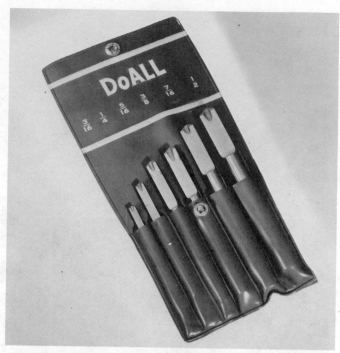

Figure 12. Set of hard steel drills (Courtesy of DoAll Company).

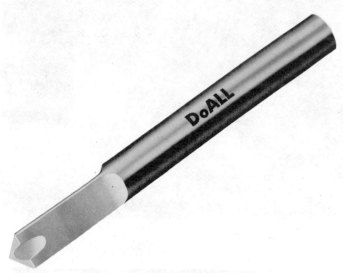

Figure 11. The hard steel drill (Courtesy of DoAll Company).

Another special drill, used for drilling very hard steel, is the Hard Steel Drill (Figure 11). A set of hard steel drills is shown in Figure 12. These drills are cast from a heat-resistant alloy, and the fluted end is ground to a triangular point. These drills work by heating the metal beneath the drill point by friction and then cutting out the softened metal as a chip. (Figures 13 a–d show this drill in use).

Drill Selection
The type of drill selected for a particular task depends upon several factors. The type of machine being used, rigidity of the workpiece, setup and size of the hole to be drilled are all

Figure 13a. Hole started in file with a hard steel drill (DeAnza College).

Figure 13*b*. Drilling the hole (DeAnza College).

important. The composition and hardness of the workpiece are especially critical. The job may require a starting drill or one for secondary operations such as counterboring or spot facing,

Figure 13*c*. Drill removed from file showing chip form (DeAnza College).

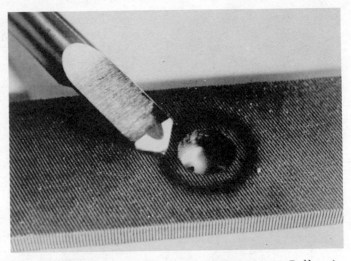

Figure 13*d*. Finished hole in file (DeAnza College).

and it might need to be a drill for a deep hole or for a shallow one. If the drilling operation is too large for the size or rigidity of the machine, there will be chatter and the work surface will be rough or distorted.

A machinist also must make the selection on the size of the drill, the most important dimension of which is the diameter. Twist drills are measured across the margins near the drill point (Figure 14). Worn drills measure slightly smaller here. Drills are normally tapered back along the margin so that they will measure a few thousandths of an inch smaller at the shank.

Drilling is basically a roughing operation. This provides a hole that must be finished by other operations such as boring or reaming. Drills almost never make a hole smaller than their measured diameter, but often make a larger hole depending on how they have been sharpened. There are four drill size series: fractional, number, letter, and metric sizes. The fractional divisions are in $\frac{1}{64}$ in. increments, while the number, letter, and metric series have drill diameters that fall between the fractional inch measures. Together, the four series make up a long-running series in decimal equivalents, as shown in Table 1.

Identification of a small drill is simple enough as long as the number or letter remains on the shank. Most shops, however, have several series of drills, and individual drills often become hard to identify since the markings become worn off by the drill chuck. The machinist must then use a decimal equivalent table such as Table 1. The drill in question is first measured by a micrometer, the decimal

Table 1
Decimal Equivalents for Drills

Decimals of an Inch	Inch	Wire Gage	Milli-meter	Decimals of an Inch	Inch	Wire Gage	Milli-meter
.0135		80		.0469	$\frac{3}{64}$		
.0145		79		.0472			1.2
				.0492			1.25
				.0512			1.3
				.0520		55	
				.0531			1.35
.0156	$\frac{1}{64}$			.0550		54	
.0157			.4	.0551			1.4
.0160		78		.0571			1.45
.0180		77		.0591			1.5
.0197			.5	.0595		53	
.0200		76		.0610			1.55
.0210		75					
.0217			.55				
.0225		74					
.0236			.6				
.0240		73					
.0250		72		.0625	$\frac{1}{16}$		
.0256			.65	.0630			1.6
.0260		71		.0635		52	
.0276			.7	.0650			1.65
.0280		70		.0669			1.7
.0293		69		.0670		51	
.0295			.75	.0689			1.75
.0310		68		.0700		50	
				.0709			1.8
				.0728			1.85
				.0730		49	
				.0748			1.9
.0313	$\frac{1}{32}$			.0760		48	
.0315			.8	.0768			1.95
.0320		67					
.0330		66					
.0335			.85				
.0350		65					
.0354			.9	.0781	$\frac{5}{64}$		
.0360		64		.0785		47	
.0370		63		.0787			2
.0374			.95	.0807			2.05
.0380		62		.0810		46	
.0390		61		.0820		45	
.0394			1	.0827			2.1
.0400		60		.0846			2.15
.0410		59		.0860		45	
.0413			1.05	.0866			2.2
.0420		58		.0886			2.25
.0430		57		.0890		43	
.0433			1.1	.0906			2.3
.0453			1.15	.0925			2.35
.0465		56		.0935		42	

Table 1 (continued)

Decimals of an Inch	Inch	Wire Gage	Milli-meter	Decimals of an Inch	Inch	Wire Gage	Milli-meter
.0938	$\frac{3}{32}$			.1563	$\frac{5}{32}$		
.0945			2.4	.1570		22	
.0960		41		.1575			4
.0966			2.45	.1590		21	
.0980		40		.1610		20	
.0984			2.5	.1614			4.1
.0995		39		.1654			4.2
.1015		38		.1660		19	
.1024			2.6	.1673			4.25
.1040		37		.1693			4.3
.1063			2.7	.1695		18	
.1065		36		.1719	$\frac{11}{64}$		
.1083			2.75	.1730		17	
				.1732			4.4
				.1770		16	
				.1772			4.5
.1094	$\frac{7}{64}$			.1800		15	
.1100		35		.1811			4.6
.1102			2.8	.1820		14	
.1110		34		.1850		13	
.1130		33		.1850			4.7
.1142			2.9	.1870			4.75
.1160		32		.1875	$\frac{3}{16}$		
.1181			3	.1890			4.8
.1200		31		.1890		12	
.1220			3.1	.1910		11	
				.1929			4.9
				.1935		10	
.1250	$\frac{1}{8}$			.1960		9	
.1260			3.2	.1969			5
.1280			3.25	.1990		8	
.1285		30		.2008			5.1
.1299			3.3	.2010		7	
.1339			3.4	.2031	$\frac{13}{64}$		
.1360		29		.2040		6	
.1378			3.5	.2047			5.2
.1405		28		.2055		5	
				.2067			5.25
				.2087			5.3
.1406	$\frac{9}{64}$			.2090		4	
.1417			3.6	.2126			5.4
.1440		27		.2130		3	
.1457			3.7	.2165			5.5
.1470		26		.2188	$\frac{7}{32}$		
.1476			3.75	.2205			5.6
.1495		25		.2210		2	
.1496			3.8	.2244			5.7
.1520		24		.2264			5.75
.1535			3.9	.2280		1	
.1540		23		.2283			5.8

Table 1 (continued)

Decimals of an Inch	Inch	Letter Sizes	Milli-meter	Decimals of an Inch	Inch	Letter Sizes	Milli-meter	Decimals of an Inch	Inch	Milli-meter
.2323			5.9	.3281	21/64			.5469	35/64	
.2340		A		.3307			8.4	.5512		14
.2344	15/64			.3320		Q		.5625	9/16	
.2362			6	.3346			8.5	.5709		14.5
.2380		B		.3386			8.6	.5781	37/64	
.2402			6.1	.3390		R		.5906		15
.2420		C		.3425			8.7	.5938	19/32	
.2441			6.2	.3438	11/32			.6094	39/64	
.2460		D		.3345			8.75	.6102		15.5
.2461			6.25	.3465			8.8	.6250	5/8	
.2480			6.3	.3480		S		.6299		16
.2500	1/4	E		.3504			8.9	.6406	41/64	
.2520			6.4	.3543			9	.6496		16.5
.2559			6.5	.3580		T		.6563	21/32	
.2570		F		.3583			9.1	.6693		17
.2598			6.6	.3594	23/64			.6719	43/64	
.2610		G		.3622			9.2	.6875	11/16	
.2638			6.7	.3642			9.25	.6890		17.5
.2656	17/64			.3661			9.3	.7031	45/64	
.2657			6.75	.3680		U		.7087		18
.2660		H		.3701			9.4	.7188	23/32	
.2677			6.8	.3740			9.5	.7283		18.5
.2717			6.9	.3750	3/8			.7344	47/64	
.2720		I		.3770		V		.7480		19
.2756			7	.3780			9.6	.7500	3/4	
.2770		J		.3819			9.7	.7656	49/64	
.2795			7.1	.3839			9.75	.7677		19.5
.2810		K		.3858			9.8	.7812	25/32	
.2812	9/32			.3860		W		.7874		20
.2835			7.2	.3898			9.9	.7969	51/64	
.2854			7.25	.3906	25/64			.8071		20.5
.2874			7.3	.3937			10	.8125	13/16	
.2900		L		.3970		X		.8268		21
.2913			7.4	.4040		Y		.8281	53/64	
.2950		M		.4063	13/32			.8438	27/32	
.2953			7.5	.4130		Z		.8465		21.5
.2969	19/64			.4134			10.5	.8594	55/64	
.2992			7.6	.4219	27/64			.8661		22
.3020		N		.4331			11	.8750	7/8	
.3031			7.7	.4375	7/16			.8858		22.5
.3051			7.75	.4528			11.5	.8906	57/64	
.3071			7.8	.4531	29/64			.9055		23
.3110			7.9	.4688	15/32			.9063	29/32	
.3125	5/16			.4724			12	.9219	59/64	
.3150			8	.4844	31/64			.9252		23.5
.3160		O		.4921			12.5	.9375	15/16	
.3189			8.1	.5000	1/2			.9449		24
.3228			8.2	.5118			13	.9531	61/64	
.3230		P		.5156	33/64			.9646		24.5
.3248			8.25	.5313	17/32			.9688	31/32	
.3268			8.3	.5315			13.5	.9843		25
								.9844	63/64	

Source: *Bendix Cutting Tool Handbook,* "Decimal Equivalents—Twist Drill Sizes,"
The Bendix Corporation, Industrial Tools Division, 1972.

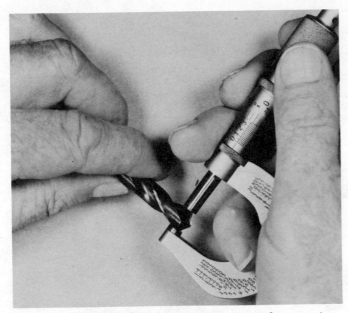

Figure 14. Drill being measured across the margins.

Figure 15. Morse taper drill sleeve (Courtesy of DoAll Company).

Figure 16. Morse taper drill socket (Courtesy of DoAll Company).

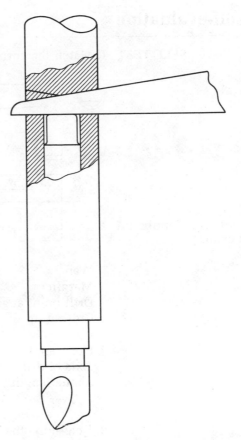

Figure 18. Cutaway of a drill and sleeve showing a drift in place.

reading is located in the table, and the equivalent fraction, number, letter, or metric size is found and noted.

Morse taper shanks on drills and Morse tapers in drill press spindles vary in size and are numbered from 1 to 6; for example, the smaller light duty drill press has a number 2 taper. Steel sleeves (Figure 15) have a Morse taper inside and outside with a slot provided at the end of the inside taper to facilitate removal of the drill shank. A sleeve is used for enlarging the taper end on a drill to fit a larger spindle taper. Steel sockets (Figure 16) function in the reverse manner of sleeves, as they adapt a smaller spindle taper to a larger drill. The drift (Figure 17) is made in several sizes and is used to remove drills or sleeves. The drift is placed round side up, flat side against the drill (Figure 18), and is struck a light blow with a hammer. A block of wood should be placed under the drill to keep it from being damaged and from being a safety hazard.

self-evaluation

SELF-TEST 1. Match the correct letter from Figure 19 to the list of drill parts.

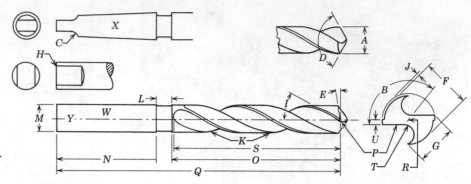

Figure 19. (Courtesy of Bendix Industrial Tools Division).

Web

Margin

Drill point angle

Cutting lip

Flute

Helix angle

Axis of drill

Shank length

Body

Lip relief angle

Land

Chisel edge angle

Body clearance

Tang

Taper Shank

Straight shank

2. Determine the letter, number, fractional, or metric equivalents of the 10 following decimal measurements of drills: .0781, .1495, .272, .159, .1969, .323, .3125, .4375, .201, and .1875.

	Decimal Diameter	Fractional Size	Number Size	Letter Size	Metric Size
a.	.0781				
b.	.1495				
c.	.272				
d.	.159				
e.	.1969				
f.	.323				
g.	.3125				
h.	.4375				
i.	.201				
j.	.1875				

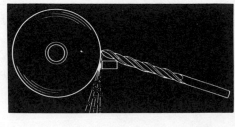

unit 3
hand grinding
of drills
on the
pedestal grinder

Hand sharpening of twist drills has been until recent times the only method used for pointing a drill. Of course, various types of sharpening machines are now in use that can give a drill an accurate point. These precision machines are not found in every shop, however, so it is still necessary for a good machinist to learn the art of off-hand drill grinding.

objective After completing this unit, you will be able to:

Properly hand sharpen a twist drill on a pedestal grinder so it will drill a hole not more than .010 in. times diameter oversize.

One of the advantages of hand grinding drills on the pedestal grinder is that special alterations of the drill point such as web thinning and rake modification can be made quickly. The greatest disadvantage to this method of drill sharpening is the possibility of producing inaccurate, oversize holes (Figure 1). If the drill has been sharpened with unequal angles, the lip with the large angle will do most of the cutting (Figure 1*a*), and will force the opposite margin to cut into the wall of the hole. If the drill has been sharpened with unequal lip lengths, both will cut with equal force, but the drill will wobble and one margin will cut into the hole wall (Figure 1*b*). When both conditions exist (Figure 1*c*), holes drilled may be out of round and oversize. When drilling with the inaccurate points, a great strain is placed on the drill and on the drill press spindle bearings. The frequent use of a drill point gage (Figure 2) during the sharp-

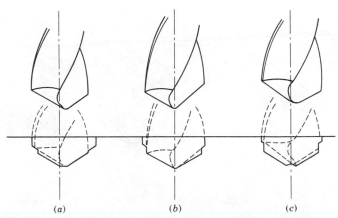

Figure 1. Causes of oversize drilling. (*a*) Drill lips ground to unequal lengths. (*b*) Drill lips ground to unequal angles. (*c*) Unequal angles and lengths.

ening process will help to keep the point accurate and avoid such drilling problems.

363

The web of a twist drill (Figure 3) is thicker near the shank. As the drill is ground shorter a thicker web results near the point. Also the dead center or chisel point of the drill is wider and requires greater pressure to force it into the workpiece, thus generating heat. Web thinning (Figure 4) is one method of narrowing the dead center in order to restore the drill to its original efficiency.

Split-point design (Figure 5) is often used for drilling crankshafts and tough alloy steels. The shape of this point is quite critical and too

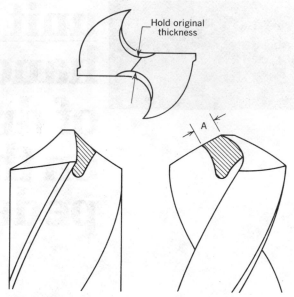

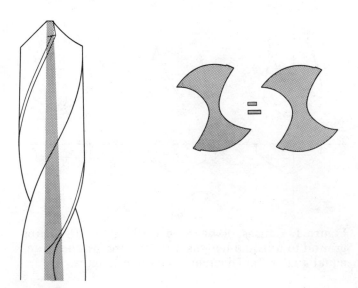

Figure 2. Using a drill point gage (DeAnza College).

Figure 4. The usual method of thinning the point on a drill. The web should not be made thinner than it was originally when the drill was new and full length (Courtesy of Bendix Industrial Tools Division).

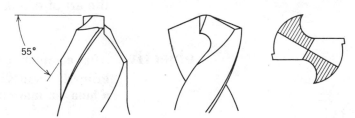

Figure 5. Split point design of a drill point (Courtesy of Bendix Industrial Tools Division).

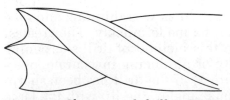

Figure 6. Sheet metal drill point.

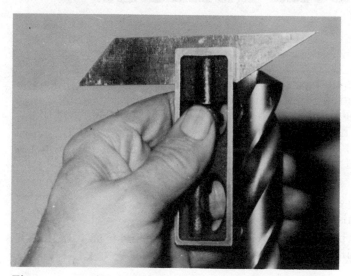

Figure 3. The tapered web of twist drills (Courtesy of Bendix Industrial Tools Division).

difficult to grind by hand; it should be done on a machine. A sheet metal drill point (Figure 6) may be ground by an experienced hand. The rake angle on a drill can be modified for drilling brass as shown in Figure 7.

The standard drill point angle is an 118 degree included angle, while for drilling hard materials point angles should be from 135 to 150 degrees. A drill point angle from 60 to 90 de-

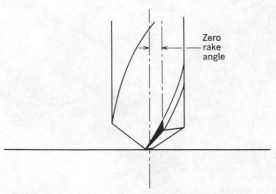

Figure 7. Modification of the rake angle for drilling brass.

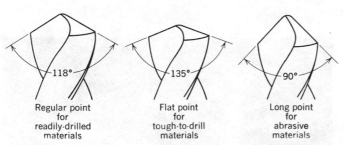

Figure 8. Drill point angles (Courtesy of Bendix Industrial Tools Division).

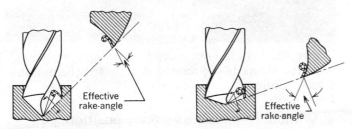

Figure 9. The effective rake angles (Courtesy of Bendix Industrial Tools Division).

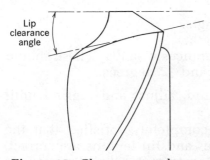

Figure 10. Clearance angles on a drill point.

grees should be used when drilling soft materials, cast iron, abrasive materials, plastics, and some nonferrous metals (Figure 8). Too great a decrease in the included point angle is not advisable, however, because it will result in an abnormal decrease in effective rake angle (Figure 9). This will increase the required feed pressure and change the chip formation and chip flow in most steels. Clearance angles (Figure 10) should be 8 to 12 degrees for most drilling.

DRILL GRINDING PROCEDURE

Check to see if both the roughing and finishing wheels are true (Figure 11). If not, move a wheel dresser across them (Figure 12). If the end of the drill is badly damaged, use the coarse wheel to remove that part. If you overheat the drill, let it cool in air; do not cool high speed steel drills in water.

The following method of grinding a drill is suggested:

1. Hold the drill shank with one hand and the drill near the point with the other hand. Rest your fingers that are near the point on the grinder tool rest. Hold the drill lightly at this point so you can manipulate it from the shank end with the other hand (Figure 13).

2. Hold the drill approximately horizontal with the cutting lip (Figure 14) that is being ground level. The axis of the drill should be at 59 degrees from the face of the wheel.

Figure 11. Grinding wheel that is dressed and ready to use (DeAnza College).

Figure 12. Trueing up a grinding wheel with a wheel dresser (DeAnza College).

Figure 14. Drill being held in the same starting position, approximately horizontal (Lane Community College).

Figure 13. Starting position showing 59 degree angle with the wheel and the cutting lip horizontal (Lane Community College).

Figure 15. Drill is now moved very slighty to the left with the shank being moved downward (Lane Community College).

3. Using the tool rest and fingers as a pivot, slowly move the shank downward and slightly to the left (Figure 15). The drill must be free to slip forward slightly to keep it against the wheel (Figure 16). Rotate the drill very slightly. It is the most common mistake of the beginner to rotate the drill until the opposite cutting edge has been ground off. Do not rotate small drills at all, only larger ones. As you continue the downward movement of the shank, crowd the drill into the wheel so that it will grind lightly all the way from the lip to the heel (Figure 17). This should all be one smooth movement. It is very important at this point to allow proper clearance (8 to 12 degrees) at the heel of the drill.

4. Without changing your body position, pull the drill back slightly and rotate 180 degrees so that the opposite lip is now in a level position. Repeat step 3.

5. Check *both cutting lips* with the drill point gage (Figure 18):

   a. For correct angle.
   b. For equal length.
   c. Check lip clearances visually. These should be between 8 and 12 degrees.

   If errors are found, adjust and regrind until they are correct.

6. When you are completely satisfied that the drill point angles and lip lengths are correct, drill a hole in a scrap metal that has been set

Figure 16. Another view of the same position as shown in Figure 15 (Lane Community College).

aside for this purpose. Consult a drill speed table so you will be able to select the correct RPM. Use cutting oil or coolant.

7. Check the condition of the hole. Did the drill chatter and cause the start of the hole to be misshapen? This could be caused by too much lip clearance. Is the hole oversize more than .005 or .010 in.? Are the lips uneven or the lip angles off or both? Running the drill too slowly for its size will cause a rough hole; too fast will burn the drill. If the hole size is more than .010 in. × diameter over the drill size, resharpen and try again.

8. When you have a correctly sharpened drill, show this and the drilled hole to your instructor for his evaluation.

Figure 17. Drill is now almost to final position of grinding . It has been rotated slightly from the starting position (Lane Community College).

Figure 18. After the sequence has been completed on both cutting lips, they are checked with a drill point gage for length and angle (DeAnza College).

self-evaluation

SELF-TEST Sharpen drills on the pedestal grinder and submit them to your instructor for approval.

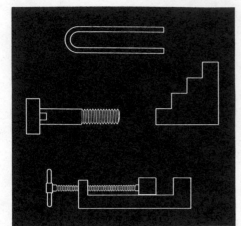

unit 4
work locating and holding devices on drilling machines

Workpieces of a great many sizes and shapes are drilled by machinists. In order to hold these parts safely and securely while they are drilled, several types of workholding devices are used. In this unit you will learn how to properly set up these devices for a machining operation.

objectives **After completing this unit, you will be able to:**
1. **Identify and explain the correct uses for several workholding and locating devices.**
2. **Set up and drill holes in two parts of a continuing project; align and start a tap using the drill press.**

Because of the great forces applied by the machines in drilling, some means must be provided to keep the workpiece from turning with the drill or from climbing up the flutes after the drill breaks through. This is not only necessary for safety's sake but also for workpiece rigidity and good workmanship.

One method of workholding is to use strap clamps (Figure 1) and T-bolts (Figure 2). The clamp must be kept parallel to the table by the use of step blocks (Figure 3) and the T-bolt should be kept as close to the workpiece as possible (Figure 4). Parallels (Figure 5) are placed under the work at the point where the clamp is holding. This provides a space for the drill to break through without making a hole in the table. A thin or narrow workpiece should not be supported too far from the drill, however, since it will spring down under the pressure of the drilling. This can cause the drill on breakthrough to suddenly "grab" more material than it can handle. The result is often a

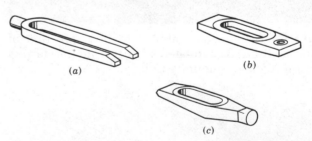

Figure 1. Strap clamps: (*a*) U-clamp, (*b*) Straight clamp, and (*c*) finger clamp.

broken drill (Figure 6). C-clamps of various sizes are used to hold workpieces on drill press tables and on angle plates (Figure 7).

Angle plates facilitate the holding of odd shaped parts for drilling. The angle plate is either bolted or clamped to the table and the work is fastened to the angle plate. For example, a gear or wheel that requires a hole to be drilled into a projecting hub could be clamped to an angle plate.

368

Figure 2. T-bolts.

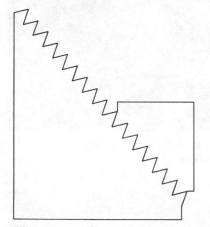

Figure 3. Adjustable step blocks.

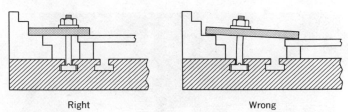

Figure 4. Right and wrong setup for strap clamps. The clamp bolt should be as close to the workpiece as possible.

Figure 5. Parallels of various sizes (Lane Community College).

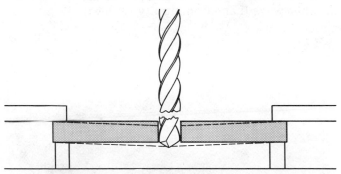

Figure 6. Thin, springy material is supported too far from the drill. Drilling pressure forces the workpiece downward until the drill breaks through, relieving the pressure. The work then springs back and the remaining "fin" of material is more than the drill can cut in one revolution. The result is drill breakage.

Drill press vises (Figure 8) are very frequently used for holding small workpieces of regular shape and size with parallel sides. Vises provide the quickest and most efficient set up method for parallel work, but should not be used if the work does not have parallel sides. The workpiece must be supported so the drill will not go into the vise. If precision parallels are used for support, they and the drill can be easily damaged since they are both hardened (Figure 9). For rough drilling, however, cold finished (CF) keystock would be sufficient for supporting the workpiece. Angular vises can pivot a workpiece to a given angle so that angular holes can be drilled (Figure 10). Another method of drilling angular holes is by tilting the drill press table. If there is no angular scale on the vise or table, a protractor head with a level may be used to set up the correct angle for drilling. Angle plates are also sometimes used for drilling angular holes (Figure 11a). The drill press table must be level (Figure 11b).

Vee-blocks come in sets of two, often with clamps for holding small size rounds (Figure 12).

Figure 7. C-clamp being used on an angle plate to hold work that would be difficult to safely support in other ways (Lane Community College).

Figure 9. Part set up in vise with parallels under it (Lane Community College).

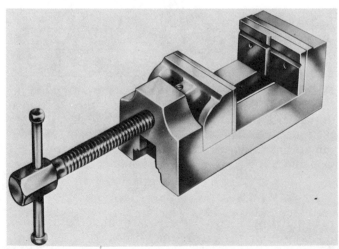

Figure 8. Drill press vise. Small parts are held for drilling and other operations with the drill press vise. (Courtesy of Wilton Corporation).

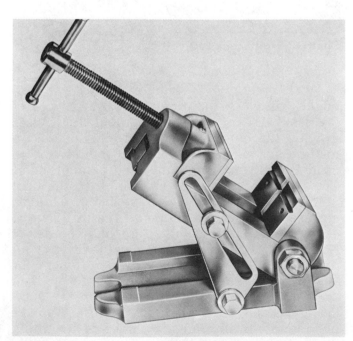

Figure 10. Angle vise. Parts that must be held at an angle to the drill press table while being drilled are held with this vise (Courtesy of Wilton Corporation).

Figure 11a. View of an adjustable angle plate on a drill press table using a protractor to set up (Lane Community College).

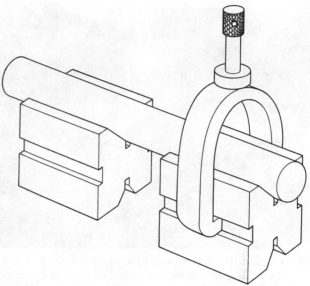

Figure 12. Set of vee-blocks with a vee-block clamp.

Figure 11b. Checking the level of the table (Lane Community College).

Larger size round stock is set up with a strap clamp between the vee-blocks (Figure 13). The hole to be cross drilled is first laid out and center punched. The workpiece is lightly clamped in the vee-blocks and the punch mark is centered as shown in Figure 14. The clamps are

tightened, and the drill is located precisely over the punch mark by means of a wiggler.

The wiggler is clamped into a drill chuck and the machine is turned on (Figure 15a). Push on the knob near the end of the pointer until it runs with no wobble (Figure 15b). With the machine still running, bring the pointer down into the punch mark. If the pointer begins to wobble again, the mark is not centered under the spindle and the workpiece will have to be shifted.

After the work is centered, use a spotting or center drill to start the hole. Then, for larger holes, use a pilot drill, which is always a little larger than the dead center of the next drill size used. Pilot drills are not used in industrial applications; only the spotting drill (if used) and the full size drill are used. Use the correct cutting speed and coolant. Chamfer both sides of the finished hole with a countersink. Round stock can also be cross drilled when held in a vise, using the same technique as with vee-blocks.

A tap may be started straight in the drill press by hand. After tap drilling the workpiece and without removing any clamps, remove the tap drill from the chuck and replace it with a straight shank center. Insert a tap in the work and attach a tap handle. Then put the center into the tap, but *do not* turn on the machine. Apply sulfurized cutting oil and start the tap by turning the tap handle a few turns with one hand while feeding down with the other hand.

Figure 13. Setup of two vee-blocks and round stock with strap clamp between the vee-blocks (Lane Community College).

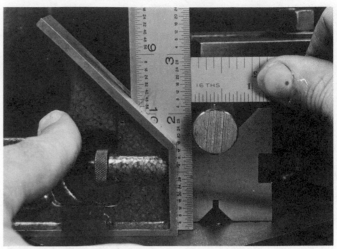

Figure 14. Round stock in vee-blocks. One method of centering layout line or punch mark using a combination square and rule (Lane Community College).

Figure 15a. Wiggler set in offset position (Lane Community College).

Figure 15b. Wiggler centered (Lane Community College).

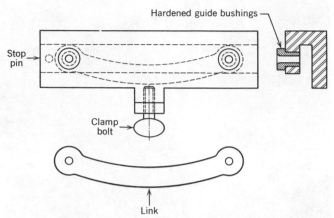

Figure 16. Simple box jig for drilling link. The link is shown below. Hardened guide bushings in the jig are used to limit wear.

Release the chuck while the tap is still in the work and finish the job of tapping the hole.

Jigs and fixtures are specially made tooling for production work. In general, a fixture holds a workpiece, while a jig guides a tool

Figure 17. Setup of C-clamp project in vise by squaring it with combination square (Lane Community College).

such as a drill (Figure 16). The use of a jig assures exact positioning of the hole pattern in duplicate and eliminates layout work on every part.

DRILLING PROCEDURE, C-CLAMP BODY

Given a combination square, wiggler, Q drill, countersink, $\frac{3}{8}$–24 tap, cutting oil, set of vee-blocks with clamps, drill press vise, the C-clamp body as it is finished up to this point, and a piece of $\frac{1}{2}$ in. diameter CF round stock $4\frac{1}{8}$ in. long:

1. Set up a workpiece square with the drill press table.
2. Locate punch mark and center drill in the mark.
3. Pilot drill and then tap drill.
4. Hand start the tap in the drill press.
5. Set up round stock in vee-blocks, locate center, then clamp in place and drill $\frac{3}{16}$ in. hole.

Drilling the clamp body:

1. Clamp the C-clamp body in the vise as shown in Figure 17 so the back side extends from the vise jaws about $\frac{1}{16}$ in. Square it with the table by using the combination square. Tighten the vise.
2. Put wiggler into the chuck and align the center as explained in this unit. Clamp the vise to the table, taking care not to move it.
3. Using the center drill or spotting drill, start the hole (Figure 18). Change to $\frac{1}{8}$ to $\frac{3}{16}$ in. pilot drill and make a hole clear through. Now change to the Q drill and enlarge the hole to this size (Figure 19). Chamfer the drilled hole with a countersink tool (Figure 20). The chamfer should measure about $\frac{3}{8}$ in. across. Use cutting oil or coolant for drilling.
4. Place a straight shank center in the chuck and tighten it. Insert a $\frac{3}{8}$–24 tap and tap handle in the tap-drilled hole and support the other end on the center. Apply sulfurized cutting oil. *Do not* turn on the machine. Feed lightly downward with one hand while hand turning the tap handle with the other hand (Figure 21). The tap will be started straight when part way into the work. Release the chuck and finish tapping with the tap handle.

Cross drilling the $\frac{1}{2}$ in. round

1. Take the $\frac{1}{2}$ in. CF round; lay out the hole location and punch. Place it in one or two vee-

Figure 18. Using a center drill to start the hole (Lane Community College).

Figure 20. Chamfering the drilled hole (Lane Community College).

Figure 19. Making the tap drill hole. Note the correct chip formation (Lane Community College).

Figure 21. Hand tapping in the drill press to assure good alignment (Lane Community College).

blocks, depending on their size. Lightly clamp with about 1 in. extended from one end.

2. Set up the punch mark so it is centered and on top by using the combination square and a rule. Refer to Figure 13.

3. With a wiggler, locate the punch mark directly under the spindle.

4. Center drill; change to a $\frac{3}{16}$ in. drill and drill through. Chamfer both sides lightly. This part is now ready for lathe work.

Complete drawings for this project may be found in the instructor's manual. Ask your instructor.

self-evaluation

SELF-TEST

1. What is the main purpose for using workholding devices on drilling machines?

2. List the names of all the workholding devices that you can remember.

3. Explain the uses of parallels for drilling setups.

4. Why should the support on a narrow or thin workpiece be as close to the drill as possible?

5. Angle drilling can be accomplished in several ways. Describe two methods. How would this be done if no angular measuring devices were mounted on the equipment?

6. What shape of material is the vee-block best suited to hold for drilling operations? What do you think its most frequent use would be?

7. What is the purpose of using a wiggler?

8. Why would you ever need an angle plate?

9. What is the purpose of starting a tap in the drill press?

10. Do you think jigs and fixtures are used to any great extent in small machine shops? Why?

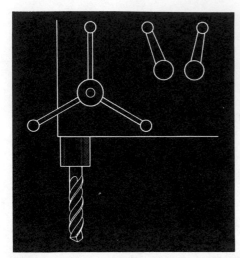

unit 5
operating drilling machines

You have already learned many things about drilling machines and tooling. You should now be ready to learn some very important facts about the use of these machines. How fast should the drill run? How much feed should be applied? Which kind of coolant should be used? These and other questions are answered in this unit.

objectives After completing this unit, you will be able to:

1. **Determine the correct drilling speeds for five given drill diameters.**
2. **Determine the correct feed in steel by chip observation.**
3. **Set up the correct feed on a machine by using a feed table.**
4. **Successfully drill several holes in a continuing project or an alternate project.**

After the workpiece is properly clamped and operator safety is assured, the most important considerations for drilling are speeds, feeds, and coolants. Of these, the control and setting of speeds will have the greatest effect on the tool and the work.

CUTTING SPEEDS

Cutting speeds (CS) are normally given for high speed steel cutting tools and are based on surface feet per minute (FPM or SFM). Since machine spindle speeds are given in revolutions per minute (RPM), this can be derived in the following manner:

$$RPM = \frac{\text{Cutting speed (in feet per minute)} \times}{\text{Diameter of cutter (in inches)} \times}$$

$$\frac{12 \text{ (inches in 1 foot)}}{\pi}$$

If you use 3 to approximate π then the formula becomes

$$\frac{CS \times 12}{D \times 3} = \frac{CS \times 4}{D}$$

This formula is certainly the most common one used in the machine shop practice and it applies to the full range of machine tool operations. The simplified formula $\frac{CS \times 4}{D}$ will be used throughout this text.

The formula is used as follows:

where D = the diameter of the drill
and CS = an assigned cutting speed for a particular material

Low carbon steel = 90 CS; aluminum = 300 CS; cast iron = 70 CS; and alloy steel SAE 4140 = 50 CS. Therefore, for a $\frac{1}{2}$ inch drill in low carbon steel the speed would be

$$\frac{90 \times 4}{\frac{1}{2}} = 720 \text{ RPM}$$

Cutting speeds/RPM tables for various materials are available in handbooks and as wall charts. Excessive speeds can cause the outer corners and margins of the drill to break down. This will in turn cause the drill to bind in the hole, even if the speed is corrected and more cutting oil is applied. The only cure is to grind the drill back to its full diameter (Figure 1) using methods discussed in Units 4 and 5 of this section.

A blue chip from steel indicates the speed is too high. The tendency with very small drills, however, is to set the RPM of the spindle too slow. This gives the drill a very low cutting speed and very little chip is formed unless the operator forces it with an excessive feed. The result is often a broken drill.

CONTROLLING FEEDS

The feed may be controlled by the "feel" of the cutting action and by observing the chip. A long, stringy chip indicates too much feed. The proper chip in soft steel should be a tightly rolled helix in both flutes (Figure 2). Some materials such as cast iron will produce a granular chip. Drilling machines that have power feeds are arranged to advance the drill a given amount for each revolution of the spindle. Therefore, .006 in. feed means that the drill advances .006 in. every time the drill makes one full turn. The amount of feed varies according to the drill size and the work material. See Table 1.

It is a better practice to start with smaller feeds than those given in tables. Materials and setups vary, so it is safer to start low and work up to an optimum feed. You should stop the feed occasionally to break the chip and allow coolant to flow to the cutting edge of the drill.

Table 1
Drilling Feed Table

Drill Size Diameter (in Inches)	Feeds per Revolution (in Inches)
Under $\frac{1}{8}$	.001 to .002
$\frac{1}{8}$ to $\frac{1}{4}$	.002 to .004
$\frac{1}{4}$ to $\frac{1}{2}$	.004 to .007
$\frac{1}{2}$ to 1	.007 to .015
over 1	.015 to .025

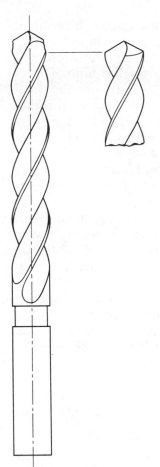

Figure 1. Broken down drill corrected by grinding back to full diameter margins and regrinding cutting lips.

Figure 2. Properly formed chip (Lane Community College).

There is generally no breakthrough problem when using power feed, but when hand feeding, the drill may catch and "grab" while coming through the last $\frac{1}{8}$ in. or so of the hole. Therefore, the operator should let up on the feed handle near this point and ease the drill through the hole. This "grabbing" tendency is especially true of brass and some plastics, but it is also a problem in steels and other materials. Large upright drill presses and radial arm drills have power feed mechanisms with feed clutch handles (Figure 3) that also can be used for hand feeding when the power feed is disengaged. Both feed and speed controls are set by levers or dials (Figure 4a). Speed and feed tables on plates are often found on large drilling machines (Figure 4b).

Tapping with small taps is often done on a sensitive drill press with a tapping attachment (Figure 5) that has an adjustable friction clutch and reverse mechanism that screws the tap out when you raise the spindle. Large size taps are power driven on upright or radial drill presses. These machines provide for spindle reversal (sometimes automatic) to screw the tap back out.

COOLANTS AND CUTTING OILS

A large variety of coolants and cutting oils are used for drilling operations on the drill press. Emulsifying or soluble oils (either mineral or synthetic) mixed in water are used for drilling

Figure 4a. Speed and feed control dials (Lane Community College).

Figure 4b. Large speed and feed plates on the front of the head of the upright drill press can be read at at a glance (Courtesy of Giddings & Lewis, Inc.).

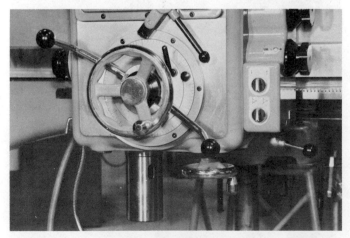

Figure 3. Feed clutch handle. The power feed is engaged by pulling the handles outward. When the power feed is disengaged, the handles may be used to hand feed the drill (Lane Community College).

Figure 5. Tapping attachment (Lane Community College).

holes where the main requirement is an inexpensive cooling medium. Operations that tend to create more friction and, hence, need more lubrication to prevent galling (abrasion due to friction), require a cutting oil. Animal or mineral oils with sulfur or chlorine added are often used. Reaming, counterboring, countersinking, and tapping all create friction and require the use of cutting oils, of which the sulfurized type is most used. Cast iron and brass are usually drilled dry, but water soluble oil can be used for both. Aluminum can be drilled with water soluble oil or kerosene for a better finish. Both soluble and cutting oils are used for steel.

DRILLING PROCEDURES

Deep hole drilling requires sufficient drill length and quill stroke to complete the needed depth. A high helix drill helps to remove the chips, but sometimes the chips bind in the flutes of the drill and, if drilling continues, will cause the drill to jam in the hole (Figure 6). A method of avoiding this problem is called "pecking"; that is, when the hole is drilled a short distance, the drill is taken out from the hole allowing the accumulated chips to fly off. The drill is again inserted into the hole, a similar amount

is drilled, and the drill is again removed. This pecking is repeated until the required depth is reached.

A depth stop is provided on drilling machines to limit the travel of the quill so that the drill can be made to stop at a predetermined depth (Figure 7). The use of a depth stop makes drilling several holes to the same depth quite easy. Spotfacing and counterboring should also be set up with the depth stop. Blind holes (holes that do not go through the piece) are measured from the edge of the drill margin to the required depth (Figure 8). Once measured, the depth can be set with the stop and drilling can proceed. One of the most important uses of a depth stop, from a maintenance standpoint, is that of setting the depth so that the machine table or drill press vise will not be drilled full of holes.

Holes that must be drilled partly into or

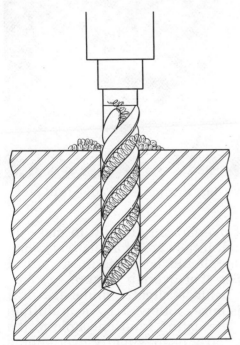

Figure 6. Drill jammed in hole because of packed chips.

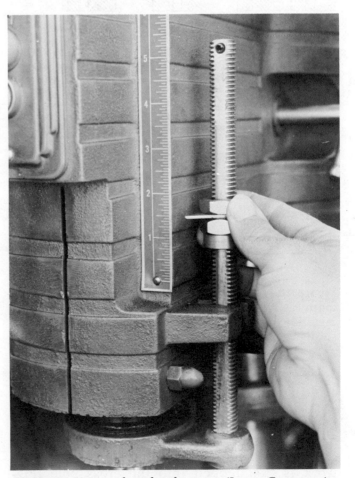

Figure 7. Using the depth stop (Lane Community College).

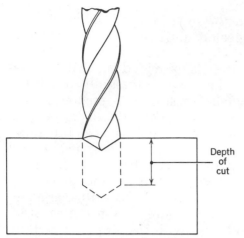

Figure 8. Measuring the depth of a drilled hole.

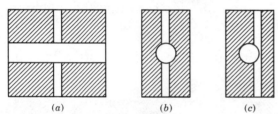

Figure 9. Hole drilled at 90 degree angle into existing hole. Cross drilling is done off center as well as on center. (*a*) Side view. (*b*) End view drilled on center. (*c*) End view drilled off center.

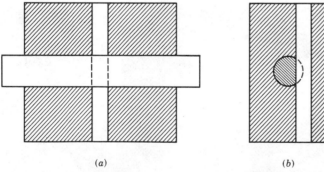

Figure 10. Existing hole is plugged to make the hole more easily drilled. (*a*) Hole is plugged with the same material that the workpiece consists of. (*b*) End view showing hole drilled through plug.

across existing holes (Figure 9) may jam or bind a drill unless special precautions are taken. A special drill with a double margin and high helix could be used directly. However, an ordinary jobbers drill will do a satisfactory job if a tight plug made of the same material as the work is first tapped into the cross hole (Figure 10). The hole may then be drilled in a normal manner and the plug removed.

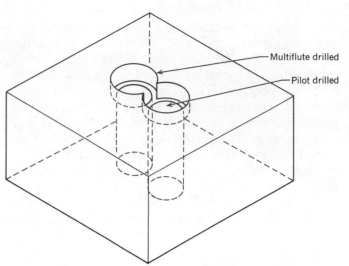

Figure 11. Deep holes that overlap are difficult to drill. The holes are drilled alternately with a smaller pilot drill and a multiflute drill for final size.

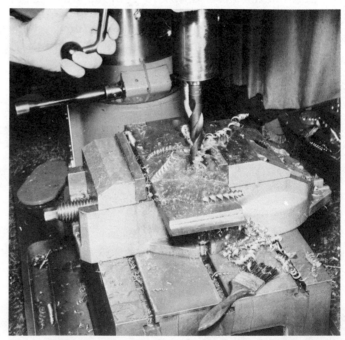

Figure 12. Heavy duty drilling on a radial drill press.

Holes that overlap are difficult to make on a drill press and, if they are of a relatively shallow depth, they should be made on a vertical mill. Deep holes that overlap (Figure 11) may be drilled by using a slightly undersize drill for a short distance, followed by a multiflute core drill. By alternating from one hole to the other in steps, fairly deep, slightly overlapping holes may be drilled. Step or subland drills can also be used for this purpose.

Heavy duty drilling should be done on an upright or radial drill press (Figure 12). The workpiece should be made very secure since high drilling forces are used with the larger drill sizes. The work should be well clamped or bolted to the work table. The head and column clamp should always be locked when drilling is being done on a radial drill press. When several operations are performed in a continuing sequence, quick change drill and toolholders are often used. Coolant is necessary for all heavy duty drilling.

self-evaluation

SELF-TEST

1. Name three important things to keep in mind when using a drill press (not including operator safety and clamping work).

2. If RPM = $\dfrac{CS \times 4}{D}$ and the cutting speed for low carbon steel is 90, what would the RPMs be for the following drills: $\frac{1}{4}$, 2, $\frac{3}{4}$, $\frac{3}{8}$, $1\frac{1}{2}$ in. diameter.

3. What are some of the results of excessive drilling speed? What corrective measures can be taken?

4. Explain what can happen to small diameter drills when the cutting speed is too slow.

5. How can an operator tell by observing the chip if the feed is about right?

6. In what way are power feeds designated?

7. Name two differing cutting fluid types.

8. Such operations as counterboring, reaming, and tapping create friction that can cause heat. This can ruin a cutting edge. How can this situation be helped?

9. How can jamming of a drill be avoided when drilling deep holes?

10. Name three uses for the depth stop on a drill press.

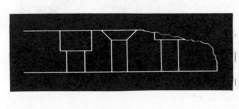

unit 6
countersinking and counterboring

In drill press work it is often necessary to make a recess that will leave a bolt head below the surface of the workpiece. These recesses are made with countersinks or counterbores. When holes are drilled into rough castings or angular surfaces, a flat surface square to these holes is needed, and spot facing is the operation used. This unit will familiarize you with these drill press operations.

objectives After completing this unit, you will be able to:
1. Identify tools for countersinking and counterboring.
2. Select speeds and feeds for countersinking and counterboring.
3. Countersink and counterbore holes.

COUNTERSINKS

A countersink is a tool used to make a conical enlargement of the end of a hole. Figures 1 and 2 illustrate two types of single flute countersinks, both of which are designed to produce smooth surfaces, free from chatter marks. A countersink is used as a chamfering or deburring tool to prepare a hole for reaming or tapping. Unless a hole needs to have a sharp edge, it should be chamfered to protect the end of the hole from nicks and burrs. A chamfer from $\frac{1}{32}$ to $\frac{1}{16}$ in. wide is sufficient for most holes.

A hole made to receive a flathead screw or rivet should be countersunk deep enough for the head to be flush with the surface or up to .015 in. below the surface. A flathead fastener should never project above the surface. The included angles on commonly available countersinks are 60, 82, 90, and 100 degrees. Most flathead fasteners used in metalworking have an 82 degree head angle, except for the aircraft industry where the 100 degree angle is prevalent. The cutting speed used when countersinking should always be slow enough to avoid chattering.

A combination drill and countersink with a 60 degree angle (Figure 3) is used to make center holes in workpieces for machining on lathes and grinders. The combination drill and countersink, also known as a center drill, also is used for spotting holes when using a drill press or milling machine, since it is extremely rigid and will not bend under pressure.

Figure 1. (*left*) Single flute countersink (Courtesy of DoAll Company).

Figure 2. (*right*) Chatterfree countersink (Courtesy of DoAll Company).

Figure 3. Center drill or combination drill and countersink (Courtesy of DoAll Company).

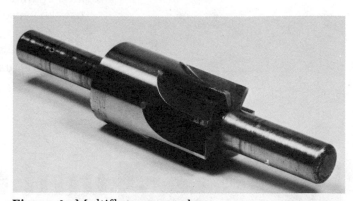

Figure 4. Multiflute counterbore.

COUNTERBORES

Counterbores are tools designed to enlarge previously drilled holes, much like countersinks, and are guided into the hole by a pilot to assure the concentricity of the two holes. A multiflute counterbore is shown in Figure 4, and a two flute counterbore is shown in Figure 5. The two flute counterbore has more chip clearance and a larger rake angle than the counterbore in Figure 4. Counterbored holes have flat bottoms, unlike the angled edges of countersunk

holes, and are often used to recess a bolt head below the surface of a workpiece. Solid counterbores, such as shown in Figures 4 and 5, are used to cut recesses for socket head cap screws or filister head screws (Figure 6). The diameter of the counterbore is usually $\frac{1}{32}$ in. larger than the head of the bolt, so that the counterbore is freer cutting and better suited for soft and ductile materials.

When a variety of counterbore and pilot sizes is necessary, a set of interchangeable pilot counterbores is available. Figure 7 shows a

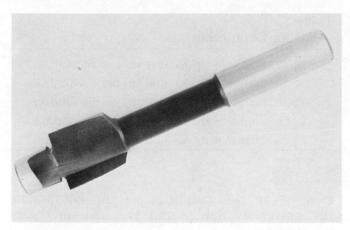

Figure 5. Two flute counterbore.

Figure 8. Pilot for interchangeable pilot counterbore.

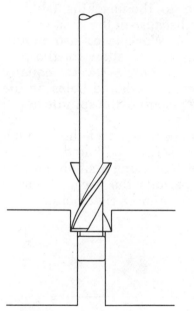

Figure 6. The counterbore is an enlargement of a hole already drilled.

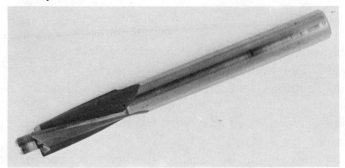

Figure 7. Interchangeable pilot counterbore.

counterbore in which a number of standard or specially made pilots can be used. A pilot is illustrated in Figure 8.

Counterbores are made with straight or tapered shanks to be used in drill presses, milling machines, or even lathes. When counterboring a recess for a hex head bolt, remember to measure the diameter of the socket wrench so the hole will be large enough to accommodate it. For most counterboring operations the pilot should have from .002 to .005 in. clearance in the hole. If the pilot is too tight in the hole, it may seize and break. If there is too much clearance between the pilot and the hole, the counterbore will be out of round and will have an unsatisfactory surface finish.

It is very important that the pilot be lubricated while counterboring. Usually this lubrication is provided if a sulfurized cutting oil or soluble oil is used. When cutting dry, which is often the case with brass and cast iron, the hole and pilot should be lubricated with a few drops of lubricating oil.

Counterbores are often used to provide a flat bearing surface for nut or bolt heads on rough castings or a raised boss (Figure 9). This operation is called spot facing. Because these rough surfaces may not be at right angles to the pilot hole, great strain is put on the pilot and counterbore, and can cause breakage of either one. To avoid breaking the tool, be very careful when starting the cut, especially when hand feeding. Prevent hogging into the work by tightening the spindle clamps slightly to remove possible backlash.

Recommended power feed rates for counterboring are shown in Table 1. The feed rate

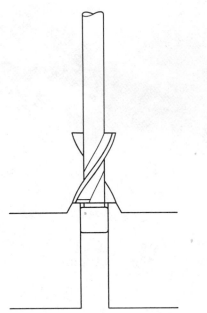

Figure 9. Spot facing on a raised boss.

Table 1
Feeds for Counterboring

$\frac{3}{8}$ in. diameter up to .004 in. per revolution
$\frac{5}{8}$ in. diameter up to .005 in. per revolution
$\frac{7}{8}$ in. diameter up to .006 in. per revolution
$1\frac{1}{4}$ in. diameter up to .007 in. per revolution
$1\frac{1}{2}$ in. diameter up to .008 in. per revolution

should be great enough to get under any surface scale quickly, thus preventing rapid dulling of the counterbore. The speeds used for counterboring are one third less than the speeds used for twist drills of corresponding diameters. The choice of speeds and feeds is very much affected by the condition of the equipment, the power available, and the material being counterbored.

Before counterboring a workpiece, it should be securely fastened to the machine table or tightly held in a vise because of the great cutting pressures encountered. Workpieces also should be supported on parallels to allow for the protrusion of the pilot. To obtain several equally deep countersunk or counterbored holes on the drill press or milling machine, the spindle depth stop can be set.

Counterbores can be used on a lathe to rough out a hole before it is finished-bored. This is often more efficient than using a single cutting edge boring bar. It permits the use of a larger diameter boring bar for a more rigid setup.

self-evaluation

SELF-TEST
1. When is a countersink used?
2. Why are countersinks made with varying angles?
3. What is a center drill?
4. When is a counterbore used?
5. What relationship exists between pilot size and hole size?
6. Why is lubrication of the pilot important?
7. As a rule, how does cutting speed compare between an equal size counterbore and twist drill?
8. What affects the selection of feed and speed when counterboring?
9. What is spot facing?
10. What important points should be considered when a counterboring setup is made?

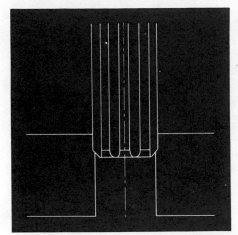

unit 7
reaming
in the drill
press

Many engineering requirements involve the production of holes having smooth surfaces, accurate location, and uniform size. In many cases, holes produced by drilling alone do not entirely satisfy these requirements. For this reason, the reamer was developed for enlarging or finishing previously formed holes. This unit will help you properly identify, select, and use machine reamers.

objectives   **After completing this unit, you will be able to:**

1. **Identify commonly used machine reamers.**
2. **Select the correct feeds and speeds for commonly used materials.**
3. **Determine appropriate amounts of stock allowance.**
4. **Identify probable solutions to reaming problems.**
5. **Properly ream two holes.**

Reamers are tools used mostly to precision finish holes, but they are also used in the heavy construction industry to enlarge or align existing holes.

COMMON MACHINE REAMERS

Machine reamers have straight or taper shanks, the taper usually is a standard Morse taper. The parts of a machine reamer are shown in Figure 1 and the cutting end of a machine reamer is shown in Figure 2.

Chucking reamers (Figures 3, 4, and 5) are efficient in machine reaming a wide range of materials and are commonly used in drill presses, turret lathes, and screw machines. Helical flute reamers have an extremely smooth cutting action that finishes holes accurately and precisely. Chucking reamers cut on the chamfer at the end of the flutes. This chamfer is usually at a 45 degree angle.

Jobber's reamers (Figure 6) are used where a longer flute length than chucking reamers is needed. The additional flute length gives added guide to the reamer, especially when reaming deep holes.

The rose reamer (Figure 7) is primarily a roughing reamer used to enlarge holes to within .003 to .005 in. of finish size. The teeth are slightly backed off, which means that the reamer diameter is smaller toward shank end by approximately .001 in./in. of flute length. The lands on these reamers are ground cylindrically without radial relief, and all cutting is done on the end of the reamer. This reamer will remove a considerable amount of material in one cut.

Shell reamers (Figure 8) are finishing reamers. They are more economically produced, especially in larger sizes, than solid reamers because a much smaller amount of tool material is used in making them. Two slots in the shank end of the reamer fit over matching driving

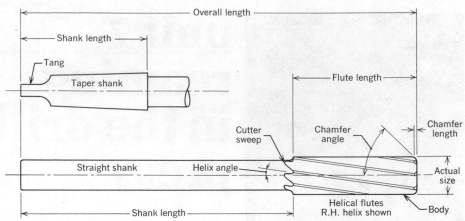

Figure 1. The parts of a machine reamer (Courtesy of Bendix Industrial Tools Division).

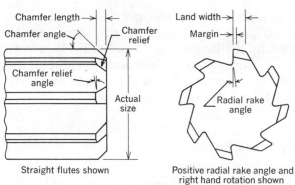

Figure 2. The cutting end of a machine reamer (Courtesy of Bendix Industrial Tools Division).

Figure 3. Straight shank straight flute chucking reamer (Courtesy of Whitman & Barnes Division, TRW Inc.).

Figure 4. Straight shank helical flute chucking reamer (Courtesy of Whitman & Barnes Division, TRW Inc.).

Figure 5. Taper shank helical flute chucking reamer (Courtesy of Whitman & Barnes Division, TRW Inc.).

Figure 6. Taper shank straight flute jobbers reamer (Courtesy of Whitman & Barnes Division, TRW Inc.).

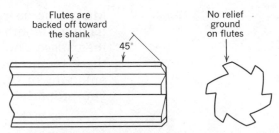

Figure 7. Rose reamer.

Figure 8. Shell reamer helical flute (Courtesy of Whitman & Barnes Division, TRW Inc.).

Figure 9. Taper shank shell reamer arbor (Courtesy of Whitman & Barnes Division, TRW Inc.).

lugs on the shell reamer or box (Figure 9). The hole in the shell reamer has a slight taper ($\frac{1}{8}$ in./ft) in it to assure exact alignment with the shell reamer arbor. Shell reamers are made with straight or helical flutes and are commonly produced in sizes from $\frac{3}{4}$ to $2\frac{1}{2}$ in. in diameter. Shell reamer arbors come with matching straight or tapered shanks and are made in designated sizes from numbers 4 to 9.

Morse taper reamers (Figure 10), with

straight or helical flutes, are used to finish ream tapered holes in drill sockets, sleeves, and machine tool spindles. Helical taper pin reamers (Figure 11) are especially suitable for machine reaming of taper pin holes. There is no packing of chips in the flutes, which reduces the possibility of breakage. These reamers have a free-cutting action that produces a good finish at high cutting speeds. Taper pin reamers have a taper of $\frac{1}{4}$ in. per foot of length and are manufactured in 18 different sizes ranging from smallest number 8/0 (eight naught) to the largest at number 10.

Taper bridge reamers (Figure 12) are used in structural iron or steel work, bridge work, and ship construction where extreme accuracy is not required. They have long tapered pilot points for easy entry in out-of-line holes often encountered in structural work. Taper bridge reamers are made with straight and helical flutes to ream holes with diameters from $\frac{1}{4}$ to $1\frac{5}{16}$ in.

Carbide tipped chucking reamers (Figure 13) are often used in production setups, particularly where abrasive materials or sand and scale as in castings are encountered. The right hand helix chucking reamer (Figure 14) is recommended for ductile materials or highly abrasive materials or when machining blind holes. The carbide tipped left-hand helix chucking reamer (Figure 15) will produce good finishes on heat treated steels and other hard materials, but should be used on through holes only. Carbide expansion reamers (Figure 16) after becoming worn can be expanded and resized by grinding. This feature offsets normal wear from abrasive materials and provides for a long tool life. These tools should not be adjusted for reaming size by loosening or tightening the expansion plug but only by grinding.

Reaming is intended to produce accurate and straight holes of uniform diameter. The

Figure 10. Morse taper reamer (Courtesy of Whitman & Barnes Division, TRW Inc.).

Figure 11. Helical taper pin reamer (Courtesy of Whitman & Barnes Division, TRW Inc.).

Figure 12. Helical flute taper bridge reamer (Courtesy of Whitman & Barnes Division, TRW Inc.).

Figure 13. Carbide tipped straight flute chucking reamer (Courtesy of Whitman & Barnes Division, TRW Inc.).

Figure 14. Carbide tipped, helical flute chucking reamer, right-hand helix (Courtesy of Whitman & Barnes Division, TRW Inc.).

Figure 15. Carbide tipped, helical flute chucking reamer, left-hand helix (Courtesy of Whitman & Barnes Division, TRW Inc.).

Figure 16. Carbide tipped expansion reamer (Courtesy of Whitman & Barnes Division, TRW Inc.).

required accuracy depends on a high degree of surface finish, tolerance on diameter, roundness, straightness, and absence of bellmouth at the ends of holes. To make an accurate hole it is necessary to use reamers with adequate support for the cutting edges; an adjustable reamer may not be adequate. Machine reamers are often made of either high speed steel or cemented carbide. Reamer cutting action is controlled to a large extent by the cutting speed and feed used.

SPEED

The most efficient cutting speed for machine reaming depends on the type of material being reamed, the amount of stock to be removed, the tool material being used, the finish required, and the rigidity of the setup. A good starting point, when machine reaming, is to use $\frac{1}{2}$ to $\frac{1}{3}$ of the cutting speed used for drilling the same materials. Table 1 may be used as a guide.

Where conditions permit the use of carbide reamers, the speeds may often be increased over those recommended for HSS (high speed steel) reamers. The limiting factor is usually an absence of rigidity in the setup. Any chatter, which is often caused by too high a speed, is likely to chip the cutting edges of a carbide reamer. Always select a speed that is slow enough to eliminate chatter. Close tolerances and fine finishes often require the use of considerably lower speeds than those recommended in Table 1.

FEEDS

Feeds in reaming are usually 2 to 3 times greater than those used for drilling. The amount of feed may vary with different materials, but a good starting point would be between .0015 and .004 in. per flute per revolution. Too low a feed may "glaze" the hole, which has the same result of work hardening the material, causing occasional chatter and excessive wear on the reamer. Too high a feed tends to reduce the accuracy of the hole and the quality of the surface finish. Generally, it is best to use as high a feed as possible to produce the required finish and accuracy.

The stock removal allowance should be sufficient to assure a good cutting action of the reamer. Too small a stock allowance results in burnishing (a slipping or polishing action) or it wedges the reamer in the hole, and causes excessive wear or breakage of the reamer. The

Table 1
Cutting Speeds in Surface Feet per Minute (FPM or SFM) for Reaming with an HSS Reamer

Aluminum and its alloys	130–200
Brass	130–200
Bronze, high tensile	50– 70
Cast iron	
Soft	70–100
Hard	50– 70
Steel	
Low carbon	50– 70
Medium carbon	40– 50
High carbon	35– 40
Alloy	35– 40
Stainless steel	
AISI 302	15– 30
AISI 403	20– 50
AISI 416	30– 60
AISI 430	30– 50
AISI 443	15– 30

condition of the hole before reaming also has an influence on the reaming allowance since a rough hole will need a greater amount of stock removed than an equal size hole with a fairly smooth finish. Commonly used stock allowances are $\frac{1}{64}$ in. for diameters up to $\frac{1}{2}$ in. and $\frac{1}{32}$ in. for diameters above $\frac{1}{2}$ in. When materials that work harden readily are reamed, it is especially important to have adequate material for reaming.

To ream a hole to a high degree of surface finish, a cutting fluid is needed. A good cutting fluid will cool the workpiece and tool and will also act as a lubricant between the chip and the tool to reduce friction and heat buildup. Cutting fluids should be applied in sufficient volume to flush the chips away. Table 2 lists

Table 2
Coolants Used for Reaming

Material	Dry	Soluble Oil	Kerosene	Sulfurized Oil	Mineral Oil
Aluminum		x	x		
Brass	x	x			
Bronze	x	x			x
Cast iron	x				
Steels					
Low carbon		x		x	
Alloy		x		x	
Stainless		x		x	

some coolants used for reaming different materials.

When a drill press that has only a hand feed is used to ream a hole, the feed rate should be estimated and converted into inches per second as shown in the following example. The objective is to ream a $\frac{1}{2}$ in. hole in a 1 in. thick mild steel plate with a 6-fluted reamer. First, calculate what RPM to use.

$$RPM = \frac{CS \times 4}{D} = \frac{60 \times 4}{\frac{1}{2}} = 480 \text{ RPM}$$

Next find the feed per revolution, which equals the feed per tooth multiplied by the number of teeth of the reamer. The feed per tooth should be between .0015 and .004 in., or roughly, .003 in. Thus, .003 × 6 = .018 inch per revolution of the spindle. This is the feed rate you would use in a power feed drill press. To get the feed rate in inches per minute multiply the inches per revolution times the revolutions per minute.

$$.018 \times 480 = 8.64 \text{ IPM}$$

For hand feeding a drill press divide this figure by 60, which gives you .144 in./sec. The 1 in. plate thickness divided by .144 = 6.9, so the total reaming time through the part should be about 7 seconds.

REAMING PROBLEMS

Chatter is often caused by the lack of rigidity in the machine, workpiece, or the reamer itself. Corrections may be made by reducing the speed, increasing the feed, putting a chamfer on the hole before reaming, using a reamer with a pilot (Figure 17), or reducing the clearance angle on the cutting edge of the reamer. Carbide tipped reamers especially cannot tolerate even a momentary chatter at the start of a hole, as such a vibration is likely to chip the cutting edges.

Oversize holes can be caused by inadequate workpiece support, worn guide bushings, worn or loose spindle bearings, or a bent reamer shank. When reamers gradually start cutting larger holes, it may be because of the work material galling or forming a built up edge on reamer cutting surfaces (Figure 18). Mild steel and some aluminum alloys are particularly troublesome in this area. Changing to a different coolant may help. Reamers with highly polished flutes, margins, and relief angles or reamers that have

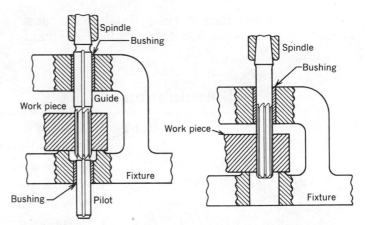

Figure 17. Use of pilots and guided bushings on reamers. Pilots are provided so that the reamer can be held in alignment and can be supported as close as possible while allowing for chip clearance (Courtesy of Bendix Industrial Tools Division).

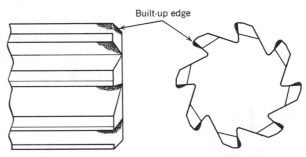

Figure 18. Reamer teeth having built-up edges.

special surface treatment may also improve the cutting action.

Bell-mouthed holes are caused by misalignment of the reamer with the hole. The use of accurate bushings or pilots may correct bellmouth, but in many cases the only solution is the use of floating holders. A floating holder will allow movement in some directions while restricting it in others. A poor finish can be improved by decreasing the feed, but this will also increase the wear and shorten the life of the reamer. A worn reamer will never leave a good surface finish as it will score or groove the finish and often produce a tapered hole.

Too fast a feed will cause a reamer to break. Too large a stock allowance for finish reaming will produce a large volume of chips with heat build-up, and will result in a poor hole finish. Too small a stock allowance will cause the reamer teeth to rub as they cut, not cut freely, which will produce a poor finish and cause rapid reamer wear. Coolant applied in insufficient quantity may also cause rough surface finishes when reaming.

self-evaluation

SELF-TEST

1. How is a machine reamer identified?
2. What is the difference between a chucking and a rose reamer?
3. What is a jobbers reamer?
4. Why are shell reamers used?
5. How does the surface finish of a hole affect its accuracy?
6. How does the cutting speed compare between drilling and reaming for the same material?
7. How does the feed rate compare between drilling and reaming?
8. How much reaming allowance will you leave on a $\frac{1}{2}$ in. hole?
9. What is the purpose of using a coolant while reaming?
10. What can be done to overcome chatter?
11. What will cause oversize holes?
12. What causes a bell-mouthed hole?
13. How can poor surface finish be overcome?
14. When are carbide tipped reamers used?
15. Why is vibration harmful to carbide tipped reamers?

section h
turning machines

No one man was the sole originator of the metal cutting lathe. Treadle or foot-operated wood turning lathes have been used for many centuries (Figure 1), and as early as 1569 Besson invented a screw cutting lathe. However, the history of modern machine tools began with the first basic machine tool (Figure 2) and the first practical all-metal screw cutting lathe built by Henry Maudslay in 1800 (Figure 3).

The power drive lathe or engine lathe is truly the father of all machine tools. With suitable attachments the engine lathe may be used for turning, threading, boring, drilling, reaming, facing, spinning, and grinding, although many of these operations are preferably done on specialized machinery. Sizes range from the smallest jeweler's or precision lathes (Figure 4) to the massive lathes used for machining huge forgings (Figure 5).

Engine lathes (Figures 6 and 7) are used by machinists to produce one-of-a-kind parts or a few pieces for a short run production. They are also used for toolmaking, machine repair, and maintenance.

Figure 1. The spring pole and treadle lathe of the thirteenth century (Courtesy of Cincinnati Milicron, Inc.).

Figure 2. The first basic machine tool: John Wilkinson Boring Mill, built in 1775 (Courtesy of DoAll Company).

391

392

Figure 3. The first all metal screw cutting lathe built by Henry Maudslay in 1800 (Courtesy of DoAll Company).

Figure 4. (*right*) Jeweler's or instrument lathe (Photograph courtesy of Louis Levin & Son, Inc.).

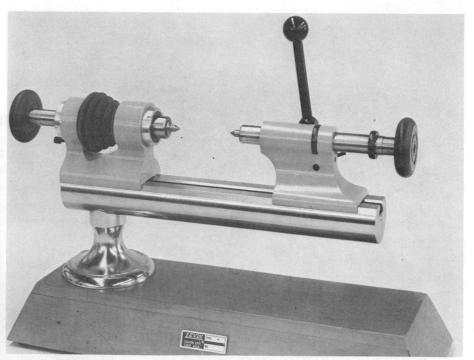

Figure 5. A massive forging being machined to exact specifications on a lathe (Courtesy of Bethlehem Steel Corporation, Bethlehem, Pa.).

Figure 6. (*right*) Heavy duty engine lathe with $11\frac{1}{2}$ in. diameter hole through the spindle. The carriage has rapid power traverse feature (Courtesy of Lodge & Shipley Company).

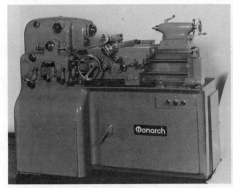

Figure 7. Ten in. toolmaker's lathe (Courtesy of The Monarch Machine Tool Company, Sidney, Ohio).

Figure 8. (*right*) Vertical boring mill (Courtesy of El-Jay Inc., Eugene, Oregon).

Some lathes have a vertical spindle instead of a horizontal one with a large rotating table on which the work is clamped. These huge machines, called Vertical Boring Mills (Figure 8), are the largest of our machine tools. A 25-foot diameter table is not unusual. Huge turbines, weighing many tons, can be placed on the table and clamped in position to be machined. The machining of such castings would be impractical on a horizontal spindle lathe.

A more versatile and higher production version of the boring mill is the vertical turret lathe (Figure 9). It does similar work to the vertical boring mill, but on a smaller scale. It is arranged with toolholders and turret with multiple tools much like that on a turret lathe, which give it flexibility and relatively high production.

Turret lathes are strictly production machines. They are designed to provide short machining time and quick tool changes. Two types of semiautomatic turret lathes require an operator in constant attendance, the ram type (Figure 10) and the saddle type (Figure 11). Small, precision, hand operated turret lathes (Figure 12) are used to produce very small parts. Automatic bar chuckers require little operator action.

Fully automatic machines such as automatic turret lathes and automatic screw machines are programmed to do a sequence of machining operations to make a completed product. Automatic turret lathes are programmed by peg board or by numerical control (NC) on punched tape. Automatic screw machines, used for high production of small parts, are typically programmed with cams, although some are numerically controlled. There are several types of automatic machines: the single spindle machine, the sliding head (or Swiss type), the multiple spindle, and the revolving head wire feed type. Multiple spindle bar and chucking machines

Figure 9. Vertical turret lathe (Courtesy of Giddings & Lewis, Inc.).

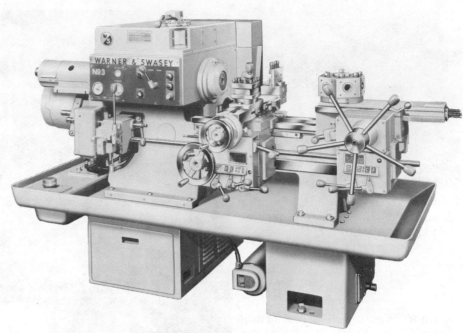

Figure 10. Ram-type turret lathe (Courtesy of The Warner & Swasey Company).

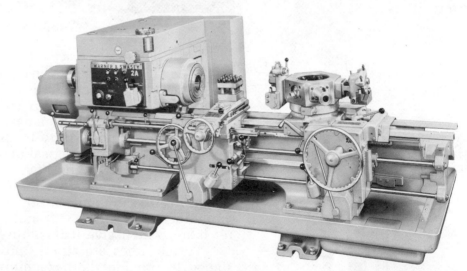

Figure 11. Saddle-type turret lathe (Courtesy of The Warner & Swasey Company).

are high production turning machines that can do a variety of operations at different stations.

Tracer lathes (Figure 13) follow a pattern or template to reproduce an exact shape on a workpiece. Tracing attachments (Figure 14) are often used on engine or turret lathes.

Lathes (Figure 15) that are numerically controlled (NC) by programming and punching tape produce workpieces such as shafts with tapers and precision diameters. NC chuckers (Figure 16) are high production automatic lathes designed for chucking operations. Similar bar feeding NC types take a full length bar through the spindle and automatically feed it in as needed. Some automatic lathes operate as either chucking machines or bar feed machines.

Figure 12. Small precision manually operated turret lathe (Photograph courtesy of Louis Levin & Sons, Inc.).

Figure 13. Tracer lathe (Lane Community College).

Lathes are also used for metal spinning. Reflectors, covers, and pans, for example, are made by this method out of aluminum, copper, and other soft metals.

Manufacturing methods for spinning heavy steel plate are entirely different. Hydraulic operated tools are used to form the steel (Figure 17). The dimensions of the part shaped by this Floturn process are shown in Figure 18.

Digital readout systems for machine tools such as the engine

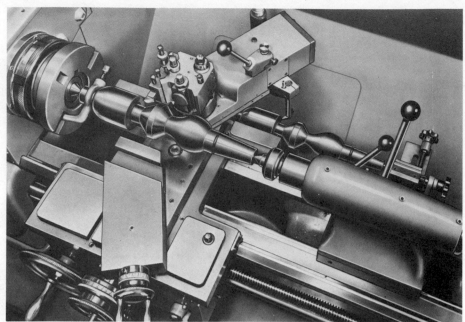

Figure 14. Tracer attachments on an engine lathe (Courtesy of Clausing Corporation).

Figure 15. Numerically controlled lathe (El-Jay Inc., Eugene, Oregon).

Figure 16. Numerically controlled chucking lathe with turret (Courtesy of The Monarch Machine Tool Company, Sidney, Ohio).

Figure 17. Floturn lathe. The photograph (*right*) shows the completion of the first two operations. Starting with the flat blank, the workpiece is shear formed (Floturn) to the shape shown on the mandrel. In the process the 1 in. thickness of the blank is reduced to .420 in. The second operation, although performed on a Florturn machine, is more properly spinning than shear forming. In this operation, the workpiece is brought to the finished shape as seen in the background (Photo courtesy of Floturn, Inc., Division of the Lodge & Shipley Co.).

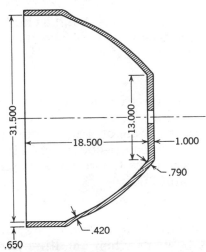

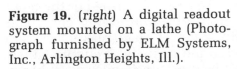

Figure 18. The principle of the Floturn process (Photo courtesy of Floturn, Inc., Division of the Lodge & Shipley Co.).

Figure 19. (*right*) A digital readout system mounted on a lathe (Photograph furnished by ELM Systems, Inc., Arlington Heights, Ill.).

lathe in Figure 19 are becoming more common. This system features a completely self-guided rack and pinion that operates on the cross slide. The direct readout resolution is .001 inch on both the diameters and the cross slide movement. These systems can also be converted to metric measure.

In ordinary turning, metal is removed from a rotating workpiece with a single point tool. The tool must be harder than the workpiece and held rigidly against it. Chips formed from the workpiece slide across the face of the tool. This essentially is the way chips are produced in all metal cutting operations. The pressures used in metal cutting can be as much as 20 tons per square inch. Tool geometry therefore is quite important to maintain the strength at the cutting edge of the tool bit.

In this section you will learn how to use common lathes, how to grind high speed tools, and how to select carbide tools for lathe work.

TURNING MACHINE SAFETY The lathe can be a safe machine only if the machinist is aware of the hazards involved in its operation. In the machine shop as anywhere, you must always keep your mind on your work in order to avoid accidents. Develop safe work habits in the use of setups, chip breakers, guards, and other protective devices. Standards for safety have been established as guidelines to help you eliminate unsafe practices and procedures on lathes. Some of the hazards are as follows:

Pinch points due to lathe movement. A finger caught in gears or between the compound rest and a chuck jaw would be an example. The rule is to keep your hands away from such dangerous positions when the lathe is operating.

Hazards associated with broken or falling components. Heavy chucks or workpieces can be dangerous when accidentally dropped. Care must be used when handling them. If a threaded spindle is suddenly reversed, the chuck can come off and fly out of the lathe. A chuck wrench left in the chuck can become a missile when the machine is turned on. Always remove the chuck wrench immediately after using it (Figures 20a and 20b).

Hazards resulting from contact with high temperature components. Burns usually result from handling hot chips (up to 800° F or even more) or a hot workpiece. Gloves may be worn when handling hot chips or workpieces, but never worn when the machine is running.

Hazards resulting from contact with sharp edges, corners, and projections. These are perhaps the most common cause of hand injuries in lathe work. Dangerous sharp edges may be found many places: on a long stringy chip, on a tool bit, or on a burred edge of a turned or threaded part. Shields should be used for protection from flying chips and coolant. These shields are usually made of clear plastic and are hinged over the chuck or clamped to the carriage of engine lathes. Stringy chips must not be removed with bare hands; wear heavy gloves and use hook tools or pliers. Always turn off the machine before attempting to remove chips. Chips should be broken and 9-shaped rather than in a stringy mass or a long wire (Figure 21). Chip breakers on tools and correct feeds will help to produce safe, easily handled chips. Burred edges must be removed before the workpiece

Figure 20a. A chuck wrench left in the chuck is a danger to everyone in the shop (Lane Community College).

Figure 20b. A safety-conscious lathe operator will remove the chuck wrench when he finishes using it (Lane Community College).

is removed from the lathe. Always remove the tool bit when setting up or removing workpieces from the lathe.

Hazards of workholding devices or driving devices. When workpieces are clamped, their components often extend beyond the outside diameter of the holding device. Guards, barriers, and warnings such as signs or verbal instructions are all used to make you aware of the hazards. On power chucking devices you should be aware of potential pinch points between workpiece and workholding device. Make certain sufficient gripping force is exerted by the jaws to safely hold the work. Never run a geared scroll chuck without having something being gripped in the jaws. Centrifugal force on the jaws can cause the scroll to unwind and the jaws to come out of the chuck. Keep tools, files, and micrometers off the machine. They may vibrate off into the revolving chuck or workpiece.

Spindle braking. The spindle or workpiece should never be slowed or stopped by hand gripping or by using a pry bar. Always use machine controls to stop or slow it.

Workpieces extending out of the lathe should be supported by a stock tube (Figure 22). If a slender workpiece is allowed to extend beyond the headstock spindle a foot or so without support, it can fly outward from centrifugal force. The piece will not only be bent, but it will present a very great danger to anyone standing near.

Other Safety Considerations Hold one end of abrasive cloth strips in each hand when polishing rotating work. Don't let either hand get closer than a few inches from the work (Figure 23). Keep rags, brushes, and fingers away from

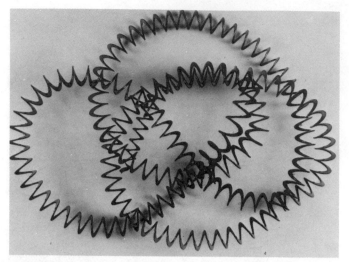

Figure 21. Unbroken lathe chips are sharp and hazardous to the operator (Lane Community College).

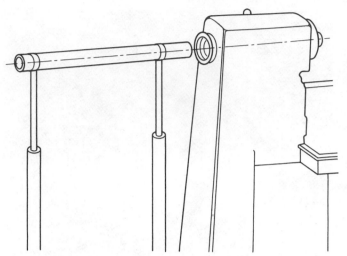

Figure 22. Stock tube is used to support long workpieces that extend out of the headstock of a lathe.

rotating work, especially when knurling. Roughing cuts tend to quickly drag in and wrap up rags, clothing, neckties, emery cloth, and hair. Move the carriage back out of the way and cover the tool with a cloth when checking boring work. When removing or installing chucks or heavy workpieces, use a board on the ways (a part of the lathe bed) so it can be slid into place. To lift a heavy chuck or workpiece (larger than an 8-inch diameter chuck) get help or use a crane (Figure 24). Remove the tool or turn it out of the way during this operation. Do not shift gears or try to take measurements while

Figure 23. Polishing in the lathe with emery cloth (Lane Community College).

Figure 24. The skyhook in use bringing a large chuck into place for mounting (Courtesy of Production Components, Inc., Ranchita, California).

the machine is running and the workpiece is in motion. Never use a file without a handle as the file tang can quickly cut your hand or wrist if the file is struck by a spinning chuck jaw or lathe dog. Left-hand filing is considered safest in the lathe; that is, the left hand grips the handle while the right hand holds the tip end of the file (Figure 25).

Figure 25. Left-hand filing in the lathe (Lane Community College).

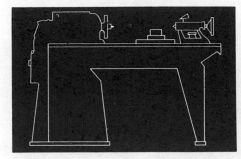

unit 1
the
engine
lathe

Modern lathes are highly accurate and complex machines capable of performing a great variety of operations. Before attempting to operate a lathe, you should familiarize yourself with its principal parts and their operation.

objective After completing this unit, you will be able to:
Identify the most important parts of a lathe and their functions.

One of the most important machine tools in the metal working industry is the lathe (Figure 1a). A lathe is a device in which the work is rotated against a cutting tool. As the cutting tool is moved lengthwise and crosswise to the axis of the workpiece, the shape of the workpiece is generated. Before you attempt to operate a lathe, you should learn the names and functions of its important parts.

Figure 1b shows a lathe and its most important parts identified. A lathe consists of the following major component groups: headstock, bed carriage, tailstock, quick-change gearbox, and a base or pedestal. The headstock is fastened on the left side of the bed. It contains the spindle that drives the various workholding devices. The spindle is supported by spindle bearings on each end. If they are sleeve-type bearings, a thrust bearing is also used to take up end play. Tapered roller spindle bearings are often used on modern lathes. Spindle speed changes are also made in the headstock, either with belts or with gears (Figure 2). The threading and feeding mechanisms of the lathe are also powered through the headstock.

Most belt driven lathes obtain a slow speed range through the use of back gears. Figure 4a

shows a back geared headstock. Slow RPM are obtained by disengaging the bull gear lockpin from the belt pulley and engaging the backgears with the backgear lever. When higher RPM are needed, the procedure is reversed by disengaging the backgears, then rotating the pulley by hand until the bull gear lockpin engages in the pulley (Figure 4b). Do not forget to close the gear and pulley guards before starting the lathe.

The spindle is hollow to allow long slender workpieces to pass through. The spindle end facing the tailstock is called the spindle nose (Figure 5). Spindle noses usually are one of three designs, a long taper key drive (Figure 5), a cam lock type (Figure 6), or a threaded spindle nose (Figure 7). Lathe chucks and other workholding devices are fastened to and driven by the spindle nose. The hole in the spindle nose typically has a standard Morse taper. The size of this taper varies with the size of the lathe.

The bed (Figure 8) is the foundation and backbone of a lathe. Its rigidity and alignment affect the accuracy of the parts machined on it. Therefore, lathe beds are constructed to withstand the stresses created by heavy machining cuts. On top of the bed are the ways, which

Figure 1*a*. The engine lathe (Courtesy of Clausing Corporation).

Figure 2. Geared headstock for heavy duty lathe (Courtesy of the Lodge & Shipley Company).

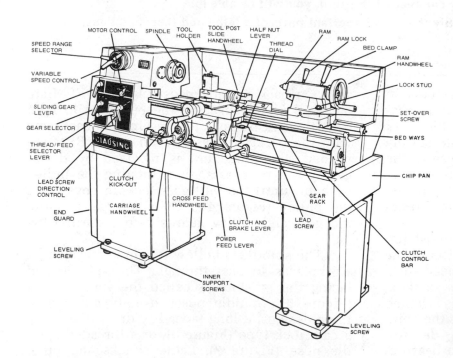

Figure 3. Spindle drive showing gears and shifting mechanism located in the headstock (Courtesy of the Lodge & Shipley Company).

Figure 1*b*. Engine lathe with the parts identified (Courtesy of Clausing Corporation).

usually consist of two inverted vees and two flat bearing surfaces. The ways of the lathes are very accurately machined by grinding or by milling and hand scraping. Wear or damage to the ways will affect the accuracy of workpieces machined on them. A gear rack is fastened below the front way of the lathe. Gears that link the carriage handwheel to this rack make possible the lengthwise movement of

the carriage by hand.

The carriage (Figure 9) is made up of the saddle and the apron. The saddle rides on top of the ways and carries the cross slide and the compound rest. The cross slide is moved perpendicular to the axis of the lathe by manually turning the cross feed screw handle or by engaging the cross feed lever for automatic power feed. The compound rest is mounted on the

Figure 4a. View of headstock showing flat belt drive, the back gear engaged, and the lock pin disengaged (Lane Community College).

Figure 6. Camlock spindle nose (DeAnza College).

Figure 4b. View of headstock showing back gear disengaged and lock pin engaged for direct belt drive (Lane Community College).

Figure 7. Threaded spindle nose (Lane Community College).

Figure 5. Long taper key drive spindle nose (Lane Community College).

cross slide and can be swiveled to any angle horizontal with the lathe axis in order to produce bevels and tapers. The compound rest can only be moved manually by turning the compound rest feed screw handle. Cutting tools are fastened on a tool post that is located on the compound rest.

The apron is the part of the carriage facing the operator. It contains the gears and feed clutches that transmit motion from the feed rod or lead screw to the carriage and cross slide. Fastened to the apron is the thread dial, which indicates the exact time to engage the half-nuts

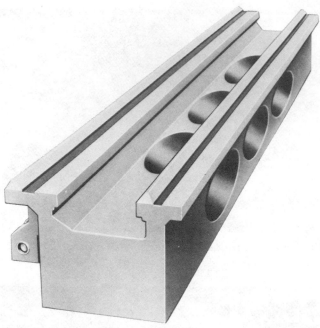

Figure 8. Lathe bed (Courtesy of Clausing Corporation).

Figure 10. Tailstock (Courtesy of The Monarch Machine Tool Company, Sidney, Ohio).

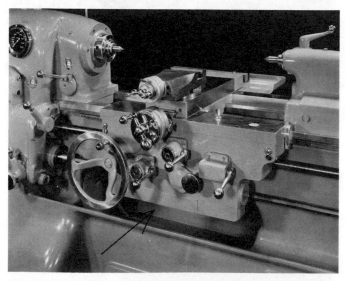

Figure 9. Lathe apron (Courtesy of The Monarch Machine Tool Company, Sidney, Ohio).

Figure 11. Quick-change gearbox showing index plate (Lane Community College).

while thread cutting. The half-nut lever is used only while cutting threads. The entire carriage can be moved along the lathe bed manually by turning the carriage handwheel or under power by engaging the power feed controls on the apron. Once in position, the carriage can be clamped to the bed by tightening the carriage lock screw.

The tailstock (Figure 10) is used to support one end of a workpiece for machining or to hold various cutting tools such as drills, reamers, and taps. The tailstock slides on the ways and can be clamped in any position along the bed. The tailstock has a sliding spindle that is operated by a handwheel and locked in position with a spindle clamp lever. The spindle is bored to receive a standard Morse taper shank. The tailstock consists of an upper and lower unit and can be adjusted to make tapered workpieces by turning the adjusting screws in the base unit.

The quick-change gearbox (Figure 11) is the link that transmits power between the spin-

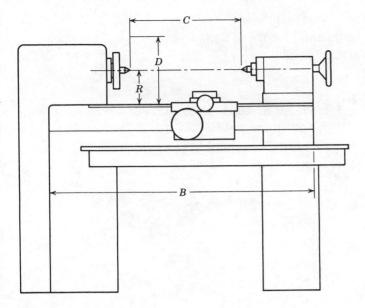

dle and the carriage. By using the gear shift levers on the quick-change gearbox, you can select different feeds. Power is transmitted to the carriage through a feed rod, or as on smaller lathes, through the leadscrew with a keyway in it. The index plate on the quick-change gearbox indicates the feed in thousandths of an inch or as threads per inch for the lever positions.

The base of the machine is used to level the lathe and to secure it to the floor. The motor of the lathe is usually mounted in the base. Figure 12 shows how the lathe is measured.

Figure 12. Measuring a lathe for size. C = maximum distance between centers; D = maximum diameter of workpiece over ways—swing of lathe; R = radius, one-half swing; B = length of bed.

self-evaluation

SELF-TEST At a lathe in the shop, identify the following parts and describe their functions. *Do not* turn on the lathe until you get permission from your instructor.

The Headstock:
1. Spindle
2. Spindle speed changing mechanism
3. Backgears
4. Bull gear lockpin
5. Spindle nose
6. What kind of spindle nose is on your lathe?
7. Feed reverse lever

The Bed:
1. The ways
2. The gear rack

The Carriage:
1. The cross slide
2. The compound rest
3. Saddle
4. Apron
5. Power feed lever
6. Feed change lever
7. Half-nut lever
8. Thread dial
9. Carriage handwheel
10. Carriage lock

The Tailstock: 1. Spindle and spindle clamping lever
2. Tapered spindle hole and the size of its taper
3. The tailstock adjusting screws

The Quick-change Gearbox: 1. The leadscrew
2. Shift the levers to obtain feeds of .005 and .013 in. per revolution. Rotating the leadscrew with your fingers aids in shifting these levers.
3. Set the levers to obtain 4 threads/in. and then 12 threads/in.
4. Measure the lathe and record its size.

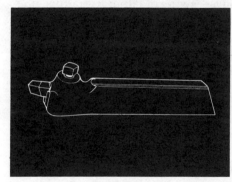

unit 2
toolholders and tool holding for the lathe

For lathe work, cutting tools must be supported and fastened securely in the proper position to machine the workpiece. There are many different types of tool holders available to satisfy this need. Anyone working with a lathe should be able to select the best tool holding device for the operation performed.

objectives **After completing this unit, you will be able to:**
1. **Identify standard, quick-change, and turret-type toolholders mounted on a lathe carriage.**
2. **Identify tool holding for the lathe tailstock.**

A cutting tool is supported and held in a lathe by a toolholder that is secured in the tool post of the lathe with a clamp screw. A common tool post found on smaller or older lathes is shown in Figure 1. Tool height adjustments are made by swiveling the rocker in the tool post ring. Making adjustments in this manner changes the effective back rake angle and also the front relief angle of the tool.

Many types of toolholders are used with the standard tool post. A straight shank turning toolholder (Figure 2) is used with high speed

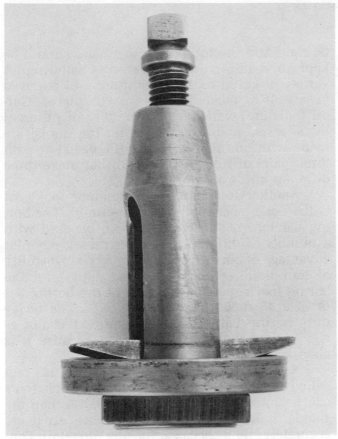

Figure 1. Standard-type tool post with ring and rocker.

Figure 2. Straight shank toolholder with built-in back rake holding a high speed right-hand tool (Lane Community College).

Figure 3. Right-hand toolholder for carbide tool bits without back rake.

Figure 4. Left-hand toolholder with right-hand tool (Lane Community College).

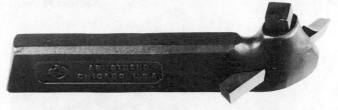

Figure 5. Right-hand toolholder with left-hand tool (Lane Community College).

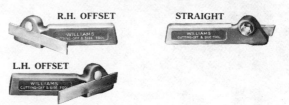

Figure 6. Three kinds of cutoff toolholders with cut-off blades (Courtesy of J. H. Williams Division of TRW, Inc.).

Figure 7. Knuckle-joint knurling tool (Lane Community College).

tool bits. The tool bit is held in the toolholder at a $16\frac{1}{2}$ degree angle, which provides a positive back rake angle for cutting. Straight shank toolholders are used for general machining on lathes. The type shown in Figure 3 is used with carbide tools.

Offset toolholders (Figures 4 and 5) allow machining close to the chuck or tailstock of a lathe without tool post interference. The left-hand toolholder is intended for use with tools cutting from right to left or toward the headstock of the lathe.

A toolholder should be selected according to the machining to be done. The setup should

Figure 8. Triple head knurling tool (Lane Community College).

Figure 9. Toolholder for small boring bars (Courtesy of J. H. Williams Division of TRW, Inc.).

Figure 10. Boring bar tool post and bars with special wrench (Courtesy of J. H. Williams Division of TRW, Inc.).

Sleeve Bar

Figure 11. Heavy duty boring bar holder (Courtesy of J. H. Williams Division of TRW, Inc.).

be rigid and the toolholder overhang should be kept to a minimum to prevent chattering. A variety of cutoff toolholders (Figure 6) are used to cut off or make grooves in workpieces. Cut-off tools are available in a number of different thicknesses and heights. Knurling tools are made with one pair of rollers (Figure 7) or with three pairs of rollers (Figure 8) that make three different kinds of knurls.

Another tool used in a standard tool post is the boring bar toolholder (Figure 9). The boring bar tool post (Figure 10) can be used with a number of different boring bar sizes. Another advantage of boring bars is the interchangeability of tool holding end caps. End caps hold the boring tool square to the axis of the boring bar or at a 45 or 60 degree angle to it. The heavy duty boring bar holder in Figure 11 is not as rigid as the holder in Figure 10 because it is clamped in the tool post.

A quick-change tool post (Figure 12), so-called because of the speed with which tools can be interchanged, is more versatile than the standard post. The toolholders used on it are accurately held because of the dovetail construction of the post. This accuracy makes for more exact repetition of setups. Tool height

Figure 12. Quick-change tool post, dovetail type (Copyright © 1975, Aloris Tool Company, Inc.).

Figure 13. Three-sided quick-change tool post (Lane Community College).

Figure 15b. Threading is accomplished with the bottom edge of the blade with the lathe spindle in reverse. This assures cutting right-hand threads without hitting the shoulders (Copyright © 1975, Aloris Tool Company, Inc.).

Figure 14. Turning toolholder in use (Copyright © 1975, Aloris Tool Company, Inc.).

Figure 15a. Threading toolholder, using the top of the blade (Copyright © 1975, Aloris Toll Company, Inc.).

Figure 16. Drill toolholder in the tool post. Mounting the drill in the tool post makes drilling with power feed possible (Copyright © 1975, Aloris Tool Company, Inc.).

adjustments are made with a micrometer adjustment collar, and the height alignment will remain constant through repeated tool changes.

A three-sided quick-change tool post (Figure 13) has the added ability to mount a tool on the tailstock side of the tool post. These tool posts are securely clamped to the compound rest. The tool post in Figure 13 uses double vees to locate the toolholders, which are clamped and released from the post by turning the top lever.

Toolholders for the quick-change tool posts include those for turning (Figure 14), threading (Figures 15a and b) and holding drills (Figure 16). The drill holder makes it possible to use the carriage power feed when drilling holes instead of the tailstock hand feed. Figure 17 shows a boring bar toolholder in use; the boring bar is very rigidly supported.

An advantage of the quick-change tool post toolholders is that cutting tools of various shank thicknesses can be mounted in the toolholders (Figure 18). Shims are sometimes used when the shank is too small for the setscrews to reach.

Figure 19. Tailstock turret used in quick-change toolholder (Courtesy of Enco Manufacturing Company).

Figure 20. Quick-change cutoff toolholder (Courtesy of Enco Manufacturing Company).

Figure 17. Boring toolholder. This setup provides good boring bar rigidity (Copyright © 1975, Aloris Tool Company, Inc.).

Figure 18. Toolholders are made with wide or narrow slots to fit tools with various shank thicknesses (Lane Community College).

Figure 21. Quick-change knurling and facing toolholder (Courtesy of Enco Manufacturing Company).

Figure 22. Facing cut with a turret-type toolholder (Courtesy of Enco Manufacturing Company).

Figure 23. Turning cut with a turret-type toolholder (Courtesy of Enco Manufacturing Company).

Figure 24. Chamfering cut with a turret-type toolholder (Courtesy of Enco Manufacturing Company).

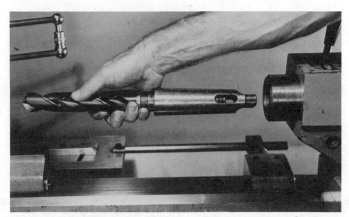

Figure 25. Taper shank just in front of tailstock spindle hole (Lane Community College).

Another example of quick-change tool post versatility is shown in Figure 19, where a tailstock turret is in use. Figure 20 shows a cutoff tool mounted in a toolholder. Figure 21 is a combination knurling tool and facing toolholder. A four-tool turret toolholder (Figure 22) can be set up with several different tools such as turning tools, facing tools, threading or boring tools. Often one tool can perform two or more operations, especially if the turret can be indexed in 30-degree intervals. A facing operation (Figure 22), a turning operation (Figure 23), and chamfering of a bored hole (Figure 24) are all performed from this turret. Tool height adjustments are made by placing shims under the tool.

The toolholders studied so far are all intended for use on the carriage of a lathe. Tool-

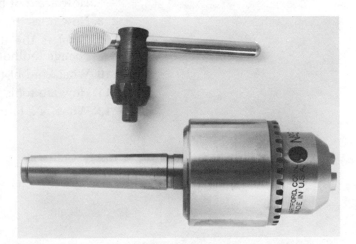

Figure 26. Drill chuck with Morse taper shank.

holding is also done on the tailstock. Figure 25 shows how the tailstock spindle is used to hold Morse taper shank tools. One of the most common toolholding devices used on a tailstock is the drill chuck (Figure 26). A drill chuck is used for holding straight shank drilling tools. When a series of operations must be performed and repeated on several workpieces, a tailstock turret (Figure 27) can be used. The illustrated tailstock turret has six tool positions, one of which is used as a workstop. The other positions are for center drilling, drilling, reaming, counterboring, and tapping. Tailstock tools are normally fed by turning the tailstock handwheel.

Figure 27. Tailstock turret (Courtesy of Enco Manufacturing Company).

self-evaluation

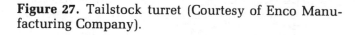

SELF-TEST At a lathe in the shop, identify various toolholders and their functions.

1. What is the purpose of a toolholder?
2. How is a standard left-hand toolholder identified?
3. What is the difference between a standard-type toolholder for high speed steel tools and for carbide tools?
4. Which standard toolholder would be best used for turning close to the chuck?
5. How are tool height adjustments made on a standard toolholder?
6. How are tool height adjustments made on a quick-change toolholder?
7. How are tool height adjustments made on a turret-type toolholder?
8. How does the toolholder overhang affect the turning operation?
9. What is the difference between a standard toolholder and a quick-change toolholder?
10. What kind of tools are used in the lathe tailstock?
11. How are tools fastened in the tailstock?
12. When is a tailstock turret used?

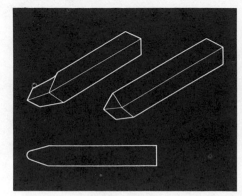

unit 3
cutting tools
for the lathe

A machinist must fully understand the purpose of tool geometry, since it is the lathe tool that removes the metal from the workpiece. Whether this is done safely, economically, and with quality finishes depends to a large extent upon the shape of the point, the rake and relief angles, and the nose radius of the tool. In this unit, you will learn this tool geometry and also how to grind a lathe tool.

objectives **After completing this unit, you will be able to:**
1. **Explain the purpose of rake and relief angles, chip breakers, and form tools.**
2. **Grind an acceptable right-hand roughing tool.**

On a lathe, metal is removed from a workpiece by turning it against a single point cutting tool. This tool must be very hard and it should not lose its hardness from the heat generated by machining. High speed steel is used for many tools as it fulfills these requirements and is easily shaped by grinding. It should be noted, however, that their use is limited since most production machining today is done with carbide tools. High speed steel tools are required for older lathes that are equipped with only low speed ranges. They are also useful for finishing operations, especially on soft metals.

The most important aspect of a lathe tool is its geometric form: the side and back rake, front and side clearance or relief angles, and chip breakers. (See Figure 14 for the nomenclature of the lathe tool.)

Grinding a tool provides both a sharp cutting edge and the shape needed for the cutting operation. When the purpose for the rake and relief angles on a tool are clearly understood, then a tool suitable to the job may be ground. Left-hand tools are shaped just the opposite to right-hand tools (Figure 1). The right-hand tool has the cutting edge on the left side and cuts to the left or toward the headstock.

Tools are given a slight nose radius to strengthen the tip. A larger nose radius will give a better finish (Figure 2), but will also promote chattering (vibration) in a nonrigid setup. A facing tool (Figure 3) for shaft ends and mandrel work has very little nose radius and an included angle of 58 degrees. This facing tool is not used for chucking work, however, as it is a relatively weak tool. A right-hand (RH) or left-hand (LH) roughing or finishing tool is often used for facing in chuck mounted workpieces.

Some useful tool shapes are shown in Figure 4. These are used for general lathe work.

Tools that have special shaped cutting edges are called form tools (Figure 5). These tools are plunged directly into the work, making the full cut in one operation.

Parting or cutoff tools are often used for necking or undercutting, but their main function is cutting off material to the correct length. The correct and incorrect ways to grind a cutoff tool are shown in Figure 6. Note that the width of the cutting edge becomes narrower than the blade as it is ground deeper, which causes the blade to bind in a groove that is deeper than the sharpened end.

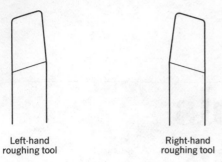

Left-hand
roughing tool

Right-hand
roughing tool

Figure 1. Left-hand and right-hand roughing tools.

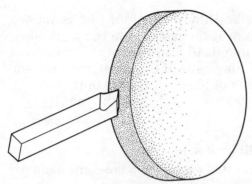

Figure 2. One method of grinding the nose radius on the point of the tool.

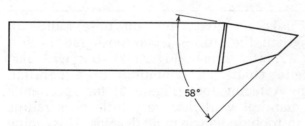

58°

Figure 3. A right-hand facing tool showing point angles. This tool is not suitable for roughing operations because of its acute point angle.

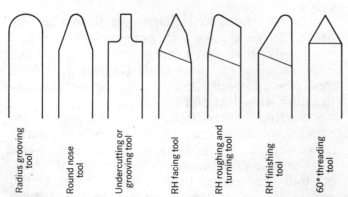

Radius grooving tool

Round nose tool

Undercutting or grooving tool

RH facing tool

RH roughing and turning tool

RH finishing tool

60° threading tool

Figure 4. Some useful tool shapes most often used. The first tool shapes needed are the three on the right, which are the roughing or general turning tool, finishing tool, and threading tool.

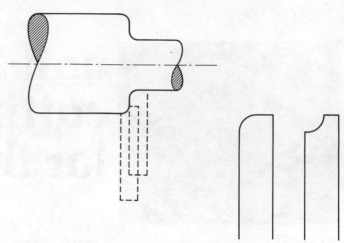

Figure 5. Form tools are used to produce the desired shape in the workpiece. External radius tools, for example, are used to make outside corners round, while fillet radius tools are used on shafts to round the inside corners on shoulders.

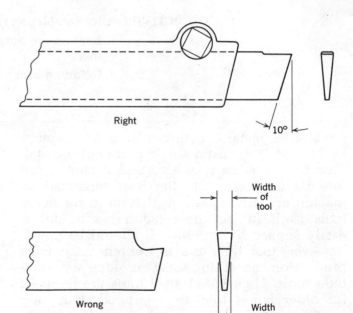

Right

10°

Width of tool

Wrong

Width of groove cut by tool

Figure 6. Correct and incorrect methods of grinding a cut-off tool for deep parting.

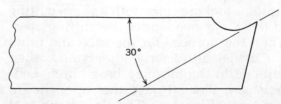

30°

Figure 7. Cut-off tools are sometimes ground with large back rake angles for aluminum and other soft metals.

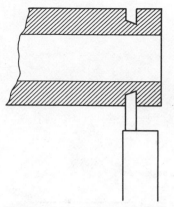

Figure 8. Parting tool ground on an angle to avoid burrs on the cut off pieces.

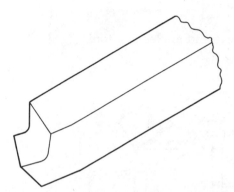

Figure 9. Misshapen tool caused by many resharpenings. The chip trap should be ground off and a new point ground on the tool.

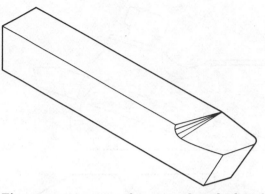

Figure 10. A properly ground right-hand roughing tool.

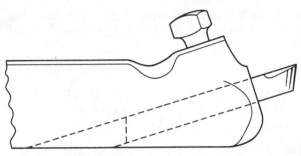

Figure 11. Toolholder with back rake.

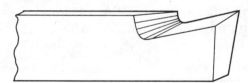

Figure 12. Right-hand roughing tool with back rake.

Tools are sometimes especially ground, however, for parting very soft metals or specially shaped grooves (Figure 7). The end is sometimes ground on a slight angle when a series of small hollow pieces is being cut off (Figure 8). This helps to eliminate the burr on small parts. This procedure is not recommended for deep parting.

Tools that have been ground back for resharpening too many times often form a "chip trap" causing the metal to be torn off or the tool to not cut at all (Figure 9). A good machinist will never allow tools to get in this condition, but will grind off the useless end and regrind a proper tool shape (Figure 10).

Although many modern lathes have toolholders that hold the tool horizontally, some lathe toolholders have built-in back rake (Figure 11) so it is not necessary to grind one into the tool as used in Figure 11. The tool in Fig-

ure 12, however, is ground with a back rake and can be used in a toolholder that does not have built-in back rake. Threading tools should have zero rake (Figure 13a). If a horizontal toolholder is used, the flat on top is unnecessary as the tool would have a zero rake. The tool should also be checked for the 60 degree angle with a center gage (Figure 13b) while grinding. Relief should be ground on each side. A slight flat should be honed on the end with an oilstone. (See Unit 11, "Cutting Unified External Thread" for this dimension.)

Tools for brass or plastics should have zero to negative rake to keep the tools from "digging in." Figure 14 shows the parts and angles of the tool according to a commonly used industrial

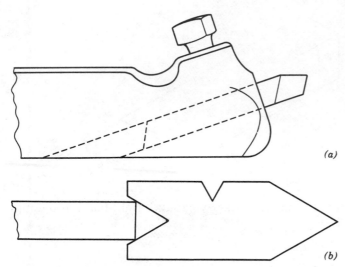

Figure 13. (*a*) Toolholder with back rake showing tool ground for zero rake. (*b*) Checking a threading tool with center gage.

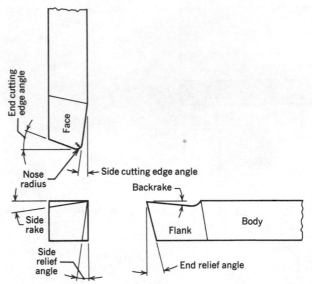

Figure 14. The parts and angles of a tool.

tool signature. The terms and definitions follow (the angles given are only examples and they could vary according to the application):

Back rake	BR	12°
Side rake	SR	12°
End relief	ER	10°
Side relief	SRF	10°
End cutting edge angle	ECEA	30°
Side cutting edge angle	SCEA	15°
Nose radius	NR	$\frac{1}{32}$ inch

Tool Signature

1. The tool shank is that part held by the toolholder.
2. Back rake is very important to smooth chip flow, which is needed to have a uniform chip and a good finish.
3. The side rake directs the chip flow away from the point of cut and it provides for a keen cutting edge.
4. The end relief angle prevents the front edge of the tool from rubbing on the work.
5. The side relief angle provides for cutting action by allowing the tool to feed into the work material.
6. The cutting edge angle may vary considerably (from 5 to 32 degrees). For roughing, it should almost be square (5 degrees off 90 degrees), but tools used for squaring shoulders or for other light machining could have angles from 15 to 32 degrees.
7. The side cutting edge angle, which is usually 10 to 20 degrees, directs the cutting forces back into a stronger section of the tool point. It helps to direct the chip flow away from the workpiece. It also affects the thickness of the cut (Figure 15).

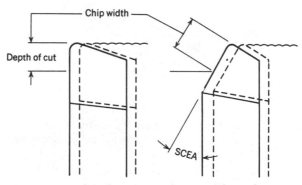

Figure 15. The change in chip width with an increase of the side cutting edge angle. A large SCEA can sometimes cause chatter (vibration of work or tool).

8. The nose radius will vary according to the finish required.

Side rake, back rake, and relief angles are given for tools in Table 1 for machining various metals.

Figure 16 shows the use of a gage for checking angles when grinding a tool. It is designed to check tool angles on any flat surface. The angles are for tools to be used in toolholders having $16\frac{1}{2}$ degree back rake. Tools for straight or horizontal toolholders should have an end relief of 10 degrees; in these cases, change the end relief angle of the gage. A protractor or optical comparator can also be used to check tool angles.

Table 1
Angle Degrees for High Speed Steel Tools

Material	End Relief	Side Relief	Side Rake	Back Rake
Aluminum	8 to 10	12 to 14	14 to 16	30 to 35
Brass, free cutting	8 to 10	8 to 10	1 to 3	0
Bronze, free cutting	8 to 10	8 to 10	2 to 4	0
Cast iron, gray	6 to 8	8 to 10	10 to 12	3 to 5
Copper	12 to 14	13 to 14	18 to 20	14 to 16
Nickel and monel	12 to 14	14 to 16	12 to 14	8 to 10
Steels, low carbon	8 to 10	8 to 10	10 to 12	10 to 12
Steels, alloy	7 to 9	7 to 9	8 to 10	6 to 8

For your safety, it is important to make tools that will produce chips that are not hazardous. Long, unbroken chips are extremely dangerous. Tool geometry, especially side and back rakes, has a considerable effect on chip formation. Side rakes with smaller angles tend to curl the chip more than those with large angles, and the curled chips are more likely to break up. Coarse feeds for roughing and maximum depth of cut also promote chip breaking. Feeds, speeds, and depth of cut will be further considered in Unit 7, "Turning Between Centers."

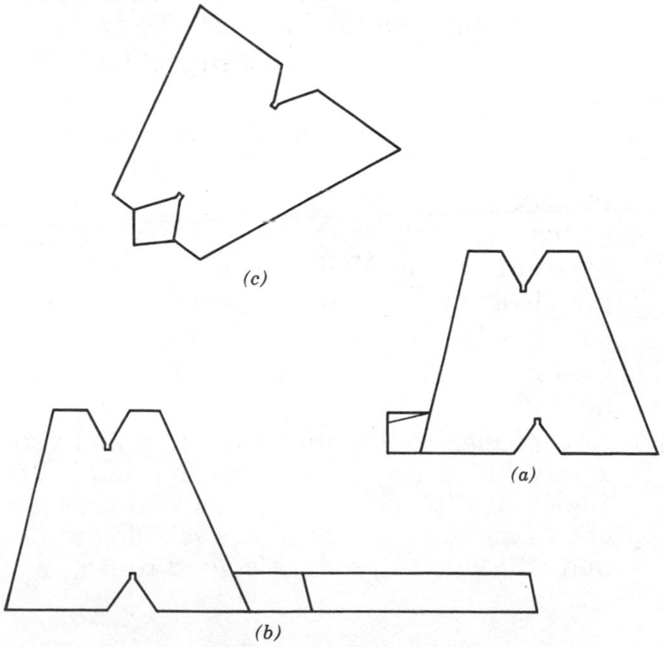

Chip breakers are extensively used on both carbide and high speed tools to curl the chip as it flows across the face of the tool. Since the chip is curled back into the work, it can go no farther and breaks (Figure 17). A C-shaped chip is often the result, but a figure-9 shaped chip is considered ideal (Figure 18). This chip should drop safely into the chip pan without flying out.

Grinding the chip breaker too deep will form a chip trap that may cause binding of the chip and tool breakage (Figure 19). The correct depth to grind a chip breaker is approximately $\frac{1}{32}$ inch. Chip breakers are typically of the parallel or angular types (Figure 20). More skill is needed to offhand grind a chip breaker on a high speed tool than is required to grind the basic tool angles. Therefore, the basic tool should be ground and an effort made to produce safe chips through the use of correct feeds and depth of cut before a chip breaker is ground.

(c)

(a)

(b)

Figure 16. Using a tool gage for checking angles. (a) Checking the side relief angle; (b) checking end relief; and (c) checking the wedge angle for steel.

Figure 17. Chip flow with a plain tool and with a chip breaker.

Figure 18. A figure-9 chip is considered the safest kind of chip to produce (DeAnza College).

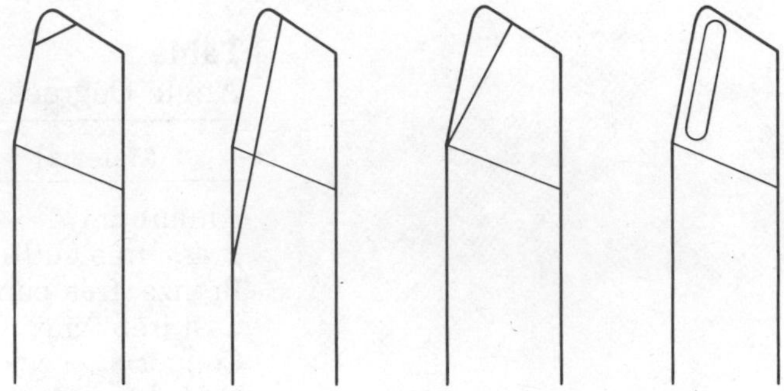

Care must be exercised while grinding on high speed steel. A glazed wheel can generate heat up to 2000°F (1093°C) at the grinder-tool interface. Do not overheat the tool edge as this will cause small surface cracks that can result in the failure of the tool. Frequent cooling in water will keep the tool cool enough to handle. Do not quench in water, however, if you have overheated it. Let it cool in air.

Since the right-hand roughing tool is the one that is most commonly used by machinists and the first one that you will need, you should begin with it. A piece of keystock the same size as the tool bit should be used for practice until you are able to grind an acceptable tool.

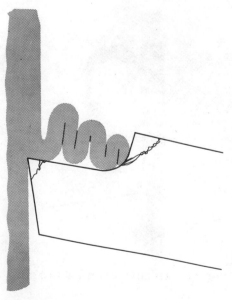

Figure 19. Crowding of the chip is caused by a chip breaker that is ground too deeply.

Figure 20. Four common types of chip breakers.

GRINDING PROCEDURE, TOOL BIT

Given a tool blank, a piece of keystock about 3 in. long, a tool gage and a tool holder.

1. Grind one acceptable practice right-hand roughing tool. Have your instructor evaluate your progress.
2. Grind one acceptable right-hand roughing tool from a high speed tool blank.

Wear goggles and make certain the tool rest on the grinder is adjusted properly (about $\frac{1}{16}$ in. from the wheel). True up the wheels with a wheel dresser, if they are grooved, glazed, or out of round.

1. Using the roughing wheel, grind the side relief angle and the side cutting angle about 10 degrees by holding the blank and supporting your hand on the tool rest (Figure 21).
2. Check the angle with a tool gage (Figure 22). Correct if needed.
3. Rough out the end relief angle about 14 degrees and the end cutting edge angle (Figure 23).
4. Check the angle with the tool gage (Figure 24). Correct if needed.
5. Rough out the side rake. Stay clear of the side cutting edge by $\frac{1}{16}$ in. (Figure 25).
6. Check for wedge angle (Figure 26). Correct if needed.
7. Now change to the finer grit wheel and very gently finish grind the side and end relief angles. Try to avoid making several facets or grinds on one surface. A side to side oscillation will help to produce a good finish.

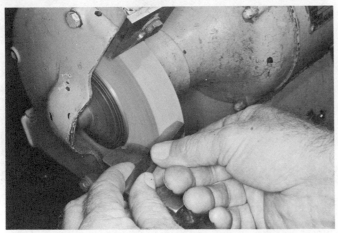

Figure 21. Roughing the side relief angle and the side cutting edge angle (Lane Community College).

Figure 24. Checking the end relief angle with a tool gage (Lane Community College).

Figure 22. Checking the side cutting edge angle with a tool gage (Lane Community College).

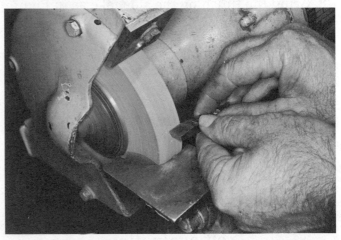

Figure 25. Roughing the side rake (Lane Community College).

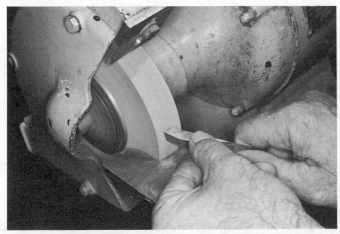

Figure 23. Roughing the end relief angle and the end cutting edge angle (Lane Community College).

Figure 26. Checking for wedge angle with a tool gage (Lane Community College).

Figure 27. One method of grinding the nose radius is on the circumference of the wheel (Lane Community College).

Figure 28. Using an oil stone to remove the burrs from the cutting edge (Lane Community College).

8. Grind the finish on the side rake as in Figure 25 and bring the ground surface just to the side cutting edge, but avoid going deeper.

9. A slight radius on the point of the tool should be ground on the circumference of the wheel (Figure 27) and all the way from the nose to the heel of the tool.

10. A medium to fine oilstone is used to remove the burrs from the cutting edge (Figure 28). The finished tool is shown in Figure 29.

Figure 29. The finished tool (Lane Community College).

self-evaluation

SELF-TEST

1. Name the advantages to using high speed steel for tools.
2. Other than hardness and toughness, what is the most important aspect of a lathe tool?
3. How do form tools work?
4. A tool that has been reground too many times on the same place can form a chip trap. Describe the problems that result from this condition.
5. Why is not always necessary to grind a back rake into the tool?
6. When should a zero or negative rake be used?
7. Explain the purpose of the side and end relief angles.
8. What is the function of the side and back rakes?
9. How can these angles be checked while grinding?
10. Why should chips be broken up?
11. In what ways can chips be broken?
12. Overheating a high speed tool bit can easily be done by using a glazed wheel that needs dressing or by exerting too much pressure. What does this cause in the tool?

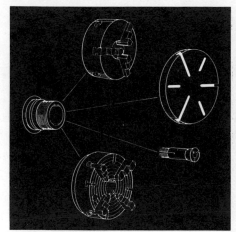

unit 4 lathe spindle tooling

Workholding and driving devices that are fastened to the spindle nose are very important to machining on lathes. Various types of these workholding devices, their uses, and proper care are detailed in this unit.

objectives After completing this unit, you will be able to:
1. Explain the uses and care of independent and universal chucks.
2. Explain the limitations and advantages of collets and describe a collet setup.
3. Explain the use of a face driver or drive center.
4. Explain the uses and differences of drive plates and face plates.

The lathe spindle nose is the carrier of a variety of workholding devices that are fastened to it in several ways. The spindle is hollow and has an internal Morse taper at the nose end, which makes possible the use of taper shank drills or drill chucks (Figure 1). This internal taper is also used to hold live centers, drive centers, or collet assemblies. The outside of the spindle nose can have either a threaded nose (Figure 2), a long taper with key drive (Figure 3), or a camlock (Figure 4).

Threaded spindle noses are mostly used on older lathes. The chuck or face plate is screwed on a coarse, right-hand thread until it is forced against a shoulder on the spindle that aligns it. Two disadvantages of the threaded spindle nose are that the spindle cannot be rotated in reverse against a load and that it is sometimes very difficult to remove a chuck or face plate (Figures 5a and 5b).

The long taper key drive spindle nose relies on the principle that a tapered fit will always repeat its original position. The key gives additional driving power. A large nut

having a right-hand thread is turned with a spanner wrench. It draws the chuck into position and holds it there.

Camlock spindle noses use a short taper for alignment. A number of studs arranged in a circle fit into holes in the spindle nose. Each stud has a notch into which a cam is turned to lock it in place.

All spindle noses and their mating parts must be carefully cleaned before assembly. Small chips or grit will cause a workholding device to run out of true and be damaged. A spring cleaner (Figure 6) is used on mating threads for threaded spindles. Brushes and cloths are used for cleaning. A thin film of light oil should be applied to threads and mating surfaces.

Figure 1. Section view of the spindle.

423

Figure 2. Threaded spindle nose (Lane Community College).

Figure 4. Camlock spindle nose (DeAnza College).

Figure 3. Long taper with key drive spindle nose (Lane Community College).

Figure 5a. One method of removing a chuck or face plate from a threaded spindle nose is to put the lathe in the lowest gear and then reverse it against a block of wood by switching it on and off to loosen the chuck. The chuck should then be screwed off by hand (Lane Community College).

Independent four-jaw and universal three-jaw chucks and, occasionally, drive or face plates are mounted on the spindle nose of engine lathes. Each of the four jaws of the independent chuck moves independently of the others, which makes it possible to set up oddly shaped pieces (Figure 7). The concentric rings on the chuck face help to set the work true before starting the machine. Very precise setups

Figure 5b. The chuck can also be removed by using a large monkey wrench on one of the chuck jaws while the spindle is locked in a low gear (Lane Community College).

Figure 7. Four-jaw independent chuck holding an offset rectangular part (DeAnza College).

Figure 6. A spring cleaner is used for cleaning internal threads on chucks.

Figure 8. Four-jaw chuck in reverse position holding a large diameter workpiece (Lane Community College).

also can be made with the four-jaw chuck by using a dial indicator, especially on round material. Each jaw of the chuck can be removed and reversed to accommodate irregular shapes. Some types are fitted with top jaws that can be reversed after removal of bolts on the jaw. Jaws in the reverse position can grip larger diameter workpieces (Figure 8). The independent chuck will hold work more securely for heavy cutting than will the three-jaw universal chuck.

Universal chucks usually have three jaws, but some are made with two jaws (Figure 9)

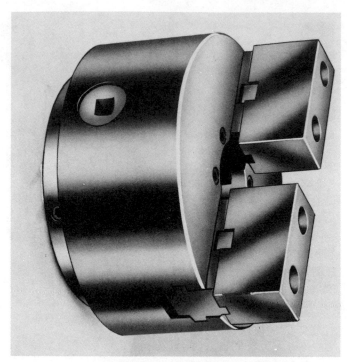

Figure 9. Two-jaw Universal chuck (Photo courtesy of Hardinge Brothers, Inc., Elmira, N.Y.).

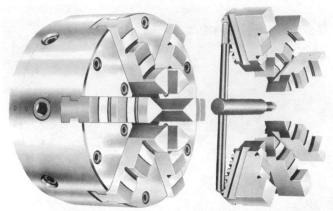

Figure 10. Six-jaw universal chuck (Courtesy of Buck Tool Company).

or six jaws (Figure 10). All the jaws are moved in or out equally in their slides by means of a scroll plate located back of the jaws. The scroll plate has a bevel gear on its reverse side that is driven by a pinion gear. This gear extends to the outside of the chuck body and is turned with the chuck wrench (Figure 11). Universal chucks provide quick and simple chucking and centering of round stock. Uneven or irregularly shaped material will damage these chucks.

The jaws of universal chucks will not re-

verse as with independent chucks, so a separate set of reverse jaws is used (Figure 12) to hold pieces with larger diameters. The chuck and each of its jaws are stamped with identification numbers. Do not interchange any of these parts with another chuck or both will be inaccurate. Also each jaw is stamped 1, 2, or 3 to correspond to the same number stamped by the slot on the chuck. The jaws are removed from the chuck in the order 3, 2, 1 and should be returned in the reverse order, 1, 2, 3.

A universal chuck with top jaws (Figure 13) is reversed by removing the bolts in the top jaws and by reversing them. They must be carefully cleaned when this is done. Soft top jaws are frequently used when special gripping problems arise. Since the jaws are machined to fit the shape of the part (Figure 14), they can grip it securely for heavy cuts (Figure 15).

One disadvantage of most universal chucks

Figure 11. Exploded view of a universal three-jaw chuck (Adjust-tru) showing the scroll plate and gear drive mechanism (Courtesy of Buck Tool Company).

Figure 12. Universal three-jaw chuck (Adjust-tru) with a set of outside jaws (Courtesy of Buck Tool Company).

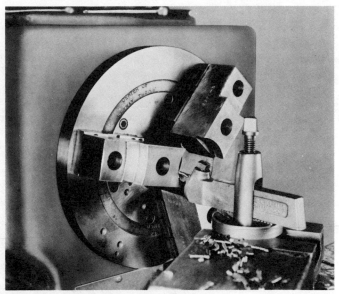

Figure 14. Machining soft jaws to fit an odd-shaped workpiece on a jaw turning fixture (Courtesy of The Warner & Swasey Company).

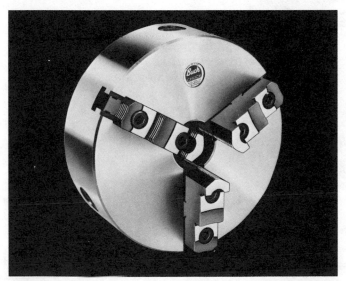

Figure 13. Universal chuck with top jaws (Courtesy of Buck Tool Company).

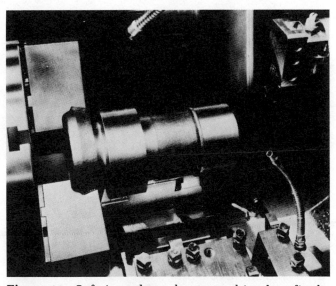

Figure 15. Soft jaws have been machined to fit the shape of this cast steel workpiece in order to hold it securely for heavy cuts (Courtesy of The Warner & Swasey Company).

is that they lose their accuracy when the scroll and jaws wear, and normally there is no compensation for wear other than regrinding the jaws. The three-jaw adjustable chuck in Figure 16 has a compensating adjustment for wear or misalignment.

Combination universal and independent chucks also provide for quick opening and closing and have the added advantage of independent adjustment on each jaw. These chucks are like the universal type since three or four

jaws move in or out equally, but each jaw can be adjusted independently as well.

Magnetic chucks (Figure 17) are sometimes used for making light cuts on ferromagnetic material. They are useful for facing thin material that would be difficult to hold in conventional workholding devices. Magnetic chucks

Figure 16. (*right*) Universal chuck (Adjust-tru) with special adjustment feature (G) makes it possible to compensate for wear (Courtesy of Buck Tool Company).

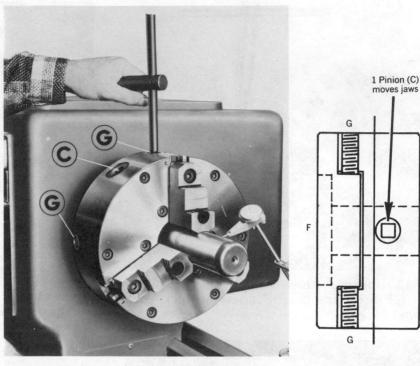

Figure 17. Magnetic chuck (Courtesy of Enco Manufacturing Company).

do not hold work very securely and so are not used much in lathe work.

All chucks need frequent cleaning of scrolls and jaws. These should be lightly oiled after cleaning and chucks with grease fittings should be pressure lubricated. Chucks come in all diameters and are made for light, medium, and heavy duty uses.

Drive plates are used together with lathe dogs to drive work mounted between centers (Figure 18). The live center fits directly into the spindle taper and turns with the spindle. A sleeve is sometimes used if the spindle taper is too large in diameter to fit the center. The live center is usually made of soft steel so the point can be machined as needed to keep it running true. Live centers are removed by means of a knockout bar (Figure 19).

Often when a machinist wants to machine the entire length of work mounted between centers without the interference of a lathe dog, specially ground and hardened drive centers are used (Figure 20). These are serrated so they will turn the work, but only light cuts can be made. Modern drive centers or face drivers (Figure 21) can also be used to machine a part without interference from a lathe dog. Quite heavy cuts are possible with these drivers, which are used especially for manufacturing purposes.

Face plates are used for mounting workpieces or fixtures. Unlike drive plates which have only slots, face plates have T slots and are more heavily built (Figure 22).

Collet chucks (Figure 23) are very accurate workholding devices and are used in producing small high precision parts. Steel spring collets are available for holding and turning hexagonal,

Figure 18. Drive plate for turning between centers (Lane Community College).

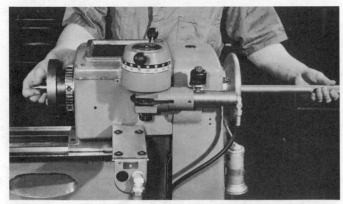

Figure 19. Knockout bar is used to remove centers (Courtesy of California Community Colleges, IMC Project).

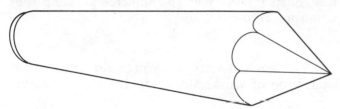

Figure 20. Hardened and serrated drive center is used for very light turning between centers.

Figure 21. Face driver is mounted in headstock spindle and work is driven by the drive pins that surround the center. (Courtesy of Madison-Kosta®, Madison Industries, a Division of Amtel, Inc.) (Madison-Kosta® is a registered trademark of Madison Industries, A Division of Amtel, Inc.)

Figure 22. T-slot face plate. Workpieces are clamped on the plate with T-bolts and strap clamps (Courtesy of The Monarch Machine Tool Company, Sidney, Ohio).

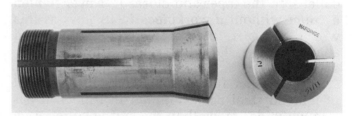

Figure 23. Side and end views of a spring collet for round work.

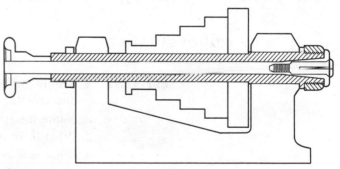

Figure 24. Cross section of spindle showing construction of draw-in collet chuck attachment.

square, and round workpieces. They are made in specific sizes (which are stamped on them) with a range of only a few thousandths of an inch. Rough and inaccurate workpieces should not be held in the collet chuck since the gripping surfaces of the chuck would form an angle with the workpiece. The contact area would then be at one point on the jaws instead of along

the entire length, and the piece would not be held firmly. If it is not held firmly, workpiece accuracy is impaired and the collet may be damaged. An adapter called a collet sleeve is fitted into the spindle taper and a draw bar is inserted into the spindle at the opposite end (Figure 24). The collet is placed in the adapter,

Figure 25. Rubber flex collet.

Figure 26. Collet handwheel attachment for rubber flex collets (Courtesy of The Monarch Machine Tool Company, Sidney, Ohio).

and the draw bar is rotated, which threads the collet into the taper and closes it. *Never tighten a collet without a workpiece in its jaws* as this will damage it. Before installing collets and adapters, they should be cleaned to insure accuracy.

The rubber flex collet (Figure 25) has a set of tapered steel bars mounted in rubber. It has a much wider range than the spring collet, each collet with a range of about $\frac{1}{8}$ in. A large handwheel is used to open and close the collets instead of a draw bar (Figure 26).

The concentricity that you could expect from each type of workholding device is as follows:

Device	Centering accuracy in inches (indicator reading difference)
Centers	Within .001
Four-jaw chuck	Within .001 (depending on the ability of machinist)
Collets	.0005 to .001
Three-jaw chuck	.001 to .003 (good condition) .005 or more (poor condition)

self-evaluation

SELF-TEST
1. Briefly describe the lathe spindle. How does the spindle support chucks and collets?
2. Name the spindle nose types.
3. What is an independent chuck and what is it used for?
4. What is a universal chuck and what is it used for?
5. What chuck types make possible the frequent adjusting of chucks so they will hold stock with minimum runout?
6. Workpieces mounted between centers are driven with lathe dogs. Which type of plate is used on the spindle nose to turn the lathe dog?
7. What is a live center made of? How does it fit in the spindle nose?
8. Describe a drive center and a face driver.
9. On which type of plate are workpieces and fixtures mounted? How is it identified?
10. Name one advantage of using steel spring collets. Name one disadvantage.

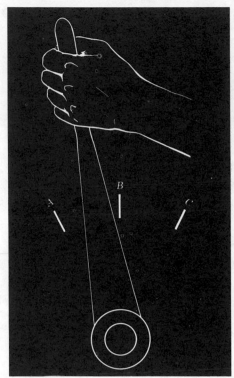

unit 5
operating the machine controls

Before using any machine, you must be able to properly use the controls, know what they are for, and how they work. You must also be aware of the potential hazards that exist for you and the machine, if it is mishandled. This unit prepares you to operate lathes.

objectives After completing this unit, you will be able to:

1. Explain drives and shifting procedures for changing speeds on lathes.
2. Describe the use of various feed control levers.
3. Explain the relationship between longitudinal feeds and cross feeds.
4. State the differences in types of cross feed screw micrometer collars.

Most lathes have similar control mechanisms and operating handles for feeds and threading. Some machines, however, have entirely different driving mechanisms as well as different speed controls.

DRIVES

Spindle speed is controlled on some lathes by a belt on a step cone pulley in the headstock (Figure 1). The speed is changed by turning the belt tension lever to loosen the belt, moving the belt to the proper step for the desired speed, and then moving the lever to its former position. Several more lower speeds are available by shifting to back gear. To do this, pull out or release the bull gear lockpin to disengage the spindle from the step cone pulley and engage the back gear lever as shown in Figure 1. It may be necessary to rotate the spindle by

Figure 1. Speeds are changed on this lathe by moving the belt to various steps on the pulley (Lane Community College).

hand slightly to bring the gear into mesh. The back gears must *never* be engaged when the spindle is turning with power.

431

Figure 2. Variable speed control and speed selector (Courtesy of Clausing Corporation).

Another drive system uses a variable speed drive (Figure 2) with a high and low range using a back gear. On this drive system, the motor must be running to change the speed on the vari-drive, but the motor must be turned off to shift the back gear lever. Geared head lathes are shifted with levers on the outside of the headstock (Figure 3). Several of these levers are used to set up the various speeds within the range of the machine. The gears will not mesh unless they are perfectly aligned, so it is sometimes necessary to rotate the spindle by hand. *Never try to shift gears with the motor running and the clutch lever engaged.*

FEED CONTROL LEVERS

The carriage is moved along the ways by means of the lead screw when threading, or by a separate feed rod when using feeds. On most small lathes, however, a lead screw-feed rod combination is used. In order to make left-hand threads and reverse the feed, the feed reverse lever is used. This lever reverses the leadscrew. It should never be moved when the machine is running.

The quick-change gearbox (Figures 4a and b) typically has two or more sliding gear shifter levers. These are used to select feeds or threads per inch. On those lathes also equipped with metric selections, the threads are expressed in pitches (measured in millimeters).

The carriage apron (Figure 5) contains the handwheel for hand feeding and a power feed

Figure 3. Speed change levers and feed selection levers on a geared head lathe (Lane Community College).

Figure 4a. Quick-change gearbox with index plate (Lane Community College).

Figure 6. Facing on a lathe (Lane Community College).

Figure 4b. Exposed quick-change gear mechanism for a large, heavy duty lathe (Courtesy of Lodge & Shipley Company).

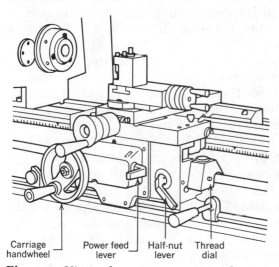

Carriage handwheel   Power feed lever   Half-nut lever   Thread dial

Figure 5. View of carriage apron with names of parts (Courtesy of Clausing Corporation).

Figure 7. Micrometer collar on the crossfeed screw that is graduated in English units. Each division represents .001 in. (Lane Community College).

lever that engages a clutch to a gear drive train in the apron.

Hand feeding should not be used for long cuts as there would be lack of uniformity and a poor finish would result. When using power feed and approaching a shoulder or the chuck jaws, disengage the power feed and hand feed the carriage for the last $\frac{1}{8}$ in. or so. When doing delicate work, always hand feed the carriage. The handwheel is used to quickly bring the tool close to the work before engaging the feed and for rapidly returning to the start of a cut after disengaging the feed. A feed change lever diverts the feed to either the carriage for longi-

tudinal movement or to the cross feed screw to move the cross slide. There is generally some slack or backlash in the cross feed and compound screws. As long as the tool is being fed in one direction against the work load, there is no problem, but if the screw is *slightly* backed

Figure 8. Crossfeed and compound screw handles with metric-English conversion collars (Courtesy of The Monarch Machine Tool Company, Sidney, Ohio).

off, the readings will be in error. To correct this problem, back off two turns and come back to the desired position.

Cross feeds are geared differently than longitudinal feeds. On most lathes the cross feed is approximately one third to one half that of the longitudinal feed, so a facing job (Figure 6) with the quick-change gearbox set at about .012 in. feed would actually only be .004 in. for facing. The cross feed ratio for each lathe is usually found on the quick-change gearbox index plate.

The half-nut or split-nut lever on the carriage engages the thread on the lead screw directly and is used only for threading. It cannot be engaged unless the feed change lever is in the neutral position.

Both the cross feed screw handle and the compound rest feed screw handle are fitted with micrometer collars (Figure 7). These collars traditionally have been graduated in English units, but metric conversion collars (Figure 8) will help in the transition to the metric system.

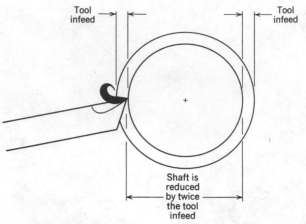

Figure 9. The diameter of the workpiece is reduced twice the amount in which the tool is moved.

Some micrometer collars are graduated to read single depth; that is, the tool moves as much as the reading shows. When turning a cylindrical object such as a shaft, tools that read single depth will remove twice as much from the diameter (Figure 9). For example, if the cross feed screw is turned in .020 in. and a cut is taken, the diameter will have been reduced by .040 in. Sometimes only the compound is calibrated in this way. Many lathes, however, are graduated on the micrometer dial to compensate for double depth on cylindrical turning. On this type of lathe if the cross feed screw is turned in until .020 in. shows on the dial and a cut is taken, the diameter will have been reduced .020 in. The tool would have actually moved into the work only .010 in.

To determine which type of graduation you are using, set a fractional amount on the dial (such as .250 in. $= \frac{1}{4}$ in.) and measure on the cross slide with a rule. The actual slide movement you measure with the rule will be either the same as the amount set on the dial, for the single depth collar, or one half that amount, for the double depth collar.

Some lathes have a brake and clutch rod that is the same length as the lead screw. A clutch lever connected to the carriage apron rides along the clutch rod (Figure 10). The spindle can be started and stopped without turning off the motor by using the clutch lever. Some types also have a spindle brake that quickly stops the spindle when the clutch lever is moved to the stop position. An adjustable automatic clutch kickout is also a feature of the clutch rod.

When starting a lathe for the first time use the following checkout list:

1. Move carriage and tailstock to the right to clear workholding device.

2. Locate feed clutches and half-nut lever and disengage before starting spindle.

3. Set up to operate at low speeds.

4. Read any machine information panels that may be located on the machine and observe precautions.

5. Note the feed direction; there are no built-in travel limits or warning devices to prevent feeding into the chuck or against the end of the slides.

6. When you are finished with a lathe, disengage all clutches, clean up chips, and remove any attachments or special setups.

Figure 10. Clutch rod is actuated by moving the clutch level. This disengages the motor from the spindle (Lane Community College).

self-evaluation

SELF-TEST
1. How can the shift to the low speed range be made on the belt drive lathes that have the step cone pulley?
2. Explain speed shifting procedure on the variable speed drive.
3. In what way can speed changes be made on gearhead lathes?
4. What lever is shifted in order to reverse the lead screw?
5. The sliding gear shifter levers on the quick-change gearbox are used for just two purposes. What are they?
6. When is the proper time to use the carriage handwheel?
7. Why will you not get the same surface finish (tool marks per inch) on the face of a workpiece as you would get on the outside diameter when on the same power feed?
8. The half-nut lever is not used to move the carriage for turning. Name its only use.
9. Micrometer collars are attached on the cross feed handle and compound handle. In what ways are they graduated?
10. How can you know if the lathe you are using is calibrated for single or double depth?

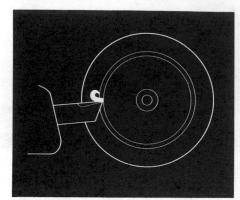

unit 6
facing and center drilling

Facing and center drilling the workpiece are often the first steps taken in a turning project to produce a stepped shaft or a sleeve from solid material. Much lathe work is done in a chuck and requires considerable facing and some center drilling. These important lathe practices will be covered in detail in this unit.

objectives After completing this unit, you will be able to:
1. Correctly set up a workpiece and face the ends.
2. Correctly center drill the ends of a workpiece.
3. Determine the proper feeds and speeds for a workpiece.
4. Explain how to set up to make facing cuts to a given depth and how to measure them.

SETTING UP FOR FACING

Facing is done to obtain a flat surface on the end of cylindrical workpieces or on the face of parts clamped in a chuck or face plate (Figures 1a and b). The work most often is held in a three- or four-jaw chuck. If the chuck is to be removed from the lathe spindle, a lathe board must first be placed on the ways. Figure 2 shows a camlock mounted chuck being removed. The correct procedure for installing a chuck on a camlock spindle nose is shown in Figures 3a to f. The cams should be tightly snugged (Figure 3f) for one or two revolutions around the spindle.

Setting up work in an independent chuck is simple, but mastering the setup procedures takes some practice. Round stock can be set up by using a dial indicator (Figure 4). Square or rectangular stock can either be set up with a dial indicator or by using a toolholder turned backwards (Figure 5).

Begin the set up by aligning two opposite jaws with the same concentric ring marked in the face of the chuck while the jaws are near the workpiece. This will roughly center the

work. Set up the other two jaws with a concentric ring also when they are near the work. Next, bring all of the jaws firmly against the work. When using the dial indicator, zero the bezel at the lowest reading. Now rotate the chuck to the opposite jaw with the high reading and tighten it half the amount of the runout. It might be necessary to loosen the jaw on the low side slightly. Always tighten the jaws at the position where the dial indicator contacts the work since any other location will give erroneous readings. When using the back of the toolholder, the micrometer dial on the cross slide will show the difference in runout. Chalk is sometimes used for setting up rough castings and other work too irregular to be measured with a dial indicator. Workpieces can either be chucked normally, internally or externally (Figures 6a to c).

FACING

The material to be machined usually has been cut off in a power saw and so the piece is not square on the end or cut to the specified length.

436

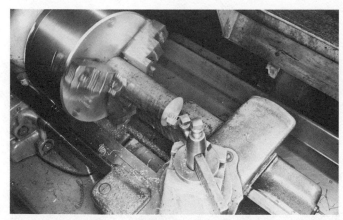

Figure 1a. Facing a workpiece in a chuck (Lane Community College).

Figure 1b. Facing the end of a shaft (Lane Community College).

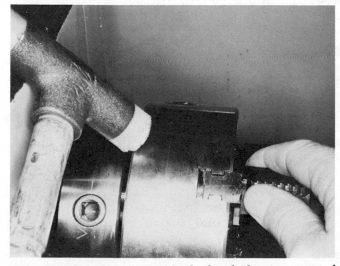

Figure 2. Removing a camlock chuck that is mounted on a lathe spindle (DeAnza College).

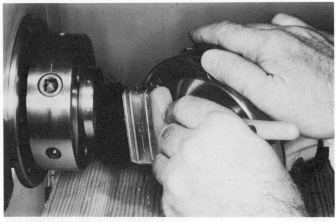

Figure 3a. Chips are cleaned from spindle nose with a brush (DeAnza College).

Figure 3b. Cleaning the chips from the chuck with a brush (DeAnza College).

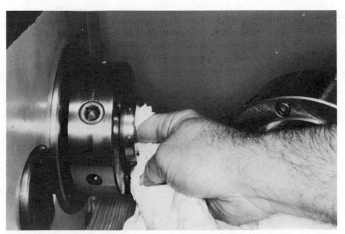

Figure 3c. Spindle nose is thoroughly cleaned with a soft cloth (DeAnza College).

Figure 3*d*. Chuck is thoroughly cleaned with a soft cloth (DeAnza College).

Figure 3*e*. Chuck is mounted on spindle nose (DeAnza College).

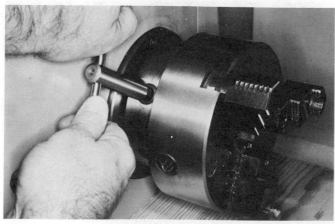

Figure 3*f*. All cams are turned clockwise until locked securely (DeAnza College).

Figure 4. Setting up round stock in an independent chuck with a dial indicator (Lane Community College).

Facing from the center out (Figure 7) produces a better finish, but it is difficult to cut on a solid face in the center. Facing from the outside (Figure 8) is more convenient since heavier cuts may be taken and it is easier to work to the scribed lines on the circumference of the work. When facing from the center out, a right-hand turning tool in a left-hand toolholder is the best arrangement, but when facing from the outside to the center, a left-hand tool in a right-hand or straight toolholder can be used. Facing or other tool machining should not be done on workpieces extending more than five diameters from the chuck jaws.

The point of the tool should be set to the center of the work (Figure 9). This is done by setting the tool to the tailstock center point or by making a trial cut to the center of the work. If the tool is off center, a small uncut stub will be left. The tool can then be reset to the center of the stub.

The carriage must be locked when taking

Figure 5a. Rectangular stock being set up by using a toolholder turned backwards. The micrometer dial is used to center the workpiece (Lane Community College).

Figure 6a. Normal chucking position (Lane Community College).

Figure 5b. Adjusting the rectangular stock at 90° from Figure 5a (Lane Community College).

facing cuts as the cutting pressure can cause the tool and carriage to move away (Figure 10), which would make the faced surface curved rather than flat. Finer feeds should be used for finishing than for roughing. Remember, the cross feed is one half to one third that of the longitudinal feed. The ratio is usually listed on the index plate of the quick-change gear box. A roughing feed could be from .005 to .015 in.

Figure 6b. Internal chucking position (Lane Community College).

Figure 6*c*. External chucking position (Lane Community College).

Figure 8. Facing from the outside toward the center of the workpiece (Courtesy of Clausing Corporation).

Figure 7. Facing from the center to the outside of a workpiece (Lane Community College).

Figure 9. Setting the tool to the center of the work using the tailstock center (Lane Community College).

Figure 10. Carriage must be locked before taking a facing cut (Lane Community College).

and a finishing feed from .003 to .005 in. Use of cutting oils will help produce better finishes on finish facing cuts.

Facing to length may be accomplished by trying a cut and measuring with a hook rule

Figure 11. Facing to length using a hook rule for measuring (Lane Community College).

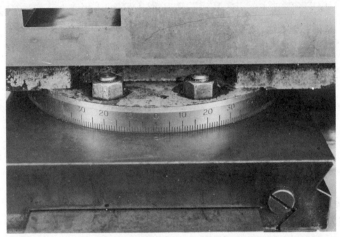

Figure 12. The compound set at 90° for facing operations (Lane Community College).

Figure 13. Setting the compound at 30° (Lane Community College).

Figure 14. Half centers make facing shaft ends easier.

(Figure 11) or by facing to a previously made layout line. A more precise method is to use the graduations on the micrometer collar of the compound. The compound is set so its slide is parallel to the ways (Figure 12). The carriage is locked in place and a trial cut is taken with the micrometer collar set on zero index. The workpiece is measured with a micrometer and the desired length is subtracted from the measurement; the remainder is the amount you should remove by facing. If more than .015 to .030 in. (depth left for finish cut) has to be removed, it should be taken off in two or more cuts by moving the compound micrometer dial the desired amount. A short trial cut (about $\frac{1}{8}$ in.) should again be taken on the finish cut and

adjustment made if necessary. Roughing cuts should be approximately .060 in. in depth.

Quite often the compound is kept at 30 degrees for threading purposes (Figure 13). It is convenient to know that at this angle, the tool feeds into the face of the work .001 in. for every .002 in. that the slide is moved. For example, if you wanted to remove .015 in. from the workpiece, you would turn in .030 in. on the micrometer dial (assuming it reads single depth).

A specially ground tool is used to face the end of a workpiece that is mounted between centers. The right-hand facing tool is shaped to fit in the angle between the center and the face of the workpiece. Half centers (Figure 14) are made to make the job easier, but they should be used only for facing and not for general turning. If the tailstock is moved off center away from the operator, the shaft end will be convex and, if it is moved toward the operator, it will be concave (Figure 15). Both right-hand and left-hand facing tools are used for facing work held on mandrels (Figure 16).

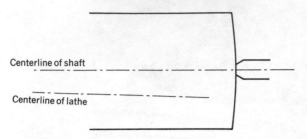

Figure 15a. Convex shaft ends caused by the tailstock being moved off center away from the operator.

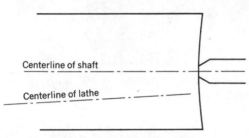

Figure 15b. Concave shaft ends caused by the tailstock being moved toward the operator.

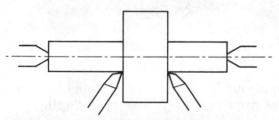

Figure 16. Work that is held between centers on a mandrel can be faced on both sides with right-hand and left-hand facing tools.

SPEEDS

Speeds (RPM) for lathe turning a workpiece are determined in essentially the same way as speeds for drilling tools. The only difference is that the diameter of the work is used instead of the diameter of the drill. For facing work, the outside diameter is always used to determine RPM. Thus:

$$RPM = \frac{CS \times 4}{D}, \text{ where}$$

D = diameter of workpiece (where machining is done)
RPM = revolutions per minute
CS = cutting speed (surface feet per minute)

EXAMPLE 1

The cutting speed for low carbon steel is 90 SFM (surface feet per minute) and the workpiece diameter to be faced is 6 inches. Find the correct RPM.

$$RPM = \frac{90 \times 4}{6} = 60$$

EXAMPLE 2

A center drill has a $\frac{1}{8}$ in. drill point. Find the correct RPM to use on low carbon steel (CS 90).

$$RPM = \frac{90 \times 4}{\frac{1}{8}} = \frac{360}{1} \times \frac{8}{1} = 2880$$

These are only approximate speeds and will vary according to the conditions. If chatter marks (vibration marks) appear on the workpiece, the RPM should be reduced. If this does not help, ask your instructor for assistance. For more information on speeds and feeds, see the next unit, "Turning Between Centers."

CENTER DRILLS AND DRILLING

When work is held and turned between centers, a center hole is required on each end of the workpiece. The center hole must have a 60 degree angle to conform to the center and have a smaller drilled hole to clear the center's point. This center hole is made with a combination drill and countersink, sometimes referred to as a center drill. These drills are available in a range of sizes from $\frac{1}{8}$ to $\frac{3}{4}$ in. body diameter and are classified by numbers from 00 to 8, which are normally stamped on the drill body. For example, a number 3 center drill has a $\frac{1}{4}$ in. body diameter and a $\frac{7}{64}$ in. drill diameter. Full listings can be found in the *Machinery's Handbook.*

Center drills are usually held in a drill chuck in the tailstock, while the workpieces are most often supported and turned in a lathe chuck for center drilling (Figure 17). A workpiece could also be laid out and supported in a vertical position for center drilling in a drill press. This method, however, is not used very often.

Round stock could be clamped in a vee-block and drill press vise (Figure 18) or on an angle plate. Crossed layout lines are scribed

Figure 17. Center drilling a workpiece held in a chuck (Lane Community College).

on round stock by means of a height gage (Figure 19), and a punch mark is made where they intersect. Layout for square or rectangular stock to be placed between centers is done simply by scribing two diagonal lines from corner to corner (Figure 20).

As a rule, center holes are drilled by rotating the work in a lathe chuck and feeding the center drill into the work by means of the tailstock spindle. Long workpieces, however, are generally faced by chucking one end and supporting the other in a steady rest (Figure 21). Since the end of stock is never sawed square, it should be center drilled only after spotting a small hole with the lathe tool. A slow feed is needed to protect the small, delicate drill end. Cutting oil should be used and the drill should be backed out frequently to remove chips. The greater the work diameter and the heavier the cut, the larger the center hole should be.

The size of the center hole can be selected by the center drill size and then regulated to some extent by the depth of drilling. You must be careful not to drill too deeply (Figure 22) as this causes the center to contact only the sharp outer edge of the hole, which is a poor bearing surface. It soon becomes loose and out of round and causes such machining problems as chatter and roughness. Center drills are broken often from feeding the drill too fast with the

lathe speed too slow or with the tailstock off center.

Center drills are often used as starting or spotting drills when a drilling sequence is to

Figure 18. Center being drilled in round stock that is held in drill press vise (Lane Community College).

Figure 19. Layout lines for center drilling are scribed on round stock by using a height gage and a vee block on a surface plate (Lane Community College).

Section H Turning Machines

be performed (Figures 23 and 24). This keeps the drill from "wandering" off the center and making the hole run eccentric. Spot drilling

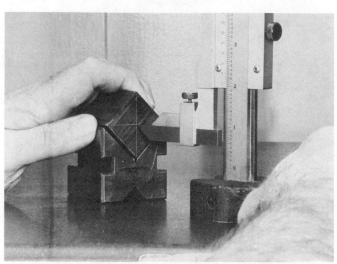

Figure 20. Layout lines for center drilling on square stock are made from corner to corner (Lane Community College).

Figure 21. Center drilling long material that is supported in a steady rest (Lane Community College).

is done when work is chucked or is supported in a steady rest. Care must be taken that the workpiece is centered properly in the steady rest or the center drill will be broken.

Figure 24. (right) Center drill is fed into work with a slow even feed (DeAnza College).

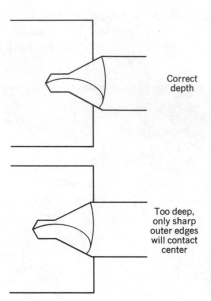

Figure 22. Correct and incorrect depth for center drilling.

Figure 23. Center drill is brought up to work and lightly fed into material (DeAnza College).

self-evaluation

1. You have a rectangular workpiece that needs a facing operation plus center drilling, and a universal chuck is mounted on the lathe spindle. What is your procedure to prepare for machining?

2. Should the point of the tool be set above, below, or at the center of the spindle axis when taking a facing cut?

3. If you set the quick-change gearbox to .012 in., would that be considered a roughing feed for facing?

4. An alignment step must be machined on a cover plate .125 in., plus or minus .003 in., in depth (Figure 25). What procedure should be taken to face to this depth? How can you check your final finish cut?

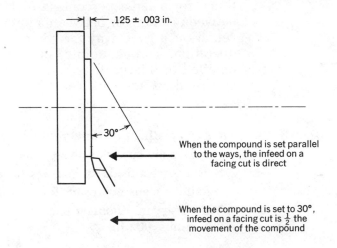

.125 ± .003 in.

30°

When the compound is set parallel to the ways, the infeed on a facing cut is direct

When the compound is set to 30°, infeed on a facing cut is ½ the movement of the compound

5. What tool is used for facing shaft ends when they are mounted between centers? In what way is this tool different from a turning tool?

6. If the cutting speed of aluminum is 300 SFM and the workpiece diameter is 4 in., what is the RPM? The formula is

$$RPM = \frac{CS \times 4}{D}$$

7. Name two reasons for center drilling a workpiece in a lathe.

8. How is laying out and drilling center holes in a drill press accomplished?

9. Name two causes for center drill breakage.

10. What happens when you drill too deeply with a center drill?

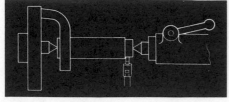

unit 7
turning
between
centers

Since for a large percentage of lathe work the workpiece is held between centers or between a chuck and a center, turning between centers is a good way for you to learn the basic principles of lathe operation. The economics of machining time and quality will be detailed in this unit, as heavy roughing cuts are compared to light cuts, and speeds and feeds for turning operations are presented.

objectives After completing this unit, you will be able to:

1. Describe the correct setup procedure for turning between centers.
2. Select correct feeds and speeds for a turning operation.
3. Detail the steps necessary for turning to size predictably.
4. Turn a $1\frac{1}{4}$-in. diameter shaft with six shoulders to a tolerance of plus 0.0, minus .002 in.

SETUP FOR TURNING BETWEEN CENTERS

To turn a workpiece between centers, it is supported between the dead center (tailstock center) and the live center in the spindle nose. A lathe dog (Figure 1) clamped to the workpiece is driven by a drive or dog plate (Figure 2) mounted on the spindle nose. Machining with a single point tool can be done anywhere on the workpiece except near or at the location of the lathe dog.

Turning between centers has some disadvantages. A workpiece cannot be cut off with a parting tool while being supported between centers as this will bind and break the parting tool and ruin the workpiece. For drilling, boring, or machining the end of a long shaft, a steady rest is normally used to support the work. But these operations cannot very well be done when the shaft is supported only by centers.

The advantages of turning between centers are many. A shaft between centers can be

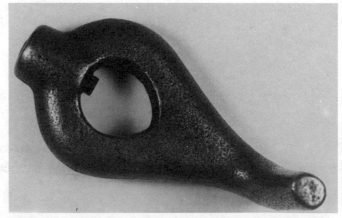

Figure 1. Lathe dog.

turned end for end to continue machining without eccentricity if the centers are in line (Figure 3). This is why shafts that are to be subsequently finish-ground between centers must be machined between centers on a lathe. If a partially threaded part is removed from between centers for checking, and everything is left the same on the lathe, the part can be returned to

Figure 2. Dog plate or drive plate on spindle nose of the lathe (Courtesy of the Monarch Machine Tool Company, Sidney, Ohio).

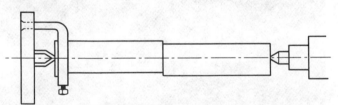

Figure 3. Eccentricity in the center of the part because of the live center being off center.

Figure 4. Work being machined between chuck and tailstock center. Note chip formation. Chip guard has been removed for clarity. (Lane Community College).

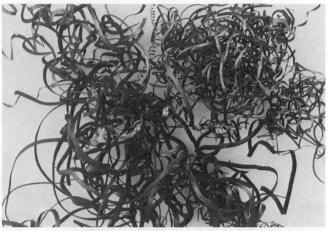

Figure 5a. A tangle of wiry chips. These chips can be hazardous to the operator (Lane Community College).

Figure 5b. A better formation of chips. This type of chip will fall into the chip pan and is more easily handled (Lane Community College).

Figure 6. Inserting the tapered spindle nose sleeve (DeAnza College).

Figure 7. Make sure the bushing is firmly seated in the taper (Lane Community College).

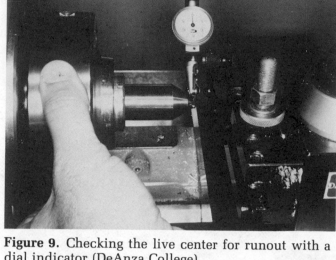

Figure 9. Checking the live center for runout with a dial indicator (DeAnza College).

Figure 8. Installing the center (DeAnza College).

Figure 10. Live center being machined in a four-jaw chuck. The lathe dog on the workpiece is driven by one of the chuck jaws (Lane Community College).

the lathe, and the threading resumed where it was left off.

A considerable amount of straight turning on shafts is done with the work held between a chuck and the tailstock center (Figure 4). The advantages of this method are quick setup and a positive drive. One disadvantage is that eccentricities in the shaft are caused by inaccuracies in the chuck jaws. Another is the tendency for the workpiece to slip endwise into the chuck jaws under a heavy cut, thus allowing the workpiece to loosen or to come out of the tailstock center.

As in other lathe operations, chip formation and handling are important to safety. Coarser feeds, deeper cuts, and smaller rake angles all tend to increase chip curl, which breaks up the chip into small, safe pieces. Fine feeds and shallow cuts, on the other hand, produce a tangle of wiry, sharp hazardous chips (Figures 5a and b) even with a chip breaker on the tool. Long strings may come off the tool, suddenly wrap in the work and be drawn back rapidly to the machine. The edges are like saws and can cause very severe cuts.

The center for the headstock spindle is called a live center and it is usually not hard-

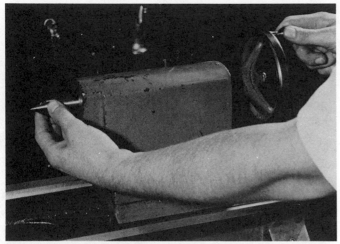

Figure 12b. Antifriction ball bearing center (Courtesy of The Monarch Machine Tool Company, Sidney, Ohio).

Figure 11. The dead center is hardened to resist wear. It is made of high speed steel or steel with a carbide insert (Courtesy of California Community Colleges, IMC Project).

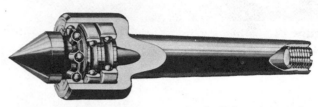

Figure 12c. Cutaway view of a ball bearing tailstock center (Courtesy of DoAll Company).

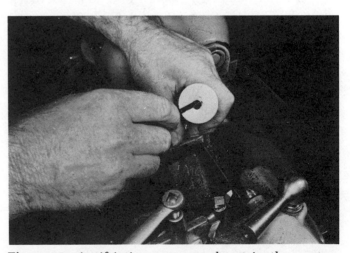

Figure 12a. Pipe center used for turning (Courtesy of The Monarch Machine Tool Company, Sidney, Ohio).

ened since its point frequently needs machining to keep it true. Thoroughly clean the inside of the spindle with a soft cloth and wipe off the live center. If the live center is too small for the lathe spindle taper, use a tapered bushing that fits the lathe (Figure 6). Seat the bushing firmly in the taper (Figure 7) and install the center (Figure 8). Set up a dial indicator on the end of the center (Figure 9) to check for runout. If there is runout, remove the center by using a knockout bar through the spindle. Be sure to catch the center with one hand. Check the outside of the center for nicks or burrs. These can be removed with a file. Check the inside of the spindle taper with your finger for nicks or grit.

Figure 13. Antifriction compound put in the center hole before setting the workpiece between centers. This step is not necessary when using an antifriction ball bearing center (Lane Community College).

If nicks are found, *do not* use a file but check with your instructor. After removing nicks, if the center still runs out more than the acceptable tolerances (usually .0001 to .0005 in.) a light cut by tool or grinding can be taken with the compound set at 30 degrees.

A live center is often machined from a short piece of soft steel mounted in a chuck (Figure 10). It is then left in place and the work-

Figure 14. Lathe dog in position (Lane Community College).

Figure 15a. Incorrect position of toolholder for roughing. If the toolholder should turn when using heavy feeds, the tool will gouge more deeply into the work (Lane Community College).

piece is mounted between it and the tailstock center. A lathe dog with the bent tail against a chuck jaw is used to drive the workpiece. This procedure sometimes saves time on large lathes where changing from the chuck to a drive plate is cumbersome and the amount of work to be done between centers is small.

The tailstock center (Figure 11) is hardened to withstand machining pressures and friction. Clean inside the taper and on the center before installing. Ball bearing, antifriction centers are often used in the tailstock as they will withstand high speed turning without the overheating problems of dead centers. Pipe centers are used for turning tubular material (Figures 12a to c).

To set up a workpiece that has been previously center drilled, slip a lathe dog on one end with the bent tail toward the drive plate. Do not tighten the dog yet. Put antifriction compound into the center hole toward the tailstock and then place the workpiece between centers (Figure 13). The tailstock spindle should not extend out too far as some rigidity in the machine would be lost and chatter or vibration may result. Set the dog in place and avoid any binding of the bent tail (Figure 14). Tighten the dog and then adjust the tailstock so there is no end play, but so the bent tail of the dog freely clicks in its slot. Tighten the tailstock binding lever. The heat of machining will expand the workpiece and cause the dead center to heat from friction. If overheated, the center may be ruined and may even be welded into your workpiece. Periodical-

Figure 15b. Correct position of toolholder for roughing. The toolholder will swing away from the cut with excessive feeds (Lane Community College).

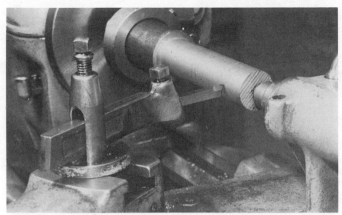

Figure 16. Tools with excessive overhang. Both tool and toolholder extend too far from the tool post for roughing operations (Lane Community College).

Figure 17. Tool and toolholder in the correct position (Lane Community College).

Figure 18. Centering a tool by means of a steel rule (Lane Community College).

ly, or at the end of each heavy cut, you should check the adjustment of the centers and reset if necessary.

When a tool post and toolholder are used, the toolholder must be positioned so it will not turn into the work when heavy cuts are taken (Figures 15a and b). The tool and toolholder should not overhang too far (Figure 16) for rough turning, but should be kept toward the tool post as far as practical (Figure 17). Tools should be set on or slightly above the center of the workpiece. The tool may be set to the dead center or to a steel rule on the workpiece (Figure 18).

SPEEDS AND FEEDS FOR TURNING

Since machining time is an important factor in lathe operations, it is necessary for you to fully understand the principles of speeds and feeds in order to make the most economical use of your machine. Speeds are determined for turning between centers by using the same formula as given for facing operations in the last unit.

$$RPM = \frac{CS \times 4}{D}$$

Where:

RPM = Revolutions per minute
$\quad D$ = Diameter of workpiece
$\quad$ CS = Cutting speed in surface feet per minute (SFM)

Cutting speeds for various materials are given in Table 1.

EXAMPLE

If the cutting speed is 40 for a certain alloy steel and the workpiece is 2 in. in diameter, find the RPM.

$$RPM = \frac{40 \times 4}{2} = 80$$

After calculating the RPM, use the nearest or next lower speed on the lathe and set the speed.

Feeds are expressed in inches per revolution (IPR) of the spindle. A .010 in. feed will move the carriage and tool .010 in. for one full

Table 1
Cutting Speeds and Feeds for High Speed Steel Tools

	Low Carbon Steel	High Carbon Steel Annealed	Alloy Steel Normalized	Aluminum Alloys	Cast Iron	Bronze
Roughing speed SFM	90	50	45	200	70	100
Finishing speed SFM	120	65	60	300	80	130
Feed IPR roughing	.010–.020	.010–.020	.010–.020	.015–.030	.010–.020	.010–.020
Feed IPR finishing	.003–.005	.003–.005	.003–.005	.005–.010	.003–.010	.003–.010

Figure 19. Index chart on the quick-change gearbox (Lane Community College).

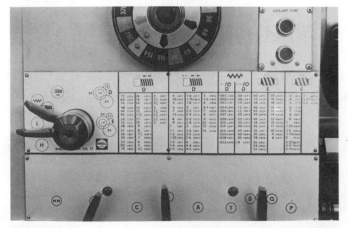

Figure 20. Index chart for feed mechanism on a modern geared head lathe with both metric and inch thread and feed selections (Lane Community College).

turn of the headstock spindle. If the spindle speed is changed, the feed ratio still remains the same. Feeds are selected by means of an index chart (Figure 19) either found on the quick-change gearbox or on the side of the headstock housing (Figure 20). The sliding gear levers are shifted to different positions to obtain the feeds indicated on the index plate. The lower decimal numbers on the plate are feeds and the upper numbers are threads per inch.

Feeds and depth of cut should be as much as the tool, workpiece, or machine can stand without undue stress. *The feed rate for roughing should be from one-fifth to one-tenth as much as the depth of cut.* A small 10 or 12 in. swing lathe should handle $\frac{1}{8}$ in. depth of cut in soft steel, but in some cases this may have to be reduced to $\frac{1}{16}$ in. If .100 in. were selected as a trial depth of cut, then the feed could be anywhere from .010 to .020 in. If the machine seems to be overloaded, reduce the feed. Finishing feeds can be from .003 to .005 in. for steel. Use a tool with a larger nose radius for finishing.

TURNING TO SIZE

The cut-and-try method of turning to size a workpiece, or making a cut and measuring how close you came to the desired result, was used in the past when calipers and rule were used for measuring work diameters. A more modern method of turning to size predictably uses the compound and cross feed micrometer collars and micrometer calipers for measurement. If the micrometer collar on the cross feed screw reads in single depth, it will remove twice the amount from the diameter of the work as the

Figure 21. A trial cut is made to establish a setting of a micrometer dial in relation to the diameter of the workpiece (Lane Community College).

Figure 22. Measuring the workpiece with a micrometer (Lane Community College).

reading shows. A micrometer collar that reads directly or double depth will remove the same amount from the diameter that the reading shows, though the tool will actually move in only half that amount.

After taking one or several roughing cuts (depending on the diameter of the workpiece), .015 to .030 in. should be left for finishing. This can be taken in one cut if the tolerance is large, such as plus or minus .003 in. If the tolerance is small (plus or minus .0005 in.), two finish cuts should be taken, but enough stock must be left for the second cut to make a chip. If insufficient material is left for machining, .001 in. for example, the tool will rub and will not cut. Between .005 and .010 in. should be left for the last finish cut.

The position of the tool is set in relation to the micrometer dial reading, and the first of the two finish cuts is made (Figure 21). *The tool is then returned to the start of the cut with-*

Figure 23. Filing in the lathe, left-handed (Lane Community College).

out moving the cross feed screw. The diameter of the workpiece is checked with a micrometer (Figure 22) and the remaining amount to be cut is dialed on the cross feed micrometer dial.

Figure 24a. Using abrasive cloth for polishing (Lane Community College).

Figure 25. Measuring the workpiece length to a shoulder with a machinist's rule (Lane Community College).

Figure 24b. Using a file for backing abrasive cloth for more uniform polishing (Lane Community College).

A short trial cut is taken (about $\frac{1}{8}$ in. long) and the lathe stopped. A final check with a micrometer is made to validate the tool setting, and then the cut is completed. If the lathe makes a slight taper, see the next unit on "Alignment of the Lathe Centers" to correct this problem.

Finishing of machined parts with a file and abrasive cloth should not be necessary if the tools are sharp and honed and if the feeds, speeds, and depth of cut are correct. A machine-finished part looks better than a part finished with a file and abrasive cloth. In the past, filing and polishing the precision surfaces of lathe workpieces were necessary because of lack of rigidity and repeatability of machines. In the same way, worn lathes are not dependable for

Figure 26. Carriage stop set to limit tool travel in order to establish a shoulder (Lane Community College).

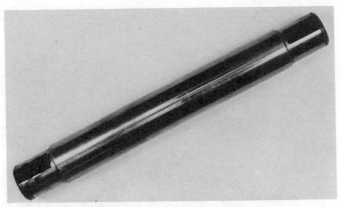

Figure 27a. Tapered mandrel or arbor.

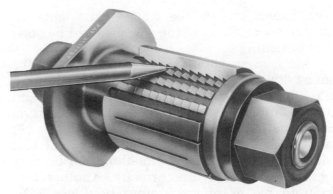

Figure 28. Operating principle of a special type of an expanding mandrel. The "Saber-tooth" design provides a uniform gripping action in the bore (Courtesy of Buck Tool Company).

Figure 27b. Tapered mandrel and workpiece set up between centers (Courtesy of Clausing Corporation).

Figure 29. Expanding mandrel and workpiece set up between centers (Lane Community College).

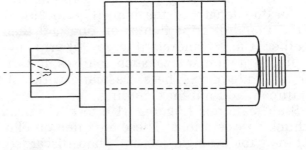

Figure 30. Many similar parts are machined at the same time with a gang mandrel.

close tolerances and so an extra allowance must be made for filing. The amount of surface material left for filing ranges from .0005 to .005 in., depending on the final finish, and diameter. If more than the tips of the tool marks are removed with a file and abrasive cloth, a wavy surface will result. For most purposes .002 in. is sufficient material to leave for finishing.

When filing on a lathe, use a low speed, long strokes and file left handed (Figure 23). For polishing with abrasive cloth, set the lathe for a high speed and move the cloth back and forth across the work. Hold an end of the cloth strip in each hand (Figures 24a and b). Abrasive cloth leaves grit on the ways of the lathe, so a thorough cleaning of the ways should be done after polishing.

Machining shoulders to specific lengths can be done in several ways. Using a machinist's rule (Figure 25) to measure workpiece length to a shoulder is a very common but semiprecision method. Preset carriage stops (Figure 26) can be used to limit carriage movement and establish a shoulder. This method can be very accurate if it is set up correctly. Another very accurate means of machining shoulders is by using special dial indicators with long travel plunger rods. These indicators show the longitudinal position of the carriage. Whichever method is used, *the power feed should be turned off one eighth in. short of the workpiece shoulder* and the tool handfed to the desired length. If the tool should be accidentally fed into an existing shoulder, the feed mechanism may jam and be very difficult to release. A broken tool, toolholder, or lathe part may be the result.

Mandrels, sometimes called lathe arbors, are used to hold work that is turned between centers (Figures 27a and b). Tapered mandrels are made in standard sizes and have a taper of only .006 in./ft. A flat is milled on one end of the mandrel for the lathe dog set screw. High pressure lubricant is applied to the bore of the workpiece and the mandrel is pressed into the workpiece with an arbor press. The assembly is mounted between centers and the workpiece is turned or faced on either side.

Expanding mandrels (Figure 28) have the advantage of providing a uniform gripping surface for the length of the bore (Figure 29). A tapered mandrel grips tighter on one end than the other. Gang mandrels (Figure 30) grip several pieces of similar size, such as discs, to turn their circumference. These are made with collars, thread, and nut for clamping.

Stub mandrels (Figures 31 a to c) are used in chucking operations. These are often quickly machined for a single job and then discarded. Expanding stub mandrels (Figures 32a and b) are used when production of many similar parts is carried out. Threaded stub mandrels are used for machining the outside surfaces of parts that are threaded in the bore.

Coolants are used for heavy duty and production turning. Oil-water emulsions and synthetic coolants are the most commonly used, while sulfurized oils usually are not used for

Figure 31a. Stub mandrel being machined to size with a slight taper, about .006 in./ft (Lane Community College).

Figure 31b. Part to be machined being affixed to the mandrel. The mandrel is oiled so that the part can be easily removed (Lane Community College).

Figure 31c. Part being machined after assembly on mandrel (Lane Community College).

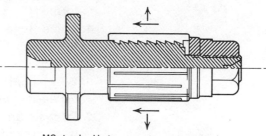

MC standard between centers nut actuated

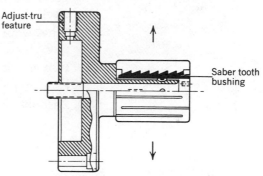

MLDR — flange mount — draw bolt actuated
stationary sleeve

Figure 32a. Special stub mandrel with adjust-tru feature (Courtesy of Buck Tool Company).

turning operations except for threading. Most job work or single piece work is done dry. Many shop lathes do not have a coolant pump and tank, so, if any coolant or cutting oil is used, it is applied with a pump oil can. Coolants and cutting oils for various materials are given in Table 2.

Figure 32b. Precision adjustment can be maintained on the expanding stub mandrels with the adjust-tru feature on the flange mount (Courtesy of Buck Tool Company).

Table 2
Coolants and Cutting Oils Used for Turning

Material	Dry	Water Soluble Oil	Synthetic Coolants	Kerosene	Sulfurized Oil	Mineral Oil
Aluminum		x	x	x		
Brass	x	x	x			
Bronze	x	x	x			x
Cast iron	x					
Steel						
Low carbon		x	x		x	
Alloy		x	x		x	
Stainless		x	x		x	

self-evaluation

SELF-TEST 1. Name two advantages and two disadvantages of turning between centers.

2. What other method besides turning between centers is extensively used for turning shafts and long workpieces that are supported by the tailstock center?

3. What factors tend to promote or increase chip curl so that safer chips are formed?

4. Name three kinds of centers used in the tailstock and explain their uses.

5. How is the dead center correctly adjusted?

6. Why should the dead center be frequently adjusted when turning between centers?

7. Why should you avoid excess overhang with the tool and toolholder when roughing?

8. Calculate the RPM for roughing a $1\frac{1}{2}$ in. diameter shaft of machine steel.

9. What would the spacing or distance between tool marks on the workpiece be with a .010 in. feed?

10. How much should the feed rate be for roughing?

11. How much should be left for finishing?

12. Describe the procedure in turning to size predictably.

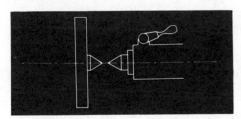

unit 8
alignment
of the lathe
centers

As a machinist, you must be able to check a workpiece for taper and properly set the tailstock of a lathe. Without these skills, you will lose much time in futile attempts to restore precision turning between centers when the workpiece has an unintentional taper. This unit will show you several ways to align the centers of a lathe.

objectives After completing this unit, you will be able to:

1. **Check for taper with a test bar and restore alignment by adjusting the tailstock.**

2. **Check for taper by taking a cut with a tool and measuring the workpiece and restore alignment by adjusting the tailstock.**

The tailstock will normally stay in good alignment on a lathe that is not badly worn. If a lathe has been used for taper turning with the tailstock offset, however, the tailstock may not have been realigned properly (Figure 1). The tailstock also could be slightly out of alignment if an improper method of adjustment was used. It is therefore a good practice to occasionally check the center alignment of the lathe you usually use and to always check the alignment before using a different lathe.

It is often too late to save the workpiece by realigning centers if a taper is discovered while making a finish cut. A check for taper should be made on the workpiece while it is still in the roughing stage. You can do this by taking a light cut for some distance along the workpiece or on each end *without resetting* the cross feed dial. Then check the diameter on each end with a micrometer; the difference between the two readings is the amount of taper in that distance.

Four methods are used for aligning centers on a lathe. In one method, the center points are brought together and visually checked for alignment (Figure 2). This is, of course, not a precision method for checking alignment.

Another method of aligning centers is by using the tailstock witness marks. Adjusting the tailstock to the witness marks (Figure 3), however, is only an approximate means of eliminating taper. The tailstock is moved by means of a screw or screws. A typical arrangement is shown in Figure 4, where one set screw is released and the opposite one is tightened to move the tailstock on its slide (Figure 5). The tailstock clamp bolt must be released before the tailstock is offset.

Figure 2. Checking alignment by matching center points (DeAnza College).

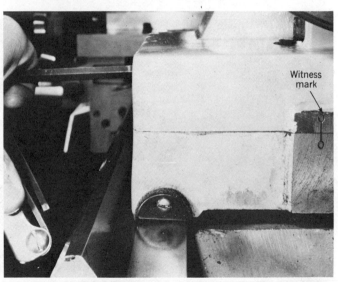

Figure 3. Adjusting the tailstock to the witness marks for alignment (DeAnza College).

Two more accurate means of aligning centers are by using a test bar and by machining and measuring. A test bar is simply a shaft that has true centers (is not off center) and has no taper. Some test bars are made with two diameters for convenience. When checking alignment with a test bar, no dog is necessary as the bar is not rotated. A dial indicator is mounted, preferably in the tool post, so it will travel with the carriage (Figure 6). Its contact point should be on the center of the test bar.

Begin with the indicator at the headstock end, and set the indicator bezel to zero (Figure 7). Now move the setup to the tailstock end

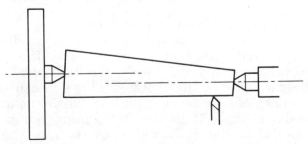

Figure 1. Tailstock out of line causing a tapered workpiece.

Figure 4. Hexagonal socket set screw which, when turned, moves the tailstock provided that the opposite one is loosened (DeAnza College).

Figure 7. Indicator is moved to measuring surface at headstock end and the bezel is set on zero (DeAnza College).

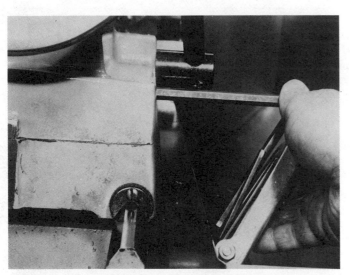

Figure 5. The opposite set screw being adjusted (DeAnza College).

Figure 8. The carriage with the dial indicator is moved to the measuring surface near the tailstock. In this case the dial indicator did not move, so the tailstock is on center (DeAnza College).

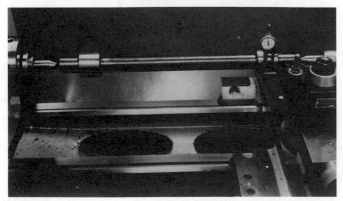

Figure 6. Test bar setup between centers with a dial indicator mounted in the tool post (DeAnza College).

of the test bar (Figure 8), and check the dial indicator reading. If no movement of the needle has occurred, the centers are in line. If the needle has moved clockwise, the tailstock is misaligned toward the operator. This will cause the workpiece to taper with the smaller end near the tailstock. If the needle has moved counterclockwise, the tailstock is away from the operator too far and the workpiece will taper with the smaller end at the headstock.

Figure 9. Checking for taper by taking a cut on a workpiece. After the cut is made for the length of the workpiece, a micrometer reading is taken at each end to determine any difference in diameter (Lane Community College).

Figure 10. Using a dial indicator to check the amount of movement of the tailstock when it is being re-aligned (Lane Community College).

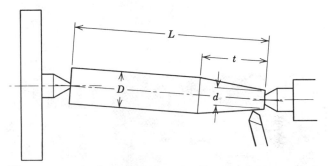

Figure 11. A short taper is made on a longer shaft.

In either case, move the tailstock until both diameters have the same reading.

Since usually only a minor adjustment is needed while a job is in progress, the most common method of aligning lathe centers is by cutting and measuring. It is also the most accurate. This method, unlike the bar test method, usually uses the workpiece while it is in the roughing stage (Figure 9). A light cut is taken along the length of the test piece and both ends are measured with a micrometer. If the diameter at the tailstock end is smaller, the tailstock is toward the operator, and if the diameter at the headstock end is smaller, the tailstock is away from the operator. Set up a dial indicator (Figure 10) and move the tailstock half the difference of the two micrometer readings. Make another light cut and check for taper.

When only a short distance of the total length of a shaft (Figure 11) can be turned to check for taper, the amount of tailstock set-over may be calculated by the following formula:

$$\text{Offset} = \frac{L \times (D - d)}{2t}$$

Where:

$L =$ total length of shaft
$t =$ length of taper
$D =$ diameter at large end
$d =$ diameter at small end

EXAMPLE

$L = 14$ inches
$t = 4$ inches
$D = 1.495$ inches
$d = 1.490$ inches

$$\text{Offset} = \frac{14 \times (1.495 - 1.490)}{2 \times 4} = .010 \text{ inch}$$

In this case, the tailstock should be moved away from the operator .010 in.

self-evaluation

1. What is the result on the workpiece when the centers are out of line?
2. What happens to the workpiece when the tailstock is offset toward the operator?
3. Name three methods of aligning the centers.
4. Which measuring instrument is used when using a test bar?
5. By what means is the measuring done when checking taper by taking a cut?

This unit has no post test.

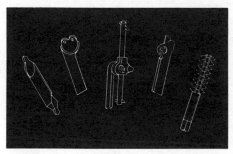

unit 9
drilling, boring, reaming, knurling, recessing, parting, and tapping in the lathe

Much of the versatility of the lathe as a machine tool is due to the variety of tools and workholding devices used. This equipment makes possible the many special operations that you will begin to do in this unit.

objectives After completing this unit, you will be able to:

1. Explain the procedures for drilling, boring, reaming, knurling, recessing, parting, and tapping in the lathe.

2. Set up to drill, ream, bore, and tap on the lathe and complete each of these operations.

3. Set up for knurling, recessing, die threading, and parting on the lathe and complete each of these operations.

DRILLING

The lathe operations of boring, tapping, and reaming usually begin with spotting and drilling a hole. The workpiece, often a solid material that requires a bore, is mounted in a chuck collet or faceplate, while the drill is typically mounted in the tailstock spindle that has a Morse taper. If there is a slot in the tailstock spindle, the drill tang must be aligned with it when inserting the drill.

Drill chucks with Morse taper shanks are used to hold straight shank drills and center drills (Figure 1a). Center drills are used for spotting or making a start for drilling (Figure 1b). When drilling with large size drills, a pilot drill should be used first. If a drill wobbles when started, place the heel of a toolholder against it near the point to steady the drill while it is starting in the hole. Taper shank drills (Figure 2) are inserted directly into the tailstock spindle. The friction of the taper is usually all that is needed to keep the drill from turning while a hole is being drilled (Figures 3 and 4), but when using larger drills, the friction is not enough. A lathe dog is sometimes clamped to the drill just above the flutes (Figure 5) with the bent tail resting on the compound. Hole depth can be measured with a rule or by means of the graduations on top of the tailstock spindle. The alignment of the tailstock with the lathe center line should be checked before drilling or reaming.

Drilled holes are not accurate enough for many applications, such as for gear or pulley bores, which should not be made over the nominal size more than .001 to .002 in. Drilling typically produces holes that are oversize and run eccentric to the center axis of the lathe (Figure 6). This is not true in the case of some manufacturing types such as gun drilling. However, truer axial alignment of holes is possible when the work is turned and the drill remains stationary, as in a lathe operation, in comparison to operations where the drill is turned and the

Figure 1a. Mounting a straight shank drill in a drill chuck in the tailstock spindle (Lane Community College).

Figure 1b. Center drilling is the first step prior to drilling, reaming, or boring (DeAnza College).

work is stationary, as on a drill press. Drilling also produces holes with rough finishes, which along with size errors can be corrected by boring or reaming. The hole must first be drilled

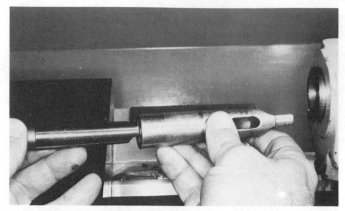

Figure 2. A drill sleeve is placed on the drill so that it will fit the taper in the tailstock spindle (DeAnza College).

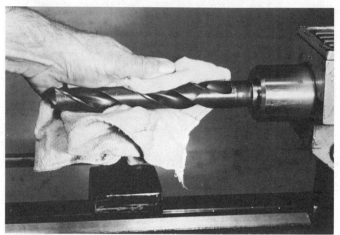

Figure 3. The drill is then firmly seated in the tailstock spindle (Lane Community College).

Figure 4. The feed pressure when drilling is usually sufficient to keep the drill seated in the tailstock spindle, thus keeping the drill from turning (DeAnza College).

Figure 5. A lathe dog is used when the drill has a tendency to turn in the tailstock spindle (Lane Community College).

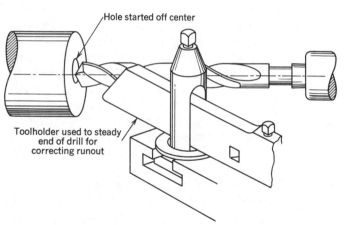

Hole started off center

Toolholder used to steady end of drill for correcting runout

Figure 6. An exaggerated view of the runout and eccentricity that is typical of drilled holes.

slightly smaller than the finish diameter in order to leave material for finishing by either of these methods.

BORING

Boring is the process of enlarging and truing an existing or drilled hole. A drilled hole for boring can be from $\frac{1}{32}$ to $\frac{1}{16}$ in. undersize, depending on the situation. Speeds and feeds for boring are determined in the same way as they are for external turning. Boring to size predictably is also done in the same way as in external turning except that the cross feed screw is turned counterclockwise to move the tool into the work.

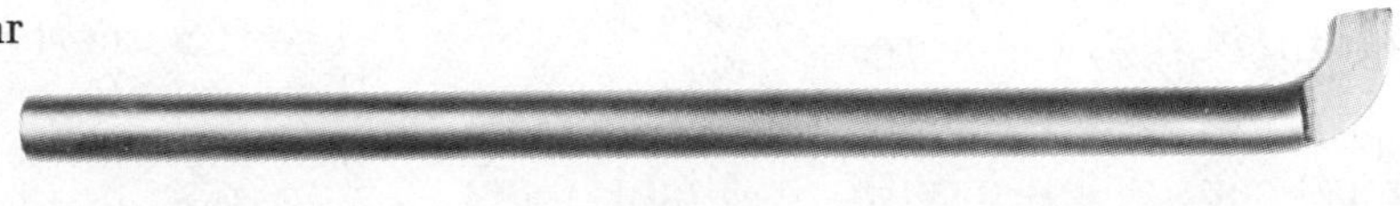

Figure 7. A small forged boring bar
made of high speed steel.

An inside spring caliper and a rule are sometimes useful for rough measurement. Vernier calipers are also used by machinists for internal measuring, though the telescope gage and outside micrometer are most commonly used for the precision measurement of small bores. Inside micrometers are used for larger bores. Other means of measurement are an inside spring caliper used with an outside micrometer, and on large bores an inside micrometer used with an outside caliper. Precision bore gages are used where many bores are checked for similar size, such as for acceptable tolerance.

A boring bar is clamped in a holder mounted on the carriage compound. Several types of boring bars and holders are used. Boring bars designed for small holes ($\frac{1}{2}$ in. and smaller) are usually the forged type (Figure 7). The forged end is sharpened by grinding. When the bar gets ground too far back, it must be reshaped or discarded. Boring bars for holes with diameters over $\frac{1}{2}$ in. (Figure 8) use high speed tool inserts, which are typically hand ground in the form of a left-hand turning tool. These tools can be removed from the bar for resharpening when needed. The cutting tool can be held at various angles to obtain different results, which makes the boring bar useful for many applications. Standard bars generally come with a tool angle of 30, 45, or 90 degrees. Some boring bars are made for carbide inserts (Figure 9).

Chatter is the rattle or vibration between a workpiece and a tool because of the lack of rigid support for the tool. Chatter is a great problem in boring operations since the bar must extend away from the support of the compound (Figure 10). For this reason boring bars should be kept back into their holders as far as practicable. Tuned boring bars can be adjusted so that their vibration is dampened (Figure 11). Stiffness of boring bars is increased by making them of solid tungsten carbide. If chatter occurs when boring, one or more of the following may help to eliminate the vibration of the boring tool.

1. Shorten the boring bar overhang, if possible.
2. Make sure the tool is on center.
3. Reduce the spindle speed.
4. Use a boring bar as large in diameter as possible without it binding in the hole.

Figure 9. A boring bar with carbide insert. When one edge is dull, a new one is selected (Photo courtesy of Kennamental Inc., Latrobe, Pa.).

Figure 10. A boring bar setup with a large overhang for making a deep bore. It is difficult to avoid chatter with this arrangement (DeAnza College).

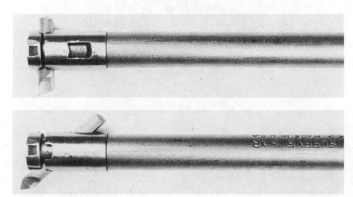

Figure 8. Two boring bars with inserted tools set at different angles.

5. Reduce the nose radius on the tool.

6. Apply cutting oil to the bore.

Boring bars sometimes spring away from the cut and cause bell-mouth, a slight taper at the front edge of a bore. One or two extra cuts taken without moving the cross feed will usually eliminate this problem.

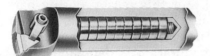

Figure 11. Tuned boring bars contain dampening slugs of heavy material that can be adjusted by applying pressure with a screw (Photo courtesy of Kennametal Inc., Latrobe, Pa.).

Figure 12. Boring tools must have sufficient side relief and side rake to be efficient cutting tools. Back rake is not normally used (Lane Community College).

A large variety of boring bar holders are used. Some types are designed for small, forged bars while others of more rigid construction are used for larger, heavier work.

Boring tools are made with side relief and end relief, but usually with zero back rake (Figure 12). Insufficient end relief will allow the heel of the tool to rub on the workpiece (Figure 13). The end relief should be between 10 and 20 degrees. The machinist must use his judgment when grinding the end relief because the larger the bore, the less end relief is required (Figure 14). If the end of the tool is relieved too much, the cutting edge will be weak and break down.

The point of the cutting tool should be positioned exactly on the center line of the workpiece (Figures 15a to c). There must be a space to allow the chips to pass between the bar and the surface being machined, or the chips will wedge and bind on the back side of the bar, forcing the cutting tool deeper into the work (Figure 16).

"Through boring" is the boring of a workpiece from one end to the other or all the way through it. For through boring, the tool is held

Figure 13. A tool with insufficient end relief will rub on the heel of the tool and will not cut (Lane Community College).

Figure 14. End relief angle varies depending on the diameter of the workpiece bore. These views are looking outward from inside the chuck (Lane Community College).

Figure 15*a*. The point of the boring tool must be positioned on the centerline of the workpiece (Lane Community College).

Figure 15*c*. If the tool is too high, the back rake becomes excessively negative and the tool point is likely to be broken off. A poor quality finish is the result of this position (Lane Community College).

Figure 15*b*. If the boring tool is too low, the heel of the tool will rub and the tool will not cut, even if the tool has the correct relief angle (Lane Community College).

Figure 16. Allowance must be made so the chips can clear the space between the bar and the surface being machined (Lane Community College).

in a bar that is perpendicular to the axis of the workpiece. A slight side cutting edge angle is often used for through boring (Figure 17). Back facing is sometimes done to true up a surface on the back side of a through bore, This is done with a straight ground right-hand tool also held in a bar perpendicular to the workpiece. The amount of facing that can be done in this way is limited to the movement of the bar in the bore.

A blind hole is one that does not go all the way through the part to be machined (Figure 18). Machining the bottom or end of a blind hole to a flat is easier when the drilled center does not need to be cleaned up. A bar with the tool set at an angle, usually 30 or 45 degrees, is used to square the bottom of a hole with a drilled center (Figure 19).

Most boring is performed on workpieces mounted in a chuck. But it is also done in the end of workpieces supported by a steady rest. Boring and other operations are infrequently done on workpieces set up on a face plate (Figure 20).

A thread relief is an enlargement of a bore at the bottom of a blind hole. The purpose of a thread relief is to allow a threading tool to disengage the work at the end of a pass (Figure 21). When the work will allow it, a hole can be drilled deeper than necessary. This will give the end of the boring bar enough space so that the tool can reach into the area to be relieved and still be held

at a 90 degree angle (Figure 22). When the work will not allow for the deeper drilling, a special tool must be ground (Figure 23).

Grooves in bores are made by feeding a

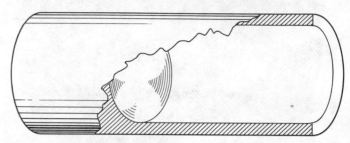

Figure 18. A blind hole machined flat in the bottom.

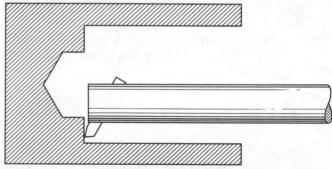

Figure 19. A bar with an angled tool used to square the bottom of a hole with a drilled center.

Figure 17. Bar and tool arrangement for through boring (Lane Community College).

Figure 20. Workpiece clamped on face plate has been located, drilled, and bored (California Community Colleges IMC Project).

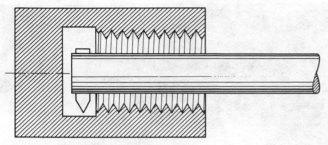

Figure 21. Ample thread relief is necessary when making internal threads.

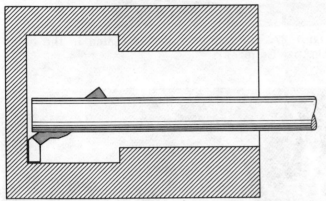

Figure 23. A special tool is needed when the thread relief must be next to a flat bottom.

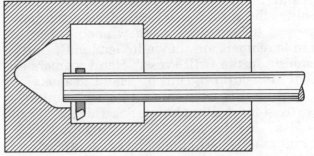

Figure 22. The hole is drilled deeper than necessary to allow room for the boring bar.

Figure 24. A tool that is ground to the exact width of the desired groove can be moved directly into the work to the correct depth (Lane Community College).

Figure 25. A square shoulder is made with a counterboring tool (Lane Community College).

form tool (Figure 24) straight into the work. Snap ring, O-ring, and oil grooves are made in this way. Cutting oil should always be used in these operations.

Counterboring in a lathe is the process of enlarging a bore for definite length (Figure 25). The shoulder that is produced in the end of the counterbore is usually made square (90 degrees) to the lathe axis. Boring and counterboring are also done on long workpieces that are supported in a steady rest. All boring work should have the edges and corners broken or chamfered.

REAMING

Reaming is done in the lathe to quickly and accurately finish drilled or bored holes to size. Machine reamers, like drills, are held in the tailstock spindle of the lathe. Floating reamer holders are sometimes used to assure alignment of the reamer, since the reamer follows the eccentricity of drilled holes. This helps eliminate bell-mouth bores that result from reamer

wobble, but does not eliminate the hole eccentricity. Only boring will remove the bore runout.

Roughing reamers (rose reamers) are often used in drilled or cored holes followed by machine or finish reamers. When drilled or cored holes have excessive eccentricity, they are bored .010 to .015 in. undersize and machine reamed. If a greater degree of accuracy is required, the hole is bored to within .003 to .005 in. of finish size and hand reamed (Figure 26). For hand reaming, the machine is shut off and the hand reamer is turned with a tap wrench. Types of machine reamers are shown in Section G, Unit 7, "Reaming in the Drill Press." Hand reamers are shown in Section B, Unit 6, "Hand Reamers."

Cutting oils used in reaming are similar to those used for drilling holes (See Table 2, "Coolants Used for Reaming," in Section G, Unit 7, "Reaming in the Drill Press"). Cutting speeds are dependent on machine and workpiece material finish requirements. A rule of thumb for reaming speeds is to use one half the speed used for drilling (See Table 1, Cutting Speeds in SFPM for Reaming with an HSS Reamer," in Section G, Unit 7, "Reaming in the Drill Press").

Feeds for reaming are about twice that used for drilling. The cutting edge should not rub without cutting as it causes glazing, work hardening, and dulling of the reamer.

A simple machine reaming sequence would be as follows:

1. Assuming that the hole has been drilled $\frac{1}{64}$ in. undersize, a taper shank machine reamer is seated in the taper by hand pressure (Figure 27).
2. Cutting oil is applied to the hole and the reamer is started into the hole by turning the tailstock handwheel (Figure 28). The hole is completed and the machine is turned off. Then the reamer is removed from the hole. Never reverse the machine when reaming.
3. The reamer is removed from the tailstock spindle and cleaned with a cloth (Figure 29).

Figure 27. The reamer must be seated in the taper (DeAnza College).

Figure 28. Starting the reamer in the hole. Kerosene is being used as a cutting fluid (DeAnza College).

Figure 26. Hand reaming in the lathe (Lane Community College).

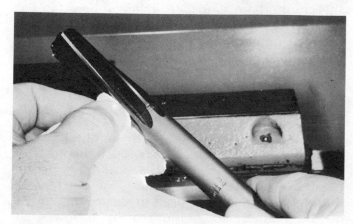

Figure 29. The reamer should be cleaned and put away after using it (DeAnza College).

The reamer is then returned to the storage rack.
4. The lathe is cleaned with a brush (Figure 30).

TAPPING

The tapping of work mounted in a chuck is a quick and accurate means of producing internal threads. Tapping in the lathe is similar to tapping in the drill press but it is generally reserved for small size holes, as tapping is the only way they can be internally threaded. Large internal threads are made in the lathe with a boring tool. A large tap requires considerable force or torque to turn, more than can be provided by hand turning. A tap that is aligned by the dead center will make a straight tapped hole that is in line with the lathe axis.

A plug tap (Figure 31) or spiral point tap (Figure 32) may be used for tapping through holes. When tapping blind holes, a plug tap could be followed by a bottoming tap (Figure 33), if threads are needed to the bottom of the hole. A good practice is to drill a blind hole deeper than the required depth of threads.

Two approaches may be taken for hand tapping. Power is not used in either case. One method is to turn the tap by means of a tap wrench or adjustable wrench with the spindle engaged in a low gear so it will not turn (Figure 34). The other method is to disengage the spindle and turn the chuck by hand while the tap wrench handle rests on the compound (Figure 35). In both cases the tailstock is clamped

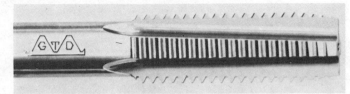

Figure 31. Plug tap (Photo courtesy of TRW Greenfield Tap & Die Division).

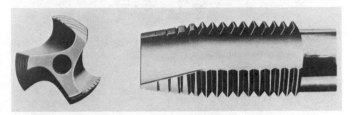

Figure 32. Spiral point tap or gun tap (Photo courtesy of TRW Greenfield Tap & Die Division).

Figure 33. Bottoming tap (Photo courtesy of TRW Greenfield Tap & Die Division).

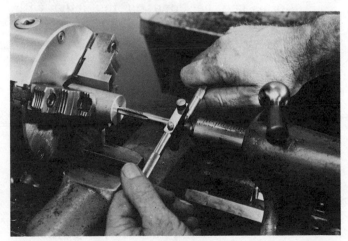

Figure 34. Hand tapping in the lathe by turning the tap wrench (Lane Community College).

to the ways, and the dead center is kept in the center of the tap by slowly turning the tailstock handwheel. The tailstock on small lathes need not be clamped to the ways for small taps, but held firmly with one hand. Cutting oil should

Figure 30. The chips are brushed into the chip pan (DeAnza College).

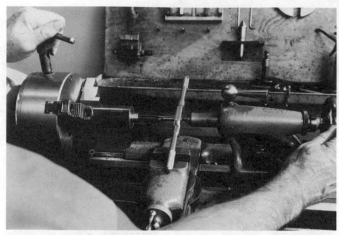

Figure 35. Tapping by turning the chuck (Lane Community College).

Figure 36. Starting die on rod to be threaded in the lathe. The tailstock spindle (without a center) is used to start the die squarely onto the work (Lane Community College).

be used and the tap backed off every one or two turns to break chips unless it is a spiral point tap, sometimes called a gun tap.

The correct tap drill size should be obtained from a tap drill chart. Drills tend to drill slightly oversize, and tapping the oversize hole can produce poor internal thread with only a small percentage of thread cut. Make sure the drill produces a correctly sized hole by drilling first with a slightly smaller drill, then use the tap drill as a reamer.

Tapping can be done on the lathe with power, but it is recommended that it be done only if the spindle rotation can be reversed, if a spiral point tap is used, and if the hole is clear through the work. The tailstock is left to move on the ways. Insert the tap in a drill chuck in the tailstock and set the lathe on a low speed. Use cutting oil and slide the tailstock so the tap engages the work. Reverse the tap and remove it from the work every $\frac{3}{8}$ to $\frac{1}{2}$ in. When reversing, apply light hand pressure on the tailstock to move it to the right until the tap is all of the way out.

External threads cut with a die should only be used for nonprecision purposes, since the die may wobble and the pitch (the distance from a point on one thread to the same point on the next) may not be uniform. The rod to be threaded extends a short distance from the chuck and a die and diestock are started on the end (Figure 36). Cutting oil is used. The handle is rested against the compound. The chuck may be turned by hand, but if power is used, the machine is set for low speed and reversed every $\frac{3}{8}$ to $\frac{1}{2}$ in.

to clear the chips. Finish the last $\frac{1}{4}$ in. by hand if approaching a shoulder. Reverse the lathe to remove the die.

RECESSING, GROOVING, AND PARTING

Recessing and grooving on external diameters (Figures 37 *a* to *c*) is done to provide grooves for thread relief, snap rings and O-rings. Special tools (Figure 38) are ground for both external and internal grooves and recesses. Parting tools are sometimes used for external grooving and thread relief.

Parting or cutoff tools (Figure 39) are designed to withstand high cutting forces, but if chips are not sufficiently cleared or cutting oil is not used, these tools can quickly jam and break. Parting tools must be set on center and square with the work (Figure 40). Lathe tools are often specially ground as parting tools for small or delicate parting jobs (Figure 41). Diagonally ground parting tools leave no burr.

Parting alloy steels and other metals is sometimes difficult, and step parting (Figure 42) may help in these cases. When deep parting difficult material, extend the cutting tool from the holder a short distance and part to that depth. Then back off the cross feed and extend the tool a bit farther; part to that depth. Repeat the process until the center is reached. Sulfurized cutting oil works best for parting unless

Figure 37a. The undercutting tool is brought to the workpiece and the micrometer dial is zeroed. Cutting oil is applied to the work (Lane Community College).

Figure 37c. The finished groove (Lane Community College).

Figure 37b. The tool is fed to the single depth of the thread or the required depth of the groove. If a wider groove is necessary, the tool is moved over and a second cut is taken as shown (Lane Community College).

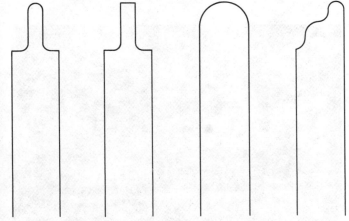

Figure 38. Recessing or grooving tools for internal and external use.

the lathe is equipped with a coolant pump and a steady flow of soluble oil is available. Parting tools are made in either straight or offset types. A right-hand offset cutoff tool is necessary when parting very near the chuck.

All parting and grooving tools have a tendency to chatter; therefore any setup must be rigid as possible. A low speed should be used for parting; if the tool chatters, reduce the speed.

Work should not extend very far from the chuck when parting or grooving, and no parting should be done in the middle of a workpiece or at the end near the dead center. A feed that is too light can cause a chatter, but a feed that is too heavy can jam the tool. The tool should always be making a chip. Hand feeding the tool is best at first.

KNURLING

A knurl is a raised impression on the surface of a workpiece produced by two hardened rolls, and is usually of two patterns, diamond or straight (Figure 43). The diamond pattern is

Figure 39. Parting tool making a cut (Lane Community College).

Figure 40. Parting tools must be set to the center of the work (Lane Community College).

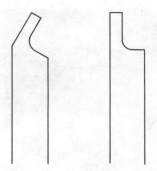

Figure 41. Special parting tools that have been ground from lathe tools for small or delicate parting jobs.

Figure 42. Step parting (Lane Community College).

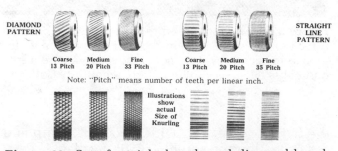

Figure 43. Set of straight knurls and diagonal knurls (Courtesy of J. H. Williams Division of TRW Inc.).

formed by a right-hand and a left-hand helix mounted in a self-centering head. The straight pattern is formed by two straight rolls. These common knurl patterns can be either fine, medium, or coarse.

Diamond knurling is used to improve the appearance of a part and to provide a good gripping surface for levers and tool handles. Straight knurling is used to increase the size of a part for press fits in light duty applications. A dis-

advantage to this use of knurls is that the fit has less contact area than a standard fit.

Three basic types of knurling toolholders are used: the knuckle-joint holder (Figure 44a),

Figure 44a. Knuckle-joint knurling toolholder.

Figure 44b. Revolving head knurling toolholder.

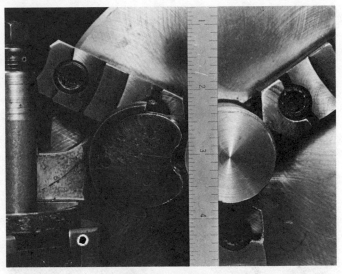

Figure 45. Knurls are centered on the workpiece (Lane Community College).

Figure 44c. Straddle knurling toolholder (Courtesy of Ralmike's Tool-A-Rama).

the revolving head holder (Figure 44b), and the straddle holder (Figure 44c). The straddle holder allows small diameters to be knurled with less distortion. This principle is used for knurling on production machines.

Knurling works best on workpieces mounted between centers. When held in a chuck and supported by a center, the workpiece tends to crawl back into the chuck and out of the supporting center with the high pressure of the knurl. This is especially true when the knurl is started at the tailstock end and the feed is toward the chuck. Long slender pieces push away from the knurl and will stay bent if the knurl is left in the work after the lathe is stopped.

Knurls do not cut, but displace the metal with high pressures. Lubrication is more important than cooling, so a lard oil or lubricating oil is satisfactory. Low speeds (about the same as for threading) and a feed of about .015 to .030 in. are used for knurling.

The knurls should be centered on the workpiece vertically (Figure 45) and the knurl toolholder should be square with the work, unless the knurl pattern is difficult to establish, as it often is in tough materials. In that case, the toolholder should be angled about 5 degrees to the work so the knurl can penetrate deeper (Figure 46).

A knurl should be started in soft metal about half depth and the pattern checked. An even diamond pattern should develop (Figure 47). But, if one roll is dull or placed too high or too low, a double impression will develop (Figure 48) because the rolls are not tracking evenly. If this happens, move the knurls to a new position along the workpiece, readjust up or down, and try again. The knurls should be cleaned with a wire brush between passes. Another way to check for even tracking of the knurls is to roll the knurl on a piece of paper (Figure 49).

Figure 46. Angling the toolholder 5° often helps establish the diamond pattern (Lane Community College).

Figure 47. The knurl is started approximately half depth. Notice the diamond pattern is not fully developed (Lane Community College).

Figure 48. Double impression on the left is the result of the rolls not tracking evenly (Lane Community College).

Figure 49. Checking the diamond pattern by using a strip of paper (Lane Community College).

Material that hardens as it is worked, such as high carbon or spring steel, should be knurled in one pass if at all possible, but in not more than two passes. Even in ordinary steel, the surface will work harden after a diamond pattern has developed to points. It is best to stop knurling just before the points are sharp (Figure 50). Metal flaking off the knurled surface is evidence that work hardening has occurred. Avoid knurling too deeply as it produces an inferior knurled finish.

Figure 50. More than one pass is usually required to bring the knurl to full depth (Lane Community College).

Figure 51. A knurling tool that cuts a knurl rather than forming it by pressure.

Knurls are also produced with a type of cutting tool (Figure 51) that is similar in appearance to a knurling tool. The serrated rolls form a chip on this edge (Figure 52). Material that is difficult to knurl by pressure rolling, such as tubing and work hardening metals, can be knurled by this cutting tool. Sulfurized cutting oil should be used when knurling steel with this kind of knurling tool.

Figure 52. A knurl being cut showing formation of chip (Lane Community College).

self-evaluation

SELF-TEST
1. Why are drilled holes not used for bores in machine parts such as pulleys, gears, and bearing fits?
2. Describe the procedure used to produce drilled holes on workpieces in the lathe with minimum oversize and runout.
3. What is the chief advantage of boring over reaming in the lathe?
4. List five ways to eliminate chatter in a boring bar.
5. Explain the differences between through boring, counterboring, and boring blind holes.
6. By what means are grooves and thread relief made in a bore?
7. Reamers will follow an eccentric drilled hole, thus producing a bell-mouth bore with runout. What device can be used to help eliminate bell-mouth? Does it help remove the runout?
8. Machine reamers produce a better finish than is obtained by boring. How can you get an even better finish with a reamer?

9. Cutting speeds for reaming are (twice, half) that used for drilling; feeds used for reaming are (twice, half) that used for drilling.

10. Are large internal threads produced with a tap or a boring tool? Explain the reason for your answer.

11. How can you avoid drilling oversize with a tap drill?

12. Standard plug or bottoming taps can be used when hand tapping in the lathe. If power is used, what kind of tap works best?

13. Why would threads cut with a hand die in a lathe not be acceptable for using on a feed screw with a micrometer collar?

14. By what means are thread relief on external grooves produced?

15. If cutting oil is not used on parting tools or chips do not clear out of the groove because of a heavy feed, what is generally the immediate result?

16. How can you avoid chatter when cutting off stock with a parting tool?

17. State three reasons for knurling.

18. Ordinary knurls do not cut. In what way do they make the diamond or straight pattern on the workpiece?

19. If a knurl is producing a double impression, what can you do to make it develop a diamond pattern?

20. How can you avoid producing a knurled surface on which the metal is flaking off?

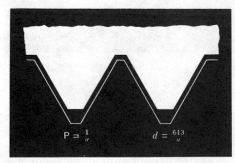

unit 10
sixty degree thread information and calculations

To cut threads, a good machinist must know more than how to set up the lathe. He must know the thread form, class of fit, and thread calculation. This unit prepares you for the actual cutting of threads, which you will do in the next unit.

objectives After completing this unit, you will be able to:

1. **Describe the several 60 degree thread forms, noting their similarities and differences.**

2. **Calculate thread depth, infeeds, and minor diameters of threads.**

THE SHARP VEE THREAD FORM

Various screw thread forms are used for fastening and for moving or transmitting parts against loads. The most widely used of these forms are the 60 degree thread types. These are mostly used for fasteners. An early form of the 60 degree thread is the sharp V (Figure 1). The sides of the thread form a 60 degree angle with each other. Theoretically, the sides and the base between two thread roots would form an equilateral triangle, but in practice this is not the case; it is necessary to make a slight flat on top of the thread in order to deburr it. Also, the tool will always round off and leave a slight flat at the thread root.The greatest drawback to this thread form is that it is so easily damaged while handling. The sharp V thread will fit closer and seal better than most threads, but is seldom used today. The depth (d) for the sharp V thread is calculated as follows:

$$d = \text{pitch} \times \cos 30 = .866 \text{ pitch} =$$

$$\frac{.866}{\text{number of threads per inch}}$$

The relationship between pitch and threads per inch should be noted (Figure 2). Pitch is the distance between a point on one screw thread and the corresponding point on the next thread, measured parallel to the thread axis. Threads per inch means the number of threads in one

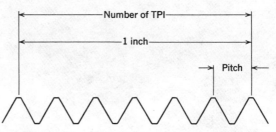

Figure 2. The difference between threads per inch and pitch.

inch. The pitch (P) may be derived by dividing the number of threads per inch (TPI) into one.

$$P = \frac{1}{\text{TPI}}$$

EXAMPLE

Find the pitch of a $\frac{1}{2}$ in. diameter-20 TPI machine screw thread.

$$P = \frac{1}{20} = .050 \text{ in.}$$

Pitch is checked on a screw thread with a screw pitch gage (Figures 3a and b). General dimensions and symbols for screw threads are shown in Figure 4.

UNIFIED AND AMERICAN NATIONAL FORMS

The American National form (Figure 5), formerly United States Standard, was used for many years for screws, bolts, and other products. These National form threads are in either the national fine (NF) or the national coarse (NC) series. Other screw threads are listed in machinist's handbooks.

Taps and dies are marked with letter symbols to designate the series of the threads they form. For example, the symbol for American Standard Taper pipe thread is NPT, for Unified coarse thread it is UNC, and for Unified fine thread the symbol is UNF.

Thread Depth

The American Standard for Unified threads (Figure 6) is very similar to the American National Standard with certain modifications. The thread forms are practically the same and the basic 60 degree angle is the same. The depth

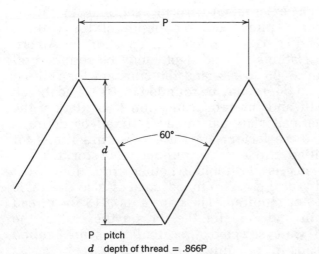

P pitch
d depth of thread = .866P

Figure 1. The 60° sharp-V thread.

Figure 3a. Screw pitch gage for inch threads.

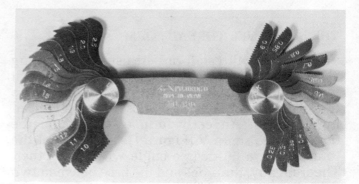

Figure 3b. Screw pitch gage for metric threads.

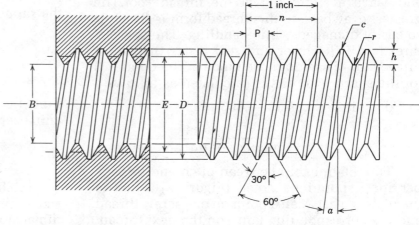

Figure 4. General dimensions of screw threads.

D	Major Diameter	*a*	Helix (or Lead) Angle
E	Pitch Diameter	*c*	Crest of Thread
B	Minor Diameter	*r*	Root of Thread
n	Number of threads per inch (TPI)	*h*	Basic Thread Height (or depth)
P	Pitch		

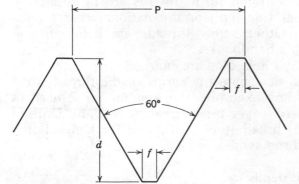

P   pitch
d   depth of thread = .6495P
f   flat at crest and root of thread = $\frac{P}{8}$

Figure 5. The American National form thread.

of an external American National thread is .6495 × pitch, and the depth of the Unified thread is .6134 × pitch. The constant for American National thread depth may be rounded off from .6495 to .65, and the constant for Unified thread depth may be rounded to .613. The thread depth and the root truncation (tool flat) of the American National thread is fixed or definite, but these factors are variable within limits for Unified threads. A rounded root for Unified threads is desirable whether from tool wear or by design. A rounded crest is also desirable but not required. The constants .613 for thread depth and .144 for the flat on the end of the tool were selected for calculations on Unified threads in this unit.

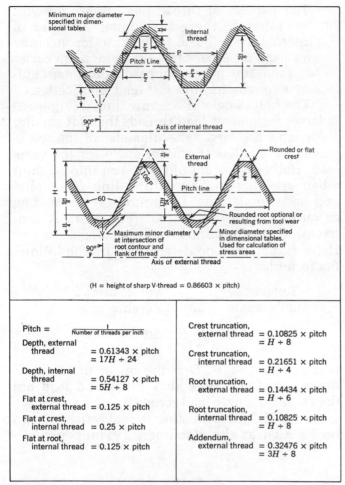

(H = height of sharp V-thread = 0.86603 × pitch)

Pitch =	$\dfrac{1}{\text{Number of threads per inch}}$	Crest truncation, external thread	= 0.10825 × pitch = H ÷ 8
Depth, external thread	= 0.61343 × pitch = 17H ÷ 24	Crest truncation, internal thread	= 0.21651 × pitch = H ÷ 4
Depth, internal thread	= 0.54127 × pitch = 5H ÷ 8	Root truncation, external thread	= 0.14434 × pitch = H ÷ 6
Flat at crest, external thread	= 0.125 × pitch	Root truncation, internal thread	= 0.10825 × pitch = H ÷ 8
Flat at crest, internal thread	= 0.25 × pitch	Addendum, external thread	= 0.32476 × pitch = 3H ÷ 8
Flat at root, internal thread	= 0.125 × pitch		

Figure 6. Unified screw threads. *Illustration:* Reprinted from ASA Bl.1—1960, Unified Screw Threads, with permission of the publisher, The American Society of Mechanical Engineers. *Data:* Source; *Machinery's Handbook*—20th Edition, Copyright © 1975. Reprinted with permission of Industrial Press, Inc., the publisher.

Thread Fit Classes and Thread Designations

Unified and American National Standard form threads are interchangeable. An NC bolt will fit an UNC nut. The principle difference between the two systems is that of tolerances. The Unified system, a modified version of the old system, allows for more tolerances of fit. Thread fit classes 1, 2, 3, 4, and 5 were used with the American National Standard; one being a very loose fit, two a free fit, three a close fit, four a snug fit, and five a jam or interference fit. The Unified system expanded this number system to include a letter, so the threads could

be identified as class 1A, 1B, 2A, 2B, and so on. "A" indicates an external thread and "B" an internal thread. Because of this expansion to the Unified system, tolerances are now possible on external threads and are 30 percent greater on internal threads. These changes make easier the manufacturer's job of controlling tolerances to insure the interchangeability of threaded parts. See the *Machinery's Handbook* for tables of Unified thread limits. Limits are the maximum and minimum allowable dimensions of a part, in this case, internal and external threads.

Threads are designated by the nominal bolt size or major diameter, the threads per inch, the letter series, the thread tolerance, and the thread direction. Thus, $1\frac{1}{4}$ in.– 12 UNF-2BLH would indicate a $1\frac{1}{4}$ in. Unified nut with 12 threads per inch, a class 2 thread fit, and a left-hand helix.

Unified screw thread systems are the American Standard for fastening types of screw threads. Manufacturing processes where V threads are produced are based on the Unified system. Many job and maintenance machine shops, on the other hand, still use the American National thread system when chasing a thread with a single point tool on an engine lathe.

Tool Flats and Infeeds for Thread Cutting

The truncation or flat on the crest of the thread is P/8 for both the Unified and American National systems. The root truncation is calculated P/8 for the American National system, but it is calculated H/6 in the Unified system, where H is the height of the sharp V thread (.866 × pitch). Therefore, for Unified threads the flat on the end of the external threading tool equals $\dfrac{.866P}{6}$; or .144P. For American National threads the flat on the end of the external threading tool is P/8 or P × .125.

To cut 60 degree form threads, the tool is fed into the work with the compound (Figure 7), which is set at 29 or 30 degrees. The infeed depth along the flank of the thread at 30 degrees is greater than the depth at 90 degrees from the work axis. This depth may be calculated for American National threads by dividing the number of threads per inch (n) into .75,

$$\text{Infeed} = \frac{.75}{n} \text{ or } P \times .75.$$

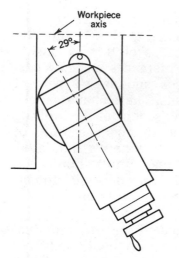

Figure 7. The compound at 29° for cutting 60° threads.

Thus, for a thread with 10 threads per inch (.100 inch pitch):

$$\text{Infeed} = \frac{.75}{10} = .075 \text{ in.}$$

or

$$\text{Infeed} = .75 \times .100 = .075 \text{ in.}$$

For external Unified threads the infeed at 29 degrees may be calculated by the formula:

$$\text{Infeed} = \frac{.708}{n} \quad \text{or} \quad .708P$$

Thus, for a thread with 10 threads per inch (.100 in. pitch):

$$\text{Infeed} = \frac{.708}{10} = .0708 \text{ in.}$$

or

$$\text{Infeed} = .708 \times .100 = .0708 \text{ in.}$$

PITCH DIAMETER, HELIX ANGLE, AND PERCENT OF THREADS

The making of external and internal threads that are interchangeable depends upon the selection of thread fit classes. The clearances and tolerances for thread fits are derived from the pitch diameter. The pitch diameter on a straight thread is the diameter of an imaginary cylinder that passes through the thread profiles at a point where the width of the groove and thread are equal. The mating surfaces are the flanks of the thread.

The percent of thread has little to do with fit, but refers to the actual minor diameter of the internal thread. The typical nut for machine screws has 75 percent threads, which are easier to tap than 100 percent threads and retain sufficient strength for most thread applications.

The helix angle of a screw thread (Figure 8) is larger for greater lead threads than for smaller leads; and the larger the diameter of the workpiece, the smaller the helix angle for the same lead. Helix angles should be taken into account when grinding tools for threading. The relief and helix angle must be ground on the leading or cutting edge of the tool (Figure 9). A protractor may be used to check this angle.

Helix angles may be determined by the following formula:

$$\frac{\text{Tangent of}}{\text{helix angle}} = \frac{\text{lead of thread}}{\text{circumference of screw}}$$

$$= \frac{\text{lead of thread}}{\pi D}$$

where $\pi = 3.1416$, $D =$ the major diameter of the screw. (Also note that pitch and lead are the same for single lead screws.) Helix angles are given for Unified and other thread series in hand books such as the *Machinery's Handbook*.

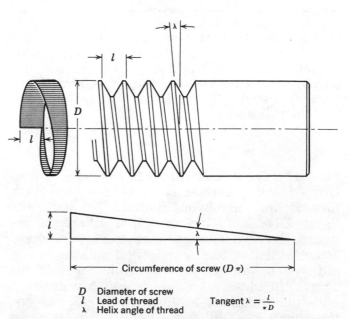

D Diameter of screw
l Lead of thread
λ Helix angle of thread

$$\text{Tangent } \lambda = \frac{l}{\pi D}$$

Figure 8. Screw thread helix angle. $D =$ diameter of screw; $l =$ lead of thread, and $\lambda =$ helix angle of thread.

$$\text{Tangent } \lambda = \frac{1}{\pi D}$$

Figure 9. Checking the relief and helix angle on the threading tool with a protractor (Lane Community College).

A taper thread is made on the internal or external surface of a cone. An example of a 60 degree taper thread is the American National Standard pipe thread (Figure 10). A line bisecting the 60 degree thread is perpendicular to the axis of the workpiece. On a taper thread the pitch diameter at a given position on the thread axis is the diameter of the pitch cone at that position.

The British Standard Whitworth thread (Figure 11) has rounded crests and roots and has an included angle of 55 degrees. This thread form has been largely replaced by the Unified and metric thread forms.

METRIC THREAD FORMS

Several metric thread systems such as the SAE standard spark plug threads and the British Standard for spark plugs are in use today. The Système Internationale (SI) thread form (Figure 12), adopted in 1898, is similar to the American National Standard. Metric bolt sizes differ slightly from one European country to the next. The British Standard for ISO (International Organization for Standardization) metric screw threads was set up to standardize metric thread forms. The basic form of the ISO metric thread (Figure 13) is similar to the Unified thread form. These and other metric thread systems are listed in the *Machinery's Handbook*. See Table 1 for ISO metric tap drill sizes.

A new metric standard was endorsed by

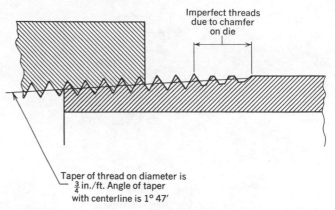

Figure 10. American National standard taper pipe thread.

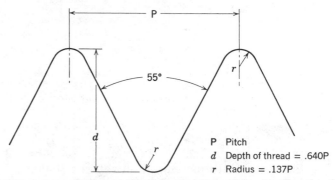

Figure 11. Whitworth thread. P = pitch; d = depth of thread, .640P; r = radius, .137P.

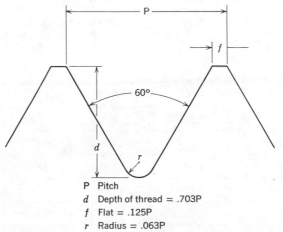

Figure 12. The SI metric thread form. P = pitch; d = depth of thread, .703P; f = flat; .125P; r = radius, .063P.

the Industrial Fasteners Institute (IFI) January 31, 1974. It is called the IFI-500 Trial Standard. For further information on the new thread sys-

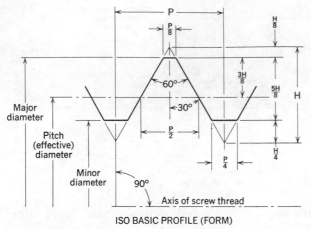

Figure 13. ISO metric thread form (Material courtesy of TRW Greenfield Tap & Die Division).

tem, refer to *Machinery's Handbook*, 20th edition, page 1327.

BASIC DESIGNATIONS

ISO Metric Threads are designated by the letter "M" followed by the *nominal size* in millimeters, and the *pitch* in millimeters, separated by the sign "×."

Example: M16 × 1.5

Above designation format is followed for all thread series except *coarse* pitch series as explained below.

Coarse Pitch ISO Metric Threads are designated by only the letter "M" and the *nominal size* in millimeters.

Example: M16

This is a 16 millimeter diameter, 2 millimeter pitch

Table 1
Metric Tap Drill Sizes

Metric Tap Size	Recommended Metric Drill				Closest Recommended Inch Drill			
	Drill Size (Millimeters)	Inch Equivalent	Probable Hole Size (Inches)	Probable Percent of Thread	Drill Size	Inch Equivalent	Probable Hole Size (Inches)	Probable Percent of Thread
M1.6 × .35	1.25	.0492	.0507	69	—	—	—	—
M1.8 × .35	1.45	.0571	.0586	69	—	—	—	—
M2 × .4	1.60	.0630	.0647	69	#52	.0635	.0652	66
M2.2 × .45	1.75	.0689	.0706	70	—	—	—	—
M2.5 × .45	2.05	.0807	.0826	69	#46	.0810	.0829	67
M3 × .5	2.50	.0984	.1007	68	#40	.0980	.1003	70
M3.5 × .6	2.90	.1142	.1168	68	#33	.1130	.1156	72
M4 × .7	3.30	.1299	.1328	69	#30	.1285	.1314	73
M4.5 × .75	3.70	.1457	.1489	74	#26	.1470	.1502	70
M5 × .8	4.20	.1654	.1686	69	#19	.1660	.1692	68
M6 × 1	5.00	.1968	.2006	70	#9	.1960	.1998	71
M7 × 1	6.00	.2362	.2400	70	15/64	.2344	.2382	73
M8 × 1.25	6.70	.2638	.2679	74	17/64	.2656	.2697	71
M8 × 1	7.00	.2756	.2797	69	J	.2770	.2811	66
M10 × 1.5	8.50	.3346	.3390	71	Q	.3320	.3364	75
M10 × 1.25	8.70	.3425	.3471	73	11/32	.3438	.3483	71
M12 × 1.75	10.20	.4016	.4063	74	Y	.4040	.4087	71

Table 1
Metric Tap Drill Sizes

Metric Tap Size	Recommended Metric Drill				Closest Recommended Inch Drill			
	Drill Size (Millimeters)	Inch Equivalent	Probable Hole Size (Inches)	Probable Percent of Thread	Drill Size	Inch Equivalent	Probable Hole Size (Inches)	Probable Percent of Thread
M12 × 1.25	10.80	.4252	.4299	67	27/64	.4219	.4266	72
M14 × 2	12.00	.4724	.4772	72	15/32	.4688	.4736	76
M14 × 1.5	12.50	.4921	.4969	71	—	—	—	—
M16 × 2	14.00	.5512	.5561	72	35/64	.5469	.5518	76
M16 × 1.5	14.50	.5709	.5758	71	—	—	—	—
M18 × 2.5	15.50	.6102	.6152	73	39/64	.6094	.6144	74
M18 × 1.5	16.50	.6496	.6546	70	—	—	—	—
M20 × 2.5	17.50	.6890	.6942	73	11/16	.6875	.6925	74
M20 × 1.5	18.50	.7283	.7335	70	—	—	—	—
M22 × 2.5	19.50	.7677	.7729	73	49/64	.7656	.7708	75
M22 × 1.5	20.50	.8071	.8123	70	—	—	—	—
M24 × 3	21.00	.8268	.8327	73	53/64	.8281	.8340	72
M24 × 2	22.00	.8661	.8720	71	—	—	—	—
M27 × 3	24.00	.9449	.9511	73	15/16	.9375	.9435	78
M27 × 2	25.00	.9843	.9913	70	63/64	.9844	.9914	70
M30 × 3.5	26.50	1.0433						
M30 × 2	28.00	1.1024						
M33 × 3.5	29.50	1.1614						
M33 × 2	31.00	1.2205			Reaming Recommended to the Drill Size Shown			
M36 × 4	32.00	1.2598						
M36 × 3	33.00	1.2992						
M39 × 4	35.00	1.3780						
M39 × 3	36.00	1.4173						

Formula for Metric Tap Drill Size:

$$\text{Basic major diameter (mm)} - \frac{\% \text{ Thread} \times \text{Pitch (mm)}}{76.980} = \text{Drilled Hole Size (mm)}$$

Formula for Percent of Thread:

$$\frac{76.980}{\text{Pitch (mm)}} \times \left[\text{Basic Major Diameter (mm)} - \text{Drilled Hole Size (mm)} \right] = \text{Percent of Thread}$$

Source. Material courtesy of TRW Greenfield Tap & Die Division, *New Greenfield Geometric ISO Metric Screw Thread Manual*, 1973.

ISO metric thread. Although the ISO standards use the above designations for coarse pitch, USA practice has been to include the pitch symbol even for the coarse pitch series. The inclusion of the pitch symbol should not create any problems, and will serve to avoid confusion.

TOLERANCE SYMBOLS

3   4   5   6̲
7   8   9

Numbers are used to define the amount of product tolerance permitted on either internal or external threads. Smaller grade numbers carry smaller tolerances, i.e. grade 4 tolerances are smaller than grade 6 tolerances, and grade 8 tolerances are larger than grade 6 tolerances.

e   H   G   g

Letters are used to designate the "position" of the product thread tolerances relative to basic diameters. Lower case letters are used for external threads, and capital letters for internal threads.

In some cases the "position" of the tolerance establishes an allowance (a definite clearance) between external and internal threads.

By combining the tolerance amount number and the tolerance position letter, the *tolerance symbol* is established which identifies the actual maximum and minimum product limits for external or internal threads. Generally the first number and letter refer to the pitch diameter symbol. The second number and letter refer to the crest diameter symbol (minor diameter of internal threads or major diameter of external threads.)

Example:

5g 6g
Pitch Diameter   Crest Diameter
Tolerance Symbol   Tolerance Symbol

Where the pitch diameter and crest diameter tolerance symbols are the same, the symbol need only be given once.

Example:

6g
Pitch Diameter and Crest
Diameter Tolerance Symbol.

It is recommended that the *coarse series* be selected whenever possible, and that *general purpose grade 6* be used for both internal and external threads.

Tolerance positions "g" for external threads and "H" for internal threads are preferred.

Other product information may also be conveyed by the ISO metric thread designations. Complete specifications and product limits may be found in the ISO Recommendations or in the B1 report "ISO Metric Screw Threads".

Some examples of ISO Metric Thread designations are as follows:

M10
M18 × 1.5
M6 - 6H
M4 - 6g
M12 × 1.25 - 6H
M20 × 2 - 6H/6g
M6 × 0.75 - 7g 6g

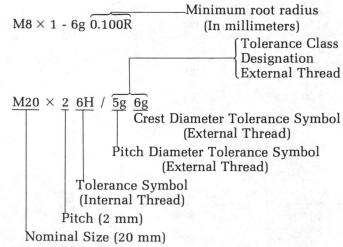

Source: (Material courtesy of TRW Greenfield Tap & Die Division, *New Greenfield* Geometric Screw ISO Metric Thread Manual, 1973).

self-evaluation

SELF-TEST   1. Name one disadvantage of the sharp V thread.

2. Explain the difference between the threads per inch and the pitch of the thread.

3. Name two similarities and two differences between American National and Unified threads.

4. What is a major reason for thread allowances and classes of fits?

5. What does $\frac{1}{2}$–20 UNC—2A describe?

6. The root truncation for unified threads is found by .144P and for American National threads it is found by .125P. What should the flat on the end of the threading tools be for both systems on a $\frac{1}{2}$–20 thread?

7. How far should the compound set at 30° move to cut a $\frac{1}{2}$–20 Unified thread? The formula is $\frac{.708}{n}$. How far should the compound move to cut a $\frac{1}{2}$–20 American National thread? The formula is $\frac{.75}{n}$.

8. Explain the difference between the fit of threads and the percent of thread.

9. Name two metric thread standard systems.

10. In metric tolerance symbols, which is a closer or smaller tolerance, a grade 3 or a grade 6?

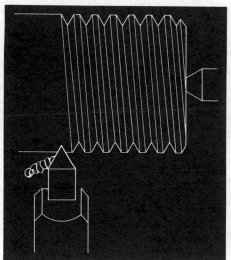

unit 11
cutting unified
external threads

A machinist is frequently called upon to cut threads of various forms on the engine lathe. The threads most commonly made are the V form, American National, or Unified. This unit will show you how to make these threads on a lathe. You will need much practice to gain confidence in your ability to make external Unified threads on any workpiece.

objectives After completing this unit, you will be able to:

1. **Detail the steps and procedures necessary to cut a Unified thread to the correct depth.**

2. **Set up a lathe for threading and cut seven different thread pitches and diameters.**

3. **Identify tools and procedures for thread measurement.**

HOW THREADING IS DONE ON A LATHE

Thread cutting on a lathe with a single point tool is done by taking a series of cuts in the same helix of the thread. This is sometimes called chasing a thread. A direct ratio exists between the headstock spindle rotation, the leadscrew rotation, and the number of threads on the lead-screw. This ratio can be altered by the quick-change gearbox to make a variety of threads. When the half-nuts are clamped on the thread of the leadscrew, the carriage will move a given distance for each revolution of the spindle. This distance is the lead of the thread.

If the infeed of a thread is made with the cross slide (Figure 1), equal size chips will be formed on both cutting edges of the tool. This causes higher tool pressures that can result in tool breakdown, and sometimes causes tearing of the threads because of insufficient chip clearance. A more accepted practice is to feed in with the compound, which is set at 29 degrees (Figure 2) toward the right of the operator, for cutting right-hand threads. This assures a cleaner cutting action than with 30° with most of the chip taken from the leading edge and a scraping cut from the following edge of the tool.

SETTING UP FOR THREADING

Begin setup by obtaining or grinding a tool for cutting Unified threads of the required thread

pitch. The only difference in tools for various pitches is the flat on the end of the tool. For Unified threads this is .144P, as discussed in the last unit. If the toolholder you are using has no back rake, no grinding on the top of the tool is necessary. If the toolholder does have back rake, the tool must be ground to provide zero rake (Figure 3).

A center gage (Figure 4) or an optical comparator may be used to check the tool angle. An adequate allowance for the helix angle on the leading edge will assure sufficient side relief.

The part to be threaded is set up between centers, in a chuck, or in a collet (Figures 5a to c). An undercut of .005 in. less than the minor diameter should be made at the end of the thread. Its width should be sufficient to clear the tool.

Figure 2. A chip is formed on the leading edge of the tool when the infeed is made with the compound set at 29° to the right of the operator to make right-hand threads (Lane Community College).

Figure 1. An equal chip is formed on each side of the threading tool when the infeed is made with the cross slide (Lane Community College).

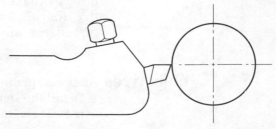

Figure 3. The tool must have zero rake and be set on the center of the work in order to produce the correct form.

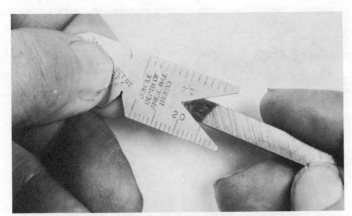

Figure 4. Checking the tool angle with a center gage.

Figure 5c. The sleeve to be threaded externally is mounted on the stub mandrel (DeAnza College).

Figure 5a. Lathe is set up for a threading project by inserting collet in collet sleeve (DeAnza College).

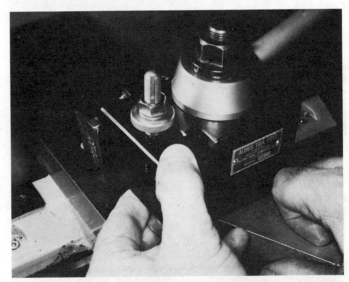

Figure 6a. The threading tool is placed in the holder and lightly clamped (DeAnza College).

Figure 5b. A stub mandrel is inserted in the collet and the collet is tightened (DeAnza College).

The tool is clamped in the holder and set on the centerline of the workpiece (Figures 6a to c). A center gage is used to align the tool to

the workpiece (Figure 7). The toolholder is clamped tightly after the tool is properly aligned.

Setting Dials on the Compound and Cross Feed
The point of the tool is brought into contact with the work by moving the cross feed handle, and the micrometer collar is set on the zero mark (Figure 8). The compound micrometer collar should also be set on zero (Figure 9), but first be sure all slack or backlash is removed by turning the compound feed handle clockwise.

Figure 6*b*. The tool is adjusted to the dead center for height. A tool that is set too high or too low will not produce a true 60° angle in the cut thread (DeAnza College).

Figure 7. The tool is properly aligned by using a center gage. The toolholder is adjusted until the tool is aligned. The toolholder is then tightened (DeAnza College).

Figure 6*c*. An alternate method of adjusting the tool for height is to use the steel rule. It will be in a vertical position when the tool is on center (DeAnza College).

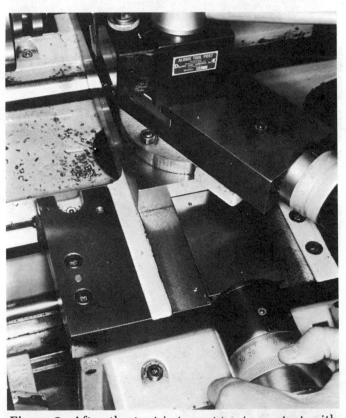

Figure 8. After the tool is brought into contact with the work, the cross feed micrometer collar is set to the zero index (DeAnza College).

Setting Apron Controls

On some lathes a feed change lever, which selects either cross or longitudinal feeds, must be moved to a neutral position for threading. This action locks out the feed mechanism so that no mechanical interference is possible.

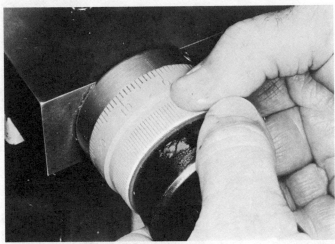

Figure 9. The operator will then set the compound micrometer collar to the zero index (DeAnza College).

Figure 10. The threads per inch selection is made on the quick-change gearbox (DeAnza College).

All lathes have some interlock mechanism to prevent interference when the half-nut lever is used. The half-nut lever causes two halves of a nut to clamp over the leadscrew. The carriage will move the distance of the lead of the thread on the leadscrew for each revolution of the leadscrew.

Threading dials operate off the leadscrew and continue to turn when the leadscrew is rotating and the carriage is not moving. When the half-nut lever is engaged, the threading dial stops turning and the carriage moves. The marks on the dial indicate when it is safe to engage the half-nut lever. If the half-nuts are engaged at the wrong place, the threading tool will not track in the same groove as before but may cut into the center of the thread and ruin it. With any even number of threads such as 4, 6, 12, and 20, the half-nut may be engaged at *any line*. Odd numbered threads such as 5, 7, 13 may be engaged at any *numbered line*. With fractional threads it is safest to engage the half-nut at the same line every time.

The Quick-Change Gearbox

The settings for the gear shift levers on the quick-change gearbox are selected according to the threads per inch desired (Figure 10). If the lathe has an interchangeable stud gear, be sure the correct one is in place.

Spindle Speeds

Spindle Speeds for thread cutting are approximately one-fourth turning speeds. The speed

Figure 11. The half-nut lever is engaged at the correct line or numbered line depending upon whether the thread is odd, even, or a fractional numbered thread (DeAnza College).

should be slow enough so you will have complete control of the thread cutting operation.

CUTTING THE THREAD

The following is the procedure for cutting right-hand threads:

Figure 12. A light scratch cut is taken for the purpose of checking the pitch (DeAnza College).

Figure 13. The pitch of the thread is being checked with a screw pitch gage (DeAnza College).

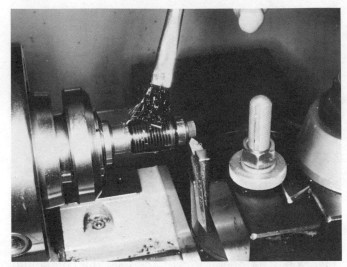

Figure 14. Cutting fluid is applied before taking the first cut (DeAnza College).

Figure 15. The second cut is taken after feeding in the compound .005 in. (DeAnza College).

1. Move the tool off the work and reset the cross feed micrometer dial to zero.
2. Feed it in .002 in. on the compound dial.
3. Turn on the lathe and engage the half-nut lever (Figure 11).
4. Take a scratch cut without using cutting oil (Figure 12). Stop the lathe at the end of the cut and back out the tool using the cross feed. Disengage the half-nut. Return the carriage to the starting position.
5. Check the thread pitch with a screw pitch gage or a rule (Figure 13). If the pitch is wrong, it can still be corrected.
6. Apply sulfurized cutting fluid to the work (Figure 14).

7. Feed the compound in .005 in. and reset the cross feed dial to zero. Make the second cut (Figure 15).
8. Continue this process until the tool is within .010 in. of the finish depth (Figure 16).
9. Brush the threads to remove the chips. Check the thread fit with a ring gage (Figure 17a), standard nut (Figure 17b), or comparison thread micrometer (Figure 17c). The work may be removed from between centers and returned without disturbing the threading setup, provided that the tail of the dog is returned to the same slot.

Figure 16. The finish cut is taken with infeed of .001 to .002 in. (DeAnza College).

Figure 17a. The thread is checked with a ring gage (DeAnza College).

Figure 17b. A standard nut is often used to check a thread (DeAnza College).

Figure 17c. A thread comparison micrometer may be used to check the threads against a known standard such as a precision thread plug gage (DeAnza College).

fingers; it should go on easily but without end play. A class 2 fit is desirable for most purposes.

11. Chamfer the end of the thread to protect it from damage.

Left-Hand Threads

The procedure for cutting left-hand threads (Figure 18a) is the same as that used for cutting right-hand threads with two exceptions. The

10. Continue to take cuts of .001 or .002 in. (as shown in Figure 16) and check the fit between each cut. Thread the nut with your

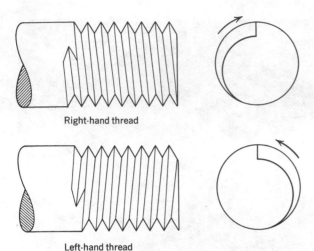

Figure 18a. The difference between right-hand and left-hand threads as seen from the side and end.

Figure 18b. Compound set for cutting a left-hand thread (Lane Community College).

compound is set at 29 degrees to the left of the operator (Figure 18b) and the leadscrew rotation is reversed so the cut is made from the left to the right. The feed reverse lever is moved to reverse the leadscrew. Sufficient undercut must be made for a starting place for the tool. Also, sufficient relief must be provided on the *right* side of the tool.

Methods of Terminating Threads

Undercuts are often used for terminating threads. They should be made the single depth of the thread plus .005 in. The undercut should have a radius to lessen the possibility of fatigue failure resulting from stress concentration in the sharp corners.

Machinists sometimes simply remove the tool quickly at the end of the thread while disengaging the half-nuts. If a machinist misjudges and waits too long, the point of the threading tool will be broken off. A dial indicator is sometimes used to locate the exact position for re-

Figure 19. A threading tool that is used for threading to a shoulder (Lane Community College).

Figure 20. By placing the blade of the threading tool in the upper position, this tool can be made to thread on the bottom side. The lathe is reversed and the thread is cut from left to right making the job easier when threading next to a shoulder (Copyright 1975, Aloris Tool Company, Inc.).

Figure 21a. Carbide threading insert for external threading. Inserts are made for many different thread forms and for boring bars (Photo courtesy of Kennametal Inc., Latrobe, Pa.)

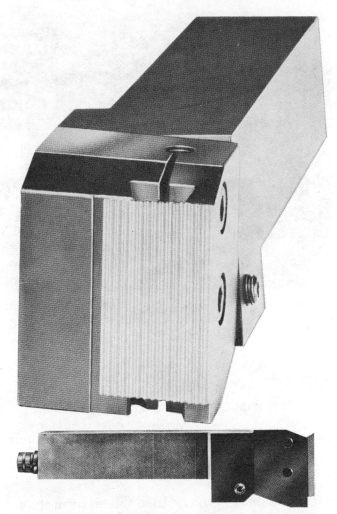

Figure 22a. Uni-chaser. A single tangent threading tool chaser that is a multiple point tool and will produce a bottom and top radius on the thread (Courtesy of Geometric Tool Division of TRW, Inc.).

Figure 21b. Threads being turned at high speed with carbide insert threading tool (Photo courtesy of Kennametal Inc., Latrobe, Pa.).

moving the tool. When this tool withdrawal method is used, an undercut is not necessary.

Terminating threads close to a shoulder requires specially ground tools (Figure 19). Sometimes it is convenient to turn the tool upside down and reverse the lathe when cutting right-hand threads to a shoulder. Some com-

Figure 22b. A right-hand thread being cut with a Uni-chaser (Courtesy of Geometric Tool Division of TRW, Inc.).

Figure 22c. Thread being cut with the bottom side of the Uni-chaser (Courtesy of Geometric Tool Division of TRW, Inc.).

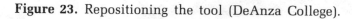

Figure 23. Repositioning the tool (DeAnza College).

mercial threading tools are made for this purpose (Figure 20).

Other Tool Types

Since the crest of the Unified thread form should be rounded, some commercial threading tools provide this and other advantages (Figures 21a and b). Another form of threading tool used in the lathe is shown in Figures 22a to c. This is a multiple point tool that will produce a full form of thread. The advantages are rapid threading, good finishes, and the ability to thread close to shoulders.

Picking Up a Thread

It sometimes becomes necessary to reset the tool when its position against the work has been changed during a threading operation. This position change may be caused by removing the threading tool for grinding, by the work slipping in the chuck or lathe dog, or by the tool moving from the pressure of the cut.

To reposition the tool the following steps may be taken:

1. Check the tool position with reference to the work by using a center gage. If necessary, realign the tool.

2. With the tool backed away from the threads, engage the half-nuts with the machine running. Turn off the machine with the half-nut still engaged and the tool located over the partially cut threads.

3. Position the tool in its original location in the threads by moving both the cross feed and compound handles (Figure 23).

4. Set the micrometer dial to zero on the cross feed collar and set the dial on the compound to the last setting used.

5. Back off the cross feed and disengage the half-nuts. Resume threading where you left off.

BASIC THREAD MEASUREMENT

The simplest method for checking a thread is to try the mating part for fit. The fit is determined solely by feel with no measurement involved. While a loose, medium or close fit may be determined by this method, the threads cannot be depended upon for interchangeability with others of the same size and pitch.

More precise methods of checking threads depend on the Go/No Go principle. The thread plug gage (Figure 24) is used for checking inter-

Figure 24. Thread plug gage (Courtesy of PMC Industries).

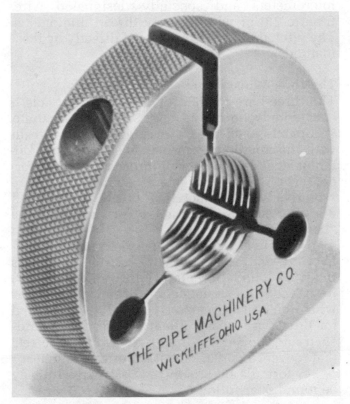

Figure 25. Thread ring gage (Courtesy of PMC Industries).

hole. If the part is within the range or tolerance of the gage, the Go end should turn in flush to the bottom of the internal thread, but the No Go end should just start into the hole and become snug with no more than three turns. The gage should never be forced into the hole.

Thread ring gages are used to check the accuracy of external threads (Figure 25). The outside of the ring gage is knurled and the No Go gage can be easily identified by a groove on the knurled surface. When these gages are used, the Go ring gage should enter the thread fully. The No Go gage should not exceed more than $1\frac{1}{2}$ turns on the thread being checked.

Thread roll snap gages are used to check the accuracy of external screw threads (Figure 26). These common measuring tools are easier and faster to use than thread micrometers or ring gages. The part size is compared to a preset dimension on the roll gage. The first set of rolls are the Go and the second the No Go rolls.

Thread roll snap gages, ring gages, and plug gages are used in production manufacturing where quick gaging methods are needed. These gaging methods depend on the operator's "feel" and the level of precision is only as good as the accuracy of the gage. The thread sizes are not measurable in any definite way.

The thread comparator micrometer (Figure 27) has two conical points. This micrometer does

nal threads. These gages are available in various sizes, which are stamped on the handle. A *male* screw thread is on each end of the handle. The longer threaded end is called the "Go" gage, while the shorter end is called the "No Go" gage. The No Go end is made to a slightly larger dimension than the pitch diameter for the class of fit that the gage tests. To test an internal thread, both the Go and No Go gages should be tried in the

Figure 26. Thread roll snap gage (Courtesy of PMC Industries).

Figure 27. Thread comparator micrometer (DeAnza College).

Figure 28. Set of "best" three-wires for various thread sizes.

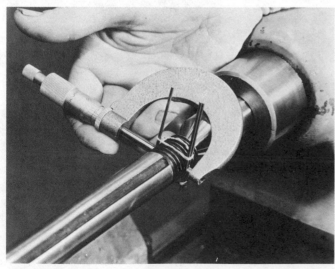

Figure 29. Measuring threads with the three-wire and micrometer method.

not measure the pitch diameter of a thread, but is used only to make a comparison with a known standard. The micrometer is first set to the threaded part and then it is compared to the reading obtained from a plug gage.

ADVANCED METHODS OF THREAD MEASUREMENT

The most accurate place to measure a screw thread is on the flank or angular surface of the thread at the pitch diameter. The outside diameter measured at the crest or the minor diameter measured at the root could vary considerably. Threads may be measured with standard micrometers and specially designated wires (Figure 28) or with a screw thread micrometer. The pitch diameter is measured directly by these methods.

The Three-Wire Method

The three-wire method of measuring threads is considered one of the best and most accurate. Figure 29 shows three wires placed in the threads with the micrometer measuring over them. Different sizes and pitches of threads require different size wires. For greatest accuracy a wire size that will contact the thread at the pitch diameter should be used. This is called the "best" wire size.

The pitch diameter of a thread can be calculated by subtracting the wire constant (which is the single depth of a sharp V thread, or $.866 \times P$) from the measurement over the three wires when the best wire size is used. The wires used for three-wire measurement of threads are hardened and lapped steel, and are available in sets that cover a large range of thread pitches.

A formula by which the best size wire may be found is as follows:

$$\text{Wire size} = \frac{.57735}{n}$$

where n = the number of threads per inch.

EXAMPLE

To find the best size wire for measuring a $1\frac{1}{4}$ in.–12 UNC screw thread:

$$\text{Wire size} = \frac{.57735}{12} = .048$$

The best wire size to use for measuring a 12 pitch thread would be .048 in.

If the best wire sizes are not available, smaller or larger wires may be used within limits. They should not be so small that they are below the major diameter of the thread, or so large that they do not contact the flank of the thread. Subtract the constant (.866 × P) for best wire size from the pitch diameter and add three times the diameter of the available wire when the best wire size is not available.

EXAMPLE

The best wire size for $1\frac{1}{4}$–12 is .048 in., but only $\frac{3}{64}$-in. diameter drill rod is available, which has a diameter of .0469 in.

1.1959	Pitch diameter of $1\frac{1}{4}$ – 12
−.0722	constant for best wire size
1.1237	
+.1407	3 × .0469 available wire size
= 1.2644	measurement over wires

The measurement over the wires will be slightly different than that of the best wire size because of the difference in wire size.

After the best size wire is found, the wires are positioned in the thread grooves as shown in Figure 29. The anvil and spindle of a standard outside micrometer are then placed against the three wires and the measurement is taken.

To calculate what the reading of the micrometer should be if a thread is the correct

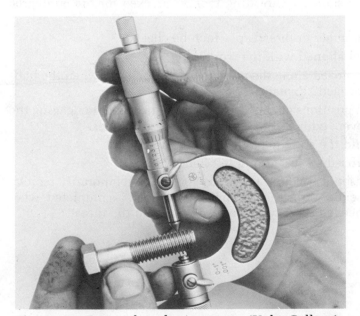

Figure 30. Screw thread micrometer (Yuba College).

finished size, use the following formula when measuring Unified coarse threads or American National threads.

$$M = D + 3W - \frac{1.5155}{n}$$

where

$M =$ micrometer measurement over wires
$D =$ diameter of the thread
$n =$ number of threads per inch
$W =$ diameter of wire used

EXAMPLE

To find M for a $1\frac{1}{4}$–12 UNC thread proceed as follows:

where

$$W = .048$$
$$D = 1.250$$
$$n = 12$$

then

$$M = 1.250 + (3 \times .048) - \frac{1.5155}{12}$$
$$= 1.250 + 1.44 - .126$$
$$M = 1.268 \text{ (micrometer measurement)}$$

When measuring a Unified fine thread, the same method and formula are used, except that the constant should be 1.732 instead of 1.5155.

The three-wire method is also used for other thread forms such as Acme and Buttress. Information and tables may be found in the *Machinery's Handbook*.

The screw thread micrometer (Figure 30) may be used to measure sharp V, Unified, and American National threads. The spindle is pointed to a 60 degree included angle. The anvil, which swivels, has a double-V shape to contact the pitch diameter. The thread micrometer measures the pitch diameter directly from the screw thread. This reading may be compared with pitch diameters given in handbook tables. Thread micrometers have interchangeable anvils that will fit a wide range of thread pitches. Some are made in sets of four micrometers that have a capacity up to one inch, and each covers a range of threads, as follows:

No. 1	8 to 13 threads per inch
No. 2	14 to 20 threads per inch
No. 3	22 to 30 threads per inch
No. 4	32 to 40 threads per inch

The optical comparator is sometimes used to check thread form, helix angle, and depth of thread on external threads (Figure 31). The part is mounted in a screw thread accessory that is adjusted to the helix angle of the thread so that the light beam will show a true profile of the thread.

Since internal threads are most often made by tapping, the pitch diameter and fit are determined by the tap used. Internal threads cut with a single point tool, however, need to be checked. A precision Go/No Go plug gage is generally sufficient in this case. If no gage is available, a shop gage can be made by cutting the required external thread to very precise dimensions. If only one threaded part of a kind is to be made and no interchangeability is required, the mating part may be used as a gage.

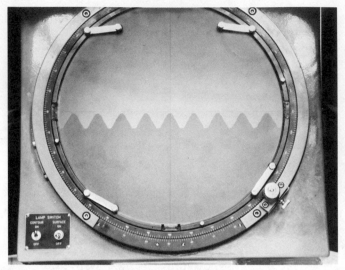

Figure 31. Profile of thread as shown on the screen of the optical comparator (Lane Community College).

self-evaluation

SELF-TEST
1. By what method are threads cut or chased with a single point tool in a lathe? How can a given helix or lead be produced?
2. The better practice is to feed the tool in with the compound set at 29 degrees rather than with the cross slide when cutting threads. Why is this so?
3. By what means should a threading tool be checked for the 60 degree angle?
4. How can the number of threads per inch be checked?
5. How is the tool aligned with the work?
6. Is the carriage moved along the ways by means of gears when the half-nut lever is engaged? Explain.
7. Explain which positions on the threading dial are used for engaging the half-nuts for even, odd, and fractional numbered threads.
8. How fast should the spindle be turning for threading?
9. What is the procedure for cutting left-hand threads?
10. If for some reason it becomes necessary for you to temporarily remove the tool or the entire threading setup before a thread is completed, what procedure is needed when you are ready to finish the thread?

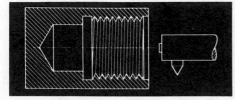

unit 12
cutting
unified
internal threads

While small internal threads are tapped, larger sizes from one inch and up are often cut in a lathe. The problems and calculations involved with cutting internal threads differ in some ways from those of cutting external threads. This unit will help you understand these differences.

objectives **After completing this unit, you will be able to:**

1. Calculate the dimensions for a given internal Unified thread.

2. Machine a nut blank and cut a 1–8 UNC internal thread to fit a plug gage.

Many of the same rules used for external threading apply to internal threading: the tool must be shaped to the exact form of the thread, and the tool must be set on the center of the workpiece. When cutting an internal thread with a single point tool, the inside diameter of the workpiece should be the *minor diameter* of the internal thread (Figure 1). On the other hand, if the thread is made by tapping in the lathe, the inside diameter of the workpiece can be varied to obtain the desired percent of thread.

Full depth or 100 percent threads are very difficult to tap in soft metals and impossible in tough materials. Tests have proven that above 60 percent of thread very little additional strength is gained. Lower percentages provide less flank surface for wear, however. Most commercial internal threads in steel are about 75 percent. Tap drill charts are generally based on 75 percent thread calculations for American National threads. The larger the tap drill, the lower the percent of thread. These figures are correct, however, only if the drilled hole is not oversize.

The probable percent of thread may be 5

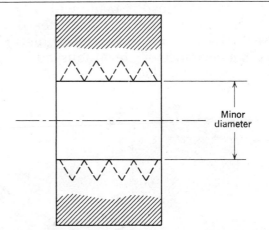

Figure 1. Diagram of internal thread showing minor diameter.

to 10 percent lower than the calculated percent since drills usually make an oversize hole. In practice, a drill slightly under the tap drill size should first be used so the tap drill will make a more accurate hole. Reaming is often done for more precise tapping.

501

The major and minor diameters of the internal thread are shown in Figure 2. For tapped internal threads, the minor diameter is varied according to the percent of thread desired. The percent of thread for American National threads is calculated: the basic major diameter minus the actual minor diameter, divided by two times the basic thread height, expressed as a percentage. Thus,

Percent of thread =

$$\frac{\text{major } D - \text{actual minor } D}{2 \times \text{basic thread height}} \times 100$$

If you should need to select a tap drill size, or determine the bore size for tapping an internal thread, use this formula in a different form. Solving for the actual minor diameter, which should be the inside diameter of the hole, the formula becomes:

Actual minor D = major $D - 2 \times$ basic thread height $\times$ Percent of thread

EXAMPLE

Determine the bore size for tapping an 80 percent 1 in.–8 NC internal thread.

Actual minor $D = 1 - \left(\dfrac{.65}{8} \times 2 \times .8\right) = .870$
or tap drill size

A bolt or external thread is made with 100 percent threads, and the outside diameter is always close to its nominal size within tolerances. There is a clearance of a few thousandths of an inch between the flank of the internal thread or nut and the flank of the bolt, which

is provided by making the major diameter of the internal thread .002 to .005 in. oversize.

The advantages of making internal threads with a single point tool are that large threads of various forms can be made and that the threads are concentric to the axis of the work. The threads may not be concentric when they are tapped. There are some difficulties encountered when making internal threads. The tool is often hidden from view and tool spring must be taken into account.

The hole to be threaded is first drilled to $\frac{1}{16}$ in. diameter less than the minor diameter. Then a boring bar is set up, and the hole is bored to the minor diameter of the thread. If the thread is to go completely through the work, no recess is necessary, but if threading is done in a blind hole, a recess must be made. The compound rest should be swiveled 29 degrees to the left of the operator for cutting right-hand threads (Figures 3a and b). A threading tool is clamped in the bar and aligned by means of a center gage (Figure 4).

The compound micrometer collar is moved to the zero index after the slack has been removed by turning the screw outwards or counterclockwise. The tool is brought to the work with the cross slide handle and its collar is set on zero. Threading may now proceed in the same manner as it is done with external threads. The compound is advanced outwards a few thousandths of an inch, a scratch cut is made and the thread pitch is checked with a screw pitch gage.

The cross slide is backed out of the cut and reset to zero before the next pass. Cutting oil is used. The compound is moved outward a few thousandths (.001 to .010 in.). The exact amount of infeed depends on how rigid the boring bar and holder are and how deep the cut has progressed. Too much infeed will cause the bar to spring away and produce a bell-mouth internal thread.

If a slender boring bar is necessary or there is more than usual overhang, lighter cuts must be used to avoid chatter. The bar may spring away from the cut causing the major diameter to be less than the calculated amount, or that amount fed in on the compound. If several passes are taken through the thread with the same setting on the compound, this problem can often be corrected.

The single depth of the Unified internal

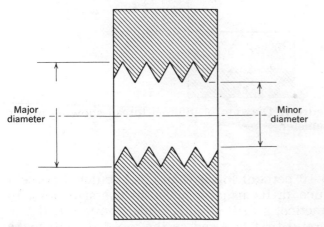

Figure 2. View of internal threads showing major and minor diameters.

Figure 3a. For right-hand internal threads, the compound is swiveled to the left (Lane Community College).

Figure 4. Aligning the threading tool with a center gage (Lane Community College).

Figure 3b. The compound rest is swiveled to the right for left-hand internal threads (Lane Community College).

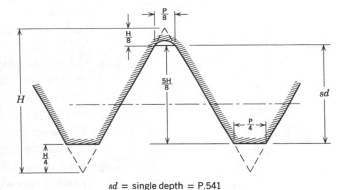

sd = single depth = P.541

Figure 5. Single depth of the Unified internal thread.

the formula is

$$d = D - (P \times .541 \times 2)$$

EXAMPLE

A $1\frac{1}{2}$–6 UNC nut must be bored and threaded to fit a stud. Find the dimension of the bore.

$$P = \tfrac{1}{6} = .1666$$
$$d = 1.500 - (.1666 \times .541 \times 2)$$
$$d = 1.500 - .180$$
$$d = 1.320$$

Thus, the bore should be made 1.320 in.

The infeed on the compound is calculated in the same way as with external threads: except for Unified internal threads, use the formula:

$$\text{Infeed} = P \times .625$$

thread (Figure 5) equals P × .541. The minor diameter is found by subtracting the double depth of the thread from the major diameter.

Thus, if

$$D = \text{Major diameter}$$
$$d = \text{Minor diameter}$$
$$P = \text{Thread pitch}$$
$$P \times .541 = \text{Single depth}$$

for the depth of cut with the compound set at 29 or 30 degrees. Using the pitch from the previous example

Infeed = .1666 × .625 = .114 in.

The single depth of the American National thread equals P × .6495 and the infeed on the compound set at 29 degrees is P × .75. These figures may be substituted in the previous calculations to determine the single

depth and infeed for a $1\frac{1}{2}$–6 NC internal thread.

Often it is necessary to realign an internal threading tool with the thread when the tool has been moved for sharpening or when the setup has moved during the cut. The tool is realigned in the same way as it is done for external threads: by engaging the half-nut and positioning the tool in a convenient place over the threads, then moving both the compound and cross slides to adjust the tool position.

self-evaluation

SELF-TEST
1. When internal threads are made with a tool, what should the bore size be?
2. In what way is percent of thread obtained? Why is this done?
3. What percent of thread are tap drill charts usually based on?
4. Drills often make an oversize hole that lowers the percent of thread that a tap will cut. How can a more precise hole be drilled?
5. Your specifications call for a 60 percent thread in a tough stainless steel casting for a $\frac{1}{2}$ – 13 tapped thread. The tap drill size = Major D – 2× basic thread height × percent of thread. What would your tap drill size be?
6. Name two advantages of making internal threads with a single point tool on the lathe.
7. When making internal right-hand threads, which direction should the compound be swiveled?
8. After a scratch cut is made, what would the most convenient method be to measure the pitch of the internal thread?
9. What does deflection or spring of the boring bar cause when cutting internal threads?
10. Using P × .541 as a constant for Unified single depth internal threads, what would the minor diameter be for a 1–8 thread?

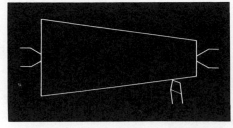

unit 13
taper turning,
taper boring,
and forming

Tapers are very useful machine elements that are used for many purposes. The machinist should be able to quickly calculate a specific taper and to set up a machine to produce it. The machinist should also be able to accurately measure tapers and determine proper fits. This unit will help you understand the various methods and principles involved in making a taper.

objectives After completing this unit, you will be able to:

1. **Describe different types of tapers and the methods used to produce and measure them.**
2. **Make a center punch, using the compound rest method of producing tapers.**
3. **Make a specific taper on a shaft using the offset tailstock method.**
4. **Make a Morse taper center and sleeve using the taper attachment.**

USES OF TAPERS

Tapers are used on machines because of their capacity to align and hold machine parts and to realign when they are repeatedly assembled and disassembled. This repeatability assures that tools such as centers in lathes, taper shank drills in drill presses, and arbors in milling machines will run in perfect alignment when placed in the machine. When a taper is slight, such as a Morse taper that is about $\frac{5}{8}$ in. taper/ft, it is called a self-holding taper since it is held in and driven by friction (Figure 1). A steep taper, such as a quick-release taper of $3\frac{1}{2}$ in./ft and used on most milling machines, must be held in place with a draw bolt (Figure 2).

A taper may be defined as a uniform increase in diameter on a round workpiece for a given length measured parallel to the axis. Internal or external tapers are expressed in taper per foot (TPF), taper per inch (TPI), or in degrees. The TPF or TPI refers to the difference in diameters in the length of one foot or one inch, respectively (Figure 3). This difference is measured in inches. Angles of taper, on the other hand, may refer to the included angles or the angles with the centerline (Figure 4).

Some machine parts that are measured in taper per foot are mandrels (.006 in./ft), taper pins and reamers ($\frac{1}{4}$ in./ft), the Jarno taper series (.600 in./ft), the Brown and Sharpe taper series ($\frac{1}{2}$ in./ft), and the Morse taper series (about $\frac{5}{8}$ in./ft). Morse tapers include eight sizes that range from size 0 to size 7. Tapers and dimensions vary slightly from size to size in both the Brown and Sharpe and the Morse series. For instance, a No. 2 Morse taper has .5944 in./ft taper and a No. 4 has .6233 in./ft taper. See Table 1 for more information on Morse tapers.

505

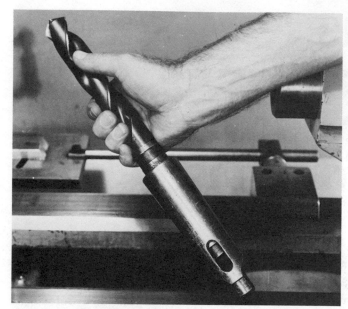

Figure 1. The Morse taper shank on this drill keeps the drill from turning when the hole is being drilled (Lane Community College).

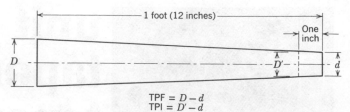

$$TPF = D - d$$
$$TPI = D' - d$$

Figure 3. The difference between taper per foot (TPF) and taper per inch (TPI).

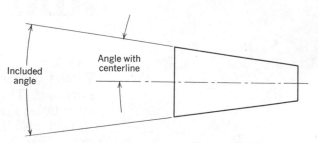

Figure 4. Included angles and angles with centerline.

Figure 2. The milling machine taper is driven by lugs and held in by a draw bolt (Lane Community College).

Figure 5. Making a taper using the compound slide (Lane Community College).

METHODS OF MAKING A TAPER

There are four methods of turning a taper on a lathe. They are the compound slide method, the offset tailstock method, the taper attachment method, and the use of a form tool. Each method has its advantages and disadvantages, so the kind of taper needed on a workpiece should be the deciding factor in the selection of the method that will be used.

The Compound Slide Method

Both internal and external short steep tapers can be turned on a lathe by hand feeding the compound slide (Figure 5). The swivel base of the compound is divided in degrees. When the compound slide is in line with the ways of the lathe, the 0 degree line will align with the index line on the cross slide (Figure 6). When the compound is swiveled off the index, which is parallel to the centerline of the lathe, a direct reading may be taken for the half angle or angle to centerline of the machined part (Figure 7). When a taper is machined on the lathe centerline, its included angle will be twice the angle

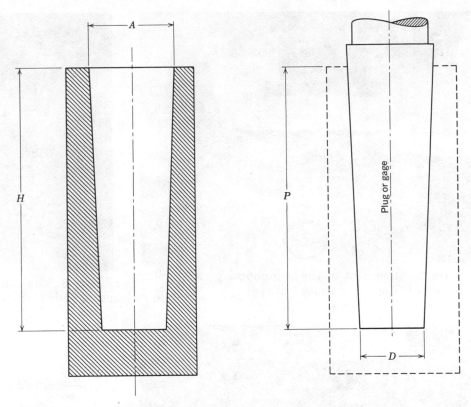

Table 1

Morse Tapers Information

Number of Taper	Taper per Foot	Taper per Inch	P Standard Plug Depth	D Diameter of Plug at Small End	A Diameter at End of Socket	H Depth of Hole
0	.6246	.0520	2	.252	.356	$2\frac{1}{32}$
1	.5986	.0499	$2\frac{1}{8}$	.396	.475	$2\frac{3}{16}$
2	.5994	.0500	$2\frac{9}{16}$	.572	.700	$2\frac{5}{8}$
3	.6023	.0502	$3\frac{3}{16}$	.778	.938	$3\frac{1}{4}$
4	.6232	.0519	$4\frac{1}{16}$	1.020	1.231	$4\frac{1}{8}$
5	.6315	.0526	$5\frac{3}{16}$	1.475	1.748	$5\frac{1}{4}$
6	.6256	.0521	$7\frac{1}{4}$	2.116	2.494	$7\frac{3}{8}$
7	.6240	.0520	10	2.750	3.270	$10\frac{1}{8}$

that is set on the compound. Not all lathes are indexed in this manner.

When the compound slide is aligned with the axis of the cross slide and swiveled off the index in either direction, an angle is directly read off the cross slide centerline (Figure 8). Since the lathe centerline is 90 degrees from the cross slide centerline, the reading on the lathe centerline index is the complementary angle. So, if the compound is set off the axis of the cross slide $14\frac{1}{2}$ degrees, the lathe centerline index reading is $90 - 14\frac{1}{2} = 75\frac{1}{2}$ degrees, as seen in Figure 8.

Tapers of any angle may be cut by this method, but the length is limited to the stroke of the compound slide. Since tapers are often given in TPF, it is sometimes convenient to consult a TPF to angle conversion table, as in

Figure 6. Alignment of the compound parallel with the ways (Lane Community College).

Figure 8. The compound set $14\frac{1}{2}°$ off the axis of the cross slide (Lane Community College).

Figure 7. An angle may be set off the axis of the lathe from this index (Lane Community College).

Table 2. A more complete table may be found in the *Machinery's Handbook.*

If a more precise conversion is desired, the following formula may be used to find the included angle.

> Divide the taper in inches per foot by 24; find the angle that corresponds to the quotient in a table of tangents and double this angle. If the angle with centerline is desired, do not double the angle.

EXAMPLE

What angle is equivalent to a taper of $3\frac{1}{2}$ in./ft?

$$\frac{3.5}{24} = .14583$$

The angle of this tangent is 8 degrees 18 minutes, and the included angle is twice this, or 16 degrees 36 minutes.

THE OFFSET TAILSTOCK METHOD

Long, slight tapers may be produced on shafts and external parts *only* between centers. Internal tapers cannot be made by this method. Power feed is used so good finishes are obtainable. The taper per foot or taper per inch must be known so the amount of offset for the tailstock can be calculated. Since tapers are of different lengths, they would not be the same TPI or TPF for the same offset (Figure 9). When the taper per inch is known, the offset calculation is as follows:

Where

> TPI = taper per inch
> L = length of workpiece

$$\text{Offset} = \frac{\text{TPF} \times L}{2}$$

When the taper per foot is known, use the following formula:

$$\text{Offset} = \frac{\text{TPF} \times L}{24}$$

Table 2
Tapers and Corresponding Angles

Taper per Foot	Included Angle		Angle with Centerline		Taper per Inch
	Degrees	Minutes	Degrees	Minutes	
$\frac{1}{8}$	0	36	0	18	.0104
$\frac{3}{16}$	0	54	0	27	.0156
$\frac{1}{4}$	1	12	0	36	.0208
$\frac{5}{16}$	1	30	0	45	.0260
$\frac{3}{8}$	1	47	0	53	.0313
$\frac{7}{16}$	2	5	1	2	.0365
$\frac{1}{2}$	2	23	1	11	.0417
$\frac{9}{16}$	2	42	1	21	.0469
$\frac{5}{8}$	3	00	1	30	.0521
$\frac{11}{16}$	3	18	1	39	.0573
$\frac{3}{4}$	3	35	1	48	.0625
$\frac{13}{16}$	3	52	1	56	.0677
$\frac{7}{8}$	4	12	2	6	.0729
$\frac{15}{16}$	4	28	2	14	.0781
1	4	45	2	23	.0833
$1\frac{1}{4}$	5	58	2	59	.1042
$1\frac{1}{2}$	7	8	3	34	.1250
$1\frac{3}{4}$	8	20	4	10	.1458
2	9	32	4	46	.1667
$2\frac{1}{2}$	11	54	5	57	.2083
3	14	16	7	8	.2500
$3\frac{1}{2}$	16	36	8	18	.2917
4	18	56	9	28	.3333
$4\frac{1}{2}$	21	14	10	37	.3750
5	23	32	11	46	.4167
6	28	4	14	2	.5000

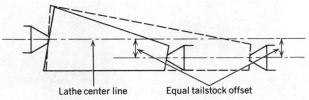

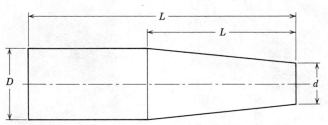

Figure 9. When tapers are of different lengths, the TPF is not the same with the same offset.

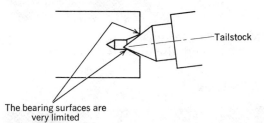

Figure 10. Long workpiece with a short taper.

Figure 11. The contact area between the center hole and the center is small.

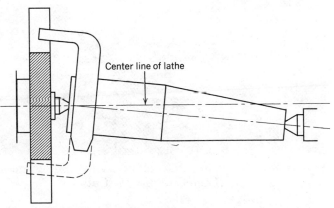

Figure 12. The bent tail of the lathe dog should have adequate clearance.

If the workpiece has a short taper in any part of its length (Figure 10) and the TPI or TPF is not given use the following formula:

$$\text{Offset} = \frac{L \times (D - d)}{2 \times L_1}$$

Where

$D =$ diameter at large end of taper
$d =$ diameter at small end of taper
$L =$ total length of workpiece
$L_1 =$ length of taper

When you set up for turning a taper between centers, remember that the contact area be-

tween the center and the center hole is limited (Figure 11). Frequent lubrication of the centers may be necessary.

You should also note the path of the lathe dog bent tail in the drive slot (Figure 12). Check to see that there is adequate clearance.

Figure 13. Measuring the offset on the tailstock by use of the centers and the rule (Lane Community College).

Figure 15. Using the dial indicator to measure the offset (Lane Community College).

Figure 14. Measuring the offset with the witness mark and a rule (Lane Community College).

To measure the offset on the tailstock, use either the centers and a rule (Figure 13) or the witness mark and a rule (Figure 14); both methods are adequate for some purposes. A more precise measurement is possible with a dial indicator as shown in Figure 15. The indicator is set on the tailstock spindle while the centers are still aligned. A slight loading of the indicator is advised. The bezel is set at zero

Figure 16a. The toolholder is brought to the tailstock spindle using a paper strip as a feeler gage. The micrometer dial is set to zero (Lane Community College).

and the tailstock is moved toward the operator the calculated amount. Clamp the tailstock

Figure 16b. The cross slide is backed off the desired amount plus one full turn (Lane Community College).

Figure 16c. The toolholder is brought forward to the desired setting and the tailstock is moved over until it contacts the paper strip (Lane Community College).

to the way. If the indicator reading changes, loosen the clamp and readjust.

Another accurate method for offsetting the tailstock is to use the cross slide (Figures 16a to c). With the centers aligned, bring the reverse end of the toolholder in contact with the tailstock spindle. A paper strip may be used as a feeler gage. Set the micrometer dial to zero. Back off the cross slide the calculated amount plus a full turn to remove backlash; then turn back in to the calculated amount. Move the tailstock until it contacts the paper strip held at the end of the toolholder.

When cutting tapered threads such as pipe threads, the tool should be square with the centerline of the workpiece, not the taper (Figure 17). When you have finished making tapers by the offset tailstock method, realign the centers to .001 in. or less in 12 in.

The Taper Attachment Method
The taper attachment features a slide external to the ways that can be angled and will move

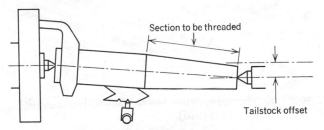

Figure 17. Adjusting the threading tool for cutting tapered threads. The tool is set square to the centerline of the work rather than the taper.

the cross slide according to the angle set. Slight to fairly steep tapers ($3\frac{1}{2}$ in./ft) may be made, but length is limited to the stroke of the taper attachment. Centers may remain in line without distortion of the center holes. Work may be held in a chuck and both external and internal tapers may be made, often with the same setting for mating parts. Power feed is used. Taper attachments are graduated in inches per foot (TPF) or in degrees.

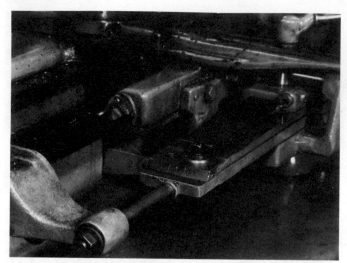

Figure 18. The plain taper attachment (Lane Community College).

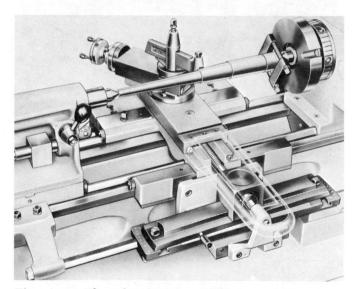

Figure 19. The telescopic taper attachment (Courtesy of Clausing Corporation).

There are two types of taper attachments, the plain taper attachment (Figure 18) and the telescoping taper attachment (Figure 19). The cross feed binding screw must be removed to free the nut when the plain type is set up. The depth of cut must then be made by using the compound feed screw handle. The cross feed may be used for depth of cut when using the telescoping taper attachment since the cross feed binding screw is not disengaged with this type.

When a workpiece is to be duplicated or an internal taper is to be made for an existing external taper, it is often convenient to set up

Figure 20. Adjusting the taper attachment to a given taper with a dial indicator (Lane Community College).

the taper attachment by using a dial indicator (Figure 20). The contact point of the dial indicator must be on the center of the workpiece. The workpiece is first set up in a chuck or between centers so there is no runout when it is rotated. With the lathe spindle stopped, the indicator is moved from one end of the taper to the other. The taper attachment is adjusted until the indicator does not change reading when moved.

The angle, the taper per foot, or the taper per inch must be known to set up the taper attachment to cut specific tapers. If none of these are known, proceed as follows.

If the end diameters (D and d) and the length of taper (L) are given in inches:

$$\text{Taper per foot} = \frac{D - d}{L} \times 12$$

If the taper per foot is given, but you want to know the amount of taper in inches for a given length, use the following formula.

$$\text{Amount of taper} = \frac{\text{TPF}}{12} \times \text{given length of tapered part}$$

When the TPF is known, to find TPI divide the TPF by 12. When the TPI is known, to find TPF multiply the TPI by 12.

To set up the taper attachment (Figure 21) proceed as follows:

1. Clean and oil the slide bar (*a*).
2. Set up the workpiece and the cutting tool on center. Bring the tool near the workpiece and to the center of the taper.

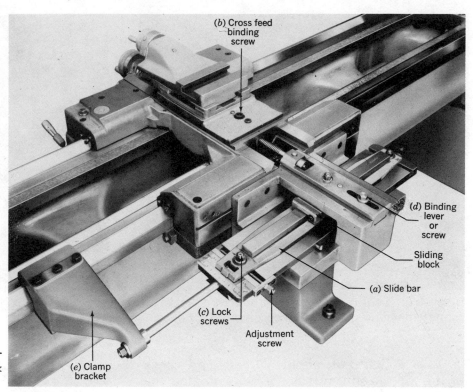

(b) Cross feed
binding
screw

(d) Binding
lever
or
screw

Sliding
block

(a) Slide bar

(c) Lock
screws

Adjustment
screw

(e) Clamp
bracket

Figure 21. The parts of the taper attachment (Courtesy of The Lodge & Shipley Company).

3. Remove the cross feed binding screw (b) that binds the cross feed screw nut to the cross slide. *Do not remove* this screw if you are using a telescoping taper attachment. The screw is removed *only* on the plain type. Put a temporary plug in the hole to keep chips out.

4. Loosen the lock screws (c) on both ends of the slide bar and adjust to the required degree of taper.

5. Tighten the lock screws.

6. Tighten the binding lever (d) on the slotted cross slide extension at the sliding block, *plain type only*.

7. Lock the clamp bracket (e) to the lathe bed.

8. Move the carriage to the right so that the tool is from $\frac{1}{2}$ to $\frac{3}{4}$ in. past the start position. This should be done on every pass to remove any backlash in the taper attachment.

9. Feed the tool in for the depth of the first cut with the cross slide unless you are using a plain-type attachment. Use the compound slide for the plain type.

10. Take a trial cut and check for diameter. Continue the roughing cut.

11. Check the taper for fit and readjust the taper attachment, if necessary.

Hole Screw

Figure 22. Internal taper being made with a taper attachment. Note that the crossfeed nut locking screw has been removed and the hole has not been plugged. This hole *must* be plugged and the screw stored in a safe place. (Lane Community College).

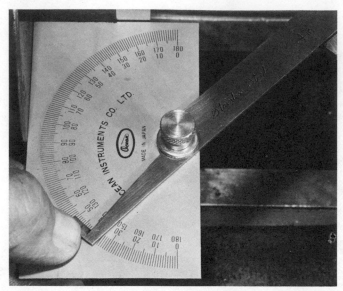

Figure 23a. Tool is set up with protractor to make an accurate chamfer or taper (Lane Community College).

Figure 23b. Making the chamfer with a tool (Lane Community College).

12. Take a light cut, about .010 in. and check the taper again. If it is correct, complete the roughing and final finish cuts.

   Internal tapers (Figure 22) are best made with the taper attachment. They are set up in the same manner as prescribed for external tapers.

Other Methods of Making Tapers
A tool may be set with a protractor to a given angle (Figures 23a and b) and a single plunge cut may be made to produce a taper. This method is often used for chamfering a workpiece to an angle such as the chamfer used for hexagonal bolt heads and nuts. Tapered form tools sometimes are used to make V-shaped grooves. Only very short tapers can be made with form tools.

   Tapered reamers are sometimes used to produce a specific taper such as a Morse taper. A roughing reamer is first used, followed by a finishing reamer. Finishing Morse taper reamers are often used to true up a badly nicked and scarred internal Morse taper.

METHODS OF MEASURING TAPERS

The most convenient and simple way of checking tapers is to use the taper plug gage (Figure 24) for internal tapers and the taper ring gage (Figure 25) for external tapers. Some taper gages have Go and No Go limit marks on them (Figure 26).

Figure 24. Taper plug gage (Lane Community College).

Figure 25. Taper ring gage (Lane Community College).

   To check an internal taper, a chalk or prussian blue mark is first made along the length of the taper plug gage (Figures 27a and b). The gage is then inserted into the internal taper and turned slightly. When the gage is taken out, the chalk mark will be partly rubbed off where contact was made. Adjustment of the taper should be made until the chalk mark is rubbed off along its full length of contact, indicating a

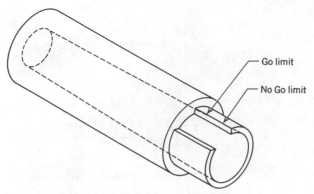

Figure 26. Go/No Go taper ring gage.

Figure 27*a*. Chalk mark is made along a taper plug gage prior to checking an internal taper (Lane Community College).

Figure 27*b*. The taper has been tested and the chalk mark has been rubbed off evenly, indicating a good fit (Lane Community College).

good fit. An external taper is marked with chalk to be checked in the same way with a taper ring gage (Figures 28*a* to *c*).

Figure 28*a*. The external taper is marked with chalk or prussian blue before being checked with a taper ring gage (Lane Community College).

Figure 28*b*. The ring gage is placed on the taper snugly and is rotated slightly (Lane Community College).

The taper per inch may be checked with a micrometer by scribing two marks one inch apart on the taper and measuring the diameters (Figures 29*a* and *b*) at these marks. The difference is the taper per inch. A more precise way of making this measurement is shown in Figures 30*a* and *b*. A surface plate is used with precision parallels and drill rods. The tapered workpiece would have to be removed from the lathe if this method is used, however.

Perhaps an even more precise method of measuring a taper is with the sine bar and gage blocks on the surface plate (Figure 31). When this

Figure 28c. The ring gage is removed and the chalk mark is rubbed off evenly for the entire length of the ring gage, which indicates a good fit. (Lane Community College).

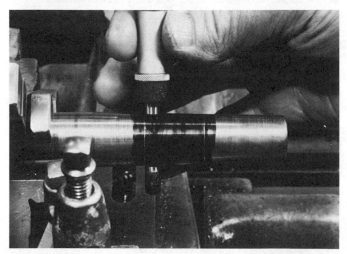

Figure 29a. Measuring the taper per inch (TPI) with a micrometer. The larger diameter is measured on the line with the edge of the spindle and the anvil of the micrometer contacting the line (Lane Community College).

Figure 29b. The second measurement is taken on the smaller diameter at the edge of the line in the same manner (Lane Community College).

Figure 30a. Checking the taper on a surface plate with precision parallels, drill rod, and micrometer. The first set of parallels is used so that the point of measurement is accessible to the micrometer (Lane Community College).

Figure 30b. When one inch wide parallels are in place, a second measurement is taken. The difference is the taper per inch (Lane Community College).

Figure 31. Using a sine bar and gage blocks with a dial indicator to measure a taper (Lane Community College).

is done, it is important to keep the centerline of the taper parallel to the sine bar and to read the indicator at the highest point.

Tapers may be measured with a taper micrometer. Refer to Section C, Unit 4 for a description of this instrument.

self-evaluation

SELF-TEST

1. State the difference in use between steep tapers and slight tapers.
2. In what three ways are tapers expressed (measured)?
3. Briefly describe the four methods of turning a taper in the lathe.
4. When a taper is produced by the compound slide method, is the reading in degrees on the compound swivel base the same as the angle of the finished workpiece? Explain.
5. If the swivel base is set to a 35 degree angle at the cross slide centerline index, what would the reading be at the lathe centerline index?
6. Calculate the offset for the taper shown in Figure 32. The formula is

$$\text{Offset} = \frac{L \times (D - d)}{2 \times L_1}$$

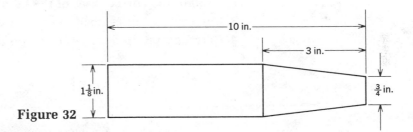

Figure 32

7. Name four methods of measuring the offset on the tailstock for making a taper.
8. What are the two types of taper attachments and what are their advantages over other means of making a taper.
9. What is the most practical and convenient way to check internal and external tapers when they are in the lathe? Name four methods of measuring tapers.
10. Describe the kinds of tapers that may be made by using a form tool or the side of a tool.

unit 14
using steady
and follower rests
in the lathe

Many lathe operations would not be possible without the use of the steady and follower rests. These valuable attachments make internal and external machining operations on long workpieces possible on a lathe.

objectives After completing this unit, you will be able to:

1. Identify the parts and explain the uses of the steady rest.
2. Explain the correct uses of the follower rest.
3. Correctly set up a steady rest on a straight shaft.
4. Correctly set up a follower rest on a prepared shaft.

THE STEADY REST

On a lathe, long shafts tend to vibrate when cuts are made, leaving chatter marks. Even light finish cuts will often produce chatter when the shaft is long and slender. To help eliminate these problems, use a steady rest to support workpieces that extend from a chuck more than four or five diameters of the workpiece for turning, facing, drilling, and boring operations.

The steady rest (Figure 1) is made of a cast iron or steel frame that is hinged so it will open to accommodate workpieces. It has three or more adjustable jaws that are tipped with bronze, plastic, or ball bearing rollers. The base of the frame is machined to fit the ways of the lathe and it is clamped to the bed by means of a bolt and crossbar.

A steady rest is also used to support long workpieces for various other machining operations such as threading, grooving, and knurling (Figure 2). Heavy cuts can be made by using one or more steady rests along a shaft.

Figure 1. The parts of the steady rest (Lane Community College).

518

Adjusting the Steady Rest

Workpieces should be mounted and centered in a chuck whether a tailstock center is used or not. If the shaft has centers and finished surfaces that turn concentric (have no runout) with the lathe centerline, setup of the steady rest is simple. The steady rest is slid to a convenient location on the shaft, which is supported in the dead center and chuck, and the base is clamped to the bed. The two lower jaws are brought up to the shaft finger tight only (Figure 3). A good high pressure lubricant is applied to the shaft and the top half of the steady rest is closed and

Figure 2. A long, slender workpiece is supported by a steady rest near the center to limit vibration or chatter (Lane Community College).

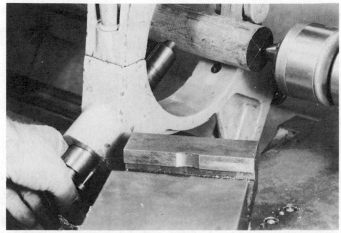

Figure 3. Adjusting the steady rest jaws to a centered shaft (Lane Community College).

clamped. The upper jaw is brought to the shaft finger tight, and then all three lockscrews are tightened. Some clearance is necessary on the upper jaw to avoid scoring of the shaft. As the shaft warms or heats up from friction during machining, readjustment of the upper jaw is necessary.

A finished workpiece can be scored if any hardness or grit is present on the jaws. To protect finishes, brass or copper strips or abrasive cloth is often placed between the jaws and the workpiece; with the abrasive cloth, the abrasive side is placed outward against the jaws.

When there is no center in a finished shaft of one diameter, setup procedure is as follows.

1. Position the steady rest near the end of the shaft with the other end lightly chucked in a three- or four-jaw chuck.
2. Scribe two cross center lines with a center head on the end of the shaft and prick punch (Figures 4a and b).
3. Bring up the dead center near to the punch mark.
4. Adjust the lower jaws of the steady rest to the shaft and lock.
5. Tighten the chuck. If it is a four-jaw chuck, check for runout with a dial indicator.
6. The steady rest may now be moved to any location along the shaft.

Stepped shafts may be set up by using a similar procedure, but the steady rest must remain on the diameter on which it is set up.

Using the Steady Rest

A frequent misconception among students is that the steady rest may be set up properly by using a dial indicator near the steady rest on a rotating shaft. This procedure would never work since the indicator would show no offset or runout, no matter where the jaws were moved.

Steady rest jaws should never be used on rough surfaces. When a forging, casting, or hot rolled bar must be placed in a steady rest, a concentric bearing with a good finish must be turned (Figure 5). Thick walled tubing or other materials that tend to be out of round also should have bearing surfaces machined on them. The usual practice is to remove no more in diameter than necessary to clean up the bearing spot.

When the piece to be set up is very irregular, such as a square or hexagonal part, a cat head

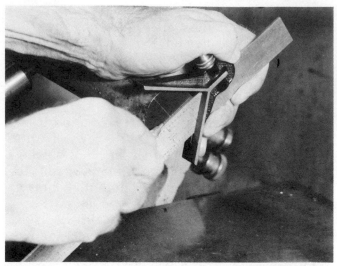

Figure 4a. Laying out the center of a shaft (Lane Community College).

Figure 5. Turning a concentric bearing surface on rough stock for the steady rest jaws (Lane Community College).

Figure 4b. Aligning the shaft center with the dead center (Lane Community College).

Figure 6. Using a cat head for supporting a square piece. Drilling and boring in the end of a heavy square bar requires the use of an external cat head (Lane Community College).

is used (Figure 6). The piece is placed in the cat head and the cat head is mounted in the steady rest while the other end of the workpiece is centered in the chuck. The workpiece is made to run true near the steady rest by adjusting screws on the cat head. In most cases the workpiece is given a center to provide more support for turning operations. A centered cat head (Figures 7a and b) is sometimes used when a permanent center is not required in the workpiece. Internal cat heads (Figure 8) are used for truing to the inside diameter of tubing that has an irregular wall thickness, so that a steady rest

bearing spot can be machined on the outside diameter. These also have adjustment screws.

THE FOLLOWER REST

Long, slender shafts tend to spring away from the tool, vary in diameter, chatter, and often climb the tool. To avoid these problems when machining a slender shaft along its entire length, a follower rest (Figure 9) is often used. Follower rests are bolted to the carriage and follow along with the tool. Most follower rests have two jaws placed to back up the work opposite to the tool thrust. Some types are made with different size bushings to fit the work.

Figure 7a. Using a centered cat head to provide a center when the end of the shaft or tube cannot be centered conveniently (Lane Community College).

Figure 7b. The cat head is adjusted over the irregular end of the shaft (Lane Community College).

Figure 8. Tubing being set up with a cat head using a dial indicator to true the inside diameter (Lane Community College).

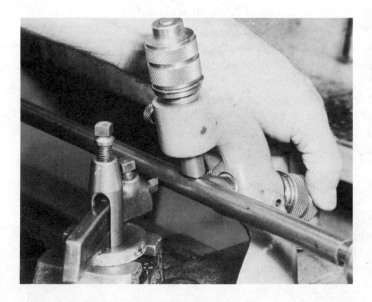

Figure 9. A follower rest is used to turn this long shaft (Lane Community College).

Figure 10. Adjusting the follower rest (Lane Community College).

Figure 11. Long, slender Acme threaded screw being machined with the aid of a follower rest (Lane Community College).

Figure 12. Both steady and follower rests being used (Lane Community College).

Using the Follower Rest

The workpiece should be one to two inches longer than the job requires to allow room for the follower rest jaws. The end is turned to smaller than the finish size. The tool is adjusted ahead of the jaws about one and one-half inches and a trial cut of two or three inches is made with the jaws backed off. Then the lower jaw is adjusted finger tight (Figure 10) followed by the upper jaw. Both locking screws are tightened.

A cutting oil should be used to lubricate the jaws.

The follower rest is often used when cutting threads on long, slender shafts, especially when cutting square or Acme threads (Figure 11). Burrs should be removed between passes to prevent them cutting into the jaws. Jaws with rolls are sometimes used for this purpose. On quite long shafts, sometimes both a steady rest and follower rest are used (Figure 12).

self-evaluation

SELF-TEST
1. When should a steady rest be used?
2. In what ways can a steady rest be useful?
3. How is the steady rest set up on a straight finished shaft when it has centers in the ends?
4. What precaution can be taken to prevent scoring of a finished shaft?
5. How can a steady rest be set up when there is no center hole in the shaft?
6. Is it possible to correctly set up a steady rest by using a dial indicator on the rotating shaft in order to watch for runout?
7. Should a steady rest be used on a rough surface? Explain.
8. How can a steady rest be used on an irregular surface such as square or hex stock?
9. When a long, slender shaft needs to be turned or threaded for its entire length, which lathe attachment could be used?
10. The jaws of the follower rest are usually one or two inches to the right of the tool on a setup. If the workpiece happens to be smaller than the dead center or tailstock spindle, how would it be possible to bring the tool to the end of the work to start a cut without interference by the follower rest jaws?

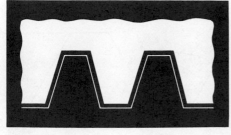

unit 15
additional thread forms

Many thread forms other than the 60 degree types are to be found on machines. Each of the forms is unique and has a special use. To be able to recognize and measure these various thread forms will be very helpful to you as an apprentice machinist.

objectives **After completing this unit, you will be able to:**
1. Identify five different thread forms and explain their uses.
2. Calculate the dimensions needed to machine the five thread forms.

Transmitting screw threads are primarily used to transmit or impart power or motion to a mechanical part. Often these transmitting screws are of multiple lead to effect rapid motion. Bench vises and house jacks are familiar applications of single lead transmitting screws. The lead screw on a lathe and the table feed screws on milling machines are examples of these screw threads being used to impart power along the axis of the screw to move a part.

The earliest type of transmitting screws were of the square thread form (Figure 1). Thrust on the flanks of the thread is fully axial, thus reducing friction to a minimum. The square thread is more difficult to produce than other types and is not now widely used. The thread has a depth and thickness that is one half the pitch. Clearance must be provided on the flanks and major diameter of the thread.

The modified square thread form (Figure 2) was designed to replace the square thread. It is easier to produce than the square thread, yet it has all of the advantages and some of the drawbacks of the square thread. It, like the square thread, is not widely used.

The Acme thread (Figure 3) is generally accepted throughout the mechanical industries

as an improved thread form over that of the square and modified square. The Acme thread

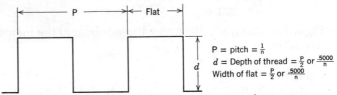

Figure 1. The square thread. $P = \text{pitch} = \dfrac{1}{n}$; $D = $ depth of thread $= \dfrac{P}{2}$ or $\dfrac{.5000}{n}$. Width of flat $= \dfrac{P}{2}$ or $\dfrac{.5000}{n}$.

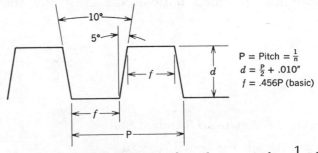

Figure 2. Modified square thread. $P = \text{pitch} = \dfrac{1}{n}$; $d = \dfrac{P}{2} + .010$ in.; $f = .456P$ (basic).

523

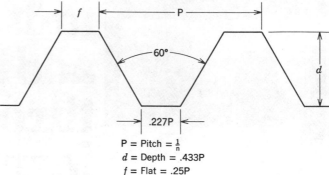

$$P = \text{Pitch} = \frac{1}{n}$$
$$d = \text{Depth} = \frac{P}{2} = \frac{0.500}{n} + 0.010''$$
$$\text{Crest} = \frac{0.3707}{n}$$
$$\text{Root} = \frac{0.3707}{n} = 0.0052''$$

Figure 3. Acme thread, $P = \text{pitch} = \frac{1}{n}$; $d = \text{depth} = \frac{P}{2} = \frac{.500}{n} + .010$ in.; crest $= \frac{.3707}{n}$; root $= \frac{.3707}{n} = .0052$ in.

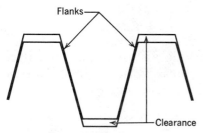

Figure 4. General purpose Acme threads bear on the flanks.

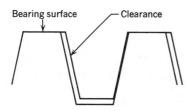

Figure 5. Centralizing Acme threads bear at the major diameter.

is easier to machine and stronger than the square form thread since the Acme root cross section is thicker than its root clearance. Acme thread screws are used on milling machines and lathes. Like the square thread, the Acme has a basic depth equal to one-half the pitch; however, clearance is added both at the crest and the

$$P = \text{Pitch} = \frac{1}{n}$$
$$d = \text{Depth} = .433P$$
$$f = \text{Flat} = .25P$$

Figure 6b. Stub Acme thread. $P = \text{pitch} = \frac{1}{n}$; $d = \text{depth} = .433P$; $f = \text{flat} = .25P$.

root of the thread for the general purpose fit. The Acme general purpose threads bear on the flanks (Figure 4). Centralizing fits bear at the major diameter (Figure 5). For more detailed information on Acme thread fits, see the *Machinery's Handbook*.

Three classes of general purpose Acme threads 2G, 3G, and 4G are used. Class 2G is preferred for general purpose assemblies. If less backlash or end play is desired, classes 3G and 4G are given. Internal threads of any class may be combined with any external class to provide other degrees of fit. The included angle of General Purpose Acme threads is 29 degrees. Depth of thread is one-half the pitch plus .010 in. for 10 TPI and coarser. For finer pitches the depth of thread is one-half the pitch plus .005 in.

Stub Acme threads (Figures 6a and b) are used where a coarse pitch thread with a shallow depth is required. The depth for the Stub Acme is only .3P and .433P for the American Standard Stub, as compared to .5P for the standard Acme threads.

The Buttress thread (Figure 7) is not usually used for translating motion. It is often used

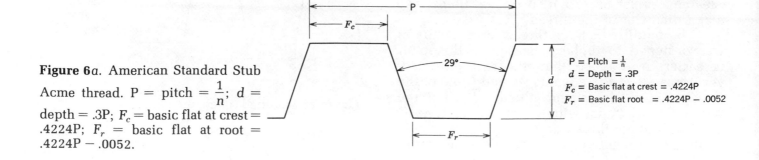

Figure 6a. American Standard Stub Acme thread. $P = \text{pitch} = \frac{1}{n}$; $d = \text{depth} = .3P$; $F_c = \text{basic flat at crest} = .4224P$; $F_r = \text{basic flat at root} = .4224P - .0052$.

$$P = \text{Pitch} = \frac{1}{n}$$
$$d = \text{Depth} = .3P$$
$$F_c = \text{Basic flat at crest} = .4224P$$
$$F_r = \text{Basic flat root} = .4224P - .0052$$

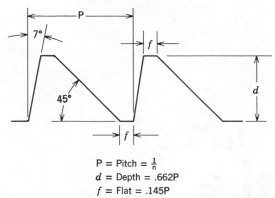

P = Pitch = $\frac{1}{n}$
d = Depth = .662P
f = Flat = .145P

Figure 7. Buttress thread. P = pitch = $\frac{1}{n}$; d = depth = .662P; f = flat = .145P.

where great pressures are applied in one direction only, such as the breech of large guns.

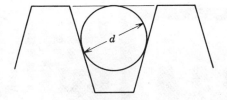

d = Diameter of wire = .48725P

Figure 8. The one-wire method for measuring Acme threads. d = diameter of wire = .48725P.

Acme threads may be measured by using the one-wire method (Figure 8). If a wire with a diameter equal to .48725 × P is placed in the groove of an Acme thread, the wire will be flush with the top of the thread. For further information on one-wire and three-wire methods for checking Acme threads, see the *Machinery's Handbook.*

self-evaluation

SELF-TEST
1. For what two major purposes are translating-type screw threads used?
2. Name five thread forms used as translating screws.
3. What would the depth of thread be for a square thread that is 4 TPI?
4. What would the depth of thread be for a general purpose Acme thread that is 4 TPI?
5. What is the main difference between general purpose Acme threads and centralizing Acme threads?
6. What is the included angle of Acme threads?
7. Explain the general use of stub Acme threads.
8. Which thread form has a 10 degree included angle?
9. Of the translating thread forms, which type is most used and is easiest to machine?
10. What are Buttress threads mostly used for?

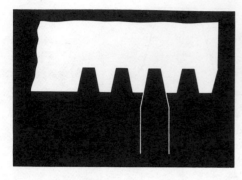

unit 16
cutting
acme threads
on the lathe

Machinists are sometimes required to cut internal and external Acme threads. These threads are, in most cases, larger and coarser than 60 degree form threads, and require greater skill to produce them well. In this unit, you will not only learn the procedure for cutting these threads, but you will also make internal and external Acme threads.

objectives **After completing this unit, you will be able to:**

1. **Describe set up and procedure for making external and internal Acme threads.**
2. **Make a $1\frac{3}{8}$–4 Acme thread plug gage.**
3. **Make a $1\frac{3}{8}$–4 Acme internal thread in a nut.**

Cutting Acme threads is similar to cutting 60 degree threads in many ways. The threads per inch and infeeds are calculated in the same way. Some calculations, tool form and relief angles, and finishes for Acme threads, however, involve different problems and procedures.

GRINDING THE TOOL FOR ACME THREADS

The cutting tool form must be checked with the Acme tool gage (Figures 1a and b) when the 29 degree included angle is ground. Side relief must be ground at the same time. The end of the tool is ground flat and perpendicular to the bisector of the angle. The flat is checked (Figure 2) with the tool gage at the number corresponding to the threads per inch you will cut. It is very important to have the flat the exact width needed for the particular thread being cut.

The relief angles on the tool (Figure 3) are of greater importance when coarse threads are

cut, since if the heel of the tool rubs on either side, a rough, inaccurate thread will result. When the helix angle has been determined, it should be added to the relief angle (8 to 12 degrees) of the tool on the leading edge and similar relief provided on the trailing edge of the tool (Figure 4).

SETTING UP TO CUT EXTERNAL ACME THREADS ON THE LATHE

The threads per inch are set up normally and the lead screw rotation is set for right- or left-hand threads. The gears, leadscrew, and carriage should be lubricated before cutting coarse threads. The compound is most often set at $14\frac{1}{2}$ degrees to the right for right-hand external threads. The workpiece must be set up and held very securely in the workholding device; a four-jaw chuck and dead center would be most secure. The tool is aligned with the work by using

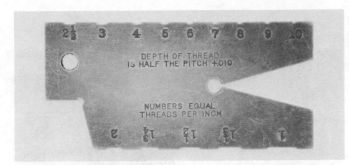

Figure 1a. Acme tool gage.

Figure 1b. Checking the tool angle with the Acme tool gage (Lane Community College).

Figure 2. Checking the flat on the end of the tool with the Acme tool gage (Lane Community College).

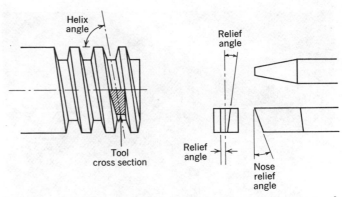

Figure 3. Acme threads showing the importance of the relief angles on both sides of the tool.

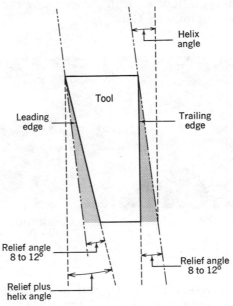

Figure 4. The relationship of the helix angle of the thread to the relief angle of the tool.

An undercut must be made at the end of the threads to clear the tool.

When Acme threads coarser than 5 TPI are cut, a square or round nose roughing tool that is smaller than the Acme tool should be used to make a first cut. The roughing tool does not cut to full depth or width and the Acme form tool is used to finish the thread. This procedure is also used when making threads in tough alloy materials.

Some machinists prefer to set the compound parallel to the ways so they can "shave" both flanks of the thread for a good finish. When this is done, the tool must be made a few thousandths of an inch narrower to allow for the "shaving" operation.

the Acme gage (Figure 5). With this setup, the tool is fed into the work by advancing the compound in small steps as with 60 degree threads.

Figure 5. Aligning the tool with the work using the Acme gage (Lane Community College).

Figure 6. Taking the scratch cut (Lane Community College).

Figure 7. Compound is set $14\frac{1}{2}°$ to the left of the operator for right-hand internal threads (Lane Community College).

Making the Cut

A scratch cut is taken (Figure 6) and measured. The cross slide is moved out and the carriage returned. The cross slide is again set on zero and the compound is advanced .005 to .010 in., depending on what the lathe and setup will handle without chatter. Use sulfurized cutting oil. The total depth of the cut is .5P + .010 in. Feed in on the cross slide for the last few thousandths of an inch so that the trailing flank will also receive a finish cut. For other Acme thread fits, see the *Machinery's Handbook*.

INTERNAL THREADS

The bore size for making an Acme General Purpose internal thread is the major diameter of the screw minus the pitch. As with Unified threads, the actual minor diameter of an Acme thread is the inside diameter of the bore; thus the minor diameter of an Acme 1–5 thread would be 1 − .200 = .800 in. The internal major diameter should be the major diameter of the screw plus .020 in. for 10 or more TPI and .010 in. for pitches less than 10 TPI.

The compound is set $14\frac{1}{2}$ degrees to the left for cutting right-hand internal threads (Figure 7). An internal Acme threading tool is ground, checked (Figure 8) and set up, then fed into the work with the compound .002 to .005 in. for each pass. An Acme screw plug gage or the mating external thread should be used to check the fit as the internal thread nears completion (Figure 9).

When internal Acme threads are too small in diameter to be cut with a boring bar and tool, an Acme tap (Figure 10) is used to make the thread. Acme taps are made in sets of two or three taps; each tap cuts more of the thread, the last tap for the finishing cut. Two taps are sometimes made on the same shank as in Figure 10. The part is drilled or bored to the minor diameter of the thread and the Acme tap is turned in by hand. Use cutting oil when tapping steel, but cut threads in bronze dry.

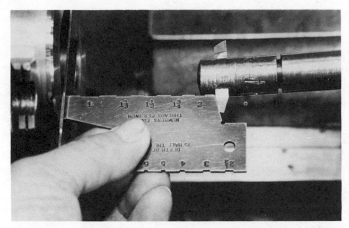

Figure 8. Aligning the Acme tool with a gage for cutting internal threads (Lane Community College).

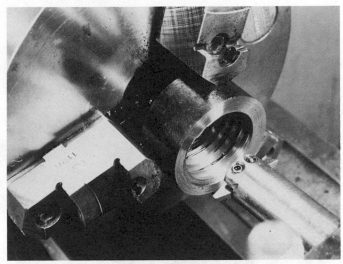

Figure 9. Completed internal Acme threads (Lane Community College).

Figure 10. An Acme tap (Lane Community College).

A problem often encountered when threading coarse threads on small lathes is that of producing a thicker than normal last thread; the plug gage will go in the nut all the way except for the final thread. This happens because the carriage is not heavy enough to provide sufficient drag to keep the tool cutting on its following side when the leading side of the tool is emerging from the cut. The result is a pitch error on the last thread. The slack in the half-nuts allows this to happen. Providing a slight drag on the carriage handwheel will eliminate this problem.

External Acme threads must often have a good finish. A final honing of the tool before the last few shaving passes will help. The setup must be very rigid and the gibs tight. Low speeds are essential. The grade of cutting oil is extremely important in this finishing operation.

The thread may be finished after it is cut by using a thin, safe edge file at low RPM and by using abrasive cloth at a higher RPM. A thin piece of wood is sometimes used to back up the abrasive cloth while each flank is being polished.

self-evaluation

SELF-TEST
1. What is the major difference between V-form threads and Acme threads?
2. When grinding an Acme threading tool, what three important parts should be carefully measured?
3. Why should the gears, ways, leadscrew, and carriage be lubricated before cutting coarse threads?
4. Where is the compound most frequently set when cutting Acme threads? What setup is preferred by some machinists?
5. What is the depth of thread for a $\frac{3}{4}$–6 external general purpose Acme thread?
6. How is the tool aligned with the workpiece?

7. Determine the bore size to make a $\frac{3}{4}$–6 general purpose internal Acme thread.
8. Which is the best way to make small internal Acme threads?
9. What can you use to check internal threads for fit?
10. Explain how a good tool finish may be obtained on an Acme thread.

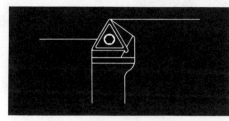

unit 17
using carbides and other tool materials on the lathe

The cutting tool materials such as carbon steels and high speed steel that served the needs of machining in the past years are not suitable in many applications today. Tougher and harder tools are required to machine the tough, hard, space age metals and new alloys. The constant demand for higher productivity led to the need for faster stock removal and quick-change tooling. You, as a machinist, must learn to achieve maximum productivity at minimum cost. Your knowledge of carbide cutting tools and ability to select them for specific machining tasks will effect your productivity directly.

objectives After completing this unit, you will be able to:

1. **List six different cutting tool materials and compare some of their machining properties.**
2. **Select a carbide tool for a job by reference to operating conditions, carbide grades, nose radii, tool style, rake angles, shank size, and insert size, shape, and thickness.**
3. **Identify carbide inserts and toolholders by number systems developed by the American Standards Association.**
4. **Control chip formation in exercises by varying feed, speed, depth of cut, and chip breakers.**

The various tool materials are high carbon steel, high speed steel, nonferrous cast alloys, cemented carbides, ceramics, and diamond.

HIGH CARBON STEELS

High carbon tool steels are used for hand tools such as files, chisels, and only to a limited extent for drilling and turning tools. They are oil or water hardening plain carbon steels with .9 to 1.4 percent carbon content. These tools maintain a keen edge and can be used for metals that produce low tool-chip interface temperatures; for example, aluminum, magnesium, copper, and brass. These tools, however, tend to soften at machining speeds above 50 feet per minute (FPM) in mild steels.

HIGH SPEED STEELS

High speed steels (HSS) may be used at higher speeds (100 FPM in mild steels) without losing their hardness. The relationship of cutting speeds to the approximate temperature of tool-chip interface is as follows:

100 FPM — 1000°F (538°C)
200 FPM — 1200°F (649°C)
300 FPM — 1300°F (704°C)
400 FPM — 1400°F (760°C)

High speed steel is sometimes used when special tool shapes are needed, especially for boring tools.

CAST ALLOYS

Cast alloys are referred to as such because they are nonferrous (not containing iron) alloys. These materials are somewhat softer than HSS at room temperature, but retain their hardness to higher temperatures. This property in tools is known as red hardness. The cast alloys can be used at speeds of nearly 200 FPM or up to 1200°F (649°C) in steels. The approximate composition of cast alloy materials is 12 to 17 percent tungsten or tantalum, 30 to 35 percent chromium, 45 to 55 percent cobalt, and 2 to 3 percent carbon.

CEMENTED CARBIDES

A carbide, generally, is a chemical compound of carbon and a metal. The term carbide is com-

monly used to refer to cemented carbides, the cutting tools composed of tungsten carbide, titanium carbide, or tantalum carbide, and cobalt in various combinations. A typical composition of cemented carbide is 85 to 95 percent carbides of tungsten and the remainder a cobalt binder for the tungsten carbide powder.

Cemented carbides are made by compressing various metal powders (Figure 1) and sintering (heating to weld particles together without melting them) the briquettes. Cobalt powder is used as a binder for the carbide powder used, either tungsten, titanium, or tantalum carbide powder or a combination of these. Increasing the percentage of cobalt binder increases the toughness of the tool material and at the same time reduces its hardness or wear resistance. Carbides have greater hardness at both high and low temperatures than high speed steel or cast alloys. At temperatures of 1400°F (760°C) and higher, carbides maintain the hardness required for efficient machining. This makes possible machining speeds of approximately 400 FPM in steels. The addition of tantalum increases the red hardness of a tool material. Cemented carbides are extremely hard tool materials (above RA90), have a high compressive strength, and resist wear and rupture.

SELECTING CARBIDE TOOLS

The following steps may be used in selecting the correct carbide tool for a job.

1. Establish the operating conditions.
2. Select the cemented carbide grade.
3. Select nose radius.
4. Select insert shape.
5. Select insert size.

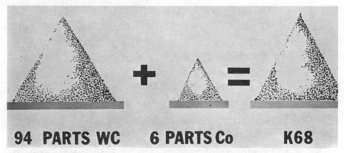

94 PARTS WC   6 PARTS Co   K68

Figure 1. The three basic materials needed to produce the straight grades of tungsten carbide (Photo courtesy of Kennametal Inc., Latrobe, Pa.).

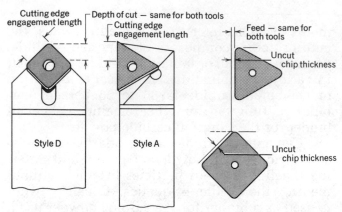

Figure 2. The difference in Style A and Style D holders for depth of cut and cutting edge engagement length (Copyright General Electric Company).

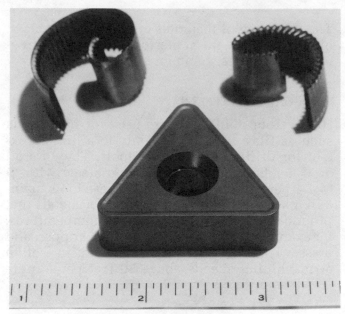

Figure 3. Large, well-formed chips were produced by this tool with built-in chip breaker (Photo courtesy of Kennametal Inc., Latrobe, Pa.)

6. Select insert thickness.
7. Select tool style.
8. Select rake angle.
9. Select shank size.

Step 1. Establishing the Operating Conditions

The tool engineer or machinist must use three variables to establish metal removal rate: speed, feed, and depth of cut. Cutting speed has the greatest effect on tool life. A 50 percent increase in cutting speed will decrease tool life by 80 percent. A 50 percent increase in feed will decrease tool life by 60 percent. The cutting edge engagement or depth of cut (Figure 2) is limited by the size and thickness of the carbide insert and the hardness of the workpiece material. Hard workpiece materials require decreased feed, speed, and depth of cut.

Figure 2 shows how the lead angle affects both cutting edge engagement length and chip thickness by comparing a Style D square insert tool, using a 45 degree lead angle, to a Style A triangular insert tool, using a 0 degree lead angle. The two tools are shown making an indentical depth of cut at an identical feed rate. The feed rate is the same as the chip thickness with the Style A tool or any tool with a 90 degree lead angle. The chip would be wider but thinner using the Style D with the same feed rate. When large lead angles are used, the Style D tool has a much greater strength than the Style A tool since the cutting forces are directed into the solid part of the holder. Large lead angles can cause chatter to develop, however, if the setup is not rigid.

Figure 4. Normal edge wear.

The depth of cut is limited by the strength and thickness of the carbide insert, the rigidity of the machine and setup, the horsepower of the machine and, of course, the amount of material to be removed. An example of a relatively large depth of cut with an insert that produced large 9-shaped chips is shown in Figure 3.

Edge wear and cratering are the most frequent tool breakdowns that occur. Edge wear (Figure 4) is simply the breaking down of the

Figure 5. Tool point breakdown caused by built-up edge.

Figure 6. Chipped or broken inserts. The triangular insert shows the typical breakage on straight tungsten carbide inserts. The square titanium-coated insert shows a stratified failure because of an edge impact.

tool relief surface caused by friction and abrasion and is considered normal wear. Edge breakdown is also caused by the tearing away of minute carbide particles by the built-up edge (Figure 5). The cutting edge is usually chipped or broken in this case. Lack of rigidity, too much feed, or too slow a speed results in chipped or broken inserts (Figure 6).

Thermal shock, caused by sudden heating and cooling, is the cracking and checking of a tool that leads to breakage (Figure 7). This condition is most likely to occur when an inadequate amount of coolant is used. It is better to machine dry if the work and tool cannot be kept flooded with coolant.

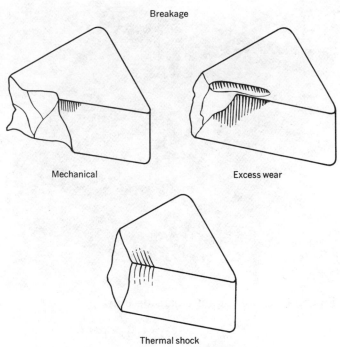

Figure 7. Three causes of tool breakage (Photo courtesy of Kennametal, Inc., Latrobe, Pa.).

If edge wear occurs:

1. Decrease machining speed.
2. Increase feed.
3. Change to a harder, more wear-resistant carbide grade.

If the cutting edge is chipped or broken:

1. Increase speed.
2. Decrease feed and/or depth of cut.
3. Change to a tougher grade carbide insert.
4. Use negative rake.
5. Hone the cutting edge before use.
6. Check the rigidity and tool overhang.

When there is a buildup on the cutting edge:

1. Increase speed.
2. Change to a positive rake tool.
3. Change to a grade containing titanium.

For cutting edge notching:

1. Increase side cutting edge angle.
2. Decrease feed.

Cratering (Figure 8) is the result of high temperatures and pressures that cause the steel chip to weld itself to the tungsten carbide and

Figure 8. Cratering on a carbide tool.

tear out small particles of the tool material. The addition of titanium carbide to the mixture of tungsten carbide and cobalt provides an anti-weld quality, but there is some loss in abrasive wear and strength in these tools.

Step 2. Selecting Cemented Carbide Grades

There are two main groups of cemented carbides from which to select most grades. First, the straight carbide grades composed of tungsten carbide and cobalt binder, which are used for cast iron, nonferrous metals and nonmetalics where resistance to edge wear is the primary factor. Second, grades composed of tungsten carbide, titanium carbide, and tantalum carbide plus cobalt binder, which are usually used for machining steels. Resistance to cratering and deformation is the major requirement for these steel grades.

Cemented carbides have been organized into grades. Properties that determine grade include hardness, toughness, and resistance to chip welding or cratering. The properties of carbide tools may be varied by the percentages of cobalt and titanium or tantalum carbides. Increasing the cobalt content increases toughness but decreases hardness. Properties may also be varied during the processing by the grain size of carbides, density, and other modifications. Some tungsten carbide inserts are given a titanium carbide coating (about .0003 in. thick) to resist cratering and edge breakdown.

The grades of carbides have been organized according to their suitable uses by the Cemented Carbide Producers Association (CCPA). It is recommended that carbides be selected by using such a table rather than by their composition. Cemented carbide grades with specific chip removal applications are:

C-1   Roughing cuts (cast iron and nonferrous materials)

C-2   General purpose (cast iron and nonferrous materials)

C-3   Light finishing (cast iron and nonferrous materials)

C-4   Precision boring (cast iron and nonferrous materials)

C-5   Roughing cuts (steel)

C-6   General purpose (steel)

C-7   Finishing cuts (steel)

C-8   Precision boring (steel)

The hardest of the nonferrous-cast iron grades would be C-4 and the hardest of the steel grades would be C-8.

This system does not specify the particular materials or alloy, and the particular machining operations are not specified. One must use the judgment of an experienced machinist in order to select a grade of carbide, for example, for turning a chromium-molybdenum steel. Factors to be considered would include the difficulty of machining such an alloy because of its toughness. Given this example, a grade of C-5 or C-6 carbide would probably be best suited to this operation. The proof of the selection would come only with the actual machining. Cemented carbide tool manufacturers often supply tables designating uses and machining characteristics of their various grades. See Table 1.

Grade classification-comparison tables that convert each manufacturer's carbide designations to CCPA "C" numbers are available. See Table 2. There is, however, one major caution. The tables are intended to correlate grades on the basis of composition and not according to tested performance. Grades from different manufacturers having the same "C" number may vary in performance. Some general guidelines to grade selection are as follows:

1. Select the grade with the highest hardness with sufficient strength to prevent breakage.

2. Select straight grades of tungsten carbide for the highest resistance to abrasion.

Table 1
Grade and Machining Applications

		Grade	Hardness R_A	Typical Machining Applications
Combined Crater and Edge-Wear Resistance	Maximum Crater Resistance	CO6	Ceramic *	The hardest of this group. For finishing most ferrous and nonferrous alloys and nonmetals as in high speed, light chip load precision machining, or for use at moderate speeds and chip loads where long tool life is desired.
		K165	93.5	Titanium carbide for finishing steels and cast irons at high to moderate speeds and light chip loads.
		K7H	93.5	For finishing steels at higher speeds and moderate chip loads.
		K5H	93.0	For finishing and light roughing steels at moderate speeds and chip loads through light interruptions.
	KC75	K45	92.5	The hardest of this group. General purpose grade for light roughing to semifinishing of steels at moderate speeds and chip loads, and for many low speed, light chip load applications.
		K4H	92.0	For light roughing to semifinishing of steels at moderate speeds and chip loads, and for form tools and tools that must dwell.
		K2S	91.5	For light to moderate roughing of steels at moderate speeds and feeds through medium interruptions.
				For general purpose use in machining of steels over a wide range of speeds in moderate roughing to semifinishing applications.
		K21	91.0	For moderate to heavy roughing of steels at moderate speeds and heavy chip loads through medium interruptions where mechanical and thermal shock are encountered.
		K42	91.3	For heavy roughing of steels at low to moderate speeds and heavy chip loads through interruptions where mechanical and severe thermal shocks are encountered.
	Maximum Edge-Wear Resistance	K11	93.0	The hardest of this group. For precision finishing of cast irons, nonferrous alloys, nonmetals at high speeds and light chip loads, and for finishing many hard steels at low speeds and light chip loads.
		K68	92.6	General purpose grade for light roughing to finishing of most high temperature alloys, refractory metals, cast irons, nonferrous alloys, and nonmetals at moderate speeds and chip loads through light interruptions.
		K6	92.0	For moderate roughing of most high temperature alloys, cast irons, nonferrous alloys, and nonmetals at moderate to low speeds and moderate to heavy chip loads through light interruptions.
		K1	90.0	The most shock resistant of this group. For heavy roughing of most high temperature alloys, cast irons, and nonferrous alloys at low speeds and heavy chip loads through heavy interruptions.

*The hardness of CO6 is 91 Rockwell 45N (or about 94R_A).
Source. *Kentrol Inserts (Supplement 5 to Catalog 73,* "Kenametal Grade Systems and Machining Applications," 1975 (Data courtesy of Kenametal, Inc., Latrobe, Pa.).

Step 3. Select Nose Radius
Selecting the nose radius can be important because of tool strength, surface finish, or perhaps the forming of a fillet or radius on the work. To determine the nose radius according to strength requirements, use the nomograph in Figure 9. Consider that the feed rate, depth of cut, and workpiece condition determine strength requirements, since a larger nose radius makes a stronger tool.

Table 2
Carbide Grade Classification–Comparison Table with CCPA "C" Numbers and Manufacturers Designations

The Grades Listed Are Those Usually Recommended by the Manufacturer for the Catagories Shown

APPLICATION		C No.	Newcomer	Adamas	Atrax	Carboloy	Carmet	Ex-cell-o	Firth Sterling	Greenleaf	Kennametal	Metal Carbides	Sandvik	Valenite	V-R Wesson	Walmet	Wendt-Sonis
Cast Irons; Nonferrous, Nonmetallic, Hi-Temperature alloys; 200 & 300 series stainless	Roughing cuts	C-1	N10	B	FA5	44A	CA3	E8	H / HB	G10	K1	C89	H20	VC-1	VR54 / 2A68	WA-1 / WA-159	CQ12 / CQ22
	General purpose	C-2	N20 / N22	A / AM	FA6 / FA-62	883 / 860	CA4 / CA443	E6 / XL620	HA / HTA	G20 / G25	K6 / K68	C91	H20	VC-2 / VC-28	2A5 / VR82	WA-69 / WA-2	CQ2 / CQ23
	Light finishing	C-3	N30	PWX	FA7	905	CA7	E5	HE / HTA	G30	K8 / K68	C93	R1P	VC-3	2A7 / VR82	WA-35 / WA3	CQ3 / CQ23
	Precision boring	C-4	N40	AAA	FA8	999 / 895	CA8	E3	HF	G40	K11	C95	H1P / HO5	VC-4	2A7	WA4	CQ4
Carbon steels; Alloy steels; 400 Series stainless	Roughing cuts	C-5	N50 / N52	499 / 434	FT-3 / FT-35	370 / 78B	CA721 / CA740	10A / 945	NTA / TXH	G50 / G55	K42 / K21	S-880	S-6	VC-55 / VC-125	VR77 / WM	WA5 / WA55	CY12 / CY17
	General purpose	C-6	N60	6X / T-60	FT-4 / FT-6	78B	CA720	BA / 606	T22 / T25	G60	K2S / K21	S-900 / S-901	S-4	VC-6	26 / VR75	WA6	CY5 / CY16
	Finishing cuts	C-7	N70 / N72	495 / 548	FT-6 / FT-62	78 / 350	CA711	6A / XL70 / 6AX	T25 / T31	G70 / G74	K45 / K5H	S-92 / S-900	SM	VC7 / VC-76	WH / VR73	WA7 / WA168	CY2 / CY14
	Precision boring	C-8	N80 / N93	490 / T-80	FT-7 / FT-71	330 / 210	CA704	6AX / XL88	T31	G80	K7H / K165	S-94	FO2	VC-8 / VC-83	VR71 / VR65	WA8 / WA800	CY31 / Ti8
	Hi-velocity	C-80	N95			0-30					CO6				VR97		

CHIP REMOVAL

Source. Newcomer Products, Inc. *Reference card.*

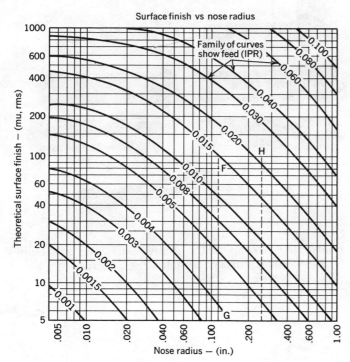

Figure 9. Surface finish vs. nose radius (Copyright General Electric Company).

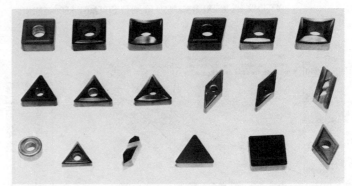

Figure 10. Insert shapes for various applications (Photo courtesy of Kannametal, Inc., Latrobe, Pa.).

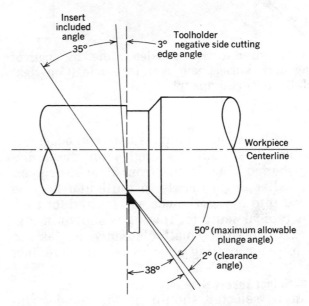

Figure 11. A 38° triangular insert used for a tracing operation (Copyright General Electric Company).

Large radii are strongest and can produce the best finishes, but they also can cause chatter between tool and workpiece. For example, the dashed line on the chart indicates that a $\frac{1}{8}$ in. radius would be required for turning with a feed rate of .015 in., and to obtain a 100 microinch finish, a $\frac{1}{4}$ in. radius would be required with a .020 in. feed rate.

Step 4. Select Insert Shapes

Indexable inserts (Figure 10), also called throw-away inserts, are clamped in toolholders of various design. These inserts provide a cutting tool with several cutting edges. After every or all edges have been used, the insert is discarded.

The round inserts have the greatest strength and, as with large radius inserts, make possible higher feed rates with equal finishes. Round inserts also have the greatest number of cutting edges possible, but are limited to workpiece configurations and operations that are not affected by a large radius. Round inserts would be ideally suited, for example, to straight turning operations.

Square inserts have lower strength and fewer possible cutting edges than round tools, but are much stronger than triangular inserts. The included angle between cutting edges (90 degrees) is greater than for triangular inserts (60 degrees), and there are eight cutting edges possible as compared to six for the triangular inserts.

Triangular inserts have the greatest versatility. They can be used, for example, for combination turning and facing operations, while round or square inserts are often not adaptable to such combinations. Because the included angle between cutting edges is less than 90 degrees, the triangular inserts are also capable of tracing operations. The disadvantages include their reduced strength and fewer cutting edges per insert.

For tracing operations where triangular inserts cannot be applied, diamond-shaped inserts

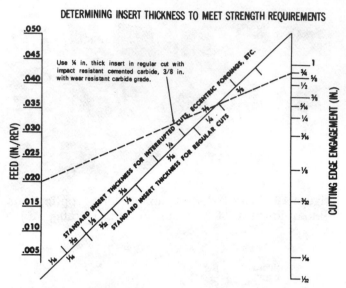

Figure 12. Insert thickness as determined by length of cutting edge engagement and feed rate (Copyright General Electric Company).

Figure 13. Several of the many tool styles available (Photo courtesy of Kennametal, Inc., Latrobe, Pa.).

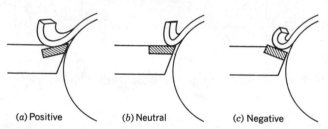

(a) Positive   (b) Neutral   (c) Negative

Figure 14. Side view of back rake angles.

with smaller included angles between edges are available. The included angles on these diamond-shaped inserts range from 35 to 80 degrees. The smaller angle inserts in particular may be plunged into the workpiece as required for tracing. A typical setup for tracing is shown in Figure 11. Note that small clearance angles are used for each cutting edge to permit plunging.

Step 5. Select Insert Size
The insert selected should be the smallest insert capable of sustaining the required depth of cut and feed rate. The depth of cut should always be as great as possible. A rule of thumb is to select an insert with cutting edges $1\frac{1}{2}$ times the length of cutting edge engagement. The feed for roughing mild steel should be approximately $\frac{1}{10}$ the depth of cut.

Step 6. Select Insert Thickness
Insert thickness is also important to tool strength. The required depth of cut and feed rate are criteria that determine insert thickness. The nomograph in Figure 12 simplifies the relationship of depth of cut and feed rate to insert thickness. Lead angle is also important in converting the depth of cut to a length of cutting edge engagement.

　　The dashed line on the nomograph in Figure 12 represents an operation with a $\frac{3}{4}$ in. length of engagement (not depth of cut) and a feed

rate of .020 IPR. Depending on the grade of carbide used (tough or hard), an insert of $\frac{1}{4}$ or $\frac{3}{8}$ inch thickness should be used.

Step 7. Select Tool Style
Tool style pertains to the configuration of toolholder for a carbide insert. To determine style, some familiarity with the particular machine tool and the operations to be performed is required. Figure 13 shows some of the styles available for toolholders.

Step 8. Select Rake Angle
When selecting the rake angles, you need to consider the machining conditions. Negative rake should be used where there is maximum rigidity of the tool and work and where high machining speeds can be maintained. More horsepower is needed when using negative rake tools. Under these conditions, negative rake tools are stronger and produce satisfactory results (Figure 14).

　　Negative rake inserts may also be used on both sides, doubling the number of cutting edges per insert. This is possible because end and side relief are provided by the angle of the tool-holder rather than by the shape of the insert.

Positive rake inserts should be used where rigidity of the tool and work is reduced and where high cutting speeds are not possible; for example, on a flexible shaft of small diameter. Positive rake tools cut with less force so deflection of the work and toolholder would be reduced. High cutting speeds (SFPM) are often not possible on small diameters because of limitations in spindle speeds. Some insert types are plain and others have built-in chip-breakers (Figure 15).

Step 9. Select Shank Size

As with insert thickness and rake angles, the rate of feed and depth of cut are important in determining shank size. Overhang of the tool shank is extremely important for the same reason. Heavy cuts at high feed rates create high downward forces on the tool. These downward forces acting on a tool with excessive overhang would cause tool deflection that would make it difficult or impossible to maintain accuracy or surface finish quality.

Having established the feed rate, depth of cut and tool overhang, use the nomograph in Figure 16 to determine the shank size. Begin by drawing a line from the depth of cut scale to the feed rate scale. From the point where this line intersects the vertical construction line, draw a line to the correct point on the tool over-

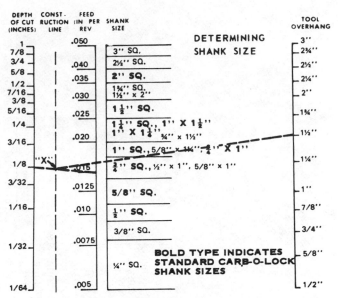

Figure 16. Determining shank size according to depth of cut, feed rate, and tool overhang (Copyright General Electric Company).

Figure 17. A boring bar with various interchangeable adjustable heads (Photo courtesy of Kennametal Inc., Latrobe, Pa.).

hang scale. Determine the shank size from where this last line crosses through the shank size scale.

In the example shown on the graph, a line has been drawn from the $\frac{1}{8}$ in. depth of cut point to the .015 IPR feed rate point. Another line has been drawn from the point of intersection on the vertical construction line to the amount of tool overhang. The second line drawn passes through the shank size scale at the $\frac{3}{4}$ square in. (cross sectional area) point. Toolholders $\frac{1}{2} \times 1$ in. or $\frac{5}{8} \times 1$ in. would meet the requirements. Throwaway carbide inserts are also used for boring bars (Figure 17) of various shapes and sizes. Brazed carbide tips are sometimes put on the end of a boring bar for special applications.

Figure 15. Chip breakers used are the adjustable chip deflector (center) with a straight insert and the type with the built-in chip control groove.

Table 3
ASA Tool Holder Identification System

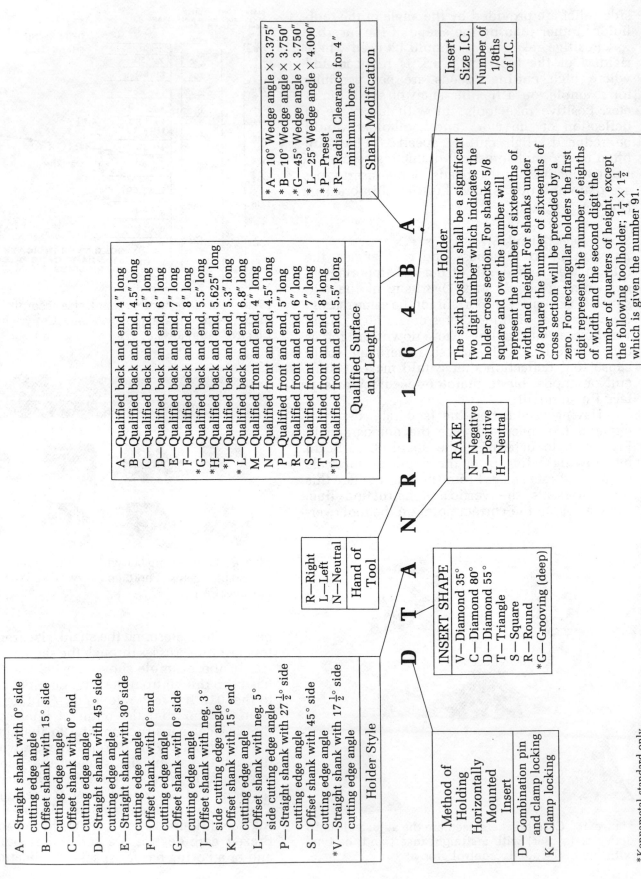

Example code: D T A N R — 1 6 4 B A

Holder Style

A—Straight shank with 0° side cutting edge angle
B—Offset shank with 15° side cutting edge angle
C—Offset shank with 0° end cutting edge angle
D—Straight shank with 45° side cutting edge angle
E—Straight shank with 30° side cutting edge angle
F—Offset shank with 0° end cutting edge angle
G—Offset shank with 0° side cutting edge angle
J—Offset shank with neg. 3° side cutting edge angle
K—Offset shank with 15° end cutting edge angle
L—Offset shank with neg. 5° side cutting edge angle
P—Straight shank with $27\frac{1}{2}$° side cutting edge angle
S—Offset shank with 45° side cutting edge angle
*V—Straight shank with $17\frac{1}{2}$° side cutting edge angle

INSERT SHAPE

V—Diamond 35°
C—Diamond 80°
D—Diamond 55°
T—Triangle
S—Square
R—Round
*G—Grooving (deep)

Method of Holding Horizontally Mounted Insert

D—Combination pin and clamp locking
K—Clamp locking

Hand of Tool

R—Right
L—Left
N—Neutral

RAKE

N—Negative
P—Positive
H—Neutral

Qualified Surface and Length

A—Qualified back and end, 4" long
B—Qualified back and end, 4.5" long
C—Qualified back and end, 5" long
D—Qualified back and end, 6" long
E—Qualified back and end, 7" long
F—Qualified back and end, 8" long
*G—Qualified back and end, 5.5" long
*H—Qualified back and end, 5.625" long
*J—Qualified back and end, 5.3" long
*L—Qualified back and end, 6.8" long
M—Qualified front and end, 4" long
N—Qualified front and end, 4.5" long
P—Qualified front and end, 5" long
R—Qualified front and end, 6" long
S—Qualified front and end, 7" long
T—Qualified front and end, 8" long
*U—Qualified front and end, 5.5" long

Holder

The sixth position shall be a significant two digit number which indicates the holder cross section. For shanks 5/8 square and over the number will represent the number of sixteenths of width and height. For shanks under 5/8 square the number of sixteenths of cross section will be preceded by a zero. For rectangular holders the first digit represents the number of eighths of width and the second digit the number of quarters of height, except the following toolholder; $1\frac{1}{4} \times 1\frac{1}{2}$ which is given the number 91.

Shank Modification

*A—10° Wedge angle × 3.375"
*B—10° Wedge angle × 3.750"
*G—45° Wedge angle × 3.750"
*L—25° Wedge angle × 4.000"
*P—Preset
*R—Radial Clearance for 4" minimum bore

Insert Size I.C.

Number of 1/8ths of I.C.

*Kennametal standard only
Source. Tool Application Handbook. (Data courtesy of Kennametal, Inc., Latrobe, Pa., 1973.

TOOLHOLDER IDENTIFICATION

The carbide manufacturers and the American Standards Association (ASA) have adopted a system of identifying toolholders for inserted carbides. This system is used to call out the toolholder geometry and for ordering tools from manufacturers or distributors. The system is shown in Table 3.

As an example, use this system to determine the geometry of a toolholder called out as TANR-8:

- T—Triangular insert shape
- A—0° side cutting edge angle
- N—Negative rake
- R—Right hand turning (from left to right)
- 8—Square shank, $\frac{8}{16}$ in. per side

CARBIDE INSERT IDENTIFICATION

As with toolholder identification, a system has been adopted by the carbide manufacturers and the American Standard Association for identifying inserts. See Table 4.

For example, use this system to determine the specifications for an insert called out as T N M G—323 E:

- T—Triangular shape
- N—0 degree relief (relief provided by holder)
- M—Plus or minus .005 in. tolerance on thickness
- G—With hole and chipbreaker
- 3—$\frac{3}{8}$ in. inscribed circle (inside square and triangular insert)
- 2—$\frac{2}{16}$ in. thickness
- 3—$\frac{3}{64}$ in. radius
- E—Unground, honed

Brazed carbide tools have been used for many years and are still used on many machining jobs. These tools can be sharpened by grinding many times, while the insert tool is thrown away after its cutting edges are dull. Grinding the tool often causes thermal shock by the sudden heating of the carbide surface. This is evident when "crazing" or several tiny checks appear on the edge or when a large crack can be seen in the tool. To prevent crazing when grinding these tools:

1. Avoid the use of aluminum oxide wheels (except to rough grind the steel shank for clearance). Use only silicon carbide wheels with a soft bond.

2. Avoid excessive grinding pressure in a small area.

3. Avoid poor cutting action of a low concentration diamond wheel.

4. Avoid dry grinding.

Chipbreakers may be ground on the edge of brazed carbide tools with diamond wheels on a surface grinder.

The side and end clearance angles are not to be confused with relief angles on carbide tools (Figure 18). Clearance refers to the increased angle ground on the shank of a carbide tipped tool. Clearance provides for a narrow flank on the carbide and for regrinding of the carbide without contacting steel.

CERAMIC TOOLS

Ceramic or "cemented oxide" tools (Figure 19) are made primarily from aluminum oxide with a binder. Some manufacturers add titanium, magnesium, or chromium oxides in quantities of 10 percent or less. The tool materials are molded at pressures over 4000 PSI (pounds per square

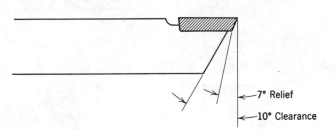

Figure 18. Relief and clearance compared.

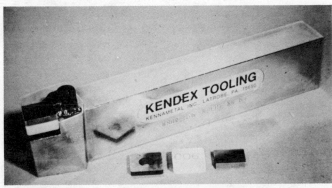

Figure 19. Ceramic tool with carbide seat and chip deflector shown assembled in toolholder behind the parts (Photo courtesy of Kennametal, Inc., Latrobe, Pa.).

Table 4
ASA Carbide Insert Identification
Throwaway Inserts Identification System

This insert identification system, developed by the American Standards Association, has been generally adopted by industry. It makes insert identification fast and accurate.

Code example: T   N   M   P — 4   3   2   □

Shape

R — Round
S — Square
T — Triangle
L — Rectangle
V — Diamond 35°
D — Diamond 55°
C — Diamond 80°
M — Diamond 86°
P — Pentagon
B — Parallelogram 82°
A — Parallelogram 85°
E — Parallelogram 55°
F — Parallelogram 70°
H — Hexagon
O — Octagon

Relief Angle

N — 0°
A — 3°
B — 5°
C — 7°
P — 10°
D — 15°
E — 20°
F — 25°
G — 30°

Tolerances

Insert	I.C.	Thickness
A	± .0002	± .001
B	± .0002	± .005
C	± .0005	± .001
D	± .0005	± .005
E	± .001	± .001
G	± .001	± .005
*M	± .002 ± .004	± .005
*U	± .005 ± .012	± .005

R = Blank with grind stock on all surfaces.

**Type

A — With hole
B — With hole and one countersink
C — With hole and two countersinks
D — Smaller than $\frac{1}{4}$" I.C. with hole
E — Smaller than $\frac{1}{4}$" I.C. without hole
F — Clamp-on type with chipbreaker
G — With hole and chipbreaker
H — With hole, one countersink and chipbreaker
J — With hole, two countersinks and chipbreaker
P — 10° Positive surface contour with hole and chipbreaker
S — 20° Positive surface countour with hole and chipbreaker

Size

Number of $\frac{1}{32}$nds on inserts less than $\frac{1}{4}$ in. I.C.

Number of $\frac{1}{8}$ths on inserts $\frac{1}{4}$ in. I.C. and over.

Rectangle and Parallelogram Inserts require two digits:
1st digit—number of $\frac{1}{8}$ths in width
2nd digit—number of $\frac{1}{4}$ths in length

Thickness

Number of $\frac{1}{32}$nds on inserts less than $\frac{1}{4}$" I.C.

Number of $\frac{1}{16}$ths on inserts $\frac{1}{4}$" I.C. and over.

Use width dimension in place of I.C. on Rectangle and Parallelogram inserts.

Cutting Point Radius, Flats

0 — Sharp corner
1 — $\frac{1}{64}$ radius
2 — $\frac{1}{32}$ radius
3 — $\frac{3}{64}$ radius
4 — $\frac{1}{16}$ radius
6 — $\frac{3}{32}$ radius
8 — $\frac{1}{8}$ radius

A — Square insert with 45° chamfer
B — Square insert with 45° chamfer 4° sweep angle, R.H. or Neg.
C — Square insert with 45° chamfer and 4° sweep angle, L.H.
D — Square insert with 30° chamfer, R.H. or Neg.
E — Square insert with 15° chamfer, R. H. or Neg.
F — Square insert with 5° chamfer, R.H. or Neg.
G — Square insert with 30° chamfer, L.H.
H — Square insert with 15° chamfer, L.H.
K — Square insert with 30° double chamfer
L — Square insert with 15° double chamfer
M — Square insert with 5° double chamfer
N — Truncated triangle insert
P — Flatted corner triangle, R.H. or Neg.
R — Flatted corner triangle, L.H.
S — Square negative insert with 10° double chamfer.
T — Square negative insert with 30° positive rake chamfer.
V — Octagon negative insert with $22\frac{1}{2}°$ corner chamfer.

Eighth position (□)

An eighth position is sometimes used in the numbering system to indicate cutting edge condition. However, cutting edge conditions such as type of hone can vary with each metal removal operation. It is therefore suggested that the size hone, or other cutting edge modification, be specified when required. (Except where noted, all KC75 and Kenloc N/P inserts are furnished hones with .002-.003 radius on all cutting edges.)

---

*Exact tolerance is determined by size of insert   **Shall be used only when required

Source. *Tool Application Handbook.* (Data courtesy of Kennametal, Inc., Latrobe, Pa., 1973.)

inch) and sintered at temperatures of approximately 3000°F (1649°C). This process partly accounts for the high density and hardness of cemented oxide tools.

Cemented oxides are brittle and require that machines and setups be rigid and free of vibration. Some machines cannot obtain the spindle speeds required to use cemented oxides at their peak capacity in terms of FPM. When practicable, however, cemented oxides can be used to machine relatively hard materials at high speeds.

Ceramic tools should be used as a replacement for carbide tools that are wearing rapidly, but not to replace carbide tools that are breaking.

DIAMOND TOOLS

Industrial diamonds are sometimes used to machine extremely hard workpieces. Only relatively small removal rates are possible with diamond tools (Figure 20a), but very high speeds are used and good finishes are obtained (Figure 20b). Nonferrous metals are turned at 2000 to 2500 FPM, for example. Sintered diamond tools (Figure 21), available in shapes similar to those of ceramic tools, are used for materials (Figure 22) that are abrasive and difficult to machine.

Each of the cutting tool materials varies in hardness. The differences in hardness tend to become more pronounced at high temperatures, as can be seen in Table 5. Hardness is related to the wear resistance of a tool and its ability to machine materials that are softer than it is. Temperature change is important since the increase of temperature during machining results in the softening of the tool material.

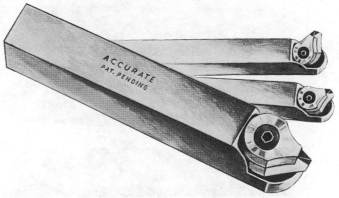

Figure 20a. Diamond tools are mounted in round or square shank holders as replaceable inserts (Courtesy of Accurate Diamond Tool Corp.). (Universal Turning and Boring Tool is a patented item. Patent is owned by: Accurate Diamond Tool Corp.)

Figure 20b. Turning at 725 s.f.m. .010 inch stock is removed on each of two passes at a rate of 5½ inches per minute (.0023 i.p.r.) along the 29 inch length of the casting. A coolant is not required for this turning (Courtesy of Accurate Diamond Tool Company).

Table 5
Hardness of Cutting Materials at High and Low Temperatures

Tool Material	Hardness at Room Temperature	Hardness at 1400°F (760°C)
High speed steel	RA 85	RA 60
Cast alloy	RA 81	RA 70
Carbide	RA 92	RA 82
Cemented oxide (ceramic)	RA 93-94	RA 84
Diamond	Hardest known substance	

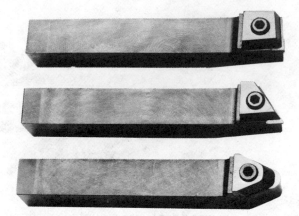

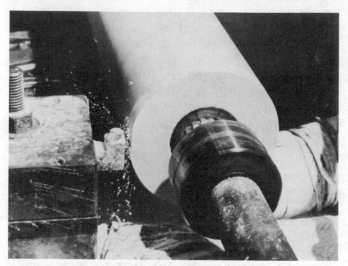

Figure 21. Sintered diamond inserts are clamped in toolholders (Courtesy of Megadiamond Industries).

Since heavy forces and high speeds are involved when using carbides, safety considerations are essential. Chip forms such as the ideal 9 or C-shape are convenient to handle, but shields, chip deflectors, or guards are required to direct chips away from workers in the shop.

Make sure the setup is secure, that centers are large enough to support the work, and that work cannot slip in the chuck. If a long, slender work is machined at high speeds and there is a possibility of it being thrown out, guards should be provided.

Figure 22. To make abrasive fused silica tubes absolutely round where they fit into tundish valves, Flo-Con turns the tubes with a sintered diamond tool insert. After limited experience, the Megadiamond sintered diamond inserts appear to give about 200 times as much wear as carbides in the same application (Courtesy of Megadiamond Industries).

self-evaluation

SELF-TEST
1. List the major materials used in "straight" cemented carbides.
2. What effect does increasing the cobalt content have on cemented carbides?
3. How can you identify normal wear on a carbide tool?
4. Is chip thickness the same as feed on a Style A tool?
5. Is the cutting edge engagement length the same as depth of cut on Style B tools?
6. What effect does the addition of titanium carbides have on tool performance?
7. What effect does the addition of tantalum carbides have on tool performance?
8. What change in tool geometry can make possible an increased rate of feed with equal surface finish quality in turning operations?
9. How does increasing the nose radius affect tool strength?
10. What is the hazard in using too large a nose radius?
11. In respect to carbide turning tools, how is clearance different from relief?

12. What are ceramic tools made of?

13. When should ceramic tools be used?

14. According to the CCPA chart, what designation of carbide would be used for finish-turning aluminum?

15. According to the CCPA chart, what designation of carbide is used for roughing cuts on steel?

16. According to the Grade Classification chart, what Carboloy designation of carbide would be used for rough cuts in cast iron?

17. According to the Grade Classification chart, what number on the CCPA table would a Kennametal K5H have?

18. Extremely hard or abrasive materials are machined with diamond tools. Is the material removal rate very high? What kind of finishes are produced?

19. Polycrystalline diamond tools are similar in some ways to ceramic inserts. What is their major advantage?

20. When brazed carbide tools are sharpened, should you use an aluminum oxide wheel or a silicon carbide wheel?

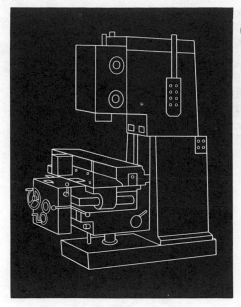

section i
vertical milling machines

The vertical milling machine is a relatively new development in comparison to the horizontal milling machine. The first vertical milling machines appeared in the 1860s (Figure 1). The vertical milling machine was a development more closely related to the drill press than to the horizontal spindle milling machine. The basic difference between drill presses and the earliest vertical milling machines was that the entire spindle assembly, pulleys and all, was moved vertically. This arrangement meant that the bearings that supported the cutting tool were always reasonably close to the tool, which permitted side thrust to be taken more readily. On the other hand, a drill press with a quill that extends some distance from the support bearing is not suitable for the side loading usually encountered in vertical milling.

The next significant step came in the mid-1880s (Figure 2) with the adaptation of the **knee and column** from the horizontal milling machine, which allowed the milling table to be raised and lowered in relation to the spindle. Also, the spindle heads on some of these machines could be tilted to an angular relationship to the table.

Just after the turn of the twentieth century, vertical milling machines began to appear with power feeds on the spindle, housed

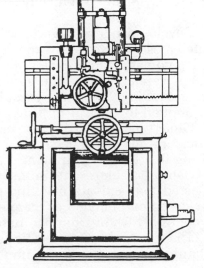 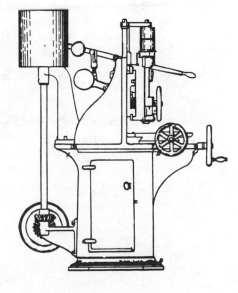

Figure 1. Vertical milling machines of 1862 (U.S. Patent) (Courtesy of The MIT Press, Massachusetts Institute of Technology).

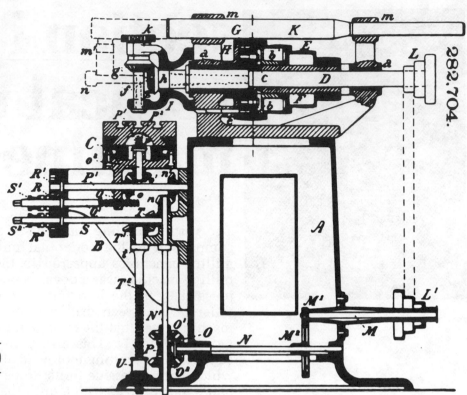

Figure 2. Cosgrove's vertical milling machine of 1883 (U.S. Patent) (Courtesy of The MIT Press, Massachusetts Institute of Technology).

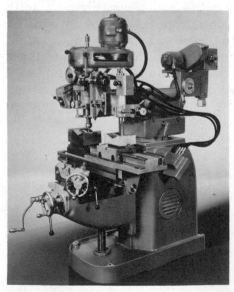

Figure 3. Hydraulic tracing controls on a vertical milling machine especially developed for die sinking (Courtesy of Cincinnati Milacron).

in a heavy duty quill. By 1906 the structural development of the vertical milling machine was essentially complete. There were various types of toolroom and production vertical milling machines, with either fixed or swiveling heads, having either fixed position or knee and column tables and either hand or power feed for longitudinal and lateral table movement. Some had vertical power feed as well.

In the first decade of the twentieth century, micrometers and vernier scales had been applied to vertical milling machines to make them suitable for precise hole locating, known as jig boring. The LeBlond Company produced a machine especially for that purpose in 1910.

Developments relative to vertical milling machines after 1910 related mostly to drive and control issues. Machines began to appear with automatic table cycles and, by 1920, electrical servomechanisms were applied to the vertical milling machine for operations like die sinking. By 1927, hydraulic tracing controls were developed and applied to vertical milling machines (Figure 3).

Other types of sensing for machine control have also been used. Optical means have been used to activate servomechanisms for milling to a line on a drawing, called line tracing (Figure 4).

Control systems, not limited to vertical milling machines, have appeared that activate machine control movements from information stored on punched or magnetic tape (Figure 5), called numerical control (NC), or from computer numerical control (CNC).

The manually operated vertical milling machine (Figure 6) is certainly one of the most important basic machine tools. It is

Figure 4. Line tracing vertical milling machine (Photo courtesy, Bridgeport Milling Machines, a Division of Textron, Inc.)

Figure 5. *(right)* Numerically controlled vertical milling machine (Courtesy of Cincinnati Milacron).

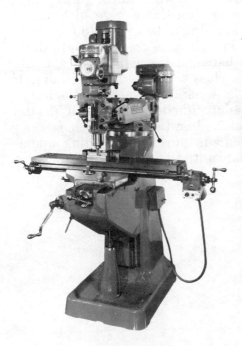

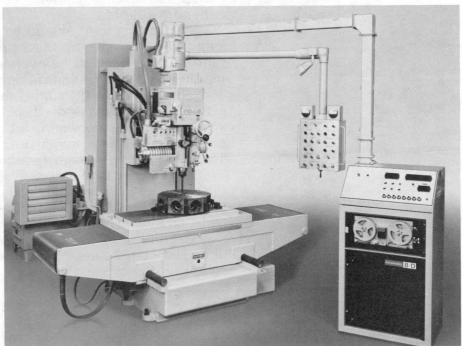

Figure 6. *(left)* A popular type of manually operated vertical milling machine with accessory slotting attachment. This is called a ram-type turret mill. (Photo courtesy, Bridgeport Milling Machines, a Division of Textron, Inc.).

convenient to use for various operations such as drilling, boring, and slotting as well as milling.

In some respects, it is even more versatile than the lathe, because in its various forms and with adaptations, it comes very close to being a machine tool that can reproduce itself.

In this section you will be called on to identify safe working practices on the vertical milling machine and to be able to name components and functions. You will be shown a variety of machine setups and operations. The cutting tools of the vertical milling machine come in a remarkable variety of forms and adaptations for specialized work. You will be shown a variety of ways to hold and locate workpieces relative to the cutting tools and how to determine reasonable speeds and feeds rates for vertical milling. Finally, you will apply the offset boring head for accurately sizing nonstandard holes.

The vertical milling machine is an interesting and challenging machine tool, capable of a large variety of work. It is important for you to be able to use this machine competently.

VERTICAL MILLING MACHINE SAFETY   Be careful when handling tools and sharp edged workpieces to avoid getting cut. Use a rag to protect your hand. Workpieces should be rigidly supported and tightly clamped to withstand the usually high cutting forces encountered in machining. When a workpiece comes loose while machining, it is usually ruined and, often, so is the cutter. The operator can also be hurt by flying particles from the cutter or workpiece.

The cutting tools should be securely fastened in the machine spindle to prevent any movement during the cutting operation. Cutting tools need to be operated at the correct revolutions per minute (RPM) and feedrate for any given material. Excessive speeds and feeds can break the cutting tools. On vertical milling machines, care has to be exercised when swivelling the workhead to make angular cuts. After loosening the clamping bolts that hold the workhead to the overarm, retighten them lightly to create a slight drag. There should be enough friction between the workhead and the overarm that the head only swivels when pressure is applied to it. If the clamping bolts are completely loosened, the weight of the heavy spindle motor will flip the workhead upside down or until it hits the table, possibly smashing the operator's hand or a workpiece.

Measurements are frequently made during machining operations. Do not make any measurements until the spindle has come to a complete stop or standstill.

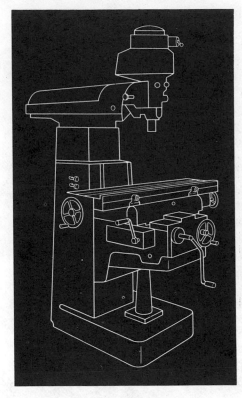

unit 1
the vertical spindle milling machine

Milling machine parts and components have names to make their identification easier. Knowing these names is helpful in locating trouble spots or in operating the machine controls.

objectives After completing this unit, you will be able to:

1. Identify the important parts of a vertical milling machine.
2. Perform routine maintenance.

The vertical milling machine is a very versatile tool in a machining operation. Figure 1 identifies many of its important parts. The *column* is the backbone of the machine; it rests on the base. The front or face of the column is accurately machined to provide a guide for the vertical travel of the knee. The upper part of the *column* is machined to provide a swivel capability to the ram. The *knee* supports and guides the saddle. The *saddle* provides cross travel for the machine and is the support and guide for the table. The *table* gives longitudinal travel to the machine and it holds the workpiece or workholding devices. The ram can be adjusted toward or away from the column to increase the working capacity of the milling machine. The *toolhead* is attached to the end of the ram. The toolhead can be swiveled on some milling machines in one or two planes. The six listed assemblies are the major components of the vertical milling machine. Most of these components have con-

trols or parts that are important to know. The toolhead (Figure 2) contains the motor, which powers the spindle. Speed changes are made with V-belts, gears, or variable speed drives. When changing speeds into the high or low speed range, the spindle has to be stopped. The same is true for V-belt or gear-driven speed changes. On variable drives the spindle has to be revolving while speed changes are made. The spindle is contained in a quill. The quill can be extended and retracted into the toolhead by a quill feed hand lever or hand wheel. The quill feed hand lever is used to position rapidly the quill or to drill holes. The quill feed hand wheels give a controlled slow manual feed as needed when boring holes.

Power feed to the quill is obtained by engaging the feed control lever. Different power feeds are available through the power feed change lever. The power feed is automatically disengaged when the quill dog contacts the adjustable

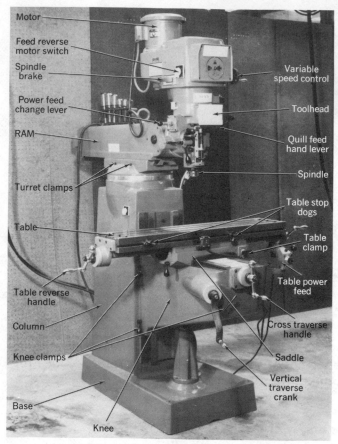

Figure 1. The important parts of a vertical milling machine (Lane Community College).

Each of these three axes of travel can be adjusted in .001 in. increments with micrometer dials. The table, saddle, and knee can be locked securely in position with the clamping levers shown in Figure 4. During machining, all axes except the moving one should be locked. This will increase the rigidity of the setup. Do not use these clamping devices to compensate for wear on the machine slides. If the machine slides become loose, make adjustments with the gib adjustment screws (Figure 5). Turning this screw in will tighten a tapered gib. Make a partial turn on the screw, then try moving the unit with the hand wheel. Repeat this operation until a free but not loose movement is obtained. Too tight an adjustment squeezes the lubricant from the slides, resulting in rapid wear.

Before operating any of the machine controls, the machine should be lubricated. The lubrication follows the cleaning of the machine. Do not use air to blow off the machine; use a rag. Lubrication is needed on all moving parts. Follow the machine manufacturer's recommendation as to the kind of lubricant required.

micrometer depth stop (Figure 3). When feeding upward, the power feed disengages when the quill reaches its upper limit. The micrometer dial allows depth stop adjustments in .001 in. increments. The quill clamp is used to lock the quill in the head to get maximum rigidity when milling. The spindle brake or spindle lock is needed to keep the spindle from rotating when installing or removing tools from it. The toolhead is swiveled on the ram by loosening the clamping nuts on the toolhead and then turning the swivel adjustment until the desired angle is obtained.

The ram is adjusted toward or away from the column by the ram positioning pinion. The ram also swivels on the column after the turret clamps are loosened.

The table is moved manually with the table traverse hand wheel. Table movement toward or away from the column is accomplished with the cross traverse hand wheel. Raising and lowering the knee is done with the vertical traverse crank.

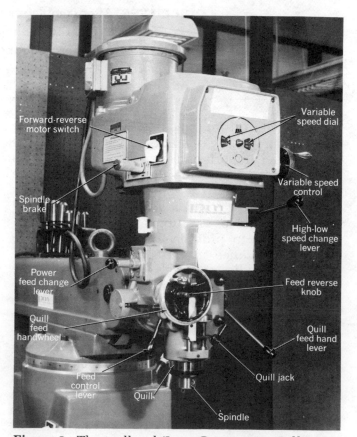

Figure 2. The toolhead (Lane Community College).

Figure 3. Quill stop (Lane Community College).

Figure 4. Clamping devices (Lane Community College).

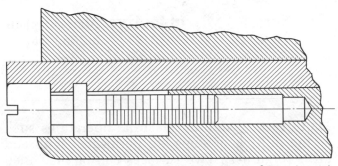

Figure 5. Gib adjusting screw (Courtesy of Cincinnati Milacron).

self-evaluation

SELF-TEST
1. Name the six major components of a vertical milling machine.
2. Which parts are used to move the table longitudinally?
3. Which parts are used to move the saddle?
4. What moves the quill manually?
5. What is the purpose of the table clamp?
6. What is the purpose of the spindle brake?
7. What is important when changing the spindle speed range from high to low?
8. Why is the toolhead fastened to a ram?
9. How is a loose table movement adjusted?
10. What is the purpose of the quill clamp?

unit 2
vertical milling machine operations

The vertical milling machine is one of the most versatile machines found in a machine shop. This unit will illustrate some of the many possible vertical milling machine operations.

objective After completing this unit, you will be able to identify and select vertical milling machine setups for a variety of different machining operations.

Many vertical milling machine operations such as the milling of steps are performed with end mills (Figure 1). Two surfaces can be machined at one time, both square to each other. The ends of workpieces can be machined square and to a given length by using the peripheral teeth of an end mill (Figure 2). Center cutting end mills make their own starting hole when used to mill a pocket or cavity (Figure 3).

Prior to making any milling cuts, the outline of the cavity should be accurately laid out on the workpiece for a guide or reference line. Only when finish cuts are made should these layout lines disappear.

Good milling practice is to rough out the cavity to within .030 in. of finish size before making any finishing cuts.

When you are milling a cavity, the direction of the feed should be against the rotation of the cutter (Figure 4). This assures positive control over the distance that the cutter travels and prevents the workpiece from being pulled into the cutter because of backlash. When you reverse the direction of table travel, you will have to compensate for the backlash in the table feed mechanism. During any milling operation, all table movements should be locked except the one that is moving to obtain the most rigid setup

possible. Spiral fluted end mills may work their way out of a split collet when deep heavy cuts are made or when the end mill gets dull. As a precaution, to warn you that this is happening, you can make a mark with a felt tip pen on the revolving end mill shank where it meets the collet face. Observing this mark during the cut will give you an early indication if the end mill is changing its position in the collet.

One operation often performed with end mills is the cutting of keyways in shafts. It is very important that such a keyway be centrally located in the shaft. A very accurate method of doing this is by positioning the cutter with the help of the machine dials. After clamping the shaft in a vise, or possibly to the table, the quill is lowered so that the cutter is along side the shaft but not touching it. Then, with the spindle motor off and the spindle rotated by hand, the table is moved until a paper feeler strip is pulled from your hand (Figure 5). At this point, the cutter is approximately .002 in. from the shaft. Zero the cross slide micrometer dial compensating for the .002-in. Raise the quill so that the cutter clears the workpiece. Move the cross slide a distance equal to half the shaft diameter plus half the cutter diameter. This will locate the cutter centrally over the shaft. Lock the cross slide table

Figure 1. Using an end mill to mill steps (Lane Community College).

Figure 2. Using an end mill to square stock (Lane Community College).

Figure 3. Using an end mill to machine a pocket (Lane Community College).

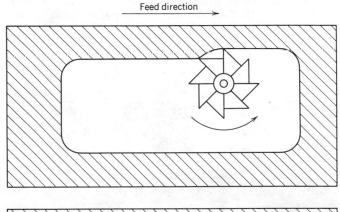

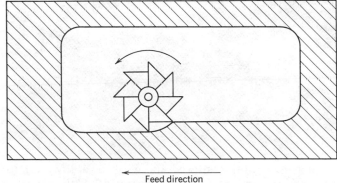

Figure 4. Feed direction is against cutter rotation.

movement into place. Raise the quill to its top position and lock it there. Move the table longitudinally to position the cutter where the keyway is to begin. Start the spindle motor and raise the knee until the cutter makes a circular mark equal to the cutter diameter (Figure 6). Zero the vertical travel micrometer dial. Move up a distance equal to half the cutter diameter plus an additional .005-in. and lock the knee. Cut the keyway the required length.

Figure 5. Setting an end mill to the side of a shaft with the aid of a paper feeler (Lane Community College).

Figure 6. Cutter centered on work and lowered to make a circular mark (Lane Community College).

Figure 7. Milling a slot and then the dovetail (Lane Community College).

Figure 8. First a slot is milled and then the T-slot cutter makes the T-slot (Lane Community College).

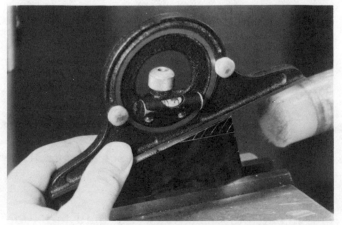

Figure 9. Setting up a workpiece for an angular cut with a protractor (Lane Community College).

To machine a T-slot or a dovetail into a workpiece, two operations are performed. First a slot is cut with a regular end mill and then a T-slot cutter or a single angle milling cutter is used to finish the contour (Figures 7 and 8). Angular cuts on workpieces can be made by tilting the workpiece in a vise with the aid of a protractor (Figure 9) and its built-in spirit level.

Figure 10. Machining an angle with an end mill (Lane Community College).

Figure 12. Cutting an angle by tilting the workhead and using the end teeth of an end mill (Lane Community College).

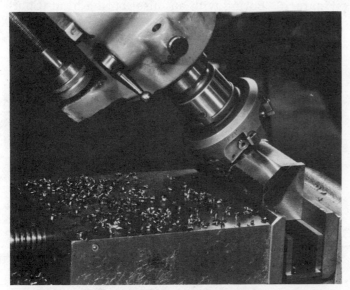

Figure 11. Using a shell mill to machine an angle (Lane Community College).

Figure 13. Cutting an angle by tilting the workhead and using the peripheral teeth of an end mill (Lane Community College).

Machining the angle can be performed with an end mill (Figure 10) or with a shell mill (Figure 11). Another possibility for machining angles is the tilting of the workhead (Figures 12 and 13).

Accurate holes can be drilled at any angle that the head can be swiveled to. These holes can be drilled by using the sensitive quill feed lever or the power feed mechanism (Figure 14) or, in the case of vertical holes, the knee can be raised.

Holes can be machine tapped by using the sensitive quill feed lever and the instant spindle reversal knob (Figure 15). When an offset boring head is mounted in the spindle, precisely located and accurately dimensioned holes can be bored (Figure 16). Circular slots can be milled when a rotary table is used (Figure 17). Precise indexing can be performed when a dividing head is mounted on the milling machine. Figure 18 shows the milling of a square on the end of a

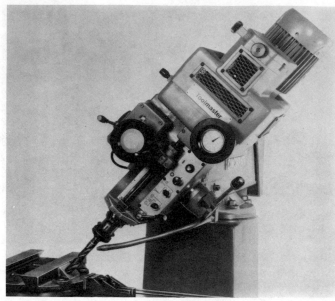

Figure 14. Drilling of accurately located holes (Courtesy of Cincinnati Milacron).

Figure 16. Boring with an offset boring head (Courtesy of Cincinnati Milacron).

Figure 15. Tapping in a vertical milling machine (Courtesy of Cincinnati Milacron).

Figure 17. (right) Using a rotary table to mill a circular slot (Courtesy of Cincinnati Milacron).

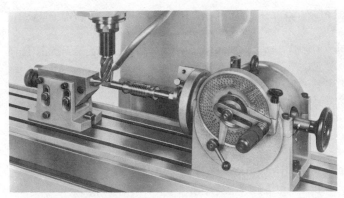

Figure 18. A dividing head in use (Courtesy of Cincinnati Milacron).

Figure 19. A shaping head used to cut a square corner hole (Lane Community College).

Figure 20. Using a right angle milling attachment (Courtesy of Cincinnati Milacron).

shaft. On many vertical milling machines a shaping attachment is mounted on the rear of the ram. This shaping attachment can be brought over the machine table by swiveling the ram 180 degrees. Shaping attachments are used to machine irregular shapes on or in workpieces such as the square corner hole shown in Figure 19. When a right angle milling attachment (Figure 20) is mounted on the spindle, it is possible to machine hard to get at cavities at often difficult angles on workpieces.

self-evaluation

SELF-TEST   1. How is an end mill centered over a shaft prior to cutting a keyway?
2. Why should the feed direction be against the cutter rotation when milling a cavity?

3. What can cause an end mill to work itself out of a collet while cutting?

4. Describe two methods of cutting angular surfaces in a vertical milling machine.

5. How are circular slots milled?

6. What milling machine attachment is used to mill a precise square or hexagon on a shaft?

7. How can a square hole in a workpiece be machined on the vertical milling machine?

8. When is a right angle milling attachment used?

9. Why should workpieces be laid out before machining starts?

10. Why are two operations necessary to mill a T-slot?

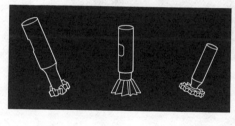

unit 3
cutting tools for the vertical milling machine

Vertical milling machines are very versatile tools. To utilize this versatility, a number of different cutting tools are available. To make an intelligent choice, a machinist needs to know the characteristics and limitations of different cutting tools.

objective After completing this unit, you will be able to identify and select from commonly used vertical milling machine cutting tools.

The most frequently used tool on a vertical milling machine is the end mill. Figure 1 is an illustration of the cutting end of a four flute end mill. End mills are made as right-hand cut or left-hand cut. Identification is made by viewing the cutter from the cutting end. A right-hand cutter rotates counterclockwise. The helix of the flutes can also be left or right hand; a right-hand helix turns to the right. Figure 1 shows a right-hand cut, right-hand helix end mill.

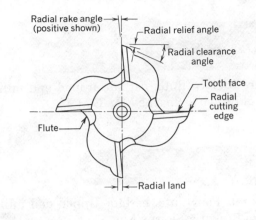

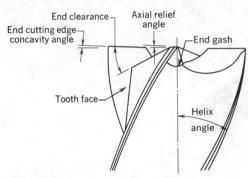

Figure 1. End mill nomenclature (Copyright © National Twist Drill & Tool Div., Lear Siegler, Inc.).

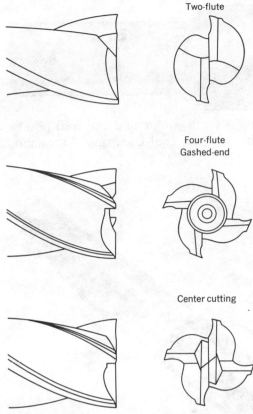

Figure 2. Types of end teeth on end mills (Copyright © National Twist Drill & Tool Div., Lear Siegler, Inc.).

The end teeth of an end mill can be different, depending on the cutting that is to be performed (Figure 2). Two flute end mills are center cutting, which means they can make their own starting hole. This is called plunge cutting. Four flute end mills may have center cutting teeth or a gashed or center drilled end. End mills with center drilled or gashed ends cannot be used to plunge cut their own starting holes. These end mills only cut with the teeth on their periphery. End mills can be single end (Figure 3) or double end (Figure 4). Double end-type end mills are usually more economical because of the savings in tool material in their production.

End mills are manufactured with two, three, four or more flutes and with straight flutes (Figure 5), slow, regular, and fast helix angles. A slow helix is approximately 12 degrees, a regular helix is 30 degrees, and the fast helix is 40 degrees or more when measured from the cutter axis. Most general purpose cutting is done with a regular helix angle cutter (Figure 6).

Aluminum is efficiently machined with a fast helix end mill with highly polished cutting faces to minimize chip adherence (Figure 7). The

Figure 3. Single end helical teeth end mill (Photo courtesy of The Weldon Tool Company, Cleveland, Ohio).

Figure 4. Two flute, double end, helical teeth end mill (Photo courtesy of The Weldon Tool Company, Cleveland, Ohio).

end mills illustrated so far are made from high speed steel. High speed steel end mills are available in great variety of styles, shapes, and sizes as stock items. High speed steel end mills are relatively low in cost when compared with carbide

Figure 5. Straight tooth, single end end mill (Photo courtesy of The Weldon Tool Company, Cleveland, Ohio).

Figure 6. Four flute, double end end mill.

Figure 7. Forty-five degree helix angle aluminum end mill (Photo courtesy of The Weldon Tool Company, Cleveland, Ohio).

tipped or solid carbide end mills. But to machine highly abrasive or hard and tough materials or in production milling, carbide tools should be considered.

Carbide tools may have carbide cutting tips brazed to a steel shank (Figure 8). This two flute carbide tipped end mill is designed to cut steel. It has a negative axial rake angle and a slow left-hand helix. Figure 9 is a carbide tipped four flute end mill designed to machine nonferrous materials such as brass, cast iron, and aluminum. This end mill has a positive radial rake with a right-hand helix. Another kind of carbide tipped end

Figure 8. Two flute, carbide tipped end mill (Courtesy of Brown & Sharpe Mfg. Co.).

Figure 9. Four flute, carbide tipped end mill (Courtesy of Brown & Sharpe Mfg. Co.).

Figure 10. Inserted blade end mill.

mill uses throwaway inserts (Figure 10). Each of the carbide inserts has three cutting edges; when all three cutting edges are dull, the insert is replaced with a new one. No sharpening is required.

Different kinds of carbide grades are available to provide the correct tool material for different work materials. When large amounts of material need to be removed, a roughing end mill (Figure 11) should be used. These end mills are also called hogging end mills and have a wavy thread form cut on their periphery. These waves form many individual cutting edges. The tip of each wave contacts the work and produces one short compact chip. Each succeeding wave tip is offset from the next one, which results in a relatively smooth surface finish. During the cutting operation, a number of teeth are in contact with the work. This reduces the possibility of vibration or chatter.

Tapered end mills (Figure 12) are used in mold making, die work, and pattern making, where precise tapered surfaces need to be made. Tapered end mills have included tapers ranging from 1 degree to over 10 degrees.

Ball-end end mills (Figure 13) have two or more flutes and form an inside radius or fillet between surfaces. Ball-end end mills are used in tracer milling and in die sinking operations. Round bottom grooves can also be machined with them. Precise convex radii can be machined on a milling machine with corner rounding end mills (Figure 14). Dovetails are machined with single angle milling cutters (Figure 15). The two commonly available angles are 45 degrees and 60 degrees. T-slots in machine tables and workholding devices are machined with T-slot cutters (Figure 16). T-slot cutters are made in sizes to fit standard T-nuts.

Figure 14. Corner rounding milling cutter (Copyright © Illinois Tool Works Inc., 1976).

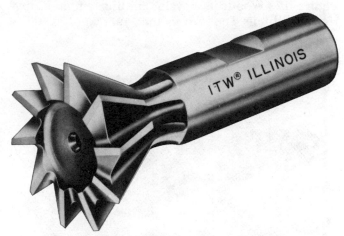

Figure 15. Single angle milling cutter (Copyright © Illinois Tool Works Inc., 1976).

Figure 16. T-slot milling cutter (Photo courtesy of The Weldon Tool Company, Cleveland, Ohio).

Figure 11. Roughing mill (Copyright © Illinois Tool Works, Inc., 1976).

Figure 12. Three flute, tapered end mill (Photo courtesy of The Weldon Tool Company, Cleveland, Ohio).

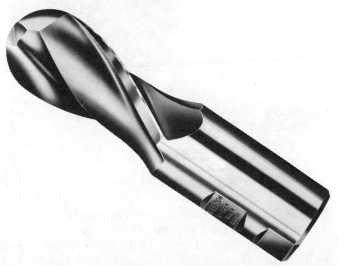

Figure 13. Two flute, single end, ball end mill (Photo courtesy of The Weldon Tool Company, Cleveland, Ohio).

Woodruff key slots are cut into shafts to retain a woodruff key as a driving and connecting member between shafts and pulleys or gears. Woodruff key slot cutters (Figure 17) come in many different standardized sizes. When larger flat surfaces need to be machined, a shell end mill (Figure 18) can be used. Shell end mills are more economical to produce because less of the costly tool material is needed to make one than for a solid shank end mill of the same size. To obtain rapid metal removal, shell end mills are made as a roughing type mill (Figure 19) with a wavy thread forming many cutting edges. Shell mills are also made with carbide inserts (Figure

Figure 17. Woodruff keyslot milling cutter (Copyright © Illinois Tool Works Inc., 1976).

Figure 19. Shell-type roughing mill (Copyright © Illinois Tool Works Inc., 1976).

Figure 18. Shell end mill (Copyright © Illinois Tool Works Inc., 1976).

Figure 20. Shell mill with carbide inserts (Lane Community College).

20). The ease with which new sharp cutting edges can be installed makes this a very practical, efficient cutting tool. The great number of different carbide grades available makes it possible to select a grade suitable for all work materials. An inexpensive face milling cutter is a fly cutter (Figure 21). A fly cutter can be made with a high speed steel or carbide tipped tool bit sharpened to have the correct clearance and rake angles.

All of these cutting tools have to be mounted and driven by the machine spindle. End mills or other straight shank tools can be held in collets. The most rigid type of collet is a solid collet (Figure 22). This collet has been precision

ground with a hole that is concentric and the exact size of the tool shank. Driving power to the tool is transmitted by one or two setscrews engaging in flats on the tool shank.

Another type of frequently used collet is the split collet (Figure 23). The shank of the tool is held by the friction created when the tapered part of the collet is pulled into the taper of the spindle nose. When a heavy side thrust is created

Figure 21. Fly cutter (Lane Community College).

Figure 22. Solid collet (Lane Community College).

Figure 23. Split collet (Lane Community College).

through a deep cut, a large feedrate, or a dull tool, helical flute end mills have a tendency to be pulled out from the collet. With the solid collet, the setscrews prevent any end movement of the cutting tool.

To speed up frequent tool changes, a quick-change tool holder (Figure 24) is used. Different tools can be mounted in their own holder, preset to a specific length, and interchanged with a

Figure 24. Quick-change adapter and tool holders (Lane Community College).

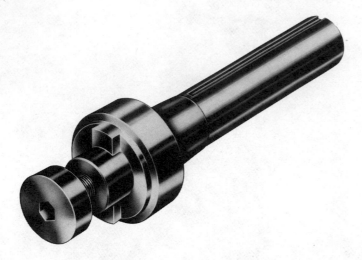

partial turn of a clamping ring. Shell end mills are mounted and driven by shell mill arbors (Figure 25). All of the tools and tool holders will perform satisfactorily only if they are cleaned and inspected for nicks and burrs and corrections are made before mounting them in the spindle.

Figure 25. Shell mill arbor (Photo courtesy of The Weldon Tool Company, Cleveland, Ohio).

self-evaluation

SELF-TEST
1. How is a right-hand cut end mill identified?
2. What is characteristic of end mills that can be used for plunge cutting?
3. What is the main difference between a general purpose end mill and one designed to cut aluminum?
4. When are carbide tipped end mills chosen over high speed steel end mills?
5. To remove a considerable amount of material, what kind of end mill is used?
6. Where are tapered end mills used?
7. Why are tools with carbide inserts used?
8. How are straight shank tools held in the machine spindle?
9. How are shell end mills driven?
10. Why are quick-change tool holders used?

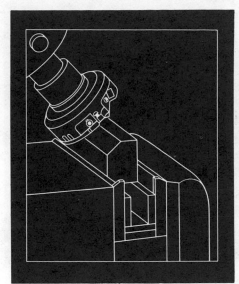

unit 4
setups on the vertical milling machine

Before any cutting takes place a milling machine has to be prepared for the operations to be performed. This preparation involves the alignment of workholding devices and the work head.

objective After completing this unit, you will be able to align vises and toolheads, locate the edges of workpieces, and find the centers of holes.

The two most commonly used workholding methods on the vertical milling machine are the fastening of workpieces to the machine table and the holding of workpieces in a machine vise. When a workpiece is fastened to the machine table, it must be aligned with the axis of the table. Machine tables are accurately machined and the table travels parallel to its outside surfaces and also parallel to its T-slots. Workpieces can be aligned by placing them against stops that fit snugly into the T-slots (Figure 1), or by measuring the distance from the edge of the table to the workpiece in a few places (Figure 2). More accurate alignments can be made when a dial indicator is used to indicate the edge of a workpiece (Figure 3). When a vise is used to hold the workpiece, the solid jaw of the vise should be indicated to assure its alignment with the axis of the table travel. For precise machining operations, the toolhead needs to be aligned squarely to the top surface of the machine table. To align the toolhead, follow the recommended procedure.

1. Fasten a dial indicator in the machine spindle (Figure 4). The dial indicator should sweep a circle slightly smaller than the width of the table.

Figure 1. Work aligned by locating against stops in T-slots (Lane Community College).

2. Lower the quill until the indicator contact point is deflected .015 to .020 in. Lock the quill in this position.

567

Figure 2. Measuring the distance from the edge of the table to the workpiece (Lane Community College).

Figure 4. Aligning the toolhead square to the table with a dial indicator (Lane Community College).

Figure 5. Offset edge finder (Lane Community College).

Figure 3. Aligning a workpiece with the aid of a dial indicator (Lane Community College).

3. Tighten the knee clamping bolts. If this is neglected, the knee will sag on the front.
4. Now set the indicator bezel to read zero.
5. Loosen the head clamping bolts one at a time and retighten each one to create a slight drag. This slight drag makes fine adjustments easier.
6. Rotate the spindle by hand until the indicator is to the left of the spindle in the center of the table, and note the indicator reading

Figure 6. Work approaches the tip of the offset edge finder (Lane Community College).

7. Rotate the spindle 180 degrees so that the indicator is to the right of the spindle, and note the indicator reading at that place. Be careful that the indicator contact does not catch and hang up when crossing the T-slots.

8. Split the difference between the left-hand reading and the right-hand reading in two by turning the head tilting screw.

9. Check and compare the indicator reading at the left side of the table. If both readings are the same, tighten the head clamping bolts. If the readings differ, repeat step 8.

10. After the head clamping bolts are tight, make another comparison on both sides. Often the tightening of the bolts changes the head location and additional adjustments need to be made.

11. Now the head needs to be aligned in relation to the width of the table. The procedure is the same as for the lengthwise alignment. A final check should be made to be sure all clamping bolts are tight.

When the workpiece edges are aligned parallel with the table travel and the toolhead is aligned square with the table top, it becomes necessary to align the spindle axis with the edges of the workpiece. A commonly used tool to locate edges on a milling machine is an offset edge finder (Figure 5). An offset edge finder consists of a shank and a tip that is held against the shank by an internal spring. The shank is usually ½ in. in diameter and the tip is either .200 or .500 in. in diameter. To use an edge finder, it is mounted in a collet. The spindle should revolve at 600 to 800 RPM. The tip should be eccentric to the shank. The workpiece is now moved slowly toward the tip of the edge finder until it just touches (Figure 6). Continue to advance the work very slowly, reducing the run out or eccentricity of the tip. Suddenly the tip will walk off sideways. At this point the spindle axis is exactly one half of the tip diameter away from the edge of the workpiece. If the tip diameter is .200 in., then the centerline of the spindle is .100 in. away from the workpiece edge. Set the micrometer dial of the just adjusted machine axis .100 in. from zero. Repeat the approach to the workpiece a few more times while observing the micrometer dial position until you feel secure in locating an edge with an edge finder. Repeat the edge finding process for the other machine axis. If your machine has any backlash, remember to locate all positions in the same direction as that used when locating with the edge finder.

If an edge finder is not available, an edge can be located with the aid of a dial indicator. The dial indicator is mounted in the spindle. Rotate the spindle by hand and set the indicator contact

Figure 7. Indicator used to locate the edge of a workpiece (Lane Community College).

point as close to the spindle centerline as possible. Lower the spindle so that the indicator contact point touches the workpiece edge and registers a .010 to .020 in. deflection (Figure 7). A slight rotating movement of the spindle forward and backward is used to locate the lowest reading on the dial indicator. Set the dial indicator to register zero. Raise the spindle so that the indicator contact point is ½ in. above the workpiece and turn the spindle 180 degrees from the way it was when the indicator was zeroed. Hold a precision parallel against the edge of the workpiece so it extends above the workpiece. Lower the spindle until the indicator contact point is against the parallel (Figure 8). Read the indicator value; use a mirror to read the indicator when it faces away from you. Now turn the table hand wheel and move the table to where the indicator pointer is halfway between the reading against the parallel and the zero on the indicator dial. Set the dial on zero and check the position of the spindle, as in Figure 7. Again, make a halfway correction until both readings are the same.

To pick up the center of an existing hole, the indicator is mounted in the spindle and swiveled so that the contact point touches the side of the hole (Figure 9). The spindle is rotated and table adjustments are made until the same reading is

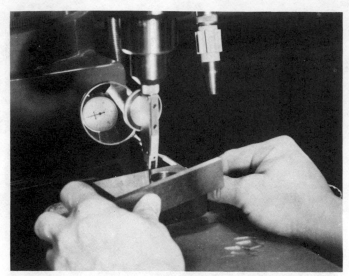

Figure 8. Indicator against parallel to locate the edge of a workpiece (Lane Community College).

obtained in a complete circle. The spindle should be centered first in one table axis and then in the other.

PROCEDURE

Machining Holes in Vise Body

1. Align the workhead square to the machine table.
2. Align and fasten a machine vise on the table so its jaw is parallel to the long axis of the table.
3. Mount the vise body in the machine vise with the bottom surface against the solid jaw of the machine vise.
4. Mount an edge finder in a spindle collet and align the spindle axis with the base surface of the vise body.
5. Move the table the required .452 in. distance and lock the table cross slide.

Figure 9. Dial indicator locating the center of a hole (Lane Community College).

6. Now pick up the outside of the solid jaw of the vise body.
7. Move to the first hole location 1.015 in. from the outside edge.
8. Center drill this hole.
9. Use a $\frac{1}{4}$ in. diameter twist drill and drill this hole $1\frac{1}{2}$ in. deep.
10. Repeat steps 8 and 9 for the remaining eight holes. Accurate positioning is done with the micrometer dials.
11. Remove the workpiece from the machine vise. Turn it over so that the just drilled holes are down and the bottom surface of the vise body is again against the solid jaw.

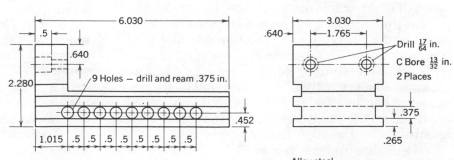

Alloy steel
Drawings for all parts of this precision vise are in the appendix

Figure 10. Machining the holes in the vise body.

Figure 11. Drilling $\frac{11}{32}$ in. diameter holes in the vise body (Lane Community College).

12. Use the edge finder to pick up the two sides, as for the first drilling operation.
13. Position the spindle over the first hole location, again with the first hole on the solid jaw side.
14. Center drill this hole.

Figure 12. Reaming $\frac{3}{8}$ in. diameter holes in the vise body (Lane Community College).

15. Drill this hole with a $\frac{1}{4}$ in. diameter drill deep enough to meet the hole from below.
16. Switch to an $\frac{11}{32}$ in. diameter drill and drill completely through the vise body (Figure 11). The $\frac{1}{4}$ in. hole acts as a pilot hole to let the $\frac{11}{32}$ in. drill come out in the correct place on the bottom side.
17. Change from the $\frac{11}{32}$ in. drill to a $\frac{3}{8}$ in. diameter machine reamer and ream this hole completely through also (Figure 12).
18. Repeat steps 14 to 17 for the remaining eight holes.

Figure 13. Drill and counterbore holes in the solid jaw of the vise body (Lane Community College).

19. Reposition the workpiece so it is upright in the machine vise with the solid jaw of the vise body up.
20. With an edge finder, pick up the edges of the workpiece.

21. Position for the two hole locations and drill the $\frac{17}{64}$ in. diameter holes with their $\frac{13}{32}$ in. diameter counterbores (Figure 13).
22. Remove all burrs.

self-evaluation

SELF-TEST

1. How can workpieces be aligned when they are clamped to the table?
2. How is a vise aligned on a machine table?
3. When is the toolhead alignment checked?
4. Why is it important that the knee clamping bolts are tight before aligning a toolhead?
5. Why does the toolhead alignment need to be checked again after all the clamping bolts are tightened?
6. How can the machine spindle be located exactly over the edge of a workpiece?
7. With a .200 in. edge finder tip, when do you know that the spindle axis is .100 in. away from the edge of the workpiece?
8. What is the recommended RPM to use with an offset edge finder?
9. When locating a number of positions on a workpiece, how can you eliminate the backlash in the machine screws?
10. How is the center of an existing hole located?

unit 5
using end mills

After the selection of a workholding device and cutting tool has been made, a decision on cutting speed, feed, depth of cut, and coolant needs to be made.

objectives After completing this unit, you will be able to:
1. Select cutting speeds and calculate RPM for end mills.
2. Select and calculate feedrates for end mills.
3. Use end mills to machine grooves and cavities.

One factor in efficient cutting with end mills requires the intelligent selection of the correct cutting speed. The cutting speed of a cutting tool is influenced by the tool material, work material, condition of the machine, rigidity of the setup, and use of coolant. Commonly used tool materials are high speed steel and cemented carbide. After the cutting speed has been selected for a job, the cutting operation should be observed carefully so that speed adjustments can be made before a job is ruined. Table 1 gives starting values for some commonly used materials. As a rule, lower speeds are used for hard materials, tough materials, abrasive materials, heavy cuts, minimum tool wear, and maximum tool life. Higher speeds are used to machine softer materials, for better surface finishes, with smaller diameter cutters, for light cuts, for frail workpieces, and delicate setups. When you calculate an RPM to use on a job, use the lower cutting speed value to start. The formula for this is:

$$RPM = \frac{CS \times 4}{D}$$

D is the diameter of the cutting tool in inches.

Cutting fluids should be used when high speed steel cutters are used. The cutting fluid dissipates the heat generated while cutting. It reduces the heat by acting as a lubricant between the tool and chip. Higher cutting speeds can be used with cutting fluids. A stream of coolant also washes the chips away. Water base coolants have very good cooling qualities and oil base coolants produce very good surface finishes. Most milling with carbide cutters is done dry unless a large enough flow of coolant at the cutting edge can be maintained to keep the cutting edge from being intermittently heated and cooled. Intermittent heating and cooling of a carbide tool usually results in thermal cracking and premature tool failure.

Some materials, such as cast iron, brass, and plastics, are commonly machined dry. A stream of compressed air can be used to cool tools and to keep the cutting area clear of chips, but precautions have to be taken to prevent flying chips from injuring anyone.

Another very important factor in efficient machining is the feed. The feed in milling is calculated by starting with a desired feed per

Table 1

Cutting Speeds and Starting Values for Some Commonly Used Materials

Work Material	Tool Material	
	High Speed Steel	Cemented Carbide
Aluminum	300-800	1000-2000
Brass	200-400	500-800
Bronze	65-130	200-400
Cast Iron	50-80	250-350
Low carbon steel	60-100	300-600
Medium carbon steel	50-80	225-400
High carbon steel	40-70	150-250
Medium alloy steel	40-70	150-350
Stainless steel	30-80	100-300

tooth. The feed per tooth determines the chip thickness. The chip thickness affects the tool life of a cutter. Very thin chips dull the cutting edges very rapidly. Commonly used feed per tooth values for end mills are given in Table 2. Usually the feed per tooth is the same for HSS and carbide end mills.

The values in Table 2 are only intended as starting points and may have to be adjusted up or down, depending on the machining conditions of the job at hand. The highest possible feed per tooth will usually give the longest tool life between sharpenings. Excessive feeds will cause tool breakage or the chipping of the cutting edges. When the feed per tooth for a cutter is selected, the feedrate can be calculated. The feedrate on a milling machine is expressed in inches per minute (IPM) and is the product of the feed per tooth (F) times the revolution per minute (RPM) times the number of teeth in the cutter (n). The formula for feedrate is:

$$Feedrate = f \times RPM \times n$$

As an example, to use the values given in Tables 1 and 2, calculate the RPM and feedrate for a $\frac{1}{2}$ in. diameter HSS two flute end mill cutting aluminum.

$$RPM = \frac{CS \times 4}{D} = \frac{300 \times 4}{1/2} = \frac{1200}{.5} = 2400$$

$Feedrate = f \times RPM \times n = .005 \times 2400 \times 2 = 24$ IPM

The third factor to be considered in using end mills is the depth of cut. The depth of cut is limited by the amount of material that needs to be removed from the workpiece, by the power available at the machine spindle, and by the rigidity of the workpiece, tool, and setup. As a rule, the depth of cut for an end mill should not exceed one half of the diameter of the tool. But if deeper cuts need to be made, the feedrate needs to be reduced to prevent tool breakage. The end

Table 2

Feeds for End Mills (Feed per Tooth in Inches)

Cutter Diameter	Aluminum	Brass	Bronze	Cast Iron	Low Carbon Steel	High Carbon Steel	Medium Alloy Steel	Stainless Steel
$\frac{1}{8}$	.002	.001	.0005	.0005	.0005	.0005	.0005	.0005
$\frac{1}{4}$	.002	.002	.001	.001	.001	.001	.0005	.001
$\frac{3}{8}$	.003	.003	.002	.002	.002	.002	.001	.002
$\frac{1}{2}$	.005	.003	.003	.0025	.003	.002	.001	.002
$\frac{3}{4}$	.006	.004	.003	.003	.004	.003	.002	.003
1	.007	.005	.004	.0035	.005	.003	.003	.004
$1\frac{1}{2}$	.008	.005	.005	.004	.006	.004	.003	.004
2	.009	.006	.005	.005	.007	.004	.003	.005

mill must be sharp and should run concentric in the end mill holder. The end mill should be mounted with no more tool overhang than necessary to do the job.

A problem that occasionally arises when using end mills to machine grooves or slots is a slot with nonperpendicular sides. Grooves with leaning sides are caused by worn spindles, excessive tool projection from the spindle, dull end mills, or excessive feedrates. The leaning slot is produced by an end mill that is deflected by high cutting forces (Figure 1). To reduce the tendency of the tool to cut a leaning slot, reduce the feedrate, use end mills with only a short projection from the spindle, and use end mills with straight or low helix angle flues.

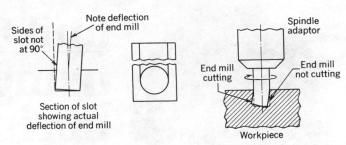

Figure 1. The causes of a leaning slot in end milling.

self-evaluation

SELF-TEST

1. When are the lower cutting speeds recommended?
2. When are the higher cutting speeds used?
3. Should you always use calculated RPM?
4. When are cutting fluids used?
5. When should machining be performed dry?
6. How is the tool life of an end mill affected by the chip thickness of a cut?
7. What is normally considered the maximum depth of cut for an end mill?
8. What are the limitations on the depth of cut?
9. Calculate the RPM for a $\frac{3}{4}$ in. diameter HSS end mill to machine brass.
10. Calculate the feed rate for a two flute $\frac{1}{4}$ in. diameter carbide end mill to machine medium alloy steel.

unit 6
using the offset boring head

The offset boring head is used on a vertical milling machine to make accurately located holes with precisely controlled diameters.

objective After completing this unit, you will be able to use an offset boring head to bore a hole pattern to the specified dimensions and within the tolerances given.

Most holes in a machine shop are made with a drill; when a higher accuracy is required, these holes are reamed. But the reaming and drilling of holes is limited to standard sized holes for which these tools are available. Also, drilled and reamed holes may wander off from a desired location during the machining process. To produce standard or nonstandard holes at specific locations, an offset boring head can be used. An offset boring head can only be used to enlarge an existing hole. The hole should be drilled to approximately $\frac{1}{16}$ in. smaller in diameter than the finished size of the hole, but it is better to leave more material in the hole, especially if the hole is rough, than to come to the finished size and have a hole that is not cleaned up completely.

The hole can be drilled on a drill press or on the vertical milling machine. The workpiece is fastened to the machine table or other workholding device. If the hole is to be bored through the workpiece it should be supported on parallels. The parallels are spaced far enough apart so they will not be interfering with the penetrating boring tool (Figure 1). The next step is to position the workpiece so that the hole to be bored is aligned with the centerline of the machine spindle by moving the machine table. Before mounting the offset boring head in the spindle, clean the spindle hole and the tool shank.

Figure 1. Workpiece supported on parallels. Note the clearance for the penetrating boring tool (Lane Community College).

The offset boring head has two main parts (Figure 2). One is the body, which is fastened in the spindle by its shank. The second part is the adjustable tool slide that holds the boring tool. The tool slide can be precisely moved with the micrometer tool adjustment screw. After tool slide adjustments are made, tightening the locking screw will prevent additional tool movements. A number of holes in the tool slide give different locations where boring tools can be

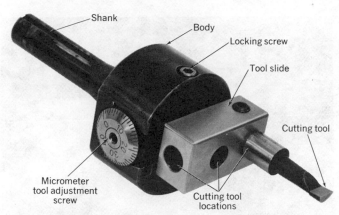

Figure 2. Offset boring head (Lane Community College).

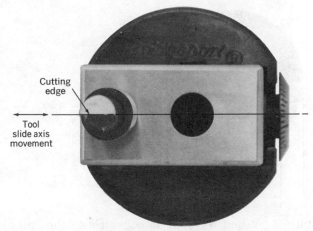

Figure 4. Boring tool cutting edge is on the centerline of the boring head (Lane Community College).

Figure 3. Set of boring tools for the offset boring head (Lane Community College).

clamped, depending on the diameter of the hole to be bored.

When using any boring head, it is very important to determine the amount of tool movement produced when the tool adjustment screw is rotated one graduation. Some boring heads are graduated to where one graduation movement will produce a .002 in. change in diameter. Others give only a .001 in. change in diameter for a single graduation movement. Boring tools are made in many different sizes and lengths (Figure 3). The best boring tool to use is the one with the largest diameter that will fit the hole to be bored and the shortest shank that will do the job.

When a boring tool is mounted in a boring head, it is very important that the cutting edge is on the centerline of the boring head and in line with the axis of the tool slide movement (Figure 4). Only in this position are the rake angles and clearance angles correct as ground on the tool. This is also the only position when the tools cutting edge moves the same distance as the tool slide when adjustments are made. Boring tools are made from high speed steel, carbide tipped or solid carbide. The kind of tool material and the workpiece material determine the cutting speed that should be used. But the rigidity of the machine spindle and the setup often require a lower than calculated RPM because the imbalance of the offset boring head creates heavy machine vibrations.

The quill feed on many vertical milling machines is limited to .0015, .003, and .006 in. of feed per spindle revolution. Roughing cuts should be made at the higher figure and finishing cuts should be made with the two lower values. Roughing cuts are usually made with the tool feeding down into the hole. Finishing cuts are made with the tool feeding down and often the tool is fed back up thru the hole by changing the feed direction at the bottom of the hole. Because of the tool deflection, a light cut will be made without resetting the tool on that second cut. When cuts are made with the tool only feeding down but not out, the spindle rotation is stopped before the tool is withdrawn from the hole. If the spindle rotates while the quill is raised, a helical groove will be cut into the wall of the just completed hole, possibly spoiling it.

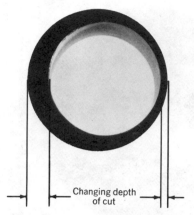

Figure 5. When the hole is eccentric to the spindle centerline, it will cause a variable depth of cut for the boring tool (Lane Community College).

Figure 6. A radius is machined on a workpiece with the offset boring head (Lane Community College).

To obtain a predictable change in hole size for a given tool slide adjustment, certain conditions have to be met. The depth of cut of the boring tool needs to be the same around the circumference of the hole and not like the varying depth of cut illustrated in Figure 5. Roughing cuts should be taken until the hole is round and concentric with the spindle centerline. The depth of cut needs to be equal for successive cuts. As an example, assume that a hole has been rough bored to be concentric with the spindle axis. The tool is now resharpened and fastened in the tool slide. The tool slide is advanced until the tool just touches the wall of the hole. After raising the tool above the work, the tool is moved 20 graduations, or a distance that should increase the hole diameter by .020 in. The spindle is turned on, and with a feed of .003 in. per revolution, the cut is made thru the hole. The spindle is stopped and the tool is withdrawn from the hole. Measuring the hole shows the diameter to have increased by only .015 in. What has happened is that the tool was deflected by the cutting pressure to produce a hole .005 in. smaller than expected. The tool is now advanced to again give an increase of .020 in. in the hole diameter. With the same feed as for the last cut, the hole is bored.

Measuring the hole again shows the hole to be .020 in. larger. Additional cuts made with the same depth of cut and the same feed will give additional .020 in. diameter increases. If the depth of cut is increased, more tool deflection will take place, resulting in a smaller than expected diameter increase. If the depth of cut is decreased, the tool will be deflected less, resulting in a larger than expected diameter.

Figure 7. A boring and facing head (Lane Community College).

When the same depth of cut is maintained but the feed per revolution is increased, higher cutting pressures will produce more tool deflection and a smaller than expected hole diameter.

With an equal depth of cut and a smaller feed per revolution, less cutting pressure will produce a larger than expected hole diameter. Another factor that affects the hole diameter with a given depth of cut is tool wear. As a tool cuts, it becomes dull. A dull tool will produce higher cutting pressures with a resultant larger tool deflection.

An offset boring head can also be used to machine a precise radius on a workpiece (Figure 6). The workpiece is positioned the specified distance from the spindle axis. A scrap piece of metal is clamped to the table opposite the workpiece. As cuts are being made with the offset boring head, the tool cuts on both the workpiece and the scrap piece. The diameter of the cuts is measured between the pieces being machined.

With a boring and facing head (Figure 7), it is possible to machine flat surfaces with the same tool that was used to bore a hole to size. The tool can be moved sideways while the spindle is rotating.

self-evaluation

SELF-TEST

1. When is an offset boring head used?
2. Why is the workpiece normally placed on parallels?
3. Why is the locking screw tightened after tool slide adjustments have been made?
4. Why does the tool slide have a number of holes to hold boring tools?
5. Why is it important to determine the amount of tool movement for each graduation on the adjustment screw?
6. What would be the best boring tool to use on a job?
7. How important is the alignment of the tool's cutting edge with the axis of the tool slide?
8. What factors affect the size of the hold obtained for a given amount of tool adjustment?
9. Name three causes for changes in boring tool deflection.
10. What determines the cutting speed in boring?

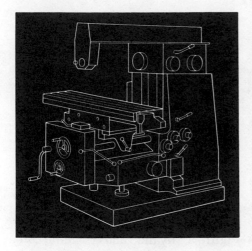

section j
horizontal milling machines

The first horizontal milling machine was probably that of Eli Whitney, invented around 1820 (Figure 1). Whitney used his machine in making parts for firearms. The milling cutter of that period was more like a rotary file than the modern coarse tooth cutter. It is interesting that this very early machine was equipped with a power feed to move the table beneath the cutter. The height of the spindle to the table surface was fixed. The table was fitted with gibs to hold the table down and to make adjustment for wear, similar to recent practices. Another milling machine utilizing a rack and pinion to move the table beneath the cutter was produced in the same area in Connecticut at about the same time by Robert Johnson. A little more than 10 years later, in Britain, James Nasmyth produced a substantial milling machine specialized for milling the flats on machine nuts using an indexing plate.

These first milling machines all had a fixed relationship between the spindle and the table. The next development appeared a few years later with a machine that permitted the cutter spindle to be raised or lowered and clamped in place. About the same time, formed cutters were devised to machine contours into metal surfaces.

By 1850, the horizontal milling machine had become more rugged and precise and had provision for moving the machine table in cross feed. By this time the milling cutter was evolving from a rotary file to a coarse-toothed cutter capable of making substantial chips.

The next major developments in the horizontal milling machine came in 1861 with Joseph Brown's universal milling machine (Figure 2). In a single machine design he included a "knee" to move the table assembly up and down in relationship to the cutter, a spiral indexing head connected to the table feed screw, and a table that could be swiveled so that spiral milling, such as the flutes of twist drills, could be machined easily.

The next 40 years showed refinement of horizontal milling machines toward convenience in control positions and improved cutter support (Figure 3). This was also a period of great cutter development with nearly all of the current forms completed by 1900.

Figure 1. One of the earliest milling machines produced by Eli Whitney (Courtesy of DoAll Company).

581

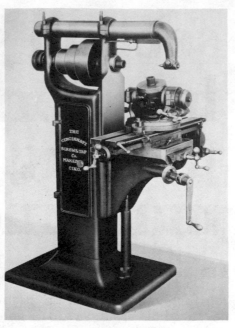

Figure 2. First universal milling machine by Brown & Sharpe in 1861 (Courtesy of the MIT Press, Massachusetts Institute of Technology).

Figure 3. By the 1880s machine controls were arranged for convenience and arbor support was provided. This is Cincinnati's first milling machine (Courtesy of Cincinnati Milacron).

Figure 4. By 1908 the horizontal milling machine was a powerful production tool (Courtesy of Cincinnati Milacron).

Scientific experimentation with cutter designs for metal removal efficiency began in the first decade of the twentieth century. With the advent of individual electric motors, massive high power milling machines began to appear with heavy support for the outboard end of the cutter arbor.

The period of the milling machine as a powerful production machine had begun (Figure 4). Machines had been developed with geared constant speed drives and independently controlled feeding rates. Machine accessories like the universal dividing head also reached a high level of development at the turn of the century. The development of the horizontal milling machine, as we know it today, was essentially complete by the time of World War I.

Figure 5. A large planer-type milling machine with two tables. A setup can be made on one table while machining is taking place on the other (Courtesy of The Ingersoll Milling Machine Company).

Since 1910, with the advent of the mass-produced automobile, highly specialized forms of milling machines have been developed that are characterized by great massiveness, high power, and multiple cutting heads, with sophisticated tooling and automatic table cycles. Hydraulic control on horizontal and vertical milling machines appeared about the same time.

The first milling machines could be termed bed-type machines because the part was moved past the cutter with only one table motion. Later machines were designed so that the spindle assembly could be raised or lowered relative to the table. A bed-type milling machine is one in which the position of the milling spindle, but not the height of the table can be changed. There are many configurations of bed-type milling machines. One type has the general appearance of a knee and column. One type of very large milling machine is known as a planer mill (Figure 5) or adjustable rail milling machine. The example shown is equipped with two separate tables, so that one table can be set up while machining is taking place on the other table. These machines often have 250 hp to their spindles. Other bed-type milling machines that employ two or more cutting heads are called duplex or triplex milling machines (Figure 6). These are commonly used in high production setups.

A particularly common form of bed milling machine is called the manufacturing milling machine (Figure 7). On this machine the spindle assembly is positioned and secured at the correct height and parts are passed, typically in a fixture, under the cutter. These machines are usually equipped with means for automatic cycling of the table; and they often have twin fixtures so that one part can be added while the other part is being machined. This is called reciprocal milling. This design of machine can also be found with hy-

Figure 6. Triplex bed-type milling machine (Courtesy of Cincinnati Milacron).

Figure 7. (Plain manufacturing-type milling machine (Courtesy of Cincinnati Milacron).

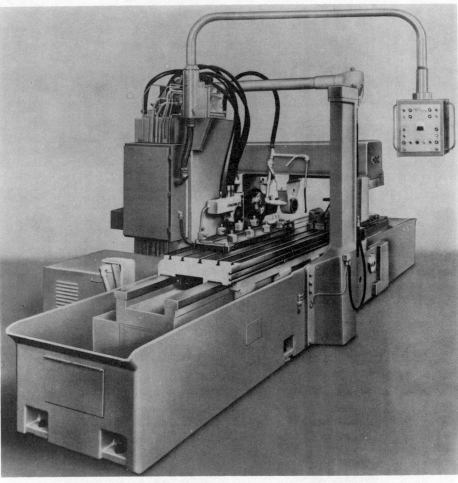

Figure 8. Plain tracer controlled milling machine (Courtesy of Cincinnati Milacron).

draulic tracer controls that move the spindle carrier and cutter vertically in response to a stylus following a cam. These are termed tracer-type manufacturing milling machines (Figure 8).

Another bed-type milling machine also has a table traverse motion, with a spindle assembly that can be moved vertically (Figure 9). This type of machine is found with either horizontal or vertical head configuration and, by general appearance, is often mistaken for a knee and column-type milling machine.

Knee and column milling machines are derived from the heritage of Joseph Brown's universal milling machines in the 1860s. Universal means that the machine table can swivel on its horizontal axis (Figure 10) so that the work can be presented to the cutter at an angle in conjunction with a suitable dividing head permitting helical milling. The plain knee and column milling machine (Figure 11) omits the table swiveling feature in the interest of greater machine rigidity. The more jointed connections, the greater the possibility of uncontrolled movement or "chatter." There is one type of knee and column milling machine used in manufacturing that is capable of vertical table positioning and longitudinal table travel, but it does not have transverse or cross feeding capability. On this type of machine (Figure 12), the spindle bearings are carried in a quill so

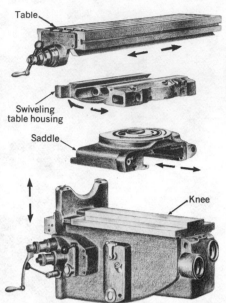

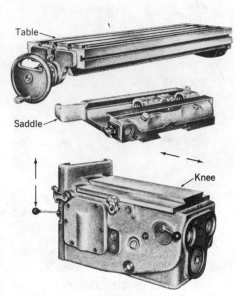

Figure 9. Bed-type horizontal milling machine with transverse table motion. This type mill is also found in a vertical spindle design (Courtesy of Cincinnati Milacron).

Figure 10. The main features of the knee, saddle, and table assembly on the universal knee and column milling machine (Courtesy of Cincinnati Milacron).

Figure 11. The feature of the knee, saddle, and column milling machine. The swiveling table housing is omitted in this design (Courtesy of Cincinnati Milacron).

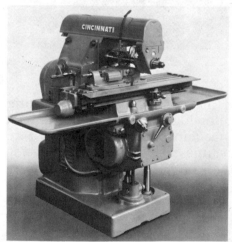

Figure 12. Small, plain, automatic knee and column milling machine (Courtesy of Cincinnati Milacron).

that the cutter can be positioned traversely over the part and locked into position. Eliminating the cross feeding saddle adds to the machine rigidity.

On the bed-type manufacturing milling machines, the table is typically moved by a hydraulic-mechanical means that includes backlash control. This control permits the cut to be made easily in either direction without the danger of having the milling cutter suddenly grab the work and take up the backlash, as can happen with most ordinary nut and screw table feeds. This sudden taking up of backlash results in many broken cutter teeth if it is not controlled by the method of milling or by special devices. One of these devices is called a backlash eliminator (Figure 13) which, when engaged, automatically takes up the backlash by applying a preload between two nuts following the leadscrew. Another method, employed mainly with numerically controlled machine tools, is a ring ball nut, commonly called ball screw, arrangement that, by design, is essentially free of backlash.

Since much of the concern in the development of the milling machine and its cutters has been toward the highest possible machine rigidity by various means such as minimum number of moving components, overarm supports, and devices like backlash eliminators, another technique should be mentioned. Machine castings often vary in their ability to absorb vibration, even with the most careful design of internal webbing. Another means to attack the problem has been the "tuning out" of vibration by special vibration dampening devices. Figure 14 shows the milling machine overarm equipped with a device to reduce the resonance of vibration

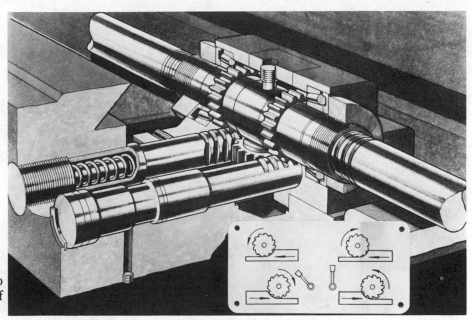

Figure 13. Backlash eliminator to permit climb milling (Courtesy of Cincinnati Milacron).

Figure 14. The overarm can reduce cutting vibration (Courtesy of Cincinnati Milacron).

passing through the casting. This capability permits increased cutting loads before chatter sets in.

A number of attachments are available to increase the capabilities of milling machines, particularly for toolroom applications where a few parts are made or for limited production where the expense of a special machine would not be warranted. The vertical milling attachment (Figure 15) is used on horizontal milling machines to obtain the capability of both vertical and angular machining. The universal milling attachments (Figure 16) permit an additional motion so that spiral milling may be done on a plain table milling machine in addition to vertical and angular cuts.

Another attachment is the independent overhead spindle with an angular swivel head (Figure 17), which is powered separately from the horizontal machine spindle. This device replaces the standard overarm and can be used in conjunction with the horizontal spindle as needed to machine angular relationships without moving the workpiece. When not needed, it can be swiveled out of the way, and the regular overarm brackets can be attached.

Figure 15. A vertical milling attachment with quill feeding capability (Courtesy of Cincinnati Milacron).

Figure 16. Universal milling attachment (Courtesy of Cincinnati Milacron).

Figure 17. Independent overhead spindle with angular head (Courtesy of Cincinnati Milacron).

Figure 18. A slotting attachment for the horizontal milling machine (Courtesy of Cincinnati Milacron).

A slotting attachment (Figure 18) is also available for horizontal milling machines to utilize a single point tool for operations like internal keyway cutting, where a vertical slotter is not available. It may be set at an angle as well as being set vertically.

Devices for tool holding, such as arbors and adapters for horizontal milling, will be studied in this section. Table-mounted attachments such as rotary tables and indexing heads and workholding devices such as clamps and vises will also be studied.

Milling machines can be equipped with accessory measuring equipment, called direct readouts (DRO), to reduce the chance of operator error when machining expensive complex parts (Figure 19). These measuring systems can be switched to present information in either inch or metric form, which is a great time-saver and eliminates the possibility of making errors in conversion between the two systems.

The horizontal milling machine, in combination with its wide array of accessories, is an extremely versatile machine tool with nearly 160 years of development in its various forms. It should be pointed out that since the advent of numerical control in 1953, many of the functions of the horizontal and vertical milling machines are being displaced, especially on the production of highly complex parts with large numbers of interrelated or repeated dimensions. When you are milling a part with more than 100 related hole posi-

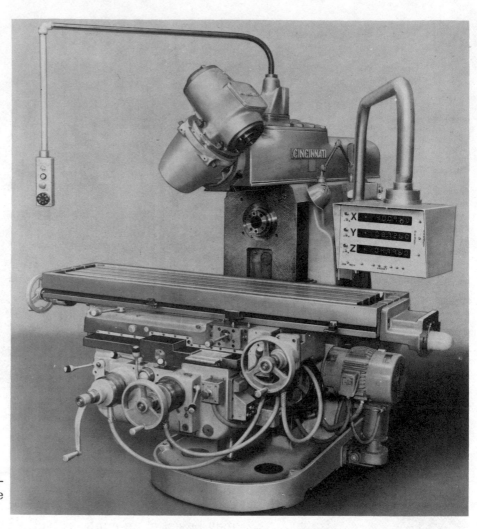

Figure 19. Direct readout (DRO) fitted to a horizontal milling machine (Courtesy of Cincinnati Milacron).

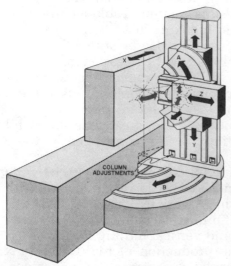

Figure 20. Diagram of five axes of machine motion for complex milling under numerical control (Courtesy of Cincinnati Milacron).

tions and depths, it is very difficult to avoid making at least one mistake. Consequently, much of the work that was formerly done by milling machines and by the accessories shown is now numerically programmed, even to make a single complex part.

It is important for you to learn how to use vertical and horizontal milling machines competently because the cutting and locating principles apply also to the most complex numerically controlled machine tool. Few companies are willing to risk damage to an expensive and complex numerically controlled machine by an operator without a background in conventional milling practice. An operator must be able to determine readily when there is something going wrong with the cutting operation and make appropriate corrections by replacing tools or manually overriding the machine control. Numerically controlled milling centers are sometimes equipped with five programmable machine axes or motions that can permit the spindle to be presented to the work at any angle between the horizontal and vertical besides moving up and down (Figure 20) under the guidance of a programmed tape or by direct computer control.

In this section you will observe specific safety precautions that relate to horizontal milling, identify components and functions of horizontal mills, and perform routine maintenance. Various mounting systems used to drive milling cutters will be studied, and you will be able to match cutters to their respective applications. You will calculate RPM and feedrates for milling cutters and set the values into the machine controls. You will learn about a variety of workholding methods and alignment procedures and how to mill a square workpiece. In addition, you will use side milling cutters in various combinations and you will use face milling cutters to machine flat surfaces.

Horizontal and vertical milling machines are as basic as the lathe, particularly where one-of-a-kind or small quantities of workpieces are involved. Both of these machine types will be in use for a long time to come. It is important for you to learn to set up and use these machines quickly, accurately, and safely.

HORIZONTAL MILLING MACHINE SAFETY

Safe operation of a machine tool requires that you think before you do something. Before starting up a machine, know the location and and operation of its controls. Operate all controls on the machine yourself; do not have another person start or stop the machine for you. Chances are good that he will turn a control at the wrong time. While operating a milling machine, observe the cutting action at all times so that you can stop the machine immediately when you see or hear something unfamiliar. Always stay within reach of the controls while the machine is running. An unexpected emergency may require quick action on your part. Never leave a running machine unattended.

Before operating the rapid traverse control on a milling machine, loosen the locking devices on the machine axis to be moved. Check that the hand wheels or hand cranks are disengaged, or they will spin and injure anyone near them when the rapid traverse is engaged. The rapid traverse control will move any machine axis that has its feed lever engaged singularly or simultaneously. Do not try to position a workpiece too close to the cutter with this control, but approach the final 2 in. by using the hand wheels or hand cranks.

A person concentrating on a machining operation should not be approached quietly from behind, since it may annoy and alarm him and he may ruin a workpiece or injure himself. Do not lean on a running machine; moving parts can hurt you. Signs posted on a machine indicating a dangerous condition or a repair in progress should only be removed by the person making the repair or by a supervisor.

Measurements should only be taken on a milling machine after the cutter has stopped rotating and after the chips have been cleared away. Milling machine chips are dangerously sharp and often hot and contaminated with cutting fluids. They should not be handled with bare hands. Chips should be removed with a brush. Compressed air should not be used to clean off chips from a machine because it will make small missiles out of chips that can injure a person, even one who's quite a distance away. A blast of air will also force small chips into the ways and sliding surfaces of the milling machine where they will cause scoring and rapid premature wear. Cleaning chips and cutting fluids from the machine or workpiece should only be done after the cutter has stopped turning. Before and during the operation of a milling machine, keep the area around the machine clean of chips, oil spills, cutting fluids, and other obstructions to prevent the operator from slipping or stumbling.

Many milling machine attachments and workpieces are heavy; use a hoist to lift them on or off the table. Do not walk under a hoisted load; the hoist may release and drop the load on you. If a hoist is not available, ask for assistance.

Injuries can be caused by improper setups or the use of wrong tools. Use the correct size wrench when loosening or tightening nuts or bolts, preferably a box wrench or a socket wrench. An oversized wrench will round off the corners on bolts and nuts and prevent sufficient tightening or loosening; a slipping wrench can cause smashed fingers or other injuries to the hands or arms. Milling machine cutters have very sharp cutting edges; handling cutters carefully and with a cloth will avoid cuts on the hands.

All machine guards should be checked to see that they are in good condition and in place to increase milling machine safety. Workpieces should be centered in a vise with only enough extending out to permit machining. Clean the working area after a job is completed. A clean machine is a safer working place than one buried under chips.

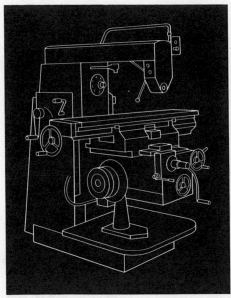

unit 1
plain and universal horizontal milling machines

Machine tool components and parts are identified by names. Anyone operating and maintaining machine tools should be familiar with these names and with the location of these parts.

objectives After completing this unit, you will be able to:
1. Identify the important parts of a horizontal milling machine.
2. Perform routine maintenance.

The major components of a horizontal milling machine are the column, knee, saddle, table, spindle, and overarm. The table of the plain horizontal milling machine does not swivel as does the universal milling machine. Figure 1 is an illustration of a horizontal milling machine with the major parts identified. The column is the main part of the milling machine. The face of the column is machined to provide an accurate guide for the vertical travel of the knee. The column also contains the main drive motor and the spindle. The spindle holds and drives the various cutting tools, chucks, and arbors. The spindle is hollow, the front end has a tapered hole with a standard milling machine taper. The front end of the spindle is called the spindle nose.

The overarm is mounted on top of the column and supports the arbor through an arbor support. The overarm slides in and out and can

be clamped securely in any position. The knee can be moved vertically on the face of the column. The knee supports the saddle and the saddle provides the sliding surface for the table. The saddle can move toward and away from the column to give crosswise movement of the machine table. The table provides the surface on which the workpieces are fastened. T-slots are machined along the length of the top surface of the table to align and hold fixtures and workpieces. Hand wheels or hand cranks are used to manually position the table. Micrometer collars make possible positioning movements as small as .001 in.

Power feed levers control automatic feeds in three axes, the feedrate being adjusted by the feed change crank (Figure 2). On many milling machines, the power feed only operates when the spindle is turning. Two safety stops at each axis

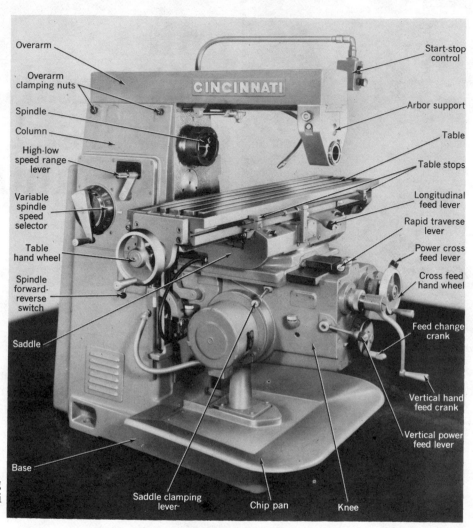

Figure 1. Horizontal milling machine (Courtesy of Cincinnati Milacron).

travel limit prevent accidental damage to the feed mechanisms by providing automatic kickout of the power feed. Two adjustable trip dogs for each axis allow the operator to preset specific power feed kickout travel distances. Rapid positioning of the table is accomplished with the rapid traverse lever (Figure 3). The direction of the rapid advance is dependent on the position of the respective feed lever. Locking devices on the table, saddle, and knee are used to prevent unwanted movements in any or all of these axes. The locking devices should be released only in the axis in which power feed is used. The spindle can rotate clockwise or counterclockwise, depending on the position of the spindle forward-reverse switch. Spindle speeds are changed to a high or low range with the speed range lever (Figure 4). The variable spindle speed selector makes any spindle speed possible between the minimum and maximum RPM available in each

Figure 2. Feed change crank (Courtesy of Cincinnati Milacron).

Figure 3. Rapid traverse lever (Courtesy of Cincinnati Milacron).

Figure 4. Speed change levers (Courtesy of Cincinnati Milacron).

speed range. The speed range lever has a neutral position between the high and low range. When the lever is in this neutral position, the spindle can easily be rotated by hand during machine setup.

The universal milling machine (Figure 5) closely resembles a plain horizontal milling machine. The main difference between these machines is that the universal machine has an additional housing that swivels on the saddle and supports the table. This allows the table to be swiveled 45 degrees in either direction in a horizontal plane. The universal milling machine is especially designed to machine helical slots or grooves as in twist drills and milling cutters. Other than these special applications, a universal mill and a plain milling machine can perform the same operations.

The size of a horizontal milling machine is usually given as the range of movement possible and the power rating of the main drive motor of the machine. An example would be a milling machine with a 28 in. longitudinal travel, 10 in. cross travel, and 16 in. vertical travel with a 5 hp main drive motor. As the physical capacity of a machine increases, more power is also available at the spindle through a larger motor.

Before any machine tool is operated, it should be lubricated. A good starting point is to

Figure 5. Universal milling machine (Courtesy of Cincinnati Milacron).

wipe clean all sliding surfaces and to apply a coat of a good way lubricant to them. Way lubricant is a specially formulated oil for sliding surfaces. Dirt, chips, and dust will act like a lapping compound between sliding members and cause excessive machine wear. Most machine tools have a lubrication chart that outlines the correct lubricants and lubrication procedures. When no lubrication chart is available, check all oil sight gages for the correct oil level and refill, if necessary. Too much oil causes leakage. Lubrication should be performed progressively, starting at the top of the machine and working down. Machine points that are hand oiled should only receive a small amount of oil at any one time, but this should be repeated at regular intervals, at least daily. Motor or pulley bearings should not get too much grease, since this may destroy the seals.

Before operating a milling machine you should be familiar with all control levers. Do not use force to engage or disengage controls or levers. Check that the locking levers are loosened on all moving slides. Check that all operating levers are in the neutral position before the machine is turned on. On a variable speed drive milling machine, change spindle speeds only while the spindle motor is running. On geared models, the spindle has to be stopped. Stop the spindle motor when shifting from one range to another. All power feed levers should be in their neutral position before feed changes are made. Spindle rotation should be reversed only after the machine has come to a complete standstill.

self-evaluation

SELF-TEST

1. Go to a plain or universal horizontal milling machine and locate the following parts.

overarm	table
column	knee
saddle clamping lever	feed change lever
speed change lever	spindle nose
powerfeed levers for	saddle
longitudinal feed,	knee clamping lever
crossfeed and	spindle forward-
vertical feed	reverse switch
rapid traverse lever	trip dogs for all
switch for spindle	three axes
ON-OFF	
arbor support	
switch for coolant pump	

2. Lubricate a plain horizontal milling machine.

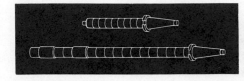

unit 2
types of spindles, arbors, and adaptors

Milling cutters, in order to be used, have to be mounted on a spindle, arbor, or in an adaptor. This unit describes different mounting methods used for milling cutters.

objective After completing this unit, you will be able to identify different mounting systems used to drive milling cutters.

The spindle of the milling machine holds and drives milling cutters. Cutters can be mounted directly on the spindle nose, as with face milling cutters (Figure 1), or by means of arbors and adaptors. These arbors and adaptors have tapered shanks that fit into the tapered hole or socket in the spindle nose. The tapers used for mounting milling cutters are divided into two general classes: (1) self-holding tapers, and (2) self-releasing or steep tapers.

Self-holding tapers have a small included angle of 5 degrees or less. When the shank is firmly seated in a socket, it will stay in place because of the high frictional forces between the contacting surfaces. Self-holding taper as-

semblies often are very difficult to take apart; that is why spindle noses are now made with a self-releasing taper.

Self-releasing tapers have a large included angle generally over 15 degrees. This steep taper permits easy and quick removal of arbors from the spindle nose. Most manufacturers have adapted the standard national milling machine taper. This taper is 3½ inches per foot (IPF) or about 16½ degrees. National milling machine tapers are available in four standard sizes, numbered 30, 40, 50, and 60, the most common being number 50. Self-releasing taper type shanks must be locked in the spindle socket with a draw-in bolt. Positive drive is obtained through two keys

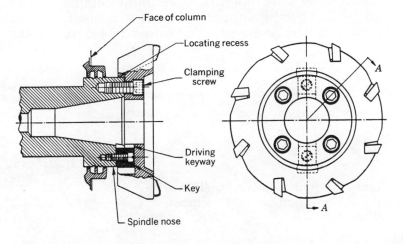

Figure 1. Mounting a face mill on the spindle nose of a milling machine (Courtesy of Cincinnati Milacron).

597

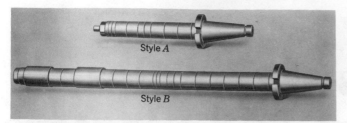

Figure 2. Arbors, styles *A* and *B* (Courtesy of Cincinnati Milacron).

in the spindle nose that engage in keyways in the flange of arbors and adaptors.

Two common arbor styles are shown in Figure 2. Style A arbor has a cylindrical pilot on the end opposite the shank. The pilot is used to support the free end of the arbor. Style *A* arbors are used mostly on small milling machines. But they are also used on larger machines when a style *B* arbor support cannot be used because of a small diameter cutter or interference between the arbor support and the workpiece.

Style *B* arbors are supported by one or more bearing collars and arbor supports. Style *B* arbors are used to obtain rigid setups in heavy duty milling operations.

Style *C* arbors are also known as shell end mill arbors or as stub arbors (Figure 3). Shell end milling cutters are face milling cutters up to 6 in. in diameter. Because of their relatively small diameter, these cutters cannot be counterbored so that they can be mounted directly on the spindle nose, as are face mills, but they are mounted on shell end mill arbors. Figure 4 shows how arbors are mounted in the milling machine. The draw-in bolt is screwed into the arbor as far as it will go, then the arbor is pulled into the spindle nose by tightening the draw-in bar lock nut. Note that the cutters are mounted close to the spindle and the first bearing support is close to the cutter. Spacing collars and a shim are used to get an

exact width in the straddle milling setup. Keys provide positive drive to the cutters and bearing collars. The bushing fit in the arbor supports can be adjusted to the bearing collars and pilot size. Bushing adjustments have to be made very carefully, because too loose a fit causes inaccuracy or chatter. Too tight a fit causes excessive friction and heat, which damage the bushing and bearing collar.

When the spindle turns at high RPM, more clearance is needed than at low RPM. The bearing collars have a larger diameter than the spacing collars for easy positioning of the arbor supports. All collars are manufactured to very close tolerances with their ends or faces being parallel and also square to the hole. It is very important that the collars and other parts fitting on the arbor are handled carefully to avoid damaging the collar faces. Any nicks, chips, or dirt between the collar faces will misalign the cutter or deflect the arbor and cause cutter run-out. The arbor nut should be tightened or loosened only with the arbor support in place. Without the arbor support, the arbor can easily be sprung and permanently bent.

Adaptors are used on milling machines to mount cutters that cannot be mounted on arbors. Adaptors can be used to hold and drive taper shank tools (Figure 5).

Collets used with these adaptors increase the range of tools that can be used in a milling machine having a given size spindle socket. The spring chuck adaptor (Figure 6), with different size removable spring collets, is used with straight shank tools such as drills and end mills. With a quick-change adaptor (Figure 7) mounted on the spindle nose, a number of milling machine operations such as drilling, end milling, and boring can be performed without changing the setup of the part being machined. The different tools are mounted on quick-change adaptors that are ready for use (Figure 8).

Figure 3. Style *C* arbor; shell end mill arbor (Courtesy of Cincinnati Milacron).

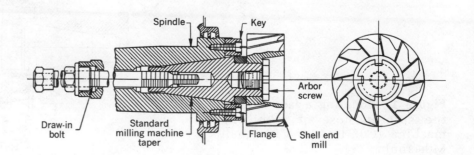

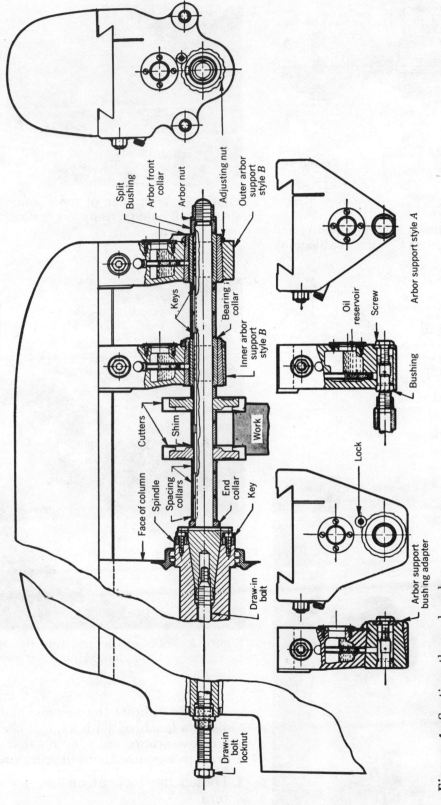

Figure 4. Section through arbor showing location of arbor collars, keys, bearing collars, and various arbor supports (Courtesy of Cincinnati Milacron).

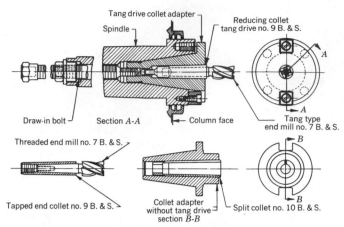

Figure 5. Adaptors and collets for self-releasing and self-holding tapers (Courtesy of Cincinnati Milacron).

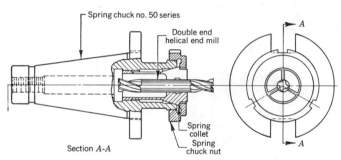

Figure 6. Spring chuck adaptor (Courtesy of Cincinnati Milacron).

Figure 7. Quick-change adaptor mounted on spindle nose (Courtesy of Cincinnati Milacron).

Figure 8. A number of tools mounted on quick-change tool holders ready to use (Courtesy of Cincinnati Milacron).

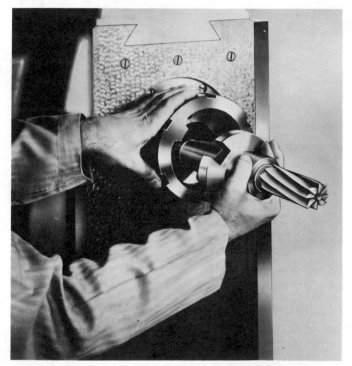

Figure 9. Tools are locked into the spindle with a partial turn of the clamp ring (Courtesy of Cincinnati Milacron).

Figure 9 shows the easy method of changing tools in a machine with a quick-change adaptor. To remove arbors and adaptors that are held with a draw-in bar, use the following procedure.

1. Loosen the locknut on the draw-in bolt one turn.
2. Tap the end of the draw-in bolt with a lead hammer. This releases the arbor shank from the spindle socket.

3. Arbors are heavy; you may need someone to hold the arbor while you unscrew the draw-in bolt from the rear of the machine.

Arbors should be stored in an upright position. Long arbors lying on their sides, if not properly supported, may bend. Rules for mounting arbors, adaptors, and cutters are:

Before inserting a tapered shank into the spindle socket, clean all mating parts and check for nicks and burrs. Nicks and burrs should be removed with a honing stone.

The cutter, spacing collars, and bearing collars should be a smooth sliding fit on the arbor. Nicks should be stoned off.

Use an arbor length that does not give much arbor overhang beyond the outer arbor support. Arbor overhang may cause vibration and chatter.

Mount cutters as close to the column as the work permits.

Cutters are sharp; handle carefully with shop towels.

When changing cutters on an arbor, place the cutter and spacers on a smooth and clean area on the worktable to avoid damage to their accurate bearing or contact surfaces.

Tighten the arbor nut with a wrench that fits accurately after the arbor support is in place. Do not use a hammer to tighten the arbor nut. Overtightening will spring or bend the arbor.

self-evaluation

SELF-TEST
1. What kind of cutters are mounted directly on the spindle nose?
2. Milling machine spindle sockets have two classes of taper. What are they?
3. What is the amount of taper on a national milling machine taper?
4. Why is it important to carefully clean the socket, shank, and arbor spaces prior to mounting them on a milling machine?
5. When is a style A arbor used?
6. Where should the arbor supports be in relation to the cutter?
7. What is a style C arbor?
8. Why should the arbor support be in place before the arbor nut is tightened or loosened?
9. Why are milling machine adaptors used?
10. If an arbor extends some distance beyond the outer arbor support, what can happen?

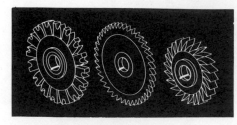

unit 3
arbor-driven milling cutters

Milling cutters are the cutting tools of the milling machines. They are made in many different shapes and sizes. These various cutters are mostly designed for a specific application. A milling machine operator should be capable of matching a cutting tool to the required application.

objective After completing this unit, you will be able to identify 10 milling cutters and list their names and common applications.

Most milling cutters are designed to perform specific kinds of operations. To make an intelligent decision as to which cutter to use, one should be able to identify milling cutters by sight and to know their capabilities and limitations. Most milling cutters are made from high speed steel; large cutters have inserted blades or teeth. More and more cutters are made with cemented carbide cutting edges. Milling cutters can be divided into profile sharpened cutters and form relieved cutters. Profile sharpened cutters are resharpened by grinding a narrow land (Figure 1) back of the cutting edges. Form relieved cutters are resharpened by grinding the face of the tooth parallel to the axis of the cutter. Cutters are classified also as being arbor driven or of the shank type. From the many different milling cutters available, this unit deals only with the more commonly used arbor-driven cutters.

Milling cutters are manufactured for either right-hand or left-hand rotation and with either right-hand or left-hand helix. The hand of a milling cutter is determined by looking at the front end of a spindle mounted cutter; a right-hand cutter requires a counterlockwise rotation (Figure 2), and a left-hand cutter rotates clockwise. The hand of the helix is determined by looking at

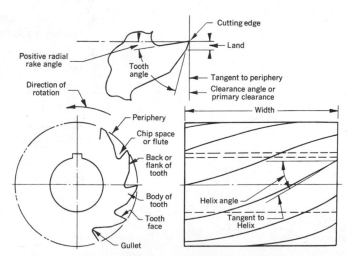

Figure 1. Nomenclature of plain milling cutter (Courtesy of Cincinnati Milacron).

the teeth or flutes from the cutter end. Flutes that turn to the right make a right-hand helix; to the left, they are a left-hand helix.

PLAIN MILLING CUTTERS

Plain milling cutters are designed for milling plain surfaces where the width of the work is

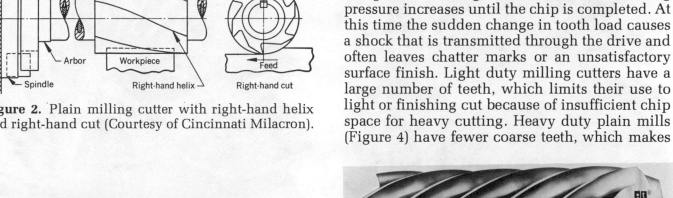

Figure 2. Plain milling cutter with right-hand helix and right-hand cut (Courtesy of Cincinnati Milacron).

narrower than the cutter (Figure 3). Plain milling cutters less than ¾ in. wide have straight teeth. On straight tooth cutters, the cutting edge will cut along its entire length at the same time. Cutting pressure increases until the chip is completed. At this time the sudden change in tooth load causes a shock that is transmitted through the drive and often leaves chatter marks or an unsatisfactory surface finish. Light duty milling cutters have a large number of teeth, which limits their use to light or finishing cut because of insufficient chip space for heavy cutting. Heavy duty plain mills (Figure 4) have fewer coarse teeth, which makes

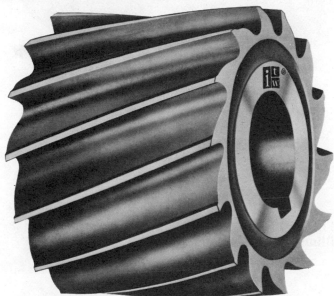

Figure 3. Light duty plain milling cutters (Copyright © Illinois Tool Works Inc., 1976.)

Figure 5. Helical plain milling cutter (Copyright © Illinois Tool Works Inc., 1976).

Figure 6. Side milling cutter (Lane Community College).

Figure 4. Heavy duty plain milling cutter (Copyright © Illinois Tool Works Inc., 1976).

Figure 7. Stagger tooth milling cutter (Copyright ©
Illinois Tool Works Inc., 1976).

Figure 8. *(top)* Half side milling cutter (Copyright ©
Illinois Tool Works Inc., 1976).

for strong teeth with ample chip clearance. The
helix angle of heavy duty mills is about 45 degrees. The helical form enables each tooth to take
a cut gradually, which reduces shock and lowers
the tendency to chatter. Plain milling cutters are
also called slab mills. Plain milling cutters with a
helix angle over 45 degrees are known as helical
mills (Figure 5). These milling cutters produce a
smooth finish when used for light cuts or on
intermittent surfaces. Plain milling cutters do not
have side cutting teeth and should not be used to
mill shoulders or steps on workpieces.

SIDE MILLING CUTTERS

Side milling cutters are used to machine steps or
grooves. These cutters are made from $\frac{1}{4}$ to 1 in. in
width. Figure 6 shows a straight tooth side milling cutter. To cut deep slots or grooves, a
staggered tooth side milling cutter (Figure 7) is
preferred because the alternate right-hand and

left-hand helical teeth reduce chatter and give
more chip space for higher speeds and feeds than
are possible with straight tooth side milling cutters. To cut slots over 1 in. wide, two or more side
milling cutters may be mounted on the arbor
simultaneously. Shims between the hubs of the
side mills can be used to get any precise width
cutter combination or to bring the cutter again to
the original width after sharpening.

Half side milling cutters are designed for
heavy duty milling where only one side of the
cutter is used (Figure 8). For straddle milling, a
right-hand and a left-hand cutter combination is
used.

Plain metal slitting saws are designed for
slotting and cutoff operations (Figure 9). Their
sides are slightly relieved or "dished" to prevent
binding in a slot. Their use is limited to a rela-

Figure 9. *(top right)* Plain metal slitting saw (Copyright © Illinois Tool Works Inc., 1976).

Figure 10. *(bottom right)* Side tooth metal slitting saw (Copyright © Illinois Tool Works Inc., 1976).

tively shallow depth of cut. These saws are made in widths from $\frac{1}{32}$ to $\frac{5}{16}$ in.

To cut deep slots or when many teeth are in contact with the work, a side tooth metal slitting saw will perform better than a plain metal slitting saw (Figure 10). These saws are made from $\frac{1}{16}$ to $\frac{3}{16}$ in. wide.

Extra deep cuts can be made with a staggered tooth metal slitting saw (Figure 11). Staggered tooth saws have greater chip carrying capacity than other saw types. All metal slitting saws have a slight clearance ground on the sides toward the hole to prevent binding in the slot and the scoring of the walls of the slot. Stagger tooth saws are made from $\frac{3}{16}$ to $\frac{5}{16}$ in. wide.

Angular milling cutters are used for angular milling such as cutting of dovetails, V-notches, and serrations. Single angle cutters (Figure 12) form an included angle of 45 or 60 degrees, with

one side of the angle at 90 degrees to the axis of the cutter.

Double angle milling cutters (Figure 13) usually have an included angle of 45, 60, or 90 degrees. Angles other than those mentioned are special milling cutters.

Convex milling cutters (Figure 14) produce concave bottom grooves or they can be used to make a radius in an inside corner. Concave milling cutters (Figure 15) make convex surfaces. Corner rounding milling cutters (Figure 16) make rounded corners. The cutters illustrated in Figures 14 to 17 are form relieved cutters.

Involute gear cutters (Figure 17) are commonly available in a set of eight cutters for a given pitch, depending on the number of teeth for which the cutter is to be used. The ranges for the individual cutters are as follows.

Figure 11. Staggered tooth metal slitting saw (Copyright © Illinois Tool Works Inc., 1976).

Figure 12. Single angle milling cutter (Copyright © Illinois Tool Works Inc., 1976).

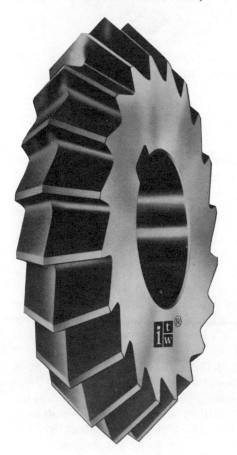

Figure 13. Double angle milling cutter (Copyright © Illinois Tool Works Inc., 1976).

Figure 14. Convex milling cutter (Copyright © Illinois Tool Works Inc., 1976.)

Figure 15. Concave milling cutter (Copyright © Illinois Tool Works Inc., 1976).

Figure 16. *(left)* Corner rounding milling cutter (Copyright © Illinois Tool Works Inc., 1976).

Figure 17. Involute gear cutter (Lane Community College).

Number of cutter	Range of teeth
1	135 to rack
2	55 to 134
3	35 to 54
4	26 to 53
5	21 to 25
6	17 to 20
7	14 to 16
8	12 and 13

These eight cutters are designed so that their forms are correct for the lowest number of teeth in each range. If an accurate tooth form near the upper end of a range is required, a special cutter is needed.

self-evaluation

SELF-TEST

1. What are the two basic kinds of milling cutters with reference to their tooth shape?
2. What is the difference between a light duty and a heavy duty plain milling cutter?
3. Why are plain milling cutters not used to mill steps or grooves?
4. What kind of cutter is used to mill grooves?
5. How does the cutting action of a straight tooth side milling cutter differ from a stagger tooth side milling cutter?
6. Give an example of an application of half side milling cutters.
7. When are metal slitting saws used?
8. Give two examples of form relieved milling cutters.
9. When are angular milling cutters used?
10. When facing the spindle, in which direction should the right-hand cutter be rotated in order to cut?

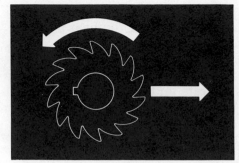

unit 4
setting speeds and feeds on the horizontal milling machine

A milling machine can be operated efficiently or inefficiently. To be an efficient machine operator, you must understand the relationships between different work materials and the cutting tools used to machine them. To use these cutting tools correctly and economically, you must know about cutting speeds and feeds.

objectives After completing this unit, you will be able to:

1. **Select cutting speeds for different materials and calculate the RPM for different milling cutters.**
2. **Select and calculate feedrates for different materials and milling cutters.**
3. **Set speeds and feeds on a horizontal milling machine.**

To get maximum use from a milling cutter, it is important that it is operated at the correct cutting speed. Cutting speed is expressed in surface feed per minute and varies for different work materials and cutting tool materials. The cutting speed of a milling cutter is the distance which the cutting edge of a cutter tooth travels in 1 min. Table 1 is a table of cutting speeds for a number of commonly used materials.

The cutting speeds given in Table 1 are only intended to be starting points. These speeds represent experience in instructional settings where cutter and machine conditions are often less than ideal. Different hardnesses within each materials group account for the wide range of cutting

Table 1
Cutting Speeds for Milling

| Material | Cutting Speed SFM | |
	High Speed Steel Cutter	Carbide Cutter
Free machining steel	100-150	400-600
Low carbon steel	60-90	300-550
Medium carbon steel	50-80	225-400
High carbon steel	40-70	150-250
Medium alloy steel	40-70	150-350
Stainless steel	30-80	100-300
Gray cast iron	50-80	250-350
Bronze	65-130	200-400
Aluminum	300-800	1000-2000

609

speeds. Generally speeds are lower for hard materials, abrasive materials, and deep cuts. Speeds are higher for soft materials, better finishes, light cuts, frail workpieces, and light setups. It is very important that the cutting speed is not too high for the material being machined or the cutting edges will dull rapidly. Too slow a cutting speed will not damage a cutter, but it will be inefficient. It is good practice to use the lower cutting speed value to start and then, if the setup allows, increase it to the higher speed. To use these cutting speed values on a milling machine, they have to be expressed in RPM. The formula used to convert cutting speed into RPM is:

$$RPM = \frac{CS \times 4}{D}$$

CS = cutting speed found in Table 1

4 = constant

D = diameter of cutter in inches

As an example of how to use the formula, calculate the RPM for a 3 in. diameter high speed steel cutter to be used on cast iron.

$$RPM = \frac{CS \times 4}{D} = RPM = \frac{50 \times 4}{3} = \frac{200}{3} = 67$$

In this example the low end of the cutting speed range was used to get a starting point. Other factors such as the use of coolant and machine rigidity also influence the selection of a cutting speed. Tool materials, other than high speed steels such as cast alloys or cemented carbides, can be used at higher cutting speeds because they retain a sharp cutting edge at elevated temperatures. As a rule, when high speed steel cutting speeds are 100 percent, then cast alloys can be 150 percent and cemented carbides can be 200 to 600 percent of that figure.

These are the general variations that affect the selection of speed for milling cutters. As you become more experienced and begin to deal with a wide range of materials, it will be very useful to refer to references such as the *Machining Data Handbook* and the general section in *Machinery's Handbook* on feeds and speeds. Determining accurately the speeds and feeds for milling is complicated by the fact that the cutting edge does not remain in the work continuously and that the chip being made varies in thickness during the cutting period. The type of milling being done (slab, face, or end milling) also makes a difference, as does the way that the heat is transferred from the cutting edge.

The second important item in efficient milling machine operation is the feed. It is expressed as a feedrate and given in inches per minute (IPM). Most milling machines have two different drive motors, one to power the spindle and one to power the feed mechanism. These two motors make independent changes of the spindle speed and the feedrate possible. The feedrate is the product of the feed per tooth times the number of teeth on the cutter times the RPM of the spindle. You have already determined how to calculate the RPM of a cutter and, by counting the number of teeth in a cutter and knowing the feed per tooth, you can determine the feedrate. Table 2 is a chart of commonly used feeds per tooth (FPT).

As you can see in Table 2, there is only a slight difference in the feed per tooth allowance between high speed steel cutters and carbide cutters. Calculate the feedrate for a 3 in. diameter six tooth helical mill cutting free machining steel, first for a high speed steel tool and then for a carbide tool. The formula for feedrate is

$$FPT \times N \times RPM$$

EXAMPLE.

FPT = feed per tooth

N = number of teeth on cutter

RPM = revolutions per minute of cutter

For a starting point, calculate the RPM (refer to Table 1).

$$RPM = \frac{CS \times 4}{D} = \frac{100 \times 4}{3} = \frac{400}{3} = 134 \; RPM$$

Table 2 gives an FPT of .002 in.

The cutter has six teeth.

The feedrate is .002 × 6 × 134 = 1.608 IPM for the high speed steel cutter.

EXAMPLE.
For a carbide cutter the RPM is

$$RPM = \frac{CS \times 4}{D} = \frac{400 \times 4}{3} = \frac{1600}{3} = 534 \; RPM$$

The FPT is .003 in.

Table 2
Feed in inches per tooth (instructional setting)

Type of Cutter	Aluminum		Bronze		Cast Iron		Free Machining Steel		Alloy Steel	
	HSS	Carbide	HSS	Carbide	HSS	Carbide	HSS	Carbide	HSS	Carbide
Face mills	.007 to .022	.007 to .020	.005 to .014	.004 to .012	.004 to .016	.006 to .020	.003 to .012	.004 to .016	.002 to .008	.003 to .014
Helical mills	.006 to .018	.006 to .016	.003 to .011	.003 to .010	.004 to .013	.004 to .016	.002 to .010	.003 to .013	.002 to .007	.003 to .001
Side cutting mills	.004 to .013	.004 to .012	.003 to .008	.003 to .007	.002 to .009	.003 to .012	.002 to .007	.003 to .009	.001 to .005	.002 to .008
End mills	.003 to .011	.003 to .010	.003 to .007	.002 to .006	.002 to .008	.003 to .010	.001 to .006	.002 to .008	.001 to .004	.002 to .007
Form relieved cutters	.002 to .007	.002 to .006	.001 to .004	.001 to .004	.001 to .005	.002 to .006	.001 to .004	.002 to .005	.001 to .003	.001 to .004
Circular saws	.002 to .005	.002 to .005	.001 to .003	.001 to .003	.001 to .004	.002 to .006	.001 to .003	.001 to .004	.005 to .002	.001 to .004

The cutter has six teeth.

The feedrate is .003 × 6 × 534 = 9.612 IPM for the carbide cutter.

To calculate the starting feedrate, use the low figure from the feed per tooth chart and, if conditions permit, increase the feedrate from there. The most economical cutting takes place when the most cubic inches of metal per minute are removed and a long tool life is obtained. The tool life is longest when a low speed and high feed rate is used. Try to avoid feedrates of less than .001 in. per tooth, because this will cause rapid dulling of the cutter. Exceptions to this limit are small diameter end mills when used on harder materials. The depth and width of cut also affect the feedrate. Wide and deep cuts require a smaller feedrate than do shallow, narrow cuts. Roughing cuts are made to remove material rapidly. The depth of cut may be ⅛ in. or more, depending on the rigidity of the machine, the setup, and the horsepower available. Finishing cuts are made to produce precise dimensions and acceptable surface finishes. The depth of cut on a finishing cut should be between .015 and .030 in. A depth of cut of .005 in. or less will cause the cutter to rub instead of cut and also results in excessive cutting edge wear.

Cutting fluids should be used when machining most metals with high speed steel cutters. A cutting fluid cools the tool and the workpiece. It lubricates, which reduces friction between the tool face and chip. Cutting fluids prevent rust and corrosion and, if applied in sufficient quantity, will flush away chips. Cutting fluids will, through these characteristics, increase production through higher speeds and produce better surface finishes. Most milling with carbide cutters is done dry unless a large constant flow of cutting fluid can be directed at the cutting edge. An interrupted coolant flow on a carbide tool causes thermal cracking and results in subsequent chipping of the tool.

self-evaluation

unit 5
workholding and locating devices on the milling machine

A very important factor in milling is the method used to hold a workpiece while it is being machined. Considerable ingenuity on the machine operator's part is required to select a workholding method suitable to the job.

objective After completing this unit, you will be able to select workholding devices for common milling machine jobs.

Large and irregularly shaped workpieces often are fastened directly to the machine table top. T-slots, which run lengthwise along the top of the table, are accurately machined and parallel to the sides of the table. These T-slots are used to retain the clamping bolts. Workpieces can also be aligned when snug fitting parallels are set into the T-slot and the workpiece is pushed against these parallels while the work is being clamped. Figure 1 shows the workpiece clamped to the table with T-slot bolts and clamps. The bolts are

Figure 3. Highly finished surfaces should be protected from clamping damage (Lane Community College).

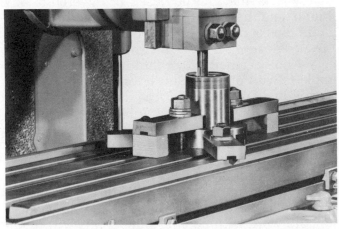

Figure 1. Work clamped to the table with T-slot bolts and clamps (Courtesy of Cincinnati Milacron).

Figure 4. Protect the machine table surface from rough workpieces (Lane Community College).

Figure 2. Clamping bolt close to the work gives effective clamping (Lane Community College).

placed close to the workpiece and the block supporting the outer end of the clamp is the same height as the shoulder being clamped. Figure 2 illustrates a good clamping arrangement. When the bolt is closer to the work than to the clamp support block, maximum leverage is obtained. The support block should never be lower than the work being clamped.

When workpieces with finished or soft surfaces are clamped, care must be taken to protect those surfaces from damage by clamping. A shim should be placed between the work surface and the clamp (Figure 3). Before placing rough castings or weldments on a machine table, protect the table surface with a shim (Figure 4). This

Figure 5. Workpiece supported under the clamp (Lane Community College).

Figure 6. Stop block prevents work slippage (Lane Community College).

shim can be paper, sheet metal, or even plywood, depending on the accuracy of the machining to be performed.

A workpiece should have a support directly underneath where a clamp exerts pressure (Figure 5). Clamping an unsupported workpiece may cause it to bend or spring, and it will bend back after clamping pressure is released. If the workpiece material is brittle, clamping pressure may break it.

Workpieces tend to move on the table from the cutting pressure against them. This movement can be prevented by clamping a stop block on the table and placing the workpiece against it (Figure 6). Different kinds of supports are used in clamping work. Figure 7 shows an assortment of screw jacks that can be raised or lowered to any height and then are locked in that position. Often solid blocks or combination of blocks are used to give the correct height of supports (Figure 8).

Another method of holding work on the table is with quick action jaws (Figure 9). These individual jaws can be located anywhere on the machine table. They are tightened by turning screws that move the jaws outward and also give a downward pull on the workpiece.

Probably the most common method of workholding on a milling machine is a vise. Vises are simple to operate and can quickly be adjusted to the size of the workpiece. A vise should be used to hold work with parallel sides if it is within the size limits of the vise, because it is the fastest and most economical workholding method. The plain vise (Figure 10) is bolted to the machine table. Alignment with the table is provided by two slots at right angles to each other on the underside of the vise. These slots are fitted with removable keys that align the vise with the table T-slots either lengthwise or crosswise. A plain vise can be converted to a swivel vise (Figure 11) by mounting it on a swivel plate. The swivel plate is graduated in degrees. This allows the upper section to be swiveled to any angle in the horizontal plane. When swivel bases are added to

Figure 7. Examples of screw jacks (Courtesy of Cincinnati Milacron).

Figure 8. Work set up and clamped on table (Courtesy of Cincinnati Milacron).

Figure 10. Plain vise (Courtesy of Cincinnati Milacron).

Figure 11. Swivel vise (Courtesy of Cincinnati Milacron).

Figure 9. Quick action jaws holding workpiece (Courtesy of Cincinnati Milacron).

a plain vise, the versatility increases, but the rigidity is lessened.

For work involving compound angles, a universal vise (Figure 12) is used. This vise can be swiveled 90 degrees in the vertical plane and 360 degrees in the horizontal plane.

The strongest setup is the one where the workpiece is clamped close to the table surface. Castings, forgings, or other rough workpieces can be securely fastened in an all-steel vise (Figure 13). The movable jaw can be set in any notch on the two bars to accommodate different workpieces. The short clamping screw makes for a very strong and rigid setup. The hardened and serrated jaws grip the workpiece securely.

Figure 12. Universal vise (Courtesy of Cincinnati Milacron).

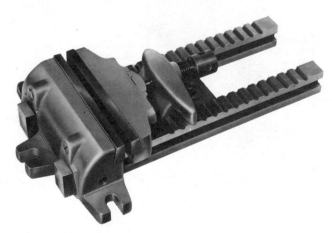

Figure 13. All-steel vise (Courtesy of Cincinnati Milacron).

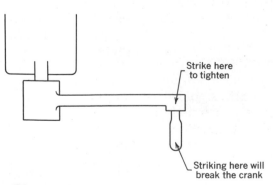

Strike here to tighten

Striking here will break the crank

Figure 14. Tightening a vise.

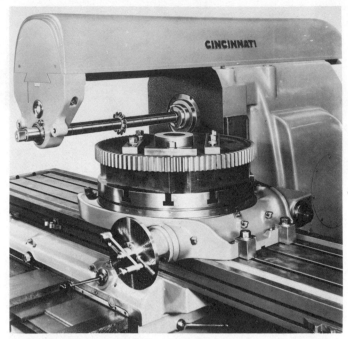

Figure 15. Rotary table (Courtesy of Cincinnati Milacron).

Figure 16. Dividing head and foot stock (Courtesy of Cincinnati Milacron).

Air or hydraulically operated vises are often used in production work, but in general toolroom work, vises are opened and closed by cranks or levers. To hold workpieces securely without slipping under high cutting forces, a vise must be tightened by striking the crank with a lead hammer (Figure 14).

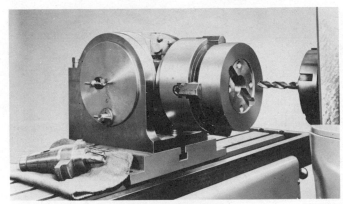

Figure 17. Dividing head used to drill equally spaced holes (Courtesy of Cincinnati Milacron).

Figure 18. Round shaft being held in vee-blocks (Lane Community College).

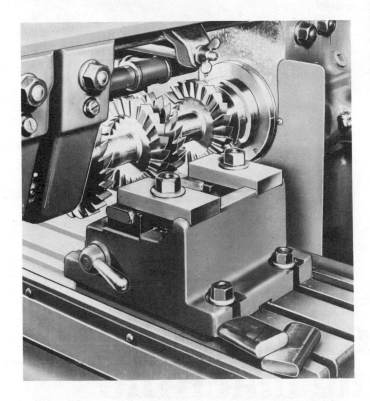

Figure 19. Milling fixture used for many identical parts (Courtesy of Cincinnati Milacron).

A rotary table or circular milling attachment (Figure 15) is used to provide rotary movement to a workpiece. The rotary table can be used for angular indexing, milling circular grooves, or to cut radii. The rotary table is shown in Figure 15 in a gear cutting operation.

The dividing head (Figure 16) is used to divide the circumference of a workpiece into any number of equally spaced divisions. Work is held between centers, in collets, or in a chuck. The supporting member opposite the dividing head is the foot stock. The dividing head can be swiveled from below a horizontal line to beyond

the vertical. The dividing head can also be used to drill equally spaced holes in workpieces held in a chuck (Figure 17). Round workpieces can be securely fastened in a set of vee blocks (Figure 18). To prevent the shaft from bending under cutting pressure, a screw jack such as those shown in Figure 7 can be used to support the shaft halfway between the vee blocks. If a number of identical workpieces are to be machined, a milling fixture (Figure 19) may be the most efficient way of holding them. A fixture is used when the savings resulting from its use are greater than the cost of making the fixture.

self-evaluation

SELF-TEST

1. In relationship to the workpiece, where should the clamping bolt be located?
2. What precautions should be taken when clamping finished surfaces?
3. When are screw jacks used?
4. What is the reason for clamping a stop block to the table?
5. What are quick action jaws?
6. What is the difference between a swivel vise and a universal vise?
7. When is an all-steel vise used?
8. When is a rotary table used?
9. When is a dividing head used?
10. When is a fixture used?

unit 6
plain milling on the horizontal milling machine

Plain milling is the operation of milling a flat surface in a plane parallel to the cutter axis. It involves the selection of a workholding device, milling cutter, speed, feed, and depth of cut.

objectives After completing this unit, you will be able to:

1. **Align workholding devices.**
2. **Mill flat surfaces to size.**
3. **Mill surfaces square to each other.**

618

Preparing a machine tool prior to machining is called setting up the machine. Before a setup can be made, the machine should be cleaned, especially all sliding surfaces such as the ways and the machine table. After wiping the table clean, use your hand to feel for nicks or burrs. If you find any, use a honing stone to remove them. Workpieces must be fastened securely for the machining operation. They can be held in a vise, clamped to an angle plate, or clamped directly to the table. Odd shaped workpieces may be held in a fixture designed for that purpose. On a universal milling machine it is good practice to check the alignment of the table before mounting a vise or fixture on it.

TABLE ALIGNMENT
ON A UNIVERSAL MILLING MACHINE

1. Clean the face of the column and the machine table.
2. Fasten a dial indicator to the table with a magnetic base or other mounting device (Figure 1).
3. Preload the indicator to approximately ½ revolution of its dial and set the bezel to zero.
4. Move the table longitudinally with the hand wheel to indicate the column.
5. If the indicator hand moves, loosen the locking bolts on the swivel table and adjust the table one half the distance of the indicated difference.

Figure 1. Aligning the universal milling machine table (Lane Community College).

Figure 2. Fixture alignment keys (Lane Community College).

6. Tighten the locking bolts and reindicate the column; make another adjustment if needed.

Never indicate the table with the indicator mounted on the column, as this would always show alignment.

A good machine vise is an accurate and dependable workholding device. When milling only the top of a workpiece, it is not necessary that the vise be square to the column or parallel to the table travel. When the job requires that the outside surface is parallel to a step or groove in the workpiece, however, the vise has to be precisely aligned and positioned on the table.

The base of the vise should be located with keys (Figure 2) that fit snugly into the T-slots on the milling machine table. This normally positions the solid jaw of the vise parallel with or square to the face of the column. Before mounting a vise or other fixture on a machine table, inspect the base carefully for small chips and nicks and remove any that you find. When the base is clean, fasten the vise to the table. Whenever possible, position the vise so that the cutting pressure will be against the solid jaw (Figure 3). Often references are made to the "solid jaw" of a vise. The solid jaw will not move or change when the vise is tightened, although the movable jaw will align itself to some degree with the work and should never be indicated. A

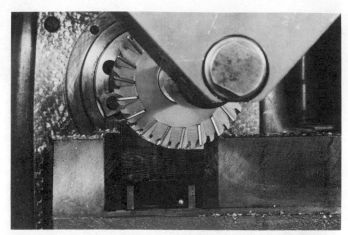

Figure 3. Cutting pressure against solid jaw (Lane Community College).

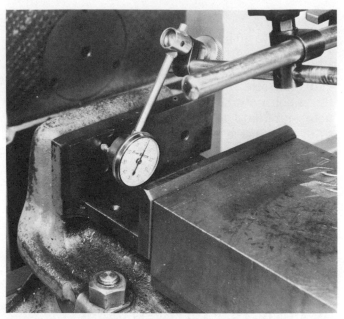

Figure 4. Aligning vise parallel to table (Lane Community College).

number of different methods of aligning a vise on a table are shown below.

ALIGNING A VISE
PARALLEL WITH THE TABLE TRAVEL

1. Fasten a dial indicator with a magnetic base to the arbor (Figure 4) and preload indicator contact point to one half revolution of the dial. Set bezel to zero.
2. Move the table so that the indicator slides along the solid jaw. Record any indicator movement.
3. Loosen the holddown bolts and lightly retighten. Lightly tap the vise with a lead or soft faced hammer to move the vise one half the distance of the indicator movement. Be sure the solid jaw moves away from the indicator contact point; tapping the jaw against the indicator may damage the indicator movement.
4. Tighten the holddown bolts securely and reindicate. Often the tightening of bolts or nuts will again move the vise.

ALIGNING A VISE
AT A RIGHT ANGLE TO THE TABLE TRAVEL

1. Fasten a dial indicator with a magnetic base to the arbor (Figure 5) and preload the indicator.
2. Move the table with the cross feed hand wheel and indicate the solid jaw.
3. Loosen the vise holddown bolts and make any necessary correction.

Figure 5. Aligning vise square to table travel (Lane Community College).

4. Indicate the solid jaw again to check the alignment. *Always* take another indicator reading after securely tightening the clamping bolts. Often the final tightening will move a vise, fixture, or workpiece.

If no indicator is available to align a vise on a table, a combination square may be used, as shown in Figure 6. The beam of the square is slid along the machined surface of the column until contact is made with the solid jaw of the vise. Two strips of paper used as feeler gages help in locating the contact point. A soft headed hammer or lead hammer is used to tap the vise into position.

ALIGNING A VISE
AT AN ANGLE OTHER THAN
90 DEGREES TO THE TABLE TRAVEL

Occasionally a vise has to be mounted on the table at an angle other than square to the table travel. This can be done with a protractor (Figure 7). Paper strips are used as feeler gages, the angular setting being correct when both strips contact the protractor blade and the vise jaw at the same time. This is not a precise method of setting an angle because of the limitations in setting an angle accurately with a protractor, maintaining the level of the protractor blade, and accurately sampling the "drag" on the paper strips.

Before machining a workpiece to size on a milling machine, several important decisions need to be made. One consideration is how to hold the workpiece while it is being machined. Large workpieces can be clamped directly to the table (Figure 8). A bar bolted to the table behind the workpiece is a safety stop that prevents the

workpiece from moving when the cutting pressure is against it. Many workpieces can be held securely in a machine vise (Figure 9). If the workpiece is high enough, seat it on the bottom of the vise. If it is not, use parallels to raise it. Remember that friction between the vise jaws and the workpiece holds the workpiece. The more contact area there is the better.

SELECTING THE CUTTER

For flat surfaces use a plain milling cutter that is wider than the surface to be machined. The diameter of the milling cutter should be as small as practical. Figure 10 illustrates the difference

Figure 7. Using a protractor to align a vise on table (Lane Community College).

Figure 8. Workpiece clamped to table (Lane Community College).

Figure 6. Using a square to align vise on table (Lane Community College).

Figure 9. Workpiece held in vise (Lane Community College).

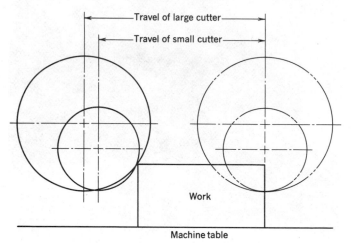

Figure 10. Different travel distances between different diameters of cutters.

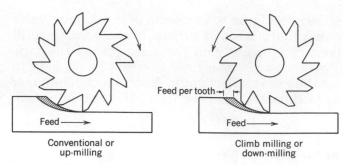

Figure 11. Conventional and climb milling.

in distance that a small diameter and a larger diameter cutter travel when machining the same length workpiece. A smaller diameter cutter is more efficient because it uses less machining time. Whatever the diameter is of the cutter used, it is important that sufficient clearance remains between the arbor support and the vise (Figure 9). Each cut taken brings the two parts closer together and, if much material is to be removed, it becomes necessary to reset the workpiece on higher parallels to avoid a collision between vise and arbor support.

Figure 11 shows one cutter operated in a conventional milling mode and the other cutter in a climb milling operation. In conventional milling modes (sometimes called up-milling), the workpiece is forced against the cutter with the teeth of the cutter trying to lift the workpiece up, especially at the beginning of a cut. In climbing (or down-milling), the cutter tends to hold the workpiece down.

Climb milling should only be performed on machines equipped with an antibacklash device. Backlash, which is play between the table drive screw and nut assembly, would let the cutter pull the workpiece under it, break the cutter, and ruin the workpiece. Remember that every cutter can be operated in an up-milling or down-milling fashion. The only difference is from which side of the workpiece the cut is started. The cutter should be located on the arbor as close to the spindlenose as the location of the workpiece permits. The cross travel of the saddle is limited. The arbor support should also be as close to the cutter as possible for a rigid setup.

Prior to assembly of the spacing collars and cutter on the arbor, all pieces should be cleaned. Keys should always be used to drive the cutter. Do not depend on the friction between the spacers and cutter. The drive keys should extend into the spacing collars on both sides of the cutter. Tighten and loosen the arbor locknut only when the arbor support is in place, and do not use a hammer on the wrench. Use only a sharp cutter to minimize cutting pressures and to get a good surface finish. Resharpen a cutter when it becomes slightly dull. A slightly dull cutter can be resharpened easily and quickly.

SETTING UP THE MACHINE

After the cutter for the job has been selected, the speed and feed can be calculated and set on the machine. The depth of cut depends largely on the amount of material that is to be removed.

Good milling practice is to take a roughing cut and then a finish cut. Better surface finishes and higher dimensional accuracy are achieved when roughing and finishing cuts are made. The depth of the roughing cut often is limited by the horsepower of the machine or the rigidity of the setup. A good starting point for roughing is .100 to .200 in. deep. The finishing cut should be .015 to .030 in. deep. Depth of cut less than .015 in. deep should be avoided, because a milling cutter, especially in conventional or up-milling, has a strong rubbing action before the cutter actually starts cutting. This rubbing action causes a cutter to dull rapidly. Assuming that a cut .100 in. deep has to be taken, the following steps outline the procedure to be used.

1. Loosen the knee locking clamp and the cross slide lock.
2. Turn on the spindle and check its rotation.
3. Position the table so that the workpiece is under the cutter.
4. Raise the knee slowly by turning the vertical hand feed crank until the cutter just touches the workpiece. If the cutter cuts a groove, you have gone too far and should try again on a different place on the workpiece.
5. Set the micrometer dial on the knee feedscrew on zero.
6. Lower the knee by approximately one half revolution of the hand feed crank. If the knee is not lowered, the cutter will leave tool marks on the workpiece in the following operation.
7. Move the table longitudinally until the cutter is clear of the workpiece. Move to the side of the workpiece, which will result in making a conventional milling cut.
8. Raise the knee past the zero mark to the 100 mark on the micrometer dial.
9. Tighten the knee lock and the cross slide lock. *Always* prior to starting a machining operation, tighten all locking clamps except the one that would restrict table movement while cutting. This aids in making a rigid chatterfree setup.
10. The machine is now ready for the cut. Turn on the coolant (Figure 12). Move the table slowly into the revolving cutter until the full depth of cut is obtained before engaging the power feed.
11. When the cut is completed, disengage the power feed, stop the spindle rotation, and turn off the coolant before returning the table to its starting position. If the revolving cutter is returned over the newly machined surface, it will leave cutter marks and mar the finish.
12. After brushing off the chips and wiping the workpiece clean, the workpiece should be measured while it is still fastened in the machine. If the workpiece is parallel at this time, additional cuts can be made if more material needs to be removed.

When machining rough castings or forgings, some thought has to be given to the setup of the workpiece, especially if the stock to be removed is limited. Figure 13 is an example of a bar that should be square but, instead, it is a parallelogram. If all the material were to be removed from sides 1 and 4 to make these sides square to sides 2 and 3, the block would be undersize between

Figure 12. Coolant cools the cutter and washes away chips (Lane Community College).

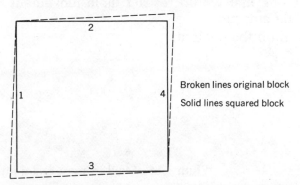

Figure 13. Machining a square from a parallelogram.

sides 1 and 4. If the block were to be shimmed so that some material would be removed from all four sides (the dotted lines in Figure 13), the resulting square would be of a larger size.

MILLING PROCEDURE, RECTANGULAR BLOCK

1. Saw off material that is between $\frac{1}{8}$ and $\frac{1}{4}$ in. larger than the finished size of the workpiece. (Figure 14)

2. Fasten the workpiece centrally in the clean vise and tap the vise screw handle with a soft headed hammer. Tightening the vise usually raises the workpiece slightly, so tap the workpiece with a soft headed (preferably lead) hammer to reseat it on the bottom of the vise.

3. Select a sharp plain milling cutter that is wider than the workpiece and put it on the arbor, taking care to have the cutter as near the spindle as possible and the arbor support as close to the cutter as feasible. Make sure the arbor key extends into the arbor spacers on each side of the cutter. Tighten the arbor nut after mounting the arbor support on the overarm.

4. Select and set the cutter speed. What RPM will you use?

$$RPM = \frac{CS \times 4}{D} = \underline{\hspace{3cm}}$$

5. Check that the spindle rotation is correct.

6. Calculate the feedrate to use and set it on the machine. Feed rate equals feed per tooth × number of teeth × RPM. What is the feedrate?

7. Set the depth of cut. Arrange your cutting plan to remove about equal amounts of material from opposite sides of the workpiece. Tighten all movement locking clamps except the one that would restrict table movement while cutting.

8. Turn on the coolant.

9. Start the cut by turning the table hand wheel; then engage the power feed. Observe the cutting operation closely so that you can turn off the power feed at the first sign of trouble. Do not remove chips or reach into the cutting area while the cutter is revolving.

10. When the cut is finished, disengage the power feed, stop the spindle, and return the table to its starting position.

11. Remove the chips with a brush and remove all burrs with a file; then remove the workpiece from the vise. Side 1 is now finished.

12. Clean the vise and workpiece and reposition the workpiece as shown in Figure 15, on parallels, so that about $\frac{1}{2}$ in. extends out from the vise. Inserting a small rod ($\frac{1}{4}$ in. diameter) between the moving jaw and the workpiece assures positive contact between side 1 and the solid jaw of the vise. The workpiece may not touch on both parallels, even when it is tapped down with a soft headed hammer, because the sides are not square to each other.

13. Set the depth of cut for side 2. Take only one half of the material that is to be removed with this cut. This side will have to be remachined if later measuring shows it not to be square to side 1.

14. Take this cut repeating procedures from steps 9 to 11.

15. Fasten the deburred workpiece in a clean vise (Figure 16). Side 1 is again against the solid jaw. Side 2 is toward the bottom of the vise. Side 3 is now to be machined.

16. The depth of cut for side 3 should be clean up cut as was side 2.

17. Repeat steps 9 to 11.

18. With a micrometer, measure dimensions "A" and "B" (Figure 17). If dimension "A" is equal to dimension "B," or within less than .004 in., the solid jaw is square to the base and you can skip step 19 and go to step 20. If the

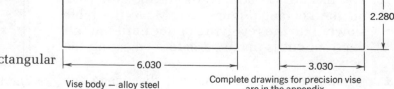

Figure 14. Machining a rectangular block.

Vise body — alloy steel

Complete drawings for precision vise are in the appendix

6.030

3.030

2.280

Figure 15. Setup to machine side 2 (Lane Community College).

Figure 16. Setup to machine side 3 (Lane Community College).

difference is more than .004 in., go to step 19.

19. If dimension "*A*" is larger than "*B*," use a thin shim and place it between the work and solid jaw at point "*C*" (Figure 18). With the shim in place and the workpiece securely fastened, take a cut .020 in. deep on side 2 and also on side 3. Measure dimensions "*A*" and "*B*" again to see if the shim corrected the prior difference. If necessary, repeat this operation with different size shims until the workpiece is parallel between sides 2 and 3.

20. Sides 1, 2, and 3 are square to each other, and only side 4 is yet to be machined (Figure 19). Set the workpiece as deep in the vise as practical. Use parallels to raise it above the vise if that is necessary. Because sides 2 and 3 are parallel, no rod is needed between the work and moving jaw. Set the depth of cut and repeat steps 9 to 11.

21. The workpiece is now square and parallel on four sides. Measure the dimensions of these sides and make additional cuts if more material has to be removed. In squaring a workpiece, do not remove all the excess material with the first cut, because additional corrective cuts often have to be made.

22. The ends of the workpiece are now machined square to the sides. Figure 20 shows how a relatively short workpiece can be set up in a vise to machine an end square to the sides. An accurate square is used to check the squareness of the workpiece in relation to the base of the vise. The vise clamps the workpiece only very lightly so that the workpiece can be aligned with the square by tapping it. Double-check the squareness by applying the square to the opposite side of the work, also. The workpiece should always be in the vise. If that cannot be done, use a spacer of equal thickness to get the vise jaws to close parallel to each other (Figure 21). The ends of a workpiece should be machined with the solid jaw parallel to the arbor axis and with the cutting pressure against the solid jaw. If the vise jaws are parallel to the table travel, there is the danger that the cutting pressure will push the workpiece out of the vise. The ends of workpieces can be safely machined when the work is clamped against an angle plate (Figure 22). The squareness of a workpiece in relation to the table surface can also be measured with a dial indicator. Fasten the dial indicator to the arbor with contact point touching the vertical work surface to be measured, then raise or lower the knee and adjust the workpiece until the indicator hand shows the workpiece to be vertical. Always test the workpiece again for squareness after the final tightening of the vise or other holding device, because the workpiece alignment is often disturbed. When both ends are to be squared, remove only enough material from the first side to clean up that end. This will leave adequate stock to machine the second side square and to the desired overall length.

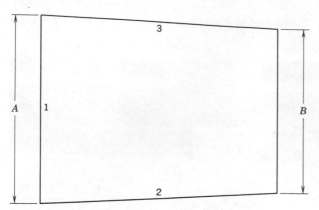

Figure 17. Measuring for squareness.

Figure 18. Location of shim to square up work (Lane Community College).

Figure 19. Setup to machine side 4 (Lane Community College).

Figure 20. Setup of a workpiece to machine an end square (Lane Community College).

Figure 21. Work clamped off center needs a spacer (Lane Community College).

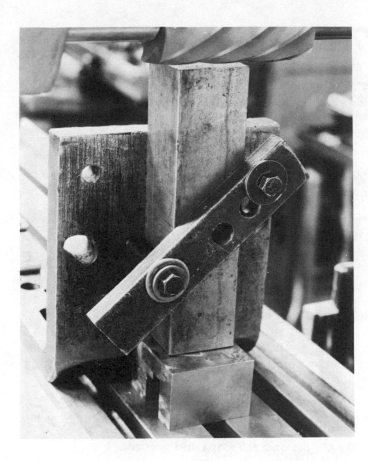

Figure 22. Use of angle plate to mill ends of work-pieces (Lane Community College).

self-evaluation

SELF-TEST
1. Why is the solid jaw used to align a vise on a milling machine table?
2. Which is more accurate, using a precision square or a dial indicator to square a vise on a machine table?
3. What is the purpose of the keys used on the base of machine vises?
4. Should you mount the indicator on the column to align the table on a universal milling machine?
5. Should the solid vise jaw be in a specific position in relation to the direction of the cut?
6. Is a large or small diameter cutter more efficient?
7. What is the difference between conventional milling and climb milling?
8. How deep should a finish cut be?
9. Why are all table movements locked except the one being used during machining?
10. Why is the cutter rotation stopped while the table is returned over the newly cut surface to its starting position?

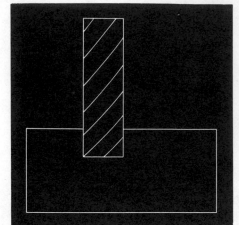

unit 7
using side milling cutters on the horizontal milling machine

Some milling cutters are used to mill only flat surfaces; others can be used to machine grooves and steps. Side milling cutters fall into this second category. A milling machine operator often makes the selection of the cutter, feed, speed, and workholding device to perform side milling operations.

objectives **After completing this unit, you will be able to:**

1. Set up side milling cutters and cut steps and grooves.

2. Use side milling cutters for straddle milling.

3. Use side milling cutters in gang milling.

information

Milling cutters with side cutting teeth are used when grooves and steps have to be machined on a workpiece. The size and kind of cutter to be used depends largely on the operation to be performed. Full side mills with cutting teeth on both sides are used when slots or grooves are cut (Figure 1).

Use only sharp cutters; they will leave a better surface finish and use less power. Cutters should be resharpened when they are only slightly dull. At this point, resharpening takes very little time. Cutters that are used until the cutting edges are worn down need an expensive reconditioning operation.

Cutters usually make a slot that is slightly wider than the nominal width of the cutter. Cutters that wobble because of dirt or chips between the arbor spacers will cut slots that are considerably oversize. A slot will become wider when more than one cut is made through it. If a slot needs to be .375 in. wide, a $\frac{3}{8}$ in. cutter probably cannot be used because it may cut a slot .3755 to .376 in. wide. A trial cut in a piece of scrap metal will tell the exact slot width. It may be necessary to use a $\frac{5}{16}$ in. wide cutter and make two or more cuts.

The width of a slot from a given cutter is also affected by the amount of feed used. A very slow feed will tend to let the cutter cut more clearance for itself. A fast feed crowds the cutter in the slot. A cutter tends to cut a wider slot in soft material than it will in harder material. The width of slots are measured with vernier calipers or with adjustable parallels. If it is a keyway, the key itself can be used as a plug gage to test the slot width.

Half side mills with cutting teeth on one side and on the periphery can be used when a step is milled where the cutter is in contact with two

628

Figure 1. Full side milling cutter machining a groove (Lane Community College).

Figure 2. Half side milling cutter machining a step (Lane Community College).

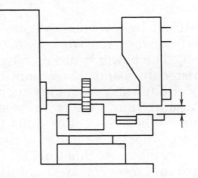

Figure 3. Check for clearance.

Figure 4. Work laid out for milling (Lane Community College).

sides only (Figure 2). The diameter of the cutter to be used on a job depends on the depth of the slot or step. As a rule, the smallest diameter cutter that will do the job should be used, as long as sufficient clearance remains between arbor and work and between arbor support and vise (Figure 3).

A good machinist will mark his workpiece with layout lines before he fastens it in a vise. The layout should be an exact outline of the part to be machined. The reason for making the layout prior to machining is that reference surfaces are often removed by machining. After the layout has been made, make diagonal lines on the portion to be machined away. This helps in identifying on which side of the layout line the cut is to be made (Figure 4). The cutter can be accurately positioned on the workpiece with the hand feed cranks and the micrometer dials. When the finished outside surface of a workpiece must not be marred or scratched by a revolving cutter, a paper strip is held between the workpiece and the cutter (Figure 5). The power is turned off and the spindle is rotated by hand. Make sure the paper strip is long enough so your hands are not near the cutter. Carefully move the table toward the revolving cutter. When the cutter pulls the paper strip from your fingers, the cutter is about .002 in. from the workpiece. At this time set the cross feed dial on zero. Lower the knee until the cutter clears the top of the workpiece. Then, by using the cross feed hand wheel, position the work where the cut is to be made. The same method will work in positioning a cutter above a workpiece and establishing the zero position for the depth of cut without actually touching the workpiece with the cutter (Figure 6).

A quicker, but not as accurate, method is illustrated in Figure 7. A steel rule is used to position a side mill a given distance from the outside edge of a workpiece. The end of the rule is held firmly against the side cutting edge of a tooth. The distance is indicated by the edge of the workpiece. The micrometer dials of the cross feed and knee controls should be zeroed when the cutter contacts the side and top of the workpiece. When these zero positions are established, additional cuts can easily be made by positioning from these points with the micrometer dials.

After the first cut is made, the distance of the side of the step or groove to the outside of the workpiece should be measured. A measurement made on both ends shows if the cut is parallel to the sides of the workpiece. If the step is not parallel, the vise needs to be aligned or, on a universal milling machine, the table may need aligning.

Figure 6. Using a paper strip to set the depth of cut (Lane Community College).

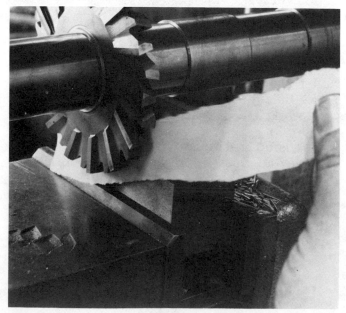

Figure 5. Setting up a cutter by using a paper strip (Lane Community College).

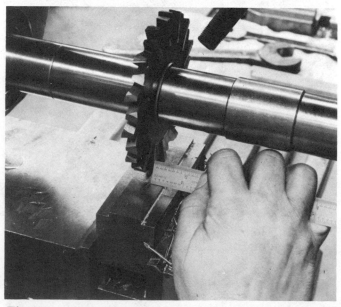

Figure 7. Positioning a cutter using a steel rule (Lane Community College).

When the table direction is reversed, compensation for backlash needs to be made. Backlash can be observed when the feed screw is turned, but the table does not start moving until all play between the drivenuts and feedscrew is removed. When possible, measurements should be made while the workpiece is still fastened in the milling machine because additional cuts can then be made without additional setup work. No measurements should be made while the cutter is revolving. Carefully remove all burrs from steps and grooves with a file prior to measuring (Figure 8).

Before making a final cut, if you are not sure of your dimensions, advance the revolving cutter until the cutter just nicks the corner of the workpiece (Figure 9). Stop the spindle and make a measurement at point "X" with a micrometer. If the location is correct, finish the cut. If the dimension is wrong, adjust the table accordingly. A small nick left on a corner is often covered up when the workpiece is chamfered.

Side milling cutters are combined to perform straddle milling operations (Figure 10). The width of the spacers between the cutters controls the width of the workpiece. It is important that the diameters of the cutters are the same if steps of equal depth are to be produced. In gang mill-

ing, a number of milling cutters are combined to cut special shapes and contours (Figure 11). The depth of the steps is determined by the difference in diameter of the various cutters. The RPM of the spindle is calculated for the largest diameter cutter in the gang.

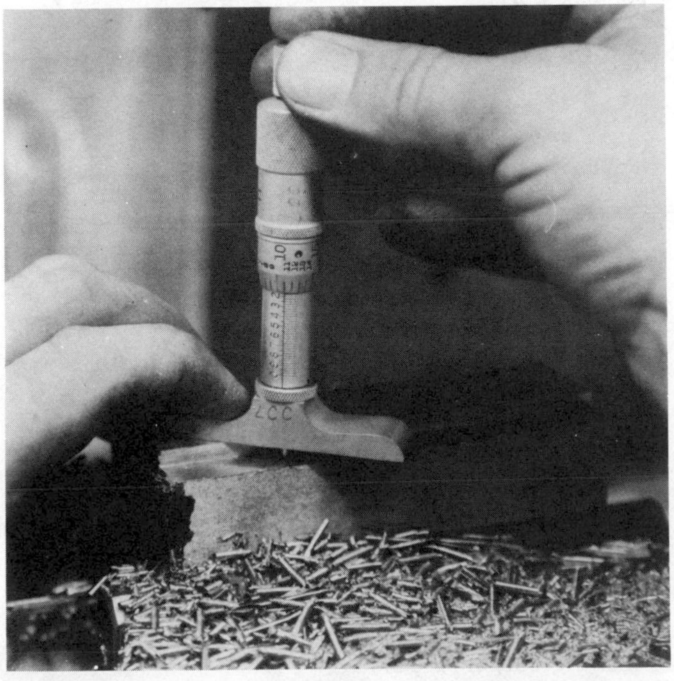

Figure 9. Taking a trial cut (Lane Community College).

Figure 10. Straddle milling (Lane Community College).

Figure 8. Measuring depth of a step (Lane Community College).

Figure 11. Gang milling (Courtesy of Cincinnati Milacron).

Figure 12. Left-hand and right-hand helical flutes on wide cuts (Courtesy of Cincinnati Milacron).

When wide cuts are made, interlocking tooth cutters with right-hand and left-hand helical flutes are used to offset the heavy side thrust (Figure 12).

Interlocking side milling cutters are used when grooves of a precise width are machined in one operation (Figure 13). Shims inserted between the individual cutters make precise adjustments possible. The overlapping teeth leave a smooth bottom groove. Cutters, which have become thinner through sharpening, can also be readjusted to their full width by adding shims.

Before any machining is started on a workpiece, you should have a plan of the sequence of operations that you will perform. This plan should include the answers to questions such as:

How is the workpiece held while it is being machined?

Do some machining operations come before others?

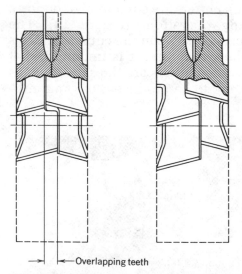

Overlapping teeth

Figure 13. Interlocking side milling cutters.

Is the setup strong enough to withstand the cutting forces?

self-evaluation

SELF-TEST 1. When are full side milling cutters used?

2. When are half side milling cutters used?

3. What diameter side milling cutter is most efficient?

4. Is a groove the same width as the cutter that produces it?

5. Why should a layout be made on workpieces?

6. How can a side milling cutter be positioned for a cut without marring the workpiece surface?

7. Why should measurements be made before removing a workpiece from the workholding device?

8. How is the width of a workpiece controlled in a straddle milling operation?

9. What determines the depth of the steps in gang milling?

10. When are interlocking side mills used?

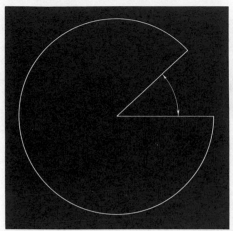

section k
indexing
devices

The precise division of the circle has been a major concern for many centuries, from the first instruments for navigation and later for clocks and instruments for surveying. A dividing engine for cutting clock gears was made in 1670. It used a plate with holes for the basis of its divisions. By the middle of the eighteenth century, it was possible to compare distances along an arc within .001 in. and to estimate to about one third that increment, or about .0003 in. The basis was set for the evolution of dividing engines into practical devices to be used for manufacturing parts requiring accurate division.

Indexing devices are used on vertical and horizontal milling machines and on other types of machine tools such as jig boring machines, planers, slotters, and shapers. These devices are also used for many types of inspection.

The indexing devices used range from quite simple to highly complex and are selected to match the requirements of the job. The use of collet fixtures and hand circular milling tables are more popular on vertical mills, while the use of dividing heads and power-fed rotary tables are more common on horizontal milling machines.

One of the simplest indexing devices is a collet index fixture (Figure 1), which can be set up either horizontally or vertically on the milling machine table. These are quick and easy to use for milling polygons and other holding and dividing operations.

Another device that is more complex is the circular milling table. The table is marked in degrees and uses either a vernier to read out minutes or a hand wheel graduated in minutes of arc (Figure 2). Large circular milling tables (Figure 3) are frequently equipped to be driven in rotation by being coupled to the machine table drive mechanism and thus can be used for dividing by hand setting or for circular milling by hand or by power feed (Figure 4).

The circular milling attachment may be equipped with an indexing attachment (Figure 5) when indexing requirements are more exacting. The indexing attachment is especially useful for the accurate division of parts that have relatively large diameters.

Devices that have been developed directly from the line of the dividing engines are known as rotary tables. These are based on either a precise worm to worm wheel relationship (Figure 6) accurate to plus or minus 2 sec of arc or, in its most refined form, on an

Figure 1. Collet index fixture (Courtesy of Hardinge Brothers, Inc.).

636

Machine Tool Practices

Figure 2. Circular milling table (Courtesy of Cincinnati Milacron)

Figure 3. A circular milling table with power feed is termed a circular milling attachment (Courtesy of Cincinnati Milacron).

optical circle read with a combination of microscope and vernier to a final reading of 1 sec of arc (Figure 7). Rotary tables are used on master machine tools such as jig boring (Figure 8) and jig grinding machines and for use in inspection (Figure 9). For calibration and inspection of these master tools, a precision index (Figure 10) is used and it is accurate to $\frac{1}{10}$ sec of arc. The second of arc is approxi-

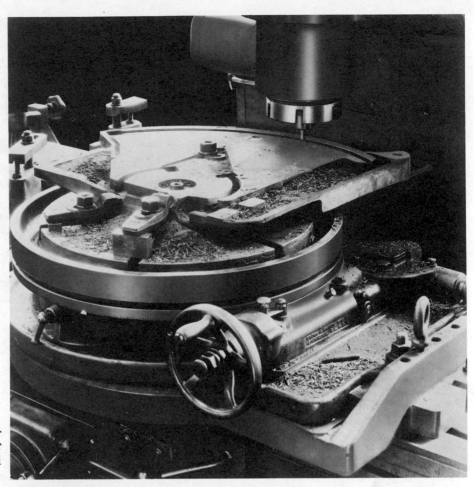

Figure 4. Circular milling attachment being used to mill a circular T-slot (Courtesy of Cincinnati Milacron).

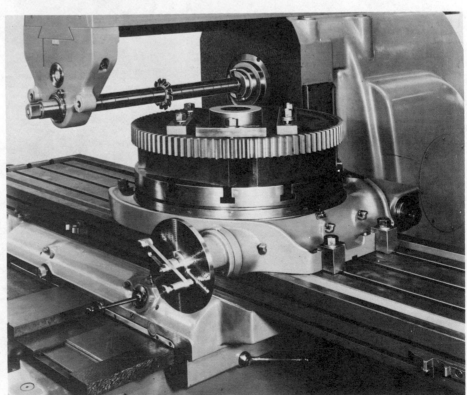

Figure 5. Circular milling attachment equipped with an indexing attachment (Courtesy of Cincinnati Milacron).

638

Figure 6. Ultraprecise rotary table (Courtesy of Moore Special Tool Co., Inc., Bridgeport, Conn.).

Figure 7. Optical dividing table (Courtesy of American SIP Corporation).

Figure 8. Using the rotary table on a jig boring machine. (Courtesy of Moore Special Tool Co., Inc., Bridgeport, Conn.).

Figure 9. Rotary table with tailstock and adjustable center being used for inspection (Courtesy of Moore Special Tool Co., Inc., Bridgeport, Conn.).

Figure 10. Precision index—a master tool for master tools (Courtesy of Moore Special Tool Co., Inc., Bridgeport, Conn.).

mately the angle subtended by the diameter of a common pencil at a range of a mile.

Many types of horizontal and universal dividing heads have been developed, since they are applied to the manufacture of parts on milling machines and other manufacturing machines. Horizontal spindle indexing devices such as the direct indexing head has the spindle directly connected with the indexing crank (Figure 11). The division equals the number of holes in hole circles, available on index plates to fit the head (50 holes maximum).

A gear cutting attachment that has the addition of 40:1 worm and worm wheel (Figure 12) gives capability of numbers up to 60, any even number, and those numbers divisible by 5 from 60 to 120.

The spiral milling head is the gear cutting attachment with the addition of a driving connection (Figure 13) to couple through gears to the leadscrew. It permits the milling of spiral forms, such as the flutes on helical reamers and slab milling cutters.

The universal spiral dividing head is an indexing device that performs the same function as the spiral milling head with provision for coupling to the leadscrew of the milling machine. In addition, the spindle assembly is carried in a swivelling block that permits the spindle to be tilted from minus 5 degrees to past the vertical by 5 to 10 degrees. A typical application of the swivel feature is shown in Figure 14. Dividing heads are also useful for cutting graduations in combination with devices like the slotting attachment (Figure 15).

Spiral indexing heads are coupled to the machine leadscrew by the use of a standard universal dividing head driving mechanism

640

Figure 11. Direct indexing head (Courtesy of Cincinnati Milacron).

Figure 12. Gear cutting attachment (Courtesy of Cincinnati Milacron).

Figure 13. Spiral milling head (Courtesy of Cincinnati Milacron.)

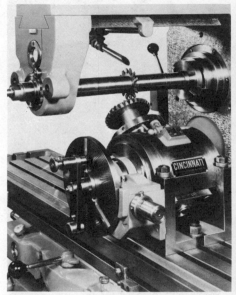

Figure 14. *(top)* Swiveling block on universal spiral dividing heads permits angular indexing (Courtesy of Cincinnati Milacron).

Figure 15. *(right)* Dividing heads are also useful for making graduations on conical surface (Courtesy of Cincinnati Milacron).

(Figure 16), which has change gears to control the direction of rotation, and the overall ratio to the dividing head. Figure 17 shows a combination of a driving mechanism and a universal spiral dividing head set up on a universal horizontal milling machine for fluting a large drill with a formed milling cutter.

The short and long lead attachment is used on universal general purpose milling machines equipped with a feature to disengage the leadscrew clutch so the table may be driven by the main shaft through gearing in the attachment gearbox (Figure 18). This permits the selection of over 13,000 leads ranging from .010 to 1000 in., or by an additional gear change from .025 to 3000 in. An example of the type of milling that can be done with this device in combination with other accessories is shown in Figure 19.

The wide range dividing head makes the regular 40:1 ratio dividing head, supplied with standard or high number index plates, adequate for the great majority of indexing work. However, there are combinations where a wide range dividing head (Figure 20), having an additional 100:1 ratio compounded with the standard 40:1 ratio, becomes necessary. This gives a minimum increment of 6 sec of arc per division. Dividing heads are also useful for inspection purposes in the same way that the rotary table is used (Figure 21).

As with the development of many specialized machine tools for production purposes, the indexing has been synchronized with the rotation of the machine spindle to create a hobbing attachment (Figure 22) for the horizontal milling machine. This type of device is rare, but it shows the basis for one type of production gear-making machine that will be shown in the next section.

Devices to divide the circle, both for the production of other instruments and for the production of gears for quiet, efficient transmission of power, are really at the core of an industrialized society. It is important for you to be able to select these devices

Figure 16. Standard universal dividing head driving mechanism (Courtesy of Cincinnati Milacron).

Figure 17. Spiral milling (Courtesy of Cincinnati Milacron).

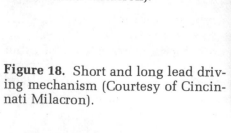

Figure 18. Short and long lead driving mechanism (Courtesy of Cincinnati Milacron).

Figure 19. A combination of attachments for the milling of a leadscrew (Courtesy of Cincinnati Milacron).

Figure 20. Dividing head with wide range divider (Courtesy of Cincinnati Milacron).

Figure 21. Inspecting a cam using a dividing head (Courtesy of Cincinnati Milacron).

Figure 22. A special hobbing attachment for the horizontal milling machine (Courtesy of Cincinnati Milacron).

unit 1
dividing
heads and
rotary tables

The dividing head and rotary table are precision milling machine attachments. They will accurately rotate a workpiece a full or partial turn. These devices are used when cutting gears, splines, keyways, or holes that must be specific angular distances apart.

objectives After completing this unit, you will be able to identify the main components of indexing devices.

information

The dividing head, which is also called an index head, is used to give rotary motion to workpieces in milling operations. Its important part is the housing, which contains the spindle (Figure 1). The spindle has a worm wheel attached to it, and this worm wheel meshes with a worm. The worm is turned thru a set of gears with the index crank or by a gear train connected to the milling machine leadscrew. The spindle axis can be swiveled to allow the holding of workpieces in a horizontal or vertical position. Once the spindle axis is adjusted, the clamping straps are tightened to lock the spindle in this position. On most dividing heads the worm can be disengaged from the worm wheel. The worm is disengaged by turning an eccentric collar. When the worm is disengaged, the spindle can easily be turned by hand while setting up or while direct indexing. In direct indexing, the plunger is used to engage the direct indexing pin into a hole of the direct indexing hole circle on the spindle nose (Figure 2). The spindle nose has a tapered hole to hold taper shank tools and chucks. On some dividing heads the spindle nose is threaded to hold screw on chucks. After the spindle has been rotated, the spindle lock is tightened to prevent any spindle movement while cutting. When the worm is en-

gaged, spindle rotation is obtained by turning the index crank. The most commonly used index ratio between the index crank and the spindle is 40:1. This means that the index crank needs to be turned 40 revolutions to get 1 revolution of the spindle. Another index ratio found on dividing heads is 5:1. The index crank has a plunger that moves an index pin in and out of a hole on the index plate. The index plate has a number of hole circles with equally spaced holes in each circle. The index plate is used to obtain precise partial revolutions of the index crank.

Often the index plate has a different set of hole circles on the reverse side. For some dividing heads, high number index plates are available to obtain a great number of different hole circles. When only a partial revolution of the index crank is made, the sector arms are set apart a distance equal to that partial turn to avoid counting the spaces for each indexing turn. The index plate stop engages in serrations on the circumference of the index plate. When it is loosened, the index plate can be rotated in small increments.

A wide range index head (Figure 3) that permits indexing of 2 to 400,000 divisions is available. The wide range divider has a large

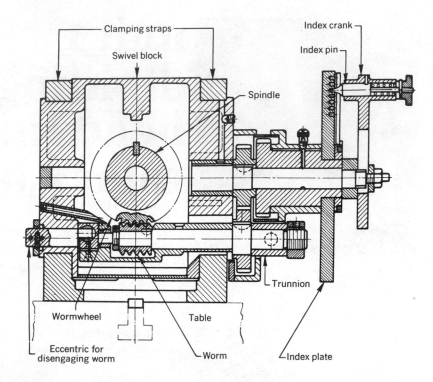

Figure 1. Section through dividing head showing worm and worm shaft (Courtesy of Cincinnati Milacron).

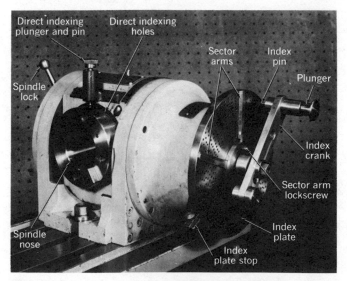

Figure 2. Indexing components of a dividing head (Lane Community College).

Figure 3. Wide range divider permits indexing of 2 to 400,000 divisions (Courtesy of Cincinnati Milacron.)

index plate, index crank, and sector arm, as does a regular dividing head, with a 40:1 ratio. In front of the large index plate a small index plate, index crank, and sector arms are mounted. The small index crank is geared to the worm of the large index crank with an additional 100:1 ratio to get a total ratio of 4000:1 (40:1 times 100:1). It takes 4000 revolutions of the small index crank to rotate the spindle one full revolution. The small

index plate has 2 hole circles of 100 and 54 holes. When the index crank is moved 1 space in the 54 hole circle on the small plate, the work is rotated through 6 sec of arc.

The footstock (Figure 4) supports one end of the workpiece. The footstock center can be adjusted toward and away from the dividing head to permit the removal of workpieces and to make up for different length workpieces. The footstock

Figure 4. Footstock (Lane Community College).

Figure 5. Adjustable center rest (Lane Community College).

center can also be adjusted vertically to allow the leveling of workpieces. Tapered workpieces are supported by swiveling the axis of the footstock center horizontally. Long or slender workpieces are supported with an adjustable center rest (Figure 5).

Another commonly used indexing device is the rotary table (Figure 6). Index ratios of rotary tables vary from 120:1, 80:1, or 40:1. Usually the table is graduated in degrees, while the hand wheel has 1 min graduations. With an index plate, very accurate spacings can be made. To determine the index ratio of an indexing device, turn the index crank 10 complete revolutions and measure the resulting spindle revolution. If the spindle has rotated $\frac{1}{4}$ turn, the index ratio is 40:1.

The index crank should be carefully adjusted so that the index pin slips easily into the holes in whatever hole circle is used. Once an indexing operation is started, the index crank should only be turned in one direction (usually

Figure 6. Rotary table (Lane Community College).

clockwise). Turning the index crank clockwise and then counterclockwise will allow the backlash between the worm and worm wheel to affect the accuracy of the indexing.

self-evaluation

SELF-TEST
1. When are indexing devices used?
2. What makes indexing devices so accurate?
3. When is the worm disengaged from the worm wheel?
4. When is the hole circle on the spindle nose used?
5. What is a commonly used index ratio on dividing heads?

6. Why does the index plate have a number of different hole circles?
7. What is the purpose of the sector arms?
8. What does the spindle lock do?
9. How can divisions be made that are not possible with a standard index plate?
10. Why should the index crank be rotated in one direction only while indexing?

unit 2
direct
and simple
indexing

Most indexing operations performed in machine shops fall into the direct and simple indexing categories. The accuracy of the indexing operation determines which dividing method to use. Examples of indexing operations are the cutting gear teeth, splines, keyways, or the machining of square or hexagon shapes.

objectives **After completing this unit, you will be able to do the calculations for direct and simple indexing.**

Direct indexing is the easiest method of dividing a workpiece into a number of equal divisions. The number of divisions obtainable by direct indexing is limited by the number of holes in the direct indexing hole circle in the spindle nose.

Hole circles available have 24, 30, or 36 holes. To perform direct indexing rapidly, the worm should be disengaged from the worm wheel to allow the spindle to be turned by hand. When a 24 hole circle is used, equal divisions of 2, 3, 4, 6,

8, 12, and 24 spaces can be made. As an example, let us assume a hexagon needs to be machined with direct indexing. The 24 holes divided by 6 equal 4 holes. This means that for each cut, the spindle has to be turned a distance of 4 more holes from the preceding cut. When making this 4 hole advance, do not count the hole the index pin is in. As soon as the index pin is seated in the hole, tighten the spindle lock to prevent any spindle movement while the machining takes place. To avoid using the wrong hole when rotating the spindle, take a felt pen or other marker to identify which holes to use prior to the machining.

Simple indexing, also known as plain indexing, involves the turning of the index crank to rotate the spindle. On most dividing heads, 40 turns of the index crank results in one revolution of the workpiece. To obtain a specific number of spaces on the circumference of a workpiece, 40 is divided by that number to get the number of whole or partial turns of the index crank for each division. To make 20 equal divisions on a workpiece, divide 40 by 20, which gives $\frac{40}{20} = 2$. The 2 represents 2 complete turns of the index crank. To cut 80 teeth on a gear, write $\frac{40}{80} = \frac{1}{2}$, or $\frac{1}{2}$ revolution of the index crank for each tooth.

When a partial turn of the index crank is needed, an index plate with a number of different hole circles is used. Index plates with the following holes are available: 24, 25, 28, 30, 34, 37, 38, 39, 41, 42, 43, 46, 47, 49, 51, 53, 54, 57, 58, 59, 62, and 66. To get a $\frac{1}{2}$ revolution of the index crank, any hole circle divisible by 2 can be used. If the 30 hole circle is used, the index pin is to be advanced 15 holes each time.

To make 27 divisions, proceed as follows. Divide 40 by 27, which is $\frac{40}{27} = 1\frac{13}{27}$. The 1 represents one complete turn of the index crank. The denominator 27 could be the hole circle to use and the numerator 13 the number of holes to be advanced. A check of the available hole circles shows that a 27 is not there, but a 54 is, and that is a multiple of 27. To keep the fraction of $\frac{13}{27}$ intact and to increase the denominator to 54, multiply both the numerator and denominator by 2. It now looks like this: $\frac{13 \times 2}{27 \times 2} = \frac{26}{54}$ and means 26 holes in the 54 hole circle. For each division then make 1 turn plus 26 holes in the 54 hole circle.

When 52 divisions are required, divide 40 by 52, $\frac{40}{52}$, and reduce it to the lowest fraction

Figure 1. Dividing head sector arms set for indexing 11 spaces (Lane Community College).

possible: $\frac{40}{52} \cdot \frac{4}{4} = \frac{10}{13}$ There is not a 13 hole circle, but a 39 is available. Now raise the fraction to a denominator of 39 by multiplying by 3, $\frac{10 \times 3}{13 \times 3} = \frac{30}{39}$, or 30 holes in a 39 hole circle. The highest indexing accuracy is achieved when the hole circle with the greatest number of holes is used that will accommodate the denominator of the fraction.

To obtain 51 divisions, write $\frac{40}{51}$. There is a 51 hole circle available, which makes no calculation necessary. It takes 40 holes in the 51 hole circle for each division.

It would be very awkward to count the number of holes for each indexing operation. This is why sector arms are found on the index plate. The sector arms can be adjusted to form different angles by loosening the lockscrew. One side of each sector arm is beveled. The number of holes of the partial turn needed are located within these beveled sides. The sector arms in Figure 1 are set for an 11 hole movement in the hole circle. The hole that the pin is in is not counted.

To obtain spacings other than those available with standard index plates, high number index plates can be used. Another choice would be the use of a wide range divider. When change gears are available for older dividing heads, compound or differential indexing can be employed to make divisions not obtainable with standard index plates. Consult *Machinery's Handbook* for the procedures used in compound and differential indexing.

self-evaluation

1. What is the difference between direct and simple indexing?
2. How can you avoid using a wrong hole in the index plate while direct indexing?
3. If the direct indexing hole circle has 24 holes, what are the different divisions you can make with it?
4. How are the sector arms used on a dividing head?
5. If both a 40 and a 60 hole circle can be used, which is the better one?
6. Calculate how to index for 6 divisions. Use the hole circles from the text.
7. Calculate how to index for 15 divisions. Use the hole circles from the text.
8. Calculate how to index for 25 divisions. Use the hole circles from the text.
9. Calculate how to index for 47 divisions. Use the hole circles from the text.
10. Calculate how to index for 64 divisions. Use the hole circles from the text.

unit 3 angular indexing

Work may be indexed to produce a given number of spaces on the circumference, or the spacing can be indicated as an angular distance measured in degrees.

objective   After completing this unit, you will be able to perform the calculations for angular indexing.

Indexing by degrees can be done on a dividing head by the direct and simple indexing method. One complete revolution of the dividing head spindle is 360 degrees. If the direct indexing hole circle has 24 holes, the angular spacing between each hole is $\frac{360}{24}$ degree = 15 degrees. Any division requiring 15 degree intervals can be made. To drill 2 holes at a 75 degree angle to each other, divide 75 by 15 $(\frac{75}{15})$ = 5. The 5 represents 5 holes on the 24 hole circle. Remember to not count the hole that the index pin is in. If angles other than 15 degrees are to be indexed, the simple indexing method is used. To obtain one 360 degree revolution of the dividing head spindle, it takes 40 turns of the index crank. One turn of the index crank produces a $\frac{360}{40}$ degrees = 9 degree movement of the dividing head spindle. Any hole circle on the index plate that is divisible by 9 can be used to index by degrees.

With a 27 hole circle it takes 3 holes for 1 degree, or 1 hole is 20 min.

With a 36 hole circle it takes 4 holes for 1 degree, or 1 hole is 15 min.

With a 45 hole circle it takes 5 holes for 1 degree, or 1 hole is 12 min.

With a 54 hole circle it takes 6 holes for 1 degree, or 1 hole is 10 min.

When indexing in degrees, the formula for the number of turns of the index crank is $\frac{\text{degrees required}}{9}$. As an example, two cuts must be made 37 degrees apart.

$\frac{37}{9} = 4\frac{1}{9}$, which is 4 complete turns and $\frac{1}{9}$ turn. The $\frac{1}{9}$ is expanded to fit a 54 hole circle $\frac{1}{9} \times \frac{6}{6} = \frac{6}{54}$. The total movement required is 4 turns and 6 holes in the 54 hole circle.

Indexing can also be done in minutes, with the formula $\frac{\text{minutes required}}{540}$. The denominator of 540 is obtained by multiplying 9 degrees by 60 min, the number of minutes in one revolution of the index crank. When using this formula, it becomes necessary to convert the degrees and partial degrees into minutes. As an example, two holes have to be drilled at 8 degrees 50 min from each other. The 8 degrees 50 min equals 530 min. Putting this value into the equation, we get a fraction of $\frac{530}{540}$ To use a 54 hole circle, this fraction has to be reduced.

$$\frac{530}{540} \div \frac{10}{10} = \frac{53}{54}$$

The result is an index crank movement of 53 holes in a 54 hole circle.

When the required minutes are not evenly divisible by 10, the required spacing may only be approximate with a slight error. As an example, calculate the index crank movement for a 1 degree 35 min spacing. Hole circles available are: 38, 39, 41, 42, 43, 46, 47, 49, 51, 53, 54, 57, 58, 59, 62, and 66. Convert the mixed number into minutes: 60 min + 35 min = 95 min. This value is entered into the equation $\frac{95}{540} = \frac{1}{5.685}$. There is no hole circle with 5.685 holes available, so use the trial and error method in expanding this fraction.

$$\frac{1}{5.685} \times \frac{7}{7} = \frac{7}{39.7949}$$

$$\frac{1}{5.685} \times \frac{8}{8} = \frac{8}{45.4799}$$

$$\frac{1}{5.685} \times \frac{9}{9} = \frac{9}{51.1649}$$

$$\frac{1}{5.685} \times \frac{10}{10} = \frac{10}{56.85}$$

It appears that the 51.1649 is the closest number to the available hole circle of 51. Each hole in the 51 hole circle is $\frac{540}{51}$, or 10.588 min spaced from the adjacent one. Moving 9 holes in the 51 hole circle gives an angle of 95.292 min or 95 min 17 sec. This is an error of only 17 sec.

High number index plates or a wide range divider will give a greater choice of accurate angular spacings.

self-evaluation

SELF-TEST **1.** If there are 24 holes in the direct indexing plate, how many degrees are between holes?

2. How many holes movement is necessary to index 45 degrees using the direct indexing method?

3. How many degrees in the movement produced by one complete turn of the index crank?

4. Which hole circles on the index plate can be used to index by whole degrees?

5. How many turns of the index crank are necessary to index 17 degrees?

6. What fraction of one degree is represented by 1 space on the 18 hole circle?

7. What fraction of one degree is represented by 1 space on the 36 hole circle?

8. What fraction of one degree is represented by 1 space on the 54 hole circle?

9. How many minutes movement is produced by one turn of the index crank?

10. How many turns of the index crank are necessary to index 54 degrees 30 min, using the 54 hole circle?

section 1
gears

Making gears for light duty purposes such as clock mechanisms goes back for several hundred years, but the history of machines that can produce gears capable of transmitting substantial power only dates back to about 1800. Since that time thousands of patents have been taken out on devices for cutting gears and gear cutting tools.

The ability to produce gears economically did not appear until about 1850 and was the basis for the development of the high speed printing press, the sewing machine, and many other useful products that began to appear in quantity after the Civil War. These early gear cutting machines (Figure 1) typically used a formed cutter made to mill out the space between the teeth to leave a tooth of correct shape. Large machines of this period often used a template to guide a tool in a slide to form the shape of the tooth.

As the demands increased for faster production of gears in the 1860s, the development of highly specialized production gear making machinery began to emerge. It was observed that a worm made as a cutter, having straight sides in the same tooth form as a rack, would generate an involute gear if the two were turned together so that the gear would move one space while the worm made one revolution (Figures 2 and 3). This is the basis for a specialized variety of milling machine known as the hobbing machine. By 1900, this was a well-developed and highly productive method of producing accurate gears. Hobbing machines come in both horizontal spindle (Figure 4) and vertical types. The larger machines usually have a vertical spindle for holding the gear blanks. Some of these machines are used for gears 16 ft in diameter and larger (Figure 5).

About 1900, another important production gear making machine appeared. It used a cutter shaped like a mating gear that was reciprocated across the gear to be cut and rotated with the blank as the cutting progressed. This machine is called a gear shaper (Figure 6). This machine can be used to generate external or internal spur or helical gears and has the advantage of being able to machine the gear form next to a shoulder, as is often the case in gear clusters in transmissions. This type machine can also be used to generate unusual forms of gears (Figure 7).

As the automobile became popular, it was also important to be able to produce quiet, accurate gears for the rear ends of cars. Machines like this hypoid gear generator (Figure 8) were developed to meet that need.

As more rapid means were needed to produce gears of accurate form for high production applications, methods such as broaching

653

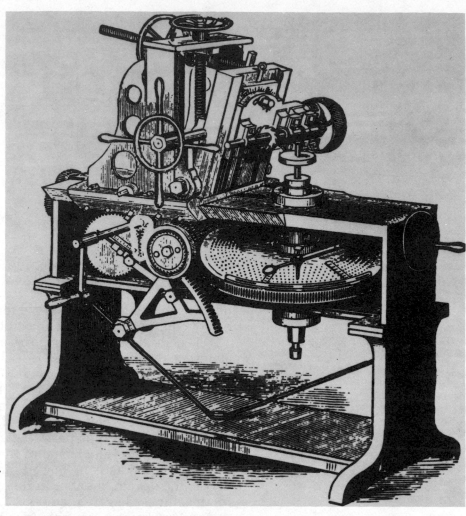

Figure 1. Joseph Brown's 1855 gear cutting machine (Courtesy of Brown & Sharpe Mfg. Co.).

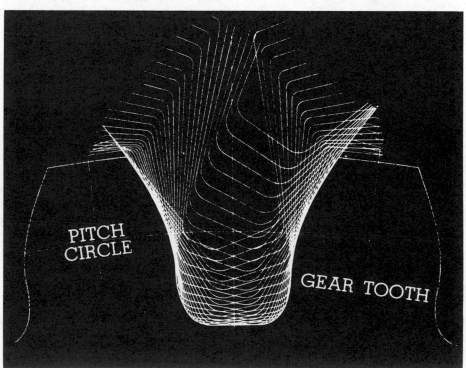

Figure 2. Schematic view of the generating action of a hob (Reprinted with permission of Barber-Colman Company).

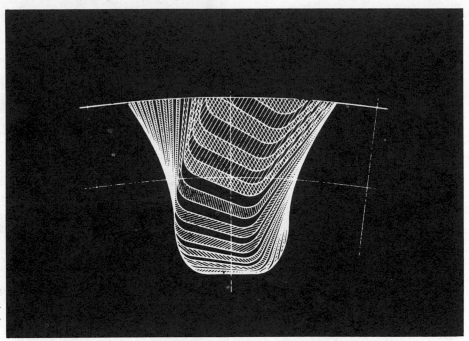

Figure 3. Schematic view of the performance of an individual hob tooth (Reprinted with permission of Barber-Colman Company).

Figure 4. Small horizontal spindle hobbing machine (DeAnza College).

were developed. This method is especially adapted to producing individual or nonclustered gears (Figure 9) with straight teeth or with helix angles under 21 degrees.

For commercial use, gear teeth are usually finished after they are cut. The type of finishing method used depends greatly on the

Figure 5. A huge vertical spindle gear hobbing machine. Notice the operator with the pendant control at the upper left (Courtesy of Ex-Cell-O Corporation).

Operator

Control console

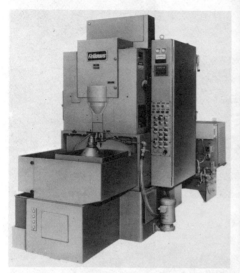

Figure 6. A gear shaper. The cutter moves up and down with the same twist as the helix angle of the gear (Courtesy of Fellows Corportion).

Figure 7. The gear shaper can produce unusual gear shapes (Courtesy of Fellows Corporation).

Figure 8. Hypoid gear generator (Courtesy of Gleason Works).

Figure 9. Pot broaching tool that produces a helical gear (Courtesy of National Broach and Machine Division, Lear Siegler, Inc.)

Figure 10. Gear shaving (Courtesy of National Broach and Machine Division, Lear Siegler, Inc.).

service condition of the gear, its accuracy requirements, noise limitation, and hardness requirements.

Gear shaving can be employed on gears that are not harder than Rockwell C30, where surface finishes in the order of 32 microinches are adequate for the application. In some cases, shaving can produce finishes of 16 microinches.

One method uses a shaving cutter rolled in mesh with the work gear at an angle (Figure 10). This is called diagonal shaving.

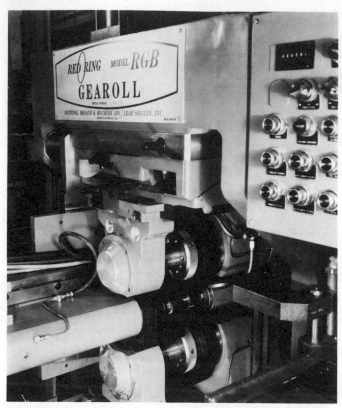

Figure 11. Roll finishing of unhardened gears (Courtesy of National Broach and Machine Division, Lear Siegler, Inc.).

Figure 12. Finishing a gear by form grinding (Courtesy of National Broach and Machine Division, Lear Siegler, Inc.).

Another method for use with gears that are not hardened is roll finishing (Figure 11). This is a cold forming process that can produce finishes as fine as 10 microinches on relatively small gears, up to about 4 in. pitch diameter.

For hardened gears, it is necessary to use more expensive processes such as gear grinding. Gear grinders either use a method in which the grinding wheel is formed to the shape of the gear space (Figure 12) or a method in which the final shape is generated by relative motion between the grinding wheel and the already cut and hardened gear (Figure 13). The surface finishes produced by grinding generally range from 20 to 30 microinches.

For still finer finishes or hardened gears in the 10 to 15 microinch range, gear honing is employed (Figure 14). The honing tool is typically an abrasive impregnated mating gear rotated with the gear in a crossed axis relationship. Honing is also readily done on internal gears.

The inspection of gears is a highly refined technology employing very complex optical mechanical and electronic devices (Figure 15).

The gears that are produced in the general machine shop utilizing formed gear cutters are typically either for emergency use or for slow speed applications where unhardened gears are satisfactory

Figure 13. Finishing a gear by generating the ground surface (Courtesy of Fellows Corporation).

Figure 14. Finishing a hardened gear by honing. The honing tool has the helical teeth (Courtesy of National Broach and Macine Division, Lear Siegler, Inc.).

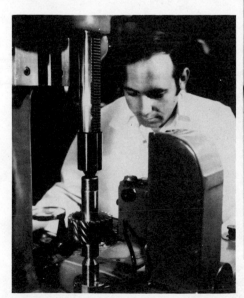

Figure 15. Inspecting gear profiles (Courtesy of Fellows Corporation).

and where accuracy and low noise requirements are not an important factor.

It is useful for machinists to know how to produce and measure spur gears for this type of service, especially in situations where mechanical maintenance is done. Producing spur gears by milling is often satisfactory, but making acceptable helical gears by spiral milling is unlikely and should probably only be done as a machine shop exercise for its instructional value.

The following units of instruction will define the common types of gears, gear tooth parts, and calculations, how to set up for milling a spur gear, and how to measure the dimensions of the resulting gear.

unit 1
introduction
to gears

Gears are used to transmit power and motion from one rotating shaft to another. Many different kinds of gears are manufactured, and this unit will explain the uses of some of the more common ones.

objective After completing this unit, you will be able to identify different gear designs and some of the materials used for gears.

Gears provide positive, no-slip power transmission and are used to increase or decrease the turning effort or speed in machine assemblies. When two gears are running together, the one with the larger number of teeth is called the gear and the one with the smaller number of teeth is called the pinion. Gears are generally used when shaft center distances are short, to provide a constant speed ratio between shafts, or to transmit high torques.

SPUR GEARS

Gears fall into several categories; the first is gears that connect parallel shafts. The best known of these gears is the spur gear. Spur gears have a cylindrical form with straight teeth cut into the periphery (Figure 1). When teeth are cut on a straight bar, a gear rack is made. A gear rack converts rotary gear motion into a linear movement. When the teeth on a cylindrical gear are at an angle to the gear axis, it is a helical gear (Figure 2). On a helical gear, several teeth are in mesh simultaneously with the mating gear, which provides a smoother operation than with spur gears. Because of the angle of the teeth, both radial and thrust loads are imposed on the gear support bearings. To offset this thrust effect, double helical gears are used. These gears are

herringbone gears (Figure 3), which consist of a right-hand and a left-hand helix. The hand of a helical gear is determined by facing the gear; if the teeth rotate to the right it is a right-hand helix.

Common reduction ratios for gears are 1:1 to 5:1 for spur gears, 1:1 to 10:1 for helical gears, and 1:1 to 20:1 for herringbone gears. These same three gear types are also manufactured as internal gears (Figure 4). An internal gear has a greater tooth strength than that of an equivalent exter-

Figure 1. Spur gear and gear rack (Lane Community College).

661

Figure 2. Helical gears (Lane Community College).

Figure 4. Helical internal gears (Lane Community College).

Figure 3. Herringbone gear (Lane Community College).

nal gear. An internal gear rotates in the same direction as its mating pinion. Internal gears permit close spacing of parallel shafts. Internal gears mesh with external pinions.

BEVEL GEARS

The second category concerns gears that connect shafts at any angle, providing the shaft's axis would intersect if extended. Gears used to transmit power between intersecting shafts are often bevel gears (Figure 5). Bevel gears are conical gears and may have straight or spiral teeth. Spiral bevel gears are smoother running and usually will transmit more power than straight bevel gears because more than one pair of teeth is in contact at all times.

Mating bevel gears with an equal number of teeth producing a 1:1 ratio are called miter gears (Figure 6). When the angle of the two shafts is other than 90 degrees, angular bevel gears are used. Bevel gears are used to provide ratios from 1:1 to 8:1. Face gears have teeth cut on the end face of a gear. The teeth on a face gear can be straight or helical. Ratios for face gears range from 3:1 to 8:1.

Figure 5. Bevel gears (Lane Community College).

Figure 6. Miter gears (Lane Community College).

Figure 7. Worm and worm gear (Lane Community College).

HELICAL GEARS

The third category is for nonintersecting, non-parallel shafts. Helical gears of the same hand, cut with a 45 degree helix angle, will mesh when two shafts are at 90 degree or right angles from each other. By changing the helix angle of the two gears, the angle of the shafts in relation to each other can be changed. This type of gear arrangement is called a crossed helical gear drive and is used for ratios of 1:1 to 100:1.

WORM GEARS

Worm and worm gears (Figure 7) are used for power transmission and speed reduction on shafts that are at 90 degrees to each other. Worm gear drives operate smoothly and quietly and give reduction ratios of 3:1 to over 100:1. In a worm gear set, the worm acts as the pinion, driving the worm gear. Two basic kinds of worm shapes are made: the single-enveloping worm gear set with a cylindrical worm, and the double-enveloping worm gear set where the worm is hourglass shaped. The worm of the dou-

ble enveloping worm gear set is wrapped around the worm gear and gives a much larger load carrying capacity to the worm gear drive. Worms are made with single thread or lead, double lead thread, triple lead thread, or other multiple lead threads. The number of leads in a worm is determined by counting the number of thread starts at the end of the worm. A single thread worm is similar to a single tooth gear when the ratio of a worm gear set is determined. A worm gear is made to run with the worm of a given number of leads. A double thread worm gear must be run with a double lead worm and cannot be interchanged with a single lead worm.

Another commonly used gear form is hypoid gears (Figure 8). Hypoid gears are similar in form to spiral bevel gears except that the pinion axis is offset from the gear axis. Hypoid gear ratios range from 1:1 to 10:1.

GEAR MATERIALS

Gear materials fall into three groups: ferrous, nonferrous, and nonmetallic materials. In the ferrous materials, steel and cast iron are used often. Steel gears, when hardened, carry the greatest load relative to their size. Steel gears can be hardened and tempered to exacting specifications. The composition of the steel can be changed. When the carbon content is increased, the wear resistance also increases. Lowering the carbon content gives better machinability. Cast iron is low in cost and is easily cast into any desired shape. Cast iron machines easily and has good wear resistance. Cast iron gears run relatively quietly and have about three quarters of the load carrying capacity of an equal size steel gear. One drawback in cast iron is its low impact strength, which prohibits its use where severe shockloads occur. Other ferrous gear materials are ductile iron, malleable iron, and sintered metals.

Nonferrous gear materials are used where corrosion resistance, light weight, and low cost production are desired. Gear bronzes are very tough and wear resistant. Gear bronzes make very good castings and have a high machinability. Lightweight gears are often made from aluminum alloys. When these alloys are anodized, a hard surface layer increases their wear resistance.

Low cost gears can be produced by die casting. Most die cast gears are completely finished when ejected from the mold with the exception of the removal of the flash on one side. Die casting materials are zinc base alloys, aluminum base alloys, magnesium base alloys, and copper base alloys.

Nonmetallic gears are used primarily because of their quiet operation at high speed. Some of these materials are layers of canvas im-

Figure 8. Hypoid gears (Lane Community College).

pregnated with phenolic resins; they are then heated and compressed to form materials such as formica or micarta. Other materials include thermoplastics, such as nylon. These nonmetallic materials exhibit excellent wear resistance. Some of these materials need very little lubrication. When plastic gears are used in gear trains, excessive temperature changes must be avoided to control damaging of dimensional changes. Good mating materials with nonmetallic gears are hardened steel and cast iron.

In many instances, gear sets are made up with different gear materials. Many worm drives use a bronze worm gear with a hardened steel worm. Cast iron gears work well with steel gears. To equalize the wear in gear sets, the pinion is made harder than the gear. Even wear in gear sets can be obtained when a gear ratio is used that allows for a "hunting tooth." For example, a gear set ratio of approximately 4:1 is needed. This is possible by using an 80 tooth gear in mesh with a 20 tooth gear. In this gear arrangement, the same tooth of the pinion will mesh with the same tooth of the gear in every revolution. If an 81 tooth gear were used, the teeth of the pinion will not equally divide into it, but each tooth of one gear will mesh with all of the mating teeth one after the other, distributing wear evenly over all teeth.

self-evaluation

SELF-TEST

1. Name two types of gears used to connect parallel shafts.
2. What are some advantages and some disadvantages of helical gears?
3. What is the direction of rotation of a pinion in relation to an internal gear when they are meshed together?
4. Two helical gears of the same hand and a 45 degree helix angle are in mesh. What relationship exists between the axis of the two shafts?
5. What is the gear reduction in a worm gear set when the worm has a double lead thread and the worm gear has 100 teeth?
6. Can a worm gear set ratio be changed by substituting a single start worm with a triple start worm?
7. What kind of gear material gives the greatest load carrying capacity for a given size?
8. What kind of material can be used for gears to run quietly at high speed?
9. How can the wear on the gear teeth be equalized when a large and a small gear are running together?
10. What kind of gear materials give corrosion resistance to gear sets?

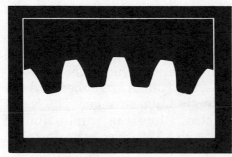

unit 2
spur gear terms and calculations

Most spur gears are mass produced on production gear making machines. Occasionally a spur gear needs to be made on a milling machine. Before a gear can be cut, its dimensions are calculated. This unit will help you to learn the names of gear tooth parts and how to calculate their sizes.

objective After completing this unit, you will be able to identify gear tooth parts and to calculate their dimensions.

Spur gear terms are illustrated in Figure 1. The definitions of these terms are as follows.

Addendum. The radial distance from the pitch circle to the outside diameter.

Dedendum. The radial distance between the pitch circle and the root diameter.

Circular thickness. The distance of the arc along the pitch circle from one side of a gear tooth to the other.

Circular pitch. The length of the arc of the pitch circle from one point on a tooth to the same point on the adjacent tooth.

Pitch diameter. The diameter of the pitch circle.

Outside diameter. The major diameter of the gear.

Root diameter. The diameter of the root circle measured from the bottom of the tooth spaces.

Chordal addendum. The distance from the top of the tooth to the chord connecting the circular thickness arc.

Chordal thickness. The thickness of a tooth on a straight line or chord on the pitch circle.

Whole depth. The total depth of a tooth space equal to the sum of the addendum and dedendum.

Working depth. The depth of engagement of two mating gears.

Clearance. The amount by which the tooth space is cut deeper than the working depth.

Backlash. The amount by which the width of a tooth space exceeds the thickness of the engaging tooth on the pitch circles.

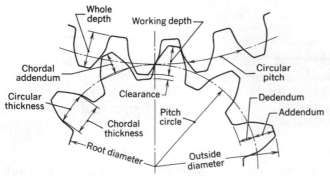

Figure 1. Spur gear terms.

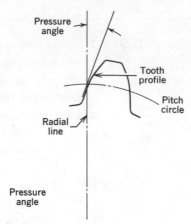

Figure 2. Pressure angle.

Diametral pitch. The number of gear teeth to each inch of pitch diameter.

Pressure angle. The angle between a tooth profile and a radial line at the pitch circle (Figure 2).

Center distance. The distance between the centers of the pitch circles.

Three spur gear tooth forms are generally used with pressure angles of $14\frac{1}{2}$, 20, and 25 degrees. The $14\frac{1}{2}$ degree tooth form is being replaced and made obsolete by the 20 and 25 degree forms. Figure 3 illustrates these three pressure angles as applied to a gear rack with all teeth being the same depth. The larger pressure angle makes teeth with a much larger base, which also makes these teeth much stronger. The larger pressure angles also allow the production of gears with fewer teeth. Any two gears in mesh with each other must be of the same pressure angle and the same diametral pitch.

Information in this unit will include constants for $14\frac{1}{2}$ degree pressure angle gear calculations because many existing gears and gear cutters are of this tooth form. When gear tooth measurements are to be made with gear tooth calipers, the chordal tooth thickness and also the chordal addendum must be calculated. Most of the gear tooth dimensions that you will calculate in this unit can be found in tables in the *Machinery's Handbook*, by Oberg and Jones.

The following symbols are used to represent gear tooth terms in spur gear calculations.

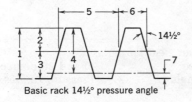

Basic rack 14½° pressure angle

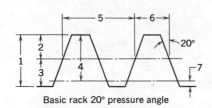

Basic rack 20° pressure angle

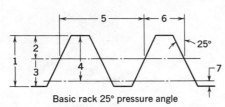

Basic rack 25° pressure angle

Figure 3. Comparison of tooth shape on gear rack with different pressure angles.

Legend for all three tooth forms

1 — Whole depth
2 — Addendum
3 — Dedendum
4 — Working depth
5 — Circular pitch
6 — Tooth thickness
7 — Clearance

P = diametral pitch
D = pitch diameter
D_o = outside diameter
N = number of teeth in gear
t = tooth thickness — circular
a = addendum
b = dedendum
c = clearance
C = center distance
h_k = working depth
h_t = whole depth
t_c = tooth thickness — chordal
a_c = addendum — chordal

Table 1 gives the formulas to calculate the gear dimensions in the following examples.

EXAMPLE 1. Determine the dimensions for a 30 tooth gear, 14½ degree pressure angle, and a 2.500 in. pitch diameter.

1. Number of teeth $N = 30$.

2. Pitch diameter $D = 2.500$ in.

3. Diametral pitch $P = \dfrac{N}{D} = \dfrac{30}{2.500} = 12$

4. Addendum $a = \dfrac{1}{P} = \dfrac{1}{12} = .083$ in.

5. Dedendum $b = \dfrac{1.157}{P} = \dfrac{1.157}{12} = .096$ in.

6. Tooth thickness
$t = \dfrac{1.5708}{P} = \dfrac{1.5708}{12} = .131$ in.

7. Clearance $c = \dfrac{.157}{P} = \dfrac{.157}{12} = .013$ in.

8. Whole depth $h_t = \dfrac{2.157}{P} = \dfrac{2.157}{12} = .179$ in.

9. Working depth $h_k = \dfrac{2}{P} = \dfrac{2}{12} = .166$ in.

10. Chordal tooth thickness

$$t_c = D \sin \left(\dfrac{90 \text{ degrees}}{N} \right) =$$

$$2.500 \times \sin \left(\dfrac{90 \text{ degrees}}{30} \right) =$$

$$2.500 \times \sin 3 \text{ degrees} =$$

$$2.5 \times .0523 = .1307 \text{ in.}$$

11. Chordal addendum $a_c = a + \dfrac{t^2}{4D} =$

$.083 + \dfrac{.131^2}{4 \times 2.5} = .083 + \dfrac{.017}{10} =$
$.083 + .0017 = .0847$ in.

12. Outside diameter $D_o = \dfrac{N + 2}{P} = \dfrac{30 + 2}{12} =$

$\dfrac{32}{12} = 2.6666$ in.

EXAMPLE 2. Determine the gear dimensions for a 45 tooth gear, 8 diametral pitch, 20 degree pressure angle.

1. Number of teeth $N = 45$

2. Diametral pitch $P = 8$

3. Pitch diameter $D = \dfrac{N}{P} = \dfrac{45}{8} = 5.625$ in.

4. Addendum $a = \dfrac{1}{P} = \dfrac{1}{8} = .125$ in.

5. Dedendum $b = \dfrac{1.250}{P} = \dfrac{1.250}{8} = .1562$ in.

Table 1
Spur Gear Formulas

To Find	Spur Gear Formulas	
	$14\frac{1}{2}$ degree Pressure Angle	20 and 25 degree Pressure Angles
Addendum, a	$a = \dfrac{1.0}{P}$	$a = \dfrac{1.0}{P}$
Dedendum, b	$b = \dfrac{1.157}{P}$	$b = \dfrac{1.250}{P}$
Pitch diameter, D	$D = \dfrac{N}{P}$	$D = \dfrac{N}{P}$
Outside diameter, D_o	$D_o = \dfrac{N + 2}{P}$	$D_o = \dfrac{N + 2}{P}$
Number of teeth, N	$N = D \times P$	$N = D \times P$
Tooth thickness, t	$t = \dfrac{1.5708}{P}$	$t = \dfrac{1.5708}{P}$
Whole depth, h_t	$h_t = \dfrac{2.157}{P}$	$h_t = \dfrac{2.250}{P}$
Clearance, c	$c = \dfrac{.157}{P}$	$c = \dfrac{.250}{P}$
Center distance, C	$C = \dfrac{N_1 + N_2}{2 \times P}$	$C = \dfrac{N_1 + N_2}{2 \times P}$
Working depth, h_k	$h_k = \dfrac{2}{P}$	$h_k = \dfrac{2}{P}$
Chordal tooth thickness, t_c	$t_c = D \sin\left(\dfrac{90 \text{ degrees}}{N}\right)$	$t_c = D \sin\left(\dfrac{90 \text{ degrees}}{N}\right)$
Chordal addendum, a_c	$a_c = a + \dfrac{t^2}{4D}$	$a_c = a + \dfrac{t^2}{4D}$
Diametral pitch, P	$P = \dfrac{N}{D}$	$P = \dfrac{N}{D}$
Center distance, C	$C = \dfrac{D_1 + D_2}{2}$	$C = \dfrac{D_1 + D_2}{2}$

6. Tooth thickness $t = \dfrac{1.5708}{P} = \dfrac{1.5708}{8}$
$= .1963$ in.

7. Clearance $c = \dfrac{.250}{P} = \dfrac{.250}{8} = .031$ in.

8. Whole depth $h_t = \dfrac{2.250}{P} = \dfrac{2.250}{8} = .281$ in.

9. Working depth $h_k = \dfrac{2}{P} = \dfrac{2}{8} = .250$ in.

10. Outside diameter $D_o = \dfrac{N + 2}{P} = \dfrac{45 + 2}{8} =$
$\dfrac{47}{8} = 5.875$ in.

11. Chordal tooth thickness
$t_c = D \sin\dfrac{90 \text{ degrees}}{N} = 5.625 \times$
$\sin\dfrac{90 \text{ degrees}}{45} = 5.625 \times \sin 2 \text{ degrees} =$
$5.625 \times .0349 = .1963$ in.

12. Chordal addendum $a_c = a + \dfrac{t^2}{4D} =$

$.125 \text{ in. } + \dfrac{.1963^2}{4 \times 5.625}$

$= .125 \text{ in. } + .0017 \text{ in. } = .1267 \text{ in.}$

The center distance between gears can be calculated when the number of teeth in the gears and the diametral pitch is known. Two gears in mesh make contact at their pitch diameters.

EXAMPLE 3. Determine the center distance between gears with 25 and 40 teeth and a diametral pitch of 14.

Center distance $C = \dfrac{N_1 + N_2}{2 \times P} = \dfrac{25 + 40}{2 \times 14} = \dfrac{65}{28} =$ 2.3214 in.

Another method of finding the center distance if the pitch diameters are known is to add both pitch diameters and divide that sum by 2.

EXAMPLE 4. What is the center distance of two gears when their pitch diameters are 2.500 and 3.000 in., respectively?

Center distance $C = \dfrac{D_1 + D_2}{2} = \dfrac{2.500 + 3.000}{2} = \dfrac{5.500}{2} = 2.750 \text{ in.}$

self-evaluation

SELF-TEST
1. What are commonly found pressure angles for gear teeth?
2. Why are larger pressure angles used on gear teeth?
3. What is the center distance between two gears with 20 and 30 teeth and a diametral pitch of 10?
4. What is the center distance between two gears with pitch diameters of 3.500 and 2.500 in.?
5. What is the difference between the whole depth of a tooth and the working depth of a tooth?
6. What relationship does the addendum and the dedendum have with the pitch diameter on a tooth?
7. What is the outside diameter and the tooth thickness on a 50 tooth gear with a diametral pitch of 5?
8. What is the diametral pitch of a gear with 36 teeth and a pitch diameter of 3.000 in.?
9. What is the outside diameter, whole depth, pitch diameter, and dedendum for a 40 tooth, 8 diametral pitch, 20 degree pressure angle gear?
10. What is the outside diameter, clearance, whole depth, tooth thickness, and pitch diameter for a 48 tooth, 6 diametral pitch, $14\frac{1}{2}$ degree pressure angle gear?

unit 3
cutting
a spur gear

When a spur gear is cut on a milling machine, specific operations must be performed. This unit deals with the order of these operations and how they are accomplished.

objective After completing this unit, you will be able to set up for and machine a spur gear on a milling machine.

Gears cut on a milling machine are limited in their possible uses because of the quality obtainable and the availability of the correct cutter. But in an emergency or for experimental purposes, satisfactory gears can be cut on a milling machine. Gears produced on gear generating machines are usually produced more economically and with higher accuracy. Before a gear is cut, the gear cutter must be selected. Involute gear cutters are available in sets numbered from 1 to 8 with pressure angles of either $14\frac{1}{2}$ or 20 degrees. These eight cutters are made with eight different tooth forms for each diametral pitch, depending on the number of teeth for which the cutter is to be used. Each cutter is designed to cut a number of teeth. See Table 1.

Each cutter is made with the correct tooth shape for the lowest number of teeth in its range. This means that a Number 5 cutter has the exact tooth shape for a 21 tooth gear, and gears with 22 to 25 teeth have only an approximate tooth

shape. Gears with 22 to 25 teeth will work when used with another gear of the same pitch, but they will not run as smoothly as a 21 tooth gear. More accurate tooth forms within each cutter's range can be obtained by half number gear tooth cutters, such as a Number $5\frac{1}{2}$ cutter, which cuts 19 and 20 teeth. Special cutters are made for a specific number of teeth in a gear or for low number of teeth in a pinion, from 6 to 11 teeth. Gear tooth cutters are marked on their side with the information as to their number, the diametral pitch, the number of teeth the cutter is designed for, the pressure angle, the whole depth the tooth is to be cut and, usually, the manufacturer's trademark.

Let us assume that a gear is to be made with 48 teeth and a 12 diametral pitch with a $14\frac{1}{2}$ degree pressure angle. The first step is to calculate the gear blank dimensions. Determine the dimensions for the:

Outside diameter $= D_o = \dfrac{N + 2}{P} = \dfrac{50}{12} =$

4.167 in.

Width of gear $= .750$ in.

Whole depth of cut $= h_t = \dfrac{2.157}{P} = \dfrac{2.157}{12} =$

.180 in.

Number of cutter to use $= 48$ teeth $=$ Number 3

Table 1

Number of Cutter	Number of Teeth Cut	Number of Cutter	Number of Teeth Cut
1	135 to rack	5	21 to 25
2	55 to 134	6	17 to 20
3	35 to 54	7	14 to 16
4	26 to 34	8	12 and 13

Then make the gear blank from the required material. You have to make it possible to hold the gear blank while cutting teeth. This can be done on a mandrel or with a hub on the gear. If more than one gear of the same size is needed, teeth can be cut on a piece of bar stock, and the individual gears are later parted off in a lathe or cutoff machine.

Prepare the milling machine for gear cutting by mounting the dividing head on the milling machine table. Make sure that the dividing head spindle axis is parallel to the table surface. Mount the footstock on the machine table. Check its axis for being parallel to the machine table.

Lubricate the hole in the gear blank before pushing the gear blank on the mandrel in the arbor press.

Fasten a driving dog at the large end of the mandrel and mount it between the dividing head and footstock centers. Clamp the dog tail in the driving slot at the dividing head spindle. With a dial indicator, test the height of the mandrel at the foot stock and at the dividing head (Figure 1). Remember to allow for the taper on the mandrel. Make footstock center height adjustments if necessary.

Move the saddle close to the column and mount the gear cutter on the machine arbor near the centerline of the gear blank. Keeping the work and cutter as close to the column as possible gives a more rigid setup. Mount the cutter so that the cutting pressure will be against the dividing head.

Align the cutter so it is centrally located over the gear blank (Figure 2). Calculate the RPM to use and adjust the machine to it. With the cutter

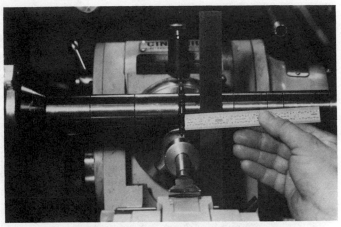

Figure 2. Aligning the cutter centrally over the gear blank (Lane Community College).

rotating, raise the knee until the cutter just touches the gear blank, then set the micrometer dial to zero. Move the table longitudinally until the cutter clears the workpiece and turn off the machine.

Calculate the dividing head movement for the number of teeth to be cut. The formula is $\frac{40}{N} = \frac{40}{48} = \frac{5}{6}$. Any hole circle divisible by 6 can be used. Using the 54 hole circle we get $\frac{5}{6} \times \frac{9}{9} = \frac{45}{54}$, or 45 holes in the 54 hole circle. Adjust the index crank to the correct hole circle. Set the sector arms for the wanted number of holes, remembering not to count the hole the index pin is in. Rotate the index crank clockwise one revolution and seat the index pin in the hole circle number. Lock the dividing head spindle. This will be the location for the first tooth space. If you develop the habit of always making the first cut with the index pin in the numbered hole of the hole circle, it will be easy to double-check your location by coming back to the original location.

Set the depth of cut, allow for a finishing cut of approximately $\frac{1}{32}$ in. Start the spindle and hand feed the table to the cutter and let the cutter just mark the edge of the gear blank. Back the cutter away from the gear blank. Index to the next position and repeat marking the gear blank around the complete circumference (Figure 3). Count the number of spaces. If you made an indexing error, it can be corrected without having ruined a gear. When the number of spaces is correct, make all the roughing cuts. Adjust the table feed trip dogs to stop the table feed at the

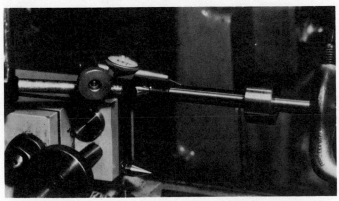

Figure 1. Checking the axis of the mandrel to be parallel with the table surface (Lane Community College).

Figure 3. Marking all the tooth spaces around the circumference of the gear blank (Lane Community College).

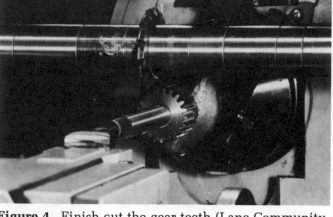

Figure 4. Finish cut the gear teeth (Lane Community College).

completion of the cut on one side and when the cutter clears the gear blank at the starting side. Adjust the feed rate to obtain an acceptable finish. Using a center rest under the gear blank during the cutting operation helps to control chatter.

Raise the knee for the finish cut. Cut two spaces and measure the tooth between these spaces for its thickness with a gear tooth vernier caliper.

Make the remaining finishing cuts (Figure 4). Remove all burrs with a file or wire brush.

Remove the mandrel from the gear and clean the machine.

This completes the cutting of a spur gear in the milling machine.

self-evaluation

SELF-TEST
1. What are two factors against cutting spur gears in the milling machine?
2. How many gear cutters are in a standard set?
3. Which cutter is used to cut a gear with 38 teeth?
4. Can a $14\frac{1}{2}$ degree pressure angle gear be in mesh with a 20 degree pressure angle gear?
5. A Number 6 gear cutter has the correct tooth shape for a gear with how many teeth?
6. What information is marked on the side of gear cutters?
7. Why should the work and cutter be mounted close to the column?
8. How many holes should be located within the sector arms?
9. Why is the gear blank marked with the cutter prior to cutting the teeth?
10. When is a center rest used?

unit 4
gear
inspection
and
measurement

When gears are made, their important dimensions must be measured to assure a useful quality product at their completion. This unit explains commonly used gear measurement methods.

objective After completing this unit, you will be able to measure gears by two methods: the gear tooth vernier caliper and the micrometer with two pins.

A machinist making spur gears in a machine shop should be able to measure gears by two common methods. These methods are with a gear tooth vernier caliper and with two pins and a micrometer. These gear measurements either are made on the pitch diameter or they involve the pitch diameter. When a gear tooth vernier caliper is used in gear measurement, the outside diameter of the gear is important because the height adjustment of the vernier caliper is set to the chordal addendum (Figure 1). Any changes in the outside diameter would affect the chordal addendum for a given pitch diameter. A gear tooth vernier caliper measures the tooth thickness on the pitch diameter, but the measurement is the chordal tooth thickness, which is less than the circular tooth thickness. The chordal addendum is larger than the addendum on a gear tooth. Dimensions of the addendum, chordal addendum, circular tooth thickness, and chordal tooth thickness can be found in the *Machinery's Handbook* or they may be calculated. As an example,

let us determine the dimensions required to measure a 6 diametral pitch gear with 12 teeth.

Outside diameter

$$D_o = \frac{N + 2}{P} = \frac{12 + 2}{6} = \frac{14}{6} = 2.3333 \text{ in.}$$

Addendum $a = \frac{1}{P} = \frac{1}{6} = .1666 \text{ in.}$

Pitch diameter $D = \frac{N}{P} = \frac{12}{6} = 2.000 \text{ in.}$

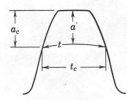

a = Addendum
a_c = Chordal addendum
t = Circular tooth thickness
t_c = Chordal tooth thickness

Figure 1. Illustration of the difference between the standard and chordal addendum and tooth thickness.

673

Circular tooth thickness

$$t = \frac{1.5708}{P} = \frac{1.5708}{6} = .2618 \text{ in.}$$

Chordal addendum

$$a_c = a + \frac{t^2}{4D} = .1666 + \frac{.2618^2}{4 \times 2}$$

$$= .1666 + \frac{.0685}{8} = .1666 + .0085 = .1751 \text{ in.}$$

Chordal tooth thickness

$$t_c = D \sin \frac{90 \text{ degrees}}{N} = 2 \times \sin \frac{90 \text{ degrees}}{12}$$

$$= 2 \times \sin 7.5 = 2 \times .1305 = .261 \text{ in.}$$

With these dimensions known, the gear can be measured with a gear tooth vernier caliper. The vertical scale is set to the chordal addendum of .175 in. and, with the vernier caliper resting on the top of the tooth (Figure 2), the tooth thickness should measure .261 in. If the outside diameter of the gear is .002 in. less than the calculated dimension, the vertical scale reading of the vernier caliper should be reduced to .174 in. This would assure that the tooth thickness measurement is made on the pitch diameter.

When any gear tooth measurements are made, especially while the gear is still on the milling machine, any burrs on the gear teeth must be carefully removed, or inaccurate measurements will result.

Another very accurate method of measuring gears is with two pins or wires and a micrometer (Figure 3). The diameter of the pins is calculated by dividing 1.728 by the diametral pitch. The constant 1.728 is the most commonly used for external spur gears. The dimension "M" (Figure 4) is found in the *Machinery's Handbook* calculated for different gear pressure angles and different teeth numbers conforming to the Van Keuren standard. In most of these tables, the dimensions are given for a diametral pitch of 1. For other diametral pitches divide the value in the table by the diametral pitch used. Internal spur gears can also be measured with this method.

As an example, determine the dimension "M" for a 45 tooth gear 10 diametral pitch, 14½ degree pressure angle P.

The pin diameter $\frac{1.728}{10} = .1728 \text{ in.}$

The dimension "M" for 1 diametral pitch is 47.4437 in.

For a 10 diametral pitch divide $\frac{47.4437}{10} =$ 4.74437 in. 4.74437 in. is the dimension for a gear without any backlash allowance. Gears need some backlash and, if .005 in. backlash is required, the dimension "M" must be recalculated. Again, we consult the *Machinery's Handbook* and find that for a 45 tooth gear, each .001 in. tooth thickness reduction at the pitch line re-

Figure 2. Measuring a spur gear with a gear tooth vernier caliper (Lane Community College).

Figure 3. Measuring a spur gear with two pins and a micrometer (Lane Community College).

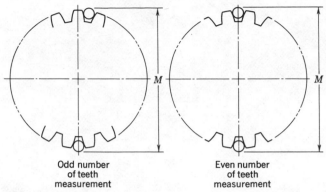

Odd number
of teeth
measurement

Even number
of teeth
measurement

Figure 4. Measuring of gear dimensions over two pins.

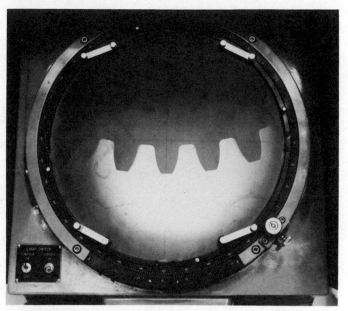

Figure 5. Using an optical comparator to check gear tooth dimensions (Lane Community College).

quires dimension "M" to be reduced by .0031 in.

To obtain .005 in. backlash, multiply .0031 in. by 5, which equals .0155 in. Then reduce the measurement over wires, which is 4.74437 in. by .0155 in., to get a dimension of 4.72887 in.

Another gear inspection method available to a machinist is an optical comparator. The optical comparator employs a light, a lens, a mirror, and a viewing screen to enlarge the shadow outline of a gear being inspected (Figure 5). An enlarged transparent drawing of the part equal to the magnification power of the lens is matched to the shadow on the screen for a direct comparison of

the enlarged shadow. Special testing and gear measuring tools are made to check the involute profile of a gear tooth, the gear tooth spacing, or the accuracy of the helix on a helical gear, or to check the meshing of a gear with a master gear to determine runout.

self-evaluation

SELF-TEST
1. What are two common gear measurements performed by a machinist?
2. Which tooth thickness is measured by a gear tooth vernier caliper?
3. The vertical scale of a gear tooth vernier caliper is set to what dimension?
4. Which is larger, the chordal addendum or the standard addendum?
5. Which is larger, the chordal tooth thickness or the circular tooth thickness?
6. Name two ways of determining the chordal thickness of a gear tooth?
7. When a gear is measured with a micrometer and two pins, can any size pin be used?
8. How is the dimension determined when measuring over pins?
9. When a dimension is given for 1 diametral pitch but the gear measured is 12 diametral pitch, how is the measurement determined?
10. How is an optical comparator used in gear measurement?

section m
shapers

The shaper cuts by passing a single point tool by the workpiece. The shaper cutting tool moves by reciprocating motion through a single axis while the workpiece moves past the tool either horizontally, vertically, or rotationally to the tool motion.

The shaper is considered by some to be an obsolete machine tool, and it is true that much of the shaper's work is now done on the milling machine. However, the shaper is a very versatile machine tool that can accomplish a wide variety of machining tasks. For this reason it is still a common sight in many machine shops.

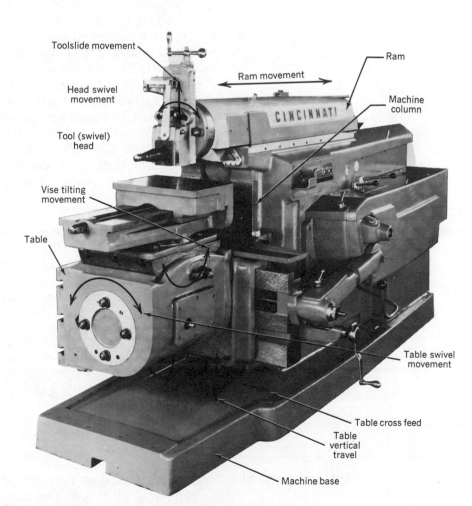

Figure 1. The horizontal shaper is a versatile machine tool capable of many motions. The model shown is a universal shaper (Courtesy of Cincinnati Incorporated).

677

THE HORIZONTAL SHAPER The horizontal shaper (Figure 1) is an extremely versatile machine tool, particularly in those situations where small numbers of parts are to be made and a variety of angles are to be machined into the part (Figure 2).

One of the more common uses of shapers is the production of dovetail assemblies (Figures 3a and 3b) for applications such as tool-slides used on machine tools like lathes, shapers, and planers.

Another interesting application of the shaper is the production of internal features in a workpiece (Figure 4). The making of internal keyways on limited production parts, particularly in sizes and shapes where the use of an ordinary keyway broach and shop press is not feasible, makes the shaper a difficult machine to do without. This internal capability is called slotting and can be combined with the use of the index head (Figure 5) to produce a variety of internal shapes. In some instances, specialized tooling is made up (Figure 6) to make unusual internal and external parts (Figure 7). Sophisticated hydraulic tracing systems have also been applied to shapers to produce internal contours on a production basis (Figures 8a, 8b, and 8c).

Contour work on shapers is by no means limited to internal work. External contouring is done on the shaper by a wide variety of means. The most basic form of contouring takes place when the operator manually controls the downfeeding of the tool and the cross feeding of the work to follow a layout line marked on the part (Figure 9). Another unique method of contouring on the shaper is the use of a strong wide template with a roller follower that moves the shaper table up and down while the cross-feeding is taking place (Figure 10). This type of system must be strongly constructed because the weight of the table and workpiece and much of the force of the cut must be taken through the template and roller. Hydraulic systems are also used in external contouring with the shaper (Figures 11 and 12). Although the shaper is most often considered to be a toolroom machine, it can be adapted to production functions by the addition of specialized components (Figures 13a, 13b, and 14).

THE VERTICAL SHAPER (SLOTTER) The vertical shaper or slotter (Figure 15) is used often in job shops and toolrooms. This machine functions much like its horizontal counterpart except that it is supplied with a rotary table as standard equipment. This rotary table can be moved both traversely and longitudinally. The ram can also be tilted 10 degrees from the vertical. This makes the machine especially versatile for making complex internal shapes. It is also much easier for the operator than an arrangement with a dividing head on a horizontal shaper, as shown in Figure 6, because the work is quite visible. These machines can also be equipped with tracing equipment for the production of contours (Figure 16).

SHAPER SAFETY The shaper has one very major hazard in its use that has to be considered at all times. The travel of the ram back and forth is very

Figure 2. The workpiece in the universal shaper can be rotated about three axes to obtain compound angles on the workpiece (Courtesy of Cincinnati Incorporated).

Figure 3a. The toolhead of the shaper can be readily swivelled to produce dovetails for machine assemblies. A toolslide is being machined.

Figure 3b. An internal dovetail is being machined (Courtesy of Cincinnati Incorporated).

Figure 4. Making of internal keyways is an important ability of the shaper. This Kennedy keyway has to be held to close size tolerance to work effectively (Courtesy of Cincinnati Incorporated).

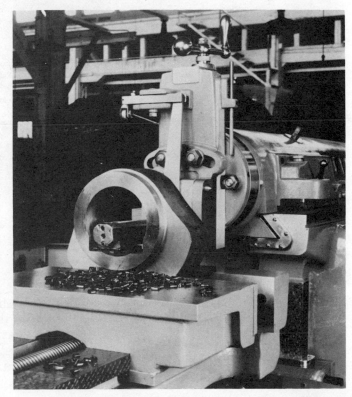

Figure 5. The dividing or indexing head in combination with the shaper can be used for making complex internal shapes (Courtesy of Cincinnati Incorporated).

Figure 6. A specially tooled shaper for complex internal parts (Courtesy of Rockford Machine Tool Company).

unyielding to objects that get in the way. You must insure that no tools are left in the way and that you keep your hands and the rest of your body away from the working area. Do not stand, or permit anyone else to stand, in front or in back of the ram. Be sure that the machine is completely stopped, including the motor, before making measurements or any adjustments that require you to place any part

Figure 7. A selection of complex parts on which the shaper was used with special tooling (Courtesy of Rockford Machine Tool Company).

Figure 8*a*. A horizontal shaper with indexing system and hydraulic tracing controls for a production application. (Courtesy of Rockford Machine Tool Company).

Figure 8*b*. Tracer stylus with master in place between centers. The master rotates at the same rate as the internal part shown in Figure 9*c* (Courtesy of Rockford Machine Tool Company).

Figure 8*c*. The internal part being machined in the hydraulic tracing set up (Courtesy of Rockford Machine Tool Company).

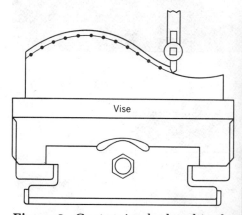

Figure 9. Contouring by hand in the shaper to a layout line on the workpiece (Courtesy of Cincinnati Incorporated).

Figure 10. Contouring can be done with this device, which moves the table up and down during cross feeding (Courtesy of Cincinnati Incorporated).

of yourself within danger of being struck by the ram. Allow no one but yourself to operate the machine controls.

Be sure that the stroke rate and stroke length are in balance. A fast stroke rate and a long stroke can cause the machine to move

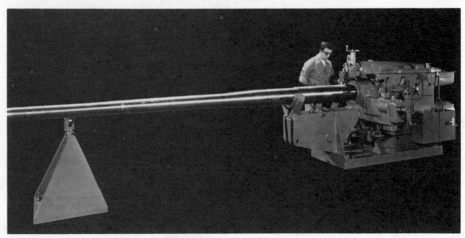

Figure 11. (Top) Two-dimensional contour cutting in the shaper by using a hydraulic tracing device with a template. Note the massive chip. Cuts like this can be made on heavy duty shapers using large, well-supported cutting tools (Courtesy of Cincinnati Incorporated).

Figure 12. (Top right) Horizontal shaper with hydraulic tracing attachment for three-dimensional duplicating. The stylus on the left contacts the master on the table. The toolhead hydraulically follows the stylus movements during the ram travel to produce a part that duplicates the master (Courtesy of Rockford Machine Tool Company).

Figure 13*a*. Special tooling for holding and working on the breech end of a gun barrel (Courtesy of Rockford Machine Tool Company).

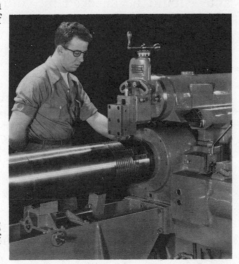

Figure 13*b*. Details of the indexing fixture and toolhead (Courtesy of Rockford Machine Tool Company).

Figure 14. Tooling applied to the shaper for making a bevel cut on both ends of precut steel plates on a production basis (Courtesy of Rockford Machine Tool Company).

about or even to be permanently damaged if the clutch is engaged under these circumstances. Engage the clutch slowly to bring the machine up to speed to be safe.

Large, powerful shapers can eject large, hot chips with great force. It is important to set up a screen of substantial fireproof material in front of the ram beyond the end of the cutting stroke to stop chips from being a hazard to yourself and to others in the shop.

It is important to review your workholding methods before making a cut. Any workholding method that is the least bit questionable should be checked by your instructor and approved before cutting begins. If you are using a vise, it is also important to check the fastening of the vise to the worktable. Some vises also have a special lock pin that fits into the machine table for added security.

It is also important to remove all handles from the stroke adjusting shaft, the ram positioning clamp, and the cross feed shaft before operating the shaper. The vise handle should be positioned downward or removed entirely.

Other hazards of the shaper are similar to those of most other machine tools. Safety glasses must be worn in the machine shop at all times to comply with federal and state safety regulations.

If there are any doubts about safety or compliance with safety regulations in your shop, the instructor must be consulted before starting the cut. In many instruction labs, the rules specify that the instructor be present the first time a student operates this specific machine tool.

Figure 15. A slotter being used for internal slotting in a casting (Courtesy of Rockford Machine Tool Company).

Figure 16. A slotter with a hydraulic tracer installed (Courtesy of Rockford Machine Tool Company).

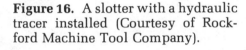

When you have finished using the shaper, the machine must be cleaned and, in the interest of safety for the next operator, the following steps should be taken.

1. Remove the cutting tool.
2. Remove the tool holder.
3. Set the toolhead to vertical and retract the toolslide (this is very important!).
4. Set the stroke length to minimum.
5. Set the stroke rate to the minimum setting.
6. Check that the clutch is disengaged.

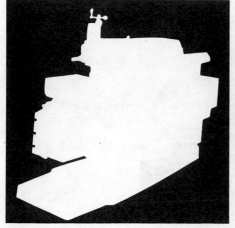

unit 1
features and tooling of the horizontal shaper

Horizontal shapers are important, not because of their great popularity, but because they are so versatile. Before operating the horizontal shaper, you should be familiar with the operating controls and components, the mechanisms of shaper operation, the workholding mechanisms, and the cutting tools commonly used on shapers.

objectives **After completing this unit, you will be able to:**

1. **Name components and operating controls on shapers.**
2. **Describe two differing types of shaper tables.**
3. **Explain the functions of two basically different shaper drive systems.**
4. **Describe mechanisms that relate to tool positioning functions.**
5. **Describe at least three differing types of shaper workholding procedures.**
6. **Explain differences between shaper tools prepared for steels and cast iron.**

OPERATING CONTROLS AND COMPONENTS OF THE HORIZONTAL SHAPER

Figure 1 provides detailed nomenclature of the components and control features of a horizontal shaper. You will refer to this figure several times in this unit.

Base and Column
The base of a shaper (Figure 2) provides a platform for the main casting, frequently referred to as the column of the shaper that contains the drive mechanisms. It also serves as a reservoir for lubricants or hydraulic oil to operate shaper mechanisms. On the front of this casting are ways machined accurately square with other ways that provide guidance for the ram of the

shaper. This relationship is critical to the accuracy of the machine.

Cross Rail
The cross rail assembly rides on a saddle that moves vertically on the face of the column driven by the vertical leadscrew. The cross rail is maintained accurately square to the column and contains drive mechanisms for moving the table assembly in a horizontal direction. On the cross rail is mounted an apron that is moved horizontally along the rail by a leadscrew.

Apron and Table
The apron is the attachment point for the table. The top and side of the table are provided with

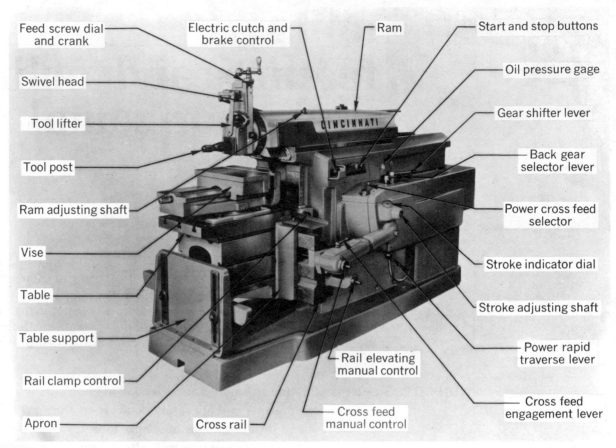

Figure 1. Detailed nomenclature of a plain horizontal shaper. Shapers are quite similar in location of basic controls despite drive differences (Courtesy of Cincinnati Incorporated).

T-slots for the attachment of vises, fixtures, and other tooling needed for shaper work. Since there are so many attachment points and such great mass and leverage relative to the column, a table support is typically provided as an **outrigger** to support the overhang of the table and to transmit machining forces back to the machine base.

Shaper tables are supplied in two designs. The shaper shown in Figure 1 has a plain table that provides for motion of the workpiece only in vertical and cross feed (other than the swiveling action of the vise). The universal table shaper has two additional motions (Figure 3) that allow for machining of complex angles, as shown in Figure 3 of the section introduction. The table is mounted to the apron with a trunion, which permits full rotation of the table, and one face of the table has a swivel plate with limited motion. When this feature is not needed for a particular job, it is often rotated to the side in the interest of easier setup and greater rigidity.

Ram
The ram (Figures 1 and 2) moves in ways on the top of the shaper. It is designed to be as rigid as possible, consistent with reasonable lightness to allow rapid reversals of direction without having the machine move along the floor.

Toolhead
On the face of the ram is mounted the toolhead or swivel head. This head is typically capable of being swiveled on the face of the ram through a full 180 degrees or more. The base of the toolhead is graduated. The toolhead consists of several components (Figure 4). It has a leadscrew that drives an accurately made toolslide. Fitted to the face of the toolslide is an apron that can be turned and secured at an angle to the slide. This apron is part of an assembly that consists of the clapper box, a hinge pin, a clapper block, and a

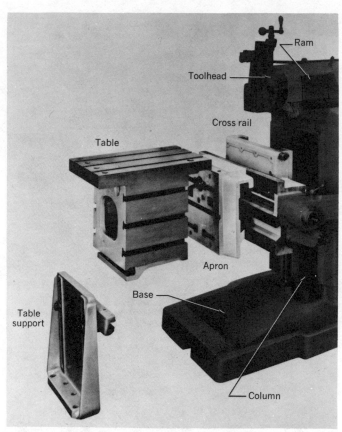

Figure 2. Components of the plain horizontal shaper table assembly (Courtesy of Cincinnati Incorporated).

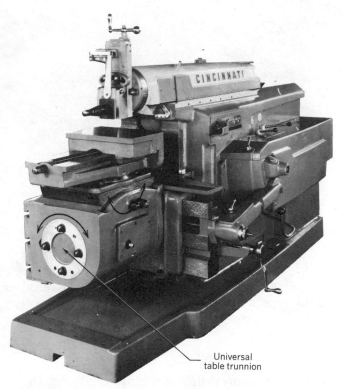

Figure 3. Additional workpiece motions are available on the universal shaper (Courtesy of Cincinnati Incorporated).

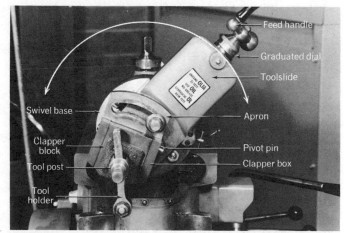

Figure 4. Shaper toolhead details (Lane Community College).

tool post for holding the cutting tool or tool holder. The function of the clapper box is to allow the cutting tool to tilt up on the return (feeding) stroke without damage to the tool. On some shapers, a tool lifter (Figure 1) is provided, programmed to lift the tool actively on the return stroke of the ram. The tool lifter is of great importance when using carbide cutting tools or for making deep, narrow slots.

TYPES OF SHAPER DRIVES

The Crank Shaper
The plain and universal shapers shown in the previous figures have been examples of crank or mechanically driven shapers. In this type of shaper the motor drives through a gearbox and clutch to a crank gear in the center of the machine (Figure 5). The crank gear (Figure 6) has a movable pivot on a crank block that, in turn, fits accurately in a sliding block on the rocker arm. The crank block is moved toward and away from

the center of the crank gear by the stroke adjustment shaft, thereby increasing or decreasing the movement of the rocker arm and its attached ram. The ram in turn can be changed in position relative to rocker arm by means of the ram adjustment shaft. Examination of the top diagram in

Figure 5. The power train of the mechanical or crank-type shaper (Courtesy of Cincinnati Incorporated).

Figure 6 shows the working relationship between the reciprocating drive components. As you can see, the return of the ram is made during the part of the cycle where the sliding block is closest to the stationary pivot point. This relationship accelerates the return stroke in relationship to the forward or cutting stroke, thus saving noncutting time. The ratio of the time of the cutting stroke to the return stroke will be about 3:2. We will go into this further in the next unit.

The Hydraulic Shaper
Hydraulically driven shapers (Figure 7) do not differ greatly in appearance from the crank-type mechanical shaper. Hydraulically operated shapers have nearly constant cutting tool velocity, and many have an arrangement for high speed ram return independent of the cutting stroke speed. Figure 8 is a schematic diagram of the hydraulic circuit, which includes the high speed return feature. Examine this diagram closely and trace the flow of oil through each circuit.

The pump supplies hydraulic oil at a constant volume to the system, which is regulated by the speed control valve. Oil not bypassed by this valve is admitted to the start-stop valve, which also distributes oil to each side of the drive piston under the command of a pilot valve, which is tripped by the ram reverse dogs. The pilot valve also activates the high return valve if it has been

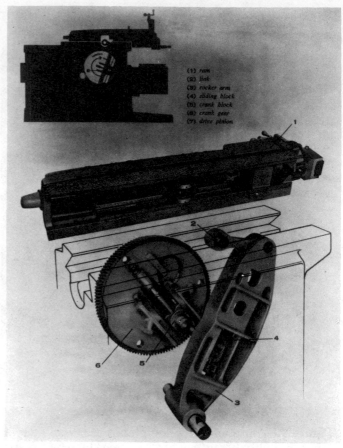

(1) ram
(2) link
(3) rocker arm
(4) sliding block
(5) crank block
(6) crank gear
(7) drive pinion

Figure 6. Details of the crank mechanism of a mechanical shaper (Courtesy of Cincinnati Incorporated).

selected by the operator to speed up the return of the ram. In addition, the pilot valve actuates the tool lifter (if fitted) and the power feed during the return stroke. Lubrication for the ram is filtered separately and passed through this valve as well.

SHAPER FEEDS

Table Feeds
Any shaper, whether simple or sophisticated, can be hand fed vertically or horizontally for table travel. Most shapers, even the simplest table top models, have a power feed for the horizontal travel of the table. On crank shapers the drive for the feed is in the form of either an adjustable eccentric (Figure 9) or a cam arrangement, which is attached to the hub of the crank gear. The reciprocating motion that results from this arrangement is passed to a ratchet mechanism connected to the cross feed

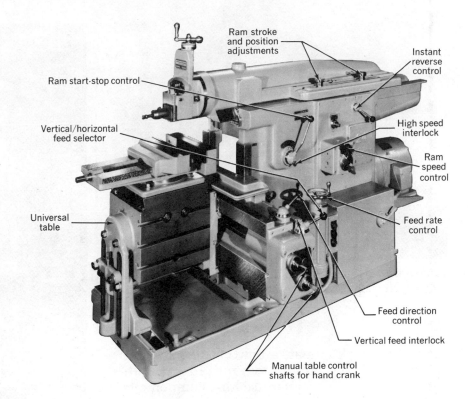

Figure 7. Twenty-four in. hydraulic shaper. Shaper size is designated by the maximum length of available ram stroke. The example shown is equipped with a universal table (Courtesy of Rockford Machine Tool Company).

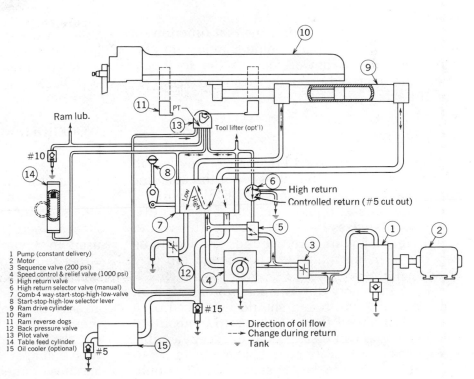

1 Pump (constant delivery)
2 Motor
3 Sequence valve (200 psi)
4 Speed control & relief valve (1000 psi)
5 High return valve
6 High return selector valve (manual)
7 Comb-4 way-start-stop-high-low-valve
8 Start-stop-high-low selector lever
9 Ram drive cylinder
10 Ram
11 Ram reverse dogs
12 Back pressure valve
13 Pilot valve
14 Table feed cylinder
15 Oil cooler (optional)

→ Direction of oil flow
- - -> Change during return
= Tank

Figure 8. Schematic diagram of the hydraulic circuit in a hydraulic shaper with high speed return (Courtesy of Rockford Machine Tool Company).

leadscrew. On many shapers, vertical feed can be selected as an alternative to cross feed. On hydraulically operated shapers the table feeds are accomplished by a hydraulic table feed cylinder (Figure 8). Most larger shapers are also equipped with a rapid traverse, which is a device to move the table either vertically or horizontally at a relatively high rate of speed. This is very useful

Figure 9. Adjustable eccentric on crank gear shaft to drive feed rachet (Lane Community College).

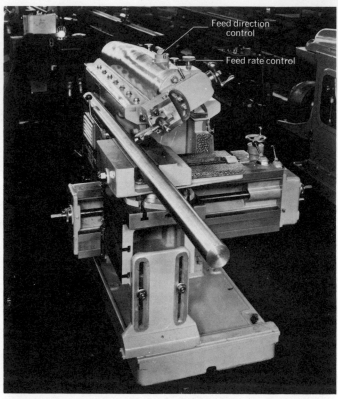

Figure 10. Shaper head swivelled at a steep angle to make a keyway cut in a long shaft. This machine is also equipped with a power feed on the toolhead (Courtesy of Rockford Machine Tool Company).

for getting back to start a new cut or for use in setup for moving rapidly to a position for indicating in the vise or other machine components.

Head Feeds

Feeding of the toolhead is a manual function on most horizontal shapers. Some shapers are also equipped with a power feed to the toolhead, often cam operated or hydraulically driven (Figure 10).

SHAPER SIZE

Shapers are size classified by the working length of the maximum cutting stroke that the machine can make. They are also classified by the working capability such as light, standard, or heavy duty. In the smaller sizes like 7 in., the machine would weigh only a few hundred pounds and would be capable of being mounted on a bench top or a sheet metal base.

A 28 in. heavy duty shaper would weigh on the order of 7500 lb and should be placed on a heavily reinforced floor. In sizes up to about 16 in., the shaper can be expected to accept a cubic workpiece of about its rated size, as 16 in. × 16 in. × 16 in. In the larger sizes, the maximum table to ram distance usually does not exceed that of a 16 in. shaper.

Heavy duty shapers are capable of making massive chips and are equipped with motors on the order of 10 hp.

WORKHOLDING DEVICES ON THE SHAPER

Vises

By far, the most common method of holding workpieces in the shaper is with a swivel base, single screw vise that is standard equipment on most shapers (Figure 11). This type of vise can be used in a variety of ways to hold workpieces either square to the travel of the ram or parallel to ram travel. Sometimes the combined arrangement of the vise and the toolhead can result in an arrangement to do what would not ordinarily be thought of as a shaper job (Figure 10). Auxiliary workholding can also be used in conjunction with a shaper vise to hold workpieces that have features of unequal length that would be difficult to hold by other methods (Figure 12). Another accessory device that can be used with a shaper vise is a holddown wedge (Figure 13), which can be used to force the workpiece against the solid or reference jaw of the vise to help insure that the

Figure 11. Swivel base single screw shaper vise (Courtesy of Cincinnati Incorporated).

Figure 12. Auxiliary workholding method to extend the usefulness of the vise (Courtesy of Rockford Machine Tool Company).

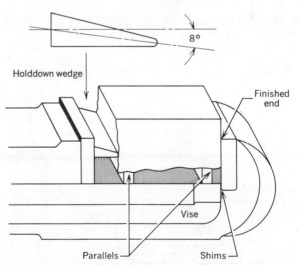

Figure 13. Holddown wedge for squaring to a finished surface (Courtesy of Cincinnati Incorporated).

surface being cut is perpendicular to the finished surface.

An important thing to remember, especially when using the standard vise with its carefully finished jaws, is that rough, ferrous castings usually have a hard, scaly outside surface before they are machined, and this surface can badly mar the vise jaws unless they are protected. Emery cloth (Figure 14) is one way that this is done. Other compressible materials also can be used for this protection.

When shaping is to be done on a side parallel to a finished surface, a cylinder of soft metal such as 1020 steel, brass, or aluminum can be used to provide line contact on the edge of the work. This prevents the movable jaw from providing a lifting movement to the work as the screw is tightened. The movable jaw on most vises has some free play that permits the jaw to tilt slightly as it is tightened. As the work is tightened with this arrangement (Figure 15) it is tapped down onto the parallels with a babbitt hammer. Then, after tightening, the parallels are checked to see that they are firmly in contact with the bottom of the workpiece. Sometimes paper strips are used on top of the parallels to check this contact; if the strip can be pulled out, you do the job over.

Since much of shaper work done in vises has the force directed toward the movable jaw, and

some of this work is quite critical, especially in operations like slotting (Figure 16), some vises are made with additional clamping for the movable jaw to reduce the tendency to tilt.

When the force of the cut is parallel to the

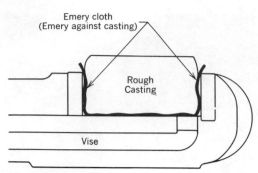

Figure 14. Protecting vise jaws from rough castings (Courtesy of Cincinnati Incorporated).

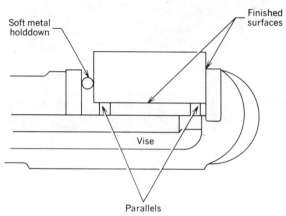

Figure 15. Method for shaping parallel to a finished surface (Courtesy of Cincinnati Incorporated).

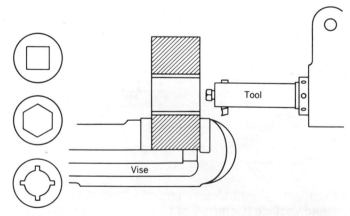

Figure 16. Slotting makes special accuracy demands on vise holding (Courtesy of Cincinnati Incorporated).

jaw as is common in making keyways on long parts (Figures 17 and 10), it is important to be sure that the vise has an adequate grip on the part. Single screw shaper vises are supplied with

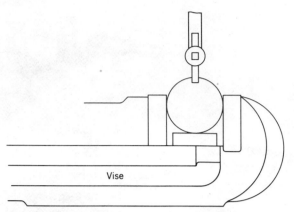

Figure 17. Keyway cutting demands adequate work-holding because of limited contact (Courtesy of Cincinnati Incorporated).

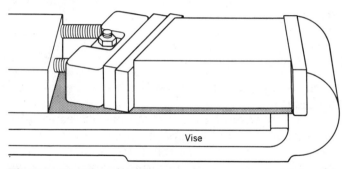

Figure 18. The double screw vise can hold work-pieces that are not parallel (Courtesy of Cincinnati Incorporated).

a handle of correct length for adequate tightening by an average sized man without using a hammer on the handle.

The double screw vise (Figure 18) is often used with both shapers and planers to hold tapered or out-of-parallel work. This type of vise can be had with a swivel base but, for heavy work, they can be supplied with a plain base and multiple screws (Figure 19). The jaws can be supplied either plain or serrated, but the serrated jaws, as shown in Figure 19, are more common.

Direct Attachment to the Table

Some shaper work is best accomplished by direct attachment of the workpiece to the table with straps and T-bolts. When this is done, it is important to make sure that the clamping force in the setup is directed more toward the work (Figure 20a) than the riser block (Figure 20b) that is used to level the clamping strap. Shapers and planers

Figure 19. Multiple screw vise with serrated jaws has exceptional holding ability (Courtesy of Cincinnati Incorporated).

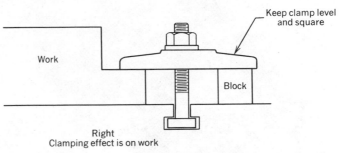

Figure 20a. Right method for direct clamping to the table (Courtesy of Cincinnati Incorporated).

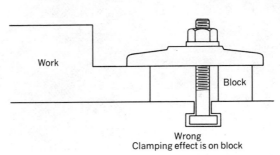

Figure 20b. Wrong method for direct clamping to the table (Courtesy of Cincinnati Incorporated).

are relatively less forgiving of poor setup than milling machines because of the high forces that are transmitted through the work upon tool impact.

It is best to strive for some excess in fastening the work than to take a chance on having the work come loose on impact. Figure 21 is an example of a secure direct clamping setup, employing adjustable table stops.

Clamping is not limited to work on the top of the table. It is useful to keep in mind that workpieces also can be attached to the side of the table (Figure 22).

On some workpieces the T-bolt can be used directly with nuts to hold down the workpiece. With recessed workholding machining can take place right up to an obstruction, as shown in the surfacing of this machine vise (Figure 23).

Thin work can be directly attached to the table with poppets and toe dogs (Figures 24 and 25) combined with the use of a stop. The poppet must be aligned straight with the toe dog, as viewed from the top, to work properly. A poppet designed to hold in a T-slot is sometimes called a bunter.

Using Fixtures with the Shaper

A fixture is a workholding device that is usually directly fastened to a machine table (Figure 26) or spindle. It typically serves as an intermediate device to locate and fasten parts more rapidly

than would be the case for individual setup. Another example of a fixture would be the indexing center (Figure 27) used in combination with the tilting base on the universal table shaper to make a forming die.

Other fixtures would include devices like angles for attaching angular workpieces for shaping, as shown in Figure 28.

SHAPER CUTTING TOOLS

The tools used on the shaper are very similar to the single point tools used in the lathe. The nomenclature for the cutting geometry is the same as that used for lathe tools. Most of the tools used in the shaper are of high speed steel, with preference toward grades that have increased amounts of vanadium for better shock resistance. Cemented carbide tools are also used on occasion, but the impact characteristic of the shaper and a cutting speed, which at a maximum is less than 150 surface feet per minute, makes the carbide less the tool of choice except on hard and abrasive materials. Carbides are brittle and do not readily stand being drawn back along the cut, even with a properly set and working clapper box, so for carbide use, the tool lifter described

Figure 21. Use of stops and straps to make secure direct table clamping (Courtesy of Cincinnati Incorporated).

Figure 23. With recessed holding, machining can be done right up to an obstruction (Courtesy of Rockford Machine Tool Company).

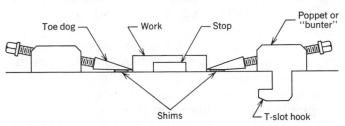

Figure 24. Using a bunter with toe dogs for a thin workpiece (Courtesy of Cincinnati Incorporated).

Figure 22. The side of the table may also be used for direct attachment (Courtesy of Rockford Machine Tool Company).

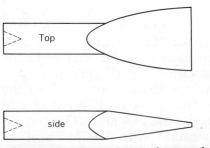

Figure 25. Two views of a toe dog.

earlier is important. In past years, forged tools were used, especially with the larger shapers. At the present time, their use is rare, replaced by tool bits in holders or by large tool bits ground for the application and directly clamped into the tool post.

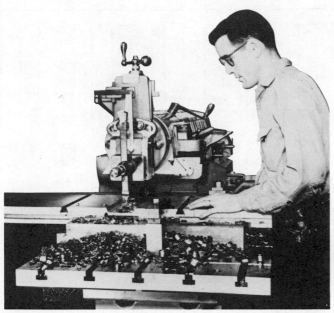

Figure 26. Using a fixture on a plain shaper (Courtesy of Cincinnati Incorporated).

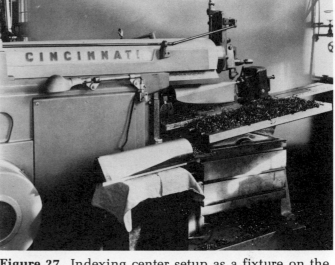

Figure 27. Indexing center setup as a fixture on the tilt table of a universal shaper (Courtesy of Cincinnati Incorporated).

Preparing Tools for Shaper Use

Tools for use in the shaper (and the planer) are only generally similar to lathe tools. They are characterized by clearances back of the cutting edges on the order of 4 degrees. The reason for these minimal clearances is to provide maximum support for the cutting edges. In addition, on the planer or shaper, feeding does not take place during the cut; hence no additional clearance is needed to allow for the feedrate, as is necessary on the lathe. Figure 29 illustrates typical cutting tools recommended for use with low carbon steels. On work in steels the roughing tool should be used to bring the work to within .010 to .015 in. (.25 to .3 mm) of final dimension. Where considerable stock is removed and where the setup is substantial, the depth of cut and feed rate should be the maximum possible consistent with the power and rigidity of the machine, the strength of the tool, and the strength of the workpiece. Where it is necessary to reduce cutting forces, the feedrate should be lessened before a reduction in depth of cut is made. Finishing the last .010 to .015 in. should be done in one cut, whenever possible, to save time.

On cast iron, the fundamental difference between shaper and planer tools prepared for steels is on the face of the tool. The side and back rake angles of the tools are less, but the clearances

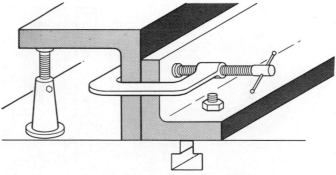

Figure 28. Angle fixture with a planer jack for shaping angles (Courtesy of Cincinnati Incorporated).

remain at about 4 degrees (Figure 30). Since dovetail work is mainly done in castings, recommendations for dovetail tools (Figure 31) are given for cast iron applications. Dovetail tools for steel should have additional rake. On cast iron, it is important to have the roughing cut completely penetrate the scale on the first cut. If the tool rides up on the scale it will dull very quickly. As with steel, you should strive for maximum depth of cut consistent with the strength of workpiece, workholding, and tool. For cast iron, the roughing cut should be carried to .005 to .010 in. (.12 to .25 mm) of finished size. Finishing cuts on flat surfaces in cast iron and free cutting steels are generally done with wide tools and a cross feed rate about three quarters of tool width. This is

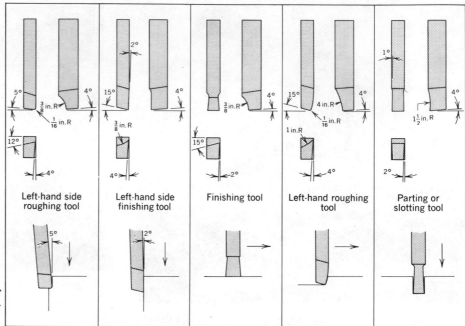

Figure 29. Shaper tool geometry for cutting low carbon steels (Courtesy of Cincinnati Incorporated).

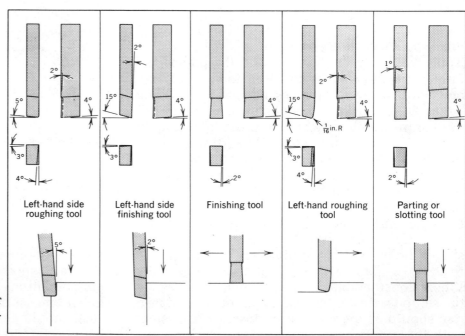

Figure 30. Shaper tool geometry for cutting cast iron (Courtesy of Cincinnati Incorporated).

one instance where finishing is usually quite fast and the time for more than one finishing cut is minimal where very accurate work is called for.

If chatter is experienced during the roughing cut in either steel or cast iron, it may be due to poor adjustment of machine gibs or to excessive overhang of the tool, toolslide, or workpiece. With regard to the cutting tool proper, this may

be reduced by limiting the radius on the point of the tool to $\frac{1}{32}$ or less. It is also important to have the tool mounted as close into the tool post as possible and the toolside overhang minimal (Figures 32a and 32b). Tools used on shapers and planers should be kept sharp and the cutting edges should be honed after grinding to give the longest possible tool life.

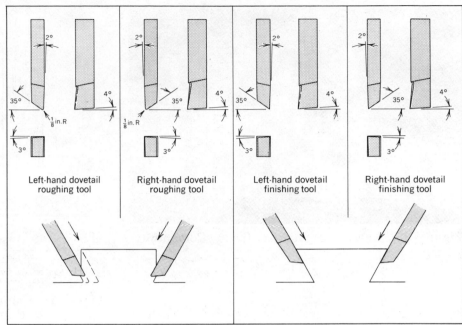

Figure 31. Shaper tool geometry for dovetail work in cast iron (Courtesy of Cincinnati Incorporated).

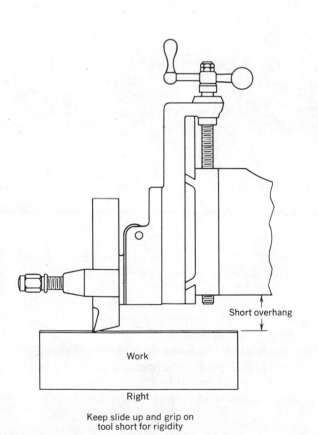

Figure 32a. Keep overhang to a minimum by using the highest possible table position for the job (Courtesy of Cincinnati Incorporated).

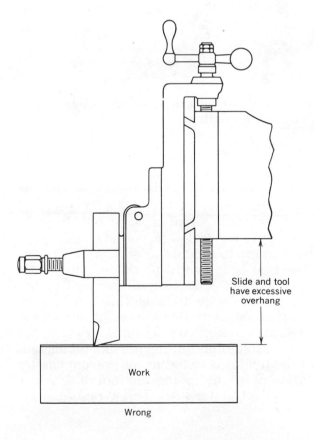

Figure 32b. Excessive overhang of slide and tool may cause chatter (Courtesy of Cincinnati Incorporated).

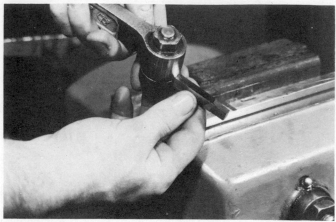

Figure 33. Small tool holder (Lane Community College).

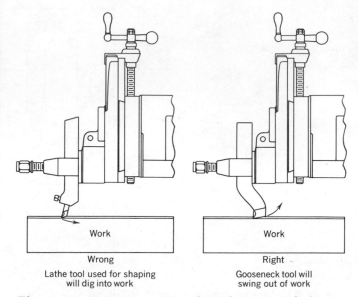

Figure 35. Using a gooseneck tool can avoid chatter (Courtesy of Cincinnati Incorporated).

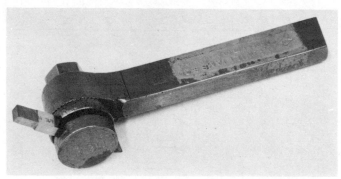

Figure 34. Tool holder can be set at various angles (Lane Community College).

Figure 36. Extension tool holder for internal shaping (Lane Community College).

Tool Holding

A common form of tool holder used with shapers and planers is shown in Figure 33. This type of tool holder permits a tool bit to be held straight as shown or to be angled (Figure 34) relative to the shank of the tool holder. For relatively heavy cuts, the tool is typically placed in the holder with the tool bit side toward the ram of the shaper. For relatively light cutting, and where maximum visibility is also important, the tool is often mounted in front, on the side away from the ram. The reason for mounting the tool on the ram side of the holder is to reduce or prevent chatter and "digging in" by giving the tool and holder room to deflect under load. This type of deflec-

tion decreases the load instead of increasing it. This is why lathe tool holders with built-in rake angles are poor for use in shapers and why gooseneck tools often relieve problems with chatter (Figure 35), particularly with worn, out-of-adjustment machines.

The extension shaper tool holder (Figure 36) is another useful tool for internal shapes. It allows the tool to be extended into existing holes for making internal keyways and other forms.

self-evaluation

SELF-TEST 1. Name the shaper features indicated in Figure 37.

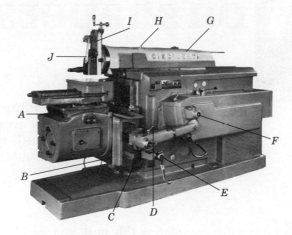

Figure 37. (Courtesy of Cincinnati Incorporated).

2. What is the function of the clapper box on the toolhead?

3. Explain the basic differences between crank and hydraulic shaper drives.

4. Describe the table feeding functions on a crank type shaper.

5. How is the size of the shaper specified? Would a 28 in. shaper hold a workpiece of the same proportion as a 12 inc. shaper?

6. What are the two basic types of vises used on shapers?

7. Explain a holding method for securing thin parts to a shaper table.

8. What advantage is gained by using a fixture over direct attachment to the machine table?

9. What is the fundamental difference between shaper tools prepared for cast iron and those prepared for steel cutting?

unit 2
cutting factors
on shapers

Maximum cutting efficiency and productivity while using a shaper can only be realized if you are aware of several important factors including selection of shaper cutting tool materials and the proper setting of speeds and feeds relative to workpiece material and machining tasks.

objectives After completing this unit, you will be able to:

1. Select a grade of high speed steel or cemented carbide suitable for use on shapers.
2. Name a source of machining data useful for interrelating machine tools, cutting tools, and workpiece characteristics.
3. State generally the machining characteristics sought for maximum productivity on shapers.
4. Determine the number of strokes per minute needed to match cutting speed on crank shapers.
5. Establish feedrates that are reasonable for both roughing and finishing on shapers.

CUTTING CHARACTERISTICS OF SHAPERS

Shapers have the common characteristic of non-continuous cutting, with shock to the tool at the start of each cutting stroke. For this reason, the cutting tools must have more adequate support beneath the cutting edges than is the case for most turning tools. As previously discussed, carbide tools are not frequently applied to shapers because of the relatively low attainable cutting velocities (below 150 SFPM maximum). Carbide tools used in shaping typically have negative back and side rake in the interests of maximum support for the cutting edges. The selection of carbide cutting tools for the shaper use is typically grade C-2 for the straight tungsten carbides and C-6 for the steel cutting grades. These are relatively less brittle grades and will

stand the impacts of interrupted cutting better than the more brittle but very hard C-4 and C-8 grades.

The selection of high speed steel tools used in shaping is also in the direction of the less brittle tools. High speed steel grades M-2 and M-3 are the most common recommendations. These grades have a relatively high vanadium content, which improves impact resistance. The high cobalt grades are to be avoided because of brittleness.

INTERRELATIONSHIPS, SPEED, FEED, DEPTH OF CUT, AND MATERIAL CONDITIONS

The generalities that are made for roughing and finishing for lathes do not readily apply to shaper operations. Speeds for roughing are generally much lower than those used in turning, and they are tied substantially to the depth of cut as well. Finishing done on lathes is generally at a higher speed than roughing, but on shapers the reverse is true. Speeds are generally cut about one third from the rate recommended for shallow roughing cuts, mainly because finishing tools for either cast iron or the softer steels are usually wide and fed nearly full width each pass. This full tool contact can lead to chatter at higher speeds.

Another major factor that affects cutting speed is the metallurgical condition of the workpiece, as usually reflected in hardness measurements.

The United States Army and Air Force made extensive studies on machinability of a large variety of workpiece materials using various cutting tool combinations on most types of machine tools. These findings are collected in the *Machining Data Handbook*, published by Metcut Research Associates in 1972.

For shaper application very few generalities can be made. For most steels, whether low or high carbon, in the Brinell hardness range of 175 to 225, the recommended cutting speed is about 60 SFPM for depth of cut of .100 in. The feedrate recommendations, however, vary with carbon content. The high carbon steel feedrate is reduced about 40 percent from the rate recommended for low and medium carbon steels. For gray cast iron in annealed condition, as compared with steels, the cutting speed recommendation is substantially higher for the same depth of cut and with a substantially larger feedrate. But, at the other extreme, some conditions of cast iron call for cutting speeds as low as 10 SFPM, pearlitic or acicular plus free carbides, for example.

As depth of cut is increased to .500 in., the cutting speed recommendation in most materials is reduced to about 60 percent of that recommended for a .100 in. depth.

There are many material factors, such as workpiece hardness and inclusions in the workpiece, that are especially important in shaping where many castings are machined. Since there are also many cutting factors, such as workholding, workpiece strength, and tool contact area, compounded by great variation in machine rigidity and power, it is helpful to have a source like the *Machining Data Handbook* for guidance. Most information in the handbook is derived from actual cutting practice and serves as a good starting point when setting up planers, shapers, and other machine tools.

A generality that can be made is that the combination of depth of cut and feedrate should be as much as can be tolerated by the weakest link of the cutting tool, workholding, part strength, or machine tool rigidity and power, consistent with the required surface characteristics required on the part such as flatness and finish. Where reductions must be made, it is better to reduce the feedrate than to reduce the depth of cut. With the increased cutting depth you have the work spread out over more cutting edge, and thinner, more easily controlled chips are produced for a given volume of metal being removed in a given time.

DETERMINING STROKE REQUIREMENTS ON A CRANK SHAPER

The duration of the cutting stroke of the crank shaper is about 220 degrees of crank rotation (Figure 1). This leaves 140 degrees for the return stroke or a ratio of 3 cutting to 2 return. The sum of this ratio is 5, so three fifths of the time is spent cutting and two fifths of the time is spent in the return strokes. Since cutting speed (CS) is given in feet per minute, the length of the stroke (L) times the number of strokes per minute (N) must also be expressed in the same unit. This is done by dividing $\frac{L \times N}{12}$ = feet per minute of tool movement.

Since the cutting stroke occupies three fifths of the time, it is necessary to allow for this fact in the equation.

$$CS = \frac{L \times N}{\frac{3}{5} \text{ of } 12} = \frac{L \times N}{7.2 \text{ (or 7)}}$$

Now, since you know the cutting speed you intend to use and the stroke length that you will be using for each operation, you can convert:

$$N = \frac{CS \times 7}{L}$$

As an example, consider the problem of machining a 10 in. long piece of medium carbon hot rolled steel with a rated cutting speed of 60 SFPM:

$CS = 60$

$L = 11$ (10 in. plus 1 in. allowance for overtravel)

therefore:

$$N = \frac{60 \times 7}{11} = \frac{420}{11} = 38.18$$

You would then set a value as close to 38 strokes per minute as available on the machine.

If your shop makes considerable use of the shaper, the formula $N = \dfrac{CS \times 7}{L}$ should be committed to memory.

Another somewhat easier approximation can be used while you are setting up. Since you know that about 40 degrees more of crank pin travel is required for the cutting stroke than the return stroke, you have a little more than 10 percent more for the time of the cutting stroke than the average that you would have if the machine had a theoretical uniform speed in both directions. If you multiply the cutting speed by 12, you will have the inches per minute of cutting speed. If you multiply the stroke length by 2, you will have the distance the tool must move in both forward and reverse in a minute to cover the distance, assuming uniform velocity.

$$N = \frac{CS \times 12}{L \times 2} \quad \text{or} \quad \frac{CS \times 6}{L}$$

Now increase the number of strokes indicated by 10 percent because of the additional crank pin travel above the average time to raise the cutting velocity to average speed.

$$N = \frac{CS \times 6}{L} + 10 \text{ percent}$$

For the example just given:

$$\frac{60 \times 6}{11} = \frac{360}{11} = 32.7 + 10 \text{ percent } (3.3) = 36$$

which reasonably approximates the more exact procedure. If this approach is easier for you, it is close enough to be useful.

For hydraulically driven shapers, the situation is very simple. The cutting speed can be entered directly into the machine controls.

ESTABLISHING FEEDRATES

For roughing in shaping, the feedrate should be about one tenth the depth of cut, where a deep roughing cut ($\frac{1}{2}$ in., for example) is made. Where a relatively shallow roughing cut is made, like .100 in., feedrates from one third to one half the depth of cut are more usual.

For finishing in shaping, the feedrate to obtain a finished surface is dependent mainly on the shape of the tool. For most steels and cast iron, the finishing tool should be essentially square on the nose and be fed up to three quarters

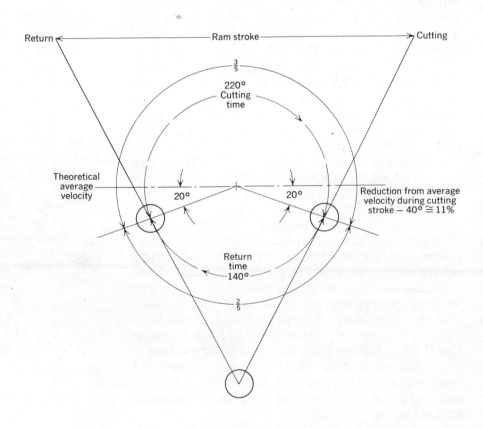

Figure 1. Timing characteristics of the crank shaper.

of the tool width on each stroke. To summarize the relationship of feed to finishing speed, the finishing speed should be somewhat faster than that recommended for deep roughing cuts and slightly slower than the recommended speed for shallow roughing cuts.

self-evaluation

SELF-TEST

1. Why are carbide shaper cutting tools usually specified with negative back and side rake?

2. What grades of high speed steel and cemented carbide tools are generally specified for use on shapers? Why are these grades specified?

3. Finishing speeds on the shaper are often lower than some roughing speeds. Why is this the case?

4. Name a source of machining information useful to determining speeds and feeds on shapers and other machine tools.

5. What general statement can be made about maching factors including depth of cut to feedrate when striving for maximum productivity on shapers.

6. If you have to choose between reducing feedrate and depth of cut to stay within other machining limitations, which would you reduce and why?

7. You have a 15 in. long workpiece of pearlitic cast iron with free carbides that has a rated deep cutting speed of 12 SFPM. What number of strokes per minute would be right for this job?

8. You have a $4\frac{1}{4}$ in. long piece of medium carbon hot rolled steel with a shallow roughing speed of 60 SFPM. What number of strokes per minute would be right for this job?

9. For roughing cut, on shapers, what would be an expected feedrate to match a .450 in. depth of cut for deep roughing and a .090 in. cut for a shallow roughing cut, assuming that the overall machining system is adequate for the jobs?

10. If you are making a .005 in. deep finishing cut on a large shaper in cast iron using a square nose finishing tool $\frac{1}{2}$ in. wide, what feedrate would you set into the machine for making this cut?

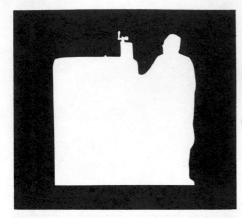

unit 3
using the shaper

Before using the shaper you must have a background of information about shaper tools, cutting speeds and feeds, and how to calculate the required strokes per minute. This unit covers operating procedures for several types of cutting operations on the shaper.

objectives After completing this unit, you will be able to:

1. Demonstrate that you have complied with the lubrication requirements specific to the shaper in your shop.
2. Set the stroke length and stroke rate, and indicate in the vise on a plain shaper (or universal shaper, as necessary).
3. Calculate and enter stroke rates for a specific cutting job.
4. Prepare worksheets for shaper operations.
5. Surface a workpiece in the shaper.
6. Square a workpiece in the shaper.
7. Make angular cuts in the shaper.

LUBRICATION

Before setting up the horizontal shaper, the lubrication requirements should be satisfied. On some larger shapers a lubrication pump is directly coupled to the main drive motor or, if the shaper is hydraulically driven, part of the hydraulic oil will be diverted into a special lubrication channel that meters oil to the ways of the ram. These larger machines are usually equipped with a sight glass to check on oil flow. On shapers without these automatic features, it is necessary to service lubrication points *before* the shaper is operated. Shaper lubrication is usually done with an *oil* (not a grease) especially compounded for sliding lubrication and fed through automotive-type fittings with a pressure lubricating gun. Check the specific requirements of the machine you will be using and follow the steps recommended for lubrication.

SETTING UP TO MAKE A CUT IN THE HORIZONTAL SHAPER

Indicating the Vise

Before the shaper is operated, you must be completely familiar with the controls of your specific machine. With Figure 1 as a guide, locate the same controls or control functions on your machine.

In the following sequences, you will see basic setups being made on the machine in Figure 1. Initially, the machine is checked to see that the toolhead is vertical, that the toolslide is retracted flush with its own base, and that cutting tools are removed. Then the plate that indicates the number of strokes per minute (Figure 2) is examined and the transmission controls are set for the fewest number of strokes per minute (Figure 3). A check is made to see that there are no tools on the machine that could interfere with

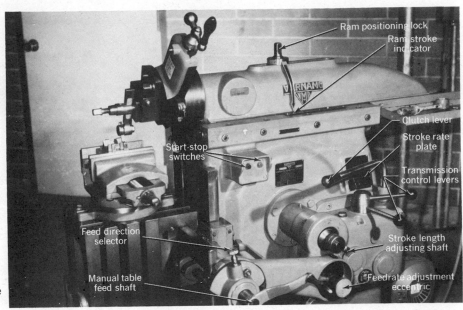

Figure 1. Shaper controls (Lane Community College).

Figure 2. Stroke rate plate (Lane Community College).

Figure 3. Transmission controls being set (Lane Community College).

Figure 4. Adjusting the stroke length. A knurled ring unlocks the adjustment shaft (Lane Community College).

ram travel, and that no wrenches remain on the stroke adjusting shaft or on the ram positioning shaft or lock. The clutch lever is also checked to see that it was not left in engagement by the last operator. The areas both in front of the ram and behind the ram should be checked for safe operation of the ram. The next step is the setting of the stroke length. The procedure varies greatly with different designs of shapers. Most large crank

Figure 5. Positioning the ram. On light machines this is done by releasing the ram and sliding it into position (Lane Community College).

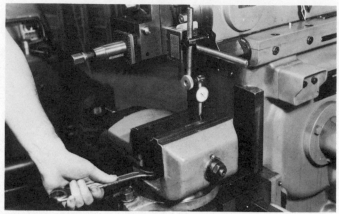

Figure 6. Indicating the shaper vise (Lane Community College).

Figure 7. Setting the clapper box angle (Lane Community College).

shapers are adjusted in the rearmost position of the ram. The shaper shown is adjusted at the center position, and some bench shapers are adjusted with the ram forward. Hydraulic shapers are arranged to set both stroke length and position by means of adjustment dogs that trip a pilot valve. The motor is then started and the clutch (or operating valve) is **drifted** to move the ram gently to the required position for adjusting stroke length (Figure 4). Since this operation is **indicating in,** the stroke length set should be about an inch less than the length of the jaw of the vise you will be using. If your machine has a universal table, the table must be indicated in on the cross feed direction before the vise is indicated.

Next, the ram is unlocked and positioned (Figure 5) so that an indicator attached to the ram or held in the tool post is over the correct part of the vise jaw. On the machine shown, the indicator would be placed central to the vise jaw. Then the table of the shaper is brought up and locked and the cross feed adjusted until the solid jaw contacts the indicator tip, with at least a quarter turn of indicator travel, below the top of the jaw. Now, with the motor started, the indicator is tracked slowly across the face of the fixed jaw by drifting the clutch to check alignment, and adjustment is made as necessary (Figure 6) until minimum indicator travel is obtained. If the machine is a universal table type, and you are working on the tilt table side, first check for parallel on the tilt table by indicating along precision parallels resting on the bottom of the vise jaw.

Surfacing a Workpiece

PREPARING THE WORKPIECE TO TOOL RELATIONSHIP. A square sectioned rectangular workpiece of hot rolled steel is being used for an example. This is mounted in the vise on parallels so that the top surface extends somewhat above the vise jaws. In mounting a rough workpiece, it may be necessary to use shims to obtain a solid hold on the part. The table is unlocked and lowered as necessary to clamp a tool in the tool post clear of the workpiece. As necessary, the stroke length control is adjusted and the ram is repositioned so that at least $\frac{1}{2}$ in. of pretravel and $\frac{1}{4}$ in. of travel are obtained beyond the workpiece. The pretravel gives the clapper box time to close fully before

Figure 8. Positioning the tool holder after setting the clapper box (Lane Community College).

Figure 10. Raising and clamping the table (Lane Community College).

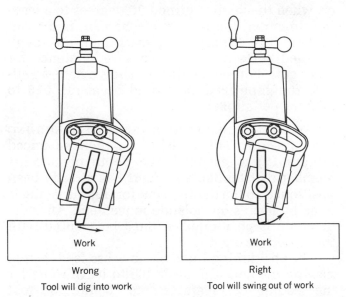

Wrong
Tool will dig into work

Right
Tool will swing out of work

Figure 9. On heavy cuts, leave room for the tool to escape (Courtesy of Cincinnati Incorporated).

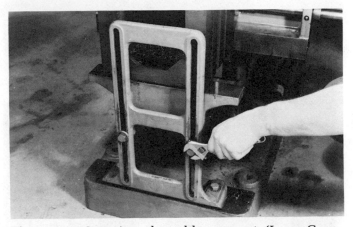

Figure 11. Securing the table support (Lane Community College).

the tool strikes the workpiece on the cutting stroke. Next, the apron carrying the clapper box is adjusted (Figure 7) to allow the tool to swing clear of the work on the return stroke. The cutting tool is then reset to vertical (Figure 8) or, in a heavy cutting situation, just a little off the vertical to allow the tool to swing away from the work if the feeding load is excessive (Figure 9). The table is then brought nearly to tool level and clamped firmly (Figure 10). Then the table support is raised and secured (Figure 11).

SETTING THE SPEED AND FEED. The number of strokes per minute must be determined to obtain

the rated cutting speed of the material for the type of cut (shallow roughing).

$$\frac{CS \times 7}{L}$$

Remember to add an inch to your workpiece length to allow for overtravel.

The calculated number of strokes are set into the machine by shifting the transmission lever to the nearest indicated setting (Figures 2 and 3). If the shaper you are using is hydraulic, the cutting speed can be set directly.

As discussed in the previous unit, the rate of feed for roughing should be as much as possible consistent with the machining factors: tool strength, workpiece strength, workholding, machining conditions, and available power. This balance must also result in usable part surface in

Figure 12. Setting the feedrate with an adjustable eccentric (Lane Community College).

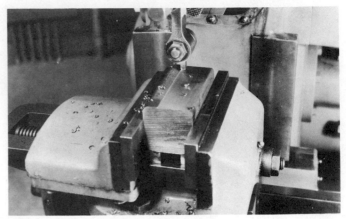

Figure 14. Making the first cut (Lane Community College).

Figure 13. Setting and locking the depth of cut (Lane Community College).

or, when the knob is turned 90 degrees to a feeding direction, the ratchet mechanism is disconnected. On the part being machined, the rough finished surface is to serve as a reference for other cuts. For the conditions shown here, a suitable roughing feedrate would be about .015 to .020 in. per stroke.

ESTABLISHING THE DEPTH OF CUT AND MACHINING THE SURFACE. The toolslide is then advanced (Figure 13) to establish the depth of cut. Remember that the table support has already been positioned so the depth of cut for this surfacing is done by means of toolslide movement. After the toolslide is positioned, it must be secured with the slide locking screw.

In positioning the table in cross feed, before making the cut, it is desirable to leave room for the feeding of several strokes before the tool reaches the work. This gives time to stop the machine if something does not seem right. When you engage the clutch for the first time, ease in the clutch gently to make sure you have not made a mistake in setting the stroke rate. The first surfacing cut is then made (Figure 14). This surface will be designated as side 1.

terms of finish and flatness for the operations that follow.

The shaper illustrated uses an adjustable eccentric (Figure 12) that increases the feedrate in increments of .005 in. as the radius from center is increased. The ratchet mechanism driven by the eccentric has a control to set the direction of feed

Squaring the Workpiece

With a completed surface from the previous cut, this surface is used as a reference for squaring the workpiece. Before this is done, the workpiece is removed and burrs are carefully removed. The vise and parallels are cleaned and the parallels are reinstalled. Then the workpiece is replaced in

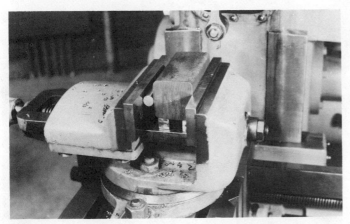

Figure 15. Round bar forces reference surfaces together (Lane Community College).

Figure 16. Turn the part over end for end to prepare for side 3 (Lane Community College).

the vise with the machined side toward the fixed jaw of the vise. A piece of unhardened round bar stock is obtained and placed between the workpiece and the movable jaw of the vise (Figure 15), and the vise is tightened. The round bar establishes line contact with the workpiece and forces the finished surface into full contact with the fixed vise jaw.

When the second cut is made it establishes a surface square to the first cut if everything was done correctly. This assumption should be checked with a square. After the part has been deburred and the vise and parallels have been recleaned, the just completed surface (2) is placed against the parallels with the original surface (1) again toward the fixed vise jaw (Figure 16). This is done by turning the part end for end in the process of turning it over. Again, the round bar is used between the movable jaw and the workpiece. As the vise is tightened, the workpiece is tapped down onto the parallels with a babbitt hammer or other soft-faced hammer with a weighted head. The parallels must be snug beneath the workpiece with the jaw fully tightened.

Now the exposed surface of side 3 is machined. The relationship between the newly machined surface (3) and the previous surface (2) should be checked with a micrometer on the machine, if possible, to insure that the two surfaces are parallel. This may also be checked off the machine and the part reset before making a final cut to the required dimension. Since the relationship of side 1 to side 2 was established as square, surface 3 should also be square to surface 1, but there is still the possibility of other errors, such as taper along the part axis.

Figure 17. Setting the vise crosswise to the ram travel (Lane Community College).

The fourth side is machined after checking, deburring, and cleaning the vise by placing the completed side 3 toward the solid vise jaw. Again, it is important that the parallels be snug as the vise jaw is tightened. Check the fit of the parallels before machining the fourth side to the required dimensions.

Squaring the End of the Workpiece
For this operation, the vise is set crosswise to the ram travel direction (Figure 17). If the work is to be done from the end of the vise, the fixed jaw of the vise needs to be indicated in by moving the vise with the table cross feed under the indicator. It is also desirable to indicate across parallels resting on the bottom of the vise opening, and then across the top edge of the vise jaw to see that this surface is also parallel.

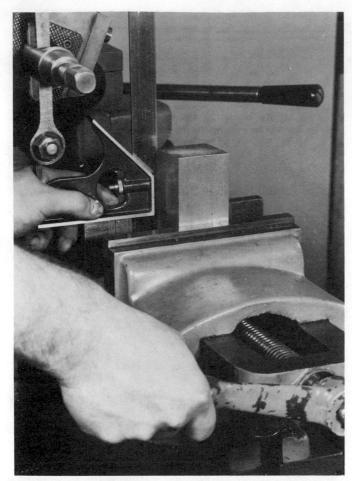

Figure 18. Setting a workpiece with the combination square (Lane Community College).

Figure 19. Squaring the end of the part (Lane Community College).

Figure 20. Setting the tool for roughing out the excess material (Lane Community College).

For relatively short workpieces, the part can be set vertically in the vise using the combination square, as shown in Figure 18. Since the reference edge is the top of the fixed vise jaw in this case, this surface must be set parallel to the table travel.

The stroke length needs to be reset and the number of strokes per minute recalculated and the ram repositioned before making the cut (Figure 19). If the part is too long to be squared in this way, it can be held out the end of the vise and cut with a side tool (Unit 1, Figure 29). If squaring is done in this fashion, vise squareness is critical.

Making Angular Cuts

Angular cuts are easy to make in the shaper. The example shown is the sequence for making a 90 degree V in the previously squared workpiece. The machine is set up and indicated as in preparing the part for surfacing. Initially a roughing tool is used to remove excess material. For this operation the toolhead is set vertically, and the apron is set to permit the tool to lift away from the cut on the return stroke. The tool and tool holder are placed with the tool mounted on the forward side (Figure 20) because for this operation visibility is quite important.

For roughing out the form, cuts are made by combinations of toolhead and table cross feed to hand contour the part just short of a layout line (Figure 21).

After the excess material has been roughed out, the toolhead is repositioned to 45 degrees (Figure 22) and the clapper box is reset to permit the tool to lift from the cut on the return stroke (Figure 23). Then the finishing cut is made along

Figure 21. Roughing out the V by hand contouring (Lane Community College).

Figure 24. Finishing one side of the V. Turn the part end for end to finish the opposite side (Lane Community College).

Figure 22. Setting the toolhead to 45 degrees (Lane Community College).

Figure 25. Vee block workpiece cut in two being finished on the ends (Lane Community College).

Figure 23. Resetting the clapper box apron (Lane Community College).

the 45 degree axis (Figure 24), using the toolhead feedscrew. Since it is important to have the V central to the workpiece, the cross feed must not be moved once the finishing cut is established. When the bottom of the V is reached and the ram is stopped, the slide is retracted and the part removed and *reversed end for end*; the finishing cut is taken on the opposite face, using the same exact machine position. In this way, the part will be symmetrical.

After the straight and angular cuts have been made on the vee block workpiece, this workpiece can be sawed in two and the ends finished as shown in Figure 25.

Securing the Machine

The last step is to secure the machine. It should be cleaned, the tooling put away, and the following steps observed for the safety of the next operator.

1. the cutting tool and holder should be removed.

2. The toolhead should be set to vertical and the toolslide retracted flush with its base.
3. The stroke length should be set to minimum.
4. The stroke rate should be set to minimum.
5. The clutch must be left disengaged.

self-evaluation

SELF-TEST
1. How are large mechanical or hydraulic shapers lubricated?
2. What should you do if you are concerned about the security of your workholding method?
3. How much additional ram travel should be allowed at both ends of the workpiece?
4. What specific, additional dial indicating should be done on a universal shaper?
5. Calculate the ram speed in strokes per minute for machining 4140 alloy steel. Assume a shallow roughing speed of 35 SPFM.
6. When squaring a workpiece standing vertically in the shaper vice, why is it better to indicate the vise rather than indicate the workpiece?
7. How does a piece of round stock aid in workholding when squaring a workpiece?
8. When shaping angles on a vee-block, why is it better to reverse the workpiece instead of swiveling the tool head to the other side?

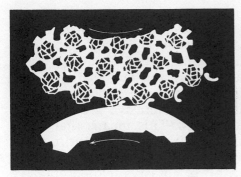

section n
grinding machines

It has been several thousand years since man first discovered that he could brighten up and sharpen his tools by rubbing them against certain stones or by plunging them into the sand several times. About the middle of the nineteenth century, with the development of mechanical holding devices and of circular, rotating abrasive wheels, this became a precision process. Grinding received another push forward around 1900 with the discovery of silicon carbide and aluminum oxide, manufactured abrasives that freed grinding from the disadvantages of natural abrasives.

Until only a few years ago, grinding still remained a polishing and finishing process, following some other primary metal cutting process. Then, in the development called **abrasive machining,** metal cutting with abrasives was rounded out to a full range of cutting, from the very fast metal removal to very fine surface finishing with extremely tight tolerances. It should be noted, of course, that rough snagging or grinding of castings and billets to remove unwanted metal, had begun very soon after the discovery of artificial abrasives. Abrasive machining, on the other hand, is usually considered to involve the removal of substantial amounts of stock to some kind of dimensional tolerance. Final finishing to exact dimensional tolerances may be done on the original grinder by changing the method of dressing the wheel or by changing wheels or, depending on the requirements for the part, by changing to another machine.

This concept of using abrasives (grinding wheels or coated abrasive belts from raw piecepart to finished part) is still relatively new and has a long way to go to reach general shop acceptance. High stock removal by abrasive machining is still generally considered to be for the high production shops with big, powerful machines and high speed wheels. However, it is also true that almost any grinder, even the little 6 × 12 in. surface grinders found in almost every machine shop, is capable, with proper selection of grinding wheel and coolant, of removing more stock than it is usually called on to do.

It cannot be overemphasized that grinding and other abrasive processes are *cutting* processes. In spite of the name abrasive, which suggests rubbing and removal of stock by friction instead of by cutting, a grinding wheel or a coated abrasive is a cutting tool similar in function to a milling cutter or a planer tool. The significant difference is that instead of taking off large chips with one tooth or a few teeth, the grinding wheel or the belt takes off tiny chips with thousands or maybe millions of tiny cutting edges (Figure 1). Some idea of the number may be gleaned from the fact that grinding wheels operate at speeds of from 1 to 3 mi/min (6,000 to 18,000 surface feet per minute) and that a single abrasive grain usually measures less than $\frac{1}{8}$ in. in any direction. This means that every minute of contact with the grinding wheel, an abrasive cutting area from 6,000 to 18,000 ft long and as wide as the wheel, is passing over the area to be cut. Even with each bit of abrasive (called abrasive grain) taking only a very small chip, the possibilities for stock removal are considerable.

DEVELOPMENT OF GRINDING MACHINES AND GRINDING WHEELS

By 1860, a grinding machine capable of grinding cylinders and tapers had been developed (Figure 2). This design of machine made it possible to get both accurate dimensions and a superior finish on cylindrical parts. It represents a major step in the development of machine tools. By 1876, the concept of the cylindrical grinding machine was well developed (Figure 3) and had essentially all of the motions of the modern universal grinder. With this machine it was possible to grind steep tapers and cylindrical and tapered workpieces.

The grinding of flat surfaces also received the attention of the machine builders of the period. While the early cylindrical grinder was an offshoot of the lathe family, the early surface grinder was a development based on the milling machine (Figure 4).

The grinding wheels of this period were quite primitive, and many experiments were under way to develop more adequate wheels. Natural abrasives like corundum (an impure form of aluminum oxide) could not be depended on to provide consistent results from batch to batch of wheels, and the problems of holding the abrasive grains in an adequate bond were huge.

In the 1880s, Charles Norton suggested that a grinding wheel a foot in diameter and an inch in thickness could be used to an advantage. For his projection, he was ridiculed by his associates but, by 1900, his concept was embodied in a production cylindrical grinding machine (Figure 5). Norton's production grinder became a reality because of the development of manufactured abrasives in the 1890s and the development of adequate bonds for holding the abrasive grains together.

In 1891, Dr. Edward Acheson heated some powdered coke and clay in a crude electric furnace and obtained crystals of silicon carbide (Figure 6), which would cut glass. At first he was able to produce only a few ounces per day, which sold for over $800 per pound, for the gemstone polishing market. Silicon carbide is now produced by the ton in large electric furnaces (Figure 7) and sells for a few cents per pound.

In 1897, Charles Jacobs was the principal in the discovery of aluminum oxide (Figure 8), which is made by fusing natural alumina or bauxite in an electric furnace and crushing the resulting glassy mass.

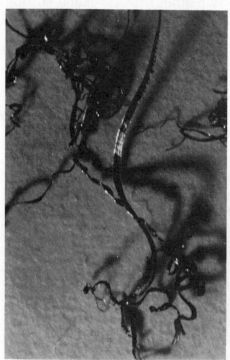

Figure 1. Chips from grinding, magnified about 350× (Courtesy of Bay State Abrasives, Dresser Industries, Inc.).

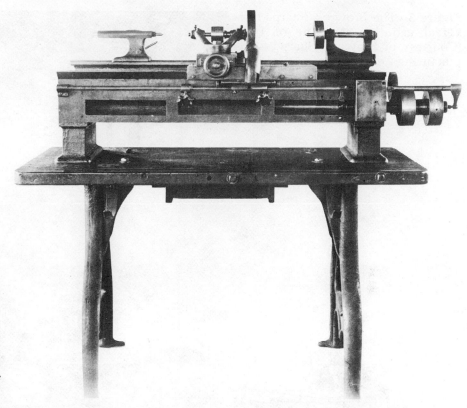

Figure 2. This is a back view of one of the earliest cylindrical grinders, made for the watchmaking industry. Note the mechanism that moves the grinding head along the workpiece. This is the same principle used with modern roll grinders (Courtesy of Norton Company).

The idea of cementing or bonding together bits of abrasive to make a **solid** wheel developed early in the nineteenth century. Rubber appears to have been the first material tried, although it was soon outdistanced by vitrified or clay-type bonds that used a furnacing technique similar to that for making dishes and that is still the major bond, certainly in precision grinding. Silicate of soda, known as silicate bond, which required only low-temperature firing, and oxychloride bond, a mixture of oxide and chloride of magnesium, which makes a cold-setting cement, were also popular in the nineteenth century, but they have gone out of use now. In 1876, indeed, oxychloride was acclaimed as the strongest and best bond of all. Shellac came along about 1880 for certain fine finishing applications, and it retains a limited popularity for finishing. Resin or resinoid bond, which has found wide use in foundry and other rough grinding applications, did not come along until about 1920. Today, it is likely that vitrified bonds and resinoid bonds share at least 90 percent of the market, with rubber and shellac taking most of the remainder. Nothing else is of much consequence. It is true, however, that there are many variations of each of these general groups, particularly in vitrified and resinoid bonds.

Some of the early grinding wheels were referred to as solid emery or corundum wheels. Possibly this was meant to indicate only that the wheel had been made without a core of steel or other such material; it certainly could not have suggested that the wheel was really solid and without pores. Every grinding wheel has some open

Figure 3. The Brown & Sharpe universal grinding machine of 1876 (Courtesy of Brown & Sharpe Manufacturing Company).

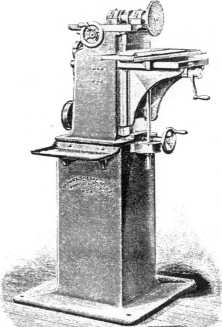

Figure 4. This surface grinder of 1887 was a close relative of the horizontal milling machine (Courtesy of Brown & Sharpe Manufacturing Company).

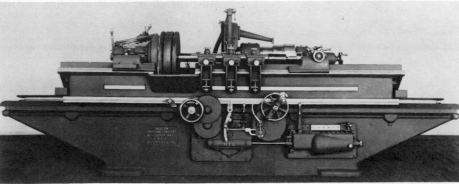

Figure 5. Production cylindrical grinding machine of 1900 (Courtesy of Norton Company).

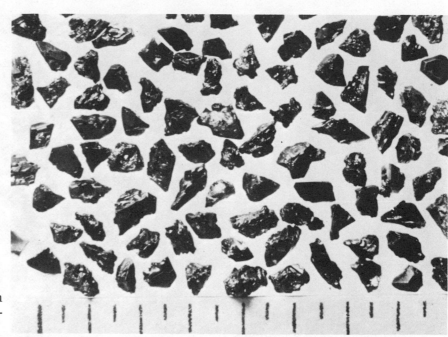

Figure 6. Large grains of silicon carbide (Courtesy of Exolon Company).

Figure 7. Silicon carbide production furnace where tons of coke, sand, sawdust, and salt are fused into the abrasive by electric heat in the range of 4000° F (Courtesy of Norton Company).

spaces, particularly vitrified bonded wheels. In fact, vitrified wheels are porous enough that one way of feeding coolant is through a hollow wheel mount and then out through the wheel to the grinding zone. This does ensure that the coolant gets right into the grinding zone. With a flood coolant, even though it may *look* as though there is plenty of coolant covering the wheel-work contact area, the fan-like action of the wheel, revolving at 6000 surface feet and more per minute, may simply be blowing the coolant away, so that you are actually "grinding dry with coolant," as one authority puts it.

Figure 8. Large grains of aluminum oxide (Courtesy of Exolon Company).

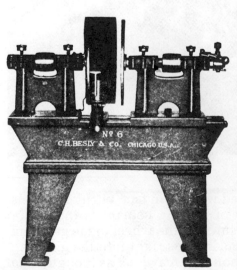

Figure 9. Besly disc grinder of 1904. A coated abrasive disc is used as a grinding surface (Courtesy of Bendix Corporation).

Once the development of consistent abrasives in commercial quantities and the problems in developing satisfactory bonding methods were solved, the development of highly productive grinding machines emerged quickly. The emergence of this class of production machine tool made the mass production of automobiles feasible. One of the early high production machines was the disc grinder (Figure 9). With this type of machine, thin parts, like piston rings, for example, could be surface ground for thickness quickly and inexpensively. Other developments of the period included magnetic workholding for surface grinding machines, with the invention of the magnetic chuck by O.S. Walker in 1896.

The need for production machines to satisfy the needs for high production of accurate highly finished parts for the emerging automotive industry led to rapid-fire developments in machine tools of all classes. One of these machines was the centerless grinding machine (Figure 10). This machine, invented about 1915 by L.R. Heim, was especially useful for the uniform sizing and finishing of automotive parts such as piston pins. The principle of the centerless grinder is quite interesting. The rubber bonded regulating wheel turns in the same direction as the grinding wheel, but at a much slower speed. The angle on top of the work rest blade causes the workpiece to be rotated (and effectively braked from spinning) by the regulating wheel so that there is a large differential speed between the work and the grinding wheel. It is also important that the center of the workpiece be above the common centers of the grinding and regulating wheels to obtain roundness on the part (Figure 11).

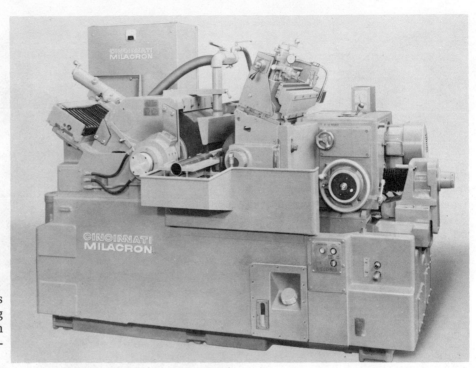

Figure 10. A modern centerless grinding machine used for grinding the OD of parts at high production rates (Courtesy of Cincinnati Milacron).

Another interesting grinding machine that emerged in the early 1900s was the vertical spindle rotary table grinder (Figure 12). With this machine, the grinding is done on the flat side of a cylinder wheel (Figure 13). In large models of this type of machine (Figure 14), the grinding is done by segments of abrasive held in a **chuck** in the form of a wheel attached to the spindle. Originally, this type of machine was used mainly for the grinding of glass optical parts, but it has developed to the point where it is one of the best ways to grind large volumes of metal on parts adapted for rotary table work-holding. Rapid metal removal, on the order of $\frac{1}{2}$ in., is common on the larger machines.

SURFACE GRINDERS Since our historical development of grinding machines and grinding wheels closes on the vertical spindle rotary table grinder, it would be a good time to deal with other types of grinders, called surface grinders. The most basic form is the horizontal spindle, reciprocating table-type grinder (Figure 15), called Type I surface grinder. This is the type that most people think of when the words surface grinder are mentioned. There are certainly more of this type in existence that the other types (other than pedestal grinders) combined. The work is usually held by a magnetic chuck and is traversed under the rotating wheel with the table. The table in turn is mounted on a saddle, which provides cross motion of the table under the wheel. In some designs, the grinding head with the wheel, is moved across the work surface instead of carrying the table on a saddle (Figure 16). The size of these machines can vary greatly from as small as 4 by 8 in. surface grinding area to 6 by 16 ft and larger (Figure 17). The great majority of this type of grinder is 6 by 12 in.

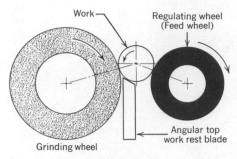

Figure 11. Principle of the center-less grinder. The grinding wheel travels at normal speeds and the regulating wheel travels at a slower speed to control the rate of spin of the workpiece (Courtesy of Bay State Abrasives, Dresser Industries, Inc.).

Figure 12. Blanchard vertical rotary surface grinder circa 1925 (Courtesy of Cone-Blanchard Machine Company, Windsor, Vermont).

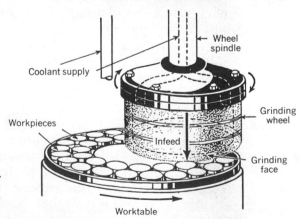

Figure 13. Principle of the vertical spindle rotary grinder (Courtesy of Bay State Abrasives, Dresser Industries, Inc.).

Most of the grinding on the Type I surface grinders is to produce a flat plane on the workpiece, usually to flatnesses less than .0002 in. overall. But this type of grinder can also be used to create contours in the work. The wheel can be dressed to the reverse form of that desired in the workpiece and the contour then ground into the part (Figure 18). This is covered in more detail in Unit 7 of this section.

Figure 14. A large vertical spindle rotary table grinding machine (Courtesy of Mattison Machine Works).

Figure 15. Type I surface grinder (Courtesy of Boyar-Schultz Corp.).

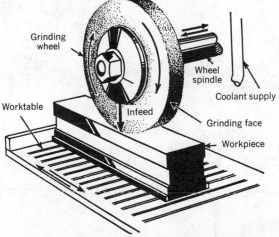

Grinding wheel

Wheel spindle

Coolant supply

Grinding face

Worktable

Infeed

Workpiece

Figure 16. Principle of the Type I surface grinder with alternative method of cross feeding motion (Courtesy of Bay State Abrasives, Dresser Industries, Inc.).

Figure 17. A large Type I surface grinder (Courtesy of Mattison Machine Works).

Figure 18. Grinding contours with the horizontal spindle surface grinder. The workpiece is directly under the wheel and the diamond-plated form dressing block is just in front of it (Courtesy of Engis Corporation).

Figure 20. Principle of the Type II surface grinder. Sometimes circular parts are centered on this grinder; the resulting concentric scratch pattern is excellent for metal-to-metal seals of mating parts (Courtesy of Bay State Abrasives, Dresser Industries, Inc.).

Figure 19. Horizontal spindle rotary table surface grinder (Type II). With the workpiece centered on the chuck, the wheel will produce a concentric scratch pattern. The table can also be tilted, making it possible to grind a part thinner in the middle than at the rim or vice versa (Courtesy of Heald Machine Division/ Cincinnati Milacron Company).

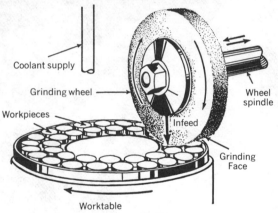

The horizontal spindle rotary table grinder is another interesting form of the surface grinder (Figure 19). This is sometimes referred to as a Type II surface grinder. The motions of this type grinder are shown in Figure 20. Not shown in this figure is that the table's rotational axis may be tilted a few degrees for operations like the hollow grinding of circular saws. When sealing joints are ground in this fashion (Figure 19), the resulting circular scratch pattern makes for exceptionally good sealing.

A third type of surface grinding machine employs a vertical spindle and a reciprocating table (Figure 21). One form of this Type III design is the way grinder (Figure 22), which is well adapted for long and norrow work like the grinding of ways on other machine tools. These grinders are typically fitted with auxiliary spindles so that the complete configuration of the ways can be completed in a single machine setup (Figure 23).

A fourth and relatively uncommon type is the face grinder, which typically employs a segmented wheel mounted on a horizontal spindle, so that the end of the spindle is presented to the part. This machine is especially suited to the surfacing of wide vertical surfaces.

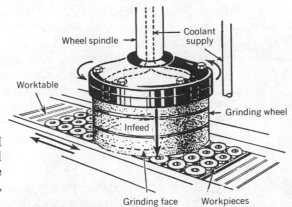

Figure 21. Principle of the Type III surface grinder, which has a vertical spindle and a reciprocating table (Courtesy of Bay State Abrasives, Dresser Industries, Inc.).

Figure 22. A combination way and surface grinder. This machine can remove a great deal of stock to good accuracy on machine ways and other long workpieces, usually eliminating hand scraping (Courtesy of Mattison Machine Works).

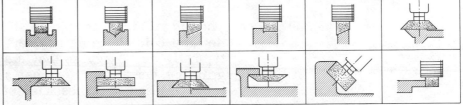

Figure 23. A great variety of surfaces may be ground with an accessory spindle by dressing or tilting the wheel (Courtesy of Mattison Machine Works).

CYLINDRICAL GRINDERS The term cylindrical grinder covers a wide assortment of grinding machine tools, including those that grind workpieces mounted on centers; extremely heavy workpieces mounted on bearing journals; centerless grinding; and internal grinding, either with the part held in a chuck or in a centerless fashion.

Center-Type Cylindrical Grinders The most basic form of cylindrical grinding is done with the workpiece mounted on centers (Figure 24). The operating principles of this type of machine are shown in Figure 25. For accuracy, the workpiece is rotated on dead centers on both ends, with the work-

Figure 24. Working area of a plain cylindrical grinder (Photo courtesy of Diamond Abrasives Corporation).

piece being driven by a plate that rotates concentric to the headstock center. The plain cylindrical grinder is capable of grinding tapered parts as well by the swiveling of the table around a vertical axis like the table of the universal milling machine. It is also capable of being used in a **plunge** mode where the workpiece and the wheel are brought together **without** table traverse.

A number of design variations exist with the center-type cylindrical grinder, the most common being the **universal** cylindrical grinder (Figure 26). With this type of grinder, the wheel head assembly can be rotated to present the grinding wheel to the work at an angle. This makes the grinding of steep tapers easy to do. You will see more details of this type of machine in the units that follow. The grinding of slender workpieces on center-type grinders of either plain or universal design can be done with the application of accessory support (Figure 27) in the form of a steady rest.

Another interesting variation of the plain center-type cylindrical grinding machine employs a grinding wheel set at an angle to the centers (Figure 28). These machines are especially suited to grinding into shoulders, especially where the relationship between the diameter and the face is critical (Figure 29). This type of grinder is also capable of traversing the table in the same way as the plain and universal types. The angular wheel-type cylindrical grinding machine is typically a production grinder and, in recent years, many production type cylindrical grinders have been supplied with automatic controls (Figure 30) to improve productivity and to reduce operator errors.

Form grinding is also possible on center-type cylindrical grinders (Figure 31). In this type of grinding, the reverse of the form to be imparted into the workpiece is dressed into the grinding wheel, and then the part is ground by a direct plunge of the wheel into the workpiece. This is a high production method widely used on complex parts like hydraulic valves.

The Roll Grinder The roll grinder is used to finish and resurface rolls used in both the hot and cold finishing of steels and other metals. These rolls are typically very heavy, so they are supported on journal bearings for grinding, just as they are when supported in the rolling mill where they are used. Also, because of the weight of the rolls, the roll

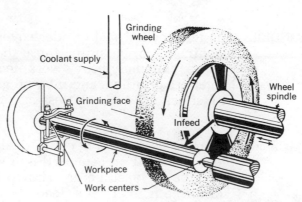

Figure 25. Principle of the cylindrical grinding machine showing the workpiece and wheel motions (Courtesy of Bay State Abrasives, Dresser Industries, Inc.).

Figure 26. A universal cylindrical grinder (Courtesy of Cincinnati Milacron).

Figure 27. The use of a steady rest for grinding slender workpieces (Courtesy of Cincinnati Milacron).

Figure 28. Angular center-type grinder equipped with in-process gaging.

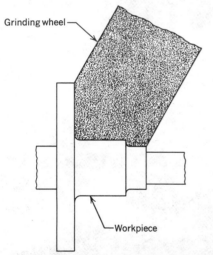

Figure 29. Typical application of the angular center-type grinding machine showing angled grinding wheel finishing both diameter and shoulder.

Figure 30. Angular center-type grinder with automatic control (Courtesy of Cincinnati Milacron).

grinding machine is designed (Figure 32) so that the roll rotates in a fixed position and the grinding head is moved along ways that are parallel to the roll. When the roll is to be used for rolling steel cold, where very high pressures are encountered, the machine is adjusted to grind a slightly convex cambered roll so the product comes out flat (Figure 33). In hot mill rolling the reverse is true and the roll is compensated to be slightly concave (Figure 34). For smaller rolling applications, some rolls are gound on centers in plain cylindrical

Figure 31. Cylindrical form grinding (Courtesy of Bendix Automation and Measurement Division).

Figure 32. Roll grinding machine used for grinding steel mill rolls or other very large workpieces (Courtesy of Landis Tool Co., Division of Litton Industries).

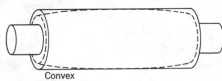

Convex

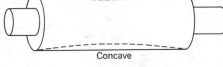

Concave

Figure 33. This steel mill roll has a convex surface (dotted lines) to compensate for the high pressures involved in cold rolling steel. The drawing is exaggerated for clarity.

Figure 34. This roll has a concave surface (dotted lines) for hot rolling steel. The drawing is exaggerated to show the slight variation.

grinders, or even sometimes on coated abrasive belt grinders that are able to grind the whole surface at one time.

The Centerless Grinder When someone speaks of the centerless grinding machine, it usually means a machine for work on the outside diameter of a cylindrical workpice, as shown in Figures 10, 11, and 35. These machines are usually used in high production applications, but they are by no means limited to simple cylindrical parts. Parts with differing diameters, like automotive valves, can be infed to a fixed stop (Figure 36). It is also possible to do tapered parts by shaping both the grinding and regulating wheels to the reverse of the shape required (Figure 37), and it is even possible to centerless grind parts with center portions larger than the ends by loading the part down from the top with special feeding apparatus (Figure 38). Even headless threaded parts, like setscrews, can have the threads formed on centerless grinders.

Figure 35. Centerless grinding machine. The regulating wheel is on the left and parts for through-feed grinding can be loaded into the hopper (Courtesy of Landis Tool Co., Division of Litton Industries).

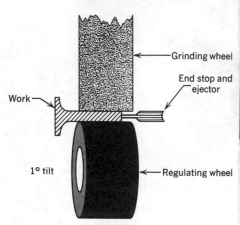

Figure 36. End feeding in the centerless grinder. The regulating wheel is set at a small tilt angle to keep the workpiece against the end stop (Courtesy of Bay State Abrasives, Dresser Industries, Inc.).

Internal Cylindrical Grinding

Internal Cylindrical Grinding

Internal cylindrical grinding (Figure 39) is usually done with a **mounted** grinding wheel having a shank that is held in the grinding spindle. The principles of this method of grinding are shown in Figure 40. The internal grinding wheel should be about three quarters of the finished hole diameter, mainly to have enough abrasive so that the interval between dressing is prolonged somewhat. With conventional wheels, dressing is frequent, sometimes even each pass in critical situations. In recent years a manufactured abrasive, cubic boron nitride (Figure 41) has been successfully applied in this type of application to reduce dressing requirements to the point where it is needed infrequently. This abrasive is close to the hardness of the diamond but, unlike the diamond, it works well in hardened steels. Mounted wheels of cubic boron nitride are also used in jig grinding machines to maintain uniform size between dressings. The use of this abrasive is not limited to internal grinding applications; it can occasionally be found in surface grinding use as well.

Internal grinding is not limited to concentric workpieces. In many instances, special fixturing is applied to machines to hold irregularly shaped workpieces (Figure 42). Internal grinding attachments are also applied to universal grinding machines to accomplish this type of work.

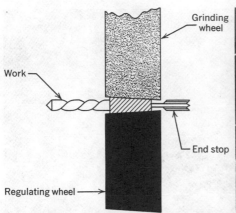

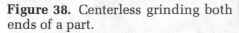

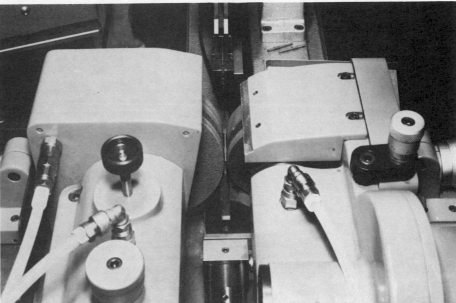

Figure 37. Centerless form grinding (Courtesy of Bay State Abrasives, Dresser Industries, Inc.).

Figure 38. Centerless grinding both ends of a part.

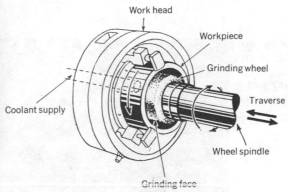

Figure 39. Internal cylindrical grinding (Courtesy of Heald Machine Division/Cincinnati Milacron Company).

Figure 40. Principles of internal cylindrical grinding (Courtesy of Bay State Abrasives, Dresser Industries, Inc.).

The Cutter and Tool Grinder This machine (Figure 43) can be considered a type of cylindrical grinder, but it also can be used for certain classes of surface grinding. It is often called the **universal** cutter and tool grinder because it can be applied in so many ways to a variety of jobs. (A number of ways of using this machine will be detailed in units in this section.) On this machine the spindle can be tilted and swiveled; straight, dish, or flaring cup wheels can be mounted. Wheels with diamonds (Figure 44) are commonly used for the sharpening of carbide tipped cutting tools (Figure 45). Even though the cost of a small diamond wheel may be several hundred dollars, the abrasive cost is ultimately lower than it would be using silicon carbide for the same application. Also, the size of the diamond wheel is maintained, which results in consistent size of the workpiece.

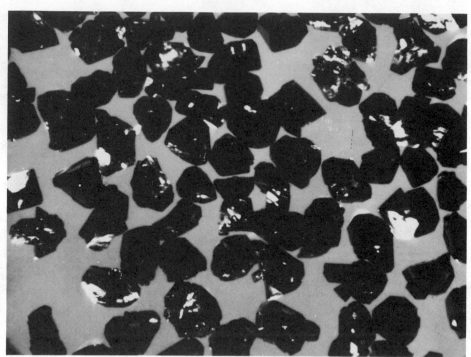

Figure 41. Cubic boron nitride (BOROZON™ CBN). This abrasive is almost as hard as diamond and very effective for grinding on tool steels (Courtesy of Specialty Materials Department, General Electric Company).

Figure 42. Internal grinding is not limited to cylindrical parts. This automotive connecting rod is an example of odd configurations that can be ground internally (Courtesy of Heald Machine Division/Cincinnati Milacron Company).

Figure 43. The cutter and tool grinder (Courtesy of Cincinnati Milacron).

Most of the sharpening work on cutters is done with ordinary free cutting aluminum oxide grinding wheels on cutters mounted between centers (Figure 46).

Some unique types of cutter and tool grinders also exist (Figure 47). This grinder is often used for the sharpening of tapered, and

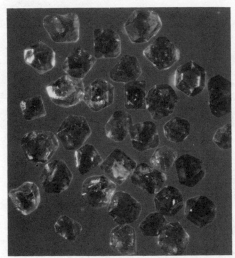

Figure 44. Manufactured diamond abrasive (Man-Made™ diamond), sometimes called synthetic diamond; it has the advantage over industrial grade natural diamond in that its mesh size can be controlled and varied to fit various materials (Courtesy of Specialty Materials Department, General Electric Company).

Figure 45. Sharpening of a large brazed carbide tipped face mill with a diamond cup wheel (Courtesy of Cincinnati Milacron).

Figure 46. Sharpening a plain milling cutter held between centers on the cutter and tool grinder (Courtesy of Cincinnati Milacron).

tapered and ball ended milling cutters. This type of cutter is used in duplicating or die sinking milling machines.

MISCELLANEOUS GRINDING MACHINES The disc grinder, which first appeared in 1904 (Figure 9), has evolved into a powerful, high production machine tool (Figure 48). This machine feeds parts between the faces of two grinding wheels. Various mehods are used for feeding parts to this type of machine. The type shown used a "ferris wheel" fixture that carries the parts between the wheels to be simultaneously ground to parallel and to a finished dimension. Other methods of feeding include the use of an oscillating arm to carry parts one at a time between the wheels or the use of guide bars to roll the parts between the wheels.

Gear grinders are another interesting type of grinding machine. There are a variety of principles employed with these machines, some of which were discussed earlier in this volume, with other gear producing machines. Basically these machines fall in two categories: form grinders (Figure 49), where the wheel is accurately dressed to the opposite shape of the tooth to be ground, and generating types (Figure 50), where the form results from the conjugate action of the wheel and workpiece.

Grinding is not always done with grinding wheels. Abrasive belt machines (Figure 51) are coming into increasing use, especially where precise dimensions are not required.

Other Abrasive Processes Some abrasive processes use relatively slow moving abrasive materials to perform their work. One type uses free abrasive grains circulated on a hardened steel plate to generate flat surfaces or workpieces (Figure 52). This type is capable of comparatively large amounts of stock removal, because the abrasive tumbles along the

Figure 47. Cutter and tool grinder capable of generating helical surfaces (Courtesy of Cincinnati Milacron).

Figure 48. Double disc production grinding machine with feed wheel to carry the parts between the two opposed discs for grinding. The parts then drop out at the bottom (Courtesy of Bendix Corporation).

Figure 49. Form-type gear grinding machine (Courtesy of Ex-Cell-O Corporation).

Figure 50. Hypoid gear grinding machine uses a cup wheel (Courtesy of Gleason Works).

hardened steel plate while it cuts the workpiece material. Another similarly appearing machine that works on a different principle is the lapping machine (Figure 53). In this machine, the cutting surface is prepared by imbedding abrasive grain into a relatively soft plate, which holds the abrasive, while relative motion is imparted to the workpiece. Exceptional flatness is obtained by this method, but the cutting rate is very slow. Temperature control is critical to accurate lapping.

Other relatively low velocity methods are used in the precise sizing of previously machined bores. The honing machine (Figure 54) can be used either for external or internal honing of workpieces within its size range, and it is often equipped with attachments (Figure 55) for mechanical movement of the workpiece along the honing mandrel.

Another relatively slow velocity finishing method is the **superfinisher** (Figure 56). This process uses a formed abrasive pad that is held in contact with the surface to be finished while it slowly turns in contact. In addition, a simultaneous slight side motion is also provided, which results in a highly finished accurate surface on the workpiece.

The vibratory deburring of parts is also a widely used abrasive process (Figure 57). The tumbling abrasive can even deburr the interior of workpieces that are inaccessible to other methods.

Finally, there is a hybrid electrochemical and abrasive machine called the electro-chemical grinder (Figure 58). This is really a plating machine operated in reverse. The electro-chemical action (Figure 59) removes the material from the workpiece (anode) but, in the process, insulating oxides are formed. The abrasive serves mainly to remove the oxides so that the process of deplating can

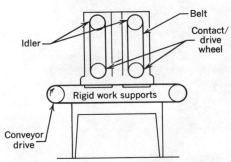

Figure 51. Diagram of a double belt coated abrasive grinder (Courtesy of 3M Company).

Figure 52. Free abrasive grinding machine has a hard, water cooled plate on which the abrasive is fed in a slurry. The abrasive grains do not become embedded as they do in lapping. Hence, the grains always roll around under the workpieces (Courtesy of Speedfam Corporation).

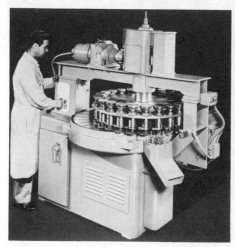

Figure 53. Lapping machine. The pressure plate on top holds the parts to be lapped by the abrasive embedded in the plate beneath (Courtesy of Lapmaster Division of Crane Packing Company).

Figure 54. Honing machine. Honing is a very popular method for finishing inside diameters of everything from bushings to cylinders of automobile engine blocks (Courtesy of Sunnen Products Company).

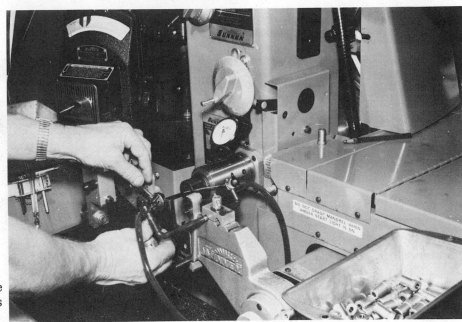

Figure 55. Using a honing machine with an attachment that reciprocates the workpiece.

Figure 56. Superfinishing an automotive crank (Courtesy of Taft-Peirce Mfg. Co.).

continue. The abrasive wheel is usually metal bonded diamond, for conductivity, and the wheels last an extremely long time. This type of machine is used frequently in the sharpening of single point carbide lathe and planer tools.

Figure 57. This high production vibratory finisher shows the parts and abrasive coming from the vibrator to the front where the parts are unleaded; then the abrasive is returned to the vibrator by the conveyor (Courtesy of UltraMatic Equipment Co.).

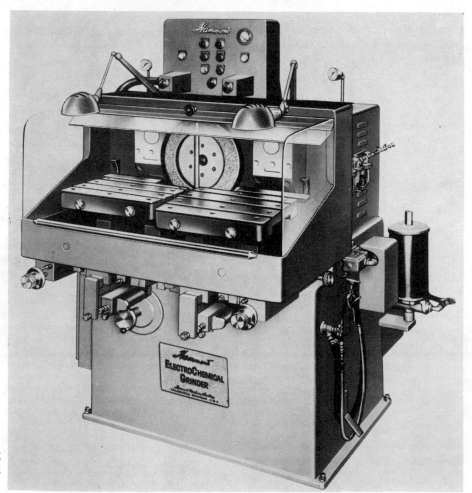

Figure 58. Electrochemical grinding machine (Courtesy of Hammond Machinery Builders Inc.).

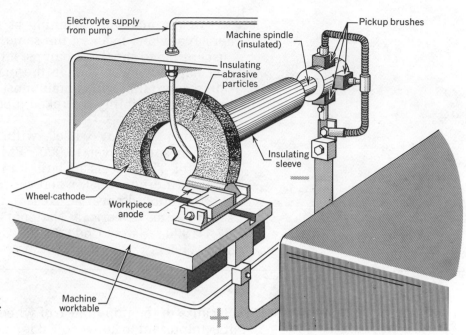

Figure 59. Principles of the electrochemical grinder (Courtesy of Hammond Machinery Builders Inc.).

GRINDING MACHINE SAFETY The potential hazards while operating any machine tool are not to be underestimated. The same general safety rules that apply to any power driven mechanical device also apply to grinding machines. However, the grinding machine is somewhat unique in that its cutting tool, the grinding wheel, presents a greater potential hazard because of its great speed. For this reason, safety considerations specific to grinding wheels must be developed in more detail.

Figure 60. Safety guard on a surface grinder. Note that the guard is somewhat squared off and covers well over half the wheel (Courtesy of DoAll Company).

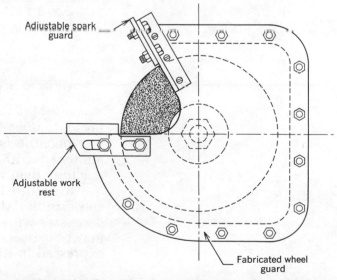

Figure 61. Safety guard for high speed wheel. The work is hand-held against the exposed peripheral grinding face of the wheel on top of the work rest. The squared corners tend to retain fragments in case of wheel breakage.

In the machine, spinning at 5000 FPM or moe, is a vitrified abrasive wheel made of the same material as dishes. The wheel is susceptible to shocks or bumps. It can easily be cracked or broken. If that happens, even though the machine has been designed with a safety guard that will contain most of the pieces (Figures 60 and 61), there is a possibility of broken pieces from the wheel flying around the shop.

Other grinding wheels with organic bonds can be operated safely at speeds over 15,000 FPM, but these are mostly for rough grinding. The wheels are built for it, but the principles are the same. Every grinding wheel, wherever used, has a safe maximum speed, and this should never be exceeded. The standard of grinding wheel safety is an American National Standards Institute publication B7.1-1970, "Safety Code for the Use, Care, and Protection of Abrasive Wheels,"

DETERMINING WHEEL SPEED Because of the importance of wheel speed in grinding wheel safety, it is important to know how it is calculated. This quality is expressed in terms of surface feet per minute (SFPM). This simply expresses the distance that a given spot on a wheel travels in a minute. It is calculated by multiplying the diameter (in inches) by 3.1416, dividing the result by 12 to convert to feet, and multiplying that result by the number of revolutions per minute (RPM) of the wheel. Thus, a 10 in. diameter wheel traveling at 2400 RPM would be rated at approximately 6283 SFPM, under the safe speed of most vitrified wheels of 6500 SFPM.

To find the safe speed in RPMs of a 10 in. diameter wheel, the formula becomes

$$\frac{SFPM}{D \times 3.1416 \div 12}$$

or,

$$\frac{6500}{10 \text{ in.} \times 3.1416 \div 12} = 2483 \text{ RPM}$$

Most machine shop-type flat surface or cylindrical grinders are preset to operate at a safe speed for the largest grinding wheel that the machine is designed to hold. As long as the machine is not tampered with and no one tries to mount a larger wheel on the machine than it is designed for, there should be no problem.

It should be clear that with a given spindle speed (RPM) the speed in SFPM increases as the wheel diameter is increased, and it decreases with a decrease in wheel diameter. Maximum safe speed may be expressed in either way, but on the wheel blotter it is usually expressed in RPM.

The Ring Test
The primary method of determining whether or not a wheel is cracked is to give it the ring test. A crack may or may not be visible; of course, if there is a visible crack you need go no further and should discard the wheel immediately.

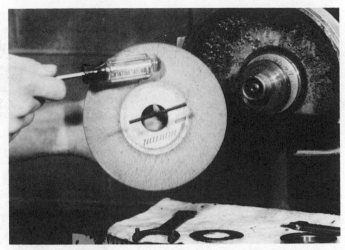

Figure 62. Making a ring test on a small wheel (Lane Community College).

The test is simple. All that is required is to hold the wheel on your finger if it is small enough (Figure 62) or rest it on a clean, hard floor if it is too big to be held (Figure 63) and preferably strike it about 45 degrees either side of the vertical centerline with a wooden mallet or a similar tool. If it is sound, it should give forth a clear ringing sound. If it is cracked, it will sound dead. The sound of a vitrified wheel is clearer than the sound of any other, but there is always a different sound between a solid wheel and a cracked one.

There are a few permissible variations. Some operators prefer to hold the wheel on a stick or a metal pin instead of a finger. Some shops prefer to suspend large wheels by a sling (chain or cable covered by a rubber hose to protect the wheel) instead of resting them on the floor (Figure 64).

The important point is that the test be done when each lot of wheels is received from the supplier, just before a wheel is mounted on a grinder, and again each time before it is remounted. If a wheel sounds cracked, or even questionable, then discard it or set it aside to be checked by the supplier, and get another wheel which, of course, must also be ring tested.

Mounting the Wheel Safely, Starting the Grinder

Any time you mount a wheel on the grinder or start the machine, good safety practice requires a set routine like the following, which is covered in detail in Unit 8.

1. Ring test the wheel as shown in Figure 62, then check the safe wheel speed printed on the blotter with the spindle speed of the machine. The spindle speed must never exceed the safe wheel speed, which is established by the wheel manufacturer after considerable research.

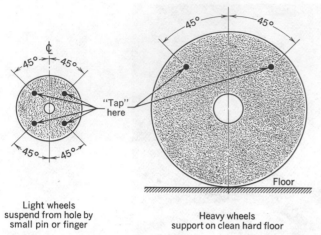

Light wheels suspend from hole by small pin or finger

Heavy wheels support on clean hard floor

Figure 63. Sketch shows where to tap wheels for ring test, 45 degrees off centerline and 1 to 2 in. in from the periphery. After first tapping, rotate wheel about 45 degrees and repeat the test (Courtesy of Bay State Abrasives, Dresser Industries Inc.).

Figure 64. Wheel being lifted by a sling for ring testing (Courtesy of DoAll Company).

2. The wheel should fit snugly on the spindle or mounting flange. Never try to force a wheel onto the spindle or enlarge the hole in the wheel. If it is too tight or too loose, get another wheel.

3. Be sure the blotters on the wheel are larger in diameter than the flanges, and be sure the flanges are flat, clean, and smooth. Smooth up any nicks or burrs on the flanges with a small abrasive stone. Do not overtighten the mounting nut. It just needs to be snug (Figure 65).

4. Always stand to one side when starting the grinder; that is, stand out of line with the wheel.

5. Before starting to grind, let the wheel run at operating speed for about a minute with the guard in place.

Steps 4 and 5 are essentially a double check on the ring test. If anything is going to happen to a wheel, it usually happens very quickly after it starts to spin.

Operator Responsibilities In grinding, there are a number of operator duties that have long been a part of company safety policy and state safety regulations, and that, with the passage of the OSHA regulations, also became a matter of federal regulations. These are mostly designed for your protection, and not, as some operators have thought, to make the job more awkward or less productive.

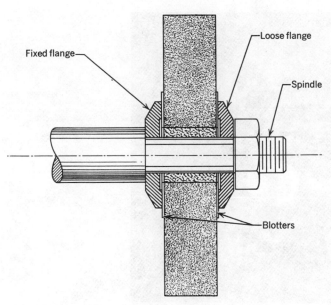

Figure 65. Typical set of flanges with flat rims and hollow centers with blotters separating the wheel and the flanges. Tightening the nut too much could spring the flanges and perhaps even crack the wheel.

Figure 66. Straight wheels are designed for grinding on the periphery. Never grind on the side of one of these, because it is not considered safe. On the other hand, cylinder wheels, cup wheels (both straight and flaring), and segments or segmental wheels (shown here without the holder) are all designed and safe for side grinding (Courtesy of Bay State Abrasives, Dresser Industries Inc.).

1. Wear approved safety glasses or other face protection when grinding. This is the first and most important rule.
2. Do not wear rings, a wrist watch, gloves, long sleeves, or anything that might catch in a moving machine.
3. Grind on the side of the wheel *only* if the wheel is designed for this purpose, such as Type 2 cylinder wheels, Types 6 and 11 cup wheels, and Type 12 dish wheels (Figure 66).
4. If you are grinding with coolant, turn off the coolant a minute or so before you stop the wheel. This prevents coolant from collecting in the bottom half of the wheel while it is stopped and throwing it out of balance.
5. Never jam work into the grinding wheel. This applies particularly in off-hand grinding on a bench grinder.

Figure 67. Storing extra wheels at the machine on pegs is often convenient and practical. The main requirement is to keep the wheels separate or protected, and off the floor (Courtesy of DoAll Company).

Wheel Storage At The Machine   General wheel storage is a responsibility of shop management, but at the machine this is usually the operator's concern. Operator carelessness can ruin an otherwise good wheel storage plan. On the other hand, care by the operator can do much to help out a substandard shop plan.

If the shop provides you with a proper storage area, all you have to do is to use it and keep it clean. If the shop does not provide such an area, you may need some ingenuity. Here are some suggestions.

1. Keep on hand only the wheels you really need. Do not allow extra wheels to accumulate around your machine. They can become cracked or broken and may get in your way. If you do not need a wheel, take it back to the general storage.
2. Store wheels above floor level either on a table or under it, on pegs in the wall (Figure 67), or in a cabinet. Be especially careful to protect wheels from each other and from metal tools or parts. Use corrugated cardboard, cloth, or even newspaper to keep wheels apart. *Keep wheels off the floor.*

unit 1
selection and identification of grinding wheels

A grinding wheel is at once both a many-toothed cutting tool and a tool holder. Selecting a grinding wheel is somewhat more complicated than selecting a milling cutter because there are more factors to be considered. These factors, including size, shape, and composition of the wheel, are expressed in a system of symbols (numbers and letters) that you must be able to interpret and apply.

objectives After completing this unit, you will be able to:

1. List the five principal abrasives with their general areas of best use.
2. List the four principal bonds with the types of applications where they are most used.
3. Identify by type number and name, from unmarked sketches or from actual wheels, the four most commonly used shapes of grinding wheels.
4. Interpret wheel shape and size markings together with the five basic symbols of a wheel specification into a description of the grinding wheel.
5. Given several standard, common grinding jobs, recommend the kind of abrasive, approximate grit size and grade, and bond.

Although six or seven factors must be considered in selecting a grinding wheel for a particular job, most of the decisions are almost automatic. For example, the size and shape of the wheel, are usually determined by the type and size of the grinder. The abrasive, of which there are five major kinds, is determined primarily by the material being ground. The abrasives are not equally efficient on all materials, and there are only a few areas where there would be much doubt about the choice. The size of the abrasive particles selected, of which the largest are perhaps $\frac{1}{8}$ to $\frac{1}{4}$ in. long and the smallest less than .001 in. long, depends on the amount of stock to be removed and the final finish desired.

Selecting the grade or hardness of the wheel, which is a measure of the force required to pull out the abrasive grains, is a difficult decision sometimes but, here again, it is typically a choice between one of two or, at most, three grades. Grain spacing or structure is most often standard, according to the grain or grit size and the grade of the wheel. Finally, there is the bond that holds the wheel together. And although there are four

745

common bonds, the choice is usually made clear from the job.

All of these factors (wheel shape, wheel size, kind of abrasive, abrasive grain size, hardness, grain spacing, and bond) are expressed in symbols consisting of numbers and letters, most of which are easy to understand and interpret.

There are many different kinds of grinding wheels in practically any toolroom or machine shop or, for that matter, in any grinding production department or foundry. Keeping them straight and knowing which wheel to use for which job can be something of a problem. Some of the ways this could be done are as follows.

Color can be useful for identificaton, and it is used to some extent. It happens that one of the best abrasives for grinding tools is white when it is manufactured, and a more commonly used abrasive is gray or brown in color. Some wheels are green, pink, or black. Among a given number of grinding wheels, there are wheels of different diameters, thicknesses, or hole sizes. Different wheels are made of different sizes of abrasive grain or with different proportions of grain and the bond that holds the grain together in the wheel.

What has been developed over the years, mostly by the grinding wheel makers, is a code of numbers or letters that provide a paragraph of information about a given wheel in just a few letters and numbers. Within a given group of symbols, the order of listing is important.

Grinding wheels are designed for grinding either on the periphery (outside diameter), which is a curved surface, or on the flat side, but rarely on both. It is not a safe practice to grind on the side of a wheel designed for peripheral grinding. The shape of the wheel determines the type of grinding performed.

The shapes of grinding wheels are designated according to a system published in full in an American National Standard, *Specifications for Shapes and Sizes of Grinding Wheels,* whose number is ANSI B74.2-1974. The various shapes have been given numbers ranging from 1 to 28, but only five are important for you now. These are described below.

Type 1 (Figure 1) is a peripheral grinding wheel, a straight wheel with three dimensions: diameter, thickness, and hole, in that order. A typical wheel for cylindrical grinding is 20 in. (diameter) × 3 in. (thickness) × 5 in. (hole). Probably most wheels are of this type.

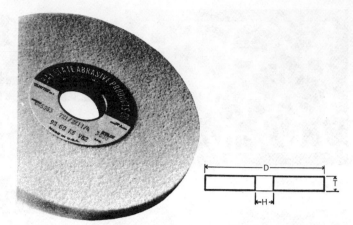

Figure 1. Straight or Type 1 wheel, whose grinding face is the periphery. Usually comes with the grinding face at right angles to the sides, in what is sometimes called an "A" face (Courtesy of Bay State Abrasives, Dresser Industries, Inc.).

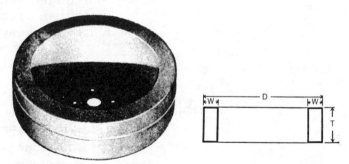

Figure 2. Cylinder or Type 2 wheel, whose grinding face is the rim or wall end of the wheel. Has three dimensions — diameter, thickness, and wall thickness (Courtesy of Bay State Abrasives, Dresser Industries, Inc.).

A *Type 2* or cylinder wheel (Figure 2) is a side grinding wheel, to be mounted for grinding on the side instead of on the periphery. This also has three dimensions, for example, 14 in. (diameter) × 5 in. (thickness) × 1½ in. (wall). This, of course, might also be called a 14 × 5 × 11 (14 in. D minus 2 times 1½ — the two wall thicknesses) but the wall thickness is more important than the hole size. Hence the change.

A *Type 6* or straight cup wheel (Figure 3) is a side grinding wheel with one side flat and the opposite side deeply recessed. It has four essential dimensions: the diameter, thickness, hole size (for mounting), and wall.

A *Type 11* or flaring cup wheel (Figure 4) is a side grinding wheel that resembles a Type 6,

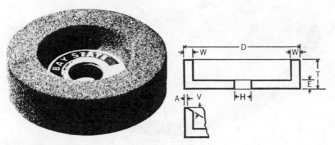

Figure 3. Straight cup or Type 6 wheel, whose grinding face is the flat rim or wall end of the cup (Courtesy of Bay State Abrasives, Dresser Industries, Inc.).

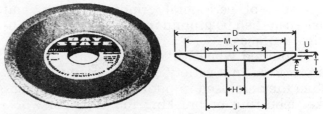

Figure 5a. Dish or Type 12 wheel, similar to Type 11, but a narrow, straight peripheral grinding face in addition to the wall grinding face. Only wheel of those shown that is considered safe for both peripheral and wall or rim grinding (Courtesy of Bay State Abrasives, Dresser Industries, Inc.).

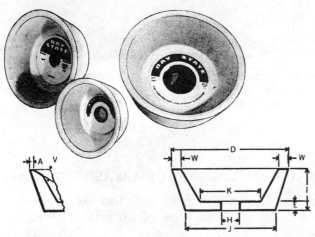

Figure 4. Flaring cup or Type 11 wheel, whose grinding face is also the flat rim or wall of the cup. Note that the wall of the cup is tapered (Courtesy of Bay State Abrasives, Dresser Industries, Inc.).

Figure 5b. An assortment of mounted wheels most often used for deburring and other odd jobs (Courtesy of Bay State Abrasives, Dresser Industries, Inc.).

except that the walls flare out from the back to the diameter and are thinner at the grinding face than at the back. This introduces a couple of new dimensions, the diameter at the back, called the "J" dimension, and the recess diameter at the back, the "K" dimension. This is mentioned only to emphasize that the "D" dimension, the diameter, is always the largest diameter of any wheel.

The *Type 12* dish wheel (Figure 5a) is essentially a very shallow Type 11 wheel, mostly for side grinding. The big difference is the dish wheel has a secondary grinding face on the periphery, the "U" dimension, so that it is an exception to the rule of grinding *only* on the side or the periphery.

This factor of grinding on the side or the periphery of a wheel is important because it affects the grade of the wheel to be chosen. The larger the area, the softer the wheel should be. In

peripheral grinding the contact is always between the arc of the wheel and either a flat (in surface grinding) or another arc (in cylindrical grinding). This makes for small areas of contact and somewhat harder wheels. On the other hand, if the flat side of the wheel is grinding the flat surface of a workpiece, then the contact area is larger and the wheel can be still softer.

It is important to understand that any grinding machine grinds either a flat surface like a planer or shaper, or a round or cylindrical surface, like a lathe or boring mill. The first group of grinders is collectively called surface grinders; the second group is called cylindrical grinders, whether the workpiece is held between centers or not, and whether the grinding is external or internal. Various forms can be cut into the grinding faces of peripheral grinding wheels, and these can then be ground into either flat or cylindrical surfaces.

Mounted wheels like the ones in Figure 5b can be used in a variety of ways around a shop. Often they are used in portable grinders for jobs like deburring or breaking the edges of workpieces where the tolerances are not too critical. They are also used in internal grinding.

STANDARD MARKING SYSTEM

The description of a grinding wheel's composition is contained in a group of symbols known as the *standard marking system*. That is, the basic symbols for the various elements are standard, but they are usually amplified by individual manufacturer's symbols, so that it does not follow that two wheels with the same basic markings, but made by two different suppliers, would act the same. However, it is a useful tool for anyone concerned with grinding wheels.

There are five basic symbols. The first is a letter indicating the kind of abrasive in the wheel, called the abrasive *type*. The second is a number to indicate the approximate size of the abrasive; this is commonly called *grit size*. In the third position, a letter symbol indicates the *grade* or relative hardness of the wheel. The fourth, *structure*, is a number describing the spacing between abrasive grains. The fifth is a letter indicating the *bond*, the material that holds the grains together as a wheel. Thus, a basic toolroom wheel specification (Figure 6) might be

A60-J8V

But, since most wheel makers have, for example, a number of different aluminum oxide or other types of abrasives and a number of different vitrified or other bonds, the symbol sometimes appears cluttered:

9A80-K7V22

This means that the wheel maker is using a particular kind of aluminum oxide (A) abrasive, which is indicated by the 9, and a particular vitrified (V) bond, which is indicated by the 22.

The wheel markings for diamond or cubic boron nitride wheels are a little different and are not standard enough for a simple explanation.

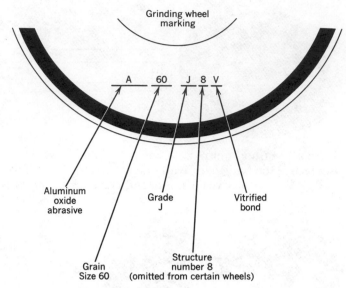

Figure 6. Sketch to illustrate the wheel specification that will be used as an illustration in following pages (McKee Editorial Assistance).

FIRST SYMBOL — TYPE OF ABRASIVE: (A60-J8V)

The symbol in the first position, as suggested above, denotes the type of abrasive. Basically, there are five:

A Aluminum oxide.

C Silicon carbide.

D Natural diamond.

MD or SD Manufactured diamond (sometimes called synthetic diamond).

B Cubic boron nitride.

The first two are inexpensive abrasives, cheap enough so that it is practical to make whole wheels of the abrasive. Both are made in electric furnaces that hold literally tons of materials, and both are crushed and graded by size for grinding wheels and other uses.

Diamond, both natural and manufactured, and cubic boron nitride are expensive enough that most wheels are made of a layer of abrasive around a core of other material. However, they have made a place for themselves because they will grind materials that no other abrasive will touch and because they stay sharp and last so long that they are actually less expensive per piece of parts ground. Natural or mined diamond that is definitely diamond, but is of less than gem quality, is crushed and sized. Both manufactured diamond and cubic boron nitride are made by a combination of high heat (in the range of 3000° F) and tremendous pressure (1 million or more PSI). This heat, incidentally, is somewhat less than that for other manufactured abrasives, which require something in the range of 4000°F for fusing or crystallization.

Each of these abrasives generally has an area in which it excels, but there is no one abrasive that is first choice for all applications. Aluminum oxide is best for grinding most steels, but on the very hard tool steel alloys it is outclassed by cubic boron nitride. However, aluminum oxide (Al_2O_3) may have close to 75 percent of the market, because it is used in foundries for grinding castings and steel mills for billet grinding, and in other high volume applications. Silicon carbide, which does poorly on most steels, is excellent for grinding nonferrous metals and nonmetallic materials. Diamond abrasive excels on cemented carbides, although green silicon carbide is occasionally used. Green silicon carbide was the recommended abrasive for cemented carbides until the introduction of diamonds. Cubic boron nitride (CBN), finally, is superior for grinding high speed steels. CBN is a hard, sharp, cool cutting, long wearing abrasive.

There is a lot still unknown about why certain abrasives grind well on some materials and not on others, and this makes for some interesting speculation. All abrasives are harder than the materials ground. But their relative hardness is apparently only one of the factors in effective ness. Aluminum oxide is three to perhaps five times harder than most steels, but it grinds them easily. Silicon carbide is harder than aluminum oxide, but not at all effective on steel; on the other hand, it grinds glass and other nonmetallics that are as hard or harder than many steels. Diamond, which is much harder than cubic boron nitride and many times harder than the hardest steel, does not grind steels well, either. The best theory is that there are chemical reactions between certain abrasives and certain materials that make some abrasives ineffective on some metals and other materials.

The basic symbol in the first position, A, C, D, or B, is often preceded and sometimes followed by a manufacturer's symbol that indicates which abrasive within the group is meant, for example, 9A, 38A, AA.

SECOND SYMBOL — GRIT SIZE (A60-J8V)

The symbol in the second position of the standard marking represents the size of the abrasive grain, usually called grit size. This is a number ranging from 4 to 8 on the coarse side to 500 or higher on the fine side. The number is derived from the approximate number of openings per linear inch in the final screen used to size the grain; the larger the number, the smaller the abrasive grain. Any standard grit size contains grain of smaller and larger sizes, whose amounts are strictly regulated by the federal government, because it would be very expensive to reduce the mix to just the size indicated.

While there is no real agreement as to what is coarse or fine, for general purposes anything from 46 to 100 might be considered medium, with everything 36 and lower considered as coarse and anything 120 and higher considered as fine. Selection of grit in any shop depends on the kind of work it is doing. Thus, where the job is to remove as much metal as fast as possible, 46 or 60 grit size would be considered very fine. On the other hand, in a shop specializing in fine finishes and close tolerances, 240 might be considered coarse.

A final point is in order about grit size. Most standard symbols end in a 0, particularly the three digit sizes 100 and finer. However, every abrasive grain manufacturer also makes some combinations that are not standard for special uses, and these usually end in a 1 or other low digit. Thus, a 240 grit is the finest that is sized by screening. Finer grits are sized by other means. On the other hand, 241 grit is a coarse 24 grit in a 1 combination.

Coarse grain is used for fast stock removal and for soft, ductile materials. Fine grain is used to obtain good finishes and for hard or brittle materials. Some materials are hard enough that fine grain removes as much stock as coarse, and neither removes very much. In a general machine shop or toolroom, most of the wheels used will be between 46 and 100 grit.

THIRD POSITION*—GRADE OR HARDNESS: (A60-J8V)

In the third position is a letter of the alphabet called grade. The later the letter, the harder the grade. Thus, a wheel graded F or G would be

Weak holding power

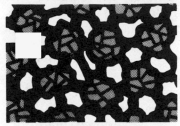

Medium holding power

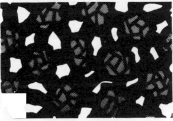

Strong holding power

Figure 7. Three sketches illustrating (from top down) a soft, a medium, and a hard wheel. This is the "grade" of the wheel. The white areas are voids with nothing but air; the black lines are the bond, and the others are the abrasive grain. The harder the wheel, the greater the proportion of bond and, usually, the smaller the voids (Courtesy of Bay State Abrasives, Dresser Industries, Inc.).

considered **soft**, and one graded R to Z would be very **hard**. What is being measured is the hold that the bond has on the abrasive grain (Figure 7), and the greater the proportion of bond to grain, the stronger the hold and the harder the wheel. Precision grinding wheels tend to be on the soft side, because it is necessary to have the grains pull out as they become dull; otherwise, the wheel glazes and its grinding face becomes shining, but the abrasive is dull. On high speed, high pressure application like foundry snagging, the pressures to pull out the grains are much greater, and a harder wheel is needed to hold in the

grains until they have lost their sharpness. Ideally, a wheel should be self-sharpening. The bond should hold each grain only long enough for it (the grain) to become dull. In practice, this is difficult to achieve.

One thing, however, should be mentioned. Grade is a much less measurable thing than type or size of abrasive or, as you will see later, than bond. Grade depends on the formula for the mix used in the wheels, but it must be checked after the wheel is finished.

FOURTH POSITION — STRUCTURE: (A60-J8V)

Following the grade letter, in the fourth position of the symbol, is a number from 1 (dense) to 15 (open); this number describes the spacing of the abrasive grain in the wheel (Figure 8). The use of

Dense spacing

Medium spacing

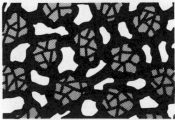

Open spacing

Figure 8. Three similar sketches showing structure. From the top down, dense, medium, and open structure or grain spacing. The proportions of bond, grain, and voids in all three sketches are about the same (Courtesy of Bay State Abrasives, Dresser Industries, Inc.).

structure is to provide chip clearance, so that the chips of ground material have some place to go and will be flung out of the wheel by centrifugal force or washed out by the coolant. If the chips remain in the wheel, then the wheel becomes loaded (Figure 9), stops cutting, and starts rubbing, and it has to be resharpened or dressed, which is the trade term.

Structure is also a result of grain size and proportion of bond, similar to grade. Quite often, large grain size wheels tend to have open structure, while smaller sized abrasive grain is often associated with dense structure. On the open side, say 11 to 12 and up, the openness is aided by the inclusion of something in the mix like ground up walnut shells, which will burn out as the wheel is fired and leave definite open spacing.

However, for many grit size and grade combinations, a best or standard structure has been worked out through experience and research, and so the structure number may be omitted.

FIFTH POSITION—BOND: (A60-J8<u>V</u>)

The fifth position of the wheel marking is a letter indicating the bond used in the wheel. This is always a letter, as follows: vitrified—V; resinoid—B (originally the bakelite bond): rubber—R (rubber was used well before resinoid): and shellac—E (originally the elastic bond, and also preceded by the now obsolete silicate bond). These are really general bond groups; each wheel maker uses extra symbols to indicate, for instance, which vitrified bond he has used in a particular wheel, and there is no standardization in these extra symbols.

The bond used has important influence on both the manufacturing process and on the final use of the wheel. Vitrified-bonded wheels are fired at temperatures between 2000 and 2500° F; for that reason, no steel inserts can be used. If such inserts are needed, they must be cemented

Figure 9. The wheel at the left is called "loaded" with small bits of metal imbedded in its grinding face. It is probably too dense in structure or perhaps has too fine an abrasive grain. Same wheel at the right has been dressed to remove all the loading (Photographs by courtesy of Desmond-Stephan Mfg. Co.).

in afterward. The others, also grouped together as organic bonds, are all baked at around 400° F, and inserts may be molded in without problems.

Vitrified, resinoid, and shellac wheels are all pressed in molds after mixing. Rubber bonded wheels, on the other hand, are mixed in a process similar to that of making dough for cookies; they are then reduced to thickness by passing the mass or grain-impregnated rubber between precisely spaced rolls. For that reason, it is possible to make much thinner wheels with rubber bond than can be made with any other bond. Thin grinding wheels with either resinoid or rubber bonds can be used in cutoff operations like very fine bandsaws, but the rubber wheels can be made thinner than the resinoid bonded wheels.

In general, vitrified wheels are used for precision grinding; resinoid wheels are used for rough grinding with high wheel speeds and heavy stock removal; and rubber or shellac wheels, are used for more specialized applications. The first two bonds monopolize over 90 percent of the market.

Bond also determines maximum safe wheel speed. Vitrified bonded wheels, with a few minor exceptions, are limited to about 6500 SFPM or a little over a mile a minute. The others run much faster, with some resinoid wheels getting up to 16,000 SFPM or more. Of course, all grinding wheels must be properly guarded to protect the operator in the unlikely case of wheel breakage. Grinding machines in general, however, are safe machines. Many will not operate unless the guards are properly in place and closed.

FACTORS IN GRINDING WHEEL SELECTION

It is generally considered that there are seven or eight factors to consider in the choice of a grinding wheel. Some people group together the elements of the amount of stock removal and of the finish required. Dividing them seems to be more logical.

These factors have also been divided into two groups — three concerned with the workpiece, which thus change frequently, and five concerned with the grinder, which are more constant.

Variable Factors
In the first group are things such as composition, hardness of the material being ground, amount of stock removal, and finish.

The composition of the material generally determines the abrasive to be used. For steel and most steel alloys, use aluminum oxide. For the very hard, high speed steels, use cubic boron nitride. For cemented carbides, use diamond. Whether this is natural or manufactured is too specialized a question to be discussed here, but the trend appears to be toward the use of manufactured diamond. For cast iron, nonferrous metals, and nonmetallics, use silicon carbide. On some steels, aluminum oxide may be used for roughing and silicon carbide for finishing. Between aluminum oxide and cubic boron nitride, because of the greatly increased cost, the latter is generally used only when it is superior by a wide margin, or if a large percentage of work done on the machine is in the very hard steel alloys like T15.

Material hardness is of concern in grit size and grade. Generally, for soft, ductile materials, the grit is coarser and the grade is harder; for hard materials, finer grit and softer grades are the rule. Of course, it is understood that for most machine shop grinding these coarse grits are mostly in the range of 36 to 60, and the finer grits are perhaps in the range of 80 to 120. Likewise, the "soft" wheels are probably something in the range of F, G, H, and perhaps I, and the "hard" wheels are in the range of J, K, or L, and maybe M or N. Too coarse a grit might leave scratches that would be difficult to remove later. And sometimes on very hard materials, coarse grit removes no more stock than fine grit, so you use fine grit. Too soft a wheel will wear too fast to be practical and economical. Too hard a wheel will glaze and not cut.

If **stock removal** is the only objective, then you can use a very coarse (30 and coarser) resinoid bonded wheel. However, in machine shop grinding, you are probably in the 36 to 60 grit sizes mentioned above, and definitely in vitrified bonds.

For **finishing** on production jobs, the wheel may be rubber, shellac, or resinoid bonded. But resinoid bonds are often softened by coolants and therefore are rarely used. And for finishing, fine grit sizes are usually preferred; however, as you will learn later when you study wheel dressing (sharpening or renewing the grinding face of a wheel), it is possible to dress a wheel so that a comparatively coarse grit like 54 or 64 will finish a surface as smoothly as 100 or 120 grit.

Fixed Factors

For any given machine the following five factors are likely to remain constant.

Horsepower of the machine, of course, is a fixed consideration. Grinding wheel manufacturers and grinding machine builders are constantly pressing for higher horsepower, because that gives the machine and the wheels the capacity to do more work, but that is a factor, usually, only in the original purchase of the machine. In wheel selection, this affects only grade. The general rule is the higher the horsepower, the harder the wheel that can be used.

The **severity of the grinding** also remains pretty constant on any given machine. This affects the choice of a particular kind of abrasive within a general group. Thus, you would probably use a regular or intermediate aluminum oxide on most jobs. But, in the toolroom, where pressures are low, you would probably want an easily fractured abrasive, white aluminum oxide. On the other end of the scale, for very severe operations like foundry snagging, you need the toughest abrasive you can get, probably an alloy of aluminum oxide and zirconium oxide. For most machine shop grinding, you will probably look first at white wheels.

The **area of grind contact** is also important but, again, it remains constant for a given machine. The rule is finer grit sizes and harder wheels for small areas of contact; and coarser grit sizes and softer wheels for larger areas of contact. All of this, of course, is within the grit size range of about 36 to 120, or 150 or 100, and within a grade range from E or F to L or N.

It is easy to understand that on a side grinding wheel, where a flat abrasive surface is grinding a flat surface, the contact area is large and the wheels are fairly coarse grained and soft (Figure 10). However, on peripheral grinding wheels, it is a different story. The smallest area is in ball grinding, where the contact area is a point, the point where the arc of the grinding wheel meets the sphere or ball. Thus ball grinding wheels are very hard and very fine grained, for instance, 400 grit, Z grade. In cylindrical grinding the contact area is the line across the thickness of the wheel, usually where the arc of the wheel meets the arc of the workpiece (Figure 11). Here grit sizes of 54, 60, or 80, with grades of K, L, or M, are common. A still larger area is in surface grinding with peripheral grinding where the line of contact is slightly wider because the wheel is cutting

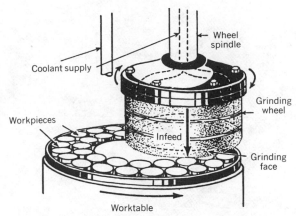

Figure 10. With the flat wall or rim of the wheel grinding a flat surface, as shown here, the wheel must be soft in grade and can have somewhat coarser abrasive grain. The area of contact between wheel and work is large (Courtesy of Bay State Abrasives, Dresser Industries, Inc.).

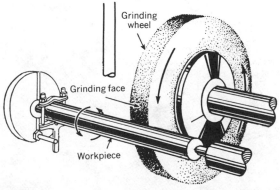

Figure 11. In center-type cylindrical grinding, as shown here, the arc of the grinding face meets the arc of the cylindrical workpiece, making the area of contact a line. This requires a harder wheel than in Figure 10 (Courtesy of Bay State Abrasives, Dresser Industries, Inc.)

into a flat surface (Figure 12). And a combination like 46 I or 46 J is not unusual. An internal grinding wheel where the OD of the wheel grinds the ID of the workpiece may have just a shade more area of grinding contact (Figure 13). And then when you get to side grinding wheels (cylinder Type 2), cup wheels, and segmental wheels, which are flats grinding flats, you get grit sizes and grades like 30 J or 46 J (see Figure 10). Of course, you have to realize that there may be other factors important enough to override contact area. For example, for grinding copper, you might use a grit size and grade like 14 J, in which

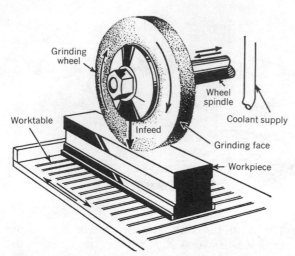

Figure 12. The contact area between the arc of the grinding face and the flat surface of the workpiece in surface grinding makes a somewhat wider line of contact than in cylindrical grinding (Courtesy of Bay State Abrasives, Dresser Industries, Inc.).

Figure 14. The blotter on the wheel, besides serving as a buffer between the flange and the rough abrasive wheel, provides information as to the dimensions and the composition of the wheel, plus its safe speed in RPMs. This wheel is 7 in. in diameter $\times \frac{1}{4}$ in. thick $\times$ $1\frac{1}{4}$ in. hole. It is a white aluminum oxide wheel, 100 grit, H grade, 8 structure, vitrified 52 bond. It can be run safely at up to 3600 RPMs. (Courtesy of Bay State Abrasives Division, Dresser Industries, 9Inc.)

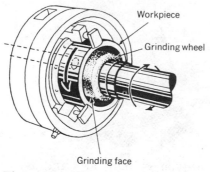

Figure 13. The contact area of the OD grinding face of the wheel and the ID surface of the workpiece creates a still larger area of contact and requires a somewhat "softer" grinding wheel than the two previous examples (Courtesy of Bay State Abrasives, Dresser Industries, Inc.).

the softness of the metal is probably the key factor.

Wheel speed is a factor that can be dealt with quickly. You must always stay within the safe speeds, which are shown on the blotter or label on every wheel of any size (Figure 14). Vitrified wheels generally have a maximum safe speed of 6500 SFPM or a little more; organic wheels (resinoid, rubber or shellac) go up to 16,000 SFPM or sometimes higher, but these speeds are generally

set by the machine designer, and they are safe speeds for the recommended wheels.

Wet or dry grinding is a factor only in that using a coolant will usually permit the use of about one grade harder wheel than would be used for dry grinding, without as much concern about burning the workpieces. Burning is a discoloration of the workpiece surface caused by overheating. The most common cause is usually the use of a wheel that is too hard.

In any shop, however, unless you are really starting from scratch, there will be some information on what wheels have been used and how they have worked. If a factor seems to need changing, it will probably be grit size or grade: You must remember that the shop probably handles a range of work, and that it does not pay to switch wheels all the time. But change only one element, either grade or grit size, at a time.

self-evaluation

SELF-TEST
1. In the course of a week's grinding you might come up with some of each of the following to grind: bronze valve bodies, steel fittings, tungsten carbide tool inserts, and high speed steel tools. If you could pick the ideal abrasive for each metal, what would you use? List four abrasives. If you were limited to three, which one of the four could be eliminated most easily?

2. Straight (Type 1) and cylinder (Type 2) wheels both have three dimensions: diameter, thickness, and a third. What is the third dimension for each and why is it stated that way?

3. Five shapes of grinding wheels are described in this unit. Four are for side grinding and two are for peripheral grinding. List the wheels in the two groups either by name or shape number.

4. Tungsten used in the points of automobile engines is very expensive, which makes it necessary to use the thinnest abrasive cutting wheels possible, $6 \times .008 \times 1$ in. What bond would be used and why?

5. Area of contact between wheel and workpiece is probably the most important factor in picking a wheel grade. Five different sets of grinding conditions are discussed, ranging from flat surfacing with a cylinder wheel to ball grinding. List the five in order by wheel grade, starting with the hardest.

6. Here are two wheel specifications, both for straight (Type 1) wheels: (a) A14-Z3 B, and (b) C14-J6 V. Describe the composition of each wheel in a sentence or two, and suggest the material to be ground by each.

7. Here are two more specifications: (a) C36-K8V and (b) C24-H9V, one for peripheral grinding and one for side grinding. From these specifications, tell which is which.

8. Here is an actual wheel specification: 32A46-H8VBE. Describe the wheel's composition, stating at least the abrasive used, the size of the abrasive, the grade, structure and bond.

9. A wheel specification for cylindrical grinding of a hard steel fitting with a straight wheel is: A54-L5 V. If you were grinding a flat piece of the same steel with a straight wheel, what elements of the specification might change? Which way? For flat grinding of the same material with a segmental or a cup wheel, what further changes might be made?

10. Write one or two sentences about each of the following to show what elements of a wheel specification it affects.
    a. Material to be ground.
    b. Hardness of the material.
    c. Amount of stock to be removed.
    d. Kind of finish required.

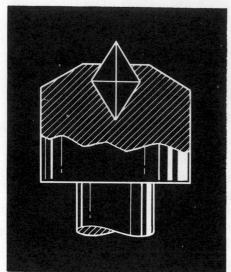

unit 2
care of abrasive wheels: trueing, dressing, and balancing

This unit focuses on the care of a grinding wheel *as a cutting tool* to keep it running true, that is, with every spot on the outside diameter (OD) of the wheel at the same distance from the center of the spindle, keeping it sharp, and keeping it in balance. All three conditions must be met if the wheel is to perform at its best as a cutting tool, producing dimensionally accurate parts with good to superior surface finish.

objectives After completing this unit, you will be able to:

1. Make a check list of steps to be followed in trueing a grinding wheel.
2. Make a similar list for dressing a wheel, including the variations required for a wheel to be used for roughing as against those for a wheel intended for finishing.
3. Define form dressing, listing at least three types of form dressers, and describe how each operates.
4. List steps to be followed in balancing a grinding wheel. Include frequency of balancing and wheels that must be balanced.

Ideally, a grinding wheel should be so well matched to the work that grains will pull out of the wheel as they become dull and bits of the work material never become lodged in the wheel. In practice, however, especially in shops where a variety of metals must be ground with as little wheel changing as possible, this rarely happens. Hence, silicon carbide and aluminum oxide wheels must be dressed, that is, sharpened. Diamond and cubic boron nitride wheels are not dressed as often, and never as severely, because of the expense of the abrasive grain and partly because they can be more accurately mounted. In addition, both kinds of grains are so much harder than aluminum oxide and silicon carbide that they do not wear as fast.

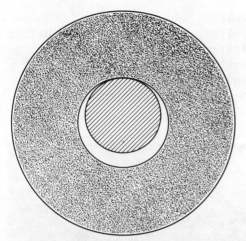

Figure 1. Sketch, exaggerated for effect, showing a grinding wheel on a spindle. Actually, the gap under the spindle may be only a few thousandths of an inch, but still enough to cause problems if the wheel is not trued to the center of the spindle.

Trueing and balancing of silicon carbide and aluminum oxide wheels are essential in the first place, because it is difficult to mount common abrasive wheels true to the centerline of the spindle. Clearance at the bottom of the center hole of the wheel is always a little greater than at the top, where there is none. Figure 1 illustrates the point, although the clearance is, of course, greatly exaggerated. The machined hole of a diamond wheel (Figure 2) makes possible a much closer fit on the spindle. Balancing is usually a factor for wheels with a 12 in. diameter and larger, because of the weight involved. Below that size, an out-of-balance condition would probably cause problems only in supercritical finishing. Wheels are in balance when they leave the factory, but not necessarily after they are mounted on the spindle with mounting flanges, as is often the case with large wheels on several types of grinders. To provide good finish, wheels must be true and in balance.

It is important to realize that while a grinding wheel is removing metal from a workpiece, the workpiece is also removing material from the wheel, although, of course, at a much slower rate. This reverse step of the operation means that the wheel eventually will have to be sharpened or dressed by removing material (the dulled abrasive grain and some bond material) from the wheel. Dressing of precision, vitrified-bonded wheels is typically done with a diamond tool, usually a single diamond in a holder (Figure 3),

Figure 2. Since the hole in the core of a diamond or CBN wheel is machined, it can be fitted to much closer tolerances (K & M Industrial Tool, Inc.).

Figure 3. Single point diamond dresser. The most important precaution in using such a dresser is to turn the diamond often to avoid grinding flats on it. This diamond is pointing to the right (Photograph courtesy of The Desmond-Stephan Manufacturing Co.).

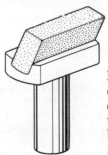

Figure 4. Cluster-type dressers have come into use mostly because several smaller diamonds are cheaper than one large diamond (Photograph by courtesy of The Desmond-Stephan Manufacturing Co.).

that can be moved back and forth across the grinding face of the rotating wheel. Sometimes a number of smaller diamonds are imbedded in a metal matrix, a comparatively soft material with the diamond chips across its face (Figure 4). This second type of dresser, sometimes called a cluster dresser, may be wide enough to cover the grinding face of the wheel without being traversed. These are used for dressing a flat grinding surface on the wheel.

Even for a flat grinding face it is usually considered good practice to dress a little radius between the grinding face and the side of the wheel, since a sharp corner could leave undesirable marks on the surface of the workpiece. This is done by swinging the dresser in a small freehand arc around the corner or, to use the shop term, breaking the corner. Holding a dressing stick lightly against the corner for a couple of seconds will probably do the job even more easily.

As a point of information, dressing of resinoid bonded grinding wheels for snagging and other rough grinding operations, where the only purpose is the quick and cheap removal of metal, is done with a Huntington dresser (Figure 5). This type is made up of a group of spurlike discs, strung on a shaft in a holder with a handle. The discs can spin freely, while the dresser is held against the surface of the rotating wheel, and the points remove dull grains and embedded metal to renew the grinding face (Figure 6a). It is worth noting, also that grinding wheels can be used to dress other grinding wheels (Figure 6b).

Vitrified grinding wheels can also grind forms (slots, grooves, and contours) in workpieces. Mostly this is done by form-dressing the wheel to the *reverse* of the form desired in the workpiece. That is, where there is a ridge in the workpiece, there is a corresponding valley in the wheel. Sometimes this is done as a finishing operation on a form started by some other operation, such as milling. More often, however, the economical way is to grind the form from the solid, or flat — in short, doing the whole job by grinding.

For example, if you need a rounded groove in the workpiece, there are radius dressers available to swing the single point diamond dresser in an arc of the correct radius. This is done mechanically and more accurately than the job referred

to earlier as breaking the corners of the grinding wheel (Figure 7). Refinements of this kind of mechanical linkage make it possible, for instance, to reproduce on the grinding face of a wheel some fairly complicated contoured forms. It should be clear by now that, for all its capabil-

Figure 6a. Huntington dresser set up for dressing. The dresser has a hook or lug that fits over the workrest of the grinder (Photo courtesy of Norton Company).

Figure 6b. Sometimes it is practical to use one grinding wheel to dress another. The abrasive wheel in the dresser is being used with a metal-bonded diamond wheel on the grinding machine (Photo courtesy of Norton Company).

Figure 5. Huntington-type dresser (sometimes called a star) used for coarse wheels intended for rough grinding (Photo courtesy of Norton Company).

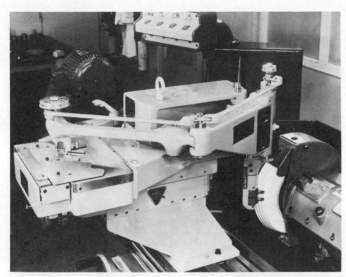

Figure 7. A dressing unit such as this can be set to dress practically any desired shape in a wheel. It is very versatile (Courtesy of Engis Co., Diamond Tool Division).

Figure 9. The crush roll must be mounted to provide support against considerable force, either as here on a center-type cylindrical grinder, or on a surface grinder (Courtesy of Bendix Automation and Measurement Division).

Figure 8. This is a crush roll, which literally crushes out the form in the wheel, instead of cutting it, as with diamond rolls or blocks (Courtesy of DoAll Company).

ity to remove material, the vitrified grinding wheel can be readily formed.

Of the several kinds of form dressers available, the crush roll and the formed diamond dresser deserve brief mention, although both are production tools. In crush forming with the crush roll, the rapidly rotating wheel is pressed with considerable force against a rotating crush roll (Figure 8), which has the form desired in the finished workpiece. Once the form is crushed into the grinding face of the wheel (Figure 9), the wheel can be used to grind the form into workpieces. This operation can be done on either flat or cylindrical surfaces with, in the same order, either a surface or a center-type cylindrical grinder. Crushing leaves the abrasive grain very sharp; this does not result in the best finished surface on the work, although it is adequate for most uses.

Diamond roll form dressers or diamond dressing blocks (Figure 10) can be made for many of the same applications as crush rolls and also for other applications. The action of this dresser involves less pressure and more cutting of the abrasive grain, so that the wheel produces a somewhat better surface finish. Both crush rolls and diamond form rolls are expensive and specialized tools that can only be justified by large production and substantial savings over alternate

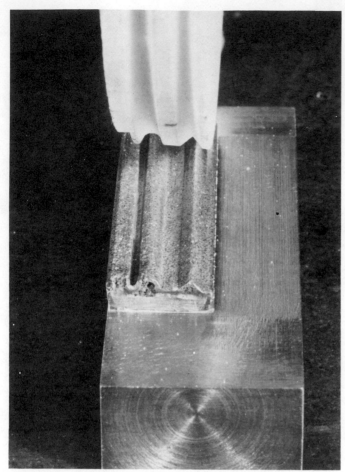

Figure 11. Overhead-type dresser for a surface grinder. The calibrated micrometer dial controls the downfeed of the diamond. The handle at the left cross feeds the diamond back and forth. In operation, of course, the wheel and the dresser would be concealed by the front piece of the safety guard (Courtesy of DoAll Company).

Figure 10. Diamond-plated dressing block, intended for use on a surface grinder. The block is held flat on the magnetic chuck, and the wheel is traversed back and forth along it. The block is formed to the shape of the finished workpiece (Courtesy of Engis Co., Diamond Tool Division).

methods. Each new form requires a new set of rolls.

Chances are that you will most often be dressing the wheel so that the peripheral surface is concentric with the center of the machine spindle and parallel to the centerline of the spindle, and with a single point diamond dresser. The dresser may be mounted above the wheel (Figure 11), on the magentic chuck that is the workholder of the machine (Figure 12), or to the side (Figure 13). In any of these positions, it is possible to move the dresser back and forth across the face of the wheel. This is called traversing the dresser.

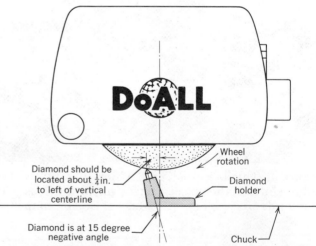

Figure 12. This is one of several ways of mounting a dresser on a surface grinder. The dresser with its diamond is simply spotted on the clean magnetic chuck. Note, however, that the diamond is slanted at a 15 degree angle and slightly past the vertical centerline of the wheel, as the wheel turns (Courtesy of DoAll Company).

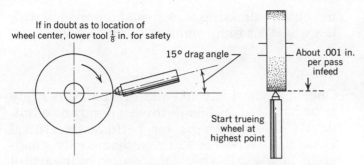

If in doubt as to location of wheel center, lower tool ⅛ in. for safety

15° drag angle

About .001 in. per pass infeed

Start trueing wheel at highest point

Figure 13. This illustrates the idea of wheel dressing instead of any specific setup. Dressing is rarely done free hand. Note that the diamond is always a little past the centerline and on an angle. (Photograph by courtesy of The Desmond-Stephan Manufacturing Co., Urbana, Ohio).

The diamond is mounted at an angle of 15 degrees, so that it contacts the wheel just after the low or high point of the wheel, or just below the centerline, depending on the location of the dresser (Figures 11, 12 and 13). Thus the wheel is cutting toward the point of the diamond. Remember that dressing with a diamond is always a two-way operation; that is, the diamond is kept sharp while it sharpens the wheel grinding face. Often there is an arrow or other indicator on the dresser to indicate its position. The diamond always points in the direction of the wheel's rotation.

Any new wheel, or a wheel that has just been reflanged, must first be trued. In trueing, the wheel and the dresser are brought together so that the dresser is touching the high point of the wheel. Otherwise, the traversing of the diamond might cause it to dig too deeply into the wheel, which could ruin the diamond. On the other hand, if you start at the high point, which is the point furthest from the center, the cross cut or traverse is short at first and gradually becomes longer until finally you are dressing the entire width of the wheel.

Infeed of the diamond into the wheel should be light, about .001 in. per pass; if dressing is being done dry, there should be frequent pauses, after every three or four passes, to allow the diamond to cool off. A hot diamond can be shattered if a drop of water or other liquid hits it. Turn the diamond frequently. This helps to keep it sharp.

Trueing the wheel is accomplished, then, by moving the dresser back and forth across the grinding face of the wheel while the wheel is rotating at operating speed. It is preferably done wet. If trueing is done wet, continuous coolant must be assured. If it is done dry, as it must be on many machine shop machines, dressing must be interrupted at intervals to allow the diamond to cool off.

The speed of traverse is probably the remaining point of concern. Generally, the faster the dresser traverses across the wheel's grinding face, the sharper the dress will be and the better suited the wheel will be for rough grinding. Slower traverse means that the diamond does more cutting on the abrasive, dulling it a little bit, so that the wheel is better suited to finishing than to roughing work.

With a new wheel or one that has just been mounted between a pair of flanges, the first step is to true the wheel. This is done as above, by traversing a diamond tool back and forth across the grinding face of the wheel until the dresser is contacting the wheel at all times.

After trueing, the wheel may or may not need to be dressed. It depends on the surface. If the wheel is to be used for roughing, then the surface should be open and the grains should be sharp. If the wheel is to be used for finishing, the grains should be a little duller. Sometimes when a wheel is intended for finishing, it is good to take a few passes across the wheel without any infeed. The point is that with a little experience you can tell the degree of sharpness that is needed in the wheel face and dress the wheel accordingly, depending on what you want to do with it afterward. As grain becomes dulled, it tends to polish more and cut less, and size becomes less of a factor in the action of the grain.

It was mentioned earlier that coarse grain is recommended for cutting and material removal and fine grain is recommended for finishing. This is true, provided both are in the same degree of sharpness. It can now be said that by proper dressing, it is possible to make a finish with comparatively coarse grain like that of a much finer grain. For example, a 46 grit wheel, dressed to dull the grain, could give the same finish as a 120 grit wheel. The reverse is not true, however.

The special situation of diamond and cubic boron nitride wheels has already been mentioned. Such wheels can be centered on the spin-

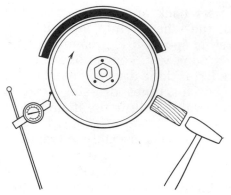

Figure 14a. Runout on a diamond wheel is checked with a dial indicator and must be within .0005 in. for resinoid wheels or .00025 in. (half as much) for metal bonded wheels. Tapping a wooden block held against the wheel to shift the wheel on the spindle is often enough to bring it within limits. Otherwise, it will have to be trued (Courtesy of Precision Diamond Tool Co.).

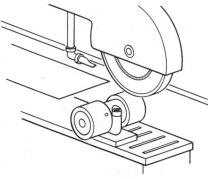

Figure 14b. Trueing a diamond wheel, when necessary, is best done with a brake-type trueing device, as shown. For resinoid wheels, the job may also be done by grinding a piece of low carbon steel (Courtesy of Precision Diamond Tool Co.).

dle with more accuracy. Furthermore, it is critical that everything else (spindle, flanges, spacers) be running true and be clean and free from burrs. If there is anything amiss, it should be corrected.

Diamond wheels are customarily trued to the bore in the manufacturing process. But if a wheel runs out more than .0005 in. (resinoid bond) or .00025 in. (metal bond), it can usually be trued by lightly tapping a wooden block held against the wheel (Figure 14a). If this does not work, then the wheel must be trued. This can be done with a brake-type trueing device (Figure 14b). Resinoid bond wheels can be trued by grinding a piece of low carbon steel.

After trueing, as with other types of wheels, it may be necessary to dress the wheel. For this,

use either a dressing stick (usually provided with the wheel) or lump pumice.

BALANCING

Once the wheel is trued and dressed, it is ready to be balanced, although this step may be eliminated (except perhaps for particularly critical work) if the wheel is 12 in. in diameter or smaller. In such small wheels the weight of the wheel is not great enough to disturb the finish if the wheel is a little out of balance. For wheels 14 in. in diameter and larger, balancing is important. It might be noted that there are no 13 in. standard wheels manufactured.

Balancing of grinding wheels is done for the same reasons, on similar equipment, and by similar methods as the balancing of automobile tires. The job is done on either balancing ways with overlapping discs (Figure 15) or on parallel ways (Figure 16), which must, of course, be perfectly level. The wheel, on a balancing arbor, is placed on the ways. The heavy point of the wheel is marked with chalk after the wheel has been allowed to come completely to rest. A horizontal

Figure 15. This type of balancing device with overlapping wheels or discs is quite common. It has an advantage in that it need not be precisely leveled (Courtesy of Bay State Abrasives, Dresser Industries, Inc.).

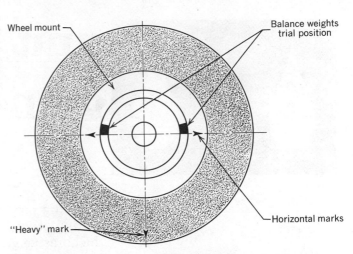

Figure 16. Balancing a wheel on two knife edges, as on this unit, is very accurate, because there is minimum friction. Of course, the unit must be perfectly level and true. Otherwise, the wheel may roll from causes other than out-of-balance (Courtesy of Bay State Abrasives, Dresser Industries, Inc.).

Figure 17. With the weights at some point between the vertical and the horizontal centerlines, the wheel should be in proper balance, stationary in any position. If not, one or two other balance weights should be used (Courtesy of Bay State Abrasives, Dresser Industries, Inc.).

line is also drawn through the center of the wheel. Then weights in the flanges (Figure 17) are shifted for a trial balance and the heavy point is turned to the top. This sequence of shifting weights and turning the wheel is continued until the wheel remains at rest in any position. This is called static balancing. Some production machines, usually some form of cylindrical grinders, are equipped so that the wheel can be automatically balanced while it is still in the machine.

self-evaluation

SELF-TEST

1. What is meant by trueing a grinding wheel, and why is it important?

2. All grinding wheels are balanced at the factory. Why and under what circumstances is it necessary to rebalance it? On what size wheels?

3. Explain the difference between dressing and trueing a wheel. How often should a wheel be dressed?

4. In your own words, define form dressing. Mention at least two methods of form dressing and one advantage and one disadvantage of each.

5. Explain why it takes longer to get a diamond or cubic boron nitride wheel ready to grind than it does an aluminum oxide wheel. Be specific as to details.

6. Explain the essentials in the placement of a single point diamond dresser.

7. List the steps in trueing an aluminum oxide or silicon carbide wheel after mounting.

8. Explain the essential differences between dressing a grinding wheel for roughing as against finishing.

9. Under what circumstances might it be necessary to true a diamond wheel and how is it done?

10. Generally, what determines whether a wheel needs to be balanced? Describe the procedure.

unit 3
grinding fluids

Grinding produces very high temperatures and is really a very hot cutting process. Temperatures at the interface (the small area where the abrasive grains are actually cutting metal) are reliably estimated to be over 2000° F, and that is enough to warp even fairly thick workpieces. Nor is it safe to assume that just because there is a lot of grinding fluid (sometimes called coolant in the shop) flowing around the grinding area, the interface is cooled. With the grinding wheel rotating at its usual 5000 SFPM plus rate, it creates enough of a fan effect to blow coolant away from the interface. This creates a condition sometimes referred to as grinding dry with water. In other words, in spite of the amount of coolant close by, the actual cutting area may be dry and hot.

Furthermore, it is not just a matter of *one* coolant. There are at least three principal groupings, many subgroups, and dozens of brand names. So there is a concern about which coolant to choose; you should know which main group to pick and why.

objectives After completing this unit, you will be able to:

1. List three principal jobs of a grinding fluid.
2. List the three principal types of grinding fluids with their major advantages and disadvantages.
3. Sketch a design of an effective nozzle for flood coolant application, and explain briefly how it works
4. Explain the advantages and disadvantages of mist coolant and through-the-wheel application of coolant.
5. Explain the advantages and disadvantages of various coolant cleaning methods.

The terms grinding fluids and coolants will be used interchangeably; the first is the engineering term and the second is the shop term. Coolants are used because they make grinding a more efficient operation. They help produce a better product the first time with less touch-up work to be done. Some specific notations about coolants are as follows.

☐ They reduce the temperature in the workpiece, thus reducing warping, especially in thin workpieces, making a more accurate product.

☐ They lubricate the interface between the wheel and the workpiece, making it more difficult for abrasive particles and bits of metal to stick in the wheel's grinding surface. Coolant also helps to soften the abrasive action and produce a better finish on the workpiece.

☐ They also flush away the bits of metal and abrasive (called swarf) into the coolant tank, where they can be filtered out and not recirculated to make random scratches on the finished surface of the workpiece.

Our discussion, then really begins from four generally accepted statements.

1. Grinding fluids or coolants are necessary in practically every type of grinding. But it would be simpler from the operator's viewpoint, for machine design, and for shop housekeeping if they were not needed.

2. Coolants vary greatly in qualities. Although there are some general rules (mostly worked out from experience) for selecting one over another, it is often puzzling as to why one works in a given situation while a similar one does not.

3. The design of the coolant and filtration system is highly important. These are not an operator's responsibility, but you should always be alert to signs that the filter is not working properly; "tramp" metal and abrasive recirculated over the grinding area can cause poor finish very quickly. Fortunately, these signs are easily recognized as random scratches without any pattern. The condition is also referred to as "fishtailing."

4. The method of application, and particularly the design and placement of the nozzle that delivers the coolant to the grinding area, are at least as important as the selection of the coolant itself. It is even possible that redesigning the nozzle and making sure that it is always located where it would do the most good is even more important than coolant selection.

METHODS OF COOLANT APPLICATION

Most grinding fluids are applied in what is called flooding. A stream of coolant under pressure is directed from a pipe, sometimes shaped but often just round, in the general direction of the grinding area (Figures 1 and 4). The fluid collects beneath the grinding area, is piped back to a tank where it is allowed to settle, and is cleaned; once more it is pumped back around the wheel (Figure 2). There is always a little waste, and periodically either more coolant concentrate or water is added

Figure 1. This is a very common method of flood coolant application. For the photograph, the volume of coolant has been reduced (Lane Community College).

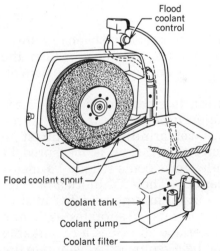

Figure 2. Fluid recirculates through the tank, piping, nozzle, and drains in flood grinding system (Courtesy of DoAll Company).

to the solution. Flooding is an effective method of applying coolant; provided the solution stays within the effective range, neither too rich nor too lean, it works very well. However, the fluid cannot just be in the vicinity of the grinding area; it must be right in that area if it is to do its job. For that purpose, particularly on high speed work, something like the nozzle shown in Figure 3 may be needed. In some cases, for example, on a small surface grinder processing small parts, the nozzle can almost be pointed at the grinding wheel-workpiece interface (Figure 4), although

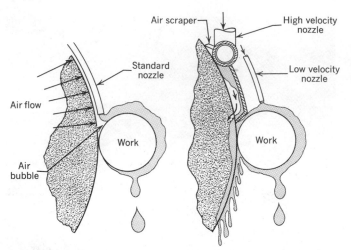

Figure 3. A specially designed nozzle like this helps to keep the fanlike effect of the rapidly rotating wheel from blowing coolant away from the wheel-work interface (Courtesy of Cincinnati Milacron, Products Division).

Figure 4. Here is another coolant application method. Under these circumstances, the pipe apparently supplies enough fluid (Courtesy of Diamond Abrasives Corporation).

here the flow of coolant may have been cut down a little when the picture was taken. Often the nozzle is lower down and closer to the wheel-work interface.

Mist lubrication is a second method of applying coolant, often used where there is an occasional need for wet grinding. In contrast to flood cooling, where fluid flow is measured in gallons per minute, mist coolant may be measured in ounces per day. The mist lubricator (Figure 5) is a small nozzle through which a mist of coolant is blown; the mist evaporates as it hits the work-wheel interface and does a remarkable job of cooling. The heat required to evaporate the coolant is considerable. The lubricating depends on the fluid used, and there is, of course, little or no removal of swarf. However, it provides an unobstructed view of the wheel-work interface; for that reason is popular with the operator who occasionally runs across the job that simply cannot be ground dry. Finally, it requires no return piping.

It was mentioned earlier that vitrified grinding wheels are porous, not solid, and this characteristic is put to work in through-the-wheel grinding fluid application. As noted in Figure 6, the wheel flange has holes leading toward the wheel. As the wheel revolves, grinding fluid is fed through the curving tube into the circular groove, through the holes into the wheel and, finally, out through the pores in the wheel to the

Figure 5. Mist grinding fluid application is sometimes used on a normally dry grinder. The cooling effect is excellent, lubrication practically nil, but no recirculating system is required (Courtesy of DoAll Company).

grinding area and to the inside of the safety guard. Centifugal force keeps the fluid moving outward. This approach has a lot of merit. It absolutely assures that the fluid will get to the wheel-work interface. It must be said, however, that at the same time it assures that fluid will be sprayed all over the inside of the guard. The fluid must be superclean. Otherwise it will clog the pores in the wheel. The guard must be very clean, or dirty fluid could drop onto the work surface.

Figure 6. Nozzle and flange design for through-the-wheel application. This allows coolant to filter through the wheel right to the cutting area. It also sprays coolant over the inside of the wheel guard. (Courtesy of DoAll Company).

As a point of interest, most machine shop-type machines have individual coolant tanks and recirculating systems that include one or more means of cleaning the fluid (Figures 7, 8, and 11). Many production grinding shops, however, have central systems that supply all the machines in the plant. Here, the coolant tanks may be larger than many swimming pools, and the cleaning system may be quite complex. However, in this discussion, we will be referring to the individual machine type, which is more in keeping with the subject matter. The principles are the same in either case.

TYPES OF FLUIDS

The three principal reasons for coolants (cooling, lubricating, and flushing away swarf) have already been mentioned. In the following discussion, each type of coolant covered will be rated in terms of these and perhaps other requirements.

Water, for example, is excellent for cooling. It carries away heat very well. However, water by itself has practically no lubricating ability, although it is good for cleaning away swarf. It also has the disadvantage of causing rust on ground surfaces. For these reasons, water by itself is not a

Figure 7. This is a settling tank for coolant and is used in connection with a single machine, in this case a cylindrical grinder. On some heavy stock removal operations, such as on a vertical spindle rotary surface grinder, the tank might have a conveyor to pull the chips off the bottom of the tank and out (Courtesy of Cincinnati Milacron).

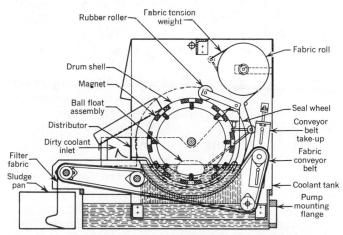

Figure 8. One of the possibilities of a combination filter for a single machine. This combines magnetic filtering to remove iron and steel with settling. Steel or iron chips picked up by the magnets are scraped off into the chute and fall down onto the dirty side of the filter fabric. Other swarf picked up by the filter goes along with the steel chips into the sludge pan (Courtesy of Barnes Drill Co., Rockford, Ill.).

good grinding fluid. However, the addition of water soluble chemicals or water soluble oils to plain water (which make two different types of water based coolants) makes up the larger share of coolants used in grinding.

Water soluble chemical grinding fluids (often called chemical fluids or coolants) are transparent and are the largest single group of grinding fluids used. Their cooling qualities are excellent and their lubricating ability is adequate in the medium to high stock removal applications where they are usually used. The additives include rust preventatives, detergents, water soluble polymers for better lubrication, and bactericides to control bacteria that cause "Monday morning odor," which is often a problem. During the workweek the agitation of the fluid helps control bacterial growth, but bacteria flourish over the weekend in the still water of the coolant tank.

Chemical fluids are good for cooling on all but the heaviest grinding jobs. They also lubricate and are satisfactory on cleaning. A critical point for the operator on a machine with an individual system, where additives come as a paste to be added to the water, is to maintain the solution at its proper richness, such as 20 parts of water to one part of paste (called a 20 to 1 solution).

Oil and water can be made to mix in the case of water soluble oil grinding fluids, by the mixture. With these agents, even though the water and oil do not mix in the common sense of the word, they are dispersed within each other to an extent that provides the effect of a mixture. The resulting milky solution reduces visibility. Otherwise, its cooling and cleaning ability are about the same as the water soluble chemical type, and its lubrication is somewhat better. This type of fluid is often used on applications involving light to medium stock removal.

Straight oils have no equal for lubrication, but they are used mainly on jobs were lubrication is the prime consideration, such as thread grinding and heavy form grinding, usually on very hard materials. The reasons are not hard to find. For all their lubricating quality, oils are much less effective than the water based fluids for cooling, and they do an average job of cleaning swarf from the work area. Furthermore, the swarf does not settle out as quickly nor as thoroughly as it does in water based coolants. Oils have two disadvantages. They cost more than other fluids

and, without or sometimes in spite of a mist-collecting system, they seem to form an objectionable oil mist over the machine and over the whole shop area. But there are some jobs where the high rating on lubrication cancels out all the minuses listed above.

Two related practices in dry grinding should be mentioned. A jet of air has both a cooling and a cleaning effect on a grinding operation, and it is occasionally used by itself. It should be noted, as was mentioned in the beginning of the unit, that the fast rotation of the wheel produces an air current that may have some cooling effect in dry grinding, but it is likely to be more of a hindrance than a help in wet grinding.

Another possibility of dry grinding, when a particularly good finish is needed, it to use a stick of a natural wax blend and hold it against the rotating wheel for a couple of seconds. As the wax heats up, it melts and forms a coating that acts as a lubricant. This will improve the finish for a few parts. This technique of loading the surface of a wheel with wax, incidentally, is much more common in coated abrasive belt grinding than it is in wheel grinding.

CLEANING THE GRINDING FLUID

It is clear that if grinding fluid is going to be recirculated, it must also have the swarf removed before it gets back to the grinding area. Clean coolant is a must in quality grinding.

There are a number of ways of doing this job. Obviously, if the tank is big enough for the coolant to become completely still, the larger bits will settle out. One way of doing this is shown in Figure 9. However, the smaller the particle, the longer it takes to settle, and many plants do not have that much time.

Next is a cloth filter or a filter of some other material. Although this loads up, it can be designed with a float switch that becomes activated when the filter is sufficiently loaded with swarf. The clean filter fabric is on a roll on one side of the tank, and there is a supporting roll on each side of the tank. When the switch is connected, a fresh section of filter is pulled into place, and the used section of filter dumps the swarf into a tote box arrangement for removal (Figure 10).

This type of filter is obviously limited in the minimum size of particle it will remove in relation to the volume of coolant flow. If the fabric will catch the smallest particles, it will at the same time slow the flow of coolant. On the other

Figure 9. A settling tank. The two requirements for cleaning coolant by settling are that the tank be big enough to allow the coolant to remain still for enough time to allow the swarf to settle out, and that there always be enough coolant in the system. This also means that the tank must be kept clean; otherwise, dirty coolant is recirculated (Courtesy of The Carborundum Company).

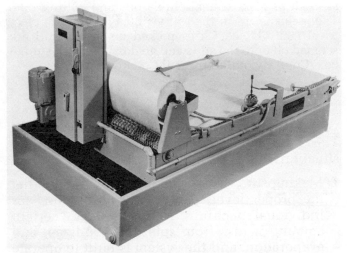

Figure 10. A widely used cleaning method is to let the fluid fall through filter fabric directly into a tank. Then, when the load on the fabric becomes heavy enough, it actuates a switch that pulls the loaded fabric into a waste container and also pulls a fresh section of fabric into place (Courtesy of The Carborundum Company).

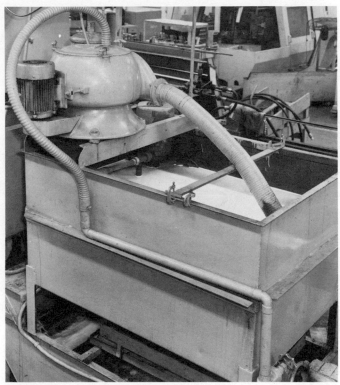

Figure 11. The centrifugal unit on top spins at high speed to remove swarf from the fluid. Dirty fluid, already partly cleaned by settling, is pumped into the unit by the large hose from the tank and recirculated through the smaller hose and piping (Courtesy of Barrett Centrifugals).

hand, the openings must be small if the filter is to be effective. While the original cost of the unit is not excessive, the cost of the filter cloth can be a problem if there is any large amount of swarf. It is a simple method, and where the demands for clean coolant are not too tight and the volume of fluid to be handled is not too great, they are popular.

Three other types of filters, at least, are in use to varying degrees. These are the centrifugal separator, the cyclonic separator, and the magnetic separator. The cyclonic separator was adapted to metal working from a type long in use in paper mills.

The centrifugal separator, as the name implies, has an inner bowl that spins at high speed. The unit is designed so that the swarf spins to the outside of the bowl and the clean coolant stays toward the center. It is very efficient where the volume of coolant is not too great, and it is most economical in space, taking up perhaps the least space of any of the possibilities. Of course, the bowl must be cleaned from time to time, and most installations have a spare bowl so that one may be cleaned while the other is in use (Figure 11).

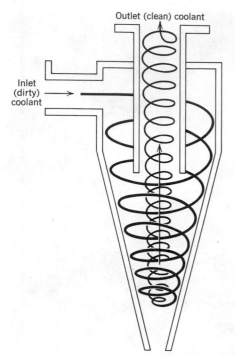

Figure 12. A cyclonic filter. The dirty coolant is fed in from the side, swirls around in the cone as the clean coolant goes out through the top, and the swarf through the bottom of the cone (Courtesy of Barnes Drill Co., Rockford, Ill.).

Figure 13. This magnetic separator removes iron and steel chips from water soluble oils and cutting oils. The magnets are in the roll visible just past the electrical control box. Chips picked up by the roll are scraped off onto the slide and go down into the sludge box (Courtesy of Barnes Drill Co., Rockford, Ill.).

The cyclonic separator also uses centrifugal force for separation of swarf from clean coolant. Figure 12 explains the principle of the operation. It must be admitted that the technical reasons for the separation may not be as clear as for the centrifugal type, although you will note that the results are similar. The swarf goes to the outside, while the clean coolant stays toward the center. Both light-duty and heavy-duty types are available, the latter with automatic sludge removal, a larger settling tank, and a larger dirt discharge opening.

The magnetic separator is, for obvious reasons, used where the grinding of iron and steel is the primary operation. Essentially, the unit is a magnetized drum that rotates into and out of the coolant flow and picks up chips of metal. As the drum rotates, a scraper, called a doctor blade, scrapes off the chips into a waste container (Figure 13).

The cleaning system or combination of systems used is not ordinarily a responsibility of an operator. However, once the system is in use, you should know in general how it works and what

your responsibilities are in connection with it. These, of course, will vary with different shop situations, but the following are some examples.

1. Making sure that the coolant is maintained at the proper level, particularly on systems for individual machines. There is always a certain amount of loss from splashing, spillage, and evaporation, and the system is built to operate on a certain amount of fluid. If this amount becomes less, the fluid circulates faster, gets warmer, as does the workpiece, and the entire heating process continues to get worse, to say nothing of the probability of recirculation of more swarf and the harmful effects that follow.

2. Maintaining the proper solution of water based fluids. It will probably not be your concern to determine the relative percentages of water and concentrate, but once these are determined, you should add water or concentrate at the times and in the amounts set up. Too rich a mixture will probably have bad effects on surface efficiency and on the effectiveness

of the fluid. Besides, it boosts fluid costs. A mixture that is too lean may seem to cut coolant costs, but it actually raises them because of the scrap rate or other difficulties.

3. Keep on the lookout for signs that lubricating oil from the grinding machine is getting into the coolant. If you are using an oil coolant, this may not be too much of a problem but, in any case, it would be difficult to detect. The addition of a few drops of lubricating oil in a water based coolant could be very bad for the finish and the coolant's efficiency.

4. Finally, always make sure that the coolant nozzle is properly adjusted and aimed. What you are doing is directing the coolant at a very specific spot.

As you gain experience, you will come to appreciate some of the fine points of coolant effectiveness. If you become a good operator in a small shop, you may very well be in the position of making the decision as to which coolant is purchased. If you are in a large shop, particularly one that is part of a larger company, your part in coolant selection may be limited to expressing an opinion to an engineer who makes the selection particularly if the plant has a central coolant system.

However, when it comes to matters like checking the cleanliness of the coolant and keeping the coolant nozzle properly adjusted and directed, there is no question about who is responsible: you are.

self-evaluation

SELF-TEST

1. Why is a knowledge of grinding fluids important?

2. Explain the three major functions of grinding fluids.

3. Explain why the design and the placement of the grinding fluid flood nozzle are so important. Use sketches if helpful.

4. For one of the three major groups of grinding fluids, evaluate it in terms of the three major functions mentioned in Problem 2.

5. Repeat the same process for another of the three major fluids.

6. Do the same for the third.

7. Explain how a mist coolant system works and under what circumstances one might be used.

8. List four methods of cleaning grinding fluids, and include advantages and disadvantages of each.

9. List reasons for one side or the other (it does not matter which) of the proposition that the operator should take part in the selection of grinding fuilds for his machine.

10. List the four major responsibilities of an operator in *using* grinding fluids.

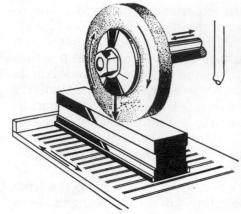

unit 4

horizontal spindle, reciprocating table surface grinders

The small horizontal spindle, reciprocating table surface grinder is considered to be basic to most general purpose machine shops. With accessories, they are extremly versatile and can do a remarkable variety of work. It is important that you learn the nomenclature and functions of this type of grinder and that you are aware of the many accessories that can be obtained to make this machine very useful in a wide variety of situations.

objectives   After completing this unit, you will be able to:

1. Name the components of the horizontal spindle surface grinder.
2. Define the functions of the various component parts of this grinder.
3. Name and describe the functions of at least four accessory devices used to increase the versatility of the surface grinder.

By any measure, the small horizontal spindle, reciprocating table surface grinder (Figure 1) must be considered the basic grinder. There is at least one in nearly every machine shop, toolroom, school shop, and anywhere else that grinding is done. It is economical both in original cost and floor space, some requiring only a 3 foot square space for the machine itself. It can be used to demonstrate the principles of grinding that apply to all kinds of grinders, such as selection of abrasive wheels and selection and application of coolants, and the results of too much or too little infeed (which on a surface grinder is called downfeed).

There are probably more machines, more different makes and models, and more companies making them than any other single type of grinder. They come in a broad range of prices; in general, the more precise the machine, the more expensive it is; but principles can be taught on an inexpensive machine as well as on a more expensive one. Operating one on straightforward work can be relatively simple. In fact, it has been said that three fourths or more of all grinding work could be done on a small 6 by 12 in. capacity surface grinder.

In a sense, of course, all grinding is done on a surface, but a surface grinder is one for producing flat surfaces (or, with these horizontal spindle machines, formed surfaces) as long as the hills and valleys of the form run parallel with the path of the wheel as the magnetic chuck moves

Figure 1. Typical toolroom-type surface grinder (Courtesy of Boyar-Schultz Corp.).

Figure 2. Grinding a simple rounded form. Note that the wheel is dressed to the reverse of the form in the finished workpiece.

the work under it. Figure 2 shows a simple form. Vertical spindle machines, however, are usually limited to flat surfaces, because the grinding is done on the flat of the wheel with the abrasive scratches making an overlapping, circular path.

It is also possible on the horizontal spindle, reciprocating table grinder, with proper fixturing, to grind surfaces that are not parallel but are either flat or formed. Finally, with accessories, it is possible to do almost any kind of grinding: center-type, cylindrical, centerless, and internal. This is usually incidental work and small workpieces. But these show that the small, hand-operated surface grinder is a very versatile machine.

The best place to start learning the part names of a hand-operated horizontal spindle, reciprocating table grinder is the wheel-work interface, the point where the action is. In this unit, you should understand that the term "grinder" means this type of machine.

The cutting element is the periphery or OD of the grinding wheel. The wheel moves up and

down under the control of handwheel A, as shown in Figure 3. The doubleheaded arrow A shows the direction of this movement. The wheel, together with its spindle, motor drive, and other necessary attachments, is often referred to as the wheelhead.

The work usually is held on a flat magnetic chuck, which is clamped to the table which, in turn, is supported on saddle ways. The magnetic chuck, which is the most common accessory for this grinder, holds iron and steel firmly enough to be ground. Using blocking of iron or steel, it will also hold nonmagnetic metals like aluminum, brass, or bronze. Only very rarely is it necessary to fasten a workpiece directly to the machine table.

So, for many workpieces, all that is needed is to clean off the chuck surface and the workpiece, place the part, turn on the current, check for firm workholding, and start grinding (assuming that the wheel on the grinder is suited to the workpiece). On all such pieces, it is imperative that the surface to be ground is parallel with the sur-

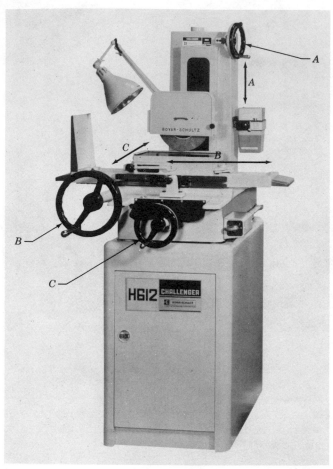

Figure 3. Surface grinder with direction and control of movements indicated by arrows. Wheel *A* controls downfeed A. Large wheel *B* controls table traverse B. Wheel *C* controls crossfeed C (Courtesy of Boyar-Schultz Corp.).

attachment called a sine dresser and then grind face resting on the magnetic chuck. The surface of the chuck and the "grinding line" of the wheel must always be parallel. For this reason part of the downfeed of the wheel is to compensate for wheel wear, and the chuck must be "ground in" when it is first mounted and periodically reground. This grinding and regrinding of the chuck will be discussed further in the next unit.

However, surfaces that are not parallel can be ground. As a simple illustration, consider a beveled edge to be ground around all four sides of the top of a rectangular block. You have two choices. One is to use a fixture that can be set to the angle you want, which is a magnetic sine chuck, as shown in Figure 13. The other is to dress the angle you want on the wheel with an

in the usual fashion. The choice is up to the operator. These applications are covered in detail in Unit 7.

The chuck moves from left to right and back again (traverse), as shown in Figure 3, arrow *B*. On many grinders, traverse and crossfeed increments are controlled hydraulically with the table moving back and forth between preset stops. When the hydraulic control is actuated, the traverse wheel is automatically disengaged. This movement is controlled by large handwheel *B*. The chuck also moves toward and away from the operator (crossfeed), as shown by arrow *C*, with the motion controlled by handwheel *C*. This is the standard arrangement, since most machines are designed for right-handed operators. In at least one make (Figure 1) the traverse handwheel can be reinstalled on the right of the crossfeed wheel to accommodate a left-handed operator. The grinding wheel, incidentally, always runs clockwise.

On a completely hand-operated machine, the operator stands in front of the machine with his left hand on the traverse wheel and his right hand on the crossfeed wheel (if he is right-handed), swinging the table and the magnetic chuck back and forth with his left hand and cross feeding with his right hand at the end of each pass across the workpiece (Figure 4). The wheel clears the workpiece at both ends of the pass, and the crossfeed is always less than the width of the wheel, so that there is an overlap. At the end of each complete pass across the entire surface to be ground, he feeds the wheel down with the downfeed wheel on the column.

The traverse handwheel is not marked, because all that is necessary is to clear both ends of the surface. However, both of the other wheels have very accurately engraved markings, so that downfeed and crossfeed can be accurately measured. The downfeed handwheel (Figure 5) has 250 marks around it, and turning the wheel from one mark to the next lowers or raises the wheel .0001 in. This is one ten-thousandth of an inch, known in the shop usually as a "tenth." The crossfeed handwheel has 100 marks, and moving the wheel from one mark to the next moves the workpiece toward or away from the operator .001 in.

With the 250 marks on the downfeed handwheel and .0001 in. movement per mark, a complete revolution of the handwheel moves the grinding wheel .025 in. (250 × .0001 in.). A

Figure 4. Operator in position at grinder. On many small surface grinders the crossfeed and traverse are hydraulic (Lane Community College).

Figure 5. Closeup of downfeed handwheel. Moving from one mark to the next lowers or raises the grinding wheel .0001 in. (Courtesy of DoAll Company).

Figure 6. Same control wheel with slip ring set to zero. Now it is simpler for operator to down feed the grinding wheel as he grinds (Courtesy of DoAll Company).

complete revolution of the crossfeed wheel, which has 100 marks, moves the table and the chuck .100 in. (100 × .001 in.).

The other feature needing mention is the zeroing slip ring on the downfeed handwheel or the crossfeed handwheel. You cannot predict in advance just where on the scale either wheel will be when the grinding wheel first contacts the workpiece, and where the control wheels ought to be when the surface is ground. These can be figured out mathematically, but with considerable chance of error. However, with the zeroing slip ring, the starting point on the scale on each wheel is simply set at zero, locked in place (Figure 6), and then ground until the required amount of stock has been removed. Allowance must be made for wheel wear, especially where considerable stock is removed by grinding.

A skilled operator develops a rhythm as he traverses the workpiece back and forth under the grinding wheel, cross feeding at the end of each pass and downfeeding when the whole surface has been covered. This is a knack you develop with experience. Using machines with hydraulic traverse and crossfeed takes out the need for physical coordination on the part of the operator, but the skill necessary to make good choices on the traverse speed and on the amount of cross-feed for each pass is very important to first-class grinding results. Generally, combinations of large crossfeed movements on the order of one half the wheel width and relatively small amounts of downfeed are preferred because wheel wear is distributed better this way.

Actually, in the wheelhead and the magnetic chuck, with the three control wheels, you have the basic parts of a surface grinder. You do not need a dressing device, which may be just a holder with the diamond mounted at the proper angle that can be mounted on the magnetic chuck or a built-in dresser (Figure 7). But everything else on the machine is either for support, for instance, the table, saddle ways, base, and column that holds the wheelhead; or an accessory that makes it possible to do something you could not do otherwise or that makes the job easier to do.

ACCESSORIES

The list of accessories for a hand and hydraulically operated toolroom-type surface grinder may be quite extensive. As mentioned before, with the proper accessories, almost any type of grinding can be done on one of these little grinders within maximum size ranges. Finally, accessories are a major point of difference between toolroom and production machines. Toolroom grinders must handle a variety of work, so accessories are needed. Production machines are used mostly for one purpose; if that purpose changes, the machine is rebuilt or modified for its new use or it is removed from service.

ATTACHMENTS

For practical purposes it probably makes very little difference whether something is called an attachment or an accessory. Both make it possible to do something with the machine that could not otherwise be done or at least could not be done so easily or quickly.

Rotary Chuck

A rotary magnetic chuck (Figure 8) can be mounted on the regular rectangular magnetic chuck and then locked in place. The rotary chuck is independently magnetized and powered. This, in effect, converts the grinder into a horizontal spindle, rotary table grinder, which is useful for

Figure 7. Built-in wheel dresser. Lever traverses dresser across wheel (Courtesy of DoAll Company).

grinding work where a circular scratch pattern around the center is desirable in the workpiece. A good example would be the grinding of custom piston ring thicknesses for use in modified engines.

Swivel Table

With some modifications, the rotating table described above can index as well as rotate for work that needs to be ground at some angle other than parallel to the front edge of the basic chuck (Figure 8).

Centerless Grinding Attachment

For the surface grinding shop that has an occasional need to do some small centerless grinding (Figure 9), there is an attachment to do just that. The part is held between the grinding wheel and two rollers. The wheel supplies the power for rotation, and the rollers act as brakes so that the part is ground and not just spun. This is one of the largest of the attachments. It can grind parts up to 5 in. in diameter and can be mounted on any surface grinder 6 × 12 in. or larger.

Figure 8. Rotary magnetic chuck turns the grinder into a rotating table type, good for grinding parts that require a circular scratch pattern; for instance, like grinding a metal-to-metal seal (Courtesy of M & M Precision Systems, Inc., Roto Grand®).

Figure 9. Centerless grinding attachment mounted on surface grinder with an assortment of parts that can be ground on it (Courtesy of Unison Corporation).

Center-Type Cylindrical Attachment

This is basically a workholder with a headstock and a tailstock (Figure 10); it is mounted cross-wise on the locked table so that the crossfeed of the grinder makes the wheel traverse end to end on the workpiece. If a flat is needed on an essentially cylindrical part, then all that is necessary is to stop the rotation of the work, reciprocate the table just a little, and cross feed as for any flat surface.

High-Speed Attachment

For incidental internal grinding, this attachment (Figure 11), driven by a belt from the grinder spindle, provides the high speed that is needed to make the small mounted wheels run at the high RPMs that are needed to make them grind efficiently. Of course, it is essential, usually, to provide an attachment for mounting the work-piece also.

Another form of the high speed spindle is the variable speed spindle. Most grinders are built to run at a constant speed in RPMs, but sometimes where a grinder is used for both regular and diamond wheels, it is worthwhile to have an attachment to vary that speed. Diamond wheels are most effective at speeds that are well below the 5000 FPM or more that is common for

Figure 10. Center-type cylindrical attachment mounted on surface grinder. Attachment can be tilted for grinding a taper, as shown here, or set level for grinding a straight cylinder (Photo of Harig Lectric-Centers, courtesy of Harig Mfg. Corp., Chicago, Ill.).

vitrified bonded aluminum oxide or silicon carbide wheels.

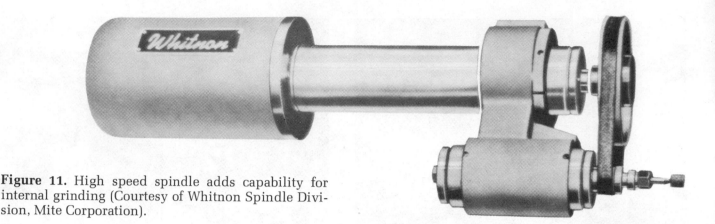

Figure 11. High speed spindle adds capability for internal grinding (Courtesy of Whitnon Spindle Division, Mite Corporation).

Vacuum Chuck

This replacement for an electromagnetic chuck (Figure 12) holds the work by exhausting the air from under it. Thus, it makes no difference whether the workpiece is magnetic or not. It is also recommended by some experts for holding pieces as thin as only a few thousandths.

Magnetic Sine Chucks

The surface of the magnetic chuck is always parallel with the line on which the wheel grinds. A sine chuck (Figure 13), usually electromagnetic, can be ajdusted so that a surface to be ground on angular work can be made parallel to the chuck. Some sine chucks are also designed so that compound angles can be ground. These devices are typically set with gage blocks in the same fashion as a sine bar of the same length.

Dressers

It makes some sense to consider all dressing devices as accessories, although a flat dresser, as discussed in Unit 3, is a necessity for any surface grinder, and when the dresser is built into the safety guard, as shown in Figure 7, then it is an integral part of the machine. Form dressing with the equipment needed will be discussed in a later unit.

The decision as to which of the attachments or accessories are justified depends mainly on production and economic factors. It is beneficial to know that there are such units available. It may be, especially in a small shop, that particular information that would widen the range of work that can be done, which is a major purpose of each attachment and accessory listed above.

Some shops, even some very large shops, find it to their advantage to have a surface grind-

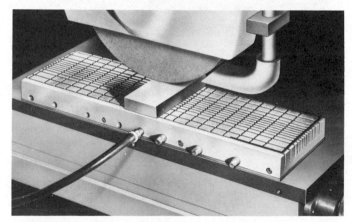

Figure 12. Vacuum chucks such as this one hold practically anything and are considered good for thin work (Courtesy of Thompson Vacuum Co., Inc.).

Figure 13. Magnetic sine chuck needed for grinding nonparallel surfaces. It is, of course, adjustable (Courtesy of Hitachi Magna-Lock Corp.).

er set up with a centerless grinding attachment, for example. It can be less expensive than a small centerless grinder; for certain work it can be adequately productive. If there comes a slack time in work for the attachment, the machine can readily be converted back for surface grinding.

self-evaluation

SELF-TEST

1. Give at least two reasons for the importance of studying the small hand or hydraulically operated horizontal spindle, reciprocating table surface grinder.

2. What three or more characteristics or qualities make the electromagnetic chuck such a valuable part of a grinder?

4. What is the principal result of these motions in terms of the wheel and the surface to be ground?

5. The handwheel that controls wheel motion (downfeed) is 10 times as precise as the one controlling crossfeed, and the traverse handwheel (if there is one) is not precise at all. Why is this?

6. What is the principal advantage of zeroing slip rings on the handwheels?

7. Why is form grinding possible only on a horizontal spindle surface grinder and not on a vertical spindle grinder? What limitations does this place on the form?

8. List at least two attachments or accessories that increase the kinds of grinding that can be done on a surface grinder. Include a sentence to tell what each unit does.

9. Make a similar list for other workholding attachments, including a sentence for each as to what it does.

10. What three reasons can you give for the popularity of these toolroom surface grinders?

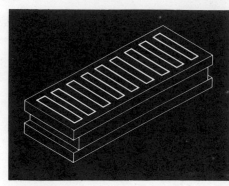

unit 5

workholding on the surface grinder

The development of the surface grinder and of the electromagnetic chuck have been so closely tied together that it is almost impossible to consider one without the other. It is true, of course, that some work must be clamped onto the surface grinder, and that the magnetic chuck is used on some machine tools other than the surface grinder, but these are both minor. Essentially, in industrial shops the electromagnetic chuck is the basic workholder for the surface grinder, but the permanent magnetic chuck is often used for instruction. There is no basic difference in principle.

This is true of all surface grinders but, in this unit, the discussion will be related to horizontal spindle, reciprocating table surface grinders. Others are different only in detail, not in principle.

Permanent magnet chucks are used to some extent; they are made up of a series of alternating plates composed of powerful alnico magnets and a nonmagnetic material. Some improved types use ceramic magnets alternating with stainless steel plates. They exert a more concentrated holding force and can be used for milling as well as grinding. Permanent magnet chucks are used for small work and are not as widely used as electromagnetic chucks for industrial purposes. Therefore the term "chuck" will be considered as meaning an electromagnetic chuck for the remainder of this unit.

objectives **After completing this unit, you will be able to:**

1. **Describe how a chuck is made and how it operates.**
2. **Detail the daily care and periodic care of a chuck.**
3. **Given a sketch or a description of a workpiece where the surface to be ground is not parallel to the side to be placed on the chuck (the chucking side), describe how to set it up for surface grinding, including accessories.**
4. **Detail the major steps in grinding a workpiece square on all six sides, including the accessories needed.**
5. **List precautions to be used in setting up thin workpieces for surface grinding.**

780

On reciprocating grinders the chuck is rectangular (Figures 1 and 2). On rotary grinders it is round. On reciprocating grinders, the dimensions of the chuck describe the size of the machine. Most of the grinders you will work on will be referred to as 6 by 12 in. or 10 by 30 in. or some other combination. Rotary grinders use the diameter of the chuck to define the machine size. Some companies, in fact, use the diameter as a model number.

Chucks operate on direct current (dc) at 24, 110, or 220 V. They are made up of alternating strips or rings of steel and some nonmagnetic metal, usually lead or brass. The holding power depends on two factors: the area of the workpiece in contact with the chuck and the amount of current that is used. Thus, a workpiece with a flat and relatively large chucking side is easily held; one with a projection of the chucking presents other problems, particularly if the surface to be ground is larger than the surface contacting the chuck. A problem like that can be handled by supporting the shoulders of the workpiece on magnetic parallels that are a little thicker than the projection is high (Figure 3).

The percentage of total available power that is used is often critical in grinding thin parts, where the full power of the chuck could pull a warped piece down to the chuck, only to have the warping recur when the power is turned off and the workpiece is removed from the chuck.

Another unique feature about the chuck is the special switch. If there were only a regular on-off switch, a large, heavy workpiece with a considerable area of chuck contact would be hard to remove after the power was turned off because of magnetism remaining in the chuck (called residual magnetism). The special switch gets rid of residual magnetism, so that any part is easy to remove. It is important to leave the part in place until the demagnetizing cycle is complete (about 20 sec), or the part will still have residual magnetism.

Mention should also be made about the weight of the workpiece and the friction that this generates between the workpiece and the chuck. This is a factor, but one not generally of much concern in most toolrooms and machine shops where the parts tend to be small anyway (Figure 4). In fact, most such workpieces have the side to be ground parallel with the chucking side. About the only precautions that need be observed are to

Figure 1. Common type of magnetic chuck for reciprocating surface grinder. The guards at the back and left side are usually adjustable and help keep work from sliding off the chuck (Courtesy of DoAll Company).

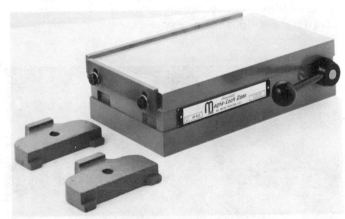

Figure 2. The permanent magentic chuck looks about the same as any other chuck of the same shape and often is used where wiring would be inconvenient (Courtesy of Hitachi Magna-Lock Corp.).

clean the chuck of chips or fluid before each piece is mounted and to spot the pieces on various parts of the chuck face so that you do not wear a depression in the center that would eventually require the regrinding of the entire chuck face. The quickness and simplicity of loading or chucking the majority of parts is one principal reason for the increasing use of the surface grinder. When producing many parts, you should load as many parts on the chuck as is practical, because for such parts you can grind several about as easily as you can grind one.

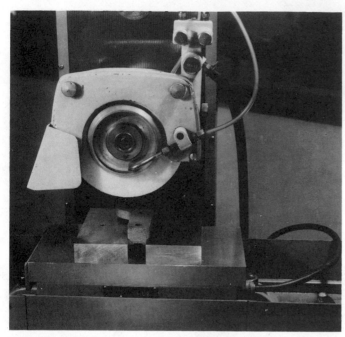

Figure 3. Chuck setup for workpiece with projection on chucking side. Work is supported on laminated magnetic parallels (Courtesy of DoAll Company).

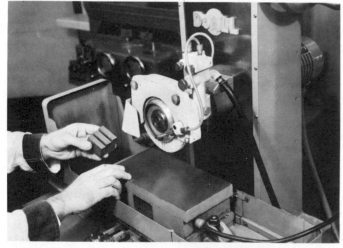

Figure 4. Much machine shop work is chucked as simply as this. If the chuck is clean, the operator can spot the workpiece on the chuck, check to see that it is firmly in place, and proceed to grind (Courtesy of DoAll Company).

ROUTINE CARE

There are a few points for routine care of the chuck that ought to be second nature with you at any time you use a surface grinder. Some of these are:

1. Clean off the top of the chuck before you place any workpiece on it. This is sometimes done with a squeegee.
2. Place the workpieces carefully on the chuck to avoid nicking, burring, or scratching the chuck. Sometimes it is even worthwhile to place a thin piece of paper between the chuck and the workpiece. However, if you are grinding more than one part at a time, it is considered good practice to place the workpiece in some kind of regular pattern. Also, particularly in placing small parts, it is important to have each part span as many of the magnetized strips in the chuck as possible to get maximum holding power. Use the same care when placing any accessory holding device on the chuck.
3. Rub the chuck occasionally with a deburring stone (Figure 5) or a medium to fine grit oilstone to remove the nicks or burrs that will come even with the best of care. When wear begins to affect work quality, you will have to regrind the chuck. (See point 6.)
4. A word of precaution: Never wear a watch while you are operating a surface grinder. It is unsafe practice and you may magnetize the watch mechanism and ruin its operation as well.
5. In case of chuck failure, a service representative of the manufacturer should be called. Any tinkering with the chuck by anyone else could void the warranty; furthermore, it could be

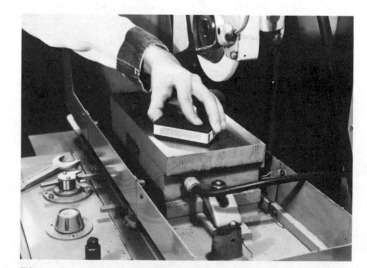

Figure 5. Periodic deburring of the chuck with a granite deburring stone like this, or with a fine grit oilstone, is a good practice (Courtesy of DoAll Company).

dangerous. A chuck failure is a serious safety hazard. DO NOT use a chuck that gives any indication of weakness; parts slipping on the chuck can lead to broken wheels, bent spindles, and impact hazards to the operator and others in the same work area.

6. Because of nicking, burring, or ordinary wear, you will occasionally have to regrind the chuck, although it would be rare for you to have to install a new chuck. Also, a chuck should be indicated in every time it is clamped on the machine. In most cases, it must be reground lightly because clamping alone is not usually accurate enough for the parallelism you need. The chuck must be mounted and indicated parallel to the table and the saddle ways under it and, most important, to the grinding line of the wheel above it as the chuck moves the work under the wheel.

Use plenty of coolant if it is available (Figure 6). Otherwise you may be limited to light cuts with time out at intervals for the chuck to cool off. The wheel specification recommended is friable (usually white) aluminum oxide, 46 grit, H grade, 8 structure, vitrified bond (friable A46-H8-V) dressed with rapid passes of a sharp diamond for open cutting. Use light downfeed, a fairly rapid table speed, and slow crossfeed. *Do not take off any more stock than you have to.* The following are some steps in regrinding the chuck.

1. Mark the entire top surface of the chuck with a thin application of layout dye or perhaps with Prussian blue. Any blue mark remaining on the chuck after the first full pass indicates a low spot.

2. Magnetize the chuck.

3. Downfeed should be about .0002 in.

4. Usually the flatness will be restored in less than .001 in. of metal removal. If more metal must be removed, you may want to redress the wheel rough before dressing it for the finish cuts.

5. Demagnetize the chuck.

6. Test the flatness of the chuck with a dial indicator fastened to the wheel guard. In appearance, the chuck surface should look dull, but almost polished.

7. If you have not done so, clean off the chuck carefully and remove any residual microburrs with a granite stone (Figure 5) or a fine India oilstone. The chuck should now be ready for use.

So far we have been discussing iron or steel workpieces on which the surface to be ground is approximately parallel to the surface that rests on the chuck. This may be 85 to 90 percent of all surface grinding work. There are other classes of work that should be covered: nonmagnetic metals such as brass and aluminum, thin work, odd-shaped work, and work to be ground square. These were mentioned briefly in the previous unit.

HOLDING NONMAGNETIC WORK

The critical point in holding nonmagnetic work on a magnetic chuck is to block it with steel bars, either with the vises that are shown separately in Figure 7 or that are shown in use in Figure 8. The surface to be ground must, of course, be higher than any of its retainers.

Figure 6. In regrinding the chuck, plenty of coolant is a necessity (Courtesy of DoAll Company).

Figure 7. The comblike teeth in these magnetically activated clamps help hold nonferrous metal work (Courtesy of DoAll Company).

Figure 8. Tooth clamps in use. Note that the toothed clamps are lower than the surface to be ground (Courtesy of DoAll Company).

Figure 9. A set of magnetic parallels and V-blocks can be very useful (Courtesy of Hitachi Magna-Lock Corp.).

The other possibility for holding nonmagnetic work is the vacuum chuck, although this is more likely to be used in production grinding of sheet metal. It has a steel bottom plate and can be used directly on top of a magnetic chuck.

HOLDING THIN WORK

Surface grinding of thin steel that has been rolled is likely to release some of the stresses in the metal caused by the rolling process; the result is that the metal may warp or twist in some unpredictable fashion. The trick here is to use only the minimum power needed to hold the work. If more power is used, the work will be flat while you are grinding, but it will spring out of flat the moment you turn off the power. Grinding with partial chuck power requires great care to be done safely.

Blocking with thin precision ground stock is often done to prevent end movement of the part. Do not grind under partial power without instructor approval for your specific setup. Then, if you have nonmagnetic thin work to grind, you may have to resort to something like double-faced tape, which is sticky on both sides. Then stick the tape to the chuck, using one or more pieces according to the size of the work, peel off the second piece of backing paper, and lay the work firmly on top of the strips of tape. For this method, you need no power for the chuck and light cuts should be taken. Do not use fluids,

since they usually cause the adhesive to become slippery.

HOLDING ODD-SHAPED WORK

There are two basic principles that apply to all surface grinding work, but particularly to odd-shaped work.

1. The surface to be ground must be parallel to the chuck surface and, hence, also parallel to the wheel's line of grinding.
2. There must be a parallel and opposite surface of sufficient area to hold in place the surface to be ground, either on the chuck itself, or on something that in turn is held on the chuck, such as magnetic bars or parallels (Figure 9) or the magnetic sine chuck (Figure 10). Sometimes you may have to grind a chucking surface on the workpiece, but keep the setup as simple as possible.

A steel vee block (Figure 11), such as the project suggested for you to finish in Unit 6, could be very useful. You must bear in mind that only specially constructed laminated vee blocks will hold a workpiece magnetically; if you use a conventional vee block, you must hold the workpiece into the vee by mechanical means while the block itself is held magnetically to the chuck surface. A steel plate with a number of drilled and tapped holes in it might also be another handy thing to have around. With a few little

Figure 10. Adjustable magnetic sine chuck is useful for holding a variety of work where the chucking side is not parallel to the side to be ground (Courtesy of Hitachi Magna-Lock Corp.).

Figure 11. A steel V-block such as this has many uses in a shop (Lane Community College).

bars of steel with similar bolt holes drilled and some miscellaneous blocking, this could be used as a base for strapping down nonmagnetic work.

Another difficult part to hold is one in which the ends to be ground are small in area compared to its length, such as a part 4 in. long by 1 in. by 1 in. Probably the best setup here is to clamp the part to an angle plate so that the end to be ground

projects above the plate. Blocking is preferable under the workpiece, but it is optional because the pressures involved are typically quite small.

There are a great many accessories to help you in holding work on the surface grinder. Probably the most important point about these for most machine shops is that the fixtures be versatile, so that they can be used on a wide variety of work.

However, as far as you are concerned, the most important need is that you have a clear idea of what you want to accomplish on the part. With that and a good knowledge of what is at hand, you can probably determine what you need to do to hold the work for finishing the job. Remember, the chuck and all the accessories that go on it are precision equipment and should be treated with great care.

self-evaluation

SELF-TEST 1. Describe the construction of an electromagnetic chuck, including the type of controls and power used, and how the construction affects the use of the chuck.
2. Explain how the chucking of a workpiece is affected by its size, shape, and material.
3. Discuss three frequent routines for caring for the magnetic chuck.
4. List at least four major points in the regrinding of a chuck.
5. List or write a paragraph about the advantages and disadvantages of electromagnetic chucks.

6. Detail at least three methods of holding metals like aluminum, brass, or bronze or a magnetic chuck.
7. What differences are there between dressing a wheel for an angular form and dressing it for a concave form?

This unit has no post-test.

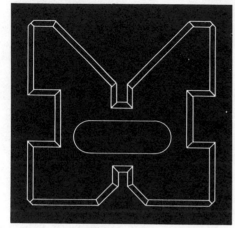

unit 6

using the surface grinder

Demonstrating any skill is, of course, the final test of whether you have learned it. The workpieces selected for this unit on using the surface grinder are two matching vee blocks of SAE 4140 or a similar alloy steel. A series of grinding steps are given in the text to detail the process for making these vee blocks. Some of these steps will include grinding operations on both blocks in the same setup at the same time. The specifications for surface grinding are often much closer than those required on other machine tools. These grinding machines, however, are capable of holding tolerances within a few tenths of a thousandth of an inch.

objectives **After completing this unit, you will be able to:**

1. **Surface grind to specification two vee blocks of SAE 4140 or similar alloy steel.**
2. **Prepare a worksheet for and grind to specification a similar part involving surface grinding.**

A SAE 4140 steel in the Rockwell C hardness range of 48 to 52 is regarded as being not too difficult to grind. It can be ground satisfactorily with a number of wheel specifications that are likely to be on hand in practically any grinding shop. As stated earlier, wheels for machine shop use have to be able to grind a wide range of materials, because the number of parts to be ground in any single lot is not likely to be large enough to warrant trials to find *exactly* the best wheel specification. Here, however, are some guidelines in the selection.

The abrasive will be aluminum oxide because this is steel, and the bond will be vitrified because this is precision grinding. Given the requirements for this particular workpiece, below are the recommendations of several specialists in the grinding field giving a range of four possible selections of wheels.

1. 9A46-H8V.
2. 9A60-K8V.
3. 32A46-I8V.
4. DA46-J9V.

All of these may be regarded as general purpose specifications for a part of this configuration, material, and hardness.

Looking at these recommendations in order, the abrasive recommendation of the first two is a white aluminum oxide, which is the most friable (brittle). The third and fourth are for a mixture of white and regular (gray) aluminum oxide, which is slightly tougher and perhaps wears a little longer. In wet grinding, as this is, probably the tougher abrasive would be preferred. However, in dry grinding, there would probably be no question about the use of white abrasive; it does not tend to burn the work as much as a tougher abrasive.

A grit size of 46 would be indicated for grinding efficiency, but 60 would provide a slightly better finish. Wheel grade provides the widest range (H to K), which could be interpreted to mean that any grade within this range would do the job. However, the wheel with the finest grit (60) also has the hardest grade (K). Such a wheel would wear a little longer than the others. Grit size and abrasive types are both likely to be fairly constant from various manufacturers.

The range of structure is only from 8 to 9, which is hardly significant. Grade and structure, however, result from manufacturing procedures, and hence are at least comparable from one wheel supplier to another.

Vitrified bond is indicated. Vitrified bonds are somewhat varied, although all wheel manufacturers have one or two general purpose or standard bonds for machine shop work.

Most general purpose grinding shops have neither the time nor the need to work out detailed comparisons and will no doubt depend on the experience of their better machinists in making such determinations. Good machinists develop opinions about which wheels perform best for them on given materials, but it is likely to be more of a process of instinct or feeling than of factual proof. The final choice of combinations of abrasive factors for particular types of work-

pieces is quite individual. The more aggressive machinist often uses a relatively harder bond because his procedures lead to more rapid wheel breakdown. Selection of the wheel to match jobs and machinist characteristics can still be considered something of an art rather than a specific science.

GRINDING THE VEE BLOCKS

The part selected for this precision grinding project is illustrated in Figure 1. Figure 2 is a sketch with the necessary dimensions and other data. These particular workpieces, a matching pair of vee blocks, follow through on a project started in a previous section on milling and shaper work. The purpose of this project is to show that you can surface grind a part flat and square to common tolerances and surface finish.

At this point, you have completed the previous steps of preparing the workpiece on other machines and completed the heat treatment of the vee blocks to RC 48-52. It is assumed that you have allowed appropriate stock for grinding .015 in. on all surfaces to be ground, except the grooves at the bottom of the vees and the $\frac{1}{2}$ by $\frac{1}{4}$ in. side grooves.

Figure 1. Finished, hardened, and ground precision vee block.

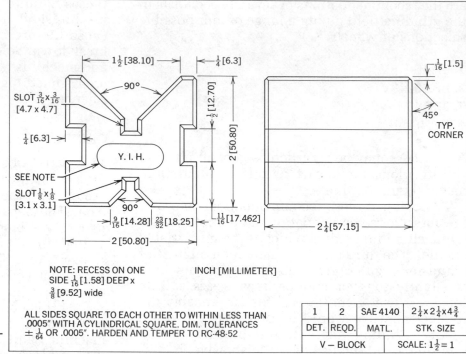

Figure 2. Dimensions and information for grinding the vee block.

STEPS IN GRINDING

1. *Select the wheel.* The previous information would help you do this.

2. *Clean the spindle (Figure 3).* Use a soft cloth to remove any grit or dirt from the spindle. Note that the chuck is protected by a cloth to prevent nicks or burrs from tools laid out on it.

3. *Ring test the wheel (Figure 4).* As indicated in Unit 2, this is a safety precaution any time a wheel is mounted.

4. *Mount the wheel (Figure 5).* The wheel should fit snugly on the spindle, and the outside flange must be of the same size as the inner flange, as shown in Figure 4. This wheel has blotters attached, but if they were damaged or there were none attached, it would be necessary to get some new blotters. Flanges should also be checked occasionally for burrs or nicks and flatness. The flange is held by a nut.

5. *Tighten the nut (Figure 6).* The nut should be tightened snugly. The blotters will take up a little extra force, but if either flange is warped or otherwise out of flat, it is possible to crack a wheel. Overtightening can also crack a wheel.

6. *Replace the safety guard (Figure 7).* This is a necessary precaution for safety.

7. *Place the diamond for dressing the wheel (Figure 8).* The wheel, as noted by the arrow on the safety guard, revolves clockwise so that the correct placement of the diamond is a little past the bottom point of the wheel, at about 6:30. This type of dresser has the correct "drag angle" of 10 to 15 degrees built into it.

8. *Dress the wheel (Figure 9).* This operation is to dress the wheel for stock removal so that the diamond can be cross fed rapidly back and forth across the wheel. Coolant is used,

Figure 3. Cleaning the wheel spindle with a soft cloth (Lane Community College).

Figure 4. Ring testing the wheel (Lane Community College).

Figure 5. Mounting the wheel, flange, and nut (Lane Community College).

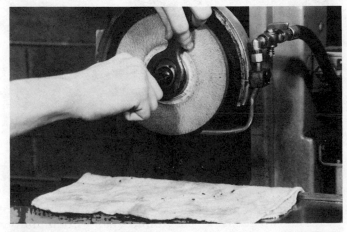

Figure 6. Tightening the nut with spanner wrenches (Lane Community College).

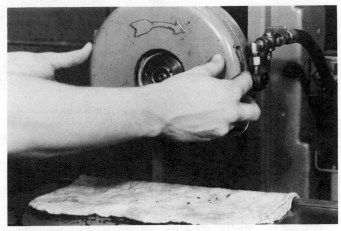

Figure 7. The safety guard being replaced (Lane Community College).

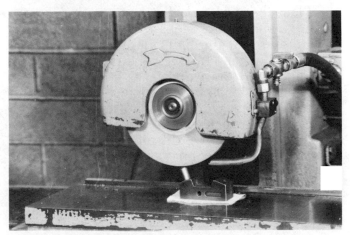

Figure 8. Diamond dresser in the correct position. The camera angle does not show the location of the diamond clearly (Lane Community College).

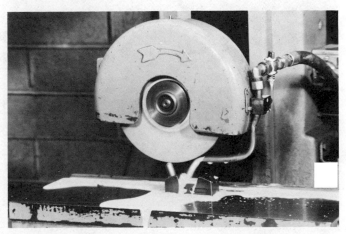

Figure 9. Dressing the wheel using coolant (Lane Community College).

since the grinding is done wet, but at greater volume than is shown in the illustration. The coolant was reduced for better visibility but, in actual dressing, the dresser should be either completely drenched with coolant or no coolant should be used at all.

9. *Wipe off the top of the chuck (Figure 10).* This is to remove any chips or bits of abrasive that may be on the chuck face. It is done with the wheel at a full stop and with the magnetic chuck turned off. A squeegee is often used before finishing the job with a shop cloth.

10. *Check the chuck for nicks and burrs (Figure 11).* The quickest and best way to do this is to run your hand over the face of the chuck. If the previous step has been done adequately, there should be no slivers or bits of metal to stick in your hand.

11. *Place the two blocks for the first grind (Figure 12).* Note that the sides with the large vees are up and that the blocks are placed near the center of the chuck. The paper protects the chuck and the workpieces from each other. For a single job like this, most operators tend to place the workpieces at the center of the chuck. However, if the grinder is in regular use, the paper would probably not be used and each group of parts would be spotted at a different place on the chuck, to equalize wear.

12. *Turn on the chuck (Figure 13).* Magnetic flux is applied on this chuck by moving a handle from left to right. On other chucks, such as

the electromagnetic type, magnetism is activated by an electrical switch.

13. *Downfeed the wheel close to the top of the vee block (Figure 14).* This is simply a matter of turning the downfeed (sometimes called the infeed) handwheel at the top of the column to lower the wheel to a point an inch or so above the workpieces.

14. *Position the workpieces for grinding (Figure*

Figure 11. Checking for nicks or burrs on the magnetic chuck (Lane Community College).

Figure 12. The rough blocks in place with the large vee side up ready to be ground (Lane Community College).

Figure 10. Cleaning the magnetic chuck with a cloth. The wheel must be completely stopped when this is done (Lane Community College).

Figure 13. Activating the magnetic chuck (Lane Community College).

Figure 14. Downfeeding the wheel (Lane Community College).

15). The large wheel, held by the operator's left hand, controls the left and right (traverse) movement of the table and chuck. The smaller wheel, held in the operator's right hand, cross feeds the table and chuck toward or away from the operator. During grinding the table is usually traversed hydraulically with the large handwheel disengaged. The large handwheel is mainly used for positioning work.

15. *Set the table stops (Figure 16).* Table stops are sometimes called trip dogs. Their purpose is to set the limits within which the table can travel, and these are usually about an inch off either end of the workpiece. All that is needed is to make sure that the wheel clears the end of the workpiece and allows time for cross feeding between traverse passes.

16. *Turn on the grinding wheel spindle.* Also turn on the hydraulic pump motion if your

Figure 15. Positioning the blocks to set the stroke length (Lane Community College).

grinder is so equipped. As a safety precaution, let the wheel run for a minute, taking care that you do not place yourself, or allow anyone else, to stand in line with the plane of rotation. The grinding wheel is then brought down close to the work (Figure 17). The high point on the work is sought by manually

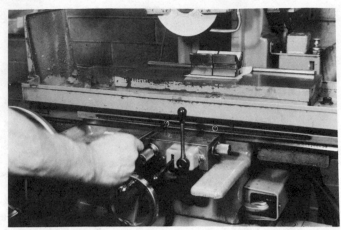

Figure 16. Setting the table stops (Lane Community College).

Figure 17. The wheel is turned on and, after running a minute for safety reasons, the rotating wheel is brought close to the workpiece (Lane Community College).

cross feeding during the table traverse to find the high spot. When it is found, set the downfeed dial to zero as a reference. Now turn on the traverse (Figure 18), if your machine is equipped with one.

17. *Turn on the coolant (Figure 19).* With the coolant flowing, downfeed about .002 in. With about .015 in. stock to be removed from each dimension and leaving about .003 to .005 in. for the finish cuts, remove about .010 to .012 in. from each of the sides and from the two ends. How much of this total comes off each side or end is usually left up to the operator's judgment, unless it is covered in a job sheet. In this case, perhaps removing a few thousandths of an inch for cleanup would be best until the squaring procedure is finished.

18. *Turn on the table crossfeed (Figure 20).* With power for both traverse and crossfeed, watch the wheel and listen for unusual sounds indicating overloading as it grinds. Downfeed .001 to .002 in. at the end of each complete pass across the two flat surfaces on either side of the large vees on both blocks.

19. *Clean up the opposite sides (Figure 21).* This step repeats steps 17 and 18 with the large vees turned down. Grinding should continue until the block is just cleaned up. (Steps 20 to 22 are done on each block singly.)

20. *Clamp either finished side of one block to a precision angle plate laid on its side on a surface plate (Figure 22).* Using a dial indicator and either a height or surface gage, clamp the vee block parallel to the surface plate. One end of the block should project

Figure 18. Starting the table traverse. The wheel is carefully adjusted to just touch the high point of the work (Lane Community College).

slightly beyond the angle plate so it may be ground. This is probably the most critical step in the procedure.

21. *Set up for grinding the end (Figure 23).* Return the angle plate and block to the grinder and place them on the magnetic chuck so that one end of the block is in position to be ground. Turn on the magnetic chuck and grind the end to clean up. The ground end should now be relatively square to the two ground sides.

22. *Grind one of the remaining sides square (Figure 24).* Use the same procedure as that explained in step 21, except that the end you have just ground on each block is clamped to

Figure 19. The coolant is turned on and an additional downfeed of .002 in. is made (Lane Community College).

Figure 22. A ground side of the vee block is clamped to a precision angle plate that is turned on its side and the top side of the vee block is being leveled with a dial indicator that is mounted on a height gage (Lane Community College).

Figure 20. Turning on the table crossfeed; both surfaces are ground on the large vee side (Lane Community College).

Figure 23. The precision angle plate and vee block setup is turned with the vee block end up on the magnetic chuck. The end of the vee block is ground square to a ground side (Lane Community College).

Figure 21. The blocks are turned over and the opposite sides (small vee) are ground (Lane Community College).

Figure 24. The other two sides being ground square to the vee block end that was previously ground (Lane Community College).

the angle plate. The side to be ground must be made parallel to the surface plate. The side to be ground must project above the end of the angle plate so there will be no interference.

23. *Grind the remaining ends at the same time (Figure 25).* Both blocks can be set on the magnetic chuck without additional support and ground separately. Grind to within .003 to .005 in. of finished dimension.

24. *Grind the remaining sides and the two cleaned up sides to .003 to .005 in. oversize (Figure 26).* This is essentially the same setup as for step 23. These steps can be done without the angle plate after two ajoining sides and one end are square with each other, because surfaces parallel to square surfaces are also square to each other.

25. *Redress the wheel for finish grinding.* This is essentially the same as step 8, except that the dresser is cross fed more slowly. Downfeed for the finish grinding should be about .0002 in., about one tenth of the downfeed used for rough grinding.

26. *Check all sides and ends for square (Figure 27).* This requires a precision cylindrical square and a .0001 in. reading dial indicator supported by a height gage on a surface plate. Correct any errors of squareness by touch-up grinding, using tissue paper shims under one side to make the faulty surface parallel with the magnetic chuck. Remember to put the tissue paper under the thick side.

27. *Check dimensions.* This can be done by using a ten-thousandths reading micrometer (Figure 28) or by using gage blocks.

28. *Grind sides and ends to finish dimensions.* If your work to this point has been done carefully, both blocks can be done at the same time, as in steps 23 and 24. Just be sure that you are grinding corresponding sides on each setup. No tissue paper is used in finish grinding.

29. *Grind one side of the large Vee (Figure 29).* Set up both blocks in a magnetic vee block, making sure that the magnetic vee block is aligned parallel with the magnetic chuck. Turn on the magnetic chuck. Turn on the

Figure 26. The sides that have not been ground are also being ground in one setup to make them parallel to the other sides (Lane Community College).

Figure 25. Grinding the opposite ends of both blocks in one setup to make them parallel (Lane Community College).

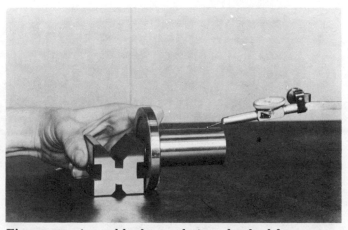

Figure 27. A vee block now being checked for square on all sides using a precision cylindrical square, a dial indicator, and a height gage on a surface plate (Lane Community College).

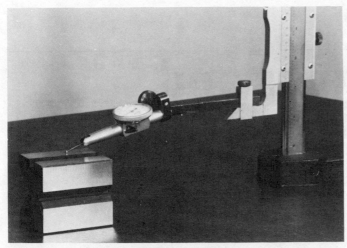

Figure 28. Dimensions being checked using a .0001 in. reading dial indicator on a height gage. Precision gage blocks may be used for comparison measurement for this operation (Lane Community College).

Figure 29. Setting up the magnetic vee block to grind the angular surfaces on the large vees (Lane Community College).

coolant and grind the side for cleanup. Use light downfeed.

30. *Reverse the vee blocks and repeat Step 29 for the other side of the large vee with the same setting (Figure 30). This will center the vee.* Check the large vee for square. The angle between the two sides of the large vee should be exactly 90 degrees. Repeat this procedure on the small vees on the opposite side.

31. *Clean up your machine.* Run the wheel for 5 min to remove coolant, which helps to maintain balance when the wheel is next used. Return the tooling to its proper place. Measure and record the finished dimensions of the vee block. Submit the vee block to your instructor for evaluation.

Figure 30. The vee blocks are repositioned and the other side of the large vees are being ground (Lane Community College).

self-evaluation

SELF-TEST Go to the surface grinder that you will be using and familiarize yourself with the controls and tooling. If someone is using the machine, observe the operation for a time. Check the supply of grinding wheels and determine their possible uses by observing their color and by reading the printed symbol on the blotter.

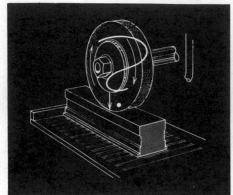

unit 7
problems and solutions in surface grinding

Producing quality work on a surface grinder bears some resemblance to driving a car. It is not too difficult when everything is going right. It is knowing what to do when something is not going right that separates the skilled from the unskilled. This unit is a discussion of some of the common problems of grinding, how to recognize them, and what to do about them. It should also give you some insight into whether you should try to do something about the problem yourself or whether you should report it to your superior or call a serviceman.

objectives After completing this unit, you will be able to:
1. Recognize common surface defects resulting from surface grinding, and suggest ways of correcting them.
2. Suggest ways of correcting situations that show up in postinspection such as work that is not flat, parallel, or to form.

Conditions causing problems in surface grinding can conveniently be divided into two groups.

1. *Mechanical.* Those problems having to do with the condition of the machine, for example, worn bearings; or those problems having to do with the electrical or hydraulic systems of the machine.
2. *Operational.* Those problems having to do with the operation of the machine; for example, the selection and condition of the grinding wheel, selection of coolant, wheel dressing, and other similar responsibilities.

Obviously, there is not a clear line that can be drawn between these two. The division of responsiblities varies between shops; in general, the mechanical condition of the machine, aside from routine daily lubrication, is a shop maintenance responsibility, while the operation of the machine is your responsibility as its operator. Still, there are some close decisions to be made; they are usually decided on the basis of shop policy or practice. For example, if the machine ways are not lubricated often enough, which causes the table to stick-slip as it traverses, shop policy or practice determines who is responsible.

Two other general observations are in order. One is that any surface grinder is limited in the degree of precision and the quality of surface finish that it will produce. A lightly built, inexpensive grinder simply will not produce the quality of finish nor the precision of a more heavily built and more expensive machine.

The second is that any machined surface, even the finest, most mirrorlike surface, is a series of scratches. It is true that on the finer finishes the scratches are finer, closer together, and follow a definite pattern, but there are scratches nonetheless, and they will show up if the surface is sufficiently magnified.

Although you have not studied surface finish in any detail in this section, it is easy to understand that on a reciprocating table surface grinder, the scratches will be parallel and running in the direction of table travel. Anything that differs significantly from this pattern can be considered a surface defect and a problem.

Dirt, heat, faulty wheel dressing, and vibration can cause problems, as indicated earlier. The condition of the wheel's grinding surface can also cause problems. A wheel whose surface is either loaded or glazed will not cut well or produce a good flat surface. *Loading* means that bits of the work material have become embedded in the wheel's face. This usually means that the wheel's structure should be more open. *Glazing* means that the wheel face has worn too smooth to cut. It may result from using a wheel with a grit that is too fine, a structure that is too dense, or a grade that is too hard.

OPERATOR'S RESPONSIBILITIES

In general, it can be said that you, as the operator, are responsible for the daily checking and running of the machine. This includes things such as selecting and dressing the wheel, selecting the coolant, checking to see that the coolant tank is full and the filters are working as they should, and that the coolant is flowing in sufficient volume. You should observe the lubrication of the machine to see whether there is too little or too much. You should be alert for signs that the wheel is not secure or if the bearings are beginning to wear too much. You should not only check the work as instructed, but you should also observe it for surface irregularities. Of course,

Table 1

Summary of Surface Grinding Defects and Possible Causes. Source: *Precision Surface Grinding*. Data courtesy of DoAll Company, 1964.

Causes	Burning or Checking	Burnishing of Work	Chatter Marks	Scratches on Work	Wheel Glazing	Wheel Loading	Work Not Flat	Work Out of Parallel	Work Sliding on Chuck
Machine Operation									
Dirty coolant				x		x			
Insufficient coolant	x						x	x	
Wrong coolant					x	x			
Dirty or burred chuck				x			x	x	
Inadequate blocking									x
Poor chuck loading							x	x	x
Sliding work off chuck				x					
Dull diamond					x				
Too fine dress	x				x	x	x		
Too long a grinding stroke								x	
Loose dirt under guard				x					
Grinding Wheel									
Too fine grain size	x				x	x			
Too dense structure					x	x			
Too hard grade	x	x	x		x	x	x		
Too soft grade			x	x					
Machine Adjustment									
Chuck out of line								x	
Loose or cracked diamond				x			x	x	
No magnetism									x
Vibration			x						
Condition of Work									
Heat treat stresses							x		
Thin							x		

you will not be working entirely on your own; much of this will be done with the advice and agreement of your instructor or supervisor, especially when you first begin to use the machine.

GOOD MACHINE CONDITION

Its an old axiom of grinding that you can not produce quality work on a machine in poor condition. It is true that you can compensate for worn bearings to some degree by substituting a softer wheel, but this is a trap to avoid if at all possible. Not that machine condition is the only factor to be considered, but it is probably the most important single factor in preventing problems.

One frequently hears of unusual causes of trouble, such as the story about a machine that had been performing satisfactorily for years on a given wheel specification. Then suddenly, with the same specification, the operator has nothing but problems. When the manufacturer checked, he found that the customer had moved the machine from the ground floor to the sixth floor, and the added vibration required a new specification. With a wheel that suited the new conditions, there were no further problems. Many problems are not so straightforward. Indeed, the ones that are most difficult are those where there is more than one condition causing the problem. However, in the discussion that follows, the best that can be done is to indicate that there is more than one condition that may cause a particular problem.

Before discussing specific problems, many of which are shown in Table 1, it is worth noting that good dressing practice is probably the best single way an operator can avoid problems. For example, if you cross feed the dresser too fast, you run the risk of causing distinctive spiral scratches on the surface of the workpiece. If you cross feed the diamond too slowly, the wheel is dressed too fine and the effect is similar to that resulting from grit size that is too fine or a wheel that is too hard. A wheel that is too hard also causes burning or burnishing of the work, or hollow spots because an area has become too hot, expanded, been ground off, and then contracted below the surface as it cooled.

Surface finish problems can occur if you have not dressed a small radius on the corner of the wheel, as recommended in Unit 2. The first corrective measure you should take, if you have surface finish problems, is to redress the wheel.

SURFACE FINISH DEFECTS

Surface defects usually show up as unwanted scratches in the finish. Figure 1 shows four slightly oversize (1.25X) pieces that have been selected to illustrate common problems of this type. Not all of those discussed in the following paragraphs lend themselves to illustration.

Chatter Marks

These are sometimes referred to as vibration marks (Figure 2) because vibration is usually the cause. The reason may be from some outside source such as a punch press operating nearby. and transmitting vibration through the machine to the wheel, so that the wheel "slips" instead of cutting for a moment. It may also come from within the machine, a wheel that is not balanced, or a wheel with one side soaked with coolant. It may be from worn wheel bearings, or even from a wheel that loads and/or glazes, and alternately slips, drops its load, and resumes cutting. This

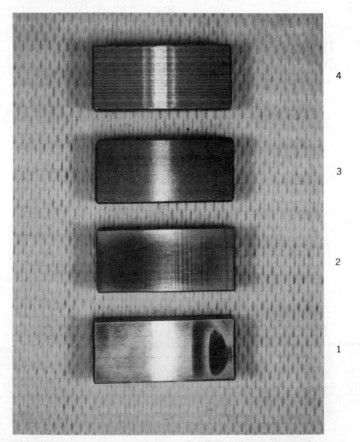

Figure 1. Four specimens of oil hardening steel, approximately 60 RC, with specific surface defects (Courtesy of Mark Drzewiecki, Surface Finishes, Inc.).

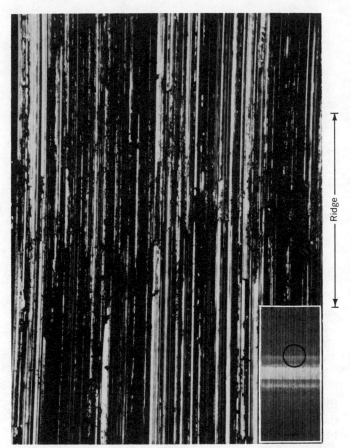

Figure 2. Chatter marks enlarged 80 ×. Marked inset (1.25 ×) shows area enlarged (Courtesy of Mark Drzewiecki, Surface Finishes, Inc.).

Figure 3. Grinding marks or "fishtails" also enlarged 80 ×. Inset (enlarged 1.25 ×) shows the damaged area (Courtesy of Mark Drzewiecki, Surface Finishes, Inc.).

slipping and cutting alternation usually produces chatter marks that are close together; the more irregular and wider spaced marks are likely to come from some other source. The remedy may be just redressing, but it could also require a change of grinding wheel, a check on the wheel bearings, or even, if none of these work, a relocation of the machine.

Irregular Scratches
These are random scratches (Figure 3), often called, for obvious reasons, "fishtails." Usually the problem is the recirculation of dirt, bits of abrasive or "tramp" metal in the coolant, or dirt falling from under the wheel guard. If there is not enough coolant, for instance, it may be recirculating too fast for the settling of swarf (grinding particles) to take place. A similar result can occur if you slide a workpiece off of a dirty chuck instead of lifting it off, which may be more difficult.

The first thing to do to remedy these problems is to clean out the inside of the wheel guard; then replace or clean the filters in the coolant tank and make sure there is enough coolant in the tank. It should be worthwhile to take some extra time to ensure that the chuck is clean before you load it again.

Discoloration
Discoloration, which is also known as burning or checking, results when a workpiece becomes overheated. It can be caused by insufficient coolant, improperly applied coolant, a wheel that is too hard or too fine, or by removing too much metal too rapidly from a small area (Figure 4). If carried on too long, it can result in expansion of the metal, probably in the middle of the surface being ground. Then, if the hump is ground off, there will be a low spot when the workpiece

cools to normal temperature. However, the burning may be considered in the first place as a surface defect that may or may not be a problem, depending on the final use of the workpiece.

Probably the first remedy is to try to get the wheel to *act* softer by speeding up table traverse, redressing the wheel rougher, or taking lighter cuts and, of course, by checking the supply and the application of the coolant. Whenever there is not enough coolant, it is recirculated before it has had a chance to cool off as it should, and so it simply becomes warmer and warmer.

Burnishing

A burnished surface (Figure 5) is one that is smoothed by abrasive rubbing instead of by cutting. It often looks good, and it may indeed not be a problem if the surface is not subject to wear. Technically, what happens when a surface is burnished is that the hills of the surface are heated enough so that they can be pushed into

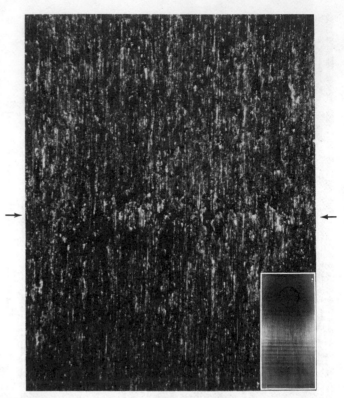

Figure 5. Burnished area enlarged 80 ×. Inset shows damaged area. (Courtesy of Mark Drzewiecki, Surface Finishes, Inc.).

the valleys of the scratches. But with any wear at all, the displaced metal breaks loose, and the surface suddenly becomes much rougher. It usually results from using a wheel that is too hard. For that reason, the usual remedy is to get a softer wheel or to change grinding conditions to make the wheel act softer by increasing work speed, redressing the wheel, or taking lighter cuts.

Miscellaneous Surface Defects

As mentioned earlier in this unit, any ground surface is a planned series of scratches, preferably of uniform depth and direction. On the grinder you have been studying, the scratches are parallel and at right angles to the direction of work traverse. Any other pattern, scratches without a pattern, or any discoloration of the work surface can be considered a defect. Some of the causes of these defects are that the wheel may be loose, the bearings may be worn, there may be vibration from some unsuspected source, the wheel may have been dressed too fast or too slow, or the wheel may be too rough or too fine. Sometimes the causes are so remote or obscure as to puzzle even experienced troubleshooters.

Figure 4. Discoloration or burning also enlarged 80 ×, with damaged area marked on inset (Courtesy of Mark Drzewiecki, Surface Finishes, Inc.).

Work Not Parallel

It has been said repeatedly that the grinding line of the wheel is parallel to the top of the chuck, and this is true if the grinder is in good condition. However, if the chuck is out of line, dirty, or burred, this parallelism may no longer exist, and any workpiece may be out of parallel. Of the three conditions mentioned above, dirt and burring of the chuck are definitely your responsibility, but the chuck alignment may or may not be. The remedies are usually obvious, as are those of other possible causes, as shown in Table 1. As you progress, you will develop a sort of routine to be followed for a given machine with this condition.

Work Out of Flat

Lack of flatness in a thick workpiece is likely to be the result of some local overheating, which causes an area of the surface to bulge. Then when the bulge is ground off and the work cools, there is a low spot.

Most flatness problems arise with thin work, and for very obvious reasons. Thin work does not have the bulk to absorb grinding heat without distortion. If it has been rolled, then stresses caused by the passage of the metal through the rolls may have been created, and the grinding may release these stresses on one side, causing the metal to warp or bow out of flat.

Correcting warpage in a thin workpiece is a matter of patience, a right start, and the minimum chuck power required to hold the workpiece in place. The procedure is as follows.

1. Using the least practical amount of chuck power, place the work on the chuck with the bowed side up so that the work rests on the ends.
2. Take a light cut with minimum downfeed. This should grind only the high spots on the work. Cutting should begin near the center.
3. Turn over the workpiece, shim the ends with paper, and take another light cut. This time cutting begins near the ends.
4. Repeat these steps, reducing the shims gradually, until the part is flat within specifications.

Finally, it should only have to be mentioned, as a reminder, that it is impossible to grind work flat if the chuck surface is not flat. When the chuck surface is between .0001 and .0002 in. out of flat, it is time to regrind it.

Perhaps the last word on solving the problems that arise in surface grinding is that care in avoiding the causes of trouble, such as careful and thorough cleaning of the chuck, frequent checking of the coolant level, the condition of the filters, and care in placing and removing workpieces from the chuck, can prevent many of the problems before they happen. It might be called preventive operation of the grinder.

self-evaluation

SELF-TEST
1. List at least three actions of an operator that will help prevent problems in surface grinding.
2. Define at least three general causes of surface grinding problems.
3. Name at least three general causes of surface grinding problems.
4. What is the principal cause of chatter marks? What are some possible remedies?
5. "Fishtails" are another common problem. What are they, what causes them, and what should you do to get rid of them?
6. Name two of the problems that can result from overheating work. What two or three things might you do if you suspect that a workpiece is getting too hot?
7. What is a burnished surface? Why is it objectionable? What are some remedies?

8. List at least two conditions that could produce out-of-parallel work.

9. List at least two conditions that could cause out-of-flat work.

10. Why is it difficult to grind thin work flat? How do you correct the condition?

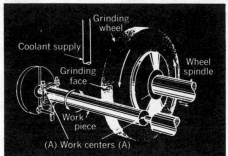

unit 8
cylindrical
grinders

Cylindrical grinding, along with surface grinding, which you have already studied, can be considered as one of the basic forms of grinding. Indeed, practically all forms of grinding with wheels or belts are some kind of variation of these two.

It was pointed out in the introduction to this section that there are several types of grinders for grinding the outside diameters of workpieces, and the internal grinder for grinding inside diameters.

The grinder that is the subject of this unit is called a universal grinder, more accurately called a universal center-type cylindrical grinder. Within its size range, it will do everything that any other center-type cylindrical grinder will do. Center type means that the workpiece is held on centers, like a lathe. The universal grinder is a toolroom-type machine, which means that it is very flexible and can do many jobs, although it does not do any one of them as efficiently as a more specialized machine. Hence, it is a very valuable machine to use to get acquainted with all the variations of center-type cylindrical grinding.

objectives After completing this unit, you will be able to:

1. Identify and explain the purpose of the major operating components of the center-type cylindrical grinder.
2. Describe the movements of the principal machine components.
3. Explain the major differences between a universal grinder, a plain grinder, an angle-head grinder, and a plunge grinder.
4. Explain when steadyrests are necessary, and how they are used.
5. List some of the major differences between surface grinding and center-type cylindrical grinding.

The center-type cylindrical grinder is the oldest type of grinder, and it is generally considered that the plain grinder, which has the grinding wheel moving toward or away from the operator on a fixed slide at right angles to the axis of the work, was the oldest. However, very shortly there were many more of the universal grinders with both the wheelhead and the table able to swivel, because there was more of a need for a finishing machine than for anything else. Furthermore, it is difficult to conceive a small 3 or 4 in. wheel, less than ½ in. thick, doing very much in the way of metal removal.

The earlier development of cylindrical grinding can probably be credited to two conditions. One was that the means for holding such work had already been pretty well developed in the lathe. In fact, the first cylindrical grinders were simply lathes with a grinding wheel mounted on the toolpost in place of the cutting tool; they were called grinding lathes. Another condition was probably the fact that the magnetic chuck had not yet been developed for the surface grinder, which made workholding more difficult.

DIFFERENCES BETWEEN SURFACE AND CYLINDRICAL GRINDING

Figure 1 shows a universal center-type cylindrical grinder. Figure 2 is a sketch of the general motions of cylindrical grinding. These provide sufficient background for some comparisons between surface grinding and cylindrical grinding.

For one thing, on a surface grinder the pressure of the wheel on the workpiece is always downward with the magnetic chuck providing the backup. The chuck obviously provides a much more stable platform for grinding than the two centers at either end of the cylindrical work-

Figure 1. Modern universal cylindrical grinder (Courtesy of Landis Tool Co., Division of Litton Industries).

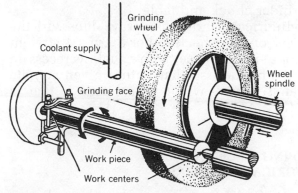

Figure 2. Sketch of center-type cylindrical grinder set up for traverse grinding. Note particularly the direction of travel of the grinding wheel and the workpiece, and the method of rotating the workpiece (Courtesy of Bay State Abrasives, Dresser Industries, Inc.).

piece, in which case the pressure is outward (toward the operator) and to some degree downward. The workpiece provides its own backup, except in the case of long, thin work when additional supports called steadyrests must be provided.

Another difference on a cylindrical grinder (Figure 2) is that the wheel-work contact area is a line at the intersection of the arc of the wheel and the smaller arc of the workpiece. Since the contact area in surface grinding is that of the arc of a wheel and a flat surface, it is slightly greater in area than that of the cylindrical grinder. The practical effect of this is that cylindrical grinding wheels tend to be just a little harder in grade than surface grinding wheels.

You will also notice in Figure 2 that the wheel cuts in only one direction (down) on the workpiece; in surface grinding the workpiece goes either with or against the direction of the abrasive cuts as it traverses.

Another significant point in terms of dimensions in cylindrical grinding is that the reduction in diameter is always approximately twice the infeed, or the reduction of the radius roughly equals the infeed. In surface grinding the reduction in the thickness of the workpiece always approximates the downfeed.

Cylindrical grinding is almost exclusively a wet operation, which aids in flushing away the swarf and dirt of grinding. This is helped by the force of gravity on the coolant. Thus, it is quite possible that the coolant on a cylindrical grinder does a better job of flushing away the grinding swarf than the same coolant would do on a surface grinder chuck and table.

Cylindrical grinding is always done with the periphery of the wheel, either a Type 1 straight wheel or a similar wheel with a shallow recess on one or both sides, so that with a given spindle you can secure a wider wheel face if that is needed.

GROUPING OF CENTER-TYPE CYLINDRICAL GRINDERS

As you will note from Figures 1 and 3, cylindrical grinders look pretty much alike from the outside. They vary in flexibility all the way from the universal (Figures 1 and 4) to the straight plunge grinder (Figure 6). On the universal, everything can swivel; on the straight plunge type, which is a specialized production grinder, the wheel moves only forward and back and the table may not even swivel. For these reasons, the plunge grinder is the most rigid, and the universal grinder is the least rigid.

Figure 3. Head-on view of a plain grinder (Courtesy of Landis Tool Co., Division of Litton Industries).

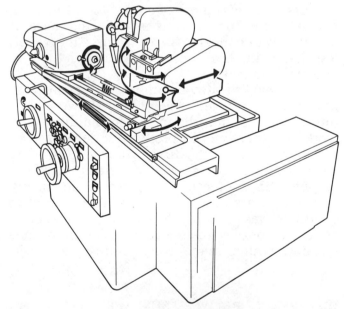

Figure 4. View of universal cylindrical grinder with arrows indicating the swiveling capabilities of the various major components (Courtesy of Cincinnati Milacron).

The type of grinding that can be done on various types of these grinders is shown in Figures 5 to 8. Traverse grinding (Figure 5) can be done on any grinder that has the ability to traverse the workpiece back and forth across the wheel face. It does not matter whether the surface is interrupted, as it is in the illustration, or continuous, as it usually is.

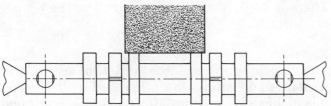

Figure 5. Sketch of traverse grinding with interrupted surfaces. Wheel should always be thick enough to span two surfaces or more at once (Courtesy of Cincinnati Milacron).

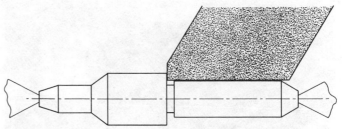

Figure 7. Angular plunge grinding with shoulder grinding. Note the dressing of the grinding wheel (Courtesy of Cincinnati Milacron).

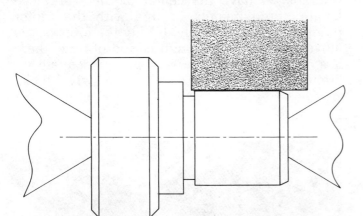

Figure 6. Straight plunge grinding, where the wheel is usually thicker than the length of the workpiece, except where the intent is to take a series of overlapping plunge cuts across a longer piece and finish with several traverses along the entire length (Courtesy of Cincinnati Milacron).

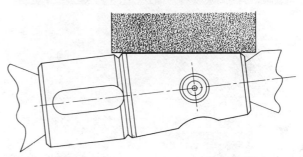

Figure 8. Taper grinding with the workpiece swiveled to the desired angle. For a steeper taper, the wheel might also have to be dressed at an angle less than 90 degrees (Courtesy of Cincinnati Milacron).

MAJOR PARTS OF THE GRINDER
Machine Bed
This is the base of the machine. It must be strong and heavy enough to support the moving elements of the grinder and rigid enough to reduce vibration to a minimum. A universal grinder has a swivel trunnion to support the wheelhead slide; on other types the slides are cast as part of the base. The base also contains ways for traversing the sliding table.

Sliding and Swivel Tables
The sliding table goes on top of the base to provide traverse motion of the work. The swivel table is mounted above the sliding table and supports both the headstock and the footstock, which actually hold the work. At one end of the table is a calibrated plate to indicate the taper both in degrees and in inches of taper per foot. Occasionally, it is desirable to make an independent check of the settings to make sure you are getting a true reading.

Headstock
The headstock (Figure 9) is generally at the left end as you face the grinder. It rotates the work

Straight plunge grinding (Figure 6) is done with a straight wheel; the only requirement, with one exception, is that the wheel be wider than the work is long. The exception is on a nominal traverse grinding job. The procedure is to make a succession of overlapping plunge cuts to produce a workpiece that is slightly oversize, and then finish it with a series of light traverse cuts. In many cases this is a faster way of doing what would normally be a traverse grinding job.

Angular plunge grinding (Figure 7) is similar, except that it uses a wheelhead at a 45 degree angle with the face dressed to the same angle. As the illustration suggests, and as you learned in Unit 7, this is an efficient and safe way of grinding a diameter and an adjoining shoulder.

Swiveling the table, as shown in Figure 8, is one way of grinding a taper or of straightening a tapered piece. This will be discussed at greater length under workholding.

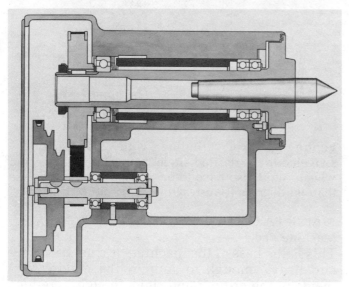

Figure 9. Typical headstock of a center-type cylindrical grinder with cutaway sketch. Note the size and complexity of the headstock in comparison with the footstock in the following figure (Courtesy of Landis Tool Co., Division of Litton Industries).

and supports one end of the work, ensuring precise alignment. Universal grinder headstocks usually have a combination live (revolving) and dead (not revolving) spindle nose. There is usually a selector lever to change from one to the other. If you select the live spindle drive, the entire spindle revolves and either a chuck or a faceplate can be mounted for work not held between centers. However, the bulk of work done on any center-type grinder is done with a dead spindle with a center that does not revolve and a workdriver plate that does. The plate has a rod or hook projection that pushes against a work driving dog clamped to the end of the workpiece.

There is usually a means of selecting the proper work speed.

Footstock

The footstock (Figure 10a and 10b) essentially supports and aligns the other end of the workpiece. Besides support, it is usually equipped with a lever, as shown, to retract the footstock center for loading or unloading. It is also smaller and easier to move than the headstock, making it convenient to move the footstock to adjust for different length workpieces. However, it is also desirable to have the center of the workpiece length approximately in line with the center point of the swivel table. If only the footstock is moved over a long enough period of time, there is a possibility of extra wear and consequent vibration on the footstock end of the swivel table.

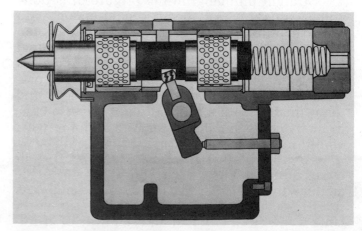

Figure 10. Typical footstock. The lever on top of the footstock retracts the work center so that the workpiece can be mounted on the grinder. The spring (right end of sketch) provides tension to hold the workpiece in place (Courtesy of Landis Tool Co., Division of Litton Industries).

Both of these conditions make it desirable to move the headstock occasionally instead of, or in addition to, moving the footstock.

Wheelhead

The wheelhead contains the grinding wheel and its driving mechanism. Both solid construction and perfect balance are essential, because any vibration from neglect in either case will be directly transmitted to the work as chatter marks.

Grinding Machine Controls

The controls for a universal center-type cylindrical grinder are quite similar in function, although they may be placed somewhat differently on different makes of machines. The controls on one make of machine will be discussed and illustrated in Unit 11.

WORKHOLDING

The basic principle of workholding on a center-type cylindrical grinder is simple. The workpiece is supported on two conical points (work centers) inserted into two matching conical center holes in either end of the blank cylinder. The work centers are supported by the headstock, which supplies the power to rotate the work, and by the footstock, which simply supports the other end. The headstock and the footstock are supported on the swivel table, which can be swung either one way or the other in order to create a tapered product or to straighten out the taper of a finished piece that is supposed to be straight. If the taper is steeper than can be produced by swiveling the table, then (on a universal grinder) the wheelhead can be swiveled. When it becomes too steep to be produced in that fashion, the face of the wheel can be dressed to whatever angle is needed.

The importance of the correct location and grinding of center holes (Figures 11, 12, and 13) cannot be too strongly emphasized. Together with painstaking care and lubrication of the work centers, this is the key to precise cylindrical grinding. No workpiece can be any better than the location and the proper angling of its center holes.

WORK POSSIBLE ON A UNIVERSAL GRINDER

Figures 14 to 21 represent a series of illustrations of the kinds of work that can be done on a universal grinder. The series starts with simple traverse

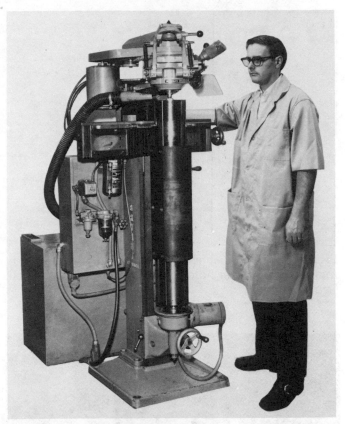

Figure 11. Center hole grinding machine (Courtesy of Bryant Grinder Corp.).

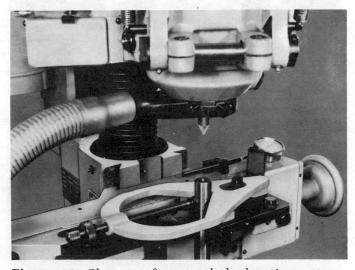

Figure 12. Closeup of center hole locating setup. Exact location of the center is a most critical step in the operation (Courtesy of Bryant Grinder Corp.).

grinding (Figure 14), progresses through OD taper grinding (Figure 15), and traverse grinding with a backrest or steadyrest (Figure 16). You will note that the workpiece in Figure 16 is long

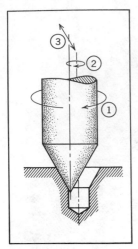

Figure 13. Sketch showing the motions of the grinding wheel in center hole grinding (Courtesy of Bryant Grinder Corp.).

Figure 15. OD taper grinding, which may be done to produce a tapered finished workpiece or to straighten up a rough piece that was previously tapered (Courtesy of Cincinnati Milacron).

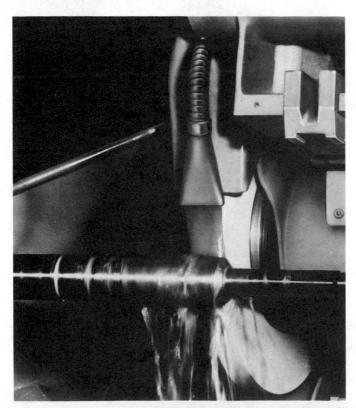

Figure 14. Traverse grinding, which is probably the most common type of cylindrical grinding (Courtesy of Cincinnati Milacron).

Figure 16. Backrest or steadyrest used to support a thin piece for grinding. If only one is used, as here, it is in the center of the workpiece. If more are used, it is always an odd number, and they are equally spaced along the workpiece (Courtesy of Cincinnati Milacron).

and thin in comparison with most of the others shown.

Steep taper grinding (Figure 17) involves swiveling the wheelhead as well as the table. Multiple diameter grinding, which can be done

by dressing the wheel in procedures like those discussed in Unit 7, is illustrated in Figure 18. Figure 19 shows simultaneous grinding of a

Figure 17. Steep taper grinding. Here the wheelhead has been swiveled to grind the workpiece taper (Courtesy of Cincinnati Milacron).

Figure 19. Angular-shoulder grinding. This is very often a production-type operation, but it is shown here on a universal grinder (Courtesy of Cincinnati Milacron).

Figure 18. Multiple-diameter or form grinding. This is usually a plunge grinding operation with the form dressed in the wheel face in the same manner as for reciprocal surface grinding (Courtesy of Cincinnati Milacron).

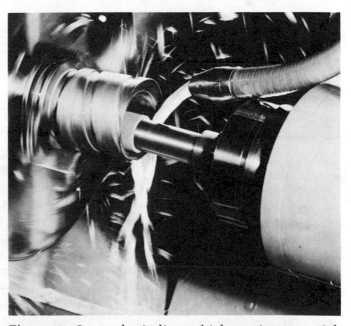

Figure 20. Internal grinding, which requires a special high speed attachment that is mounted on the wheelhead so that it can be swung up out of the way when not in use. It also requires either a chuck or a face plate on the headstock (Courtesy of Cincinnati Milacron).

diameter and a shoulder. Figures 20 and 21 illustrate internal grinding.

These are not necessarily the most efficient ways of doing these jobs. The sequence is only

Figure 21. ID taper grinding. This is the same sort of operation as shown in Figure 20, but with the workpiece swiveled at the required angle (Courtesy of Cincinnati Milacron).

designed to illustrate the range of the machine. On the other hand, it does emphasize that the universal grinder is very flexible, and that a shop equipped with a universal grinder and a small surface grinder can handle a great variety of work.

ATTACHMENTS AND ACCESSORIES

Accessories for the universal grinder can be grouped as those for headstocks, those for table mounting, and all others designed for a variety of uses. For the headstock, the units are generally three- or four-jawed chucks for mounting the small percentage of parts for which drilling and grinding a center hole would be impractical or impossible.

For the table-mounted group, there are diamond dressers, either for straight or contour dressing, that have the same purpose but a different mounting as the form dressers. Others, like the center rest, can supply extra support for very thin work, and the dial indicator for table swivel adjustment simply provides greater accuracy in tapering or removing taper than the standard machine.

self-evaluation

SELF-TEST   1. Explain the basic difference in construction between the universal center-type grinder and other types of center-type cylindrical grinders.
2. Name the most important reason that the center-type cylindrical grinder was the first to be developed.
3. List at least three differences between center-type cylindrical grinding and surface grinding.
4. List and discuss in a sentence or two the different setups for grinding cylindrical tapers.
5. What would be the major factors to consider in setups to be used?
6. How do the functions of the headstock differ from those of the tailstock? Which is moved more often? Why?
7. Describe the basic method by which a workpiece is held in a center-type cylindrical grinder.
8. What is a steadyrest and when is it used? List some of the factors that determine how many will be used on a given workpiece.
9. What is the relationship between the variety of work possible on a center-type cylindrical grinder and the rigidity of the grinder?
10. Why are the center holes so important in this type of cylindrical grinding?

unit 9
using the cylindrical grinder

The workpiece selected for the cylindrical grinder in this unit is a lathe mandrel, slightly tapered, which requires the application of many of the techniques that you studied in the previous unit. The material is SAE 4140 or a similar alloy steel, which is a widely used metal. The unit includes a step-by-step procedure for grinding a lathe mandrel. Some of the steps are not discussed because they are sufficiently clear from the illustrations.

objectives After completing this unit, you will be able to:

1. **Grind the tapered mandrel of SAE 4140 or similar alloy steel on a universal center-type cylindrical grinder to your instructor's satisfaction.**
2. **Explain any part of the operation to your instructor on request.**
3. **Prepare a worksheet for and actually grind a similar tapered part involving center-type cylindrical grinding to your instructor's satisfaction.**

The part selected for this demonstration project is sketched in Figure 1 and shows the necessary dimensions. In this unit it is also assumed that you have mastered the machining skills necessary to machine the part to allow about .025 to .030 in. oversize for grinding.

The selection of the grinding wheel, which was discussed in Unit 8, will only be reviewed here, since the only difference is that this is OD grinding instead of flat surface grinding. Hence, it will be a general purpose wheel of friable or semifriable aluminum oxide (because it is for grinding steel), a little finer (60 or 54 instead of 46) and definitely harder grit (K, L, or M grade instead of H), a shade more dense in structure (if this is mentioned, as it is frequently not mentioned in such wheels), and with a general purpose vitrified bond.

The testing and mounting of the wheel

(steps 2 to 6 of Unit 8) are very similar, and the work centers on the headstock and the footstock should be wiped off with a clean, soft cloth.

STEPS IN GRINDING

1. *Set up and adjust the diamond for dressing (Figure 2).* The footstock work center in Figure 2 provides a reference point, but there are times in actual practice when it would be desirable to move the footstock out of the way after the diamond has been located in relation to the wheel.
2. *Turn on the coolant and dress the wheel (Figure 3).* The first dressing of a wheel in a situation like this is usually for rough grinding for stock removal, and the diamond is traversed back and forth across the wheel rather rapidly.

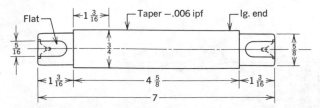

Figure 1. Sketch of tapered lathe mandrel.

Figure 2. Setting up the diamond for wheel dressing (Lane Community College).

3. *Set up the parallel test bar (Figure 4).* The test bar is mounted between centers and a dial indicator with a magnetic base is mounted at a convenient spot on the wheelhead. The indictator is zeroed at the right side of the test bar. (This is at the operator's left.)

4. *Read the indicator on the other end, 12 inches from the original position (Figure 5).* For the required taper of .006 IPF, the indicator should read .003 in. counterclockwise on the radius. You will remember that in one revolution of the workpiece the wheel removes about *twice* the amount of infeed of the wheel.

5. *Adjust the swivel table (Figure 6).* In a sense this is a sort of interim step with steps 3 and 4. This is the adjustment that has to be made to obtain the two readings called for in those steps. The scale gives an approximation, but you should follow the readings from the dial indicator.

6. *Lubricate the mandrel ends (Figures 7 a and b).* The center hole at the ends of each workpiece must be lubricated with center lube, since both the headstock and the footstock centers are dead; that is, they do not rotate. The work dog should also be mounted on the mandrel at this point.

Figure 3. Dressing the wheel with the coolant on (Lane Community College).

Figure 4. Setting up the parallel test bar (Lane Community College).

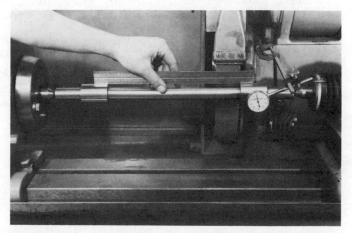

Figure 5. Second reading of the dial indicator (Lane Community College).

Figure 6. Adjusting the swivel table (Lane Community College).

Figure 7. Lubricating the center holes in the two ends of the workpiece (Lane Community College).

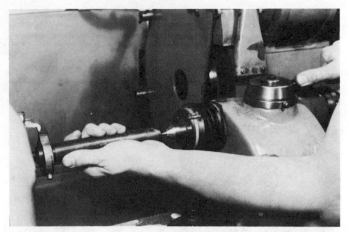

Figure 8. Clamping the footstock (Lane Community College).

Figure 9. Checking the work dog clearance and drive pin contact (Lane Community College).

7. *Clamp the footstock to the table (Figure 8).* The distance between the headstock and the footstock should be slightly less than the workpiece length, so that when the workpiece is mounted on centers there will be a little tension on it from the slightly compressed spring in the footstock.

8. *Check the work dog (Figure 9).* The work dog on the workpiece must have clearance to turn freely without binding. Also, it must contact the drive pin. This combination is what rotates the workpiece.

9. *Set stops for wheel overrun (Figure 10).* To make sure that the ends of the workpiece are ground, the wheel must definitely travel a little farther than the end of the work. About one third of the wheel thickness is recommended.

Figure 10. Adjusting the wheel overrun (Lane Community College).

Figure 12. Adjusting the wheel to the workpiece (Lane Community College).

Figure 11. Checking wheel-work dog clearance (Lane Community College).

Figure 13. Turning on the wheel (Lane Community College).

10. *Check wheel-work dog clearance (Figure 11).* Obviously, the wheel should not run into the work dog on the headstock end. On short pieces with this kind of grinder, clearance may be a problem. Incidentally, this illustration shows the best view of the contact between the work dog and the drive pin mentioned in Step 8.

11. *Adjust the wheel to the work (Figure 12).* Remove the workpiece and advance the wheel, backing it off if necessary to assure that it will clear the workpiece. Replace the workpiece. The wheel is at what will be the smaller end of the mandrel.

12. *Turn on the wheel (Figure 13).* For safety's sake, stand out of line with the wheel for a minute or so. After starting the wheel, feed it in close to the workpiece.

13. *Turn on the spindle (Figure 14).* This starts the workpiece rotating. Set the work speed. Infeed the wheel slowly toward the work until sparks begin to show.

14. *Turn on the coolant (Figure 15).*

15. *Turn on the table feed and adjust the speed (Figure 16).* For roughing, the traverse rate will be about one-half to two-thirds of the wheel width for each revolution of the work. Wheel infeed should be about .001 to .002 in. Take a roughing cut or two.

16. *Check taper (Figures 17 a and b).* Retract the wheel and make two marks 4 in. apart. Measure the differences in *diameter* with a micrometer to determine whether the taper per foot is correct. The difference in a 4 in. length should be .002 in. Adjust the swivel table as necessary.

Figure 14. Turning on the spindle (Lane Community College).

Figure 16. Adjusting the table speed (Lane Community College).

Figure 15. Turning on the coolant (Lane Community College).

Figure 17. Checking the taper (Lane Community College).

17. *Rough and finish grind the mandrel to size* (Figures 18 and 19). If your grinder has an automatic infeed, set it. If the setting is properly done, the machine will stop when the workpiece reaches finished size. If the grinder does not have an automatic infeed, calculate approximately the number of passes you will need, figuring about .001 to .002 in. for each roughing infeed (which will reduce the diameter by .002 to .004 in.) and about .0002 in. per finishing pass (which will reduce the diameter about .0004 in.).

Stop and check diameter and taper periodically. The main point to avoid is grinding undersize, because then you have only scrap.

Traverse rate in finish grinding is no more than ½ in. per work revolution, and the work speed ranges up to 100 SFPM. On a center-type cylindrical grinder, the wheel, at 12, 14, or 16 in. in diameter, may wear enough that wheel speed

Figure 18. Grinding the mandrel to size (Lane Community College).

Figure 19. The finished mandrel (Lane Community College).

(SFPM) will be affected. If this happens, it may be advisable to reduce the work speed, also. This will be further discussed in Unit 12.

Clean up the machine and return the tooling to proper storage. Show the finished mandrel to your instructor for his evaluation.

self-evaluation

SELF-TEST Your instructor will assign you a center-type grinding project similar to the one you have just studied, which will involve setting up the machine from scratch, perhaps setting and checking a taper, starting up the grinder, and roughing and finishing the workpiece.

Completing the test satisfactorily will involve writing a worksheet indicating the steps needed to complete the job and grinding the workpiece to specifications.

unit 10
cutter and tool grinder

By now, you are well aware of the variety of cutting tools used in the machine shop. Like any cutting tools these become dull or occasionally broken during normal use. No production machine shop could function at peak efficiency unless it can keep all of its cutting tools sharp. The cutter and tool grinder is an essential tool for this purpose. In a large machine shop several cutter and tool grinders will be kept continuously busy sharpening and reconditioning a variety of cutting tools. At least one tool and cutter grinder is almost essential to any machine shop that expects its machinists to turn out quality workmanship.

The operation of the cutter and tool grinder in a large machine shop is often delegated to specialists in tool grinding. These individuals may be assigned to the shop's tool making and tool grinding department. Even though tool grinding is a somewhat specialized area of machining, any well-rounded machinist should be familiar with the machines and processes used.

objectives After completing this unit, you will be able to:
1. Identify a cutter and tool grinder.
2. Briefly describe the function of this machine tool.
3. Under guidance from your instructor, sharpen common cutting tools.

The cutter and tool grinder may well be the most versatile grinding machine made. It is principally used for the sharpening of rotary cutting tools such as reamers and milling cutters but, with various accessories, it can do a remarkable number of grinding operations. These include internal and external cylindrical and taper grinding, surface grinding, cutting off operations and single point tool grinding. The standard machine (Figure 1) comes with a variety of components as standard equipment, including a workhead that may be swiveled on two axes and tailstock centers for mounting tools that have centers in each end. Also supplied are wheel guards, tooth rests, supports, and mounting equipment for positioning tooth rests.

The cutter and tool grinder may also be equipped with power table traverse and with a motor drive for the workhead, called a cylindrical grinding attachment (Figure 2). This combination makes internal and external cylindrical grinding easy to do. The motor driven workhead

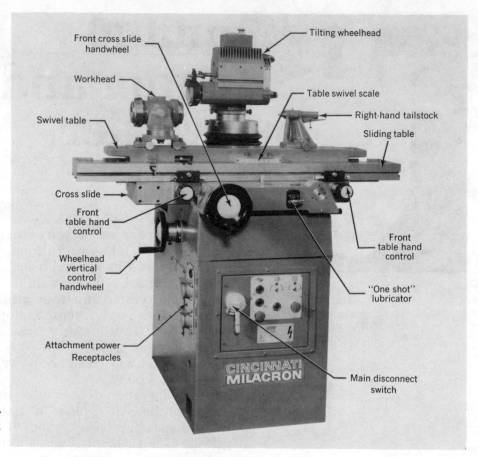

Tilting wheelhead

Front cross slide
handwheel

Workhead

Table swivel scale

Swivel table

Right-hand tailstock

Sliding table

Cross slide

Front
table hand
control

Front
table hand
control

Wheelhead
vertical
control
handwheel

"One shot"
lubricator

Attachment power
Receptacles

CINCINNATI
MILACRON

Main disconnect
switch

Figure 1. Components of the cutter and tool grinder (Courtesy of Cincinnati Milacron).

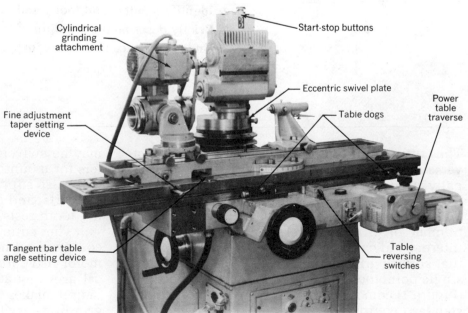

Cylindrical
grinding
attachment

Start-stop buttons

Eccentric swivel plate

Fine adjustment
taper setting
device

Table dogs

Power
table
traverse

Tangent bar table
angle setting device

Table
reversing
switches

Figure 2. Cutter and tool grinder equipped with power table traverse and with a cylindrical grinding attachment mounted on the workhead (Industrial Plastics Products, Inc.).

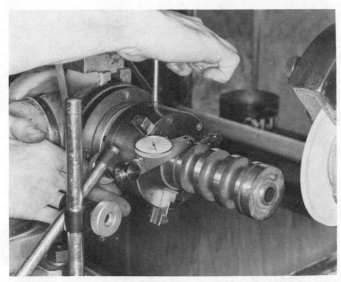

Figure 3. Cylindrical grinding attachment being used with adjustable scroll chuck to set up part for cylindrical grinding (Industrial Plastics Products, Inc.).

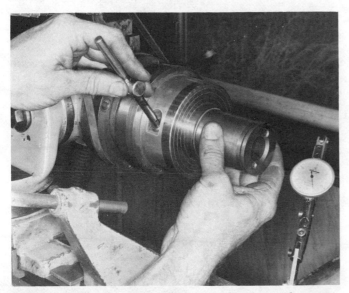

Figure 4. A permanent magnet chuck is also very useful for cylindrical grinding (Industrial Plastics Products, Inc.).

Figure 5. Internal grinding attachment set up for use with a mounted grinding wheel. Only the forward portion of the wheel is used; the rest is dressed away for clearance (Industrial Plastics Products, Inc.).

Figure 6. Another use for the cylindrical grinding attachment is the reconditioning of centers (Industrial Plastics Products, Inc.).

can be fitted with a scroll chuck with an adjustable backing plate that can be used for setting up work for cylindrical grinding (Figure 3).

A magnetic chuck (Figure 4) is also very useful as a cylindrical grinding attachment accessory, especially when external cylindrical grinding is preceded or followed by internal grinding (Figure 5). The cylindrical grinding attachment also makes it easy to recondition centers (Figure 6). This is done to restore centers for other machine tools such as lathes, but it is also done, when necessary, before mounting work between centers to ensure accuracy when parts are cylindrically ground on centers (Figure 7). The workhead is also equipped with a lock so that the center in the workhead may be held stationary (as a "dead" center) while the workpiece is driven, in the same way that it is done on regular universal and plain cylindrical grinders.

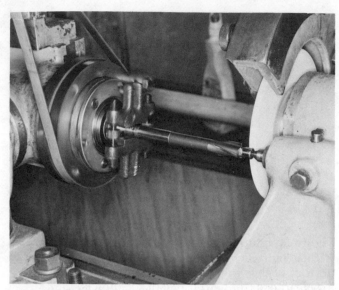

Figure 7. Resizing the pilot portion of a counterbore by cylindrical grinding (Industrial Plastics Products, Inc.).

Figure 8. Tailstock centers are basic to a large portion of cutter grinding (Industrial Plastics Products, Inc.).

Figure 9. Cutter grinding arbor and components (Lane Community College).

Figure 10. Slitting saw being mounted on grinding arbor (Lane Community College).

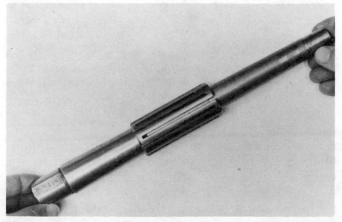

Figure 11. Adjustable grinding mandrel. The slotted bushing is moved along the mandrel to adjust for the ID of the tool to be mounted (Lane Community College).

The greatest amount of work on the cutter and tool grinder is done between tailstock centers (Figure 8) or, as will be seen later, from a manually operated workhead. Cutters such as formed cutters and plain milling cutters are usually mounted between centers on accurately made hardened and ground cutter grinding arbors (Figure 9) with precise centers in each end. This type of arbor can also be used for mounting slitting saws for resharpening (Figure 10).

Another type of holding device that can accommodate variations in internal diameter is the adjustable mandrel (Figure 11). This type of

Figure 12. Various designs of tooth rest blades (Lane Community College).

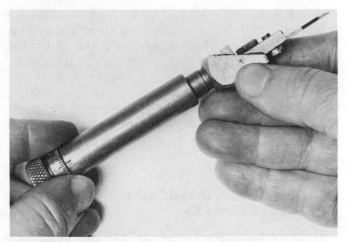

Figure 14. Micrometer-type tooth rest support. This example also has provision for spring loading of the finger to permit ratcheting of the cutter tooth, called a "flicker finger."

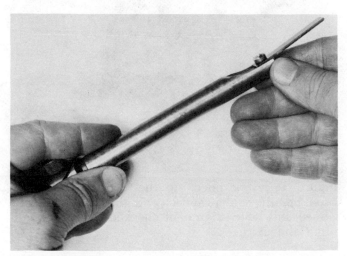

Figure 13. Plain-type tooth rest support (Lane Community College).

Figure 15. Offset tooth rest blade for use in grinding helical milling cutters.

holding device requires special care to get accurate results.

Most of the cutter grinding operations require a support for the tooth being ground. This is usually done by mounting specially shaped tooth rest blades (Figure 12) in either plain tooth rest supports (Figure 13) or in a support with a micrometer adjustment (Figure 14). The micrometer adjustment is especially useful for making small adjustments in setup and, as will be seen later, as a feeding device for finish grinding of form relieved tools.

For grinding helical milling cutters, an offset tooth rest (Figure 15) is obtained or made up to

match the helix direction and angle of the cutter to be sharpened. Other shapes of tooth rests will be seen in action later.

GENERAL SETUP PROCEDURES FOR MOST CUTTER SHARPENING

After centers are mounted and secured at the correct distance for the arbor to be mounted, it is important to check grinding arbors both for runout on centers and, in the case of cylindrical cutters like plain milling cutters, the table alignment should also be checked (Figures 16a and

16b). If the table is not aligned, it should be adjusted until the indicator hand remains stationary as the table is tracked past the indicator tip. The table should then be locked in alignment (Figure 17) and reindicated to insure that it did not shift during the securing process. When using tailstock centers, it is desirable to position a T-bolt between the centers suitable for mounting either a tooth rest plate or a dressing diamond.

ESTABLISHING AND CHECKING CUTTER CLEARANCES

For tools that are sharpened on the periphery (profile ground tools) such as plain milling cutters, stagger tooth cutters, and saws, it is necessary to establish the needed relief and clearance angles to match the cutting requirements (Figure 18). For sharpening form relieved tools, like gear milling cutters, where form and relief are manufactured into the tool, the sharpening takes place on the face of the tool, and special procedures are used that will be covered later.

Correct reliefs and clearances are of critical importance to efficient cutting, satisfactory tool life, and to avoid cutter breakage (Figure 19).

For profile ground cutters, the following angles are recommended (Table 1) for the outside diameter (peripheral clearance).

MEASURING OF CUTTER CLEARANCE ANGLES

One method of checking peripheral relief and clearance angles may be easily done right on the machine while the tool is still between centers. This is called the **indicator drop** method (Figure 20). By using a table that takes into account radial relief angle, land width, and cutter diameter, the amount of drop can easily be determined (Table 2).

The cutter illustrated in Figure 21 is 3 in. OD. The radial movement is approximately .028 in. and the indicator drop approximately .003 in. The table shows that this tool has a primary radial relief angle midway between 5 and 8 degrees, or about $6\frac{1}{2}$ degrees.

This type of checking is done mainly for demonstration purposes, since it would not be economical to do this kind of checking where high productivity is critical. A much faster and more convenient method for checking clearances than the indicator drop method is the use of a cutter clearance gage (Figure 22).

Figure 16. After checking the grinding arbor for runout on the centers at each end, the bezel should be zeroed and table alignment checked (Lane Community College).

Figure 17. Adjusting and locking the swivel table in alignment (Lane Community College).

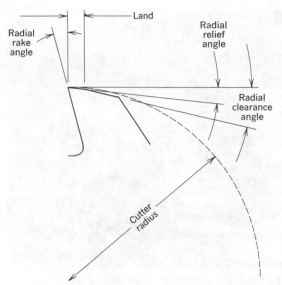

Figure 18. Relief and clearance angles.

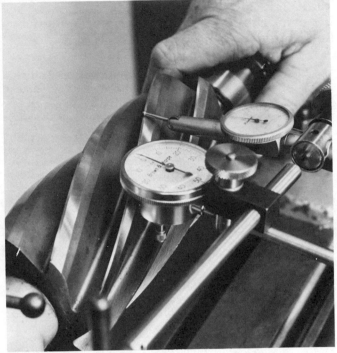

Figure 20. Setting up indicator for the indicator drop method of checking clearances (Lane Community College).

For example, if you wish a 5 degree primary relief angle on a 3 in. OD cutter, then:

$$\text{Sine of 5 degrees} =$$
$$.08715 \times 1.5 \text{ (cutter radius)} = .131 \text{ in.}$$

This is the distance to **lower** the wheelhead to obtain this angle.

Figure 19. This cutter was weakened by incorrect excessive clearance.

Most experienced cutter grinding specialists use a chart (based on the following formula) for guidance in setting relief and clearances. For flat grinding of relief or clearance with a flaring cup wheel, the amount to lower the wheel with the tooth rest attached to the wheelhead to obtain a specified primary relief angle is:

Sine of the desired angle × radius of the cutter

Table 1
Peripheral (OD) Relief and Clearance Angles Suggested for High Speed Steel Cutters

Material	Primary Relief, Degrees	Secondary Clearance, Degrees
Carbon steels	3—5	8—10
Gray cast iron	4—7	9—12
Bronze	4—7	7—12
Brasses and other copper alloys	5—8	10—13
Stainless steels	5—7	11—15
Titanium	8—12	14—18
Aluminum and magnesium alloys	10—12	15—17

Table 2

Indicator Drop Method of Determining Radial Relief Angles

Diameter of Cutter, Inches	Average Range of Radial Relief, Degrees	Indicator Drop for Range of Radial Relief Shown		Radial Movement for Checking
		Minimum	Maximum	
$\frac{1}{16}$	20–25	.0018	.0027	.010
$\frac{1}{8}$	15–19	.0021	.0032	.015
$\frac{3}{16}$	12–16	.0020	.0034	.020
$\frac{1}{4}$	10–14	.0019	.0033	.020
$\frac{5}{16}$	10–13	.0020	.0033	.020
$\frac{7}{16}$	9–12	.0025	.0038	.025
$\frac{1}{2}$	9–12	.0027	.0040	.025
$\frac{5}{8}$	8–11	.0028	.0045	$\frac{1}{32}$
$\frac{7}{8}$	8–11	.0033	.0049	$\frac{1}{32}$
1	7–10	.0028	.0045	$\frac{1}{32}$
$1\frac{1}{4}$	6–9	.0025	.0042	$\frac{1}{32}$
$1\frac{1}{2}$	6–9	.0026	.0043	$\frac{1}{32}$
$1\frac{3}{4}$	6–9	.0027	.0044	$\frac{1}{32}$
2	6–9	.0028	.0045	$\frac{1}{32}$
$2\frac{1}{2}$	5–8	.0024	.0040	$\frac{1}{32}$
3	5–8	.0024	.0041	$\frac{1}{32}$
4	5–8	.0025	.0042	$\frac{1}{32}$
5	4–7	.0020	.0037	$\frac{1}{32}$
6	4–7	.0021	.0037	$\frac{1}{32}$
8	4–7	.0021	.0037	$\frac{1}{32}$

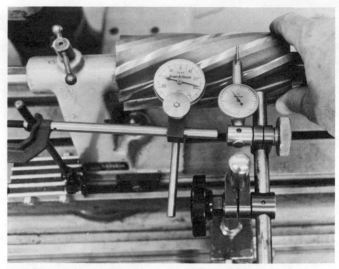

Figure 21. Checking the primary radial relief by the indicator drop method (Lane Community College).

The additional drop for the secondary clearance is done in the same way. If you are seeking a secondary clearance angle of 9 degrees, then lower the wheelhead and its attached tooth rest an additional 4 degrees.

Sine of 4 degrees = .06976 × cutter radius = .06976 × 1.5 = .105 in.

This is the additional distance for the secondary clearance.

If you have a grinding machine with a tilting head, all that is necessary is to tilt the spindle for the required primary relief and secondary clearance angles when using a flaring cup wheel. In this case, the cutter tooth tip remains level instead of being lowered as in the previous flaring cup or flat grinding method.

Another method of sharpening cutters on the periphery is by using a relatively large diameter straight (Type 1) grinding wheel (Figure 23). Using this type of wheel produces a slight hollow grind (Figure 24) behind the cutting edge. The larger the wheel, of course, the less the amount of hollow. This method is popular with cutters sharpening job shops; with the larger wheel there

Figure 22. Checking the primary relief of a stagger tooth milling cutter with a Starrett Cutter clearance gage (K & M Tool, Inc.).

Figure 23. A plain grinding wheel can be used for cutter sharpening (K & M Tool, Inc.).

Figure 24. Tooth form produced by hollow grinding with a relatively large grinding wheel.

is more available abrasive for cutting without losing dimensions from tooth to tooth, and more cutters can be ground between wheel dressings.

This method is quite similar to the flat grind method and involves either lowering the wheelhead with finger attached to produce the relief or clearance angles, or raising the wheelhead for clearance when the tooth rest is attached to the table, the cutter tooth remains level and the wheelhead is raised to produce the relief or clearances. The formula when the wheel is raised is similar to the flat grind method, except that the **grinding wheel radius** instead of the cutter diameter is used in the computation.

Amount to raise the wheelhead **with the cutter tooth maintained level with work centers =**

Sine of the angle desired × radius of the grinding wheel

For example, if a 6 degree primary relief was desired and the machine is fitted with a 7 in. diameter grinding wheel:

Sine 6 degrees = 0.1453 × 3.5 (wheel radius)
= .366 in.

the amount to raise the wheelhead.

If the finger is attached to the wheelhead, then the problem is treated in the same way as in using a flaring cup wheel.

Sine of desired angle × *cutter* radius = *drop* required

SHARPENING OF PLAIN MILLING CUTTERS USING THE CUP WHEEL (FLAT GRINDING METHOD)

The following steps demonstrate the basic method of using a flaring cup wheel to sharpen a plain milling cutter. Look over the machine you have available while you are studying this unit, and determine what differences you must allow for when you do the same operation in your own shop.

1. Mount the tailstock on the table to match the length of the correct size grinding mandrel.
2. Lubricate the centers of the grinding mandrel and fit it between centers so that the mandrel is free to rotate, but completely free of end play.
3. Indicate the mandrel for runout on centers and then also indicate to check swivel table alignment. Adjust table angle to zero indicator travel as necessary.
4. Set the wheelhead square to table and install a flaring cup wheel. Install wheel guard.
5. Install offset tooth rest, and set tooth rest,

Figure 25. Use the center gage to set wheel and tooth rest blade height (Lane Community College).

Figure 26. Recheck tooth level and adjust tooth rest blade to level if necessary (Lane Community College).

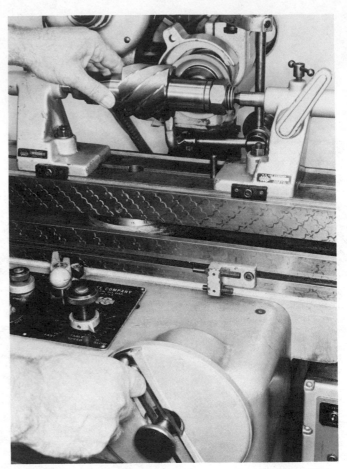

Figure 27. Make the grinding passes (Lane Community College).

centers, and wheel axis in the same plane with the center gage (Figure 25).

6. Unlock and rotate the wheelhead counterclockwise about 1 degree so that the trailing edge of the cup wheel will clear the cutter on the right side.

7. With cutter mounted on the grinding arbor and with the tooth to be ground resting on the tooth rest blade, use the center gage to check the level of the tooth. Readjust the tooth rest height if necessary to level the tooth to the center height (Figure 26).

8. Slip the calibrated ring on the wheelhead vertical control handwheel to zero, without changing the elevation.

9. Calculate the necessary drop to produce the required primary clearance. (Drop = sine of clearance angle × radius of cutter.)

10. Adjust the wheelhead vertical control handwheel to lower the grinding head the required value.

11. With the cutter just clear of the wheel, start the wheel spindle and let the machine run for at least a minute before starting to grind. This is not only a safety measure, but it allows the bearings to warm up and can result in better finish.

12. Traverse the cutter past the grinding wheel while guiding the cutter gently along the top of the tooth rest blade (Figure 27); add small increments of cross feed until a very narrow land is produced. Note the cross slide handwheel reading.

13. With the cutter clear of the tooth rest blade and of the wheel, rotate the cutter to the opposing tooth (180 degrees).

14. Grind the diametrically opposite tooth to the same cross slide handwheel reading.

15. Stop the spindle and bring the cutter clear of the wheel in order to check for taper with a micrometer.

16. Check the cutter for taper. If an unacceptable amount of taper exists (over .001 in. would be considered poor practice), remove the cutter from the grinding arbor and recheck the alignment procedure (step 3).

17. If the taper check is correct, grind the remaining teeth with additional infeed and make the final sparking-out pass on all the teeth with no additional infeed.

18. Set the additional clearance angle required for the secondary clearance.

19. Grind the secondary clearance until a primary land of .030 to .040 in. remains.

20. Check the clearances with a cutter clearance gage or by the indicator drop method, if you do not have a cutter clearance gage.

SHARPENING OF PLAIN MILLING CUTTERS USING THE STRAIGHT WHEEL (HOLLOW GRINDING METHOD)

This type of grind will result in slight hollow grinding. To reduce the amount of hollow, use the largest straight wheel consistent with the spindle RPM and with an available wheel guard. A grinding wheel of 6 in. diameter is usual. If your grinder has wheels larger than 6 in., a collet or wheel flanges having a minimum diameter of 3 in. must be used to meet safety standards (ANSI Safety Code 5.8.1-1964). It is also of *critical importance* that the spindle speed be checked and adjusted (usually by changing the spindle drive belts or pulley positions) to reduce the RPM to or below the speed marked on the grinding wheel blotter.

1. Mount the grinding wheel. If the wheel is a collet mount type, trueing may not be necessary.

Figure 28. Set the wheelhead and tooth rest blade height to center (K & M Tool, Inc.).

2. Mount the tailstock to match the length of the grinding arbor for the cutter to be ground.

3. Indicate the arbor for runout and for table alignment.

4. Mount a tooth rest and blade consistent with the helix angle of the cutter.

5. Using the center gage, adjust the wheelhead to center height and position the height of the tooth rest blade to center (Figure 28).

6. Mount the wheel guard.

7. Mount the cutter on the arbor and place between centers, free to rotate but without end play.

8. Recheck the center height of the tooth to be ground and adjust the tooth rest blade height to level the tooth tip, if required (Figure 29).

9. Set the elevation handwheel collar to zero.

10. Calculate the amount of wheelhead movement necessary to produce a clearance of 6 degrees. Will the wheelhead be raised or lowered in the situation shown in Figures 28 and 29?

11. Since the tooth rest is attached to the wheelhead, the wheelhead will be lowered by:

Sine of clearance angle × radius of the cutter

Lower the wheelhead. If the tooth rest had been attached to the table, you would raise the wheelhead by:

Sine of clearance × radius of the grinding wheel.

12. With the cutter clear of the grinding wheel, you are ready to start the spindle. For safety,

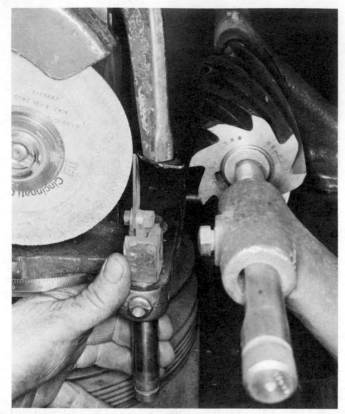

Figure 29. Establish the tooth tip level to the center (K & M Tool, Inc.).

Figure 30. Traverse the table while maintaining cutter contact with the support finger (K & M Tool, Inc.).

let the spindle run for approximately 1 min. before making the grind (Figure 30).

The completed appearance of the cutter ground surfaces should be like that shown in Figure 31.

CONVENTIONAL SHARPENING OF THE SIDE OF AN END MILL

For the grinding of end mills, an accessory spindle is usually used that has a very freely moving spindle that both rotates and slides axially with ease. In some cases, this spindle is floating in an air bearing, as will be seen later in this unit. These fixtures also have the capability of being rocked sideways to clear the end mill from contact with the grinding wheel. The fundamental practices for producing primary relief and secondary clearances are the same as for sharpening plain milling cutters but, as will be seen, there are some substantial detail differences. See Table 3 for recommended radial relief and clearance angles for end mills.

Figure 31. The completed grind should have this appearance (K & M Tool, Inc.).

Since the spindle that carries the milling cutter provides its own alignment, the table of the cutter and tool grinder can be swung at an angle for

Table 3
Radial Relief Angles Suggested for High Speed
Steel End Mills

Workpiece Material	End Mill Diameter, inches							
	⅓	¼	⅜	½	¾	1	1½	2
Carbon steels*	16°	12°	11°	10°	9°	8°	7°	6°
Nonferrous metals	19°	15°	13°	13°	12°	10°	8°	7°

*For tool steels, decrease indicated values about 20 percent. For
secondary clearance angle, increase value by about ⅓.

Figure 33. The wheelhead is lowered with the tooth
rest blade to establish the primary relief (K & M Tool,
Inc.).

Figure 32. The height of the wheelhead and the tooth
rest blade are adjusted to the center of the end mill
mounted in the special grinding fixture (K & M Tool,
Inc.).

Figure 34. The fixture is rocked away from the wheel
and the cutter is moved forward (K & M Tool, Inc.).

the operator's convenience, as seen in Figure 34.

The steps to sharpen end mills with this type
of fixture are as follows:

1. Mount a straight wheel, true if necessary, and
   narrow the wheel with a dressing stick to a
   land of about 1/16 in. width and then stop the
   spindle.
2. Mount the accessory grinding spindle on the
   table and install a tooth rest on the wheelhead with a narrow blade shaped to fit the
   helix of the end mill.
3. Install the end mill and adjust the wheelhead
   and tooth rest blade to the height of the
   centerline of the mounted end mill (Figure
   32). The tooth tip must be horizontal.

4. Zero set the elevation handwheel micrometer
   collar.
5. Calculate the required wheelhead drop.

   **Sine of the primary relief angle × the
   radius of the cutter**

6. Lower the wheelhead with tooth rest for the
   primary relief angle (Figure 33).
7. Start the spindle and allow it to run for a
   minute for safety reasons.

Figure 35. The grinding pass is started at the shank end of the tooth (K & M Tool, Inc.).

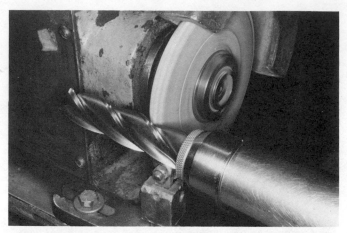

Figure 36. Passes are made with additional infeed until the primary relief is complete (K & M Tool, Inc.).

8. Rock the fixture away from the grinding wheel to clear the cutter (Figure 34) and move the cutter forward along the tooth rest blade.

9. Release the fixture gently when the tooth adjacent to the shank end is on the tooth rest blade (Figure 35). The reason for starting at the shank end is to prevent accidental contacts between the wheel and the end of the cutter. The end may not need sharpening or it may have been sharpened before the side. It is important for beginners to start this way until coordination is developed. With experience, the operation can start from either end of the tooth.

10. Make additional passes with increasing infeed increments (Figure 36) until all evidence of cutter damage has been ground away and the primary land is complete.

11. Lower the head an additional amount, using the same formula as before, to provide for secondary clearance.

12. Grind the secondary clearance (Figure 37) with added infeed until a primary land of correct width remains (about $\frac{1}{32}$ in. for the example shown) (Figure 38).

13. Inspect the cutter and remove.

Figure 37. The secondary clearance is ground (K & M Tool, Inc.).

SHARPENING THE END OF AN END MILL USING THE UNIVERSAL WORKHEAD

1. Install a relatively small diameter flaring cup wheel. Start, true if necessary, and dress to a

Figure 38. Correct appearance of the completed side grinding of the end mill (K & M Tool, Inc.).

Figure 39. A small flaring cup wheel is installed and prepared (K & M Tool, Inc.).

Figure 40. Workhead is squared to table allowing for center relief (K & M Tool, Inc.).

narrow edge ($\frac{1}{8}$ to $\frac{3}{16}$ in.) (Figure 39) and then stop the spindle.

2. Set the workhead square to the table, then turn it slightly counterclockwise (2 degrees is

Figure 41. The primary relief angle is set in (K & M Tool, Inc.).

Figure 42. Flicker tooth rest support with a micrometer base is used for "ratchet" indexing (K & M Tool, Inc.).

recommended, 3 degrees at the most) to make the center a little lower than the outside edge (Figure 40).

3. Set in the amount of axial relief angle required by tilting up the workhead spindle and securing it in place (Figure 41). The amount of axial relief for an end mill should be about 5 degrees for cutting most nonferrous metals.

4. Attach a flicker-type micrometer tooth rest support to the workhead to provide for rachet indexing (Figure 42). A regular indexing attachment may be used, but it tends to be slower both in use and in setup.

5. Install the end mill with adapters and level the tooth using a spirit level and the microm-

Figure 43. The tooth is leveled (K & M Tool, Inc.).

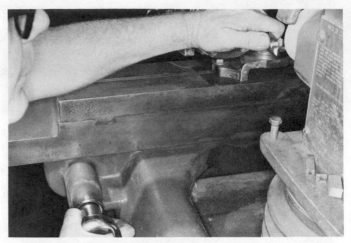

Figure 45. The primary relief is ground on all the teeth (K & M Tool, Inc.).

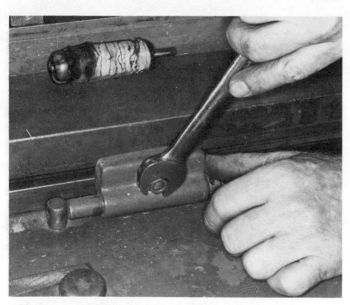

Figure 44. A stop is set to limit cross travel to the center of the cutter (K & M Tool, Inc.).

eter base of the tooth rest support (Figure 43).

6. Traverse the end mill in the workhead so that the center of the cutter is just across from the edge of the wheel and set the table stop (Figure 44).

7. Start the spindle and allow it to run for a minute.

8. Infeed the cutter about .003 in. and traverse across the face of the wheel (Figure 45), then ratchet the spindle to the next tooth.

9. Continue the infeed and grinding routine until all damage has been ground away and the cutter is sharp. The last grind around the

teeth is made as a spark-out with no additional infeed.

10. Raise the workhead spindle an additional amount for the secondary clearance. This should be about twice the value of the axial relief angle that you used. Adjust the spindle head height to compensate and grind the secondary clearance (Figure 46).

11. Examine the cutter and remove (Figure 47).

ADDITIONAL END MILL PREPARATION

For reconditioning of end mills that have been damaged excessively on the end by breakage or by excessive speed resulting in burning of the corners, it is necessary to do a cutting off operation, followed by a reshaping of the end of the tool. The tool is then resharpened and gashed carefully to center to recondition. To do these operations, take the following steps:

1. Mount the damaged end mill in a workhead set square to the table. Mount a narrow (about $\frac{1}{16}$ in.) reinforced cutting off wheel and a wheel guard.

2. Start the spindle and dress the edge of the wheel with a dressing stick.

3. Set the required distance to remove the damaged portion and lock the cross slide, if your machine is so equipped.

4. Rotate the workhead by hand to do the cutting off. Hand rotation is recommended because the usual helix on the flutes of an end mill tend to deflect the wheel initially during the cutting

Figure 46. Grinding the secondary clearance (K & M Tool, Inc.).

Figure 47. Appearance of the completed end grinding of the end mill.

Figure 48. Cutting off the damaged end (K & M Tool, Inc.).

Figure 49. Gashing across the face of each tool (K & M Tool, Inc.).

off process and, with hand turning, you can be sensitive to crowding of the wheel. If you should employ a cylindrical grinding attach-

ment for cutting off (Figure 48), use a reinforced wheel of $\frac{1}{8}$ in. thickness.

5. Using the same reinforced wheel, **gash** across the face of each tooth (Figure 49), and then

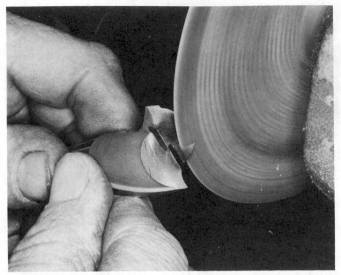

Figure 50. Rough shaping the end of the tool (K & M Tool, Inc.).

remove the excess material behind each tooth (Figure 50). This procedure is followed for multiple fluted end mills as well as the two flute type shown.

6. Mount the end mill for regular primary and secondary grinding as described in the unit on sharpening end mills (Figure 51).

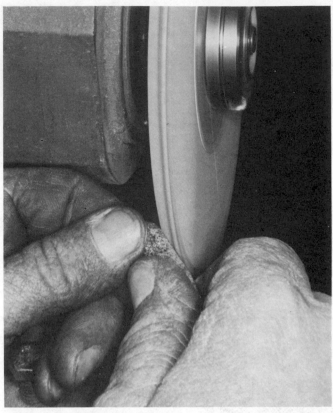

Figure 52. Dressing a straight wheel to a sharp bevel (K & M Tool, Inc.).

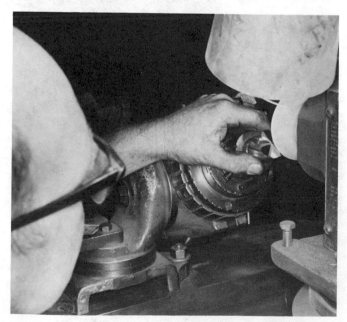

Figure 51. Sharpening the end mill in a conventional way (K & M Tool, Inc.).

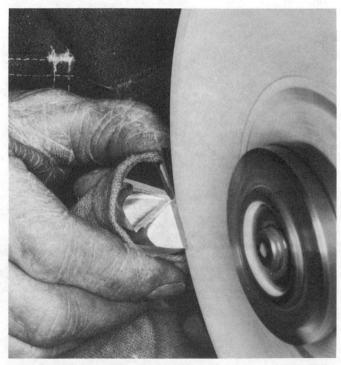

Figure 53. Grinding across the face of the tooth to center (K & M Tool, Inc.).

Figure 54. Remove the burrs—note the completed appearance (K & M Tool, Inc.).

Figure 55. The same method is used for two and four fluted end mills (K & M Tool, Inc.).

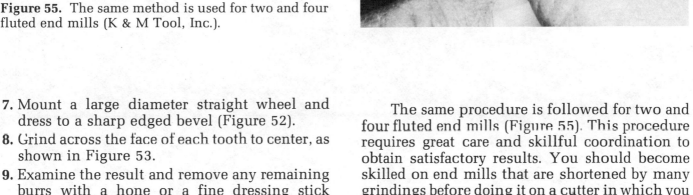

7. Mount a large diameter straight wheel and dress to a sharp edged bevel (Figure 52).
8. Grind across the face of each tooth to center, as shown in Figure 53.
9. Examine the result and remove any remaining burrs with a hone or a fine dressing stick (Figure 54). Note the completed appearance.

The same procedure is followed for two and four fluted end mills (Figure 55). This procedure requires great care and skillful coordination to obtain satisfactory results. You should become skilled on end mills that are shortened by many grindings before doing it on a cutter in which you have a good deal of sharpening time invested.

section o
numerical
control
of machine tools

Numerical control of machine tools is of major importance in modern machining technology. With numerical control (N/C), machining productivity is increased. This is an important consideration in an age where manufacturing is faced with rising costs. The N/C machine tool is helping to keep production costs down. Manufacturers have seen the value and versatility of these machines. They are appearing in machine shops of every size. Numerical control can be productively applied to almost any machining task. An individual beginning machinist training today does so in a numerical control environment.

As you begin a career in machining, N/C will play an ever increasing part. The field of N/C machining technology is constantly expanding. What was new in numerical control only a short time ago is becoming rapidly obsolete. In this section you will have an opportunity to investigate N/C technology in a general way. As you proceed into further training and possibly to a career in machining, you may find the area of N/C to be of great interest. In any case, the field is sure to be expanding and changing. For this reason, the numerical control branch of machining technology presents a number of opportunities for specialized study.

INTRODUCTION TO NUMERICAL CONTROL

Machine Control by Numerical Instructions

Numerical control is a method and a system of controlling a machine or process by instructions in the form of numbers. On a manually operated machine tool, the operator turns cranks in order to move milling machine tables or lathe cross slides and carriages. On the N/C machine tool, cranks are replaced by drive motors or hydraulic mechanisms. These are controlled from an external machine control unit (MCU).

Machine control functions previously provided by the operator are translated into numeric instructions that can be understood by the machine control unit. These control functions include positioning tables and spindles, setting milling feedrates, setting spindle speeds, cycling drill press quills, changing cutting tools, and turning coolant on and off. A simple on/off control function provided by

837

an electric switch might be used to start and stop an electric motor or open and close a hydraulic valve. This might be expressed as a simple numeric instruction of **one** to open a valve or start a motor. A numeric instruction of **zero** might close a valve or stop a motor. If the motor or hydraulic mechanism is connected to the leadscrew of a lathe or the table screw of a milling machine, the simple on off, one zero numerical control function can be used to position the component of the machine tool. The distance traveled and the direction can be controlled as well. In actual practice, numerical control is somewhat more complicated than a simple on/off function. However, this example will give you some idea of control by a numeric instruction.

Historical Development of Numerical Control The development of true numerical control dates from about 1945. However, research and development in automatic control had been done as early as 1725 (Figure 1).

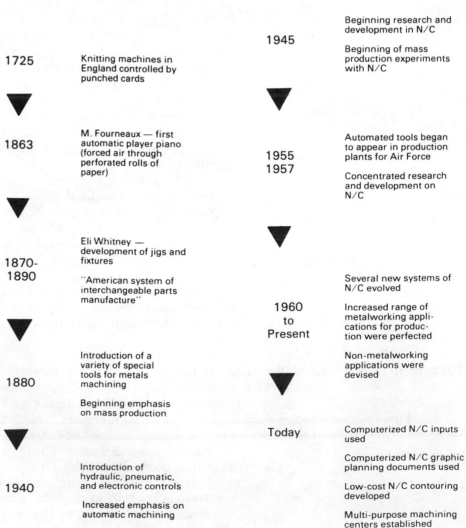

Figure 1. Time line for numerical control (Courtesy of the Superior Electric Company.)

Construction Features of N/C Machine Tools

A numerical control system can be added to an existing machine tool (Figure 2). This is called a retrofit and presents a less expensive way to gain N/C capability in the machine shop. Retrofit numerical

The many advantages of N/C have brought about extensive development of machine tools designed specifically for N/C operations. In conventional machining, a complex workpiece may have to be set up on several machine tools in order to complete all required machining tasks. The development of numerical control technology has brought about the concept of a **machining center** on which a wide variety of machining tasks can be accomplished on the same machine tool. The N/C machines used in machining centers are patterned after their conventional counterparts. However, the construction of an N/C machine tool is often quite different.

N/C machine tools are designed for long hours of continuous production. This necessitates that they be built so that accuracy will be retained as long as possible. Wear is a problem associated with

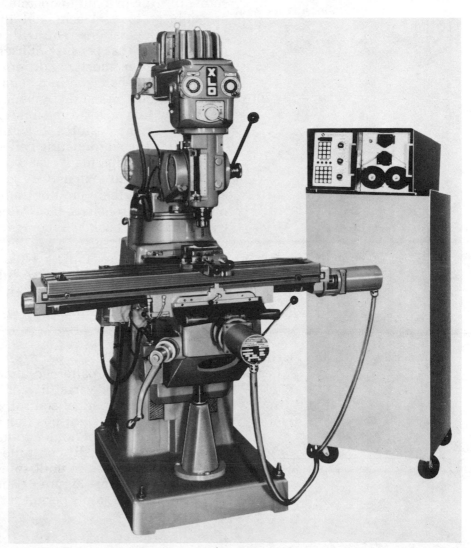

Figure 2. A numerical control system added to an existing milling machine (Courtesy of the Superior Electric Company).

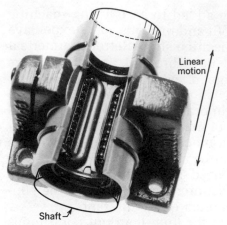

Figure 3. The linear motion bearing permits a close fit along with low wear and nearly friction free movement (California State University at Fresno).

any mechanical device. Wear in a machine tool is particularly significant in that the accuracy of the machine is directly affected. A possible advantage of a manually operated machine tool over an N/C machine is that a good machinist can know his machine and compensate for wear. However, the increased productivity and versatility of the N/C machine tool far outweighs this factor. N/C machines, like all machines, are subject to wear. Because of this, manufacturers have introduced many features designed to prolong the machine's accuracy holding capabilities.

Sliding components have given way to rolling components in order to provide nearly friction-free movement and still retain the alignment and close fit of a slide. Several types of linear motion bearings are used on N/C machines. These bearings permit the rolling advantages of the ball bearing to be used in a linear motion application (Figure 3).

Conventional machine tools such as milling machines use a screw turning in a nut to operate the table saddle and knee. Threads are subject to backlash, and this places certain limitations on machining operations. You will remember from your study of milling machines that because of backlash, climb milling is not recommended under most conditions. You will also remember that backlash must be taken into consideration when positioning a mill table or lathe cross slide by hand. The N/C machine tool cannot be limited by this factor if maximum productivity is to be realized.

One of the features designed to overcome the problem of backlash is the recirculating ball screw. The principle of this device has been used for many years in automobile steering gears. Ball screws are used extensively on N/C machine tools. Machines equipped with them are not limited by a backlash problem. Where numerical controls have been added to existing machine tools, electromechanical backlash compensation has been built into the machine control unit.

Hydraulics are employed to overcome the limitations of threads. Some N/C machine tools use full hydraulic positioning systems, thus eliminating the use of any thread mechanism for moving machine components.

N/C Machine Tools and Machining Centers

VERTICAL SPINDLE N/C MACHINING CENTERS. Vertical spindle N/C machining centers patterned after the vertical milling machine are very popular. The versatility of the vertical mill is mated to the advantages of numerical control. Vertical spindle machines may be equipped with an eight-position turret tool holder (Figure 4) or a drum or carousel tool holder containing many tools (Figure 5). Tool changing is numerically controlled. The vertical spindle design also includes large capacity multispindle milling machines such as the bridge-type profiler (Figure 6) and the gantry-type profiler (Figure 7). These machine tools permit several workpieces to be machined at the same time.

HORIZONTAL SPINDLE N/C MACHINE TOOLS. Horizontal spindle N/C machining centers, patterned after the horizontal mill and horizontal boring machine, are also very popular and versatile. These machines may be equipped with side mounted tool drums (Figure 8)

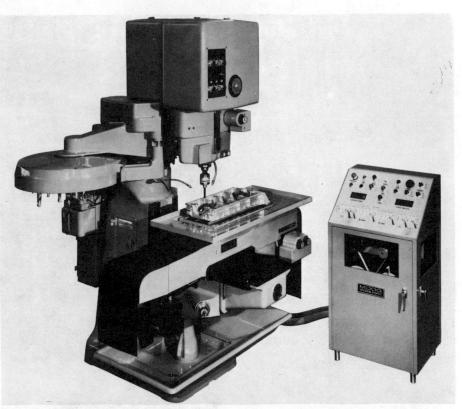

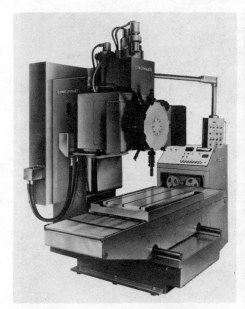

Figure 4. Vertical spindle N/C milling machine with eight position turret tool holder (Courtesy of Cincinnati Milacron).

Figure 5. Vertical N/C milling machine with side mounted tool drum (Courtesy of Hydra-Point Division, Moog Inc.).

Figure 6. Bridge-type multiple spindle N/C profiler (Courtesy of Cincinnati Milacron).

or top mounted drums (Figure 9). Tool changing is numerically controlled. The workpiece may be mounted on a rotary table, enabling both sides to be machined (Figure 10).

Large capacity horizontal spindle machines include the traveling column profiler (Figure 11). The machine tool shown is equipped with a conveyor system to remove chips. Horizontal multiple spindle profilers may be used to machine several workpieces at the same time (Figure 12).

Figure 7. Gantry-type multiple spindle N/C profiler (Courtesy of Cincinnati Milacron).

Figure 8. Horizontal spindle N/C machining center with side mounted tool drum (Courtesy of Cincinnati Milacron).

Figure 9. Horizontal spindle N/C machining center with top mounted tool drum (Courtesy of Cincinnati Milacron).

Figure 10. Rotary table on a horizontal spindle machining center (Courtesy of Cincinnati Milacron).

N/C LATHES. A sophisticated N/C lathe or turning center (Figure 13) may have numerically controlled turret tool holders for inside diameter turning (Figure 14). With the workpiece held in a chuck, a

Figure 11. Traveling column N/C profiler with chip conveyor (Courtesy of Cincinnati Milacron).

Figure 12. Horizontal N/C profiler with three spindles (Courtesy of Cincinnati Milacron).

Figure 13. Dual turret N/C turning center (Courtesy of Cincinnati Milacron).

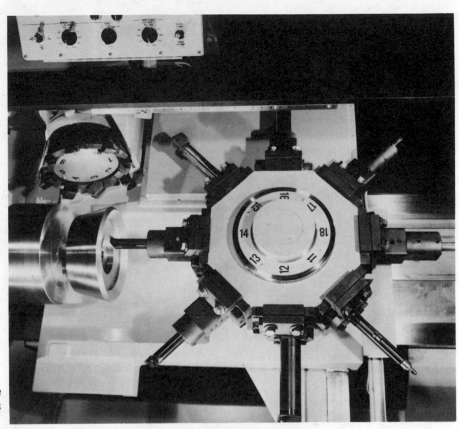

Figure 14. Inside and outside diameter N/C turning center turrets (Courtesy of Cincinnati Milacron).

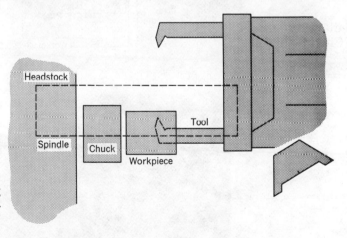

Figure 15. Inside diameter turning loop (Courtesy of Cincinnati Milacron).

variety of cutting tools can be applied in the inside diameter turning loop (Figure 15). The outside diameter turret (Figure 16) operates in the outside diameter turning loop (Figure 17). Numerically controlled lathes are also used in shaft turning operations (Figure 18). The shaft turning loop requires use of a tailstock center (Figure 19).

The Input Medium of Numerical Control Once a control function has been translated into numeric form, the information must be provided to the machine tool. This can be done through a control medium such as punched cards, magnetic tape, or

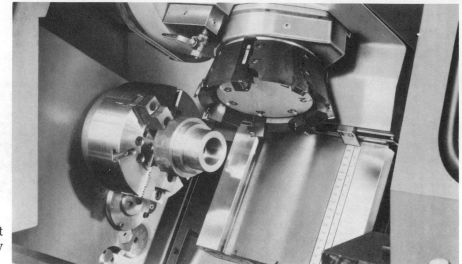

Figure 16. Outside diameter turret on the N/C turning center (Courtesy of Cincinnati Milacron).

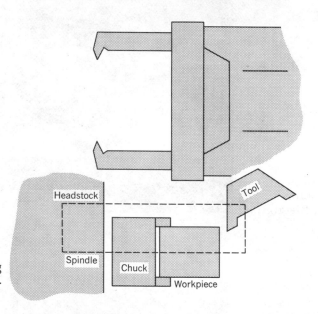

Figure 17. Outside diameter turning loop (Courtesy of Cincinnati Milacron).

Figure 18. Shaft turning on the N/C turning center (Courtesy of Cincinnati Milacron).

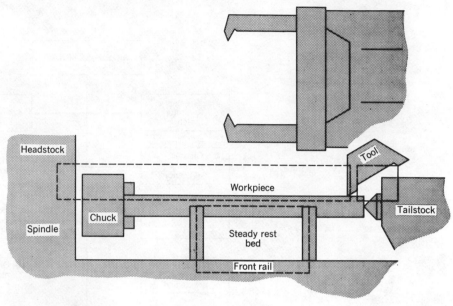

Figure 19. Shaft turning loop (Courtesy of Cincinnati Milacron).

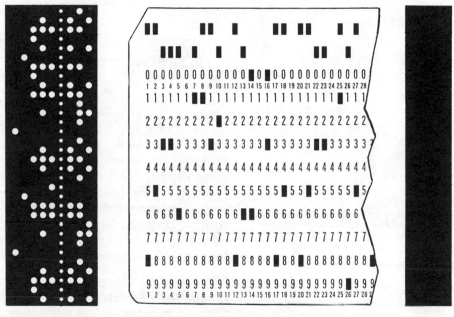

Perforated tape   Punched cards   Magnetic tape

Figure 20. N/C input program media (Courtesy of the Superior Electric Company).

punched tape (Figure 20). The punched tape medium is very popular at the present stage of the technology. Instructions are translated into numeric form and appear as specific patterns of holes punched into a tape. The tape is read by the tape reader in the machine control unit. The MCU then interprets the numerical information on the control tape and, through electric and mechanical means, controls components of the machine tool. As you study numerical control, you may hear terms such as tape machine and tape control. These are references to machine control by a tape medium.

Advantages of Numerical Control Machining REPEATABILITY. The N/C machine tool can produce 1, 10, or 10,000 parts with unvarying accuracy. This is important to the pro-

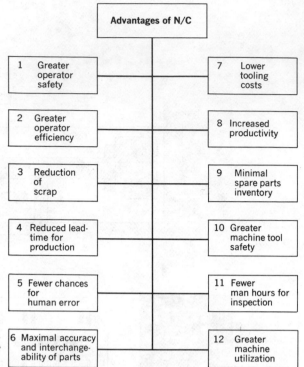

Figure 21. Advantages of numerical control (Courtesy of the Superior Electric Company).

duction of duplicate parts that are within tolerance. A set of tape instructions does not vary. The machine tool does not become fatigued or bored, as does the operator of its manually operated counterpart. The machine will repeat precisely during each machining cycle. Workpieces need only be removed and replaced with new stock. Except for downtime due to resharpening of cutting tools or routine maintenance, the N/C machine tool can function 24 hours a day, year round, if necessary. In this respect, it has a decided advantage over its manually controlled counterpart (Figure 21).

VERSATILITY. The machining function of the N/C tool can easily be changed by inserting a new tape. With computer control, N/C programs can be stored in computer memory and the machine tool can be operated directly, thus bypassing the tape. Tape N/C programs are permanent and can be stored for future use. A long production run can be stopped and a short run can be inserted. The machine tool can then be returned to its long run in a very short time.

MACHINING CAPABILITY. With the numerically controlled tool changer, the N/C machine tool or machining center can accomplish a wide variety of machining tasks. Tool changing may be a two-step operation where the drum is mounted on the side of the machine (Figure 22). The drum rotates the required tool into position, where it is pivoted down and grasped by the changing arm. The tool is then placed in the spindle (Figure 23). At the same time, the tool already in the spindle is returned to the pivot arm and replaced in the drum. Where the drum is on the top of the machine, the tool in the drum and the tool in the spindle are removed at the same time and exchanged (Figures 24*a* to 24*d*). Tools are secured in a self-releasing

Figure 22. Pivoting presenter arm on a side mounted tool changer (Courtesy of Heald Machine Division/Cincinnati Milacron).

Figure 23. Tool changing arm (Courtesy of Heald Machine Division/Cincinnati Milacron).

taper shank tool holder. The holder is secured in the machine by a locking mechanism contained within the spindle (Figure 25).

Drilling (Figure 26) is a common N/C machining capability. Spindle speed, feedrate, and depth can be controlled from tape instructions. A "peck drilling" cycle can be used for deep hole or small hole drilling. When the peck drill cycle is initiated, the drill feeds partway into the workpiece, where it dwells for a short time. The drill is then automatically withdrawn in order to clear chips. The cycle is automatically repeated and the drill is permitted to feed further into the workpiece. The cycle continues until the final depth is reached. Dwell time during each "peck" can be varied manually.

Milling an enclosed feature or "pocket milling" (Figure 27) is another very common and useful machining capability. Spindle speed, milling feedrates, and direction and distance of the cuts are controlled from tape instructions.

Close tolerance boring (Figure 28) is frequently done on the N/C machine tool. Spindle speeds and bore feedrates are tape controlled. The boring bar may be withdrawn from the bore at the same feedrate used during boring. This reduces tool marks in the workpiece. Progressive boring bars may be used where close tolerances must be maintained. The hole is first bored with a roughing bar and finished with a finishing bar. Boring bars must be preset for correct diameter before they are used.

Tapping is another machining operation that is well suited to N/C operations. One method is leadscrew tapping (Figure 29). The leadscrew causes the machine spindle to feed or lead the same amount as the tap lead. Leadscrews are changed to correspond to different tap leads.

Figure 24. Tool changing sequence from a top mounted tool drum (Courtesy of Cincinnati Milacron).

Probably the greatest machining capability of the N/C machine tool is that of contouring or continuous path machining (Figure 30). This includes circles, angles, and radius cuts, as well as irregular shapes in two or three dimensions. In fact, the N/C machine tool can

Figure 25. Spindle tool holding mechanism (Courtesy of Hydra-Point Division, Moog Inc.).

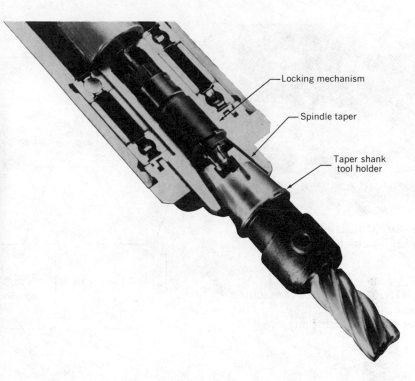

Locking mechanism

Spindle taper

Taper shank tool holder

Figure 26. N/C drilling (Courtesy of Hydra-Point Division, Moog Inc.).

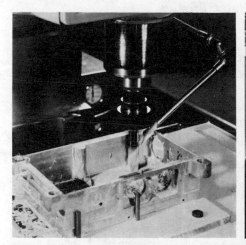

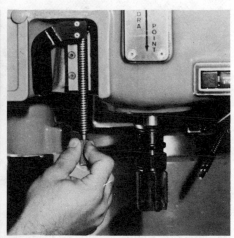

Figure 27. N/C pocket milling (Courtesy of Hydra-Point Division, Moog Inc.).

Figure 28. N/C boring (Courtesy of Hydra-Point Division, Moog Inc.).

Figure 29. N/C leadscrew tapping (Courtesy of Hydra-Point Division, Moog Inc.).

produce shapes that would be quite impossible to machine by manual means. This capability has opened new avenues of study for the designer of mechanical hardware.

Numerical Control and the Machinist

The N/C machine tool, with its amazing capabilities, seldom requires a fully qualified machinist as an operator. In fact, the talent of the machinist would be wasted if time were spent changing workpieces on an N/C machine. Most N/C machine tools can be operated

Figure 30. Continuous path N/C machining (Yuba College).

Figure 31. Role of the N/C machine tool operator (Courtesy of the Superior Electric Company).

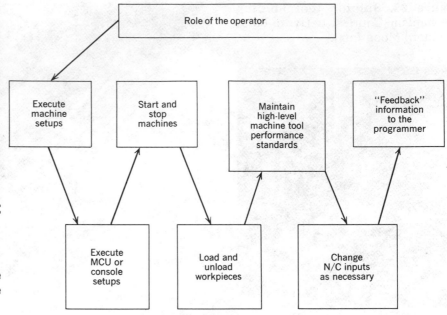

by a competent machine operator. However, the N/C operator has several important functions to perform (Figure 31). He must know the machine and its operating characteristics.

Numerical Control Systems

CLOSED LOOP SYSTEMS. In a closed loop N/C system, a signal is fed back to the machine control unit confirming the specific instruction (Figure 32). For example, if the MCU instructs a milling machine table to move 10 in., a signal would be fed back to the MCU from a sensor on the drive motor indicating that the table has moved the instructed distance. Closed loop provides the MCU with a check on the accuracy of machine movement.

OPEN LOOP SYSTEMS. No feedback signal is used in an open loop N/C system. The open loop system may use an electric stepping motor to control the movement of machine components. The stepping motor is used on many numerical controls that are added to existing machine tools.

A stepping motor operates on a pulse of electric current supplied by the MCU. Each current pulse causes the motor rotor to turn or "step" a fraction of a revolution. When the motor is coupled to a mill table, lathe cross slide, or lathe leadscrew, it can act to move the screw specific amounts according to the number of pulses received from the MCU. Stepping motors are often designed to move machine tool tables a distance of .001 in. per pulse. They are reversible and can move a component in either direction.

Numerical Control and the Computer

The computer is an important and valuable component of the modern numerical control system. The computer can do mathematics with great speed and accuracy. This has made it a valuable tool in

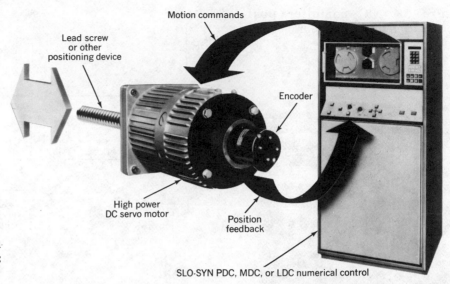

Motion commands

Lead screw
or other
positioning device

Encoder

High power
DC servo motor

Position
feedback

SLO-SYN PDC, MDC, or LDC numerical control

Figure 32. Closed loop N/C system (Courtesy of the Superior Electric Company).

the N/C programming of complex machining operations. Mathematical computation necessary for continuous path machining would be difficult and time consuming if it had to be done by hand. The computer can accommodate this kind of calculation easily.

A computer can understand direct descriptions of workpiece geometry, machine control functions, and machining operations. This permits the N/C programmer to program much in the same way that he would verbally describe the machining task to be done. The computer can understand direct statements such as GO TO, MILL, DRILL, or BORE. These direct descriptions are translated by the computer into appropriate instructions for a specific N/C machine tool.

COMPILER POST-PROCESSOR SYSTEM. Large capacity computers that can accommodate complex N/C programming are expensive. For this reason, they must be put to productive use. A manufacturing concern may use its computer facility for many purposes in addition to N/C programming. Where N/C programming is only a portion of a computer's total work, the "compiler post-processor" system may be used (Figure 33).

A computer programmer first enters a special set of instructions called a "compiler." The compiler is used to direct computer operations when N/C program instructions are entered (Figure 33). N/C instructions to the computer can now be entered in the form of an N/C computer language consisting of abbreviated "words" and punctuation. Workpiece geometry, machine control functions, and machining operations can be described directly. The computer acts on the N/C program instructions and produces an output in the form of magnetic tape or punched computer cards. This information must then be translated into a format suitable for a specific N/C machine tool.

Different types of N/C machine tools use different program formats. An additional set of instructions called a "post processor" is used to translate the computer output into specific instructions in the format required by a specific machine tool. Manufacturers of N/C

Figure 33. N/C complier post-processor computer system.

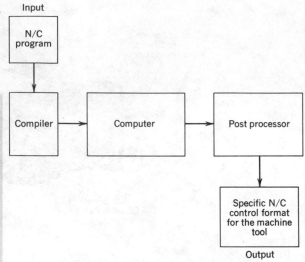

Figure 34. MCU computer numerical control (Courtesy of Cincinnati Milacron).

machine tools and numerical controls often provide post processors for their specific machines. Punched N/C tape may be produced directly by the computer and post processor. However, as this is a time consuming operation as compared to the computer time required for the programming operation, actual tape preparation may be done on a separate tape preparation unit.

DIRECT COMPUTER NUMERICAL CONTROL. Modern numerical controls are incorporating the computer into the machine control unit (Figure 34). This permits direct computer control of one or more N/C machine tools. The advantages of direct computer N/C, or C N/C, include machine tool operation directly from programs stored in the computer memory. The tape and tape reader are bypassed. Computer numerical controls also permit tapes to be verified at the machine tool location.

One versatile feature of C N/C is tape editing. With the tape editor, a tape error can be corrected or a change can be made in machining operations or the location of a part feature.

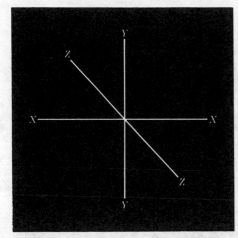

unit 1
numerical control dimensioning

The numerical control programmer studies a drawing of the workpiece and determines the direction and distance that the cutting tool must travel. The programmer then directs the machine movements along these paths by indicating the appropriate numerical instructions. In order to do this, the programmer must be able to define and identify the travel direction of a specific machine component. He must also be able to differentiate travel directions of different machine components. This is the purpose of machine tool axis identifications and N/C dimensioning.

objectives After completing this unit, you will be able to:

1. Describe the application of rectangular coordinates to machine tools.
2. Describe the purpose of rotational axes.
3. Describe the direction of machine spindle movements in a coordinate system.

MACHINE TOOL AXES

Basic Axes

The rectangular coordinate system consists of the two perpendicular axes of X and Y (Figure 1). The X and Y axes lie in the same plane and are known as coordinate axes. With the addition of a third axis, Z, that is perpendicular to the X-Y plane, a three-dimensional volume of space can be described and identified (Figure 2). The point at which the axes intersect is called the origin and has a numeric value of zero.

These notations are applied to N/C machine tools in order to identify the basic machine axes. The Z axis is always the spindle axis, even though the machine spindle may be horizontal or vertical. On a typical vertical spindle machine tool, such as a vertical mill, the spindle axis is Z.

The X axis is the table and the Y axis is the saddle (Figure 3). The knee is in the Z axis.

On a horizontal spindle machine tool, Z remains the spindle axis while Y becomes vertical and X remains horizontal (Figure 4). The N/C lathe is also a horizontal spindle machine tool. The spindle axis is Z and the cross slide is X, or the horizontal axis perpendicular to Z (Figure 5). Since the lathe tool holder is not moved vertically, the Y axis is not used.

Rotational Axes

Rotational axes define numerically controlled motion around the X, Y, and Z basic axes. An N/C machine tool may have a rotary table or part indexer (Figure 6). These accessories may operate from tape instructions, rotationally around the basic axes. The direction of rotation, as well

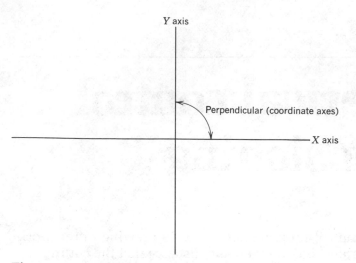

Figure 1. Coordinate axes X and Y.

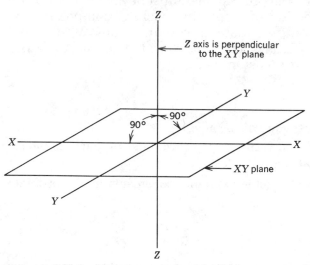

Figure 2. The Z axis is perpendicular to the XY plane.

as the basic axes around which rotation occurs, must be identified. Rotational axes are identified as a, b, and c (Figure 7). Discussions about four and five axis N/C machine tools refer to the basic axes of X, Y, and Z and the rotational axes of a, b, and c.

QUADRANTS

The perpendicular coordinate axes X and Y form four quadrants (Figure 8). Quadrants are numbered in a counterclockwise direction beginning at the upper right. The point of axial intersection or origin has a numeric value of zero. All points to the right of zero along the X axis have positive

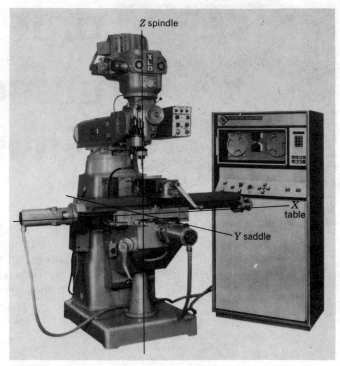

Figure 3. Basic axes of a vertical spindle N/C machine tool (Courtesy of the Superior Electric Company).

Figure 4. Basic axes of a horizontal spindle N/C machine tool (Courtesy of Cincinnati Milacron).

value. All points to the left of zero have negative values. Points on the Y axis above zero are positive and points below zero are negative.

If the X-Y plane is horizontal, as it is on all

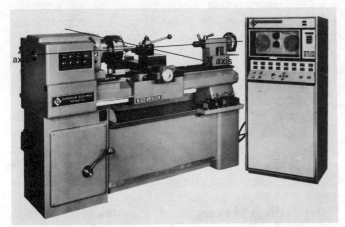

Figure 5. Basic axes of an N/C lathe (Courtesy of the Superior Electric Company).

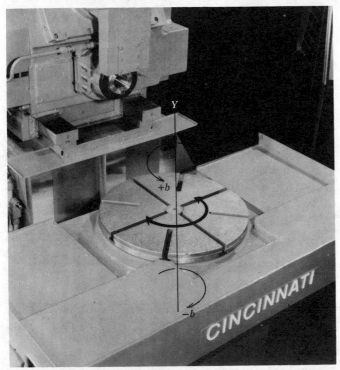

Figure 6. Rotational axis b defining rotary table motion (Courtesy of Cincinnati Milacron).

vertical spindle machine tools, there is no geographic location of above and below zero. In this case, points from the origin away from you are positive, while points toward you from zero are negative.

Quadrant Point Values
Point values in the four quadrants are as follows.

Quadrant I: X positive, Y positive.

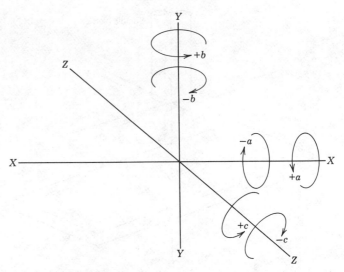

Figure 7. Rotational axes.

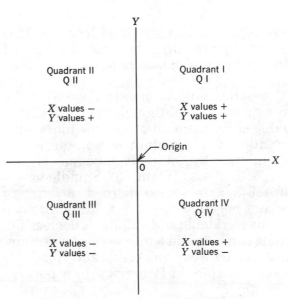

Figure 8. Quadrants formed by the X and Y coordinate axes.

Quadrant II: X negative, Y positive.
Quadrant III: X negative, Y negative.
Quadrant IV: X positive, Y negative.

DIRECTIONS OF MACHINE TOOL SPINDLE TRAVEL

Understanding quadrant point values is important to the preparation of certain types of N/C tape instructions. The MCU must tell the machine tool spindle or worktable to move in a

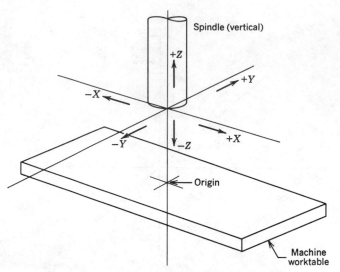

Figure 9. Directions of vertical spindle movement.

certain direction or to a specified location. This may be done by providing tape instructions that indicate positive and negative movement directions.

N/C programming is always done as if the tool were moving, even though the worktable is the moving component on a machine tool with a fixed spindle. The direction of spindle movement is expressed by noting its direction of travel along a specified axis (Figure 9). Spindle movement in the Z axis is also defined in terms of positive and negative directions. If the distance between the worktable and spindle is decreasing, the spindle is moving in a minus $(-)Z$ direction. If the distance between worktable and spindle is increasing, the spindle is moving in a positive $(+)Z$ direction.

SPINDLE POSITIONING BY INCREMENTAL MEASUREMENT

Certain numerical control programs instruct the machine tool to position the spindle by incremental measurement. This means that the spindle measures the distance to its next location from the position at which it was last located. Incremental positioning requires positive and negative travel directions.

EXAMPLE
A certain workpiece is to be drilled on an N/C machine tool positioning by incremental measurement. The workpiece is set up so that the

spindle start point is over one corner and the edges of the part are parallel with the coordinate axes. The N/C program instructs the machine spindle to move in a $+X$ direction a distance of 1 in. and in a $+Y$ direction a distance of 4 in. (Figure 10). This will position the spindle over drilling location 1.

To reach the second location, the spindle moves in the $+X$ direction an additional distance of 2 in. However, it must move in a $-Y$ direction a distance of 1 in. to reach location 2. The machine uses location 1 as a new origin from which to measure its movement to location 2. After drilling is completed, it is desired to return the spindle to the start point. This requires a $-X$ move of 3 in. and a $-Y$ move of 3 in. Once again, location 2 becomes a new origin from which to measure the distance back to the start point.

SPINDLE POSITIONING BY ABSOLUTE OR COORDINATE MEASUREMENT

Other types of N/C machine tools position the spindle by absolute or coordinate measurement. Two systems are used.

With fixed zero absolute positioning, all measurements are taken from the same zero reference point located at the lower left corner of the workpiece or worktable. With fixed zero, the need for positive and negative moves is eliminated. For practical purposes, the spindle is always operating in Quadrant I, where all points have positive value. All points are specified as coordinate locations in terms of the distance from the coordinate axes.

EXAMPLE
The same workpiece is to be drilled on a machine positioning by absolute measurement from fixed zero. Each drilling location is expressed as a dimension from the absolute zero point (Figure 11). Coordinate dimensions are measured parallel to the coordinate axes.

Drilling location 1 is at point $(1X, 4Y)$ from zero. This is a coordinate location. Location 2 is at point $(3X, 3Y)$ from zero. The N/C program instructs the spindle to position to coordinate location $(1X, 4Y)$. After drilling the first hole, the spindle is instructed to position to location 2 at coordinate $(3X, 3Y)$. Since this second location is measured from zero and not from the previous location, the machine positions to the new location without the need of movement in a negative direction. Return to the start point is ac-

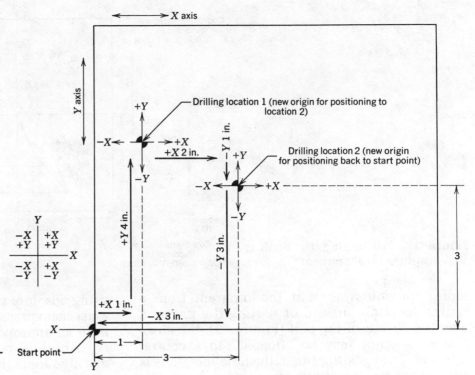

Figure 10. Positioning by incremental measurement.

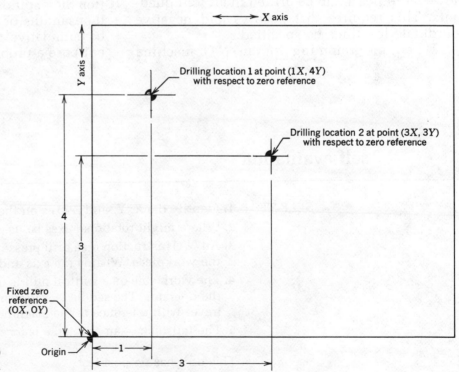

Figure 11. Positioning by fixed zero absolute or coordinate measurement.

complished by instructing the spindle to position to the coordinate location (0X, 0Y).

Some N/C machines permit any point to be established as absolute zero. This is known as

floating zero and can be used to make certain programming easier.

EXAMPLE

When drilling a symmetric pattern, it might be

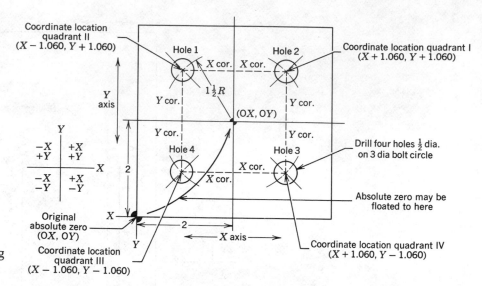

Figure 12. Positioning by floating zero absolute measurement.

more convenient to start the program from a central location. Instead of starting the program over the corner of the part (Figure 12), the absolute zero point may be "floated" to a central location. Tool positioning to the hole locations is still done by specifying coordinate locations. However, holes are to be drilled in all four quadrants. This requires that positive and negative coordinate locations be specified.

When programming for an N/C machine using absolute positioning, drawing dimensions must be expressed in terms of absolute measure from an appropriate zero reference point.

Absolute positioning may be used on N/C machine tools that are designed to position with extreme accuracy. Many N/C machines can position their spindles or worktables within a few ten thousandths of an inch. Small errors that might be cumulative with incremental positioning, do not pose a problem in absolute positioning.

self-evaluation

SELF-TEST
1. Identify the X, Y, and Z axes on the machine tool outlines (Figure 13).
2. Where might rotational axes be used?
3. An N/C instruction on a drill press instructs the spindle to move toward the workpiece. What is the axis and direction of travel?
4. The worktable on a vertical mill is moving the workpiece to the right of the operator. The saddle is stationary. What is the direction and axis of travel with reference to the machine spindle?
5. The tailstock of an N/C lathe is located in the _____ axis.

This unit has no post-test.

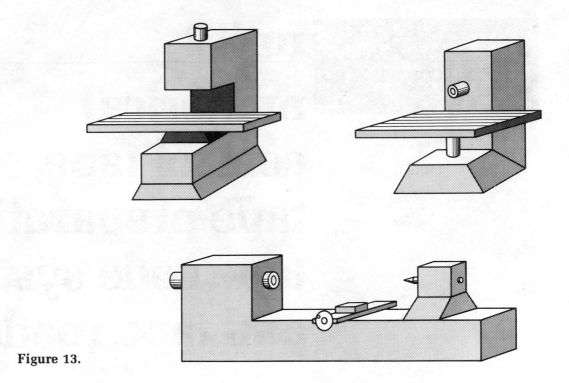

Figure 13.

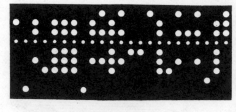

unit 2

numerical control tape, tape preparation, tape code systems, and tape readers

Before advancing to actual N/C programming, you should be familiar with N/C tape materials, tape preparation, tape code systems, and tape readers. N/C programming is sure to involve you in operating a typewriter tape punch unit to produce new tapes, verify programs, duplicate tapes, and correct tape errors. An understanding of tape readers will be useful to your overall understanding of the entire numerical control system.

objectives After completing this unit, you will be able to:

1. Describe and cite advantages of tape materials.
2. Describe the various functions of the tape preparation unit.
3. Name standard tape codes.
4. Recognize types of tape readers appearing on N/C machine tools.

NUMERICAL CONTROL TAPES

Tape Materials

Eight channel punched tape (Figure 1) is a very common and popular means by which control instructions are given to an N/C machine tool. N/C tape materials include paper or paper-plastic and aluminum-plastic laminates. Blank tape may be supplied in 1000 or 2000 ft. rolls, depending on thickness. The plastic and aluminum lami-nate materials are generally more durable than paper. They are more expensive, but they will withstand many trips through the tape reader without wear or damage. The plastic materials are also less subject to damage by oils and grease found around the machine and shop. Manufacturing tolerances are held closely (Figure 2) so that tape may be standardized between different manufacturers.

Paper tape is available in several different

862

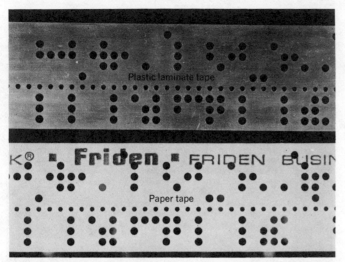

Figure 1. Paper and plastic laminate tape materials.

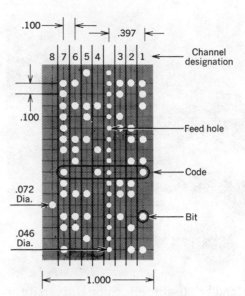

Figure 2. Dimensions of N/C tape features (Courtesy of the Superior Electric Company).

colors, including pink, yellow, and black. The color has no significance except that certain colors are more suitable for use in photoelectric tape readers. In these readers, light passes through the tape punches and activates photoelectric cells. A photoelectric tape reader using paper tape may require a black or dark color so that light will not pass through the tape material and activate improper photocells.

Standard N/C tape is 1 in. wide and contains eight channels for information (Figure 2). A row of sprocket or feed holes also appears on the tape. These are necessary for transport through certain types of tape readers. The line of feed holes is punched off center to eliminate confusion as to how the tape is to be placed in the tape reader.

TAPE PREPARATION

Control information is placed on the N/C tape by punching a specific pattern of holes. This is accomplished on a special typewriter tape punch machine (Figure 3). The typewriter keyboard operates in a similar manner as a standard typewriter. The tape punch typewriter has the same letters and numbers found on a standard typewriter. In addition, several extra symbols are included, as well as control keys for the tape punch.

The tape punch is activated as each typewriter key is depressed. This produces a pattern of holes in the tape that is unique to that typewriter symbol. As the tape is punched, a printed record is typed on paper in the typewriter car-

Figure 3. Typewriter tape punch (California State University at Fresno).

riage. The tape feed key (Figure 4) causes blank tape to feed through the punch. Feed holes are produced during this operation (Figure 5). Blank tape is run out to provide a leader that can be wound on the machine tool tape reader reels. Blank tape is run out to provide a leader that can be wound on the machine tool tape reader reels.

The typewriter tape punch also has a tape reading head. The reading head is not unlike the tape reader on the N/C machine tool. The function of the typewriter reader is to operate the typewriter from the punched tape. The reader is used to obtain a printed record of a punched tape. This is useful for verifying tape accuracy.

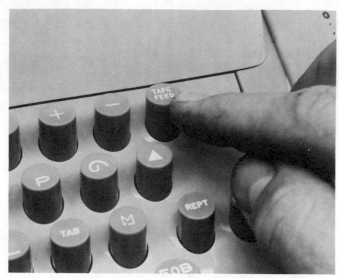

Figure 4. Tape feed key (California State University at Fresno).

Figure 5. Blank tape with feed holes (California State University at Fresno).

After a tape has been prepared, it may be inserted in the typewriter reader (Figure 6). The tape will now activate the typewriter and a record of the information will be typed out. If there is an error in the tape information, it can be detected and corrected. The typewriter reader is also used to duplicate tapes.

Correcting Tape Errors

If an error in typing is made and detected at that time by the tape punch operator, a correction may be made by pressing the **delete** key. The delete key causes all seven rows on the tape to be punched out. The operator may then retype the correct information and continue with the remainder of the program.

If an error is not detected until the tape is completed and read back using the typewriter reader, a new tape will have to be produced. This can be done by inserting the incorrect tape in the reader and duplicating a new tape to the point of the error. The operator will, of course, have to watch the typed printout and stop the duplicating process when it reaches the last correct entry. The correct information is then typed from the keyboard. The incorrect information on the original tape is advanced through the reader by hand and the duplicating process is resumed.

Punched tape may also be corrected by inserting a splice at the appropriate point. This requires that the feed holes and tape perforations be precisely aligned.

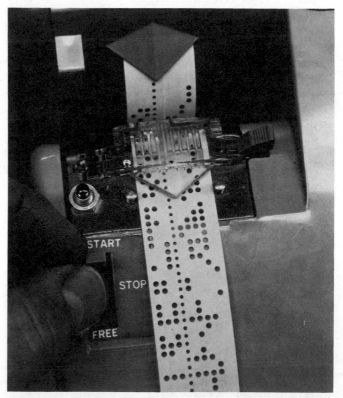

Figure 6. Typewriter tape reader (California State University at Fresno).

TAPE CODE SYSTEMS

Standard systems of tape codes are used throughout industry. One system is the Elec-

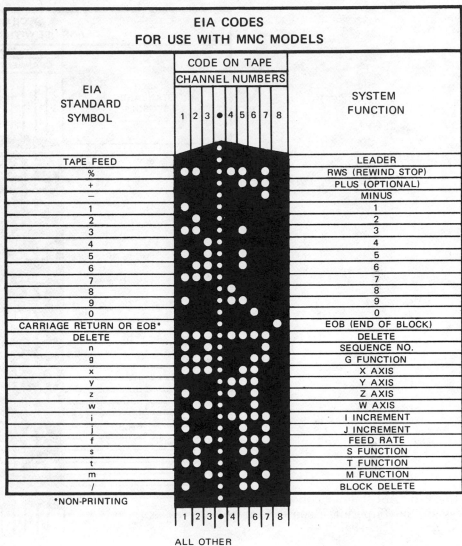

Figure 7. EIA tape codes (Courtesy of the Superior Electric Company).

tronics Industries Association code, known as EIA (Figure 7). Another system is the U.S.A. Standard Code for Information Interchange, known as ASCII (Figure 8). Note that each typewriter symbol has a specific pattern of holes in the tape. Typewriter keys such as "tab" and "carriage return" also produce a specific tape punch code. However, these codes do not produce a printed symbol.

TAPE READERS

The tape reader is usually found in the machine control unit. The MCU may be attached to the machine tool, or it may be freestanding and connected to the machine by appropriate wiring. Tape readers may be electromechanical, photoelectric, or pneumatic. A tape reader consists of the tape reading head and tape transport system. The transport system includes the drive sprocket, tension arms, and tape reels. The tension arms maintain a taut tape as it passes by the reading head. In photoelectric tape readers, the tape may be transported through the reader by a pinch roll instead of by a drive sprocket.

Electromechanical Readers
The electromechanical reader uses electrical contacts that operate through the tape punches. Electromechanical readers are quite fast reading.

ASCII CODES FOR USE WITH MNC MODELS

ASCII STANDARD SYMBOLS	CODE ON TAPE (CHANNEL NUMBERS 1 2 3 • 4 5 6 7 8)	SYSTEM FUNCTION
HERE IS		LEADER
EOT*		RWS (REWIND STOP)
D		RWS (REWIND STOP)
/		BLOCK DELETE
N		SEQUENCE NO. ADDRESS
G		G FUNCTION ADDRESS
X		X AXIS ADDRESS
Y		Y AXIS ADDRESS
Z		Z AXIS ADDRESS
W		W AXIS ADDRESS
I		I INCREMENT ADDRESS
J		J INCREMENT ADDRESS
F		F ADDRESS
S		S FUNCTION ADDRESS
T		T FUNCTION ADDRESS
M		M FUNCTION ADDRESS
LINE FEED*		EOB (END OF BLOCK)
+		PLUS (OPTIONAL)
−		MINUS
1		1
2		2
3		3
4		4
5		5
6		6
7		7
8		8
9		9
0		0
DELETE*		DELETE

*NON-PRINTING

Figure 8. ASCII tape codes (Courtesy of the Superior Electric Company).

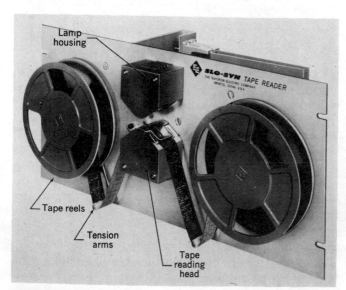

Figure 9. Photoelectric MCU tape reader (Courtesy of the Superior Electric Company).

Photoelectric Readers

The photoelectric reader (Figure 9) uses a concentrated light source that beams light through the tape punches. The light beam activates photoelectric cells. Photoelectric readers are fast reading and are used on numerical controls designed for continuous path machine tools. Continuous path machining may involve many small cuts to approximate a radius, angle, or irregular shape. Instructions for each move must be provided from tape instructions. If the machine tool has to wait for the tape reader to read a tape instruction, time will be lost, resulting in lowered productivity.

To reduce waiting time while instructions are read from the tape, the tape reader reads and stores one instruction ahead. The MCU may be equipped with a memory for this purpose. This is called a buffer storage. While the machine is performing its cut, the tape reader is reading and storing the next instruction. When the cut is completed, the next instruction is instantly available from buffer storage. In this way, the machine does not have to wait for the tape reader to read the next instruction from the tape.

Pneumatic Readers
The pneumatic tape reader uses air flowing through the tape punches to activate electromechanical switches. Pneumatic readers are slower reading than the photoelectric or electromechanical types. They also depend on a precise alignment of the tape over the reader air jets.

self-evaluation

SELF-TEST
1. Name two common tape materials.
2. What are the advantages of the laminate materials?
3. What are the various functions of the typewriter tape punch?
4. Name two standard tape code systems.
5. Name three types of tape readers.

This unit has no post-test.

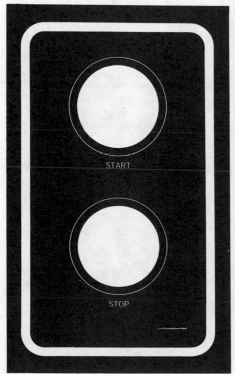

unit 3

operating the n/c machine tool

Before operating the N/C machine tool from tape instructions, you must first become totally familiar with its manual operation. The N/C machine is somewhat different than its conventional counterpart in that control functions are accomplished from the MCU for both manual and tape operation. To protect yourself from injury and to protect an expensive machine tool from damage, it is most important to know ahead of time exactly what will happen when an MCU console control is actuated.

objectives After completing this unit, you will be able to:
1. Describe safe operating procedure for an N/C machine tool.
2. Describe the purpose of tool length gaging equipment.

N/C MACHINE SAFETY

The same safety rules apply to N/C machine tools as apply to all machine tools. Always wear appropriate eye protection and short sleeves. Remove rings and watches before operating the machine. See that the workpiece is properly secured in a vise or by suitable clamps. Be sure that the cutter is clamped securely in its arbor or collet. Use proper speeds and feeds as you would for any machining task.

MANUAL OPERATION FROM THE MCU CONSOLE

Many N/C machine tools, especially those that have been retrofitted with a numerical control system, may have some controls on the MCU console and other controls on the machine itself. Because of this, you must exercise additional caution when operating the machine manually.

The MCU console may have an emergency stop control that will stop worktable or spindle positioning in case a problem should develop during a cut. However, the emergency stop control may not stop spindle rotation. This may be a separate control attached to the machine tool and may be quite far removed from the MCU console.

When positioning the worktable or spindle manually during the setup of the workpiece, be sure that the cutter and machine spindle are clear of vises and clamps before entering positioning data on the MCU console (Figure 1). Check clearances before lowering the quill on a vertical mill or drill press. It is good practice to set up the workpiece on a milling machine so that the quill must be lowered a small amount before beginning a milling cut. The quill should not be extended excessively as this will affect rigidity. If a problem should develop, the cutter can be quickly raised clear of the workpiece by actuating the quill up control on the MCU. However, if the setup has been made such that the quill is already in the full up position during milling, it cannot be raised clear of the workpiece. In case of a problem, the cutter, machine, or workpiece may be damaged.

If the N/C machine has a manually indexed turret to control quill travel (Figure 2), do not forget to index the turret at the appropriate points during the machining cycle. Be precise in adjusting the rapid traverse of the quill as it ap-

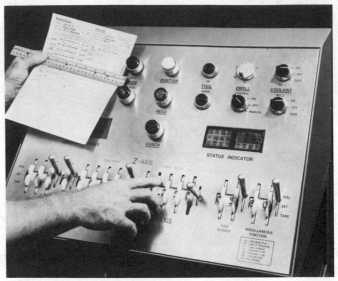

Figure 1. Entering positioning data on the MCU console (Courtesy of Hydra-Point Division, Moog Inc.).

Figure 2. Manually operated turret depth stop on an N/C milling machine (California State University at Fresno).

proaches the workpiece. If the cutter should run into the work during rapid traverse, damage can result to the machine or cutter in addition to the hazard of flying metal.

USING THE N/C MACHINE AS A CONVENTIONAL TOOL

The N/C machine tool may be used as a substitute for its conventional counterpart. This will generally not be done in industry, since the reason for having the N/C machine is to realize its increased productivity and other advantages. However, in the school shop, the N/C machine tool may be used to supplement conventional machines. In fact, the accurate positioning of an N/C machine tool often makes it very effective for a routine machining task.

N/C TAPE OPERATION

After a tape has been prepared, it should be read back at the typewriter tape punch to determine if any errors are present (Figure 3). The printout can be checked against the program manuscript. If the tape is correct, it should be verified by a "dry run" on the machine tool.

Be sure that the spindle and cutter are clear of all obstructions. On a milling machine this can be insured by moving the knee below the maximum extension of the spindle. Insert the tape in the MCU reader and observe the machine as it completes all programmed instructions. N/C machine tools will have a feature that permits tape blocks to be read one at a time. Each sequence can be initiated from the MCU. The machine will read and execute one tape block and stop. Positioning and miscellaneous functions can be observed and checked for accuracy. A dry tape run will safely verify the tape, thus preventing possible damage to the machine, cutter, or workpiece.

GAGING TOOL LENGTHS FOR N/C

An N/C machining task often requires a number of different tools. For example, a drilling operation may require center drilling, drilling, and reaming. If the holes are through, the drill and reamer will have to extend from the tool holder the appropriate distance. Since the center drill is probably shorter, the machine spindle will have to extend further or the worktable raised accordingly.

Figure 3. After punching a tape, obtain a printout using the typewriter reader (Courtesy of Hydra-Point Division, Moog Inc.).

Figure 4. Micrometer tool length gage (Courtesy of Hydra-Point Division, Moog Inc.).

In an industrial setting, an entire set of tools required for a specific job may be preset for length and stored for future use. When a machining job comes to the shop, the N/C machine operator obtains the complete set of preadjusted tools and places them in the machine tool changer or in a tool rack.

The tool length gage is used to measure the projection of cutting tools from their tool holders. Common length gages consist of a series of accurately spaced rings mounted on a column (Figure 4). The instrument is not unlike a precision height gage. Ring spacing is usually 1 in. A

Figure 5. Dial indicator gage for adjusting length and diameter (Courtesy of Cincinnati Milacron).

Figure 6. Electronic digital tool length gage (Courtesy of Cincinnati Milacron).

Figure 7. Setting a tool length with the electronic gage (Courtesy of Cincinnati Milacron).

micrometer head with 1 in. travel spans the distance between the rings and can be placed at any desired height within the range of the gage. The cutting tool to be set is placed in its holder and the amount of projection is adjusted according to the job requirements. Tool length gages may also use dial indicators (Figure 5). This type of gage can be used to set boring bars and insert tooth cutters for specific diameters.

High discrimination electronic tool length gages are also used. These instruments are equipped with digital readouts (Figure 6). They can be used for tool length adjustments (Figure 7) and cutter diameter adjustments (Figure 8). One advantage of the electronic gage is its ability to read in inch and metric dimensions.

Figure 8. Setting an insert tooth milling cutter for diameter using the electronic gage (Courtesy of Cincinnati Milacron).

self-evaluation

SELF-TEST 1. After you have punched a tape, what should you do before placing the tape in the machine tool for a dry run?

2. What is the purpose of a dry tape run?

3. What is an important precaution that must be considered when operating the machine tool from the MCU console?

4. What is the function of the tool length gage?

5. Name two types of tool length gages.

This unit has no post-test.

appendix I

Table 1
Natural Trigonometric Functions

NATURAL TRIGONOMETRIC FUNCTIONS

	0°						1°						2°						3°						
′	sin	cos	tan	cot	sec	cosec	sin	cos	tan	cot	sec	cosec	sin	cos	tan	cot	sec	cosec	sin	cos	tan	cot	sec	cosec	′
0	.00000	1.0000	.00000	Infinite	1.0000	Infinite	.01745	.99985	.01745	57.290	1.0001	57.299	.03490	.99939	.03492	28.636	1.0006	28.654	.05234	.99863	.05241	19.081	1.0014	19.107	60
1	.00029	.0000	.00029	3437.7	.0000	3437.7	.01774	.99984	.01775	56.350	.0001	56.359	.03519	.99938	.03521	28.399	.0006	28.417	.05263	.99861	.05270	18.975	.0014	19.002	59
2	.00058	.0000	.00058	1718.9	.0000	1718.9	.01803	.99984	.01804	55.441	.0001	55.450	.03548	.99937	.03550	28.166	.0006	28.184	.05292	.99860	.05299	18.871	.0014	18.897	58
3	.00087	.0000	.00087	1145.9	.0000	1145.9	.01832	.99983	.01833	54.561	.0002	54.570	.03577	.99936	.03579	27.937	.0006	27.955	.05321	.99858	.05328	18.768	.0014	18.794	57
4	.00116	.0000	.00116	859.44	.0000	859.44	.01861	.99983	.01862	53.708	.0002	53.718	.03606	.99935	.03608	27.712	.0006	27.730	.05350	.99857	.05357	18.665	.0014	18.692	56
5	.00145	1.0000	.00145	687.55	.0000	687.55	.01891	.99982	.01891	52.882	.0002	52.891	.03635	.99934	.03638	27.490	.0007	27.508	.05379	.99855	.05387	18.564	.0014	18.591	55
6	.00174	.0000	.00174	572.96	.0000	572.96	.01920	.99981	.01920	52.081	.0002	52.090	.03664	.99933	.03667	27.271	.0007	27.290	.05408	.99854	.05416	18.464	.0015	18.491	54
7	.00204	.0000	.00204	491.11	.0000	491.11	.01949	.99981	.01949	51.303	.0002	51.313	.03693	.99932	.03696	27.056	.0007	27.075	.05437	.99852	.05445	18.365	.0015	18.393	53
8	.00233	.0000	.00233	429.72	.0000	429.72	.01978	.99980	.01978	50.548	.0002	50.558	.03722	.99931	.03725	26.845	.0007	26.864	.05466	.99850	.05474	18.268	.0015	18.295	52
9	.00262	.0000	.00262	381.97	.0000	381.97	.02007	.99980	.02007	49.816	.0002	49.826	.03751	.99930	.03754	26.637	.0007	26.655	.05495	.99849	.05503	18.171	.0015	18.198	51
10	.00291	0.99999	.00291	343.77	.0000	343.77	.02036	.99979	.02036	49.104	.0002	49.114	.03781	.99928	.03783	26.432	.0007	26.450	.05524	.99847	.05532	18.075	.0015	18.103	50
11	.00320	.99999	.00320	312.52	.0000	312.52	.02065	.99979	.02066	48.412	.0002	48.422	.03810	.99927	.03812	26.230	.0007	26.249	.05553	.99846	.05562	17.980	.0015	18.008	49
12	.00349	.99999	.00349	286.48	.0000	286.48	.02094	.99978	.02095	47.739	.0002	47.750	.03839	.99926	.03842	26.031	.0007	26.050	.05582	.99844	.05591	17.886	.0016	17.914	48
13	.00378	.99999	.00378	264.44	.0000	264.44	.02123	.99977	.02124	47.085	.0002	47.096	.03868	.99925	.03871	25.835	.0007	25.854	.05611	.99842	.05620	17.793	.0016	17.821	47
14	.00407	.99999	.00407	245.55	.0000	245.55	.02152	.99977	.02153	46.449	.0002	46.460	.03897	.99924	.03900	25.642	.0008	25.661	.05640	.99841	.05649	17.701	.0016	17.730	46
15	.00436	.99999	.00436	229.18	.0000	229.18	.02181	.99976	.02182	45.829	.0002	45.840	.03926	.99923	.03929	25.452	.0008	25.471	.05669	.99839	.05678	17.610	.0016	17.639	45
16	.00465	.99999	.00465	214.86	.0000	214.86	.02210	.99975	.02211	45.226	.0002	45.237	.03955	.99922	.03958	25.264	.0008	25.284	.05698	.99837	.05707	17.520	.0016	17.549	44
17	.00494	.99999	.00494	202.22	.0000	202.22	.02240	.99975	.02240	44.638	.0002	44.650	.03984	.99921	.03987	25.080	.0008	25.100	.05727	.99836	.05737	17.431	.0016	17.460	43
18	.00524	.99999	.00524	190.98	.0000	190.99	.02269	.99974	.02269	44.066	.0002	44.077	.04013	.99919	.04016	24.898	.0008	24.918	.05756	.99834	.05766	17.343	.0017	17.372	42
19	.00553	.99998	.00553	180.93	.0000	180.93	.02298	.99974	.02298	43.508	.0003	43.520	.04042	.99918	.04045	24.718	.0008	24.739	.05785	.99832	.05795	17.256	.0017	17.285	41
20	.00582	.99998	.00582	171.88	.0000	171.89	.02326	.99973	.02327	42.964	.0003	42.976	.04071	.99917	.04075	24.542	.0008	24.562	.05814	.99831	.05824	17.169	.0017	17.198	40
21	.00611	.99998	.00611	163.70	.0000	163.70	.02356	.99972	.02357	42.433	.0003	42.445	.04100	.99916	.04104	24.367	.0008	24.388	.05843	.99829	.05853	17.084	.0017	17.113	39
22	.00640	.99998	.00640	156.26	.0000	156.26	.02385	.99971	.02386	41.916	.0003	41.928	.04129	.99915	.04133	24.196	.0008	24.216	.05872	.99827	.05883	16.999	.0017	17.028	38
23	.00669	.99998	.00669	149.46	.0000	149.47	.02414	.99971	.02415	41.410	.0003	41.423	.04158	.99913	.04162	24.026	.0009	24.047	.05902	.99826	.05912	16.915	.0017	16.944	37
24	.00698	.99997	.00698	143.24	.0000	143.24	.02443	.99970	.02444	40.917	.0003	40.930	.04187	.99912	.04191	23.859	.0009	23.880	.05931	.99824	.05941	16.832	.0018	16.861	36
25	.00727	.99997	.00727	137.51	.0000	137.51	.02472	.99969	.02473	40.436	.0003	40.448	.04217	.99911	.04220	23.694	.0009	23.716	.05960	.99822	.05970	16.750	.0018	16.779	35
26	.00756	.99997	.00756	132.22	.0000	132.22	.02501	.99969	.02502	39.965	.0003	39.978	.04246	.99910	.04249	23.532	.0009	23.553	.05989	.99820	.05999	16.668	.0018	16.698	34
27	.00785	.99997	.00785	127.32	.0000	127.32	.02530	.99968	.02531	39.506	.0003	39.518	.04275	.99908	.04279	23.372	.0009	23.393	.06018	.99819	.06029	16.587	.0018	16.617	33
28	.00814	.99997	.00814	122.77	.0000	122.78	.02559	.99967	.02560	39.057	.0003	39.069	.04304	.99907	.04308	23.214	.0009	23.235	.06047	.99817	.06058	16.507	.0018	16.538	32
29	.00843	.99996	.00844	118.54	.0000	118.54	.02589	.99966	.02589	38.618	.0003	38.631	.04333	.99906	.04337	23.058	.0009	23.079	.06076	.99815	.06087	16.428	.0018	16.459	31
30	.00873	.99996	.00873	114.59	.0000	114.59	.02618	.99966	.02618	38.188	.0003	38.201	.04362	.99905	.04366	22.904	.0009	22.925	.06105	.99813	.06116	16.350	.0019	16.380	30
31	.00902	.99996	.00902	110.89	.0000	110.90	.02647	.99965	.02648	37.769	.0003	37.782	.04391	.99903	.04395	22.752	.0010	22.774	.06134	.99812	.06145	16.272	.0019	16.303	29
32	.00931	.99996	.00931	107.43	.0000	107.43	.02676	.99964	.02677	37.358	.0003	37.371	.04420	.99902	.04424	22.602	.0010	22.624	.06163	.99810	.06175	16.195	.0019	16.226	28
33	.00960	.99995	.00960	104.17	.0000	104.17	.02705	.99963	.02706	36.956	.0004	36.969	.04449	.99901	.04453	22.454	.0010	22.476	.06192	.99808	.06204	16.119	.0019	16.150	27
34	.00989	.99995	.00989	101.11	.0000	101.11	.02734	.99963	.02735	36.563	.0004	36.576	.04478	.99900	.04483	22.308	.0010	22.330	.06221	.99806	.06233	16.043	.0019	16.075	26
35	.01018	.99995	.01018	98.218	.0000	98.223	.02763	.99962	.02764	36.177	.0004	36.191	.04507	.99898	.04512	22.164	.0010	22.186	.06250	.99804	.06262	15.969	.0019	16.000	25
36	.01047	.99994	.01047	95.489	.0000	95.495	.02792	.99961	.02793	35.800	.0004	35.814	.04536	.99897	.04541	22.022	.0010	22.044	.06279	.99803	.06291	15.894	.0020	15.926	24
37	.01076	.99994	.01076	92.908	.0000	92.914	.02821	.99960	.02822	35.431	.0004	35.445	.04565	.99896	.04570	21.881	.0010	21.904	.06308	.99801	.06321	15.821	.0020	15.853	23
38	.01105	.99994	.01105	90.463	.0001	90.469	.02850	.99959	.02851	35.069	.0004	35.084	.04594	.99894	.04599	21.742	.0010	21.765	.06337	.99799	.06350	15.748	.0020	15.780	22
39	.01134	.99993	.01134	88.143	.0001	88.149	.02879	.99958	.02880	34.715	.0004	34.729	.04623	.90893	.04628	21.606	.0011	21.629	.06366	.99797	.06379	15.676	.0020	15.708	21
40	.01163	.99993	.01164	85.940	.0001	85.946	.02908	.99958	.02910	34.368	.0004	34.382	.04652	.99892	.04657	21.470	.0011	21.494	.06395	.99795	.06408	15.605	.0020	15.637	20
41	.01193	.99993	.01193	83.843	.0001	83.849	.02937	.99957	.02939	34.027	.0004	34.042	.04681	.99890	.04687	21.337	.0011	21.360	.06424	.99793	.06437	15.534	.0021	15.566	19
42	.01222	.99992	.01222	81.847	.0001	81.853	.02967	.99956	.02968	33.693	.0004	33.708	.04711	.99889	.04716	21.205	.0011	21.228	.06453	.99791	.06467	15.464	.0021	15.496	18
43	.01251	.99992	.01251	79.943	.0001	79.950	.02996	.99955	.02997	33.366	.0004	33.381	.04740	.99888	.04745	21.075	.0011	21.098	.06482	.99790	.06496	15.394	.0021	15.427	17
44	.01280	.99992	.01280	78.126	.0001	78.133	.03025	.99954	.03026	33.045	.0004	33.060	.04769	.99886	.94774	20.946	.0011	20.970	.06511	.99788	.06525	15.325	.0021	15.358	16
45	.01309	.99991	.01309	76.390	.0001	76.396	.03054	.99953	.03055	32.730	.0005	32.745	.04798	.99885	.04803	20.819	.0011	20.843	.06540	.99786	.06554	15.257	.0021	15.290	15
46	.01338	.99991	.01338	74.729	.0001	74.736	.03083	.99952	.03084	32.421	.0005	32.437	.04827	.99883	.04832	20.693	.0012	20.717	.06569	.99784	.06583	15.189	.0022	15.222	14
47	.01367	.99991	.01367	73.139	.0001	73.146	.03112	.99951	.03113	32.118	.0005	32.134	.04856	.99882	.04862	20.569	.0012	20.593	.06598	.99782	.06613	15.122	.0022	15.155	13
48	.01396	.99990	.01396	71.615	.0001	71.622	.03141	.99951	.03143	31.820	.0005	31.836	.04885	.99881	.04891	20.446	.0012	20.471	.06627	.99780	.06642	15.056	.0022	15.089	12
49	.01425	.99990	.01425	70.153	.0001	70.160	.03170	.99950	.03172	31.528	.0005	31.544	.04914	.99879	.04920	20.325	.0012	20.350	.06656	.99778	.06671	14.990	.0022	15.023	11
50	.01454	.99989	.01454	68.750	.0001	68.757	.03199	.99949	.03201	31.241	.0005	31.257	.04943	.99878	.04949	20.205	.0012	20.230	.06685	.99776	.06700	14.924	.0022	14.958	10
51	.01483	.99989	.01484	67.402	.0001	67.409	.03228	.99948	.03230	30.960	.0005	30.976	.04972	.99876	.04978	20.087	.0012	20.112	.06714	.99774	.06730	14.860	.0023	14.893	9
52	.01512	.99988	.01513	66.105	.0001	66.113	.03257	.99947	.03259	30.683	.0005	30.699	.05001	.99875	.05007	19.970	.0012	19.995	.06743	.99772	.06759	14.795	.0023	14.829	8
53	.01542	.99988	.01542	64.858	.0001	64.866	.03286	.99946	.03288	30.411	.0005	30.428	.05030	.99873	.05037	19.854	.0013	19.880	.06772	.99770	.06788	14.732	.0023	14.765	7
54	.01571	.99988	.01571	63.657	.0001	63.664	.03315	.99945	.03317	30.145	.0005	30.161	.05059	.99872	.05066	19.740	.0013	19.766	.06801	.99768	.06817	14.668	.0023	14.702	6
55	.01600	.99987	.01600	62.499	.0001	62.507	.03344	.99944	.03346	29.882	.0005	29.899	.05088	.99870	.05095	19.627	.0013	19.653	.06830	.99766	.06846	14.606	.0023	14.640	5
56	.01629	.99987	.01629	61.383	.0001	61.391	.03374	.99943	.03375	29.624	.0006	29.641	.05117	.99869	.05124	19.515	.0013	19.541	.06859	.99764	.06876	14.544	.0024	14.578	4
57	.01658	.99987	.01658	60.306	.0001	60.314	.03403	.99942	.03405	29.371	.0006	29.388	.05146	.99867	.05153	19.405	.0013	19.431	.06888	.99762	.06905	14.482	.0024	14.517	3
58	.01687	.99986	.01687	59.266	.0001	59.274	.03432	.99941	.03434	29.122	.0006	29.139	.05175	.99866	.05182	19.296	.0013	19.322	.06918	.99760	.06934	14.421	.0024	14.456	2
59	.01716	.99985	.01716	58.261	.0001	58.270	.03461	.99940	.03463	28.877	.0006	28.894	.05204	.99864	.05212	19.188	.0013	19.214	.06947	.99758	.06963	14.361	.0024	14.395	1
60	.01745	.99985	.01745	57.290	.0001	57.299	.03490	.99939	.03492	28.636	.0006	28.654	.05234	.99863	.05241	19.081	.0014	19.107	.06976	.99756	06993	14.301	.0024	14.335	0
′	cos	sin	cot	tan	cosec	sec	cos	sin	cot	tan	cosec	sec	cos	sin	cot	tan	cosec	sec	cos	sin	cot	tan	cosec	sec	′
	89°						88°						87°						86°						

* *Courtesy - Bethlehem Steel Corp.*

874 Machine Tool Practices

NATURAL TRIGONOMETRIC FUNCTIONS

	4°						5°						6°						7°						
′	sin	cos	tan	cot	sec	cosec	sin	cos	tan	cot	sec	cosec	sin	cos	tan	cot	sec	cosec	sin	cos	tan	cot	sec	cosec	′
0	.06976	.99756	.06993	14.301	1.0024	14.335	.08715	.99619	.08749	11.430	1.0038	11.474	.10453	.99452	.10510	9.5144	1.0055	9.5668	.12187	.99255	.12278	8.1443	1.0075	8.2055	60
1	.07005	.99754	.07022	14.241	.0025	14.276	.08744	.99617	.08778	11.392	.0038	11.436	.10482	.99449	.10540	.4878	.0055	.5404	.12216	.99251	.12308	.1248	.0075	.1861	59
2	.07034	.99752	.07051	14.182	.0025	14.217	.08773	.99614	.08807	11.354	.0039	11.398	.10511	.99446	.10569	.4614	.0056	.5141	.12245	.99247	.12337	.1053	.0076	.1668	58
3	.07063	.99750	.07080	14.123	.0025	14.159	.08802	.99612	.08837	11.316	.0039	11.360	.10540	.99443	.10599	.4351	.0056	.4880	.12273	.99244	.12367	.0860	.0076	.1476	57
4	.07092	.99748	.07110	14.065	.0025	14.101	.08831	.99609	.08866	11.279	.0039	11.323	.10568	.99440	.10628	.4090	.0056	.4620	.12302	.99240	.12396	.0667	.0076	.1285	56
5	.07121	.99746	.07139	14.008	.0025	14.043	.08860	.99607	.08895	11.242	.0039	11.286	.10597	.99437	.10657	9.3831	.0057	9.4362	.12331	.99237	.12426	8.0476	.0077	8.1094	55
6	.07150	.99744	.07168	13.951	.0026	13.986	.08889	.99604	.08925	11.205	.0040	11.249	.10626	.99434	.10687	.3572	.0057	.4105	.12360	.99233	.12456	.0285	.0077	.0905	54
7	.07179	.99742	.07197	13.894	.0026	13.930	.08918	.99601	.08954	11.168	.0040	11.213	.10655	.99431	.10716	.3315	.0057	.3850	.12389	.99229	.12485	.0095	.0078	.0717	53
8	.07208	.99740	.07226	13.838	.0026	13.874	.08947	.99599	.08983	11.132	.0040	11.176	.10684	.99428	.10746	.3060	.0057	.3596	.12418	.99226	.12515	7.9906	.0078	.0529	52
9	.07237	.99738	.07256	13.782	.0026	13.818	.08976	.99596	.09013	11.095	.0040	11.140	.10713	.99424	.10775	.2806	.0058	.3343	.12447	.99222	.12544	.9717	.0078	.0342	51
10	.07266	.99736	.07285	13.727	.0026	13.763	.09005	.99594	.09042	11.059	.0041	11.104	.10742	.99421	.10805	9.2553	.0058	9.3092	.12476	.99219	.12574	7.9530	.0079	8.0156	50
11	.07295	.99733	.07314	13.672	.0027	13.708	.09034	.99591	.09071	11.024	.0041	11.069	.10771	.99418	.10834	.2302	.0058	.2842	.12504	.99215	.12603	.9344	.0079	7.9971	49
12	.07324	.99731	.07343	13.617	.0027	13.654	.09063	.99588	.09101	10.988	.0041	11.033	.10800	.99415	.10863	.2051	.0059	.2593	.12533	.99211	.12633	.9158	.0079	.9787	48
13	.07353	.99729	.07373	13.563	.0027	13.600	.09092	.99586	.09130	10.953	.0041	10.998	.10829	.99412	.10893	.1803	.0059	.2346	.12562	.99208	.12662	.8973	.0080	.9604	47
14	.07382	.99727	.07402	13.510	.0027	13.547	.09121	.99583	.09159	10.918	.0042	10.963	.10858	.99409	.10922	.1555	.0059	.2100	.12591	.99204	.12692	.8789	.0080	.9421	46
15	.07411	.99725	.07431	13.457	.0027	13.494	.09150	.99580	.09189	10.883	.0042	10.929	.10887	.99406	.10952	9.1309	.0060	9.1855	.12620	.99200	.12722	7.8606	.0080	7.9240	45
16	.07440	.99723	.07460	13.404	.0028	13.441	.09179	.99578	.09218	10.848	.0042	10.894	.10916	.99402	.10981	.1064	.0060	.1612	.12649	.99197	.12751	.8424	.0081	.9059	44
17	.07469	.99721	.07490	13.351	.0028	13.389	.09208	.99575	.09247	10.814	.0043	10.860	.10944	.99399	.11011	.0821	.0060	.1370	.12678	.99193	.12781	.8243	.0081	.8879	43
18	.07498	.99718	.07519	13.299	.0028	13.337	.09237	.99572	.09277	10.780	.0043	10.826	.10973	.99396	.11040	.0579	.0061	.1129	.12706	.99189	.12810	.8062	.0082	.8700	42
19	.07527	.99716	.07548	13.248	.0028	13.286	.09266	.99570	.09306	10.746	.0043	10.792	.11002	.99393	.11069	.0338	.0061	.0890	.12735	.99186	.12840	.7882	.0082	.8522	41
20	.07556	.99714	.07577	13.917	.0029	13.235	.09295	.99567	.09335	10.712	.0043	10.758	.11031	.99390	.11099	9.0098	.0061	9.0651	.12764	.99182	.12869	7.7703	.0082	7.8344	40
21	.07585	.99712	.07607	13.146	.0029	13.184	.09324	.99564	.09365	10.678	.0044	10.725	.11060	.99386	.11128	8.9860	.0062	.0414	.12793	.99178	.12899	.7525	.0083	.8168	39
22	.07614	.99710	.07636	13.096	.0029	13.134	.09353	.99562	.09394	10.645	.0044	10.692	.11089	.99383	.11158	.9623	.0062	.0179	.12822	.99174	.12928	.7348	.0083	.7992	38
23	.07643	.99707	.07665	13.046	.0029	13.084	.09382	.99559	.09423	10.612	.0044	10.659	.11118	.99380	.11187	.9387	.0062	8.9944	.12851	.99171	.12958	.7171	.0084	.7817	37
24	.07672	.99705	.07694	12.996	.0029	13.034	.09411	.99556	.09453	10.579	.0044	10.626	.11147	.99377	.11217	.9152	.0063	.9711	.12879	.99167	.12988	.6996	.0084	.7642	36
25	.07701	.99703	.07724	12.947	.0030	12.985	.09440	.99553	.09482	10.546	.0045	10.593	.11176	.99373	.11246	8.8918	.0063	8.9479	.12908	.99163	.13017	7.6821	.0084	7.7469	35
26	.07730	.99701	.07753	12.898	.0030	12.937	.09469	.99551	.09511	10.514	.0045	10.561	.11205	.99370	.11276	.8686	.0063	.9248	.12937	.99160	.13047	.6646	.0085	.7296	34
27	.07759	.99698	.07782	12.849	.0030	12.888	.09498	.99548	.09541	10.481	.0045	10.529	.11234	.99367	.11305	.8455	.0064	.9018	.12966	.99156	.13076	.6473	.0085	.7124	33
28	.07788	.99696	.07812	12.801	.0030	12.840	.09527	.99545	.09570	10.449	.0046	10.497	.11262	.99364	.11335	.8225	.0064	.8790	.12995	.99152	.13106	.6300	.0085	.6953	32
29	.07817	.99694	.07841	12.754	.0031	12.793	.09556	.99542	.09599	10.417	.0046	10.465	.11291	.99360	.11364	.7996	.0064	.8563	.13024	.99148	.13136	.6129	.0086	.6783	31
30	.07846	.99692	.07870	12.706	.0031	12.745	.09584	.99540	.09629	10.385	.0046	10.433	.11320	.99357	.11393	8.7769	.0065	8.8337	.13053	.99144	.13165	7.5957	.0086	7.6613	30
31	.07875	.99689	.07899	12.659	.0031	12.698	.09613	.99537	.09658	10.354	.0046	10.402	.11349	.99354	.11423	.7542	.0065	.8112	.13081	.99141	.13195	.5787	.0087	.6444	29
32	.07904	.99687	.07929	12.612	.0031	12.652	.09642	.99534	.09688	10.322	.0047	10.371	.11378	.99350	.11452	.7317	.0065	.7888	.13110	.99137	.13224	.5617	.0087	.6276	28
33	.07933	.99685	.07958	12.566	.0032	12.606	.09671	.99531	.09717	10.291	.0047	10.340	.11407	.99347	.11482	.7093	.0066	.7665	.13139	.99133	.13254	.5449	.0087	.6108	27
34	.07962	.99682	.07987	12.520	.0032	12.560	.09700	.99528	.09746	10.260	.0047	10.309	.11436	.99344	.11511	.6870	.0066	.7444	.13168	.99129	.13284	.5280	.0088	.5942	26
35	.07991	.99680	.08016	12.474	.0032	12.514	.09729	.99525	.09776	10.229	.0048	10.278	.11465	.99341	.11541	8.6648	.0066	8.7223	.13197	.99125	.13313	7.5113	.0088	7.5776	25
36	.08020	.99678	.08046	12.429	.0032	12.469	.09758	.99523	.09805	10.199	.0048	10.248	.11494	.99337	.11570	.6427	.0067	.7004	.13226	.99121	.13343	.4946	.0089	.5611	24
37	.08049	.99675	.08075	12.384	.0032	12.424	.09787	.99520	.09834	10.168	.0048	10.217	.11523	.99334	.11600	.6208	.0067	.6786	.13254	.99118	.13372	.4780	.0089	.5446	23
38	.08078	.99673	.08104	12.339	.0033	12.379	.09816	.99517	.09864	10.138	.0048	10.187	.11551	.99330	.11629	.5989	.0067	.6569	.13283	.99114	.13402	.4615	.0089	.5282	22
39	.08107	.99671	.08134	12.295	.0033	12.335	.09845	.99514	.09893	10.108	.0049	10.157	.11580	.99327	.11659	.5772	.0068	.6353	.13312	.99110	.13432	.4451	.0090	.5119	21
40	.08136	.99668	.08163	12.250	.0033	12.291	.09874	.99511	.09922	10.078	.0049	10.127	.11609	.99324	.11688	8.5555	.0068	8.6138	.13341	.99106	.13461	7.4287	.0090	7.4957	20
41	.08165	.99666	.08192	12.207	.0033	12.248	.09903	.99508	.09952	10.048	.0049	10.098	.11638	.99320	.11718	.5340	.0068	.5924	.13370	.99102	.13491	.4124	.0090	.4795	19
42	.08194	.99664	.08221	12.163	.0034	12.204	.09932	.99505	.09981	10.019	.0050	10.068	.11667	.99317	.11747	.5126	.0069	.5711	.13399	.99098	.13520	.3961	.0091	.4634	18
43	.08223	.99661	.08251	12.120	.0034	12.161	.09961	.99503	.10011	9.9893	.0050	10.039	.11696	.99314	.11777	.4913	.0069	.5499	.13427	.99094	.13550	.3800	.0091	.4474	17
44	.08252	.99659	.08280	12.077	.0034	12.118	.09990	.99500	.10040	9.9601	.0050	10.010	.11725	.99310	.11806	.4701	.0069	.5289	.13456	.99090	.13580	.3639	.0092	.4315	16
45	.08281	.99656	.08309	12.035	.0034	12.076	.10019	.99497	.10069	9.9310	.0050	9.9812	.11754	.99307	.11836	8.4489	.0070	8.5079	.13485	.99086	.13609	7.3479	.0092	7.4156	15
46	.08310	.99654	.08339	11.992	.0035	12.034	.10048	.99494	.10099	9.9021	.0051	9.9525	.11783	.99303	.11865	.4279	.0070	.4871	.13514	.99083	.13639	.3319	.0092	.3998	14
47	.08339	.99652	.08368	11.950	.0035	11.992	.10077	.99491	.10128	9.8734	.0051	9.9239	.11811	.99300	.11895	.4070	.0070	.4663	.13543	.99079	.13669	.3160	.0093	.3840	13
48	.08368	.99649	.08397	11.909	.0035	11.950	.10106	.99488	.10158	9.8448	.0051	9.8955	.11840	.99296	.11924	.3862	.0071	.4457	.13571	.99075	.13698	.3002	.0093	.3683	12
49	.08397	.99647	.08426	11.867	.0035	11.909	.10134	.99485	.10187	9.8164	.0052	9.8672	.11869	.99293	.11954	.3655	.0071	.4251	.13600	.99071	.13728	.2844	.0094	.3527	11
50	.08426	.99644	.08456	11.826	.0036	11.868	.10163	.99482	.10216	9.7882	.0052	9.8391	.11898	.99290	.11983	8.3449	.0071	8.4046	.13629	.99067	.13757	7.2687	.0094	7.3372	10
51	.08455	.99642	.08485	11.785	.0036	11.828	.10192	.99479	.10246	9.7601	.0052	9.8112	.11927	.99286	.12013	.3244	.0072	.3843	.13658	.99063	.13787	.2531	.0094	.3217	9
52	.08484	.99639	.08514	11.745	.0036	11.787	.10221	.99476	.10275	9.7322	.0053	9.7834	.11956	.99283	.12042	.3040	.0072	.3640	.13687	.99059	.13817	.2375	.0095	.3063	8
53	.08513	.99637	.08544	11.704	.0036	11.747	.10250	.99473	.10305	9.7044	.0053	9.7558	.11985	.99279	.12072	.2837	.0073	.3439	.13716	.99055	.13846	.2220	.0095	.2909	7
54	.08542	.99634	.08573	11.664	.0037	11.707	.10279	.99470	.10334	9.6768	.0053	9.7283	.12014	.99276	.12101	.2635	.0073	.3238	.13744	.99051	.13876	.2066	.0096	.2757	6
55	.08571	.99632	.08602	11.625	.0037	11.668	.10308	.99467	.10363	9.6493	.0053	9.7010	.12042	.99272	.12131	8.2434	.0073	8.3039	.13773	.99047	.13906	7.1912	.0096	7.2604	5
56	.08600	.99629	.08632	11.585	.0037	11.628	.10337	.99464	.10393	9.6220	.0054	9.6739	.12071	.99269	.12160	.2234	.0074	.2840	.13802	.99043	.13935	.1759	.0097	.2453	4
57	.08629	.99627	.08661	11.546	.0037	11.589	.10366	.99461	.10422	9.5949	.0054	9.6469	.12100	.99265	.12190	.2035	.0074	.2642	.13831	.99039	.13965	.1607	.0097	.2302	3
58	.08658	.99624	.08690	11.507	.0038	11.550	.10395	.99458	.10452	9.5679	.0054	9.6200	.12129	.99262	.12219	.1837	.0074	.2446	.13860	.99035	.13995	.1455	.0097	.2152	2
59	.08687	.99622	.08719	11.468	.0038	11.512	.10424	.99455	.10481	9.5411	.0055	9.5933	.12158	.99258	.12249	.1640	.0075	.2250	.13888	.99031	.14024	.1304	.0098	.2002	1
60	.08715	.99619	.08749	11.430	.0038	11.474	.10453	.99452	.10510	9.5144	.0055	9.5668	.12187	.99255	.12278	8.1443	.0075	8.2055	.13917	.99027	.14054	7.1154	.0098	7.1853	0
′	cos	sin	cot	tan	cosec	sec	cos	sin	cot	tan	cosec	sec	cos	sin	cot	tan	cosec	sec	cos	sin	cot	tan	cosec	sec	′
	85°						84°						83°						82°						

	8°						9°						10°						11°						
′	sin	cos	tan	cot	sec	cosec	sin	cos	tan	cot	sec	cosec	sin	cos	tan	cot	sec	cosec	sin	cos	tan	cot	sec	cosec	′
0	.13917	.99027	.14054	7.1154	1.0098	7.1853	.15643	.98769	.15838	6.3137	1.0215	6.3924	.17365	.98481	.17633	5.6713	1.0154	5.7588	.19081	.98163	.19438	5.1445	1.0187	5.2408	60
1	.13946	.99023	.14084	.1004	.0099	.1704	.15672	.98764	.15868	.3019	.0125	.3807	.17393	.98476	.17663	.6616	.0155	.7493	.19109	.98157	.19468	.1366	.0188	.2330	59
2	.13975	.99019	.14113	.0854	.0099	.1557	.15701	.98760	.15898	.2901	.0125	.3690	.17422	.98471	.17693	.6520	.0155	.7398	.19138	.98152	.19498	.1286	.0188	.2252	58
3	.14004	.99015	.14143	.0706	.0099	.1409	.15730	.98755	.15928	.2783	.0126	.3574	.17451	.98465	.17723	.6425	.0156	.7304	.19166	.98146	.19529	.1207	.0189	.2174	57
4	.14032	.99010	.14173	.0558	.0100	.1263	.15758	.98750	.15958	.2665	.0126	.3458	.17479	.98460	.17753	.6329	.0156	.7210	.19195	.98140	.19559	.1128	.0189	.2097	56
5	.14061	.99006	.14202	7.0410	.0100	7.1117	.15787	.98746	.15987	6.2548	.0127	6.3343	.17508	.98455	.17783	5.6234	.0157	5.7117	.19224	.98135	.19589	5.1049	.0190	5.2019	55
6	.14090	.99002	.14232	.0264	.0101	.0972	.15816	.98741	.16017	.2432	.0217	.3228	.17537	.98450	.17813	.6140	.0157	.7023	.19252	.98129	.19619	.0970	.0191	.1942	54
7	.14119	.98998	.14262	.0117	.0101	.0827	.15844	.98737	.16047	.2316	.0128	.3113	.17565	.98445	.17843	.6045	.0158	.6930	.19281	.98124	.19649	.0892	.0191	.1865	53
8	.14148	.98994	.14291	6.9972	.0102	.0683	.15873	.98732	.16077	.2200	.0128	.2999	.17594	.98440	.17873	.5951	.0158	.6838	.19309	.98118	.19680	.0814	.0192	.1788	52
9	.14176	.98990	.14321	.9827	.0102	.0539	.15902	.98727	.16107	.2085	.0129	.2885	.17622	.98435	.17903	.5857	.0159	.6745	.19338	.98112	.19710	.0736	.0192	.1712	51
10	.14205	.98986	.14351	6.9682	.0102	7.0396	.15931	.98723	.16137	6.1970	.0129	6.2772	.17651	.98430	.17933	5.5764	.0159	5.6653	.19366	.98107	.19740	5.0658	.0193	5.1636	50
11	.14234	.98982	.14380	.9538	.0103	.0254	.15959	.98718	.16167	.1856	.0130	.2659	.17680	.98425	.17963	.5670	.0160	.6561	.19395	.98101	.19770	.0581	.0193	.1560	49
12	.14263	.98978	.14410	.9395	.0103	.0112	.15988	.98714	.16196	.1742	.0130	.2546	.17708	.98419	.17993	.5578	.0160	.6470	.19423	.98095	.19800	.0504	.0194	.1484	48
13	.14292	.98973	.14440	.9252	.0104	6.9971	.16017	.98709	.16226	.1628	.0131	.2434	.17737	.98414	.18023	.5485	.0161	.6379	.19452	.98090	.19831	.0427	.0195	.1409	47
14	.14320	.98969	.14470	.9110	.0104	.9830	.16045	.98704	.16256	.1515	.0131	.2322	.17766	.98409	.18053	.5393	.0162	.6288	.19480	.98084	.19861	.0350	.0195	.1333	46
15	.14349	.98965	.14499	6.8969	.0104	6.9690	.16074	.98700	.16286	6.1402	.0132	6.2211	.17794	.98404	.18083	5.5301	.0162	5.6197	.19509	.98078	.19891	5.0273	.0196	5.1258	45
16	.14378	.98961	.14529	.8828	.0105	.9550	.16103	.98695	.16316	.1290	.0132	.2100	.17823	.98399	.18113	.5209	.0163	.6107	.19537	.98073	.19921	.0197	.0196	.1183	44
17	.14407	.98957	.14559	.8687	.0105	.9411	.16132	.98690	.16346	.1178	.0133	.1990	.17852	.98394	.18143	.5117	.0163	.6017	.19566	.98067	.19952	.0121	.0197	.1109	43
18	.14436	.98952	.14588	.8547	.0106	.9273	.16160	.98685	.16376	.1066	.0133	.1880	.17880	.98388	.18173	.5026	.0164	.5928	.19595	.98061	.19982	.0045	.0198	.1034	42
19	.14464	.98948	.14618	.8408	.0106	.9135	.16189	.98681	.16405	.0955	.0134	.1770	.17909	.98383	.18203	.4936	.0164	.5838	.19623	.98056	.20012	4.9969	.0198	.0960	41
20	.14493	.98944	.14648	6.8269	.0107	6.8998	.16218	.98676	.16435	6.0844	.0134	6.1661	.17937	.98378	.18233	5.4845	.0165	5.5749	.19652	.98050	.20042	4.9894	.0199	5.0886	40
21	.14522	.98940	.14677	.8131	.0107	.8861	.16246	.98671	.16465	.0734	.0135	.1552	.17966	.98373	.18263	.4755	.0165	.5660	.19680	.98044	.20073	.9819	.0199	.0812	39
22	.14551	.98936	.14707	.7993	.0107	.8725	.16275	.98667	.16495	.0624	.0135	.1443	.17995	.98368	.18293	.4665	.0166	.5572	.19709	.98039	.20103	.9744	.0200	.0739	38
23	.14579	.98931	.14737	.7856	.0108	.8589	.16304	.98662	.16525	.0514	.0136	.1335	.18023	.98362	.18323	.4575	.0166	.5484	.19737	.98033	.20133	.9669	.0201	.0666	37
24	.14608	.98927	.14767	.7720	.0108	.8454	.16333	.98657	.16555	.0405	.0136	.1227	.18052	.98357	.18353	.4486	.0167	.5396	.19766	.98027	.20163	.9594	.0201	.0593	36
25	.14637	.98923	.14796	6.7584	.0109	6.8320	.16361	.98652	.16585	6.0296	.0136	6.1120	.18080	.98352	.18383	5.4396	.0167	5.5308	.19794	.98021	.20194	4.9520	.0202	5.0520	35
26	.14666	.98919	.14826	.7448	.0109	.8185	.16390	.98648	.16615	.0188	.0137	.1013	.18109	.98347	.18413	.4308	.0168	.5221	.19823	.98016	.20224	.9446	.0202	.0447	34
27	.14695	.98914	.14856	.7313	.0110	.8052	.16419	.98643	.16644	.0080	.0137	.0906	.18138	.98341	.18444	.4219	.0169	.5134	.19851	.98010	.20254	.9372	.0203	.0375	33
28	.14723	.98910	.14886	.7179	.0110	.7919	.16447	.98638	.16674	5.9972	.0138	.0800	.18166	.98336	.18474	.4131	.0169	.5047	.19880	.98004	.20285	.9298	.0204	.0302	32
29	.14752	.98906	.14915	.7045	.0111	.7787	.16476	.98633	.16704	.9865	.0138	.0694	.18195	.98331	.18504	.4043	.0170	.4960	.19908	.97998	.20315	.9225	.0204	.0230	31
30	.14781	.98901	.14945	6.6911	.0111	6.7655	.16505	.98628	.16734	5.9758	.0139	6.0588	.18223	.98325	.18534	5.3955	.0170	5.4874	.19937	.97992	.20345	4.9151	.0205	5.0158	30
31	.14810	.98897	.14975	.6779	.0111	.7523	.16533	.98624	.16764	.9651	.9139	.0483	.18252	.98320	.18564	.3868	.0171	.4788	.19965	.97987	.20375	.9078	.0205	.0087	29
32	.14838	.98893	.15004	.6646	.0112	.7392	.16562	.98619	.16794	.9545	.9140	.0379	.18281	.98315	.18594	.3780	.0171	.4702	.19994	.97981	.20406	.9006	.0206	.0015	28
33	.14867	.98889	.15034	.6514	.0112	.7262	.16591	.98614	.16824	.9439	.0140	.0274	.18309	.98309	.18624	.3694	.0172	.4617	.20022	.97975	.20436	.8933	.0207	4.9944	27
34	.14896	.98884	.15064	.6383	.0113	.7132	.16619	.98609	.16854	.9333	.0141	.0170	.18338	.98304	.18654	.3607	.0172	.4532	.20051	.97969	.20466	.8860	.0207	.9873	26
35	.14925	.98880	.15094	6.6252	.0113	6.7003	.16648	.98604	.16884	5.9228	.0141	6.0066	.18366	.98299	.18684	5.3521	.0173	5.4447	.20079	.97963	.20497	4.8788	0208	4.9802	25
36	.14953	.98876	.15123	.6122	.0114	.6874	.16677	.98600	.16914	.9123	.0142	5.9963	.18395	.98293	.18714	.3434	.0174	.4362	.20108	.97957	.20527	.8716	.0208	.9732	24
37	.14982	.98871	.15153	.5992	.0114	.6745	.16705	.98595	.16944	.9019	.0142	.9860	.18424	.98288	.18745	.3349	.0174	.4278	.20136	.97952	.20557	.8644	.0209	.9661	23
38	.15011	.98867	.15183	.5863	.0115	.6617	.16734	.98590	.16973	.8915	.0143	.9758	.18452	.98283	.18775	.3263	.0175	.4194	.20165	.97946	.20588	.8573	.0210	.9591	22
39	.15050	.98862	.15213	.5734	.0115	.6490	.16763	.98585	.17003	.8811	.0143	.9655	.18481	.98277	.18805	.3178	.0175	.4110	.20193	.97940	.20618	.8501	.0210	.9521	21
40	.15068	.98858	.15243	6.5605	.0115	6.6363	.16791	.98580	.17033	5.8708	.0144	5.9554	.18509	.98272	.18835	5.3093	.0176	5.4026	.20222	.97934	.20648	4.8430	.0211	4.9452	20
41	.15097	.98854	.15272	.5478	.0116	.6237	.16820	.98575	.17063	.8605	.0144	.9452	.18538	.98267	.18865	.3008	.0176	.3943	.20250	.97928	.20679	.8359	.0211	.9382	19
42	.15126	.98849	.15302	.5350	.0116	.6111	.16849	.98570	.17093	.8502	.0145	.9351	.18567	.98261	.18895	.2923	.0177	.3860	.20279	.97922	.20709	.8288	.0212	.9313	18
43	.15155	.98845	.15332	.5223	.0117	.5985	.16878	.98565	.17123	.8400	.0145	.9250	.18595	.98256	.18925	.2839	.0177	.3777	.20307	.97916	.20739	.8217	.0213	.9243	17
44	.15183	.98840	.15362	.5097	.0117	.5860	.16906	.98560	.17153	.8298	.0146	.9150	.18624	.98250	.18955	.2755	.0178	.3695	.20336	.97910	.20770	.8147	.0213	.9175	16
45	.15212	.98836	.15391	6.4971	.0118	6.5736	.16935	.98556	.17183	5.8196	.0146	5.9049	.18652	.98245	.18985	5.2671	.0179	5.3612	.20364	.97904	.20800	4.8077	.0214	4.9106	15
46	.15241	.98832	.15421	.4845	.0118	.5612	.16964	.98551	.17213	.8095	.0147	.8950	.18681	.98240	.19016	.2588	.0179	.3530	.20393	.97899	.20830	.8007	.0215	.9037	14
47	.15270	.98827	.15451	.4720	.0119	.5488	.16992	.98546	.17243	.7994	.0147	.8850	.18709	.98234	.19046	.2505	.0180	.3449	.20421	.97893	.20861	.7937	.0215	.8969	13
48	.15298	.98823	.15481	.4596	.0119	.5365	.17021	.98541	.17273	.7894	.0148	.8751	.18738	.98229	.19076	.2422	.0180	.3367	.20450	.97887	.20891	.7867	.0216	.8901	12
49	.15328	.98818	.15511	.4472	.0119	.5243	.17050	.98536	.17303	.7794	.0148	.8652	.18767	.98223	.19106	.2339	.0181	.3286	.20478	.97881	.20921	.7798	.0216	.8833	11
50	.15356	.98814	.15540	6.4348	.0120	6.5121	.17078	.98531	.17333	5.7694	.0149	5.8554	.18795	.98218	.19136	5.2257	.0181	5.3205	.20506	.97875	.20952	4.7728	.0217	4.8765	10
51	.15385	.98809	.15570	.4225	.0120	.4999	.17107	.98526	.17363	.7594	.0150	.8456	.18824	.98212	.19166	.2174	.0182	.3124	.20535	.97869	.20982	.7659	.0218	.8967	9
52	.15413	.98805	.15600	.4103	.0121	.4878	.17136	.98521	.17393	.7495	.0150	.8358	.18852	.98207	.19197	.2092	.0182	.3044	.20563	.97863	.21012	.7591	.0218	.8630	8
53	.15442	.98800	.15630	.3980	.0121	.4757	.17164	.98516	.17423	.7396	.0151	.8261	.18881	.98201	.19227	.2011	.0183	.2963	.20592	.97857	.21043	.7522	.0219	.8563	7
54	.15471	.98796	.15659	.3859	.0122	.4637	.17193	.98511	.17453	.7297	.0151	.8163	.18909	.98196	.19257	.1929	.0184	.2883	.20620	.97851	.21073	.7453	.0220	.8496	6
55	.15500	.98791	.15689	6.3737	.0122	6.4517	.17221	.98506	.17483	5.7199	.0152	5.8067	.18938	.98190	.19287	5.1848	.0184	5.2803	.20649	.97845	.21104	4.7385	.0220	4.8429	5
56	.15528	.98787	.15719	.3616	.0123	.4398	.17250	.98501	.17513	.7101	.0152	.7970	.18967	.98185	.19317	.1767	.0185	.2724	.20677	.97839	.21134	.7317	.0221	.8362	4
57	.15557	.98782	.15749	.3496	.0123	.4279	.17279	.98496	.17543	.7004	.0153	.7874	.18995	.98179	.19347	.1686	.0185	.2645	.20706	.97833	.21164	.7249	.0221	.8296	3
58	.15586	.98778	.15779	.3376	.0124	.4160	.17307	.98491	.17573	.6906	.0153	.7778	.19024	.98174	.19378	.1606	.0186	.2566	.20734	.97827	.21195	.7181	.0222	.8229	2
59	.15615	.98773	.15809	.3257	.0124	.4042	.17336	.98486	.17603	.6809	.0154	.7683	.19052	.98168	.19408	.1525	.0186	.2487	.20763	.97821	.21225	.7114	.0223	.8163	1
60	.15643	.98769	.15838	6.3137	.0125	6.3924	.17365	.98481	.17633	5.6713	.0154	5.7588	.19081	.98163	.19438	5.1445	.0187	5.2408	.20791	.97815	.21256	4.7046	.0223	4.8097	0
′	cos	sin	cot	tan	cosec	sec	cos	sin	cot	tan	cosec	sec	cos	sin	cot	tan	cosec	sec	cos	sin	cot	tan	cosec	sec	′
	81°						80°						79°						78°						

876 Machine Tool Practices

	12°						13°						14°						15°						
′	sin	cos	tan	cot	sec	cosec	sin	cos	tan	cot	sec	cosec	sin	cos	tan	cot	sec	cosec	sin	cos	tan	cot	sec	cosec	′
0	.20791	.97815	.21256	4.7046	1.0223	4.8097	.22495	.97437	.23087	4.3315	1.0263	4.4454	.24192	.97029	.24933	4.0108	1.0306	4.1336	.25882	.96592	.26795	3.7320	1.0353	3.8637	60
1	.20820	.97809	.21286	.6979	.0224	.8032	.22523	.97430	.23117	.3257	.0264	.4398	.24220	.97022	.24964	.0058	.0307	.1287	.25910	.96585	.26826	.7277	.0353	.8595	59
2	.20848	.97803	.21316	.6912	.0225	.7966	.22552	.97424	.23148	.3200	.0264	.4342	.24249	.97015	.24995	.0009	.0308	.1239	.25938	.96577	.26857	.7234	.0354	.8553	58
3	.20876	.97797	.21347	.6845	.0225	.7901	.22580	.97417	.23179	.3143	.0265	.4287	.24277	.97008	.25025	3.9959	.0308	.1191	.25966	.96570	.26888	.7191	.0355	.8512	57
4	.20905	.97790	.21377	.6778	.0226	.7835	.22608	.97411	.23209	.3086	.0266	.4231	.24305	.97001	.25056	.9910	.0309	.1144	.25994	.96562	.26920	.7147	.0356	.8470	56
5	.20933	.97784	.21408	4.6712	.0226	4.7770	.22637	.97404	.23240	4.3029	.0266	4.4176	.24333	.96994	.25087	3.9861	.0310	4.1096	.26022	.96555	.26951	3.7104	.0357	3.8428	55
6	.20962	.97778	.21438	.6646	.0227	.7706	.22665	.97398	.23270	.2972	.0267	.4121	.24361	.96987	.25118	.9812	.0311	.1048	.26050	.96547	.26982	.7062	.0358	.8387	54
7	.20990	.97772	.21468	.6580	.0228	.7641	.22693	.97391	.23301	.2916	.0268	.4065	.24390	.96980	.25149	.9763	.0311	.1001	.26078	.96540	.27013	.7019	.0358	.8346	53
8	.21019	.97766	.21499	.6514	.0228	.7576	.22722	.97384	.23332	.2859	.0268	.4011	.24418	.96973	.25180	.9714	.0312	.0953	.26107	.96532	.27044	.6976	.0359	.8304	52
9	.21047	.97760	.21529	.6448	.0229	.7512	.22750	.97378	.23363	.2803	.0269	.3956	.24446	.96966	.25211	.9665	.0313	.0906	.26135	.96524	.27076	.6933	.0360	.8263	51
10	.21076	.97754	.21560	4.6382	.0230	4.7448	.22778	.97371	.23393	4.2747	.0270	4.3901	.24474	.96959	.25242	3.9616	.0314	4.0859	.26163	.96517	.27107	3.6891	.0361	3.8222	50
11	.21104	.97748	.21590	.6317	.0230	.7384	.22807	.97364	.23424	.2691	.0271	.3847	.24502	.96952	.25273	.9568	.0314	.0812	.26191	.96509	.27138	.6848	.0362	.8181	49
12	.21132	.97741	.21621	.6252	.0231	.7320	.22835	.97358	.23455	.2635	.0271	.3792	.24531	.96944	.25304	.9520	.0315	.0765	.26219	.96502	.27169	.6806	.0362	.8140	48
13	.21161	.97735	.21651	.6187	.0232	.7257	.22863	.97351	.23485	.2579	.0272	.3738	.24559	.96937	.25335	.9471	.0316	.0718	.26247	.96494	.27201	.6764	.0363	.8100	47
14	.21189	.97729	.21682	.6122	.0232	.7193	.22892	.97344	.23516	.2524	.0273	.3684	.24587	.96930	.25366	.9423	.0317	.0672	.26275	.96486	.27232	.6722	.0364	.8059	46
15	.21218	.97723	.21712	4.6057	.0233	4.7130	.22920	.97338	.23547	4.2468	.0273	4.3630	.24615	.96923	.25397	3.9375	.0317	4.0625	.26303	.96479	.27263	3.6679	.0365	3.8018	45
16	.21246	.97717	.21742	.5993	.0234	.7067	.22948	.97331	.23577	.2413	.0274	.3576	.24643	.96916	.25428	.9327	.0318	.0579	.26331	.96471	.27294	.6637	.0366	.7978	44
17	.21275	.97711	.21773	.5928	.0234	.7004	.22977	.97324	.23608	.2358	.0275	.3522	.24672	.96909	.25459	.9279	.0319	.0532	.26359	.96463	.27326	.6596	.0367	.7937	43
18	.21303	.97704	.21803	.5864	.0235	.6942	.23005	.97318	.23639	.2303	.0276	.3469	.24700	.96901	.25490	.9231	.0320	.0486	.26387	.96456	.27357	.6554	.0367	.7897	42
19	.21331	.97698	.21834	.5800	.0235	.6879	.23033	.97311	.23670	.2248	.0276	.3415	.24728	.96894	.25521	.9184	.0320	.0440	.26415	.96448	.27388	.6512	.0368	.7857	41
20	.21360	.97692	.21864	4.5736	.0236	4.6817	.23061	.97304	.23700	4.2193	.0277	4.3362	.24756	.96887	.25552	3.9136	.0321	4.0394	.26443	.96440	.27419	3.6470	.0369	3.7816	40
21	.21388	.97686	.21895	.5673	.0237	.6754	.23090	.97298	.23731	.2139	.0278	.3309	.24784	.96880	.25583	.9089	.0322	.0348	.26471	.96433	.27451	.6429	.0370	.7776	39
22	.21417	.97680	.21925	.5609	.0237	.6692	.23118	.97291	.23762	.2084	.0278	.3256	.24813	.96873	.25614	.9042	.0323	.0302	.26499	.96425	.27482	.6387	.0371	.7736	38
23	.21445	.97673	.21956	.5546	.0238	.6631	.23146	.97284	.23793	.2030	.0279	.3203	.24841	.96865	.25645	.8994	.0323	.0256	.26527	.96417	.27513	.6346	.0371	.7697	37
24	.21473	.97667	.21986	.5483	.0239	.6569	.23175	.97277	.23823	.1976	.0280	.3150	.24869	.96858	.25676	.8947	.0324	.0211	.26556	.96409	.27544	.6305	.0372	.7657	36
25	.21502	.97661	.22017	4.5420	.0239	4.6507	.23202	.97271	.23854	4.1921	.0280	4.3098	.24897	.96851	.25707	3.8900	.0325	4.0165	.26584	.96402	.27576	3.6263	.0373	3.7617	35
26	.21530	.97655	.22047	.5357	.0240	.6446	.23231	.97264	.23885	.1867	.0281	.3045	.24925	.96844	.25738	.8853	.0326	.0120	.26612	.96394	.27607	.6222	.0374	.7577	34
27	.21559	.97648	.22078	.5294	.0241	.6385	.23260	.97257	.23916	.1814	.0282	.2993	.24953	.96836	.25769	.8807	.0327	.0074	.26640	.96386	.27638	.6181	.0375	.7538	33
28	.21587	.97642	.22108	.5232	.0241	.6324	.23288	.97250	.23946	.1760	.0283	.2941	.24982	.96829	.25800	.8760	.0327	.0029	.26668	.96378	.27670	.6140	.0376	.7498	32
29	.21615	.97636	.22139	.5169	.0242	.6263	.23316	.97244	.23977	.1706	.0283	.2838	.25010	.96822	.25831	.8713	.0328	3.9984	.26696	.96371	.27701	.6100	.0376	.7459	31
30	.21644	.97630	.22169	4.5107	.0243	4.6201	.23344	.97237	.24008	4.1653	.0284	4.2836	.25038	.96815	.25862	3.8667	.0329	3.9939	.26724	.96363	.27732	3.6059	.0377	3.7420	30
31	.21672	.97623	.22200	.5045	.0243	.6142	.23373	.97230	.24039	.1600	.0285	.2785	.25066	.96807	.25893	.8621	.0330	.9894	.26752	.96355	.27764	.6018	.0378	.7380	29
32	.21701	.97617	.22230	.4983	.0244	.6081	.23401	.97223	.24069	.1546	.0285	.2733	.25094	.96800	.25924	.8574	.0330	.9850	.26780	.96347	.27795	.5977	.0379	.7341	28
33	.21729	.97611	.22261	.4921	.0245	.6021	.23429	.97216	.24100	.1493	.0286	.2681	.25122	.96793	.25955	.8528	.0331	.9805	.26808	.96340	.27826	.5937	.0380	.7302	27
34	.21757	.97604	.22291	.4860	.0245	.5961	.23458	.97210	.24131	.1440	.0287	.2630	.25151	.96785	.25986	.8482	.0332	.97[illegible]	.26836	.96332	.27858	.5896	.0381	.7263	26
35	.21786	.97598	.22322	4.4799	.0246	4.5901	.23486	.97203	.24162	4.1388	.0288	4.2579	.25179	.96778	.26017	3.8436	.0333	3.9716	.26864	.96324	.27889	3.5856	.0382	3.7224	25
36	.21814	.97592	.22353	.4737	.0247	.5841	.23514	.97196	.24192	.1335	.0288	.2527	.25207	.96771	.26048	.8390	.0334	.9672	.26892	.96316	.27920	.5816	.0382	.7186	24
37	.21843	.97575	.22383	.4676	.0247	.5782	.23542	.97189	.24223	.1282	.0289	.2476	.25235	.96763	.26079	.8345	.0334	.9627	.26920	.96308	.27952	.5776	.0383	.7147	23
38	.21871	.97579	.22414	.4615	.0248	.5722	.23571	.97182	.24254	.1230	.0290	.2425	.25263	.96756	.26110	.8299	.0335	.9583	.26948	.96301	.27983	.5736	.0384	.7108	22
39	.21899	.97573	.22444	.4555	.0249	.5663	.23599	.97175	.24285	.1178	.0291	.2375	.25291	.96749	.26141	.8254	.0336	.9539	.26976	.96293	.28014	.5696	.0385	.7070	21
40	.21928	.97566	.22475	4.4494	.0249	4.5604	.23627	.97169	.24316	4.1126	.0291	4.2324	.25319	.96741	.26172	3.8208	.0337	3.9495	.27004	.96285	.28046	3.5656	.0386	3.7031	20
41	.21956	.97560	.22505	.4434	.0250	.5545	.23655	.97162	.24346	.1073	.0292	.2273	.25348	.96734	.26203	.8163	.0338	.9451	.27032	.96277	.28077	.5616	.0387	.6993	19
42	.21985	.97553	.22536	.4373	.0251	.5486	.23684	.97155	.24377	.1022	.0293	.2223	.25376	.96727	.26234	.8118	.0338	.9408	.27060	.96269	.28109	.5576	.0387	.6955	18
43	.22013	.97547	.22566	.4313	.0251	.5428	.23712	.97148	.24408	.0970	.0293	.2173	.25404	.96719	.26266	.8073	.0339	.9364	.27088	.96261	.28140	.5536	.0388	.6917	17
44	.22041	.97541	.22597	.4253	.0252	.5369	.23740	.97141	.24439	.0918	.0294	.2122	.25432	.96712	.26297	.8027	.0340	.9320	.27116	.96253	.28171	.5497	.0389	.6878	16
45	.22070	.97534	.22628	4.4194	.0253	4.5311	.23768	.97134	.24470	4.0867	.0295	4.2072	.25460	.96704	.26328	3.7983	.0341	3.9277	.27144	.96245	.28203	3.5457	.0390	3.6840	15
46	.22098	.97528	.22658	.4134	.0253	.5253	.23797	.97127	.24501	.0815	.0296	.2022	.25488	.96697	.26359	.7938	.0341	.9234	.27172	.96238	.28234	.5418	.0391	.6802	14
47	.22126	.97521	.22689	.4074	.0254	.5195	.23825	.97120	.24531	.0764	.0296	.1972	.25516	.96690	.26390	.7893	.0342	.9190	.27200	.96230	.28266	.5378	.0392	.6765	13
48	.22155	.97515	.22719	.4015	.0255	.5137	.23853	.97113	.24562	.0713	.0297	.1923	.25544	.96682	.26421	.7848	.0343	.9147	.27228	.96222	.28297	.5339	.0393	.6727	12
49	.22183	.97508	.22750	.3956	.0255	.5079	.23881	.97106	.24593	.0662	.0298	.1873	.25573	.96675	.26452	.7804	.0344	.9104	.27256	.96214	.28328	.5300	.0393	.6689	11
50	.22211	.97502	.22781	4.3897	.0256	4.5021	.23910	.97099	.24624	4.0611	.0299	4.1824	.25601	.96667	.26483	3.7759	.0345	3.9061	.27284	.96206	.28360	3.5261	.0394	3.6651	10
51	.22240	.97495	.22811	.3838	.0257	.4964	.23938	.97092	.24655	.0560	.0299	.1774	.25629	.96660	.26514	.7715	.0345	.9018	.27312	.96198	.28391	.5222	.0395	.6614	9
52	.22268	.97489	.22842	.3779	.0257	.4907	.23966	.97086	.24686	.0509	.0300	.1725	.25657	.96652	.26546	.7671	.0346	.8976	.27340	.96190	.28423	.5183	.0396	.6576	8
53	.22297	.97483	.22872	.3721	.0258	.4850	.23994	.97079	.24717	.0458	.0301	.1676	.25685	.96645	.26577	.7627	.0347	.8933	.27368	.96182	.28454	.5144	.0397	.6539	7
54	.22325	.97476	.22903	.3662	.0259	.4793	.24023	.97072	.24747	.0408	.0302	.1627	.25713	.96638	.26608	.7583	.0348	.8890	.27396	.96174	.28486	.5105	.0398	.6502	6
55	.22353	.97470	.22934	4.3604	.0260	4.4736	.24051	.97065	.24778	4.0358	.0302	4.1578	.25741	.96630	.26639	3.7539	.0349	3.8848	.27424	.96166	.28517	3.5066	.0399	3.6464	5
56	.22382	.97463	.22964	.3546	.0260	.4679	.24079	.97058	.24809	.0307	.0303	.1529	.25769	.96623	.26670	.7495	.0349	.8805	.27452	.96158	.28549	.5028	.0399	.6427	4
57	.22410	.97457	.22995	.3488	.0261	.4623	.24107	.97051	.24840	.0257	.0304	.1481	.25798	.96615	.26701	.7451	.0350	.8763	.27480	.96150	.28580	.4989	.0400	.6390	3
58	.22438	.97450	.23025	.3430	.0262	.4566	.24136	.97044	.24871	.0207	.0305	.1432	.25826	.96608	.26732	.7407	.0351	.8721	.27508	.96142	.28611	.4951	.0401	.6353	2
59	.22467	.97443	.23056	.3372	.0262	.4510	.24164	.97037	.24902	.0157	.0305	.1384	.25854	.96600	.26764	.7364	.0352	.8679	.27536	.96134	.28643	.4912	.0402	.6316	1
60	.22495	.97437	.23087	4.3315	.0263	4.4454	.24192	.97029	.24933	4.0108	.0306	4.1336	.25882	.96592	.26795	3.7320	.0353	3.8637	.27564	.96126	.28674	3.4874	.0403	3.6279	0
′	cos	sin	cot	tan	cosec	sec	cos	sin	cot	tan	cosec	sec	cos	sin	cot	tan	cosec	sec	cos	sin	cot	tan	cosec	sec	′
	77°						76°						75°						74°						

Source. (Courtesy of Bethlehem Steel.

NATURAL TRIGONOMETRIC FUNCTIONS

	16°						17°						18°						19°						
′	sin	cos	tan	cot	sec	cosec	sin	cos	tan	cot	sec	cosec	sin	cos	tan	cot	sec	cosec	sin	cos	tan	cot	sec	cosec	′
0	.27564	.96126	.28674	3.4874	1.0403	3.6279	.29237	.95630	.30573	3.2708	1.0457	3.4203	.30902	.95106	.32492	3.0777	1.0515	3.2361	.32557	.94552	.34433	2.9042	1.0576	3.0715	60
1	.27592	.96118	.28706	.4836	.0404	.6243	.29265	.95622	.30605	.2674	.0458	.4170	.30929	.95097	.32524	.0746	.0516	.2332	.32584	.94542	.34465	.9015	.0577	.0690	59
2	.27620	.96110	.28737	.4798	.0405	.6206	.29293	.95613	.30637	.2640	.0459	.4138	.30957	.95088	.32556	.0716	.0517	.2303	.32612	.94533	.34498	.8987	.0578	.0664	58
3	.27648	.96102	.28769	.4760	.0406	.6169	.29321	.95605	.30668	.2607	.0460	.4106	.30985	.95079	.32588	.0686	.0518	.2274	.32639	.94523	.34530	.8960	.0579	.0638	57
4	.27675	.96094	.28800	.4722	.0406	.6133	.29348	.95596	.30700	.2573	.0461	.4073	.31012	.95070	.32621	.0655	.0519	.2245	.32667	.94514	.34563	.8933	.0580	.0612	56
5	.27703	.96086	.28832	3.4684	.0407	3.6096	.29376	.95588	.30732	3.2539	.0461	3.4041	.31040	.95061	.32653	3.0625	.0520	3.2216	.32694	.94504	.34595	2.8905	.0581	3.0586	55
6	.27731	.96078	.28863	.4646	.0408	.6060	.29404	.95579	.30764	.2505	.0462	.4009	.31068	.95051	.32685	.0595	.0521	.2188	.32722	.94495	.34628	.8878	.0582	.0561	54
7	.27759	.96070	.28895	.4608	.0409	.6024	.29432	.95571	.30796	.2472	.0463	.3977	.31095	.95042	.32717	.0565	.0522	.2159	.32749	.94485	.34661	.8851	.0584	.0535	53
8	.27787	.96062	.28926	.4570	.0410	.5987	.29460	.95562	.30828	.2438	.0464	.3945	.31123	.95033	.32749	.0535	.0523	.2131	.32777	.94476	.34693	.8824	.0585	.0509	52
9	.27815	.96054	.28958	.4533	.0411	.5951	.29487	.95554	.30859	.2405	.0465	.3913	.31150	.95024	.32782	.0505	.0524	.2102	.32804	.94466	.34726	.8797	.0586	.0484	51
10	.27843	.96045	.28990	3.4495	.0412	3.5915	.29515	.95545	.30891	3.2371	.0466	3.3881	.31178	.95015	.32814	3.0475	.0525	3.2074	.32832	.94457	.34758	2.8770	.0587	3.0458	50
11	.27871	.96037	.29021	.4458	.0413	.5879	.29543	.95536	.30923	.2338	.0467	.3849	.31206	.95006	.32846	.0445	.0526	.2045	.32859	.94447	.34791	.8743	.0588	.0433	49
12	.27899	.96029	.29053	.4420	.0413	.5843	.29571	.95528	.30955	.2305	.0468	.3817	.31233	.94997	.32878	.0415	.0527	.2017	.32887	.94438	.34824	.8716	.0589	.0407	48
13	.27927	.96021	.29084	.4383	.0414	.5807	.29598	.95519	.30987	.2271	.0469	.3785	.31261	.94988	.32910	.0385	.0528	.1989	.32914	.94428	.34856	.8689	.0590	.0382	47
14	.27955	.96013	.29116	.4346	.0415	.5772	.29626	.95511	.31019	.2238	.0470	.3754	.31289	.94979	.32943	.0356	.0529	.1960	.32942	.94418	.34889	.8662	.0591	.0357	46
15	.27983	.96005	.29147	3.4308	.0416	3.5736	.29654	.95502	.31051	3.2205	.0471	3.3722	.31316	.94970	.32975	3.0326	.0530	3.1932	.32969	.94409	.34921	2.8636	.0592	3.0331	45
16	.28011	.95997	.29179	.4271	.0417	.5700	.29682	.95493	.31083	.2172	.0472	.3690	.31344	.94961	.33007	.0296	.0531	.1904	.32996	.94399	.34954	.8609	.0593	.0306	44
17	.28039	.95989	.29210	.4234	.0418	.5665	.29710	.95485	.31115	.2139	.0473	.3659	.31372	.94952	.33039	.0267	.0532	.1876	.33024	.94390	.34987	.8582	.0594	.0281	43
18	.28067	.95980	.29242	.4197	.0419	.5629	.29737	.95476	.31146	.2106	.0474	.3627	.31399	.94942	.33072	.0237	.0533	.1848	.33051	.94380	.35019	.8555	.0595	.0256	42
19	.28094	.95972	.29274	.4160	.0420	.5594	.29765	.95467	.31178	.2073	.0475	.3596	.31427	.94933	.33104	.0208	.0534	.1820	.33079	.94370	.35052	.8529	.0596	.0231	41
20	.28122	.95964	.29305	3.4124	.0420	3.5559	.29793	.95459	.31210	3.2041	.0476	3.3565	.31454	.94924	.33136	3.0178	.0535	3.1792	.33106	.94361	.35085	2.8502	.0598	3.0206	40
21	.28150	.95956	.29337	.4087	.0421	.5523	.29821	.95450	.31242	.2008	.0477	.3534	.31482	.94915	.33169	.0149	.0536	.1764	.33134	.94351	.35117	.8476	.0599	.0181	39
22	.28178	.95948	.29368	.4050	.0422	.5488	.29848	.95441	.31274	.1975	.0478	.3502	.31510	.94906	.33201	.0120	.0537	.1736	.33161	.94341	.35150	.8449	.0600	.0156	38
23	.28206	.95940	.29400	.4014	.0423	.5453	.29876	.95433	.31306	.1942	.0478	.3471	.31537	.94897	.33233	.0090	.0528	.1708	.33189	.94332	.35183	.8423	.0601	.0131	37
24	.28234	.95931	.29432	.3977	.0424	.5418	.29904	.95424	.31338	.1910	.0479	.3440	.31565	.94888	.33265	.0061	.0539	.1681	.33216	.94322	.35215	.8396	.0602	.0106	36
25	.28262	.95923	.29463	3.3941	.0425	3.5383	.29932	.95415	.31370	3.1877	.0480	3.3409	.31592	.94878	.33298	3.0032	.0540	3.1653	.33243	.94313	.35248	2.8370	.0603	3.0081	35
26	.28290	.95915	.29495	.3904	.0426	.5348	.29959	.95407	.31402	.1845	.0481	.3378	.31620	.94869	.33330	.0003	.0541	.1625	.33271	.94303	.35281	.8344	.0604	.0056	34
27	.28318	.95907	.29526	.3868	.0427	.5313	.29987	.95398	.31434	.1813	.0482	.3347	.31648	.94860	.33362	2.9974	.0542	.1598	.33298	.94293	.35314	.8318	.0605	.0031	33
28	.28346	.95898	.29558	.3832	.0428	.5279	.30015	.95389	.31466	.1780	.0483	.3316	.31675	.94851	.33395	.9945	.0543	.1570	.33326	.94283	.35346	.8291	.0606	.0007	32
29	.28374	.95890	.29590	.3795	.0428	.5244	.30043	.95380	.31498	.1748	.0484	.3286	.31703	.94841	.33427	.9916	.0544	.1543	.33353	.94274	.35379	.8265	.0607	2.9982	31
30	.28401	.95882	.29621	3.3759	.0429	3.5209	.30070	.95372	.31530	3.1716	.0485	3.3255	.31730	.94832	.33459	2.9887	.0545	3.1515	.33381	.94264	.35412	2.8239	.0608	2.9957	30
31	.28429	.95874	.29653	.3723	.0430	.5175	.30098	.95363	.31562	.1684	.0486	.3224	.31758	.94823	.33492	.9858	.0546	.1488	.33408	.94254	.35445	.8213	.0609	.9933	29
32	.28457	.95865	.29685	.3687	.0431	.5140	.30126	.95354	.31594	.1652	.0487	.3194	.31786	.94814	.33524	.9829	.0547	.1461	.33435	.94245	.35477	.8187	.0611	.9908	28
33	.28485	.95857	.29716	.3651	.0432	.5106	.30154	.95345	.31626	.1620	.0488	.3163	.31813	.94805	.33557	.9800	.0548	.1433	.33463	.94235	.35510	.8161	.0612	.9884	27
34	.28513	.95849	.29748	.3616	.0433	.5072	.30181	.95337	.31658	.1588	.0489	.3133	.31841	.94795	.33589	.9772	.0549	.1406	.33490	.94225	.35543	.8135	.0613	.9859	26
35	.28541	.95840	.29780	3.3580	.0434	3.5037	.30209	.95328	.31690	3.1556	.0490	3.3102	.31868	.94786	.33621	2.9743	.0550	3.1379	.33518	.94215	.35576	2.8109	.0614	2.9835	25
36	.28569	.95832	.29811	.3544	.0435	.5003	.30237	.95319	.31722	.1524	.0491	.3072	.31896	.94777	.33654	.9714	.0551	.1352	.33545	.94206	.35608	.8083	.0615	.9810	24
37	.28597	.95824	.29843	.3509	.0436	.4969	.30265	.95310	.31754	.1492	.0492	.3042	.31923	.94767	.33686	.9686	.0552	.1325	.33572	.94196	.35641	.8057	.0616	.9786	23
38	.28624	.95816	.29875	.3473	.0437	.4935	.30292	.95301	.31786	.1460	.0493	.3011	.31951	.94758	.33718	.9657	.0553	.1298	.33600	.94186	.35674	.8032	.0617	.9762	22
39	.28652	.95807	.29906	.3438	.0438	.4901	.30320	.95293	.31818	.1429	.0494	.2981	.31978	.94749	.33751	.9629	.0554	.1271	.33627	.94176	.35707	.8006	.0618	.9738	21
40	.28680	.95799	.29938	3.3402	.0438	3.4867	.30348	.95284	.31850	3.1397	.0495	3.2951	.32006	.94740	.33783	2.9600	.0555	3.1244	.33655	.94167	.35739	2.7980	.0619	2.9713	20
41	.28708	.95791	.29970	.3367	.0439	.4833	.30375	.95275	.31882	.1366	.0496	.2921	.32034	.94730	.33816	.9572	.0556	.1217	.33682	.94157	.35772	.7954	.0620	.9689	19
42	.28736	.95782	.30001	.3332	.0440	.4799	.30403	.95266	.31914	.1334	.0497	.2891	.32061	.94721	.33848	.9544	.0557	.1190	.33709	.94147	.35805	.7929	.0622	.9665	18
43	.28764	.95774	.30033	.3296	.0441	.4766	.30431	.95257	.31946	.1303	.0498	.2861	.32089	.94712	.33880	.9515	.0558	.1163	.33737	.94137	.35838	.7903	.0623	.9641	17
44	.28792	.95765	.30065	.3261	.0442	.4732	.30459	.95248	.31978	.1271	.0499	.2831	.32116	.94702	.33913	.9487	.0559	.1137	.33764	.94127	.35871	.7878	.0624	.9617	16
45	.28820	.95757	.30096	3.3226	.0443	3.4698	.30486	.95239	.32010	3.1240	.0500	3.2801	.32144	.94693	.33945	2.9459	.0560	3.1110	.33792	.94118	.35904	2.7852	.0625	2.9593	15
46	.28847	.95749	.30128	.3191	.0444	.4665	.30514	.95231	.32042	.1209	.0501	.2772	.32171	.94684	.33978	.9431	.0561	.1083	.33819	.94108	.35936	.7827	.0626	.9569	14
47	.28875	.95740	.30160	.3156	.0445	.4632	.30542	.95222	.32074	.1177	.0502	.2742	.32199	.94674	.34010	.9403	.0562	.1057	.33846	.94098	.35969	.7801	.0627	.9545	13
48	.28903	.95732	.30192	.3121	.0446	.4598	.30569	.95213	.32106	.1146	.0503	.2712	.32226	.94665	.34043	.9375	.0563	.1030	.33874	.94088	.36002	.7776	.0628	.9521	12
49	.28931	.95723	.30223	.3087	.0447	.4565	.30597	.95204	.32138	.1115	.0504	.2683	.32254	.94655	.34075	.9347	.0565	.1004	.33901	.94078	.36035	.7751	.0629	.9497	11
50	.28959	.95715	.30255	3.3052	.0448	3.4532	.30625	.95195	.32171	3.1084	.0505	3.2653	.32282	.94646	.34108	2.9319	.0566	3.0977	.33928	.94068	.36068	2.7725	.0630	2.9474	10
51	.28987	.95707	.30287	.3017	.0448	.4498	.30653	.95186	.32203	.1053	.0506	.2624	.32309	.94637	.34140	.9291	.0567	.0951	.33956	.94058	.36101	.7700	.0632	.9450	9
52	.29014	.95698	.30319	.2983	.0449	.4465	.30680	.95177	.32235	.1022	.0507	.2594	.32337	.94627	.34173	.9263	.0568	.0925	.33983	.94049	.36134	.7675	.0633	.9426	8
53	.29042	.95690	.30350	.2948	.0450	.4432	.30708	.95168	.32267	.0991	.0508	.2565	.32364	.94618	.34205	.9235	.0569	.0898	.34011	.94039	.36167	.7650	.0634	.9402	7
54	.29070	.95681	.30382	.2914	.0451	.4399	.30736	.95159	.32299	.0960	.0509	.2535	.32392	.94608	.34238	.9208	.0570	.0872	.34038	.94029	.36199	.7625	.0635	.9379	6
55	.29098	.95673	.30414	3.2879	.0452	3.4366	.30763	.95150	.32331	3.0930	.0510	3.2506	.32419	.94599	.34270	2.9180	.0571	3.0846	.34065	.94019	.36232	2.7600	.0636	2.9355	5
56	.29126	.95664	.30446	.2845	.0453	.4334	.30791	.95141	.32363	.0899	.0511	.2477	.32447	.94590	.34303	.9152	.0572	.0820	.34093	.94009	.36265	.7575	.0637	.9332	4
57	.29154	.95656	.30478	.2811	.0454	.4301	.30819	.95132	.32395	.0868	.0512	.2448	.32474	.94580	.34335	.9125	.0573	.0793	.34120	.93999	.36298	.7550	.0638	.9308	3
58	.29181	.95047	.30509	.2777	.0455	.4268	.30846	.95124	.32428	.0838	.0513	.2419	.32502	.94571	.34368	.9097	.0574	.0767	.34147	.93989	.36331	.7525	.0639	.9285	2
59	.29209	.95639	.30541	.2742	.0456	.4236	.30874	.95115	.32400	.0807	.0514	.2390	.32529	.94561	.34400	.9069	.0575	.0741	.34175	.93979	.36364	.7500	.0641	.9261	1
60	.29237	.95630	.30573	3.2708	.0457	3.4203	.30902	.95106	.32492	3.0777	.0515	3.2361	.32557	.94552	.34433	2.9042	.0576	3.0715	.34202	.93969	.36397	2.7475	.0042	2.9238	0
′	cos	sin	cot	tan	cosec	sec	cos	sin	cot	tan	cosec	sec	cos	sin	cot	tan	cosec	sec	cos	sin	cot	tan	cosec	sec	′
	73°						72°						71°						70°						

878 Machine Tool Practices

NATURAL TRIGONOMETRIC FUNCTIONS

	20°						21°						22°						23°						
′	sin	cos	tan	cot	sec	cosec	sin	cos	tan	cot	sec	cosec	sin	cos	tan	cot	sec	cosec	sin	cos	tan	cot	sec	cosec	′
0	.34202	.93969	.36397	2.7475	1.0642	2.9238	.35837	.93358	.38386	2.6051	1.0711	2.7904	.37461	.92718	.40403	2.4751	1.0785	2.6695	.39073	.92050	.42447	2.3558	1.0864	2.5593	60
1	.34229	.93959	.36430	.7450	.0643	.9215	.35864	.93348	.38420	.6028	.0713	.7883	.37488	.92707	.40436	.4730	.0787	.6675	.39100	.92039	.42482	.3539	.0865	.5575	59
2	.34257	.93949	.36463	.7425	.0644	.9191	.35891	.93337	.38453	.6006	.0714	.7862	.37514	.92696	.40470	.4709	.0788	.6656	.39126	.92028	.42516	.3520	.0866	.5558	58
3	.34284	.93939	.36496	.7400	.0645	.9168	.35918	.93327	.38486	.5983	.0715	.7841	.37541	.92686	.40504	.4689	.0789	.6637	.39153	.92016	.42550	.3501	.0868	.5540	57
4	.34311	.93929	.36529	.7376	.0646	.9145	.35945	.93316	.38520	.5960	.0716	.7820	.37568	.92675	.40538	.4668	.0790	.6618	.39180	.92005	.42585	.3482	.0869	.5523	56
5	.34339	.93919	.36562	2.7351	.0647	2.9122	.35972	.93306	.38553	2.5938	.0717	2.7799	.37595	.92664	.40572	2.4647	.0792	2.6599	.39207	.91993	.42619	2.3463	.0870	2.5506	55
6	.34366	.93909	.36595	.7326	.0648	.9098	.36000	.93295	.38587	.5916	.0719	.7778	.37622	.92653	.40606	.4627	.0793	.6580	.39234	.91982	.42654	.3445	.0872	.5488	54
7	.34393	.93899	.36628	.7302	.0650	.9075	.36027	.93285	.38620	.5893	.0720	.7757	.37649	.92642	.40640	.4606	.0794	.6561	.39260	.91971	.42688	.3426	.0873	.5471	53
8	.34421	.93889	.36661	.7277	.0651	.9052	.36054	.93274	.38654	.5871	.0721	.7736	.37676	.92631	.40673	.4586	.0795	.6542	.39287	.91959	.42722	.3407	.0874	.5453	52
9	.34448	.93879	.36694	.7252	.0652	.9029	.36081	.93264	.38687	.5848	.0722	.7715	.37703	.92620	.40707	.4565	.0797	.6523	.39314	.91948	.42757	.3388	.0876	.5436	51
10	.34475	.93869	.36727	2.7228	.0653	2.9006	.36108	.93253	.38720	2.5826	.0723	2.7694	.37730	.92609	.40741	2.4545	.0798	2.6504	.39341	.91936	.42791	2.3369	.0877	2.5419	50
11	.34502	.93859	36760	.7204	.0654	.8983	.36135	.93243	.38754	.5804	.0725	.7674	.37757	.92598	.40775	.4525	.0799	.6485	.39367	.91925	.42826	.3350	.0878	.5402	49
12	.34530	.93849	.36793	.7179	.0655	.8960	.36162	.93232	.38787	.5781	.0726	.7653	.37784	.92587	.40809	.4504	.0801	.6466	.39394	.91913	.42860	.3332	.0880	.5384	48
13	.34557	.93839	.36826	.7155	.0656	.8937	.36189	.93222	.38821	.5759	.0727	.7632	.37811	.92576	.40843	.4484	.0802	.6447	.39421	.91902	.42894	.3313	.0881	.5367	47
14	.34584	.93829	.36859	.7130	.0658	.8915	.36217	.93211	.38854	.5737	.0728	.7611	.37838	.92565	.40877	.4463	.0803	.6428	.39448	.91891	.42929	.3294	.0882	.5350	46
15	.34612	.93819	.36892	2.7106	.0659	2.8892	.36244	.93201	.38888	2.5715	.0729	2.7591	.37865	.92554	.40911	2.4443	.0804	2.6410	.39474	.91879	.42963	2.3276	.0884	2.5333	45
16	.34639	.93809	.36925	.7082	.0660	.8869	.36271	.93190	.38921	.5693	.0731	.7570	.37892	.92543	.40945	.4423	.0806	.6391	.39501	.91868	.42998	.3257	.0885	.5316	44
17	.34666	.93799	.36958	.7058	.0661	.8846	.36298	.93180	.38955	.5671	.0732	.7550	.37919	.92532	.40979	.4403	.0807	.6372	.39528	.91856	.43032	.3238	.0886	.5299	43
18	.34693	.93789	.36991	.7033	.0662	.8824	.36325	.93169	.38988	.5649	.0733	.7529	.37946	.92521	.41013	.4382	.0808	.6353	.39554	.91845	.43067	.3220	.0888	.5281	42
19	.34721	.93779	.37024	.7009	.0663	.8801	.36352	.93158	.39022	.5627	.0734	.7509	.37972	.92510	.41047	.4362	.0810	.6335	.39581	.91833	.43101	.3201	.0889	.5264	41
20	.34748	.93769	.37057	2.6985	.0664	2.8778	.36379	.93148	.39055	2.5605	.0736	2.7488	.37999	.92499	.41081	2.4342	.0811	2.6316	.39608	.91822	.43136	2.3183	.0891	2.5247	40
21	.34775	.93758	.37090	.6961	.0666	.8756	.36406	.93137	.39089	.5583	.0737	.7468	.38026	.92488	.41115	.4322	.0812	.6297	.39635	.91810	.43170	.3164	.0892	.5230	39
22	.34803	.93748	.37123	.6937	.0667	.8733	.36433	.93127	.39122	.5561	.0738	.7447	.38053	.92477	.41149	.4302	.0813	.6279	.39661	.91798	.43205	.3145	.0893	.5213	38
23	.34830	.93738	.37156	.6913	.0668	.8711	.36460	.93116	.39156	.5539	.0739	.7427	.38080	.92466	.41183	.4282	.0815	.6260	.39688	.91787	.43239	.3127	.0895	.5196	37
24	.34857	.93728	.37190	.6889	.0669	.8688	.36488	.93105	.39189	.5517	.0740	.7406	.38107	.92455	.41217	.4262	.0816	.6242	.39715	.91775	.43274	.3109	.0896	.5179	36
25	.34884	.93718	.37223	2.6865	.0670	2.8666	.36515	.93095	.39223	2.5495	.0742	2.7386	.38134	.92443	.41251	2.4242	.0817	2.6223	.39741	.91764	.43308	2.3090	.0897	2.5163	35
26	.34912	.93708	.37256	.6841	.0671	.8644	.36542	.93084	.39257	.5473	.0743	.7366	.38161	.92432	.41285	.4222	.0819	.6205	.39768	.91752	.43343	.3072	.0899	.5146	34
27	.34939	.93698	.37289	.6817	.0673	.8621	.36569	.93074	.39290	.5451	.0744	.7346	.38188	.92421	.41319	.4202	.0820	.6186	.39795	.91741	.43377	.3053	.0900	.5129	33
28	.34966	.93687	.37322	.6794	.0674	.8599	.36596	.93063	.39324	.5430	.0745	.7325	.38214	.92410	.41353	.4182	.0821	.6168	.39821	.91729	.43412	.3035	.0902	.5112	32
29	.34993	.93677	.37355	.6770	.0675	.8577	.36623	.93052	.39357	.5408	.0747	.7305	.38241	.92399	.41387	.4162	.0823	.6150	.39848	.91718	.43447	.3017	.0903	.5095	31
30	.35021	.93667	.37388	2.6746	.0676	2.8554	.36650	.93042	.39391	2.5386	.0748	2.7285	.38268	.92388	.41421	2.4142	.0824	2.6131	.39875	.91706	.43481	2.2998	.0904	2.5078	30
31	.35048	.93657	.37422	.6722	.0677	.8532	.36677	.93031	.39425	.5365	.0749	.7265	.38295	.92377	.41455	.4122	.0825	.6113	.39901	.91694	.43516	.2980	.0906	.5062	29
32	.35075	.93647	.37455	.6699	.0678	.8510	.36704	.93020	39458	.5343	.0750	.7245	.38322	.92366	.41489	.4102	.0826	.6095	.39928	.91683	.43550	.2962	.0907	.5045	28
33	.35102	.93637	.37488	.6675	.0679	.8488	.36731	.93010	.39492	.5322	.0751	.7225	.38349	.92354	.41524	.4083	.0828	.6076	.39955	.91671	.43585	.2944	.0908	.5028	27
34	.35130	.93626	.37521	.6652	.0681	.8466	.36758	.92999	.39525	.5300	.0753	.7205	.38376	.92343	.41558	.4063	.0829	.6058	.39981	.91659	.43620	.2925	.0910	.5011	26
35	.35157	.93616	.37554	2.6628	.0682	2.8444	.36785	.92988	.39559	2.5278	.0754	2.7185	.38403	.92332	.41592	2.4043	.0830	2.6040	.40008	.91648	.43654	2.2907	.0911	2.4995	25
36	.35184	.93606	.37587	.6604	.0683	.8422	.36812	.92978	.39593	.5257	.0755	.7165	.38429	.92321	.41626	.4023	.0832	.6022	.40035	.91636	.43689	.2889	.0913	.4978	24
37	.35211	.93596	.37621	.6581	.0684	.8400	.36839	.92967	.39626	.5236	.0756	.7145	.38456	.92310	.41660	.4004	.0833	.6003	.40061	.91625	.43723	.2871	.0914	.4961	23
38	.35239	.93585	.37654	.6558	.0685	.8378	.36866	.92956	.39660	.5214	.0758	.7125	.38483	.92299	.41694	.3984	.0834	.5985	.40088	.91613	.43758	.2853	.0915	.4945	22
39	.35266	.93575	.37687	.6534	.0686	.8356	.36893	.92945	.39694	.5193	.0759	.7105	.38510	.92287	.41728	.3964	.0836	.5967	.40115	.91601	.43793	.2835	.0917	.4928	21
40	.35293	.93565	.37720	2.6511	.0688	2.8334	.36921	.92935	.39727	2.5171	.0760	2.7085	.38537	.92276	.41762	2.3945	.0837	2.5949	.40141	.91590	.43827	2.2817	.0918	2.4912	20
41	.35320	.93555	.37754	.6487	.0689	.8312	.36948	.92924	.39761	.5150	.0761	.7065	.38564	.92265	.41797	.3925	.0838	.5931	.40168	.91578	.43862	.2799	.0920	.4895	19
42	.35347	.93544	.37787	.6464	.0690	.8290	.36975	.92913	.39795	.5129	.0763	.7045	.38591	.92254	.41831	.3906	.0840	.5913	.40195	.91566	.43897	.2781	.0921	.4879	18
43	.35375	.93534	.37820	.6441	.0691	.8269	.37002	.92902	.39828	.5108	.0764	.7026	.38617	.92242	.41865	.3886	.0841	.5895	.40221	.91554	.43932	.2763	.0922	.4862	17
44	.35402	.93524	.37853	.6418	.0692	.8247	.37029	.92892	.39862	.5086	.0765	.7006	.38644	.92231	.41899	.3867	.0842	.5877	.40248	.91543	.43966	.2745	.0924	.4846	16
45	.35429	.93513	.37887	2.6394	.0694	2.8225	.37056	.92881	.39896	2.5065	.0766	2.6986	.38671	.92220	.41933	2.3847	.0844	2.5958	.40205	.91531	.44001	2.2727	.0925	2.4829	15
46	.35456	.93503	.37920	.6371	.0695	.8204	.37083	.92870	.39930	.5044	.0768	.6967	.38698	.92209	.41968	.3828	.0845	.5841	.40301	.91519	.44036	.2709	.0927	.4813	14
47	.35483	.93493	.37953	.6348	.0696	.8182	.37110	.92859	.39963	.5023	.0769	.6947	.38725	.92197	.42002	.3808	.0846	.5823	.40328	.91508	.44070	.2691	.0928	.4797	13
48	.35511	.93482	.37986	.6325	.0697	.8160	.37137	.92848	.39997	.5002	.0770	.6927	.38751	.92186	.42036	.3789	.0847	.5805	.40354	.91496	.44105	.2673	.0929	.4780	12
49	.35538	.93472	.38020	.6302	.0698	.8139	.37164	.92838	.40031	.4981	.0771	.6908	.38778	.92105	.42070	.3770	.0849	.5787	.40381	.91484	.44140	.2655	.0931	.4764	11
50	.35565	.93462	.38053	2.6279	.0699	2.8117	.37191	.92827	.40065	2.4960	.0773	2.6888	.38805	.92164	.42105	2.3750	.0850	2.5770	.40408	.91472	.44175	2.2637	.0932	2.4748	10
51	.35592	.93451	.38086	.6256	.0701	.8096	.37218	.92816	.40098	.4939	.0774	.6869	.38832	.92152	.42139	.3731	.0851	.5752	.40434	.91461	.44209	.2619	.0934	.4731	9
52	.35619	.93441	.38120	.6233	.0702	.8074	.37245	.92805	.40132	.4918	.0775	.6849	.38859	.92141	.42173	.3712	.0853	.5734	.40461	.91449	.44244	.2602	.0935	.4715	8
53	.35647	.93431	.38153	.6210	.0703	.8053	.37272	.92794	.40166	.4897	.0776	.6830	.38886	.92130	.42207	.3692	.0854	.5716	.40487	.91437	.44279	.2584	.0936	.4699	7
54	.35674	.93420	.38186	.6187	.0704	.8032	.37299	.92784	.40200	.4876	.0778	.6810	.38912	.92118	.42242	.3673	.0855	.5699	.40514	.91425	.44314	.2566	.0938	.4683	6
55	.35701	.93410	.38220	2.6164	.0705	2.8010	.37326	.92773	.40233	2.4855	.0779	2.6791	.38939	.92107	.42276	2.3654	.0857	2.5681	.40541	.91414	.44349	2.2548	.0939	2.4666	5
56	.35728	.93400	.38253	.6142	.0707	.7989	.37353	.92762	.40267	.4834	.0780	.6772	.38966	.92096	.42310	.3635	.0858	.5663	.40567	.91402	.44383	.2531	.0941	.4650	4
57	.35755	.93389	.38286	.6119	.0708	.7968	.37380	.92751	.40301	.4813	.0781	.6752	.38993	.92084	.42344	.3616	.0859	.5646	.40594	.91390	.44418	.2513	.0942	.4634	3
58	.35782	.93379	.38320	.6096	.0709	.7947	.37407	.92740	.40335	.4792	.0783	.6733	.39019	.92073	.42379	.3597	.0861	.5628	.40620	.91378	.44453	.2495	.0943	.4618	2
59	.35810	.93368	.38353	.6073	.0710	.7925	.37434	.92729	.40369	.4772	.0784	.6714	.39046	.92062	.42413	.3577	.0862	.5610	.40647	.91366	.44488	.2478	.0945	.4602	1
60	.35837	.93358	.38386	2.6051	.0711	2.7904	.37461	.92718	.40403	2.4751	.0785	2.6695	.39073	.92050	.42447	2.3558	.0864	2.5593	.40674	.91354	.44523	2.2460	.0946	2.4586	0
′	cos	sin	cot	tan	cosec	sec	cos	sin	cot	tan	cosec	sec	cos	sin	cot	tan	cosec	sec	cos	sin	cot	tan	cosec	sec	′
	69°						68°						67°						66°						

	24°						25°						26°						27°						
′	sin	cos	tan	cot	sec	cosec	sin	cos	tan	cot	sec	cosec	sin	cos	tan	cot	sec	cosec	sin	cos	tan	cot	sec	cosec	′
0	.40674	.91354	.44523	2.2460	1.0946	2.4586	.42262	.90631	.46631	2.1445	1.1034	2.3662	.43837	.89879	.48773	2.0503	1.1126	2.2812	.45399	.89101	.50952	1.9626	1.1223	2.2027	60
1	.40700	.91343	.44558	.2443	.0948	.4570	.42288	.90618	.46666	.1429	.1035	.3647	.43863	.89867	.48809	.0488	.1127	.2798	.45425	.89087	.50989	.9612	.1225	.2014	59
2	.40727	.91331	.44593	.2425	.0949	.4554	.42314	.90606	.46702	.1412	.1037	.3632	.43889	.89854	.48845	.0473	.1129	.2784	.45451	.89074	.51026	.9598	.1226	.2002	58
3	.40753	.91319	.44627	.2408	.0951	.4538	.42341	.90594	.46737	.1396	.1038	.3618	.43915	.89841	.48881	.0458	.1131	.2771	.45477	.89061	.51062	.9584	.1228	.1989	57
4	.40780	.91307	.44662	.2390	.0952	.4522	.42367	.90581	.46772	.1380	.1040	.3603	.43942	.89828	.48917	.0443	.1132	.2757	.45503	.89048	.51099	.9570	.1230	.1977	56
5	.40806	.91295	.44697	2.2373	.0953	2.4506	.42394	.90569	.46808	2.1364	.1041	2.3588	.43968	.89815	.48953	2.0427	.1134	2.2744	.45528	.89034	.51136	1.9556	.1231	2.1964	55
6	.40833	.91283	.44732	.2355	.0955	.4490	.42420	.90557	.46843	.1348	.1043	.3574	.43994	.89803	.48989	.0412	.1135	.2730	.45554	.89021	.51172	.9542	.1233	.1952	54
7	.40860	.91271	.44767	.2338	.0956	.4474	.42446	.90544	.46879	.1331	.1044	.3559	.44020	.89790	.49025	.0397	.1137	.2717	.45580	.89008	.51209	.9528	.1235	.1939	53
8	.40886	.91260	.44802	.2320	.0958	.4458	.42473	.90532	.46914	.1315	.1046	.3544	.44046	.89777	.49062	.0382	.1139	.2703	.45606	.88995	.51246	.9514	.1237	.1927	52
9	.40913	.91248	.44837	.2303	.0959	.4442	.42499	.90520	.46950	.1299	.1047	.3530	.44072	.89764	.49098	.0367	.1140	.2690	.45632	.88981	.51283	.9500	.1238	.1914	51
10	.40939	.91236	.44872	2.2286	.0961	2.4426	.42525	.90507	.46985	2.1283	.1049	2.3515	.44098	.89751	.49134	2.0352	.1142	2.2676	.45658	.88968	.51319	1.9486	.1240	2.1902	50
11	.40966	.91224	.44907	.2268	.0962	.4418	.42552	.90495	.47021	.1267	.1050	.3501	.44124	.89739	.49170	.0338	.1143	.2663	.45684	.88955	.51356	.9472	.1242	.1889	49
12	.40992	.91212	.44942	.2251	.0963	.4395	.42578	.90483	.47056	.1251	.1052	.3486	.44150	.89726	.49206	.0323	.1145	.2650	.45710	.88942	.51393	.9458	.1243	.1877	48
13	.41019	.91200	.44977	.2234	.0965	.4379	.42604	.90470	.47092	.1235	.1053	.3472	.44177	.89713	.49242	.0308	.1147	.2636	.45736	.88928	.51430	.9444	.1245	.1865	47
14	.41045	.91188	.45012	.2216	.0966	.4363	.42630	.90458	.47127	.1219	.1055	.3457	.44203	.89700	.49278	.0293	.1148	.2623	.45761	.88915	.51466	.9430	.1247	.1852	46
15	.41072	.91176	.45047	2.2199	.0968	2.4347	.42657	.90445	.47163	2.1203	.1056	2.3443	.44229	.89687	.49314	2.0278	.1150	2.2610	.45787	.88902	.51503	1.9416	.1248	2.1840	45
16	.41098	.91164	.45082	.2182	.0969	.4332	.42683	.90433	.47199	.1187	.1058	.3428	.44255	.89674	.49351	.0263	.1151	.2596	.45813	.88888	.51540	.9402	.1250	.1828	44
17	.41125	.91152	.45117	.2165	.0971	.4316	.42709	.90421	.47234	.1171	.1059	.3414	.44281	.89661	.49387	.0248	.1153	.2583	.45839	.88875	.51577	.9388	.1252	.1815	43
18	.41151	.91140	.45152	.2147	.0972	.4300	.42736	.90408	.47270	.1155	.1061	.3399	.44307	.89649	.49423	.0233	.1155	.2570	.45865	.88862	.51614	.9375	.1253	.1803	42
19	.41178	.91128	.45187	.2130	.0973	.4285	.42762	.90396	.47305	.1139	.1062	.3385	.44333	.89636	.49459	.0219	.1156	.2556	.45891	.88848	.51651	.9361	.1255	.1791	41
20	.41204	.91116	.45222	2.2113	.0975	2.4269	.42788	.90383	.47341	2.1123	.1064	2.3371	.44359	.89623	.49495	2.0204	.1158	2.2543	.45917	.88835	.51687	1.9347	.1257	2.1778	40
21	.41231	.91104	.45257	.2096	.0976	.4254	.42815	.90371	.47376	.1107	.1065	.3356	.44385	.89610	.49532	.0189	.1159	.2530	.45942	.88822	.51724	.9333	.1258	.1766	39
22	.41257	.91092	.45292	.2079	.0978	.4238	.42841	.90358	.47412	.1092	.1067	.3342	.44411	.89597	.49568	.0174	.1161	.2517	.45968	.88808	.51761	.9319	.1260	.1754	38
23	.41284	.91080	.45327	. 2062	.0979	.4222	.42867	.90346	.47448	.1076	.1068	.3328	.44437	.89584	.49604	.0159	.1163	.2503	.45994	.88795	.51798	.9306	.1262	.1742	37
24	.41310	.91068	.45362	.2045	.0981	.4207	.42893	.90333	.47483	.1060	.1070	.3313	.44463	.89571	.49640	.0145	.1164	.2490	.46020	.88781	.51835	.9292	.1264	.1730	36
25	.41337	.91056	.45397	2.2028	.0982	2.4191	42920	.90321	.47519	2.1044	.1072	2.3299	.44489	.89558	.49677	2.0130	.1166	2.2477	.46046	.88768	.51872	1.9278	.1265	2.1717	35
26	.41363	.91044	.45432	.2011	.0984	.4176	.42946	.90308	.47555	.1028	.1073	.3285	.44516	.89545	.49713	.0115	.1167	.2464	.46072	.88755	.51909	.9264	.1267	.1705	34
27	.41390	.91032	.45467	.1994	.0985	.4160	.42972	.90296	.47590	.1013	.1075	.3271	.44542	.89532	.49749	.0101	.1169	.2451	.46097	.88741	.51946	.9251	.1269	.1693	33
28	.41416	.91020	.45502	.1977	.0986	.4145	.42998	.90283	.47626	.0997	.1076	.3256	.44568	.89519	.49785	.0086	.1171	.2438	.46123	.88728	.51983	.9237	.1270	.1681	32
29	.41443	.91008	.45537	.1960	.0988	.4130	.43025	.90271	.47662	.0981	.1078	.3242	.44594	.89506	.49822	.0071	.1172	.2425	.46149	.88714	.52020	.9223	.1272	.1669	31
30	.41469	.90996	.45573	2.1943	.0989	2.4114	.43051	.90258	.47697	2.0965	.1079	2.3228	.44620	.89493	.49858	2.0057	.1174	2.2411	.46175	.88701	.52057	1.9210	.1274	2.1657	30
31	.41496	.90984	.45608	.1926	.0991	.4099	.43077	.90246	.47733	.0950	.1081	.3214	.44646	.89480	.49894	.0042	.1176	.2398	.46201	.88688	.52094	.9196	.1275	.1645	29
32	.41522	.90972	.45643	.1909	.0992	.4083	.43104	.90233	.47769	.0934	.1082	.3200	.44672	.89467	.49931	.0028	.1177	.2385	.46226	.88674	.52131	.9182	.1277	.1633	28
33	.41549	.90960	.45678	.1892	.0994	.4068	.43130	.90221	.47805	.0918	.1084	.3186	.44698	.89454	.49967	.0013	.1179	.2372	.46252	.88661	.52168	.9169	.1279	.1620	27
34	.41575	.90948	.45713	.1875	.0995	.4053	.43156	.90208	.47840	.0903	.1085	.3172	.44724	.89441	.50003	1.9998	.1180	.2359	.46278	.88647	.52205	.9155	.1281	.1608	26
35	.41602	.90936	.45748	2.1859	.0997	2.4037	.43182	.90196	.47876	2.0887	.1087	2.3158	.44750	.89428	.50040	1.9984	.1182	2.2346	.46304	.88634	.52242	1.9142	.1282	2.1596	25
36	.41628	.90924	.45783	.1842	.0998	.4022	.43208	.90183	.47912	.0872	.1088	.3143	.44776	.89415	.50076	.9969	.1184	.2333	.46330	.88620	.52279	.9128	.1284	.1584	24
37	.41654	.90911	.45819	.1825	.1000	.4007	.43235	.90171	.47948	.0856	.1090	.3129	.44802	.89402	.50113	.9955	.1185	.2320	.46355	.88607	.52316	.9115	.1286	.1572	23
38	.41681	.90899	.45854	.1808	.1001	.3992	.43261	.90158	.47983	.0840	.1092	.3115	.44828	.89389	.50149	.9940	.1187	.2307	.46381	.88593	.52353	.9101	.1287	.1560	22
39	.41707	.90887	.45889	.1792	.1003	.3976	.43287	.90145	.48019	.0825	.1093	.3101	.44854	.89376	.50185	.9926	.1189	.2294	.46407	.88580	.52390	.9088	.1289	.1548	21
40	.41734	.90875	.45924	2.1775	.1004	2.3961	.43313	.90133	.48055	2.0809	.1095	2.3087	.44880	.89363	.50222	1.9912	.1190	2.2282	.46433	.88566	.52427	1.9074	.1291	2.1536	20
41	.41760	.90863	.45960	.1758	.1005	.3946	.43340	.90120	.48091	.0794	.1096	.3073	.44906	.89350	.50258	.9897	.1192	.2269	.46458	.88553	.52464	.9061	.1293	.1525	19
42	.41787	.90851	.45995	.1741	.1007	.3931	.43366	.90108	.48127	.0778	.1098	.3059	.44932	.89337	.50295	.9883	.1193	.2256	.46484	.88539	.52501	.9047	.1294	.1513	18
43	.41813	.90839	.46030	.1725	.1008	.3916	.43392	.90095	.48162	.0763	.1099	.3046	.44958	.89324	.50331	.9868	.1195	.2243	.46510	.88526	.52538	.9034	.1296	.1501	17
44	.41839	.90826	.46065	.1708	.1010	.3901	.43418	.90082	.48198	.0747	.1101	.3032	.44984	.89311	.50368	.9854	.1197	.2230	.46536	.88512	.52575	.9020	.1298	.1489	16
45	.41866	.90814	.46101	2.1692	.1011	2.3886	.43444	.90070	.48234	2.0732	.1102	2.3018	.45010	.89298	.50404	1.9840	.1198	2.2217	.46561	.88499	.52612	1.9007	.1299	2.1477	15
46	.41892	.90812	.46136	.1675	.1013	.3871	.43471	.90057	.48270	.0717	.1104	.3004	.45036	.89285	.50441	.9825	.1200	.2204	.46587	.88485	.52650	.8993	.1301	.1465	14
47	.41919	.90790	.46171	.1658	.1014	.3856	.43497	.90044	.48306	.0701	.1106	.2990	.45062	.89272	.50477	.9811	.1202	.2192	.46613	.88472	.52687	.8980	.1303	.1453	13
48	.41945	.90778	.46206	.1642	.1016	.3841	.43523	.90032	.48342	.0686	.1107	.2976	.45088	.89258	.50514	.9797	.1203	.2179	.46639	.88458	.52724	.8967	.1305	.1441	12
49	.41972	.90765	.46242	.1625	.1017	.3826	.43549	.90019	.48378	.0671	.1109	.2962	.45114	.89245	.50550	.9782	.1205	.2166	.46664	.88444	.52761	.8953	.1306	.1430	11
50	.41998	.90753	.46277	2.1609	.1019	2.3811	.43575	.90006	.48414	2.0655	.1110	2.2949	.45140	.89232	.50587	1.9768	.1207	2.2153	.46690	.88431	.52798	1.8940	.1308	2.1418	10
51	.42024	.90741	.46312	.1592	.1020	.3796	.43602	.89994	.48449	.0640	.1112	.2935	.45166	.89219	.50623	.9754	.1208	.2141	.46716	.88417	.52836	.8927	.1310	.1406	9
52	.42051	.90729	.46348	.1576	.1022	.3781	.43628	.89981	.48485	.0625	.1113	.2921	.45191	.89206	.50660	.9739	.1210	.2128	.46741	.88404	.52873	.8913	.1312	.1394	8
53	.42077	.90717	.46383	.1559	.1023	.3766	.43654	.89968	.48521	.0609	.1115	.2907	.45217	.89193	.50696	.9725	.1212	.2115	.46767	.88390	.52910	.8900	.1313	.1382	7
54	.42103	.90704	.46418	.1543	.1025	.3751	.43680	.89956	.48557	.0594	.1116	.2894	.45243	.89180	.50733	.9711	.1213	.2103	.46793	.88376	.52947	.8887	.1315	.1371	6
55	.42130	.90692	.46454	2.1527	.1026	2.3736	.43706	.89943	.48593	2.0579	.1118	2.2880	.45269	.89166	.50769	1.9697	.1215	2.2090	.46819	.88363	.52984	1.8873	.1317	2.1359	5
56	.42156	.90680	.46489	.1510	.1028	.3721	.43732	.89930	.48629	.0564	.1120	.2866	.45295	.89153	.50806	.9683	.1217	.2077	.46844	.88349	.53022	.8860	.1319	.1347	4
57	.42183	.90668	.46524	.1494	.1029	.3706	.43759	.89918	.48665	.0548	.1121	.2853	.45321	.89140	.50843	.9668	.1218	.2065	.46870	.88336	.53059	.8847	.1320	.1335	3
58	.42209	.90655	.46560	.1478	.1031	.3691	.43785	.89905	.48701	.0533	.1123	.2839	.45347	.89127	.50879	.9654	.1220	.2052	.46896	.88322	.53096	.8834	.1322	.1324	2
59	.42235	.90643	.46595	.1461	.1032	.3677	.43811	.89892	.48737	.0518	.1124	.2825	.45373	.89114	.50916	.9640	.1222	.2039	.46921	.88308	.53134	.8820	.1324	.1312	1
60	.42262	.90631	.46631	2.1445	.1034	2.3662	.43837	.89879	.48773	2.0503	.1126	2.2812	.45399	.89101	.50952	1.9626	.1223	2.2027	.46947	.88295	.53171	1.8807	.1326	2.1300	0
′	cos	sin	cot	tan	cosec	sec	cos	sin	cot	tan	cosec	sec	cos	sin	cot	tan	cosec	sec	cos	sin	cot	tan	cosec	sec	′
	65°						64°						63°						62°						

	28°						29°						30°						31°						
′	sin	cos	tan	cot	sec	cosec	sin	cos	tan	cot	sec	cosec	sin	cos	tan	cot	sec	cosec	sin	cos	tan	cot	sec	cosec	′
0	.46947	.88295	.53171	1.8807	1.1326	2.1300	.48481	.87462	.55431	1.8040	1.1433	2.0627	.50000	.86603	.57735	1.7320	1.1547	2.0000	.51504	.85717	.60086	1.6643	1.1666	1.9416	60
1	.46973	.88281	.53208	.8794	.1327	.1289	.48506	.87448	.55469	.8028	.1435	.0616	.50025	.86588	.57774	.7309	.1549	1.9990	.51529	.85702	.60126	.6632	.1668	.9407	59
2	.46998	.88267	.53245	.8781	.1329	.1277	.48532	.87434	.55507	.8016	.1437	.0605	.50050	.86573	.57813	.7297	.1551	.9980	.51554	.85687	.60165	.6621	.1670	.9397	58
3	.47024	.88254	.53283	.8768	.1331	.1266	.48557	.87420	.55545	.8003	.1439	.0594	.50075	.86559	.57851	.7286	.1553	.9970	.51578	.85672	.60205	.6610	.1672	.9388	57
4	.47050	.88240	.53320	.8754	.1333	.1254	.48583	.87405	.55583	.7991	.1441	.0583	.50101	.86544	.57890	.7274	.1555	.9960	.51603	.85657	.60244	.6599	.1674	.9378	56
5	.47075	.88226	.53358	1.8741	.1334	2.1242	.48608	.87391	.55621	1.7979	.1443	2.0573	.50126	.86530	.57929	1.7262	.1557	1.9950	.51628	.85642	.60284	1.6588	.1676	1.9369	55
6	.47101	.88213	.53395	.8728	.1336	.1231	.48633	.87377	.55659	.7966	.1445	.0562	.50151	.86515	.57968	.7251	.1559	.9940	.51653	.85627	.60324	.6577	.1678	.9360	54
7	.47127	.88199	.53432	.8715	.1338	.1219	.48659	.87363	.55697	.7954	.1446	.0551	.50176	.86500	.58007	.7239	.1561	.9930	.51678	.85612	.60363	.6566	.1681	.9350	53
8	.47152	.88185	.53470	.8702	.1340	.1208	.48684	.87349	.55735	.7942	.1448	.0540	.50201	.86486	.58046	.7228	.1562	.9920	.51703	.85597	.60403	.6555	.1683	.9341	52
9	.47178	.88171	.53507	.8689	.1341	.1196	.48710	.87335	.55774	.7930	.1450	.0530	.50226	.86471	.58085	.7216	.1564	.9910	.51728	.85582	.60443	.6544	.1685	.9332	51
10	.47204	.88158	.53545	1.8676	.1343	2.1185	.48735	.87320	.55812	1.7917	.1452	2.0519	.50252	.86457	.58123	1.7205	.1566	1.9900	.51753	.85566	.60483	1.6534	.1687	1.9322	50
11	.47229	.88144	.53582	.8663	.1345	.1173	.48760	.87306	.55850	.7905	.1454	.0508	.50277	.86442	.58162	.7193	.1568	.9890	.51778	.85551	.60522	.6523	.1689	.9313	49
12	.47255	.88130	.53619	.8650	.1347	.1162	.48786	.87292	.55888	.7893	.1456	.0498	.50302	.86427	.58201	.7182	.1570	.9880	.51803	.85536	.60562	.6512	.1691	.9304	48
13	.47281	.88117	.53657	.8637	.1349	.1150	.48811	.87278	.55926	.7881	.1458	.0487	.50327	.86413	.58240	.7170	.1572	.9870	.51827	.85521	.60602	.6501	.1693	.9295	47
14	.47306	.88103	.53694	.8624	.1350	.1139	.48837	.87264	.55964	.7868	.1459	.0476	.50352	.86398	.58279	.7159	.1574	.9860	.51852	.85506	.60642	.6490	.1695	.9285	46
15	.47332	.88089	.53732	1.8611	.1352	2.1127	.48862	.87250	.56003	1.7856	.1461	2.0466	.50377	.86383	.58318	1.7147	.1576	1.9850	.51877	.85491	.60681	1.6479	.1697	1.9276	45
16	.47357	.88075	.53769	.8598	.1354	.1116	.48887	.87235	.56041	.7844	.1463	.0455	.50402	.86369	.58357	.7136	.1578	.9840	.51902	.85476	.60721	.6469	.1699	.9267	44
17	.47383	.88061	.53807	.8585	.1356	.1104	.48913	.87221	.56079	.7832	.1465	.0444	.50428	.86354	.58396	.7124	.1580	.9830	.51927	.85461	.60761	.6458	.1701	.9258	43
18	.47409	.88048	.53844	.8572	.1357	.1093	.48938	.87207	.56117	.7820	.1467	.0434	.50453	.86339	.58435	.7113	.1582	.9820	.51952	.85446	.60801	.6447	.1703	.9248	42
19	.47434	.88034	.53882	.8559	.1359	.1082	.48964	.87193	.56156	.7808	.1469	.0423	.50478	.86325	.58474	.7101	.1584	.9811	.51977	.85431	.60841	.6436	.1705	.9239	41
20	.47460	.88020	.53919	1.8546	.1361	2.1070	.48989	.87178	.56194	1.7795	.1471	2.0413	.50503	.86310	.58513	1.7090	.1586	1.9801	.52002	.85416	.60881	1.6425	.1707	1.9230	40
21	.47486	.88006	.53957	.8533	.1363	.1059	.49014	.87164	.56232	.7783	.1473	.0402	.50528	.86295	.58552	.7079	.1588	.9791	.52026	.85400	.60920	.6415	.1709	.9221	39
22	.47511	.87992	.53995	.8520	.1365	.1048	.49040	.87150	.56270	.7771	.1474	.0392	.50553	.86281	.58591	.7067	.1590	.9781	.52051	.85385	.60960	.6404	.1712	.9212	38
23	.47537	.87979	.54032	.8507	.1366	.1036	.49065	.87136	.56309	.7759	.1476	.0381	.50578	.86266	.58630	.7056	.1592	.9771	.52076	.85370	.61000	.6393	.1714	.9203	37
24	.47562	.87965	.54070	.8495	.1368	.1025	.49090	.87121	.56347	.7747	.1478	.0370	.50603	.86251	.58670	.7044	.1594	.9761	.52101	.85355	.61040	.6383	.1716	.9193	36
25	.47588	.87951	.54107	1.8482	.1370	2.1014	.49116	.87107	.56385	1.7735	.1480	2.0360	.50628	.86237	.58709	1.7033	.1596	1.9752	.52126	.85340	.61080	1.6372	.1718	1.9184	35
26	.47613	.87937	.54145	.8469	.1372	.1002	.49141	.87093	.56424	.7723	.1482	.0349	.50653	.86222	.58748	.7022	.1598	.9742	.52151	.85325	.61120	.6361	.1720	.9175	34
27	.47639	.87923	.54183	.8456	.1373	.0991	.49166	.87078	.56462	.7711	.1484	.0339	.50679	.86207	.58787	.7010	.1600	.9732	.52175	.85309	.61160	.6350	.1722	.9166	33
28	.47665	.87909	.54220	.8443	.1375	.0980	.49192	.87064	.56500	.7699	.1486	.0329	.50704	.86192	.58826	.6999	.1602	.9722	.52200	.85294	.61200	.6340	.1724	.9157	32
29	.47690	.87895	.54258	.8430	.1377	.0969	.49217	.87050	.56539	.7687	.1488	.0318	.50729	.86178	.58865	.6988	.1604	.9713	.52225	.85279	.61240	.6329	.1726	.9148	31
30	.47716	.87882	.54295	1.8418	.1379	2.0957	.49242	.87035	.56577	1.7675	.1489	2.0308	.50754	.86163	.58904	1.6977	.1606	1.9703	.52250	.85264	.61280	1.6318	.1728	1.9139	30
31	.47741	.87868	.54333	.8405	.1381	.0946	.49268	.87021	.56616	.7663	.1491	.0297	.50779	.86148	.58944	.6965	.1608	.9693	.52275	.85249	.61320	.6308	.1730	.9130	29
32	.47767	.87854	.54371	.8392	.1382	.0935	.49293	.87007	.56654	.7651	.1493	.0287	.50804	.86133	.58983	.6954	.1610	.9683	.52299	.85234	.61360	.6297	.1732	.9121	28
33	.47792	.87840	.54409	.8379	.1384	.0924	.49318	.86992	.56692	.7639	.1495	.0276	.50829	.86118	.59022	.6943	.1612	.9674	.52324	.85218	.61400	.6286	.1734	.9112	27
34	.47818	.87826	.54446	.8367	.1386	.0912	.49343	.86978	.56731	.7627	.1497	.0266	.50854	.86104	.59061	.6931	.1614	.9664	.52349	.85203	.61440	.6276	.1737	.9102	26
35	.47844	.87812	.54484	1.8354	.1388	2.0901	.49369	.86964	.56769	1.7615	.1499	2.0256	.50879	.86089	.59100	1.6920	.1616	1.9654	.52374	.85188	.61480	1.6265	.1739	1.9093	25
36	.47869	.87798	.54522	.8341	.1390	.0890	.49394	.86949	.56808	.7603	.1501	.0245	.50904	.86074	.59140	.6909	.1618	.9645	.52398	.85173	.61520	.6255	.1741	.9084	24
37	.47895	.87784	.54559	.8329	.1391	.0879	.49419	.86935	.56846	.7591	.1503	.0235	.50929	.86059	.59179	.6898	.1620	.9635	.52423	.85157	.61560	.6244	.1743	.9075	23
38	.47920	.87770	.54597	.8316	.1393	.0868	.49445	.86921	.56885	.7579	.1505	.0224	.50954	.86044	.59218	.6887	.1622	.9625	.52448	.85142	.61601	.6233	.1745	.9066	22
39	.47946	.87756	.54635	.8303	.1395	.0857	.49470	.86906	.56923	.7567	.1507	.0214	.50979	.86030	.59258	.6875	.1624	.9616	.52473	.85127	.61641	.6223	.1747	.9057	21
40	.47971	.87742	.54673	1.8291	.1397	2.0846	.49495	.86892	.56962	1.7555	.1508	2.0204	.51004	.86015	.59297	1.6864	.1626	1.9606	.52498	.85112	.61681	1.6212	.1749	1.9048	20
41	.47997	.87728	.54711	.8278	.1399	.0835	.49521	.86877	.57000	.7544	.1510	.0194	.51029	.86000	.59336	.6853	.1628	.9596	.52522	.85096	.61721	.6202	.1751	.9039	19
42	.48022	.87715	.54748	.8265	.1401	.0824	.49546	.86863	.57039	.7532	.1512	.0183	.51054	.85985	.59376	.6842	.1630	.9587	.52547	.85081	.61761	.6191	.1753	.9030	18
43	.48048	.87701	.54786	.8253	.1402	.0812	.49571	.86849	.57077	.7520	.1514	.0173	.51079	.85970	.59415	.6831	.1632	.9577	.52572	.85066	.61801	.6181	.1756	.9021	17
44	.48073	.87687	.54824	.8240	.1404	.0801	.49596	.86834	.57116	.7508	.1516	.0163	.51104	.85955	.59454	.6820	.1634	.9568	.52597	.85050	.61842	.6170	.1758	.9013	16
45	.48099	.87673	.54862	1.8227	.1406	2.0790	.49622	.86820	.57155	1.7496	.1518	2.0152	.51129	.85941	.59494	1.6808	.1636	1.9558	.52621	.85735	.61882	1.6160	.1760	1.9004	15
46	.48124	.87659	.54900	.8215	.1408	.0779	.49647	.86805	.57193	.7484	.1520	.0142	.51154	.85926	.59533	.6797	.1638	.9549	.52646	.85020	.61922	.6149	.1762	.8995	14
47	.48150	.87645	.54937	.8202	.1410	.0768	.49672	.86791	.57232	.7473	.1522	.0132	.51179	.85911	.59572	.6786	.1640	.9539	.52671	.85004	.61962	.6139	.1764	.8986	13
48	.48175	.87631	.54975	.8190	.1411	.0757	.49697	.86776	.57270	.7461	.1524	.0122	.51204	.85896	.59612	.6775	.1642	.9530	.52695	.84989	.62003	.6128	.1766	.8977	12
49	.48201	.87617	.55013	.8177	.1413	.0746	.49723	.86762	.57309	.7449	.1526	.0111	.51229	.85881	.59651	.6764	.1644	.9520	.52720	.84974	.62043	.6118	.1768	.8968	11
50	.48226	.87603	.55051	1.8165	.1415	2.0735	.49748	.86748	.57348	1.7437	.1528	2.0101	.51254	.85866	.59691	1.6753	.1646	1.9510	.52745	.84959	.62083	1.6107	.1770	1.8959	10
51	.48252	.87588	.55089	.8152	.1417	.0725	.49773	.86733	.57386	.7426	.1530	.0091	.51279	.85851	.59730	.6742	.1648	.9501	.52770	.84943	.62123	.6097	.1772	.8950	9
52	.48277	.87574	.55127	.8140	.1419	.0714	.49798	.86719	.57425	.7414	.1531	.0081	.51304	.85836	.59770	.6731	.1650	.9491	.52794	.84928	.62164	.6086	.1775	.8941	8
53	.48303	.87560	.55165	.8127	.1421	.0703	.49823	.86704	.57464	.7402	.1533	.0071	.51329	.85821	.59809	.6720	.1652	.9482	.52819	.84912	.62204	.6076	.1777	.8932	7
54	.48328	.87546	.55203	.8115	.1422	.0692	.49849	.86690	.57502	.7390	.1535	.0061	.51354	.85806	.59849	.6709	.1654	.9473	.52844	.84897	.62244	.6066	.1779	.8924	6
55	.48354	.87532	.55241	1.8102	.1424	2.0681	.49874	.86675	.57541	1.7379	.1537	2.0050	.51379	.85791	.59888	1.6698	.1656	1.9463	.52868	.84882	.62285	1.6055	.1781	1.8915	5
56	.48379	.87518	.55279	.8090	.1426	.0670	.49899	.86661	.57580	.7367	.1539	.0040	.51404	.85777	.59928	.6687	.1658	.9454	.52893	.84866	.62325	.6045	.1783	.8906	4
57	.48405	.87504	.55317	.8078	.1428	.0659	.49924	.86646	.57619	.7355	.1541	.0030	.51429	.85762	.59967	.6676	.1660	.9444	.52918	.84851	.62366	.6034	.1785	.8897	3
58	.48430	.87490	.55355	.8065	.1430	.0648	.49950	.86632	.57657	.7344	.1543	.0020	.51454	.85747	.60007	.6665	.1662	.9435	.52942	.84836	.62406	.6024	.1787	.8888	2
59	.48455	.87476	.55393	.8053	.1432	.0637	.49975	.86617	.57696	.7332	.1545	.0010	.51479	.85732	.60046	.6654	.1664	.9425	.52967	.84820	.62446	.6014	.1790	.8879	1
60	.48481	.87462	.55431	1.8040	.1433	2.0627	.50000	.86603	.57735	1.7320	.1547	2.0000	.51504	.85717	.60086	1.6643	.1666	1.9416	.52992	.84805	.62487	1.6003	.1792	1.8871	0
′	cos	sin	cot	tan	cosec	sec	cos	sin	cot	tan	cosec	sec	cos	sin	cot	tan	cosec	sec	cos	sin	cot	tan	cosec	sec	′
	61°						60°						59°						58°						

NATURAL TRIGONOMETRIC FUNCTIONS

	32°						33°						34°						35°						
′	sin	cos	tan	cot	sec	cosec	sin	cos	tan	cot	sec	cosec	sin	cos	tan	cot	sec	cosec	sin	cos	tan	cot	sec	cosec	′
0	.52992	.84805	.62487	1.6003	1.1792	1.8871	.54464	.83867	.64941	1.5399	1.1924	1.8361	.55919	.82904	.67451	1.4826	1.2062	1.7883	.57358	.81915	.70021	1.4281	1.2208	1.7434	60
1	.53016	.84789	.62527	.5993	.1794	.8862	.54488	.83851	.64982	.5389	.1926	.8352	.55943	.82887	.67493	.4816	.2064	.7875	.57381	.81898	.70064	.4273	.2210	.7427	59
2	.53041	.84774	.62568	.5983	.1796	.8853	.54513	.83835	.65023	.5379	.1928	.8344	.55967	.82871	.67535	.4807	.2067	.7867	.57405	.81882	.70107	.4264	.2213	.7420	58
3	.53066	.84758	.62608	.5972	.1798	.8844	.54537	.83819	.65065	.5369	.1930	.8336	.55992	.82855	.67578	.4798	.2069	.7860	.57429	.81865	.70151	.4255	.2215	.7413	57
4	.53090	.84743	.62649	.5962	.1800	.8836	.54561	.83804	.65106	.5359	.1933	.8328	.56016	.82839	.67620	.4788	.2072	.7852	.57453	.81848	.70194	.4246	.2218	.7405	56
5	.53115	.84728	.62689	1.5952	.1802	1.8827	.54586	.83788	.65148	1.5350	.1935	1.8320	.56040	.82822	.67663	1.4779	.2074	1.7844	.57477	.81832	.70238	1.4237	.2220	1.7398	55
6	.53140	.84712	.62730	.5941	.1805	.8818	.54610	.83772	.65189	.5340	.1937	.8311	.56064	.82806	.67705	.4770	.2076	.7837	.57500	.81815	.70281	.4228	.2223	.7391	54
7	.53164	.84697	.62770	.5931	.1807	.8809	.54634	.83756	.65231	.5330	.1939	.8303	.56088	.82790	.67747	.4761	.2079	.7829	.57524	.81798	.70325	.4220	.2225	.7384	53
8	.53189	.84681	.62811	.5921	.1809	.8801	.54659	.83740	.65272	.5320	.1942	.8295	.56112	.82773	.67790	.4751	.2081	.7821	.57548	.81781	.70368	.4211	.2228	.7377	52
9	.53214	.84666	.62851	.5910	.1811	.8792	.54683	.83724	.65314	.5311	.1944	.8287	.56136	.82757	.67832	.4742	.2083	.7814	.57572	.81765	.70412	.4202	.2230	.7369	51
10	.53238	.84650	.62892	1.5900	.1813	1.8783	.54708	.83708	.65355	1.5301	.1946	1.8279	.56160	.82741	.67875	1.4733	.2086	1.7806	.57596	.81748	.70455	1.4193	.2233	1.7362	50
11	.53263	.84635	.62933	.5890	.1815	.8775	.54732	.83692	.65397	.5291	.1948	.8271	.56184	.82724	.67917	.4724	.2088	.7798	.57619	.81731	.70499	.4185	.2235	.7355	49
12	.53288	.84619	.62973	.5880	.1818	.8766	.54756	.83676	.65438	.5282	.1951	.8263	.56208	.82708	.67960	.4714	.2091	.7791	.57643	.81714	.70542	.4176	.2238	.7348	48
13	.53312	.84604	.63014	.5869	.1820	.8757	.54781	.83660	.65480	.5272	.1953	.8255	.56232	.82692	.68002	.4705	.2093	.7783	.57667	.81698	.70586	.4167	.2240	.7341	47
14	.53337	.84588	.63055	.5859	.1822	.8749	.54805	.83644	.65521	.5262	.1955	.8246	.56256	.82675	.68045	.4696	.2095	.7776	.57691	.81681	.70629	.4158	.2243	.7334	46
15	.53361	.84573	.63095	1.5849	.1824	1.8740	.54829	.83629	.65563	1.5252	.1958	1.8238	.56280	.82659	.68087	1.4687	.2098	1.7768	.57714	.81664	.70673	1.4150	.2245	1.7327	45
16	.53386	.84557	.63136	.5839	.1826	.8731	.54854	.83613	.65604	.5234	.1960	.8230	.56304	.82643	.68130	.4678	.2100	.7760	.57738	.81647	.70717	.4141	.2248	.7319	44
17	.53411	.84542	.63177	.5829	.1828	.8723	.54878	.83597	.65646	.5233	.1962	.8222	.56328	.82626	.68173	.4669	.2103	.7753	.57762	.81630	.70760	.4132	.2250	.7312	43
18	.53435	.84526	.63217	.5818	.1831	.8714	.54902	.83581	.65688	.5223	.1964	.8214	.56353	.82610	.68215	.4659	.2105	.7745	.57786	.81614	.70804	.4123	.2253	.7305	42
19	.53460	.84511	.63258	.5808	.1833	.8706	.54926	.83565	.65729	.5214	.1967	.8206	.56377	.82593	.68258	.4650	.2107	.7738	.57809	.81597	.70848	.4115	.2255	.7298	41
20	.53484	.84495	.63299	1.5798	.1835	1.8697	.54951	.83549	.65771	1.5204	.1969	1.8198	.56401	.82577	.68301	1.4641	.2110	1.7730	.57833	.81580	.70891	1.4106	.2258	1.7291	40
21	.53509	.84479	.63339	.5788	.1837	.8688	.54975	.83533	.65813	.5195	.1971	.8190	.56425	.82561	.68343	.4632	.2112	.7723	.57857	.81563	.70935	.4097	.2260	.7284	39
22	.53533	.84464	.63380	.5778	.1839	.8680	.54999	.83517	.65854	.5185	.1974	.8182	.56449	.82544	.68386	.4623	.2115	.7715	.57881	.81546	.70979	.4089	.2263	.7277	38
23	.53558	.84448	.63421	.5768	.1841	.8671	.55024	.83501	.65896	.5175	.1976	.8174	.56473	.82528	.68429	.4614	.2117	.7708	.57904	.81530	.71022	.4080	.2265	.7270	37
24	.53583	.84433	.63462	.5757	.1844	.8663	.55048	.83485	.65938	.5166	.1978	.8166	.56497	.82511	.68471	.4605	.2119	.7700	.57928	.81513	.71066	.4071	.2268	.7263	36
25	.53607	.84417	.63503	1.5747	.1846	1.8654	.55002	.83469	.65980	1.5156	.1980	1.8158	.56521	.82495	.68514	1.4595	.2122	1.7693	.57952	.81496	.71110	1.4063	.2270	1.7256	35
26	.53632	.84402	.63543	.5737	.1848	.8646	.55090	.83453	.66021	.5147	.1983	.8150	.56545	.82478	.68557	.4586	.2124	.7685	.57975	.81479	.71154	.4054	.2273	.7249	34
27	.53656	.84386	.63584	.5727	.1850	.8637	.55121	.83437	.66063	.5137	.1985	.8142	.56569	.82462	.68600	.4577	.2127	.7678	.57999	.81462	.71198	.4045	.2276	.7242	33
28	.53681	.84370	.63625	.5717	.1852	.8629	.55145	.83421	.66105	.5127	.1987	.8134	.56593	.82445	.68642	.4568	.2129	.7670	.58023	.81445	.71241	.4037	.2278	.7234	32
29	.53705	.84355	.63666	.5707	.1855	.8620	.55169	.83405	.66147	.5118	.1990	.8126	.56617	.82429	.68685	.4559	.2132	.7663	.58047	.81428	.71285	.4028	.2281	.7227	31
30	.53730	.84339	.63707	1.5697	.1857	1.8611	.55194	.83388	.56188	1.5108	.1992	1.8118	.56641	.82413	.68728	1.4550	.2134	1.7655	.58070	.81411	.71329	1.4019	.2283	1.7220	30
31	.53754	.84323	.63748	.5687	.1859	.8603	.55218	.83372	.66230	.5099	.1994	.8110	.56664	.82396	.68771	.4541	.2136	.7648	.58094	.81395	.71373	.4011	.2286	.7213	29
32	.53779	.84308	.63789	.5677	.1861	.8595	.55242	.83356	.66272	.5089	.1997	.8102	.56688	.82380	.68814	.4532	.2139	.7640	.58118	.81378	.71417	.4002	.2288	.7206	28
33	.53803	.84292	.63830	.5667	.1863	.8586	.55266	.83340	.66314	.5080	.1999	.8094	.56712	.82363	.68857	.4523	.2141	.7633	.58141	.81361	.71461	.3994	.2291	.7199	27
34	.53828	.84276	.63871	.5657	.1866	.8578	.55291	.83324	.66356	.5070	.2001	.8086	.56736	.82347	.68899	.4514	.2144	.7625	.58165	.81344	.71505	.3985	.2293	.7192	26
35	.53852	.84261	.63912	1.5646	.1868	1.8569	.55315	.83308	.66398	1.5061	.2004	1.8078	.56760	.82330	.68942	1.4505	.2146	1.7618	.58189	.81327	.71549	1.3976	.2296	1.7185	25
36	.53877	.84245	.63953	.5636	.1870	.8561	.55339	.83292	.66440	.5051	.2006	.8070	.56784	.82314	.68985	.4496	.2149	.7610	.58212	.81310	.71593	.3968	.2298	.7178	24
37	.53901	.84229	.63994	.5626	.1872	.8552	.55363	.83276	.66482	.5042	.2008	.8062	.56808	.82297	.69028	.4487	.2151	.7603	.58236	.81293	.71637	.3959	.2301	.7171	23
38	.53926	.84214	.64035	.5616	.1874	.8544	.55388	.83260	.66524	.5032	.2010	.8054	.56832	.82280	.69071	.4478	.2153	.7596	.58259	.81276	.71681	.3951	.2304	.7164	22
39	.53950	.84198	.64076	.5606	.1877	.8535	.55412	.83244	.66566	.5023	.2013	.8047	.56856	.82264	.69114	.4469	.2156	.7588	.58283	.81259	.71725	.3942	.2306	.7157	21
40	.53975	.84182	.64117	1.5596	.1879	1.8527	.55436	.83228	.66608	1.5013	.2015	1.8039	.56880	.82247	.69157	1.4460	.2158	1.7581	.58307	.81242	.71769	1.3933	.2309	1.7151	20
41	.53999	.84167	.64158	.5586	.1881	.8519	.55460	.83211	.66650	.5004	.2017	.8031	.56904	.82231	.69200	.4451	.2161	.7573	.58330	.81225	.71813	.3925	.2311	.7144	19
42	.54024	.84151	.64199	.5577	.1883	.8510	.55484	.83195	.66692	.4994	.2020	.8023	.56928	.82214	.69243	.4442	.2163	.7566	.58354	.81208	.71857	.3916	.2314	.7137	18
43	.54048	.84135	.64240	.5567	.1886	.8502	.55509	.83179	.66734	.4985	.2022	.8015	.56952	.82198	.69286	.4433	.2166	.7559	.58378	.81191	.71901	.3908	.2316	.7130	17
44	.54073	.84120	.64281	.5557	.1888	.8493	.55533	.83163	.66776	.4975	.2024	.8007	.56976	.82181	.69329	.4424	.2168	.7551	.58401	.81174	.71945	.3899	.2319	.7123	16
45	.54097	.84104	.64322	1.5547	.1890	1.8485	.55557	.83147	.66818	1.4966	.2027	1.7999	.57000	.82165	.69372	1.4415	.2171	1.7544	.58425	.81157	.71990	1.3891	.2322	1.7116	15
46	.54122	.84088	.64363	.5537	.1892	.8477	.55581	.83131	.66860	.4957	.2029	.7992	.57023	.82148	.69415	.4406	.2173	.7537	.58448	.81140	.72034	.3882	.2324	.7109	14
47	.54146	.84072	.64404	.5527	.1894	.8468	.55605	.83115	.66902	.4947	.2031	.7984	.57047	.82131	.69459	.4397	.2175	.7529	.58472	.81123	.72078	.3874	.2327	.7102	13
48	.54171	.84057	.64446	.5517	.1897	.8460	.55629	.83098	.66944	.4938	.2034	.7976	.57071	.82115	.69502	.4388	.2178	.7522	.58496	.81106	.72122	.3865	.2329	.7095	12
49	.54195	.84041	.64487	.5507	.1899	.8452	.55654	.83082	.66986	.4928	.2036	.7968	.57095	.82098	.69545	.4379	.2180	.7514	.58519	.81089	.72166	.3857	.2332	.7088	11
50	.54220	.84025	.64528	1.5497	.1901	1.8443	.55678	.83066	.67028	1.4919	.2039	1.7960	.57119	.82082	.69588	1.4370	.2183	1.7507	.58543	.81072	.72211	1.3848	.2335	1.7081	10
51	.54244	.84009	.64569	.5487	.1903	.8435	.55702	.83050	.67071	.4910	.2041	.7953	.57143	.82065	.69631	.4361	.2185	.7500	.58566	.81055	.72255	.3840	.2337	.7075	9
52	.54268	.83993	.64610	.5477	.1906	.8427	.55726	.83034	.67113	.4900	.2043	.7945	.57167	.82048	.69674	.4352	.2188	.7493	.58590	.81038	.72299	.3831	.2340	.7068	8
53	.54293	.83978	.64652	.5467	.1908	.8418	.55750	.83017	.67155	.4891	.2046	.7937	.57191	.82032	.69718	.4343	.2190	.7485	.58614	.81021	.72344	.3823	.2342	.7061	7
54	.54317	.83962	.64693	.5458	.1910	.8410	.55774	.83001	.67197	.4881	.2048	.7929	.57214	.82015	.69761	.4335	.2193	.7478	.58637	.81004	.72388	.3814	.2345	.7054	6
55	.54342	.83946	.64734	1.5448	.1912	1.8402	.55799	.82985	.67239	1.4872	.2050	1.7921	.57238	.81998	.69804	1.4326	.2195	1.7471	.58661	.80987	.72432	1.3806	.2348	1.7047	5
56	.54366	.83930	.64775	.5438	.1915	.8394	.55823	.82969	.67282	.4863	.2053	.7914	.57262	.81982	.69847	.4317	.2198	.7463	.58684	.80970	.72477	.3797	.2350	.7040	4
57	.54391	.83914	.64817	.5428	.1917	.8385	.55847	.82952	.67324	.4853	.2055	.7906	.57286	.81965	.69891	.4308	.2200	.7456	.58708	.80953	.72521	.3789	.2353	.7033	3
58	.54415	.83899	.64858	.5418	.1919	.8377	.55871	.82936	.67366	.4844	.2057	.7898	.57310	.81948	.69934	.4299	.2203	.7449	.58731	.80936	.72565	.3781	.2355	.7027	2
59	.54439	.83883	.64899	.5408	.1921	.8369	.55895	.82920	.67408	.4835	.2060	.7891	.57334	.81932	.69977	.4290	.2205	.7442	.58755	.80919	.72610	.3772	.2358	.7020	1
60	.54464	.83867	.64941	1.5399	.1922	1.8361	.55919	.82904	.67451	1.4826	.2062	1.7883	.57358	.81915	.70021	1.4281	.2208	1.7434	.58778	.80902	.72654	1.3764	.2361	1.7013	0
′	cos	sin	cot	tan	cosec	sec	cos	sin	cot	tan	cosec	sec	cos	sin	cot	tan	cosec	sec	cos	sin	cot	tan	cosec	sec	′
	57°						56°						55°						54°						

NATURAL TRIGONOMETRIC FUNCTIONS

	36°						37°						38°						39°						
′	sin	cos	tan	cot	sec	cosec	sin	cos	tan	cot	sec	cosec	sin	cos	tan	cot	sec	cosec	sin	cos	tan	cot	sec	cosec	′
0	.58778	.80902	.72654	1.3764	1.2361	1.7013	.60181	.79863	.75355	1.3270	1.2521	1.6616	.61566	.78801	.78128	1.2799	1.2690	1.6243	.62932	.77715	.80978	1.2349	1.2867	1.5890	60
1	.58802	.80885	.72699	.3755	.2363	.7006	.60205	.79846	.75401	.3262	.2524	.6610	.61589	.78783	.78175	.2792	.2693	.6237	.62955	.77696	.81026	.2342	.2871	.5884	59
2	.58825	.80867	.72743	.3747	.2366	.6999	.60228	.79828	.75447	.3254	.2527	.6603	.61612	.78765	.78222	.2784	.2696	.6231	.62977	.77678	.81075	.2334	.2874	.5879	58
3	.58849	.80850	.72788	.3738	.2368	.6993	.60251	.79811	.75492	.3246	.2530	.6597	.61635	.78747	.78269	.2776	.2699	.6224	.63000	.77660	.81123	.2327	.2877	.5873	57
4	.58873	.80833	.72832	.3730	.2371	.6986	.60274	.79793	.75538	.3238	.2532	.6591	.61658	.78729	.78316	.2769	.2702	.6218	.63022	.77641	.81171	.2320	.2880	.5867	56
5	.58896	.80816	.72877	1.3722	.2374	1.6979	.60298	.79776	.75584	1.3230	.2535	1.6584	.61681	.78711	.78363	1.2761	.2705	1.6212	.63045	.77623	.81219	1.2312	.2883	1.5862	55
6	.58920	.80799	.72921	.3713	.2376	.6972	.60320	.79758	.75629	.3222	.2538	.6578	.61703	.78693	.78410	.2753	.2707	.6206	.63067	.77605	.81268	.2305	.2886	.5856	54
7	.58943	.80782	.72966	.3705	.2379	.6965	.60344	.79741	.75675	.3214	.2541	.6572	.61726	.78675	.78457	.2746	.2710	.6200	.63090	.77586	.81316	.2297	.2889	.5850	53
8	.58967	.80765	.73010	.3697	.2382	.6959	.60367	.79723	.75721	.3206	.2543	.6565	.61749	.78657	.78504	.2738	.2713	.6194	.63113	.77568	.81364	.2290	.2892	.5845	52
9	.58990	.80747	.73055	.3688	.2384	.6952	.60390	.79706	.75767	.3198	.2546	.6559	.61772	.78640	.78551	.2730	.2716	.6188	.63135	.77549	.81413	.2283	.2895	.5839	51
10	.59014	.80730	.73100	1.3680	.2387	1.6945	.60413	.79688	.75812	1.3190	.2549	1.6552	.61795	.78622	.78598	1.2723	.2719	1.6182	.63158	.77531	.81461	1.2276	.2898	1.5833	50
11	.59037	.80713	.73144	.3672	.2389	.6938	.60437	.79670	.75858	.3182	.2552	.6546	.61818	.78604	.78645	.2715	.2722	.6176	.63180	.77513	.81509	.2268	.2901	.5828	49
12	.59060	.80696	.73189	.3663	.2392	.6932	.60460	.79653	.75904	.3174	.2554	.6540	.61841	.78586	.78692	.2708	.2725	.6170	.63203	.77494	.81558	.2261	.2904	.5822	48
13	.59084	.80679	.73234	.3655	.2395	.6925	.60483	.79635	.75950	.3166	.2557	.6533	.61864	.78568	.78739	.2700	.2728	.6164	.63225	.77476	.81606	.2254	.2907	.5816	47
14	.59107	.80662	.73278	.3647	.2397	.6918	.60506	.79618	.75996	.3159	.2560	.6527	.61886	.78550	.78786	.2692	.2731	.6159	.63248	.77458	.81655	.2247	.2910	.5811	46
15	.59131	.80644	.73323	1.3638	.2400	1.6912	.60529	.79600	.76042	1.3151	.2563	1.6521	.61909	.78532	.78834	1.2685	.2734	1.6153	.63270	.77439	.81703	1.2239	.2913	1.5805	45
16	.59154	.80627	.73368	.3630	.2403	.6905	.60552	.79582	.76088	.3143	.2565	.6514	.61932	.78514	.78881	.2677	.2737	.6147	.63293	.77421	.81752	.2232	.2916	.5799	44
17	.59178	.80610	.73412	.3622	.2405	.6898	.60576	.79565	.76134	.3135	.2568	.6508	.61955	.78496	.78928	.2670	.2739	.6141	.63315	.77402	.81800	.2225	.2919	.5794	43
18	.59201	.80593	.73457	.3613	.2408	.6891	.60599	.79547	.76179	.3127	.2571	.6502	.61978	.78478	.78975	.2662	.2742	.6135	.63338	.77384	.81849	.2218	.2922	.5788	42
19	.59225	.80576	.73502	.3605	.2411	.6885	.60622	.79530	.76225	.3119	.2574	.6496	.62001	.78460	.79022	.2655	.2745	.6129	.63360	.77365	.81898	.2210	.2926	.5783	41
20	.59248	.80558	.73547	1.3597	.2413	1.6878	.60645	.79512	.76271	1.3111	.2577	1.6489	.62023	.78441	.79070	1.2647	.2748	1.6123	.63383	.77347	.81946	1.2203	.2929	1.5777	40
21	.59272	.80541	.73592	.3588	.2416	.6871	.60668	.79494	.76317	.3103	.2579	.6483	.62046	.78423	.79117	.2639	.2751	.6117	.63405	.77329	.81995	.2196	.2932	.5771	39
22	.59295	.80524	.73637	.3580	.2419	.6865	.60691	.79477	.76364	.3095	.2582	.6477	.62069	.78405	.79164	.2632	.2754	.6111	.63428	.77310	.82043	.2189	.2935	.5766	38
23	.59318	.80507	.73681	.3572	.2421	.6858	.60714	.79459	.76410	.3087	.2585	.6470	.62092	.78387	.79212	.2625	.2757	.6105	.63450	.77292	.82092	.2181	.2938	.5760	37
24	.59342	.80489	.73726	.3564	.2424	.6851	.60737	.79441	.76456	.3079	.2588	.6464	.62115	.78369	.79259	.2617	.2760	.6099	.63473	.77273	.82141	.2174	.2941	.5755	36
25	.59365	.80472	.73771	1.3555	.2427	1.6845	.60761	.79424	.76502	1.3071	.2591	1.6458	.62137	.78351	.79306	1.2609	.2763	1.6093	.63495	.77255	.82190	1.2167	.2944	1.5749	35
26	.59389	.80455	.73816	.3547	.2429	.6838	.60784	.79406	.76548	.3064	.2593	.6452	.62160	.78333	.79354	.2602	.2766	.6087	.63518	.77236	.82238	.2160	.2947	.5743	34
27	.59412	.80437	.73861	.3539	.2432	.6831	.60807	.79388	.76594	.3056	.2596	.6445	.62183	.78315	.79401	.2594	.2769	.6081	.63540	.77218	.82287	.2152	.2950	.5738	33
28	.59435	.80420	.73906	.3531	.2435	.6825	.60830	.79371	.76640	.3048	.2599	.6439	.62206	.78297	.79449	.2587	.2772	.6075	.63563	.77199	.82336	.2145	.2953	.5732	32
29	.59459	.80403	.73951	.3522	.2437	.6818	.60853	.79353	.76686	.3040	.2602	.6433	.62229	.78279	.79496	.2579	.2775	.6070	.63585	.77181	.82385	.2138	.2956	.5727	31
30	.59482	.80386	.73996	1.3514	.2440	1.6812	.60876	.79335	.76733	1.3032	.2605	1.6427	.62251	.78261	.79543	1.2572	.2778	1.6064	.63608	.77162	.82434	1.2131	.2960	1.5721	30
31	.59506	.80368	.74041	.3506	.2443	.6805	.60899	.79318	.76779	.3024	.2607	.6421	.62274	.78243	.79591	.2564	.2781	.6058	.63630	.77144	.82483	.2124	.2963	.5716	29
32	.59529	.80351	.74086	.3498	.2445	.6798	.60922	.79300	.76825	.3016	.2610	.6414	.62297	.78224	.79639	.2557	.2784	.6052	.63653	.77125	.82531	.2117	.2966	.5710	28
33	.59552	.80334	.74131	.3489	.2448	.6792	.60945	.79282	.76871	.3009	.2613	.6408	.62320	.78206	.79686	.2549	.2787	.6046	.63675	.77107	.82580	.2109	.2969	.5705	27
34	.59576	.80316	.74176	.3481	.2451	.6785	.60968	.79264	.76918	.3001	.2616	.6402	.62342	.78188	.79734	.2542	.2790	.6040	.63697	.77088	.82629	.2102	.2972	.5699	26
35	.59599	.80299	.74221	1.3473	.2453	1.6779	.60991	.79247	.76964	1.2993	.2619	1.6396	.62365	.78170	.79781	1.2534	.2793	1.6034	.63720	.77070	.82678	1.2095	.2975	1.5694	25
36	.59622	.80282	.74266	.3465	.2456	.6772	.61014	.79229	.77010	.2985	.2622	.6389	.62388	.78152	.79829	.2527	.2795	.6029	.63742	.77051	.82727	.2088	.2978	.5688	24
37	.59646	.80264	.74312	.3457	.2459	.6766	.61037	.79211	.77057	.2977	.2624	.6383	.62411	.78134	.79876	.2519	.2798	.6023	.63765	.77033	.82776	.2081	.2981	.5683	23
38	.59669	.80247	.74357	.3449	.2461	.6759	.61061	.79193	.77103	.2970	.2627	.6377	.62433	.78116	.79924	.2512	.2801	.6017	.63787	.77014	.82825	.2074	.2985	.5677	22
39	.59692	.80230	.74402	.3440	.2464	.6752	.61084	.79176	.77149	.2962	.2630	.6371	.62456	.78097	.79972	.2504	.2804	.6011	.63810	.76996	.82874	.2066	.2988	.5672	21
40	.59716	.80212	.74447	1.3432	.2467	1.6746	.61107	.79158	.77196	1.2954	.2633	1.6365	.62479	.78079	.80020	1.2497	.2807	1.6005	.63832	.76977	.82923	1.2059	.2991	1.5666	20
41	.59739	.80195	.74492	.3424	.2470	.6739	.61130	.79140	.77242	.2946	.2636	.6359	.62501	.78061	.80067	.2489	.2810	.6000	.63854	.76958	.82972	.2052	.2994	.5661	19
42	.59762	.80177	.74538	.3416	.2472	.6733	.61153	.79122	.77289	.2938	.2639	.6352	.62524	.78043	.80115	.2482	.2813	.5994	.63877	.76940	.83022	.2045	.2997	.5655	18
43	.59786	.80160	.74583	.3408	.2475	.6726	.61176	.79104	.77335	.2931	.2641	.6346	.62547	.78025	.80163	.2475	.2816	.5988	.63899	.76921	.83071	.2038	.3000	.5650	17
44	.59809	.80143	.74628	.3400	.2478	.6720	.61199	.79087	.77382	.2923	.2644	.6340	.62570	.78007	.80211	.2467	.2819	.5982	.63921	.76903	.83120	.2031	.3003	.5644	16
45	.59832	.80125	.74673	1.3392	.2480	1.6713	.61222	.79069	.77428	1.2915	.2647	1.6334	.62592	.77988	.80258	1.2460	.2822	1.5976	.63944	.76884	.83169	1.2024	.3006	1.5639	15
46	.59856	.80108	.74719	.3383	.2483	.6707	.61245	.79051	.77475	.2907	.2650	.6328	.62615	.77970	.80306	.2452	.2825	.5971	.63966	.76865	.83218	.2016	.3010	.5633	14
47	.59879	.80090	.74764	.3375	.2486	.6700	.61268	.79033	.77521	.2900	.2653	.6322	.62638	.77952	.80354	.2445	.2828	.5965	.63989	.76847	.83268	.2009	.3013	.5628	13
48	.59902	.80073	.74809	.3367	.2488	.6694	.61290	.79015	.77568	.2892	.2656	.6316	.62660	.77934	.80402	.2437	.2831	.5959	.64011	.76828	.83317	.2002	.3016	.5622	12
49	.59926	.80056	.74855	.3359	.2491	.6687	.61314	.78998	.77614	.2884	.2659	.6309	.62683	.77915	.80450	.2430	.2834	.5953	.64033	.76810	.83366	.1995	.3019	.5617	11
50	.59949	.80038	.74900	1.3351	.2494	1.6681	.61337	.78980	.77661	1.2876	.2661	1.6303	.62706	.77897	.80498	1.2423	.2837	1.5947	.64056	.76791	.83415	1.1988	.3022	1.5611	10
51	.59972	.80021	.74946	.3343	.2497	.6674	.61360	.78962	.77708	.2869	.2664	.6297	.62728	.77879	.80546	.2415	.2840	.5942	.64078	.76772	.83465	.1981	.3025	.5606	9
52	.59995	.80003	.74991	.3335	.2499	.6668	.61383	.78944	.77754	.2861	.2667	.6291	.62751	.77861	.80594	.2408	.2843	.5936	.64100	.76754	.83514	.1974	.3029	.5600	8
53	.60019	.79986	.75037	.3327	.2502	.6661	.61405	.78926	.77801	.2853	.2670	.6285	.62774	.77842	.80642	.2400	.2846	.5930	.64123	.76735	.83563	.1967	.3032	.5595	7
54	.60042	.79968	.75082	.3319	.2505	.6655	.61428	.78908	.77848	.2845	.2673	.6279	.62796	.77824	.80690	.2393	.2849	.5924	.64145	.76716	.83613	.1960	.3035	.5590	6
55	.60065	.79951	.75128	1.3311	.2508	1.6648	.61451	.78890	.77895	1.2838	.2676	1.6273	.62819	.77806	.80738	1.2386	.2852	1.5919	.64167	.76698	.83662	1.1953	.3038	1.5584	5
56	.60088	.79933	.75173	.3303	.2510	.6642	.61474	.78873	.77941	.2830	.2679	.6267	.62841	.77788	.80786	.2378	.2855	.5913	.64189	.76679	.83712	.1946	.3041	.5579	4
57	.60112	.79916	.75219	.3294	.2513	.6636	.61497	.78855	.77988	.2822	.2681	.6261	.62864	.77769	.80834	.2371	.2858	.5907	.64212	.76660	.83761	.1939	.3044	.5573	3
58	.60135	.79898	.75264	.3286	.2516	.6629	.61520	.78837	.78035	.2815	.2684	.6255	.62887	.77751	.80882	.2364	.2861	.5901	.64234	.76642	.83811	.1932	.3048	.5568	2
59	.60158	.79881	.75310	.3278	.2519	.6623	.61543	.78819	.78082	.2807	.2687	.6249	.62909	.77733	.80930	.2356	.2864	.5896	.64256	.76623	.83860	.1924	.3051	.5563	1
60	.60181	.79863	.75355	1.3270	.2521	1.6616	.61566	.78801	.78128	1.2799	.2690	1.6243	.62932	.77715	.80978	1.2349	.2867	1.5890	.64279	.76604	.83910	1.1917	.3054	1.5557	0
′	cos	sin	cot	tan	cosec	sec	cos	sin	cot	tan	cosec	sec	cos	sin	cot	tan	cosec	sec	cos	sin	cot	tan	cosec	sec	′
	53°						52°						51°						50°						

	40°						41°						42°						43°						
′	sin	cos	tan	cot	sec	cosec	sin	cos	tan	cot	sec	cosec	sin	cos	tan	cot	sec	cosec	sin	cos	tan	cot	sec	cosec	′
0	.64279	.76604	.83910	1.1917	1.3054	1.5557	.65606	.75471	.86929	1.1504	1.3250	1.5242	.66913	.74314	.90040	1.1106	1.3456	1.4945	.68200	.73135	.93251	1.0724	1.3673	1.4663	60
1	.64301	.76586	.83959	.1910	.3057	.5552	.65628	.75452	.86980	.1497	.3253	.5237	.66935	.74295	.90093	.1100	.3460	.4940	.68221	.73115	.93306	.0717	.3677	.4658	59
2	.64323	.76567	.84009	.1903	.3060	.5546	.65650	.75433	.87031	.1490	.3257	.5232	.66956	.74276	.90146	.1093	.3463	.4935	.68242	.73096	.93360	.0711	.3681	.4654	58
3	.64345	.76548	.84059	.1896	.3064	.5541	.65672	.75414	.87082	.1483	.3260	.5227	.66978	.74256	.90198	.1086	.3467	.4930	.68264	.73076	.93415	.0705	.3684	.4649	57
4	.64368	.76530	.84108	.1889	.3067	.5536	.65694	.75394	.87133	.1477	.3263	.5222	.66999	.74236	.90251	.1080	.3470	.4925	.68285	.73056	.93469	.0699	.3688	.4644	56
5	.64390	.76511	.84158	1.1882	.3070	1.5530	.65716	.75375	.87184	1.1470	.3267	1.5217	.67021	.74217	.90304	1.1074	.3474	1.4921	.68306	.73036	.93524	1.0692	.3692	1.4640	55
6	.64412	.76492	.84208	.1875	.3073	.5525	.65037	.75356	.87235	.1463	.3270	.5212	.67043	.74197	.90357	.1067	.3477	.4916	.68327	.73016	.93578	.0686	.3695	.4635	54
7	.64435	.76473	.84257	.1868	.3076	.5520	.65759	.75337	.87287	.1456	.3274	.5207	.67064	.74178	.90410	.1061	.3481	.4911	.68349	.72996	.93633	.0680	.3699	.4631	53
8	.64457	.76455	.84307	.1861	.3080	.5514	.65781	.75318	.87338	.1450	.3277	.5202	.67086	.74158	.90463	.1054	.3485	.4906	.68370	.72976	.93687	.0674	.3703	.4626	52
9	.64479	.76436	.84357	.1854	.3083	.5509	.65803	.75299	.87389	.1443	.3280	.5197	.67107	.74139	.90515	.1048	.3488	.4901	.68391	.72956	.93742	.0667	.3707	.4622	51
10	.64501	.76417	.84407	1.1847	.3086	1.5503	.65285	.75280	.87441	1.1436	.3284	1.5192	.67129	.74119	.90568	1.1041	.3492	1.4897	.68412	.72937	.93797	1.0661	.3710	1.4617	50
11	.64523	.76398	.84457	.1840	.3089	.5498	.65847	.75261	.87492	.1430	.3287	.5187	.67150	.74100	.90621	.1035	.3495	.4892	.68433	.72917	.93851	.0655	.3714	.4613	49
12	.64546	.76380	.84506	.1833	.3092	.5493	.65869	.75241	.87543	.1423	.3290	.5182	.67172	.74080	.90674	.1028	.3499	.4887	.68455	.72897	.93906	.0649	.3718	.4608	48
13	.64568	.76361	.84556	.1826	.3096	.5487	.65891	.75222	.87595	.1416	.3294	.5177	.67194	.74061	.90727	.1022	.3502	.4882	.68476	.72877	.93961	.0643	.3722	.4604	47
14	.64590	.76342	.84606	.1819	.3099	.5482	.65913	.75203	.87646	.1409	.3297	.5171	.67215	.74041	.90780	.1015	.3506	.4877	.68497	.72857	.94016	.0636	.3725	.4599	46
15	.64612	.76323	.84656	1.1812	.3102	1.5477	.65934	.75184	.87698	1.1403	.3301	1.5166	.67237	.74022	.90834	1.1009	.3509	1.4873	.68518	.72837	.94071	1.0630	.3729	1.4595	45
16	.64635	.76304	.84706	.1805	.3105	.5471	.65956	.75165	.87749	.1396	.3304	.5161	.67258	.74002	.90887	.1003	.3513	.4868	.68539	.72817	.94125	.0624	.3733	.4590	44
17	.64657	.76286	.84056	.1798	.3109	.5466	.65978	.75146	.87801	.1389	.3307	.5156	.67280	.73983	.90940	.0996	.3517	.4863	.68561	.72797	.94180	.0618	.3737	.4586	43
18	.64679	.76267	.84806	.1791	.3112	.5461	.66000	.75126	.87852	.1383	.3311	.5151	.67301	.73963	.90993	.0990	.3520	.4858	.68582	.72777	.94235	.0612	.3740	.4581	42
19	.64701	.76248	.84856	.1785	.3115	.5456	.66022	.75107	.87904	.1376	.3314	.5146	.67323	.73943	.91046	.0983	.3524	.4854	.68603	.72757	.94290	.0605	.3744	.4577	41
20	.64723	.76229	.84906	1.1778	.3118	1.5450	.66044	.75088	.87955	1.1369	.3318	1.5141	.67344	.73924	.91099	1.0977	.3527	1.4849	.68624	.72737	.94345	1.0599	.3748	1.4572	40
21	.64745	.76210	.84956	.1771	.3121	.5445	.66066	.75069	.88007	.1363	.3321	.5136	.67366	.73904	.91153	.0971	.3531	.4844	.68645	.72717	.94400	.0593	.3752	.4568	39
22	.64768	.76191	.85006	.1764	.3125	.5440	.66087	.75049	.88058	.1356	.3324	.5131	.67387	.73885	.91206	.0964	.3534	.4839	.68666	.72697	.94455	.0587	.3756	.4563	38
23	.64790	.76173	.85056	.1757	.3128	.5434	.66109	.75030	.88110	.1349	.3328	.5126	.67409	.73865	.91259	.0958	.3538	.4835	.68688	.72677	.94510	.0581	.3759	.4559	37
24	.64812	.76154	.85107	.1750	.3131	.5429	.66131	.75011	.88162	.1343	.3331	.5121	.67430	.73845	.91312	.0951	.3542	.4830	.68709	.72657	.94565	.0575	.3763	.4554	36
25	.64834	.76135	.85157	1.1743	.3134	1.5424	.66153	.74992	.88213	1.1336	.3335	1.5116	.67452	.73826	.91366	1.0945	.3545	1.4825	.68730	.72637	.94620	1.0568	.3767	1.4550	35
26	.64856	.76116	.85207	.1736	.3138	.5419	.66175	.74973	.88265	.1329	.3338	.5111	.67473	.73806	.91419	.0939	.3549	.4821	.68751	.72617	.94675	.0562	.3771	.4545	34
27	.64878	.76097	.85257	.1729	.3141	.5413	.66197	.74953	.88317	.1323	.3342	.5106	.67495	.73787	.91473	.0932	.3552	.4816	.68772	.72597	.94731	.0556	.3774	.4541	33
28	.64900	.76078	.85307	.1722	.3144	.5408	.66218	.74934	.88369	.1316	.3345	.5101	.67516	.73767	.91526	.0926	.3556	.4811	.68793	.72577	.94786	.0550	.3778	.4536	32
29	.64923	.76059	.85358	.1715	.3148	.5403	.66240	.74915	.88421	.1309	.3348	.5096	.67537	.73747	.91580	.0919	.3560	.4806	.68814	.72557	.94841	.0544	.3782	.4532	31
30	.64945	.76041	.85408	1.1708	.3151	1.5398	.66262	.74895	.88472	1.1303	.3352	1.5092	.67559	.73728	.91633	1.0913	.3563	1.4802	.68835	.72537	.94896	1.0538	.3786	1.4527	30
31	.64967	.76022	.85458	.1702	.3154	.5392	.66284	.74876	.88524	.1296	.3355	.5087	.67580	.73708	.91687	.0907	.3567	.4797	.68856	.72517	.94952	.0532	.3790	.4523	29
32	.64989	.76003	.85509	.1695	.3157	.5387	.66305	.74857	.88576	.1290	.3359	.5082	.67602	.73688	.91740	.0900	.3571	.4792	.68878	.72497	.95007	.0525	.3794	.4518	28
33	.65011	.75984	.85559	.1688	.3161	.5382	.66327	.74838	.88628	.1283	.3362	.5077	.67623	.73669	.91794	.0894	.3574	.4788	.68899	.72477	.95062	.0519	.3797	.4514	27
34	.65033	.75965	.85609	.1681	.3164	.5377	.66349	.74818	.88680	.1276	.3366	.5072	.67645	.73649	.91847	.0888	.3578	.4783	.68920	.72457	.95118	.0513	.3801	.4510	26
35	.65055	.75946	.85660	1.1674	.3167	1.5371	.66371	.74799	.88732	1.1270	.3369	1.5067	.67666	.73629	.91901	1.0881	.3581	1.4778	.68941	.72437	.95173	1.0507	.3805	1.4505	25
36	.65077	.75927	.85710	.1667	.3170	.5366	.66393	.74780	.88784	.1263	.3372	.5062	.67688	.73610	.91955	.0875	.3585	.4774	.68962	.72417	.95229	.0501	.3809	.4501	24
37	.65100	.75908	.85761	.1660	.3174	.5361	.66414	.74760	.88836	.1257	.3376	.5057	.67709	.73590	.92008	.0868	.3589	.4769	.68983	.72397	.95284	.0495	.3813	.4496	23
38	.65121	.75889	.85811	.1653	.3177	.5356	.66436	.74741	.88888	.1250	.3379	.5052	.67730	.73570	.92062	.0862	.3592	.4764	.69004	.72377	.95340	.0489	.3816	.4492	22
39	.65144	.75870	.85862	.1647	.3180	.5351	.66458	.74722	.88940	.1243	.3383	.5047	.67752	.73551	.92116	.0856	.3596	.4760	.69025	.72357	.95395	.0483	.3820	.4487	21
40	.65166	.75851	.85912	1.1640	.3184	1.5345	.66479	.74702	.88992	1.1237	.3386	1.5042	.67773	.73531	.92170	1.0849	.3600	1.4755	.69046	.72337	.95451	1.0476	.3824	1.4483	20
41	.65188	.75832	.85963	.1633	.3187	.5340	.66501	.74683	.89044	.1230	.3390	.5037	.67794	.73511	.92223	.0843	.3603	.4750	.69067	.72317	.95506	.0470	.3828	.4479	19
42	.65210	.75813	.86013	.1626	.3190	.5335	.66523	.74664	.89097	.1224	.3393	.5032	.67816	.73491	.92277	.0837	.3607	.4746	.69088	.72297	.95562	.0464	.3832	.4474	18
43	.65232	.75794	.86064	.1619	.3193	.5330	.66545	.74644	.89149	.1217	.3397	.5027	.67837	.73472	.92331	.0830	.3611	.4741	.69109	.72277	.95618	.0458	.3836	.4470	17
44	.65254	.75775	.86115	.1612	.3197	.5325	.66566	.74625	.89201	.1211	.3400	.5022	.67859	.73452	.92385	.0824	.3614	.4736	.69130	.72256	.95673	.0452	.3839	.4465	16
45	.65276	.75756	.86165	1.1605	.3200	1.5319	.66588	.74606	.89253	1.1204	1.3404	1.5018	.67880	.73432	.92439	1.0818	.3618	1.4732	.69151	.72236	.95729	1.0446	.3843	1.4461	15
46	.65298	.75737	.86216	.1599	.3203	.5314	.66610	.74586	.89306	.1197	.3407	.5013	.67901	.73412	.92493	.0812	.3622	.4727	.69172	.72216	.95785	.0440	.3847	.4457	14
47	.65320	.75718	.86267	.1592	.3207	.5309	.66631	.74567	.89358	.1191	.3411	.5008	.67923	.73393	.92547	.0805	.3625	.4723	.69193	.72196	.95841	.0434	.3851	.4452	13
48	.65342	.75700	.86318	.1585	.3210	.5304	.66653	.74548	.89410	.1184	.3414	.5003	.67944	.73373	.92601	.0799	.3629	.4718	.69214	.72176	.95896	.0428	.3855	.4448	12
49	.65364	.75680	.86368	.1578	.3213	.5299	.66675	.74528	.89463	.1178	.3418	.4998	.67965	.73353	.92655	.0793	.3633	.4713	.69235	.72156	.95952	.0422	.3859	.4443	11
50	.65386	.75661	.86419	1.1571	.3217	1.5294	.66697	.74509	.89515	1.1171	.3421	1.4993	.67987	.73333	.92709	1.0786	.3636	1.4709	.69256	.72136	.96008	1.0416	.3863	1.4439	10
51	.65408	.75642	.86470	.1565	.3220	.5289	.66718	.74489	.89567	.1165	.3425	.4988	.68008	.73314	.92763	.0780	.3640	.4704	.69277	.72115	.96064	.0410	.3867	.4435	9
52	.65430	.75623	.86521	.1558	.3223	.5283	.66740	.74470	.89620	.1158	.3428	.4983	.68029	.73294	.92817	.0774	.3644	.4699	.69298	.72095	.96120	.0404	.3870	.4430	8
53	.65452	.75604	.86572	.1551	.3227	.5278	.66762	.74450	.89672	.1152	.3432	.4979	.68051	.73274	.92871	.0767	.3647	.4695	.69319	.72075	.96176	.0397	.3874	.4426	7
54	.65474	.75585	.86623	.1544	.3230	.5273	.66783	.74431	.89725	.1145	.3435	.4974	.68072	.73254	.92926	.0761	.3651	.4690	.69340	.72055	.96232	.0391	.3878	.4422	6
55	.65496	.75566	.86674	1.1537	.3233	1.5268	.66805	.74412	.89777	1.1139	.3439	1.4969	.68093	.73234	.92980	1.0755	.3655	1.4686	.69361	.72035	.96288	1.0385	.3882	1.4417	5
56	.65518	.75547	.86725	.1531	.3237	.5263	.66826	.74392	.89830	.1132	.3442	.4964	.68115	.73215	.93034	.0749	.3658	.4681	.69382	.72015	.96344	.0379	.3886	.4413	4
57	.65540	.75528	.86775	.1524	.3240	.5258	.66848	.74373	.89882	.1126	.3446	.4959	.68136	.73195	.93088	.0742	.3662	.4676	.69403	.71994	.96400	.0373	.3890	.4408	3
58	.65562	.75509	.86826	.1517	.3243	.5253	.66870	.74353	.89935	.1119	.3449	.4954	.68157	.73175	.93143	.0736	.3666	.4672	.69424	.71974	.96456	.0367	.3894	.4404	2
59	.65584	.75490	.86878	.1510	.3247	.5248	.66891	.74334	.89988	.1113	.3453	.4949	.68178	.73155	.93197	.0730	.3669	.4667	.69445	.71954	.96513	.0361	.3898	.4400	1
60	.65606	.75401	.86929	1.1504	.3250	1.5242	.66913	.74314	.90040	1.1106	.3456	1.4945	.68200	.73135	.93251	1.0724	.3673	1.4663	.69466	.71934	.96569	1.0355	.3902	1.4395	0
′	cos	sin	cot	tan	cosec	sec	cos	sin	cot	tan	cosec	sec	cos	sin	cot	tan	cosec	sec	cos	sin	cot	tan	cosec	sec	′
	49°						48°						47°						46°						

Machine Tool Practices

44°

′	sin	cos	tan	cot	sec	cosec	′
0	.69466	.71934	.96569	1.0355	1.3902	1.4395	60
1	.69487	.71914	.96625	.0349	.3905	.4391	59
2	.69508	.71893	.96681	.0343	.3909	.4387	58
3	.69528	.71873	.96738	.0337	.3913	.4382	57
4	.69549	.71853	.96794	.0331	.3917	.4378	56
5	.69570	.71833	.96850	1.0325	.3921	1.4374	55
6	.69591	.71813	.96907	.0319	.3925	.4370	54
7	.69612	.71792	.96963	.0313	.3929	.4365	53
8	.69633	.71772	.97020	.0307	.3933	.4361	52
9	.69654	.71752	.97076	.0301	.3937	.4357	51
10	.69675	.71732	.97133	1.0295	.3941	1.4352	50
11	.69696	.71711	.97189	.0289	.3945	.4348	49
12	.69716	.71691	.97246	.0283	.3949	.4344	48
13	.69737	.71671	.97302	.0277	.3953	.4339	47
14	.69758	.71650	.97359	.0271	.3957	.4335	46
15	.69779	.71630	.97416	1.0265	.3960	1.4331	45
16	.69800	.71610	.97472	.0259	.3964	.4327	44
17	.69821	.71589	.97529	.0253	.3968	.4322	43
18	.69841	.71569	.97586	.0247	.3972	.4318	42
19	.69862	.71549	.97643	.0241	.3976	.4314	41
20	.69883	.71529	.97700	1.0235	.3980	1.4310	40
21	.69904	.71508	.97756	.0229	.3984	.4305	39
22	.69925	.71488	.97813	.0223	.3988	.4301	38
23	.69945	.71468	.97870	.0218	.3992	.4297	37
24	.69966	.71447	.97927	.0212	.3996	.4292	36
25	.69987	.71427	.97984	1.0206	.4000	1.4288	35
26	.70008	.71406	.98041	.0200	.4004	.4284	34
27	.70029	.71386	.98098	.0194	.4008	.4280	33
28	.70049	.71366	.98155	.0188	.4012	.4276	32
29	.70070	.71345	.98212	.0182	.4016	.4271	31
30	.70091	.71325	.98270	1.0176	.4020	1.4267	30
31	.70112	.71305	.98327	.0170	.4024	.4263	29
32	.70132	.71284	.98384	.0164	.4028	.4259	28
33	.70153	.71264	.98441	.0158	.4032	.4254	27
34	.70174	.71243	.98499	.0152	.4036	.4250	26
35	.70194	.71223	.98556	1.0146	.4040	1.4246	25
36	.70215	.71203	.98613	.0141	.4044	.4242	24
37	.70236	.71182	.98671	.0135	.4048	.4238	23
38	.70257	.71162	.98728	.0129	.4052	.4233	22
39	.70277	.71141	.98786	.0123	.4056	.4229	21
40	.70298	.71121	.98843	1.0117	.4060	1.4225	20
41	.70319	.71100	.98901	.0111	.4065	.4221	19
42	.70339	.71080	.98958	.0105	.4069	.4217	18
43	.70360	.71059	.99016	.0099	.4073	.4212	17
44	.70381	.71039	.99073	.0093	.4077	.4208	16
45	.70401	.71018	.99131	1.0088	.4081	1.4204	15
46	.70422	.70998	.99189	.0082	.4085	.4200	14
47	.70443	.70977	.99246	.0076	.4089	.4196	13
48	.70463	.70957	.99304	.0070	.4093	.4192	12
49	.70484	.70936	.99362	.0064	.4097	.4188	11
50	.70505	.70916	.99420	1.0058	.4101	1.4183	10
51	.70525	.70895	.99478	.0052	.4105	.4179	9
52	.70546	.70875	.99536	.0047	.4109	.4175	8
53	.70566	.70854	.99593	.0041	.4113	.4171	7
54	.70587	.70834	.99651	.0035	.4117	.4167	6
55	.70608	.70813	.99709	1.0029	.4122	1.4163	5
56	.70628	.70793	.99767	.0023	.4126	.4159	4
57	.70649	.70772	.99826	.0017	.4130	.4154	3
58	.70669	.70752	.99884	.0012	.4134	.4150	2
59	.70690	.70731	.99942	.0006	.4138	.4146	1
60	.70711	.70711	1.00000	1.0000	.4142	1.4142	0
′	cos	sin	cot	tan	cosec	sec	′

45°

Table 2
Sine-bar constants

Constants for Setting a 5-inch Sine-bar

Min.	0°	1°	2°	3°	4°	5°	6°	7°
0	0.00000	0.08725	0.17450	0.26170	0.34880	0.43580	0.52265	0.60935
1	.00145	.08870	.17595	.26315	.35025	.43720	.52410	.61080
2	.00290	.09015	.17740	.26460	.35170	.43865	.52555	.61225
3	.00435	.09160	.17885	.26605	.35315	.44010	.52700	.61370
4	.00580	.09310	.18030	.26750	.35460	.44155	.52845	.61510
5	0.00725	0.09455	0.18175	0.26895	0.35605	0.44300	0.52985	0.61655
6	.00875	.09600	.18320	.27040	.35750	.44445	.53130	.61800
7	.01020	.09745	.18465	.27185	.35895	.44590	.53275	.61945
8	.01165	.09890	.18615	.27330	.36040	.44735	.53420	.62090
9	.01310	.10035	.18760	.27475	.36185	.44880	.53565	.62235
10	0.01455	0.10180	0.18905	0.27620	0.36330	0.45025	0.53710	0.62380
11	.01600	.10325	.19050	.27765	.36475	.45170	.53855	.62520
12	.01745	.10470	.19195	.27910	.36620	.45315	.54000	.62665
13	.01890	.10615	.19340	.28055	.36765	.45460	.54145	.62810
14	.02035	.10760	.19485	.28200	.36910	.45605	.54290	.62955
15	0.02180	0.10905	0.19630	0.28345	0.37055	0.45750	0.54435	0.63100
16	.02325	.11055	.19775	.28490	.37200	.45895	.54580	.63245
17	.02475	.11200	.19920	.28635	.37345	.46040	.54725	.63390
18	.02620	.11345	.20065	.28780	.37490	.46185	.54865	.63530
19	.02765	.11490	.20210	.28925	.37635	.46330	.55010	.63675
20	0.02910	0.11635	0.20355	0.29070	0.37780	0.46475	0.55155	0.63820
21	.03055	.11780	.20500	.29220	.37925	.46620	.55300	.63965
22	.03200	.11925	.20645	.29365	.38070	.46765	.55445	.64110
23	.03345	.12070	.20795	.29510	.38215	.46910	.55590	.64255
24	.03490	.12215	.20940	.29655	.38360	.47055	.55735	.64400
25	0.03635	0.12360	0.21085	0.29800	0.38505	0.47200	0.55880	0.64540
26	.03780	.12505	.21230	.29945	.38650	.47345	.56025	.64685
27	.03925	.12650	.21375	.30090	.38795	.47490	.56170	.64830
28	.04070	.12800	.21520	.30235	.38940	.47635	.56315	.64975
29	.04220	.12945	.21665	.30380	.39085	.47780	.56455	.65120
30	0.04365	0.13090	0.21810	0.30525	0.39230	0.47925	0.56600	0.65265
31	.04510	.13235	.21955	.30670	.39375	.48070	.56745	.65405
32	.04655	.13380	.22100	.30815	.39520	.48210	.56890	.65550
33	.04800	.13525	.22245	.30960	.39665	.48355	.57035	.65695
34	.04945	.13670	.22390	.31105	.39810	.48500	.57180	.65840
35	0.05090	0.13815	0.22535	0.31250	0.39955	0.48645	0.57325	0.65985
36	.05235	.13960	.22680	.31395	.40100	.48790	.57470	.66130
37	.05380	.14105	.22825	.31540	.40245	.48935	.57615	.66270
38	.05525	.14250	.22970	.31685	.40390	.49080	.57760	.66415
39	.05670	.14395	.23115	.31830	.40535	.49225	.57900	.66560
40	0.05820	0.14540	0.23265	0.31975	0.40680	0.49370	0.58045	0.66705
41	.05965	.14690	.23410	.32120	.40825	.49515	.58190	.66850
42	.06110	.14835	.23555	.32265	.40970	.49660	.58335	.66995
43	.06255	.14980	.23700	.32410	.41115	.49805	.58480	.67135
44	.06400	.15125	.23845	.32555	.41260	.49950	.58625	.67280
45	0.06545	0.15270	0.23990	0.32700	0.41405	0.50095	0.58770	0.67425
46	.06690	.15415	.24135	.32845	.41550	.50240	.58915	.67570
47	.06835	.15560	.24280	.32990	.41695	.50385	.59060	.67715
48	.06980	.15705	.24425	.33135	.41840	.50530	.59205	.67860
49	.07125	.15850	.24570	.33280	.41985	.50675	.59345	.68000
50	0.07270	0.15995	0.24715	0.33425	0.42130	0.50820	0.59490	0.68145
51	.07415	.16140	.24860	.33570	.42275	.50960	.59635	.68290
52	.07565	.16285	.25005	.33715	.42420	.51105	.59780	.68435
53	.07710	.16430	.25150	.33865	.42565	.51250	.59925	.68580
54	.07855	.16580	.25295	.34010	.42710	.51395	.60070	.68720
55	0.08000	0.16725	0.25440	0.34155	0.42855	0.51540	0.60215	0.68865
56	.08145	.16870	.25585	.34300	.43000	.51685	.60355	.69010
57	.08290	.17015	.25730	.34445	.43145	.51830	.60500	.69155
58	.08435	.17160	.25875	.34590	.43290	.51975	.60645	.69300
59	.08580	.17305	.26028	.34735	.43435	.52120	.60790	.69445
60	0.08725	0.17450	0.26170	0.34880	0.43580	0.52265	0.60935	0.69585

Constants for Setting a 5-inch Sine-bar

Min.	8°	9°	10°	11°	12°	13°	14°	15°
0	0.69585	0.78215	0.86825	0.95405	1.0395	1.1247	1.2096	1.2941
1	.69730	.78360	.86965	.95545	.0410	.1261	.2110	.2955
2	.69875	.78505	.87110	.95690	.0424	.1276	.2124	.2969
3	.70020	.78650	.87255	.95835	.0438	.1290	.2138	.2983
4	.70165	.78790	.87395	.95975	.0452	.1304	.2152	.2997
5	0.70305	0.78935	0.87540	0.96120	1.0466	1.1318	1.2166	1.3011
6	.70450	.79080	.87685	.96260	.0481	.1332	.2181	.3025
7	.70595	.79225	.87825	.96405	.0495	.1346	.2195	.3039
8	.70740	.79365	.87970	.96545	.0509	.1361	.2209	.3053
9	.70885	.79510	.88115	.96690	.0523	.1375	.2223	.3067
10	0.71025	0.79655	0.88255	0.96830	1.0538	1.1389	1.2237	1.3081
11	.71170	.79795	.88400	.96975	.0552	.1403	.2251	.3095
12	.71315	.79940	.88540	.97115	.0566	.1417	.2265	.3109
13	.71460	.80085	.88685	.97260	.0580	.1431	.2279	.3123
14	.71600	.80230	.88830	.97405	.0594	.1446	.2293	.3137
15	0.71745	0.80370	0.88970	0.97545	1.0609	1.1460	1.2307	1.3151
16	.71890	.80515	.89115	.97690	.0623	.1474	.2322	.3165
17	.72035	.80660	.89260	.97830	.0637	.1488	.2336	.3179
18	.72180	.80800	.89400	.97975	.0651	.1502	.2350	.3193
19	.72320	.80945	.89545	.98115	.0665	.1516	.2364	.3207
20	0.72465	0.81090	0.89685	0.98260	1.0680	1.1531	1.2378	1.3221
21	.72610	.81230	.89830	.98400	.0694	.1545	.2392	.3235
22	.72755	.81375	.89975	.98545	.0708	.1559	.2406	.3250
23	.72900	.81520	.90115	.98685	.0722	.1573	.2420	.3264
24	.73040	.81665	.90260	.98830	.0737	.1587	.2434	.3278
25	0.73185	0.81805	0.90405	0.98970	1.0751	1.1601	1.2448	1.3292
26	.73330	.81950	.90545	.99115	.0765	.1615	.2462	.3306
27	.73475	.82095	.90690	.99255	.0779	.1630	.2477	.3320
28	.73615	.82235	.90830	.99400	.0793	.1644	.2491	.3334
29	.73760	.82380	.90975	.99540	.0808	.1658	.2505	.3348
30	0.73905	0.82525	0.91120	0.99685	1.0822	1.1672	1.2519	1.3362
31	.74050	.82665	.91260	.99825	.0836	.1686	.2533	.3376
32	.74190	.82810	.91405	.99970	.0850	.1700	.2547	.3390
33	.74335	.82955	.91545	1.0011	.0864	.1714	.2561	.3404
34	.74480	.83100	.91690	.0026	.0879	.1729	.2575	.3418
35	0.74625	0.83240	0.91835	1.0039	1.0893	1.1743	1.2589	1.3432
36	.74770	.83385	.91975	.0054	.0907	.1757	.2603	.3446
37	.74910	.83530	.92120	.0068	.0921	.1771	.2617	.3460
38	.75055	.83670	.92260	.0082	.0935	.1785	.2631	.3474
39	.75200	.83815	.92405	.0096	.0949	.1799	.2645	.3488
40	0.75345	0.83960	0.92545	1.0110	1.0964	1.1813	1.2660	1.3502
41	.75485	.84100	.92690	.0125	.0978	.1828	.2674	.3516
42	.75630	.84245	.92835	.0139	.0992	.1842	.2688	.3530
43	.75775	.84390	.92975	.0153	.1006	.1856	.2702	.3544
44	.75920	.84530	.93120	.0168	.1020	.1870	.2716	.3558
45	0.76060	0.84675	0.93260	1.0182	1.1035	1.1884	1.2730	1.3572
46	.76205	.84820	.93405	.0196	.1049	.1898	.2744	.3586
47	.76350	.84960	.93550	.0210	.1063	.1912	.2758	.3600
48	.76495	.85105	.93690	.0225	.1077	.1926	.2772	.3614
49	.76635	.85250	.93835	.0239	.1091	.1941	.2786	.3628
50	0.76780	0.85390	0.93975	1.0253	1.1106	1.1955	1.2800	1.3642
51	.76925	.85535	.94120	.0267	.1120	.1969	.2814	.3656
52	.77070	.85680	.94260	.0281	.1134	.1983	.2828	.3670
53	.77210	.85820	.94405	.0296	.1148	.1997	.2842	.3684
54	.77355	.85965	.94550	.0310	.1162	.2011	.2856	.3698
55	0.77500	0.86110	0.94690	1.0324	1.1176	1.2025	1.2870	1.3712
56	.77645	.86250	.94835	.0338	.1191	.2039	.2884	.3726
57	.77785	.86395	.94975	.0353	.1205	.2054	.2899	.3740
58	.77930	.86540	.95120	.0367	.1219	.2068	.2913	.3754
59	.78075	.86680	.95260	.0381	.1233	.2082	.2927	.3768
60	0.78215	0.86825	0.95405	1.0395	1.1247	1.2096	1.2941	1.3782

Source. *Machinery's Handbook*—20th Edition, Copyright © 1975. Reprinted with the permission of the Industrial Press, Inc., the publisher.

Table 2 (continued)

Constants for Setting a 5-inch Sine-bar

Min.	16°	17°	18°	19°	20°	21°	22°	23°
0	1.3782	1.4618	1.5451	1.6278	1.7101	1.7918	1.8730	1.9536
1	.3796	.4632	.5464	.6292	.7114	.7932	.8744	.9550
2	.3810	.4646	.5478	.6306	.7128	.7945	.8757	.9563
3	.3824	.4660	.5492	.6319	.7142	.7959	.8771	.9576
4	.3838	.4674	.5506	.6333	.7155	.7972	.8784	.9590
5	1.3852	1.4688	1.5520	1.6347	1.7169	1.7986	1.8797	1.9603
6	.3865	.4702	.5534	.6361	.7183	.8000	.8811	.9617
7	.3879	.4716	.5547	.6374	.7196	.8013	.8824	.9630
8	.3893	.4730	.5561	.6388	.7210	.8027	.8838	.9643
9	.3907	.4743	.5575	.6402	.7224	.8040	.8851	.9657
10	1.3921	1.4757	1.5589	1.6416	1.7237	1.8054	1.8865	1.9670
11	.3935	.4771	.5603	.6429	.7251	.8067	.8878	.9683
12	.3949	.4785	.5616	.6443	.7265	.8081	.8892	.9697
13	.3963	.4799	.5630	.6457	.7278	.8094	.8905	.9710
14	.3977	.4813	.5644	.6471	.7292	.8108	.8919	.9724
15	1.3991	1.4827	1.5658	1.6484	1.7306	1.8122	1.8932	1.9737
16	.4005	.4841	.5672	.6498	.7319	.8135	.8946	.9750
17	.4019	.4855	.5686	.6512	.7333	.8149	.8959	.9764
18	.4033	.4868	.5699	.6525	.7347	.8162	.8973	.9777
19	.4047	.4882	.5713	.6539	.7360	.8176	.8986	.9790
20	1.4061	1.4896	1.5727	1.6553	1.7374	1.8189	1.8999	1.9804
21	.4075	.4910	.5741	.6567	.7387	.8203	.9013	.9817
22	.4089	.4924	.5755	.6580	.7401	.8217	.9026	.9830
23	.4103	.4938	.5768	.6594	.7415	.8230	.9040	.9844
24	.4117	.4952	.5782	.6608	.7428	.8244	.9053	.9857
25	1.4131	1.4966	1.5796	1.6622	1.7442	1.8257	1.9067	1.9870
26	.4145	.4980	.5810	.6635	.7456	.8271	.9080	.9884
27	.4159	.4993	.5824	.6649	.7469	.8284	.9094	.9897
28	.4173	.5007	.5837	.6663	.7483	.8298	.9107	.9911
29	.4187	.5021	.5851	.6676	.7496	.8311	.9120	.9924
30	1.4201	1.5035	1.5865	1.6690	1.7510	1.8325	1.9134	1.9937
31	.4214	.5049	.5879	.6704	.7524	.8338	.9147	.9951
32	.4228	.5063	.5893	.6718	.7537	.8352	.9161	.9964
33	.4242	.5077	.5906	.6731	.7551	.8365	.9174	.9977
34	.4256	.5091	.5920	.6745	.7565	.8379	.9188	.9991
35	1.4270	1.5104	1.5934	1.6759	1.7578	1.8392	1.9201	2.0004
36	.4284	.5118	.5948	.6772	.7592	.8406	.9215	.0017
37	.4298	.5132	.5961	.6786	.7605	.8419	.9228	.0031
38	.4312	.5146	.5975	.6800	.7619	.8433	.9241	.0044
39	.4326	.5160	.5989	.6813	.7633	.8447	.9255	.0057
40	1.4340	1.5174	1.6003	1.6827	1.7646	1.8460	1.9268	2.0070
41	.4354	.5188	.6017	.6841	.7660	.8474	.9282	.0084
42	.4368	.5201	.6030	.6855	.7673	.8487	.9295	.0097
43	.4382	.5215	.6044	.6868	.7687	.8501	.9308	.0110
44	.4396	.5229	.6058	.6882	.7701	.8514	.9322	.0124
45	1.4410	1.5243	1.6072	1.6896	1.7714	1.8528	1.9335	2.0137
46	.4423	.5257	.6085	.6909	.7728	.8541	.9349	.0150
47	.4437	.5271	.6099	.6923	.7742	.8555	.9362	.0164
48	.4451	.5285	.6113	.6937	.7755	.8568	.9376	.0177
49	.4465	.5298	.6127	.6950	.7769	.8582	.9389	.0190
50	1.4479	1.5312	1.6141	1.6964	1.7782	1.8595	1.9402	2.0204
51	.4493	.5326	.6154	.6978	.7796	.8609	.9416	.0217
52	.4507	.5340	.6168	.6991	.7809	.8622	.9429	.0230
53	.4521	.5354	.6182	.7005	.7823	.8636	.9443	.0244
54	.4535	.5368	.6196	.7019	.7837	.8649	.9456	.0257
55	1.4549	1.5381	1.6209	1.7032	1.7850	1.8663	1.9469	2.0270
56	.4563	.5395	.6223	.7046	.7864	.8676	.9483	.0283
57	.4577	.5409	.6237	.7060	.7877	.8690	.9496	.0297
58	.4591	.5423	.6251	.7073	.7891	.8703	.9510	.0310
59	.4604	.5437	.6264	.7087	.7905	.8717	.9523	.0323
60	1.4618	1.5451	1.6278	1.7101	1.7918	1.8730	1.9536	2.0337

Constants for Setting a 5-inch Sine-bar

Min.	24°	25°	26°	27°	28°	29°	30°	31°
0	2.0337	2.1131	2.1918	2.2699	2.3473	2.4240	2.5000	2.5752
1	.0350	.1144	.1931	.2712	.3486	.4253	.5012	.5764
2	.0363	.1157	.1944	.2725	.3499	.4266	.5025	.5777
3	.0376	.1170	.1958	.2738	.3512	.4278	.5038	.5789
4	.0390	.1183	.1971	.2751	.3525	.4291	.5050	.5802
5	2.0403	2.1197	2.1984	2.2764	2.3538	2.4304	2.5063	2.5814
6	.0416	.1210	.1997	.2777	.3550	.4317	.5075	.5826
7	.0430	.1223	.2010	.2790	.3563	.4329	.5088	.5839
8	.0443	.1236	.2023	.2803	.3576	.4342	.5100	.5851
9	.0456	.1249	.2036	.2816	.3589	.4355	.5113	.5864
10	2.0469	2.1262	2.2049	2.2829	2.3602	2.4367	2.5126	2.5876
11	.0483	.1276	.2062	.2842	.3614	.4380	.5138	.5889
12	.0496	.1289	.2075	.2855	.3627	.4393	.5151	.5901
13	.0509	.1302	.2088	.2868	.3640	.4405	.5163	.5914
14	.0522	.1315	.2101	.2881	.3653	.4418	.5176	.5926
15	2.0536	2.1328	2.2114	2.2893	2.3666	2.4431	2.5188	2.5938
16	.0549	.1341	.2127	.2906	.3679	.4444	.5201	.5951
17	.0562	.1354	.2140	.2919	.3691	.4456	.5214	.5963
18	.0575	.1368	.2153	.2932	.3704	.4469	.5226	.5976
19	.0589	.1381	.2166	.2945	.3717	.4482	.5239	.5988
20	2.0602	2.1394	2.2179	2.2958	2.3730	2.4494	2.5251	2.6001
21	.0615	.1407	.2192	.2971	.3743	.4507	.5264	.6013
22	.0628	.1420	.2205	.2984	.3755	.4520	.5276	.6025
23	.0642	.1433	.2218	.2997	.3768	.4532	.5289	.6038
24	.0655	.1447	.2232	.3010	.3781	.4545	.5301	.6050
25	2.0668	2.1460	2.2245	2.3023	2.3794	2.4558	2.5314	2.6063
26	.0681	.1473	.2258	.3036	.3807	.4570	.5327	.6075
27	.0695	.1486	.2271	.3048	.3819	.4583	.5339	.6087
28	.0708	.1499	.2284	.3061	.3832	.4596	.5352	.6100
29	.0721	.1512	.2297	.3074	.3845	.4608	.5364	.6112
30	2.0734	2.1525	2.2310	2.3087	2.3858	2.4621	2.5377	2.6125
31	.0748	.1538	.2323	.3100	.3870	.4634	.5389	.6137
32	.0761	.1552	.2336	.3113	.3883	.4646	.5402	.6149
33	.0774	.1565	.2349	.3126	.3896	.4659	.5414	.6162
34	.0787	.1578	.2362	.3139	.3909	.4672	.5427	.6174
35	2.0801	2.1591	2.2375	2.3152	2.3922	2.4684	2.5439	2.6187
36	.0814	.1604	.2388	.3165	.3934	.4697	.5452	.6199
37	.0827	.1617	.2401	.3177	.3947	.4709	.5464	.6211
38	.0840	.1630	.2414	.3190	.3960	.4722	.5477	.6224
39	.0853	.1643	.2427	.3203	.3973	.4735	.5489	.6236
40	2.0867	2.1656	2.2440	2.3216	2.3985	2.4747	2.5502	2.6249
41	.0880	.1670	.2453	.3229	.3998	.4760	.5514	.6261
42	.0893	.1683	.2466	.3242	.4011	.4773	.5527	.6273
43	.0906	.1696	.2479	.3255	.4024	.4785	.5539	.6286
44	.0920	.1709	.2492	.3268	.4036	.4798	.5552	.6298
45	2.0933	2.1722	2.2505	2.3280	2.4049	2.4811	2.5564	2.6310
46	.0946	.1735	.2518	.3293	.4062	.4823	.5577	.6323
47	.0959	.1748	.2531	.3306	.4075	.4836	.5589	.6335
48	.0972	.1761	.2544	.3319	.4087	.4848	.5602	.6348
49	.0986	.1774	.2557	.3332	.4100	.4861	.5614	.6360
50	2.0999	2.1787	2.2570	2.3345	2.4113	2.4874	2.5627	2.6372
51	.1012	.1801	.2583	.3358	.4126	.4886	.5639	.6385
52	.1025	.1814	.2596	.3371	.4138	.4899	.5652	.6397
53	.1038	.1827	.2609	.3383	.4151	.4912	.5664	.6409
54	.1052	.1840	.2621	.3396	.4164	.4924	.5677	.6422
55	2.1065	2.1853	2.2634	2.3409	2.4177	2.4937	2.5689	2.6434
56	.1078	.1866	.2647	.3422	.4189	.4949	.5702	.6446
57	.1091	.1879	.2660	.3435	.4202	.4962	.5714	.6459
58	.1104	.1892	.2673	.3448	.4215	.4975	.5727	.6471
59	.1117	.1905	.2686	.3460	.4228	.4987	.5739	.6483
60	2.1131	2.1918	2.2699	2.3473	2.4240	2.5000	2.5752	2.6496

Table 2 (continued)

Constants for Setting a 5-inch Sine-bar

Min.	32°	33°	34°	35°	36°	37°	38°	39°
0	2.6496	2.7232	2.7959	2.8679	2.9389	3.0091	3.0783	3.1466
1	.6508	.7244	.7971	.8690	.9401	.0102	.0794	.1477
2	.6520	.7256	.7984	.8702	.9413	.0114	.0806	.1488
3	.6533	.7268	.7996	.8714	.9424	.0125	.0817	.1500
4	.6545	.7280	.8008	.8726	.9436	.0137	.0829	.1511
5	2.6557	2.7293	2.8020	2.8738	2.9448	3.0149	3.0840	3.1522
6	.6570	.7305	.8032	.8750	.9460	.0160	.0852	.1534
7	.6582	.7317	.8044	.8762	.9471	.0172	.0863	.1545
8	.6594	.7329	.8056	.8774	.9483	.0183	.0874	.1556
9	.6607	.7341	.8068	.8786	.9495	.0195	.0886	.1567
10	2.6619	2.7354	2.8080	2.8798	2.9507	3.0207	3.0897	3.1579
11	.6631	.7366	.8092	.8809	.9518	.0218	.0909	.1590
12	.6644	.7378	.8104	.8821	.9530	.0230	.0920	.1601
13	.6656	.7390	.8116	.8833	.9542	.0241	.0932	.1612
14	.6668	.7402	.8128	.8845	.9554	.0253	.0943	.1624
15	2.6680	2.7414	2.8140	2.8857	2.9565	3.0264	3.0954	3.1635
16	.6693	.7427	.8152	.8869	.9577	.0276	.0966	.1646
17	.6705	.7439	.8164	.8881	.9589	.0288	.0977	.1658
18	.6717	.7451	.8176	.8893	.9600	.0299	.0989	.1669
19	.6730	.7463	.8188	.8905	.9612	.0311	.1000	.1680
20	2.6742	2.7475	2.8200	2.8916	2.9624	3.0322	3.1012	3.1691
21	.6754	.7487	.8212	.8928	.9636	.0334	.1023	.1703
22	.6767	.7499	.8224	.8940	.9647	.0345	.1034	.1714
23	.6779	.7512	.8236	.8952	.9659	.0357	.1046	.1725
24	.6791	.7524	.8248	.8964	.9671	.0369	.1057	.1736
25	2.6803	2.7536	2.8260	2.8976	2.9682	3.0380	3.1069	3.1748
26	.6816	.7548	.8272	.8988	.9694	.0392	.1080	.1759
27	.6828	.7560	.8284	.8999	.9706	.0403	.1091	.1770
28	.6840	.7572	.8296	.9011	.9718	.0415	.1103	.1781
29	.6852	.7584	.8308	.9023	.9729	.0426	.1114	.1792
30	2.6865	2.7597	2.8320	2.9035	2.9741	3.0438	3.1125	3.1804
31	.6877	.7609	.8332	.9047	.9753	.0449	.1137	.1815
32	.6889	.7621	.8344	.9059	.9764	.0461	.1148	.1826
33	.6902	.7633	.8356	.9070	.9776	.0472	.1160	.1837
34	.6914	.7645	.8368	.9082	.9788	.0484	.1171	.1849
35	2.6926	2.7657	2.8380	2.9094	2.9799	3.0495	3.1182	3.1860
36	.6938	.7669	.8392	.9106	.9811	.0507	.1194	.1871
37	.6951	.7681	.8404	.9118	.9823	.0519	.1205	.1882
38	.6963	.7694	.8416	.9130	.9834	.0530	.1216	.1893
39	.6975	.7706	.8428	.9141	.9846	.0542	.1228	.1905
40	2.6987	2.7718	2.8440	2.9153	2.9858	3.0553	3.1239	3.1916
41	.7000	.7730	.8452	.9165	.9869	.0565	.1251	.1927
42	.7012	.7742	.8464	.9177	.9881	.0576	.1262	.1938
43	.7024	.7754	.8476	.9189	.9893	.0588	.1273	.1949
44	.7036	.7766	.8488	.9200	.9904	.0599	.1285	.1961
45	2.7048	2.7778	2.8500	2.9212	2.9916	3.0611	3.1296	3.1972
46	.7061	.7790	.8512	.9224	.9928	.0622	.1307	.1983
47	.7073	.7802	.8523	.9236	.9939	.0634	.1319	.1994
48	.7085	.7815	.8535	.9248	.9951	.0645	.1330	.2005
49	.7097	.7827	.8547	.9259	.9963	.0657	.1341	.2016
50	2.7110	2.7839	2.8559	2.9271	2.9974	3.0668	3.1353	3.2028
51	.7122	.7851	.8571	.9283	.9986	.0680	.1364	.2039
52	.7134	.7863	.8583	.9295	.9997	.0691	.1375	.2050
53	.7146	.7875	.8595	.9307	3.0009	.0703	.1387	.2061
54	.7158	.7887	.8607	.9318	.0021	.0714	.1398	.2072
55	2.7171	2.7899	2.8619	2.9330	3.0032	3.0725	3.1409	3.2083
56	.7183	.7911	.8631	.9342	.0044	.0737	.1421	.2095
57	.7195	.7923	.8643	.9354	.0056	.0748	.1432	.2106
58	.7207	.7935	.8655	.9365	.0067	.0760	.1443	.2117
59	.7220	.7947	.8667	.9377	.0079	.0771	.1454	.2128
60	2.7232	2.7959	2.8679	2.9389	3.0091	3.0783	3.1466	3.2139

Constants for Setting a 5-inch Sine-bar

Min.	40°	41°	42°	43°	44°	45°	46°	47°
0	3.2139	3.2803	3.3456	3.4100	3.4733	3.5355	3.5967	3.6567
1	.2150	.2814	.3467	.4110	.4743	.5365	.5977	.6577
2	.2161	.2825	.3478	.4121	.4754	.5376	.5987	.6587
3	.2173	.2836	.3489	.4132	.4764	.5386	.5997	.6597
4	.2184	.2847	.3499	.4142	.4774	.5396	.6007	.6607
5	3.2195	3.2858	3.3510	3.4153	3.4785	3.5406	3.6017	3.6617
6	.2206	.2869	.3521	.4163	.4795	.5417	.6027	.6627
7	.2217	.2879	.3532	.4174	.4806	.5427	.6037	.6637
8	.2228	.2890	.3543	.4185	.4816	.5437	.6047	.6647
9	.2239	.2901	.3553	.4195	.4827	.5448	.6058	.6657
10	3.2250	3.2912	3.3564	3.4206	3.4837	3.5458	3.6068	3.6666
11	.2262	.2923	.3575	.4217	.4848	.5468	.6078	.6676
12	.2273	.2934	.3586	.4227	.4858	.5478	.6088	.6686
13	.2284	.2945	.3597	.4238	.4868	.5489	.6098	.6696
14	.2295	.2956	.3607	.4248	.4879	.5499	.6108	.6706
15	3.2306	3.2967	3.3618	3.4259	3.4889	3.5509	3.6118	3.6716
16	.2317	.2978	.3629	.4269	.4900	.5519	.6128	.6726
17	.2328	.2989	.3640	.4280	.4910	.5529	.6138	.6736
18	.2339	.3000	.3650	.4291	.4921	.5540	.6148	.6745
19	.2350	.3011	.3661	.4301	.4931	.5550	.6158	.6755
20	3.2361	3.3022	3.3672	3.4312	3.4941	3.5560	3.6168	3.6765
21	.2373	.3033	.3683	.4322	.4952	.5570	.6178	.6775
22	.2384	.3044	.3693	.4333	.4962	.5581	.6188	.6785
23	.2395	.3054	.3704	.4344	.4973	.5591	.6198	.6795
24	.2406	.3065	.3715	.4354	.4983	.5601	.6208	.6805
25	3.2417	3.3076	3.3726	3.4365	3.4993	3.5611	3.6218	3.6814
26	.2428	.3087	.3736	.4375	.5004	.5621	.6228	.6824
27	.2439	.3098	.3747	.4386	.5014	.5632	.6238	.6834
28	.2450	.3109	.3758	.4396	.5024	.5642	.6248	.6844
29	.2461	.3120	.3769	.4407	.5035	.5652	.6258	.6854
30	3.2472	3.3131	3.3779	3.4417	3.5045	3.5662	3.6268	3.6864
31	.2483	.3142	.3790	.4428	.5056	.5672	.6278	.6873
32	.2494	.3153	.3801	.4439	.5066	.5683	.6288	.6883
33	.2505	.3163	.3811	.4449	.5076	.5693	.6298	.6893
34	.2516	.3174	.3822	.4460	.5087	.5703	.6308	.6903
35	3.2527	3.3185	3.3833	3.4470	3.5097	3.5713	3.6318	3.6913
36	.2538	.3196	.3844	.4481	.5107	.5723	.6328	.6923
37	.2550	.3207	.3854	.4491	.5118	.5734	.6338	.6932
38	.2561	.3218	.3865	.4502	.5128	.5744	.6348	.6942
39	.2572	.3229	.3876	.4512	.5138	.5754	.6358	.6952
40	3.2583	3.3240	3.3886	3.4523	3.5149	3.5764	3.6368	3.6962
41	.2594	.3250	.3897	.4533	.5159	.5774	.6378	.6972
42	.2605	.3261	.3908	.4544	.5169	.5784	.6388	.6981
43	.2616	.3272	.3918	.4554	.5180	.5795	.6398	.6991
44	.2627	.3283	.3929	.4565	.5190	.5805	.6408	.7001
45	3.2638	3.3294	3.3940	3.4575	3.5200	3.5815	3.6418	3.7011
46	.2649	.3305	.3950	.4586	.5211	.5825	.6428	.7020
47	.2660	.3316	.3961	.4596	.5221	.5835	.6438	.7030
48	.2671	.3326	.3972	.4607	.5231	.5845	.6448	.7040
49	.2682	.3337	.3982	.4617	.5242	.5855	.6458	.7050
50	3.2693	3.3348	3.3993	3.4628	3.5252	3.5866	3.6468	3.7060
51	.2704	.3359	.4004	.4638	.5262	.5876	.6478	.7069
52	.2715	.3370	.4014	.4649	.5273	.5886	.6488	.7079
53	.2726	.3381	.4025	.4659	.5283	.5896	.6498	.7089
54	.2737	.3391	.4036	.4670	.5293	.5906	.6508	.7099
55	3.2748	3.3402	3.4046	3.4680	3.5304	3.5916	3.6518	3.7108
56	.2759	.3413	.4057	.4691	.5314	.5926	.6528	.7118
57	.2770	.3424	.4068	.4701	.5324	.5936	.6538	.7128
58	.2781	.3435	.4078	.4712	.5335	.5947	.6548	.7138
59	.2792	.3445	.4089	.4722	.5345	.5957	.6558	.7147
60	3.2803	3.3456	3.4100	3.4733	3.5355	3.5967	3.6567	3.7157

Table 2 (continued)

Constants for Setting a 5-inch Sine-bar

Min.	48°	49°	50°	51°	52°	53°	54°	55°
0	3.7157	3.7735	3.8302	3.8857	3.9400	3.9932	4.0451	4.0957
1	.7167	.7745	.8311	.8866	.9409	.9940	.0459	.0966
2	.7176	.7754	.8321	.8875	.9418	.9949	.0468	.0974
3	.7186	.7764	.8330	.8884	.9427	.9958	.0476	.0982
4	.7196	.7773	.8339	.8894	.9436	.9967	.0485	.0991
5	3.7206	3.7783	3.8349	8.8903	3.9445	3.9975	4.0493	4.0999
6	.7215	.7792	.8358	.8912	.9454	.9984	.0502	.1007
7	.7225	.7802	.8367	.8921	.9463	.9993	.0510	.1016
8	.7235	.7811	.8377	.8930	.9472	4.0001	.0519	.1024
9	.7244	.7821	.8386	.8939	.9481	.0010	.0527	.1032
10	3.7254	3.7830	3.8395	3.8948	3.9490	4.0019	4.0536	4.1041
11	.7264	.7840	.8405	.8958	.9499	.0028	.0544	.1049
12	.7274	.7850	.8414	.8967	.9508	.0036	.0553	.1057
13	.7283	.7859	.8423	.8976	.9516	.0045	.0561	.1066
14	.7293	.7869	.8433	.8985	.9525	.0054	.0570	.1074
15	3.7303	3.7878	3.8442	3.8994	3.9534	4.0062	4.0578	4.1082
16	.7312	.7887	.8451	.9003	.9543	.0071	.0587	.1090
17	.7322	.7897	.8460	.9012	.9552	.0080	.0595	.1099
18	.7332	.7906	.8470	.9021	.9561	.0089	.0604	.1107
19	.7341	.7916	.8479	.9030	9570	.0097	.0612	.1115
20	3.7351	3.7925	3.8488	3.9039	3.9579	4.0106	4.0621	4.1124
21	.7361	.7935	.8498	.9049	.9588	.0115	.0629	.1132
22	.7370	.7944	.8507	.9058	.9596	.0123	.0638	.1140
23	.7380	.7954	.8516	.9067	.9605	.0132	.0646	.1148
24	.7390	.7963	.8525	.9076	.9614	.0141	.0655	.1157
25	3.7399	3.7973	3.8535	3.9085	3.9623	4.0149	4.0663	4.1165
26	.7409	.7982	.8544	.9094	.9632	.0158	.0672	.1173
27	.7419	.7992	.8553	.9103	.9641	.0167	.0680	.1181
28	.7428	.8001	.8562	.9112	.9650	.0175	.0689	.1190
29	.7438	.8011	.8572	.9121	.9659	.0184	.0697	.1198
30	3.7448	3.8020	3.8581	3.9130	3.9667	4.0193	4.0706	4.1206
31	.7457	.8029	.8590	.9139	.9676	.0201	.0714	.1214
32	.7467	.8039	.8599	.9148	.9685	.0210	.0722	.1223
33	.7476	.8048	.8609	.9157	.9694	.0219	.0731	.1231
34	.7486	.8058	.8618	.9166	.9703	.0227	.0739	.1239
35	3.7496	3.8067	3.8627	3.9175	3.9712	4.0236	4.0748	4.1247
36	.7505	.8077	.8636	.9184	.9720	.0244	.0756	.1255
37	.7515	.8086	.8646	.9193	.9729	.0253	.0765	.1264
38	.7525	.8096	.8655	.9202	.9738	.0262	.0773	.1272
39	.7534	.8105	.8664	.9212	.9747	.0270	.0781	.1280
40	3.7544	3.8114	3.8673	3.9221	3.9756	4.0279	4.0790	4.1288
41	.7553	.8124	.8683	.9230	.9765	.0288	.0798	.1296
42	.7563	.8133	.8692	.9239	.9773	.0296	.0807	.1305
43	.7573	.8143	.8701	.9248	.9782	.0305	.0815	.1313
44	.7582	.8152	.8710	.9257	.9791	.0313	.0823	.1321
45	3.7592	3.8161	3.8719	3.9266	3.9800	4.0322	4.0832	4.1329
46	.7601	.8171	.8729	.9275	.9809	.0331	.0840	.1337
47	.7611	.8180	.8738	.9284	.9817	.0339	.0849	.1346
48	.7620	.8190	.8747	.9293	.9826	.0348	.0857	.1354
49	.7630	.8199	.8756	.9302	.9835	.0356	.0865	.1362
50	3.7640	3.8208	3.8765	3.9311	3.9844	4.0365	4.0874	4.1370
51	.7649	.8218	.8775	.9320	.9853	.0374	.0882	.1378
52	.7659	.8227	.8784	.9329	.9861	.0382	.0891	.1386
53	.7668	.8236	.8793	.9338	.9870	.0391	.0899	.1395
54	.7678	.8246	.8802	.9347	.9879	.0399	.0907	.1403
55	3.7687	3.8255	3.8811	3.9355	3.9888	4.0408	4.0916	4.1411
56	.7697	.8265	.8820	.9364	.9896	.0416	.0924	.1419
57	.7707	.8274	.8830	.9373	.9905	.0425	.0932	.1427
58	.7716	.8283	.8839	.9382	.9914	.0433	.0941	.1435
59	.7726	.8293	.8848	.9391	.9923	.0442	.0949	.1443
60	3.7735	3.8302	3.8857	3.9400	3.9932	4.0451	4.0957	4.1452

Table 3
Inch/Metric Conversion Table

Drill No. or Letter		Inch	mm
		.001	0,0254
		.002	0,0508
		.003	0,0762
		.004	0,1016
		.005	0,1270
		.006	0,1524
		.007	0,1778
		.008	0,2032
		.009	0,2286
		.010	0,2540
		.011	0,2794
		.012	0,3048
80	.0135	.013	0,3302
79	.0145	.014	0,3556
		.015	0,3810
	1/64	.0156	0,3969
78		.016	0,4064
		.017	0,4318
77		.018	0,4572
		.019	0,4826
76		.020	0,5080
75		.021	0,5334
		.022	0,5588
74	.0225	.023	0,5842
73		.024	0,6096
72		.025	0,6350
71		.026	0,6604
		.027	0,6858
70		.028	0,7112
69	.0292	.029	0,7366
		.030	0,7620
68		.031	0,7874
	1/32	.0312	0,7937
67		.032	0,8128
66		.033	0,8382
		.034	0,8636
65		.035	0,8890
64		.036	0,9144
63		.037	0,9398
62		.038	0,9652
61		.039	0,9906
		.0394	1,0000
60		.040	1,0160
59		.041	1,0414
58		.042	1,0668
57		.043	1,0922
		.044	1,1176
		.045	1,1430
56	.0465	.046	1,1684
	3/64	.0469	1,1906
		.047	1,1938
		.048	1,2192
		.049	1,2446
		.050	1,2700
		.051	1,2954
55		.052	1,3208
		.053	1,3462
		.054	1,3716
54		.055	1,3970
		.056	1,4224
		.057	1,4478
		.058	1,4732
		.059	1,4986
53	.0595	.060	1,5240
		.061	1,5494
		.062	1,5748
	1/16	.0625	1,5875
		.063	1,6002
52	.0635	.064	1,6256
		.065	1,6510
		.066	1,6764
51		.067	1,7018
		.068	1,7272
		.069	1,7526
50		.070	1,7780
		.071	1,8034
		.072	1,8288
49		.073	1,8542
		.074	1,8796
		.075	1,9050
48		.076	1,9304
		.077	1,9558
47	.0785	.078	1,9812
	5/64	.0781	1,9844
		.0787	2,0000
		.079	2,0066
		.080	2,0320
46		.081	2,0574
45		.082	2,0828
		.083	2,1082
		.084	2,1336
		.085	2,1590
44		.086	2,1844
		.087	2,2098
		.088	2,2352
43		.089	2,2606
		.090	2,2860
		.091	2,3114
		.092	2,3368
42	.0935	.093	2,3622
	3/32	.0937	2,3812
		.094	2,3876
		.095	2,4130
41		.096	2,4384
		.097	2,4638
40		.098	2,4892
		.099	2.5146
39	.0995	.100	2.5400

Drill No. or Letter		Inch	mm
38	.1015	.101	2,5654
		.102	2,5908
		.103	2,6162
37		.104	2,6416
		.105	2,6670
36	.1065	.106	2,6924
		.107	2,7178
		.108	2,7432
		.109	2,7686
	7/64	.1094	2,7781
35		.110	2,7940
34		.111	2,8194
		.112	2,8448
33		.113	2,8702
		.114	2,8956
		.115	2,9210
32		.116	2,9464
		.117	2,9718
		.118	2,9972
		.1181	3,0000
		.119	3,0226
31		.120	3,0480
		.121	3,0734
		.122	3,0988
		.123	3,1242
		.124	3,1496
	1/8	.125	3,1750
		.126	3,2004
		.127	3,2258
		.128	3,2512
30	.1285	.129	3,2766
		.130	3,3020
		.131	3,3274
		.132	3,3528
		.133	3,3782
		.134	3,4036
		.135	3,4290
29		.136	3,4544
		.137	3,4798
		.138	3,5052
		.139	3,5306
28	.1405	.140	3,5560
	9/64	.1406	3,5719
		.141	3,5814
		.142	3,6068
		.143	3,6322
27		.144	3,6576
		.145	3,6830
		.146	3,7084
26		.147	3,7338
		.148	3,7592
		.149	3,7846
25	.1495	.150	3,8100
		.151	3,8354
24		.152	3,8608
		.153	3,8862
23		.154	3,9116
		.155	3,9370
		.156	3,9624
	5/32	.1562	3,9687
22		.157	3,9878
		.1575	4,0000
		.158	4,0132
21		.159	4,0386
		.160	4,0640
20		.161	4,0894
		.162	4,1148
		.163	4,1402
		.164	4,1656
		.165	4,1910
19		.166	4,2164
		.167	4,2418
		.168	4,2672
		.169	4,2926
18	.1635	.170	4,3180
		.171	4,3434
	11/64	.1719	4,3656
		.172	4,3688
17		.173	4,3942
		.174	4,4196
		.175	4,4450
		.176	4,4704
16		.177	4,4958
		.178	4,5212
		.179	4,5466
15		.180	4,5720
		.181	4,5974
14		.182	4,6228
		.183	4,6482
		.184	4,6736
13		.185	4,6990
		.186	4,7244
		.187	4,7498
	3/16	.1875	4,7625
		.188	4,7752
12		.189	4,8006
		.190	4,8260
11		.191	4,8514
		.192	4,8768
		.193	4,9022
10	.1935	.194	4,9276
		.195	4,9530
9		.196	4,9784
		.1969	5,0000
		.197	5,0038
		.198	5,0292
		.199	5,0546
8		.200	5,0800

Drill No. or Letter		Inch	mm
7		.201	5,1054
		.202	5,1308
		.203	5,1562
	13/64	.2031	5,1594
6		.204	5,1816
5	.2055	.205	5,2070
		.206	5,2324
		.207	5,2578
		.208	5,2832
4		.209	5,3086
		.210	5,3340
		.211	5,3594
		.212	5,3848
3		.213	5,4102
		.214	5,4356
		.215	5,4610
		.216	5,4864
		.217	5,5118
		.218	5,5372
	7/32	.2187	5,5562
		.219	5,5626
		.220	5,5880
2		.221	5,6134
		.222	5,6388
		.223	5,6642
		.224	5,6896
		.225	5,7150
		.226	5,7404
		.227	5,7658
1		.228	5,7912
		.229	5,8166
		.230	5,8420
		.231	5,8674
		.232	5,8928
		.233	5,9182
A		.234	5,9436
	15/64	.2344	5,9531
		.235	5,9690
		.236	5,9944
		.2362	6,0000
		.237	6,0198
B		.238	6 0452
		.239	6,0706
		.240	6,0960
		.241	6,1214
C		.242	6,1468
		.243	6,1727
		.244	6,1976
		.245	6,2230
D		.246	6,2484
		.247	6,2738
		.248	6,2992
		.249	6,3246
E	1/4	.250	6,3500
		.251	6,3754
		.252	6,4008
		.253	6,4262
		.254	6,4516
		.255	6,4770
		.256	6,5024
F		.257	6,5278
		.258	6,5532
		.259	6,5786
		.260	6,6040
G		.261	6,6294
		.262	6,6548
		.263	6,6802
		.264	6,7056
		.265	6,7310
	17/64	.2656	6,7469
H		.266	6,7564
		.267	6,7818
		.268	6,8072
		.269	6,8326
		.270	6,8580
		.271	6,8834
I		.272	6,9088
		.273	6,9342
		.274	6,9596
		.275	6,9850
		.2756	7,0000
		.276	7,0104
J		.277	7,0358
		.278	7,0612
		.279	7,0866
		.280	7,1120
K		.281	7,1374
	9/32	.2812	7,1437
		.282	7,1628
		.283	7,1882
		.284	7,2136
		.285	7,2390
		.286	7,2644
		.287	7,2898
		.288	7,3152
		.289	7,3406
L		.290	7,3660
		.291	7,3914
		.292	7,4168
		.293	7,4422
		.294	7,4676
M		.295	7,4930
		.296	7,5184
	19/64	.2969	7,5406
		.297	7,5438
		.298	7,5692
		.299	7,5946
		.300	7,6200

Drill No. or Letter		Inch	mm
		.301	7,6454
N		.302	7,6708
		.303	7,6962
		.304	7,7216
		.305	7,7470
		.306	7,7724
		.307	7,7978
		.308	7,8232
		.309	7,8486
		.310	7,8740
		.311	7,8994
		.312	7,9248
	5/16	.3125	7,9375
		.313	7,9502
		.314	7,9756
		.3150	8,0000
		.315	8,0010
O		.316	8,0264
		.317	8,0518
		.318	8,0772
		.319	8,1026
		.320	8,1280
		.321	8,1534
		.322	8,1788
P		.323	8,2042
		.324	8,2296
		.325	8,2550
		.326	8,2804
		.327	8,3058
		.328	8,3312
	21/64	.3281	8,3344
		.329	8,3566
		.330	8,3820
		.331	8,4074
Q		.332	8,4328
		.333	8,4582
		.334	8,4836
		.335	8,5090
		.336	8,5344
		.337	8,5598
		.338	8,5852
R		.339	8,6106
		.340	8,6360
		.341	8,6614
		.342	8,6868
		.343	8,7122
	11/32	.3437	8,7312
		.344	8,7376
		.345	8,7630
		.346	8,7884
		.347	8,8138
S		.348	8,8392
		.349	8,8646
		.350	8,8900
		.351	8,9154
		.352	8,9408
		.353	8,9662
		.354	8,9916
		.3543	9,0000
		.355	9,0170
		.356	9,0424
		.357	9,0678
T		.358	9,0932
		.359	9,1186
	23/64	.3594	9,1281
		.360	9,1440
		.361	9,1694
		.362	9,1948
		.363	9,2202
		.364	9,2456
		.365	9,2710
		.366	9,2964
		.367	9,3218
U		.368	9,3472
		.369	9,3726
		.370	9,3980
		.371	9,4234
		.372	9,4488
		.373	9,4742
		.374	9,4996
	3/8	.375	9,5250
		.376	9,5504
V		.377	9,5758
		.378	9,6012
		.379	9,6266
		.380	9,6520
		.381	9,6774
		.382	9,7028
		.383	9,7282
		.384	9,7536
		.385	9,7790
W		.386	9,8044
		.387	9,8298
		.388	9,8552
		.389	9,8806
		.390	9,9060
	25/64	.3906	9,9219
		.391	9,9314
		.392	9,9568
		.393	9,9822
		.3937	10,0000
		.394	10,0076
		.395	10,0330
		.396	10,0584
X		.397	10,0838
		.398	10,1092
		.399	10,1346
		.400	10,1600

Drill No. or Letter		Inch	mm
		.401	10,1854
		.402	10,2108
		.403	10,2362
Y		.404	10,2616
		.405	10,2870
		.406	10,3124
	13/32	.4062	10,3187
		.407	10,3378
		.408	10,3632
		.409	10,3886
		.410	10,4140
		.411	10,4394
		.412	10,4648
Z		.413	10,4902
		.414	10,5156
		.415	10,5410
		.416	10,5664
		.417	10,5918
		.418	10,6172
		.419	10,6426
		.420	10,6680
		.421	10,6934
	27/64	.4219	10,7156
		.422	10,7188
		.423	10,7442
		.424	10,7696
		.425	10,7950
		.426	10,8204
		.427	10,8458
		.428	10,8712
		.429	10,8966
		.430	10,9220
		.431	10,9474
		.432	10,9728
		.433	10,9982
		.4331	11,0000
		.434	11,0236
		.435	11,0490
		.436	11,0744
		.437	11,0998
	7/16	.4375	11,1125
		.438	11,1252
		.439	11,1506
		.440	11,1760
		.441	11,2014
		.442	11,2268
		.443	11,2522
		.444	11,2776
		.445	11,3030
		.446	11,3284
		.447	11,3538
		.448	11,3792
		.449	11,4046
		.450	11,4300
		.451	11,4554
		.452	11,4808
		.453	11,5062
	29/64	.4531	11,5094
		.454	11,5316
		.455	11,5570
		.456	11,5824
		.457	11,6078
		.458	11,6332
		.459	11,6586
		.460	11,6840
		.461	11,7094
		.462	11,7348
		.463	11,7602
		.464	11,7856
		.465	11,8110
		.466	11,8364
		.467	11,8618
		.468	11,8872
	15/32	.4687	11,9062
		.469	11,9126
		.470	11,9380
		.471	11.9634
		.472	11,9888
		.4724	12,0000
		.473	12,0142
		.474	12,0396
		.475	12,0650
		.476	12,0904
		.477	12,1158
		.478	12,1412
		.479	12,1666
		.480	12,1920
		.481	12,2174
		.482	12,2428
		.483	12,2682
		.484	12,2936
	31/64	.4844	12,3031
		.485	12,3190
		.486	12,3444
		.487	12,3698
		.488	12,3952
		.489	12,4206
		.490	12,4460
		.491	12,4714
		.492	12,4968
		.493	12,5222
		.494	12,5476
		.495	12,5730
		.496	12,5984
		.497	12,6238
		.498	12,6492
		.499	12,6746
	1/2	.500	12,7000

Source. Courtesy of the Standard Gage Company.

890

Table 3 (continued)

Inch	mm
.501	12,7254
.502	12,7508
.503	12,7762
.504	12,8016
.505	12,8270
.506	12,8524
.507	12,8778
.508	12,9032
.509	12,9286
.510	12,9540
.511	12,9794
.5118	13,0000
.512	13,0048
.513	13,0302
.514	13,0556
.515	13,0810
33/64 .5156	13,0968
.516	13,1064
.517	13,1318
.518	13,1572
.519	13,1826
.520	13,2080
.521	13,2334
.522	13,2588
.523	13,2842
.524	13,3096
.525	13,3350
.526	13,3604
.527	13,3858
.528	13,4112
.529	13,4366
.530	13,4620
.531	13,4874
17/32 .5312	13,4937
.532	13,5128
.533	13,5382
.534	13,5636
.535	13,5890
.536	13,6144
.537	13,6398
.538	13,6652
.539	13,6906
.540	13,7160
.541	13,7414
.542	13,7668
.543	13,7922
.544	13,8176
.545	13,8430
.546	13,8684
35/64 .5469	13,8906
.547	13,8938
.548	13,9192
.549	13,9446
.550	13,9700
.551	13,9954
.5512	14,0000
.552	14,0208
.553	14,0462
.554	14,0716
.555	14,0970
.556	14,1224
.557	14,1478
.558	14,1732
.559	14,1986
.560	14,2240
.561	14,2494
.562	14,2748
9/16 .5625	14,2875
.563	14,3002
.564	14,3256
.565	14,3510
.566	14,3764
.567	14,4018
.568	14,4272
.569	14,4526
.570	14,4780
.571	14,5034
.572	14,5288
.573	14,5542
.574	14,5796
.575	14,6050
.576	14,6304
.577	14,6558
.578	14,6812
37/64 .5781	14,6844
.579	14,7066
.580	14,7320
.581	14,7574
.582	14,7828
.583	14,8082
.584	14,8336
.585	14,8590
.586	14,8844
.587	14,9098
.588	14,9352
.589	14,9606
.590	14,9860
.5906	15,0000
.591	15,0114
.592	15,0368
.593	15,0622
19/32 .5937	15,0812
.594	15,0876
.595	15,1130
.596	15,1384
.597	15,1638
.598	15,1892
.599	15,2146

Inch	mm
.600	15,2400
.601	15,2654
.602	15,2908
.603	15,3162
.604	15,3416
.605	15,3670
.606	15,3924
.607	15,4178
.608	15,4432
.609	15,4686
39/64 .6094	15,4781
.610	15,4940
.611	15,5194
.612	15,5448
.613	15,5702
.614	15,5956
.615	15,6210
.616	15,6464
.617	15,6718
.618	15,6972
.619	15,7226
.620	15,7480
.621	15,7734
.622	15,7988
.623	15,8242
.624	15,8496
5/8 .625	15,8750
.626	15,9004
.627	15,9258
.628	15,9512
.629	15,9766
.6299	16,0000
.630	16,0020
.631	16,0274
.632	16,0528
.633	16,0782
.634	16,1036
.635	16,1290
.636	16,1544
.637	16,1798
.638	16,2052
.639	16,2306
.640	16,2560
41/64 .6406	16,2719
.641	16,2814
.642	16,3068
.643	16,3322
.644	16,3576
.645	16,3830
.646	16,4084
.647	16,4338
.648	16,4592
.649	16,4846
.650	16,5100
.651	16,5354
.652	16,5608
.653	16,5862
.654	16,6116
.655	16,6370
.656	16,6624
21/32 .6562	16,6687
.657	16,6878
.658	16,7132
.659	16,7386
.660	16,7640
.661	16,7894
.662	16,8148
.663	16,8402
.664	16,8656
.665	16,8910
.666	16,9164
.667	16,9418
.668	16,9672
.669	16,9926
.6693	17,0000
.670	17,0180
.671	17,0434
43/64 .6719	17,0656
.672	17,0688
.673	17,0942
.674	17,1196
.675	17,1450
.676	17,1704
.677	17,1958
.678	17,2212
.679	17,2466
.680	17,2720
.681	17,2974
.682	17,3228
.683	17,3482
.684	17,3736
.685	17,3990
.686	17,4244
.687	17,4498
11/16 .6875	17,4625
.688	17,4752
.689	17,5006
.690	17,5260
.691	17,5514
.692	17,5768
.693	17,6022
.694	17,6276
.695	17,6530
.696	17,6784
.697	17,7038
.698	17,7292
.699	17,7546
.700	17,7800

Inch	mm
.701	17,8054
.702	17,8308
.703	17,8562
45/64 .7031	17,8594
.704	17,8816
.705	17,9070
.706	17,9324
.707	17,9578
.708	17,9832
.7087	18,0000
.709	18,0086
.710	18,0340
.711	18,0594
.712	18,0848
.713	18,1102
.714	18,1356
.715	18,1610
.716	18,1864
.717	18,2118
.718	18,2372
23/32 .7187	18,2562
.719	18,2626
.720	18,2880
.721	18,3134
.722	18,3388
.723	18,3642
.724	18,3896
.725	18,4150
.726	18,4404
.727	18,4658
.728	18,4912
.729	18,5166
.730	18,5420
.731	18,5674
.732	18,5928
.733	18,6182
.734	18,6436
47/64 .7344	18,6532
.735	18,6690
.736	18,6944
.737	18,7198
.738	18,7452
.739	18,7706
.740	18,7960
.741	18,8214
.742	18,8468
.743	18,8722
.744	18,8976
.745	18,9230
.746	18,9484
.747	18,9738
.748	18,9992
.7480	19,0000
.749	19,0246
3/4 .750	19,0500
.751	19,0754
.752	19,1008
.753	19,1262
.754	19,1516
.755	19,1770
.756	19,2024
.757	19,2278
.758	19,2532
.759	19,2786
.760	19,3040
.761	19,3294
.762	19,3548
.763	19,3802
.764	19,4056
.765	19,4310
49/64 .7656	19,4469
.766	19,4564
.767	19,4818
.768	19,5072
.769	19,5326
.770	19,5580
.771	19,5834
.772	19,6088
.773	19,6342
.774	19,6596
.775	19,6850
.776	19,7104
.777	19,7358
.778	19,7612
.779	19,7866
.780	19,8120
.781	19,8374
25/32 .7812	19,8433
.782	19,8628
.783	19,8882
.784	19,9136
.785	19,9390
.786	19,9644
.787	19,9898
.7874	20,0000
.788	20,0152
.789	20,0406
.790	20,0660
.791	20,0914
.792	20,1168
.793	20,1422
.794	20,1676
.795	20,1930
.796	20,2184
51/64 .7969	20,2402
.797	20,2438
.798	20,2692
.799	20,2946

Inch	mm
.800	20,3200
.801	20,3454
.802	20,3708
.803	20,3962
.804	20,4216
.805	20,4470
.806	20,4724
.807	20,4978
.808	20,5232
.809	20,5486
.810	20,5740
.811	20,5994
.812	20,6248
13/16 .8125	20,6375
.813	20,6502
.814	20,6756
.815	20,7010
.816	20,7264
.817	20,7518
.818	20,7772
.819	20,8026
.820	20,8280
.821	20,8534
.822	20,8788
.823	20,9042
.824	20,9296
.825	20,9550
.826	20,9804
.8268	21,0000
.827	21,0058
.828	21,0312
.829	21,0566
.830	21,0820
.831	21,1074
.832	21,1328
.833	21,1582
.834	21,1836
.835	21,2090
.836	21,2344
.837	21,2598
.838	21,2852
.839	21,3106
.840	21,3360
.841	21,3614
.842	21,3868
.843	21,4122
27/32 .8437	21,4312
.844	21,4376
.845	21,4630
.846	21,4884
.847	21,5138
.848	21,5392
.849	21,5646
.850	21,5900
.851	21,6154
.852	21,6408
.853	21,6662
.854	21,6916
.855	21,7170
.856	21,7424
.857	21,7678
.858	21,7932
.859	21,8186
55/64 .8594	21,8281
.860	21,8440
.861	21,8694
.862	21,8948
.863	21,9202
.864	21,9456
.865	21,9710
.866	21,9964
.8661	22,0000
.867	22,0218
.868	22,0472
.869	22,0726
.870	22,0980
.871	22,1234
.872	22,1488
.873	22,1742
.874	22,1996
7/8 .875	22,2250
.876	22,2504
.877	22,2758
.878	22,3012
.879	22,3266
.880	22,3520
.881	22,3774
.882	22,4028
.883	22,4282
.884	22,4536
.885	22,4790
.886	22,5044
.887	22,5298
.888	22,5552
.889	22,5806
.890	22,6060
57/64 .8906	22,6219
.891	22,6314
.892	22,6568
.893	22,6822
.894	22,7076
.895	22,7330
.896	22,7584
.897	22,7838
.898	22,8092
.899	22,8346
.900	22,8600

Inch	mm
.901	22,8854
.902	22,9108
.903	22,9362
.904	22,9616
.905	22,9870
.9055	23,0000
.906	23,0124
29/32 .9062	23,0187
.907	23,0378
.908	23,0632
.909	23,0886
.910	23,1140
.911	23,1394
.912	23,1648
.913	23,1902
.914	23,2156
.915	23,2410
.916	23,2664
.917	23,2918
.918	23,3172
.919	23,3426
.920	23,3680
.921	23,3934
59/64 .9219	23,4156
.922	23,4188
.923	23,4442
.924	23,4696
.925	23,4950
.926	23,5204
.927	23,5458
.928	23,5712
.929	23,5966
.930	23,6220
.931	23,6474
.932	23,6728
.933	23,6982
.934	23,7236
.935	23,7490
.936	23,7744
.937	23,7998
15/16 .9375	23,8125
.938	23,8252
.939	23,8506
.940	23,8760
.941	23,9014
.942	23,9268
.943	23,9522
.944	23,9776
.9449	24,0000
.945	24,0030
.946	24,0284
.947	24,0538
.948	24,0792
.949	24,1046
.950	24,1300
.951	24,1554
.952	24,1808
.953	24,2062
61/64 .9531	24,2094
.954	24,2316
.955	24,2570
.956	24,2824
.957	24,3078
.958	24,3332
.959	24,3586
.960	24,3840
.961	24,4094
.962	24,4348
.963	24,4602
.964	24,4856
.965	24,5110
.966	24,5364
.967	24,5618
.968	24,5872
31/32 .9687	24,6062
.969	24,6126
.970	24,6380
.971	24,6634
.972	24,6888
.973	24,7142
.974	24,7396
.975	24,7650
.976	24,7904
.977	24,8158
.978	24,8412
.979	24,8666
.980	24,8920
.981	24,9174
.982	24,9428
.983	24,9682
.984	24,9936
63/64 .9843	25,0000
.985	25,0190
.986	25,0444
.987	25,0698
.988	25,0952
.989	25,1206
.990	25,1460
.991	25,1714
.992	25,1968
.993	25,2222
.994	25,2476
.995	25,2730
.996	25,2984
.997	25,3238
.998	25,3492
.999	25,3746
1.000	25,4000

appendix II
answers to self-tests

section a unit 1 shop safety

SELF-TEST ANSWERS

1. Eye protection equipment.
2. Wear a safety goggle or full face shield. Prescription glasses may be made as safety glasses.
3. Shoes, short sleeves, short or properly secured hair, no rings and wristwatches, shop apron or shop coat with short sleeves.
4. Use of cutting fluids and vacuum dust collectors.
5. They may cause skin rashes or infections.
6. Bend knees and squat, lift with your legs, keeping your back straight.
7. Compressed air can propel chips through the air, implant dirt into skin, and possibly injure ear drums.
8. Good houskeeping includes cleaning oil spills, keeping material off the floor, and keeping aisles clear of obstructions.
9. In the vertical position or with a person on each end.
10. Do I know how to operate this machine?
    What are the potential hazards involved?
    Are all guards in place?
    Are my procedures safe?
    Am I doing something I probably should not do?
    Have I made all proper adjustments and tightened all locking bolts and clamps?
    Is the workpiece secured properly?
    Do I have proper safety equipment?
    Do I know where the stop switch is?
    Do I think about safety in everything I do?

section a unit 2 introduction to mechanical hardware

SELF-TEST ANSWERS

1. A bolt goes through parts being assembled and is tightened with a nut. A screw is used where a part is internally threaded and no nut is needed.
2. The minimum recommended thread engagement for a screw in an assembly is as much as the screw diameter; a better assembly will result when $1\frac{1}{2}$ times the screw diameter is used.
3. Class 2 threads are found on most screws, nuts, and bolts used in the manufacturing industry. Cars and machines tools would be good examples.
4. Machine bolts are not machined to the precise dimensions of cap screws. Machine bolts have coarse threads where cap screws may have coarse or fine threads. Machine bolts have many uses in the construction industry, and cap screws are usually used in precision assemblies.

5. The formula is $D=$ number of the machine screw times .013 in. plus .060 in. $D=8\times$ .013 in. plus .060$=$
6. Set screws are used to secure gears or pulleys
7. Stud bolts can be used instead of long bolts. Stud bolts are used to aid in the assembly of heavy parts by acting as guide pins.
8. Thread forming screws form threads by displacing material. Thread cutting screws produce threads by actually cutting grooves and making chips.
9. Castle nuts can be secured on a bolt with a cotter pin to prevent their accidental loosening.
10. Cap nuts are used because of their neat appearance. They also protect projecting threads from damage.
11. Flat washers protect the surface of parts from being marred by the tightening of screws or nuts. Flat washers also provide a larger contact area than nuts and screw heads to distribute the clamping pressure over a larger area.
12. A helical spring lock washer prevents the unplanned loosening of nut and bolt or screw assemblies. Spring lock washers will also provide for a limited amount of take up when expansion or contraction takes place.
13. Internal-external tooth lock washers are used on oversized holes or to provide a large bearing surface.
14. Dowel pins are used to achieve accurate alignment between two or more parts.
15. Taper pins give accurate alignment to parts that have to be disassembled frequently.
16. Roll pins are used to align parts. Holes to receive roll pins do not have to be reamed, which is necessary for dowel pins and taper pins.
17. Retaining rings are used to hold bearings or seals in bearing housings or on shafts. Retaining rings have a spring action and are usually seated in grooves.
18. Keys transmit the driving force between a shaft and pulley.
19. Woodruff keys are used where only light loads are transmitted.
20. Gib head keys are used to transmit heavy loads. These keys are installed and removed from the same side of a hub and shaft assembly.

section b unit 1 workholding for hand operations

SELF-TEST ANSWERS

1. The vise should be positioned so that a long piece may be held vertically in the jaws without interference from the workbench.
2. The solid base and the swivel-base types.
3. By the width of the jaws.
4. Pin vise or toolmaker vise.
5. Insert jaws are hardened and have diamond pattern or criss-cross serrations.
6. Copper, soft metal, or wood may be used to protect finishes from insert jaw serrations.
7. Vises used for sheet metal work have smooth, deep jaws.
8. Vises are used for holding work for assembly and disassembly; for filing, hacksawing, and bending light metal.
9. "Cheater" bars should never be used on the handle. The movable jaw slide bar should never be hammered upon, and excessive heat should never be applied to the jaws.
10. The vise should be taken apart, cleaned, and the screw and nut cleaned in solvent. A heavy grease should be packed on the screw and thrust collars before reassembly.

section b unit 2 arbor and shop presses

SELF-TEST ANSWERS

1. To use the arbor press without instruction is unsafe for the operator; also very expensive equipment and materials may be damaged or ruined.
2. Arbor presses are hand powered and can be mechanical or hydraulic. Large power driven presses do not provide the "feel" needed when pressing delicate parts.
3. The arbor press is used for installing and removing mandrels, bushings, and ball bearings. The hydraulic shop press is also used for straightening and bending.

4. The shaft has seized or welded in the bore because it had not been lubricated with pressure lubricant.
5. A loose and rounded ram could cause a bushing to tilt or twist sideways while pressing and thus be ruined. In any case, the operator should always check to see if a bushing is going in straight. A pressing plug with a pilot would be helpful here.
6. No. Thirty tons would deform the extended end of the bushing. Just enough pressure should be applied to press in the bushing and when it contacts the press plate at the bottom, more pressure will be sensed. At that point it is time to stop.
7. Pressing on the inner race with the outer race supported will damage or break the bearing.
8. Ordinary shafts with press fits are not tapered but have the same dimension along the pressing length. Mandrels taper .006 in./ft, which causes them to tighten in the bore somewhere along their length.
9. The two most important steps, assuming that the dimensions are all correct, are to
   a. Make sure the bore has a good chamfer and the bushing should also have a chamfer or "start."
   b. Apply high pressure lubricant to the bore and the bushing.
10. Five ways to avoid tool breakage and other problems when broaching keyways in the arbor press:
   a. Make sure the press ram is not loose and check to see that the proper hole in the press plate is under the work so that the broach has clearance to go through the work.
   b. Clean and lubricate the broach, especially the back edge between each cut.
   c. Do not use a broach on hard materials (over Rc 35).
   d. Use the right size bushing for the bore and broach.
   e. Make sure at least two teeth are continuously engaged in the work.

section b unit 3 noncutting hand tools

SELF-TEST ANSWERS

1. True.
2. The purpose of false jaws is to protect finished work from being damaged by the hard serrated jaws. Also, when gripping soft metals, "soft" jaws of leather or a softer metal are used.
3. False. C-clamps are used for heavy clamping work. Some heavy duty types can hold many hundreds of pounds.
4. False. This practice can quickly ruin good machinery so that the proper tool can never be used.
5. The principal advantage of the lever jawed wrench is its great holding power. Most types have hard serrated jaws and so should not be used on nuts and bolt heads.
6. No. Hammers used for layout work range from 4 to 10 ounces in weight. A smaller one should be used with a prick punch and for delicate work.
7. Soft hammers and mallets are made for this purpose. When setting down work in a drill press or milling machine vise, for instance, a lead hammer is best because it has no rebound.
8. The box type, either socket or end wrench, would be best as it provides contact on the six points of the capscrew, thus avoiding the damage and premature wear that would be caused by using an adjustable wrench.
9. The hard serrated jaws will damage machine parts. Pipe wrenches should be used on pipe and pipe fittings only.
10. Standard screwdrivers should have the right width blade to fit the screw head. They should be shaped correctly and, if worn, reground and shaped properly.

section b unit 4 hacksaws

SELF-TEST ANSWERS

1. The kerf is the groove produced in the work by a saw blade.
2. The set on a saw blade is the width of the teeth that are bent out from the blade back.
3. The pitch of a hacksaw blade refers to the number of teeth per inch on a saw blade.
4. The first consideration in the selection of a saw blade is the kind of material being cut. For soft materials use a coarse tooth blade and for harder materials use a fine pitch blade. The second point to watch is that at least three teeth should be cutting at the same time.

5. The two basic kinds of saw blades are the all-hard blade and the flexible blade.
6. Generally a speed between 40 and 60 strokes per minute is suggested. It is best to use long and slow strokes utilizing the full length of the blade.
7. Excessive dulling of saw blades is caused by pressure on the saw blade on the return stroke, sawing too fast, letting the saw slide over the workpiece without any cutting pressure, or applying too much pressure.
8. Saw blades break if too much pressure is used or if the blade is not sufficiently tightened in the saw frame.
9. When the saw blade is used, the set wears and makes the kerf cut narrower with a used blade than the kerf cut with a new blade. If a cut started with a used blade can't be finished with that blade, but has to be completed with a new blade, the workpiece should be turned over and a new cut started from the opposite end from the original cut. A new blade, when used in a kerf started with a used blade, would lose its set immediately and start binding in the groove.
10. When a blade breaks, it shatters causing blade particles to fly quite a distance with the possibility of injuring someone. Should the blade break while sawing, it may catch the operator off balance and cause him to push his hand into the workpiece while following through with his sawing stroke. Serious cuts or abrasions can be the result of this action.

section b unit 5 files and off-hand grinding

SELF-TEST ANSWERS

1. By its length, shape, cut, and coarseness.
2. Single cut, double cut, curved cut, and rasp.
3. Four out of these: rough, coarse, bastard, second cut, smooth, and dead smooth.
4. The double cut file.
5. To make it possible to file a flat surface by offsetting the tendency to rock a file. To compensate for the slight downward deflection when pressure is applied while filing. To concentrate pressure on fewer teeth for deeper penetration.
6. A blunt file has the same cross-sectional area from heel to point, where a tapered file is larger at the heel than at the point.
7. A mill file is thinner than a comparable size flat file.
8. A warding file is rectangular in shape and it tapers to a small point, making filing possible in narrow slots and grooves.
9. Swiss pattern files are more precise in construction; they are more slender and have teeth to the extreme edges.
10. Coarseness is identified by numbers from 00 fine to 6 coarse.
11. Where files touch each other, teeth break off or become dull.
12. Too much pressure will break teeth off a file. It also will cause pinning and scratching of the work surface.
13. Files in contact with each other, files rubbing over the work without any pressure being applied, filing too fast, or filing on hardened materials cause files to dull.
14. As a safety precaution; an unprotected tang can cause serious injury.
15. Measuring the workpiece for flatness and size assures the craftsman that the filing is done in the right place and that the filing is stopped before a piece is undersize. Measuring often is not a waste of time.
16. Touching a workpiece is just like lubricating the workpiece or the file so that it slips over the work without cutting. This also causes a file to dull quickly.
17. A soft workpiece requires a file with coarser teeth because there is less resistance to tooth penetration. A fine toothed file would clog up on soft materials. For harder materials use a fine toothed file in order to have more teeth making smaller chips.
18. The drawfiling stroke should be short enough that the file never slips over the ends of the workpiece. Care should be taken that no hollow surface is created through too short of a stroke.
19. Pressure is only applied on the forward stroke, which is the cutting stroke.
20. Rotating a round file clockwise while filing makes the file cut better and improves the surface finish.

section b unit 6 hand reamers

SELF-TEST ANSWERS

1. Hand reamers have a square on the shank and a long starting taper on the fluted end.
2. A reamer does its cutting on the tapered portion. A long taper will help in keeping a reamer aligned with the hole.
3. Spiral fluted reamers cut with a shearing action. They will also bridge over keyways and grooves without chattering.
4. The shank diameter is usually a few thousandths of an inch smaller than the nominal size of the reamer. This allows the reamer to pass through the hole without marring it.
5. Expansion reamers are useful to increase hole sizes by a very very small amount.
6. Expansion reamers can only be adjusted a small amount by moving a tapered internal plug. Adjustable reamers have a larger range of adjustments, from $\frac{1}{32}$ in. on small diameters to $\frac{5}{16}$ in. on large reamers. Adjustable reamers have removable blades. Size changes are made by moving these blades with nuts in external tapered slots.
7. Coolants are used to dissipate the heat generated by the reaming process, but in reaming coolants are more important in obtaining a high quality surface finish of the hole.
8. Reamers dull rapidly if they should be rotated backwards.
9. The hand reaming allowance is rather small, only between .001 and .005 in.

section b unit 7 taps: identification and application

SELF-TEST ANSWERS

1. A set of taps consists of three taps with equal pitch diameters and major diameters with the difference being the number of chamfered threads on the cutting end. Serial taps have different pitch and major diameters within a nominal size designation. The smallest tap in the series is marked with a single ring on the shank near the square. The next larger tap has two rings on the shank and the tap that cuts the thread to its full size is marked with three rings.
2. Spiral pointed taps are used on through holes or blind holes with sufficient chip space at the bottom of a hole.
3. Fluteless spiral pointed taps are especially useful to tap holes in sheet metal or on soft, stringy materials where the thickness is no greater than one tap diameter.
4. Spiral fluted taps draw the chips out of the hole and are useful when tapping a hole that has a keyway in it bridged by the helical flutes.
5. Thread forming or fluteless taps do not produce chips because they don't cut threads. Their action can be compared to thread rolling in that material is being displaced in grooves to form ridges shaped in the precise form of a thread.
6. Taper pipe taps are being identified by the taper of the body of the tap, which is $\frac{3}{4}$ inch per foot of length; also by the size marked on the shank.
7. Pulley taps and nut taps are both extended length hand taps. The pulley tap has a shank diameter equal to the nominal size of the thread tapped where a nut tap has a shank diameter slightly smaller than the minor diameter of the nut tapped.
8. When an Acme thread is cut, the tap is required to cut too much material in one pass. To obtain a quality thread, a roughing pass and then a finishing pass are needed.
9. Rake angles vary on tools depending on the kind of work material machined. In general, we can say that softer, more ductile materials require larger rake angles than do harder, less ductile materials.
10. Friction is reduced by back tapering, eccentric or con-eccentric relief on the pitch diameter, concave groove land relief, making the tap in interrupted design, and in various surface treatments such as oxides and flash chrome plating.

section b unit 8 tapping procedures

SELF-TEST ANSWERS

1. Taps are driven with tap wrenches or T-handle tap wrenches.
2. A hand tapper is a fixture used to hold a tap in precise alignment while hand tapping holes.
3. A tapping attachment is used when tapping holes in a machine or for production tapping.
4. The strength of a tapped hole is determined by the kind of material being tapped, the percentage of thread used, and the length or depth of thread engagement.
5. Holes should be tapped deep enough to provide 1 to $1\frac{1}{2}$ times the tap diameter of usable thread.
6. Tap drilled holes should be reamed when close control over the percentage of thread produced is necessary and when fine pitches of thread are produced, because a small change in tap drilled hole size would mean a large change in the percentage of thread cut.
7. Taps break because holes are drilled too shallow, chips are packed tight in the flutes, hard materials or hard spots are encountered, inadequate or the wrong kind of lubricant is used, or the cutting speed used is too great.
8. Tapped holes that are rough and torn are often caused by dull taps, chips clogging the flutes, insufficient lubrication, wrong kind of lubrication, or already rough holes being tapped.
9. Oversize tapped holes can be caused by a loose machine spindle or a worn tap holder, misaligned spindle, oversized tap, a dull tap, chips packed in flutes, and buildup on cutting edges.
10. Broken taps can be removed by drilling them out after annealing the tap, or by using an electrical discharge machine to erode the tap.

section b unit 9 thread cutting dies and their uses

SELF-TEST ANSWERS

1. A die is used to cut external threads.
2. A die stock is used to hold the die when hand threading. A special die holder is used when machine threading.
3. The size of thread cut can only be changed a very small amount on round adjustable dies. Too much expansion or contraction may break the die.
4. The purpose of the guide is to align the die square to the workpiece to be threaded.
5. When assembling a two-piece die collet, be sure both die halves are marked with the same serial number and that the starting chamfer on the dies are toward the guide.
6. Hexagon rethreading dies are used to clean and recut slightly damaged or rusty threads. Only in emergencies should they be used to cut new threads.
7. The chamfer on the cutting end of a die distributes the cutting force over a number of threads and aids in starting the thread cutting operation.
8. Cutting fluids are very important in threading to achieve thread with a good surface finish, close tolerance, and to give long tool life.
9. Before a rod is threaded, it should be measured to assure its size is no larger than its nominal size. Preferably, it is .002 to .005 in. undersize.
10. The chamfer on a rod before threading makes it easy to start a die. It also protects the starting thread on a finished bolt.

section C unit 1 systems of measurement

SELF-TEST ANSWERS

1. To find in. knowing mm, multiply mm by .03937: $35 \times .03937 = 1.377$ in.
2. To find mm knowing in., multiply in. by 25.4: $.125 \times 25.4 = 3.17$ mm.
3. To find cm knowing in., multiply in. by 2.54: $6.273 \times 2.54 = 159.33$ cm.
4. To find mm knowing in., multiply in. by 25.4: $.050 \times 25.4 \pm 1.27$ mm.

5. 10 mm = 1 cm; therefore, to find cm knowing mm, divide mm by 10.

6. To find in. knowing mm, multiply mm by .03937: .02 × .03937 = .007 in. The tolerance would be ± .007 in.

7. SI refers to the International System of Units.

8. Conversions between metric and inch systems may be accomplished by mathematical procedures, conversion charts, and direct converting calculators.

9. The yard is presently defined in terms of the meter. 1 yard = $\frac{3600}{3937}$ meter.

10. Yes, by the use of appropriate conversion dials.

section C unit 2 using steel rules

SELF-TEST ANSWERS

Fractional Inch Rules

Figure 22 $A = 1\frac{1}{4}$ in.

Figure 22 $B = 2\frac{1}{8}$ in.

Figure 22 $C = \frac{15}{16}$ in.

Figure 22 $D = 2\frac{5}{16}$ in.

Figure 22 $E = \frac{15}{32}$ in.

Figure 22 $F = 2\frac{25}{32}$ in.

Figure 22 $G = \frac{63}{64}$ in.

Figure 22 $H = 1\frac{59}{64}$ in.

Decimal Inch Rules

Figure 25 $A = $ .300 in.

Figure 25 $B = $ .510 in.

Figure 25 $C = 1.020$ in.

Figure 25 $D = 1.200$ in.

Figure 25 $E = 1.260$ in.

Metric Rules

Figure 28 $A = 11$ mm or 1.1 cm

Figure 28 $B = 27$ mm or 2.7 cm

Figure 28 $C = 52$ mm or 5.2 cm

Figure 28 $D = 7.5$ mm or .75 cm

Figure 28 $E = 20.5$ mm or 2.05 cm

Figure 28 $F = 45.5$ mm or 4.55 cm

section C unit 3 using vernier calipers and vernier depth gages

SELF-TEST ANSWERS

Reading Inch Vernier Calipers

Figure 10*a*. 1.304 in.

Figure 10*b*. .492 in.

Figure 10*c*. .532 in.

Figure 10*d*. .724 in.

Reading Metric Vernier Calipers

Figure 12*a*. 20.26 mm

Figure 12*b*. 14.50 mm

Figure 12c. 29.84 mm
Figure 12d. 35.62 mm

Reading Inch Vernier Depth Gages
Figure 17a. .943 in.
Figure 17b. 1.326 in.
Figure 17c. 2.436 in.
Figure 17d. 3.768 in.

section c unit 4 using micrometer instruments

SELF-TEST ANSWERS

1. Anyone taking pride in his tools usually takes pride in his workmanship. The quality of a product produced depends to a large extent on the accuracy of the measuring tools used. A skilled craftsman protects his tools because he guarantees his product.
2. Moisture between the contact faces can cause corrosion.
3. Even small dust particles will change a dimension. Oil or grease attract small chips and dirt. All of these can cause incorrect readings.
4. A measuring tool is no more discriminatory than the smallest division marked on it. This means that a standard micrometer can discriminate to the nearest thousandth. A vernier scale on a micrometer will make it possible to discriminate a reading to one ten thousandths of an inch under controlled conditions.
5. The reliability of a micrometer depends on the inherent qualities built into it by its maker. Reliability also depends upon the skill of the user and the care the tool receives.
6. The sleeve is stationary in relation to the frame and is engraved with the main scale, which is divided into 40 equal spaces each equal to .025 in. The thimble is attached to the spindle and rotates with it. The thimble circumference is graduated with 25 equal divisions, each representing a value of .001 in.
7. There is less chance of accidentally moving the thimble when reading a micrometer while it is still in contact with the workpiece.
8. Measurement should be made at least twice. On critical measurements, checking the dimensions additional times will assure that the size measurement is correct.
9. As the temperature of a part is increased, the size of the part will increase. When a part is heated by the machining process, it should be permitted to cool down to room temperature before being measured.
    Holding a micrometer by the frame for an extended period of time will transfer body heat through the hand and affect the accuracy of the measurement taken.
10. The purpose of the ratchet stop or friction thimble is to enable equal pressure to be repeatedly applied between the measuring faces and the object being measured. Use of the ratchet stop or friction thimble will minimize individual differences in measuring pressure applied by different persons using the same micrometer.

Exercise Answers
Outside micrometer readings (Figure 39a to e):

Figure 39a. .669 in.
Figure 39b. .787 in.
Figure 39c. .237 in.
Figure 39d. .994 in.
Figure 39e. .072 in.

Inside micrometer readings (Figure 46a to e):
Figure 46a. 1.617
Figure 46b. 2.000
Figure 46c. 2.254
Figure 46d. 2.562
Figure 46e. 2.784

Depth micrometer readings (Figure 52a to e):
Figure 52a. .535 in.
Figure 52b. .815 in.
Figure 52c. .732 in.
Figure 52d. .535 in.
Figure 52e. .647 in.
Metric micrometer readings (Figure 57a to e):
Figure 57a. 21.21 mm
Figure 57b. 13.27 mm
Figure 57c. 9.94 mm
Figure 57d. 5.59 mm
Figure 57e. 4.08 mm
Vernier micrometer readings (Figure 59a to e):
Figure 59a. .3749 in.
Figure 59b. .5377 in.
Figure 59c. .3123 in.
Figure 59d. .2498 in.
Figure 59e. .1255 in.

section C unit 5 using comparison measuring instruments

SELF-TEST ANSWERS

1. Comparison measurement is measurement where an unknown dimension is compared to a known dimension. This often involves a transfer device that represents the unknown and is then transferred to the known where the reading can be determined.
2. Most comparison instruments do not have the capability to show measurement directly.
3. Cosine error is error incurred when misalignment exists between the axis of measurement and the axis of the measuring instrument.
4. Cosine error can be reduced by making sure that the axis of the measuring instrument is exactly in line with the axis of measurement.
5. Adjustable parallel (c).
6. Dial test indicator in conjunction with a height transfer micrometer. Height transfer measurements can also be accomplished with the test indicator and planer gage (f) and (n).
7. Optical comparator (p).
8. A combination square can be checked against a precision square or, if the actual amount of deviation is required, the cylindrical square or micrometer square can be used: (i) (j) and (l).
9. Telescope gage and outside micrometer (b).
10. Thickness gage or, in the case of a thick shim or chock, the adjustable parallel: (c) and (e).

section C unit 6 using gage blocks

SELF-TEST ANSWERS

1. The wringing interval in the space or interface between wrung gage blocks.
2. Wear blocks are made from very hard material such as tungsten carbide. Wear blocks are used in applications where direct contact with gage blocks might damage them.
3. If gage blocks should become heated or cooled above or below room temperature, normalizing is the process of returning them to room temperature.
4. AA grade — ± .000002 in.
   A+   — + .000004 in.
             — − .000002 in.
   B   Tolerance of B grade blocks is not specified.

5. The conditioning stone is a highly finished piece of granite or ceramic material and is used to remove burrs from the wringing surface of a gage block.

6. A microinch is one millionth of an inch. On surface finish it refers to the deviation of a surface from a uniform plane.

7. Gage block accuracy depends on the following factors:

    Extreme cleanliness
    No burrs
    Minimum use of the conditioning stone
    Leaving stacks assembled only for minimum amounts of time
    Cleaning before storage
    Application of gage block preservative

<table>
<tr><td>8.</td><td>3.0213</td><td>9.</td><td>1.9643</td><td></td></tr>
<tr><td></td><td>.1003</td><td></td><td>.100</td><td>(Wear blocks 2 × .050)</td></tr>
<tr><td></td><td>2.9210</td><td></td><td>1.8643</td><td></td></tr>
<tr><td></td><td>.121</td><td></td><td>1.003</td><td></td></tr>
<tr><td></td><td>2.800</td><td></td><td>1.7643</td><td></td></tr>
<tr><td></td><td>.800</td><td></td><td>.114</td><td></td></tr>
<tr><td></td><td>2.000</td><td></td><td>1.650</td><td></td></tr>
<tr><td></td><td>2.000</td><td></td><td>.650</td><td></td></tr>
<tr><td></td><td>0.0000</td><td></td><td>1.000</td><td></td></tr>
<tr><td></td><td></td><td></td><td>1.000</td><td></td></tr>
<tr><td></td><td></td><td></td><td>0.0000</td><td></td></tr>
</table>

10. Gage blocks can be used: to check other measuring instruments, to set sine bars for angles, as precision height gages for layout, in direct gaging applications, and for setting machine and cutting tool positions.

section C unit 7 using angular measuring instruments

SELF-TEST ANSWERS

1. Plate protractor and machinist's combination set bevel protractor.
2. Five minutes of arc.
3. The sine bar becomes the hypotenuse of a right triangle. Angles are measured or established by elevating the bar a specified amount or calculating the amount of bar elevation, knowing the angle.
4. 50° (Figure 19)
5. 96° 15′ (Figure 20)
6. 34° 30′ (Figure 21)
7. 61° 45′ (Figure 22)
8. 56° 25′ (Figure 23)
9. Bar elevation = bar length × sine of angle desired
    $$= 5 \text{ in.} \times \sin 37°$$
    $$= 5 \times .6018$$
    $$= 3.0090 \text{ in.}$$
10. Sine of the angle desired = elevation / bar length
    $$= 2.750 / 5$$

     Sin of angle   = .550

     Angle   = 33° 22′

This unit has no post-test.

section d unit 1 selection and identification of steels

SELF-TEST ANSWERS

1. Carbon and alloy steels are designated by the numerical SAE or AISI system.
2. The three basic types of stainless steels are: martensitic (hardenable) and ferritic (nonhardenable), both magnetic and of the 400 series, and austenitic (nonmagnetic and nonhardenable, except by work hardening) of the 300 series.
3. The identification for each piece would be as follows:
    a. AISI C1020 CF is a soft, low carbon steel with a dull metallic luster surface finish. Use the observation test, spark test, and file test for hardness.
    b. AISI B1140 (G and P) is a medium carbon, resulfurized, free machining steel with a shiny finish. Use the observation test, spark test, and machinability test.
    c. AISI C4140 (G and P) is a chromium-molybdenum alloy, medium carbon content with a polished, shiny finish. Since an alloy steel would be harder than a similar carbon or low carbon content steel, a hardness test should be used such as the file or scratch test to compare with known samples. The machinability test would be useful as a comparison test.
    d. AISI 8620 HR is a tough low carbon steel used for carburizing purposes. A hardness test and a machinability test will immediately show the difference from low carbon hot rolled steel.
    e. AISI B1140 (ebony) is the same as the resulfurized steel in **b**, only the finish is different. The test would be the same as for **b**.
    f. AISI C1040 is a medium carbon steel. The spark test would be useful here as well as the hardness and machinability tests.
4. A magnetic test can quickly determine whether it is a ferrous metal or perhaps nickel. If the metal is white in color, a spark test will be needed to determine whether it is a nickel casting or one of white cast iron, since they are similar in appearance. If a small piece can be broken off, the fracture will show whether it is white or grey cast iron. Grey cast iron will leave a black smudge on the finger. If it is cast steel, it will be more ductile than cast iron and a spark test should reveal a smaller carbon content.
5. O1 refers to an alloy-type oil hardening (oil quench) tool steel. W1 refers to a water hardening (water quench) tool steel.
6. The 40 in. long, $2\frac{7}{16}$ in. diameter shaft weighs 1.322 lbs/in. The cost is $.30/lb.
    $1.322 \times 40 \times .30 = \$15.86$ cost of the shaft.
7. **a.** No.
    b. Hardened tool steel or case hardened steel.
8. Austenitic (having a face centered cubic unit cell in its lattice structure). Examples are chromium, nickel, stainless steel, and high manganese alloy steel.
9. Nickel is a nonferrous metal that has magnetic properties. Some alloy combinations of nonferrous metals make strong permanent magnets; for example, the well-known Alnico magnet, an alloy of aluminum, nickel and cobalt.
10. Some properties of steel to be kept in mind when ordering or planning for a job would be:
    Strength
    Machinability
    Hardness
    Weldability (if welding is involved)
    Fatigue resistance
    Corrosion resistance (especially if the piece is to be exposed to a corrosive atmosphere)

section d unit 2 selection and identification of nonferrous metals

SELF-TEST ANSWERS

1. Since aluminum is about one-third lighter than steel, it is used extensively in aircraft. It also forms an oxide on the surface that resists further corrosion. The initial cost is much greater. Higher strength aluminum alloys cannot be welded.

2. The letter "H" following the four digit number always designates strain on work hardening. The letter "T" refers to heat treatment.

3. Magnesium weighs approximately one-third less than aluminum and is approximately one-quarter the weight of steel. Magnesium will burn in air when finely divided.

4. Copper is most extensively used in the electrical industries because of its low resistance to the passage of current when it is unalloyed with other metals. Copper can be strain hardened or work hardened and certain alloys may be hardened by a solution heat treat and aging process.

5. Bronze is basically copper and tin. Brass is basically copper and zinc.

6. Nickel is used to electroplate surfaces of metals for corrosion resistance, and as an alloying element with steels and nonferrous metals.

7. All three resist deterioration from corrosion.

8. Alloy.

9. Tin, lead, and cadmium.

10. Die cast metals, sometimes called "pot metal."

section d unit 3 hardening, tempering, and case hardening

SELF-TEST ANSWERS

1. No hardening would result as 1200°F (649°C) is less than the lower critical point and no dissolving of carbon has taken place.

2. There would be almost no change. For all practical purposes in the shop these low carbon steels are not considered hardenable.

3. They are shallow hardening, and liable to distortion and quench cracking because of the severity of the water quench.

4. Air and oil hardening steels are not so subject to distortion and cracking as W1 steels and are deep hardening.

5. 1450°F (788°C). 50°F (10°C) above the upper critical limit.

6. Tempering is done to remove the internal stresses in hard martensite, which is very brittle. The temperature used gives the best compromise between hardness and toughness or ductility.

7. Tempering temperature should be specified according to the hardness, strength, and ductility desired. Mechanical properties charts give this data.

8. 525°F (274°C). Purple.

9. 600°F (315°C). It would be too soft for any cutting tool.

10. Immediately. If you let it set for any length of time, it may crack from internal stresses.

11. The low carbon steel core does not harden when quenched from 1650°F (899°C), so it remains soft and tough, but the case becomes very hard. No tempering is therefore required as the piece is not brittle all the way through as a fully hardened carbon steel would be.

12. A deep case can be made by pack carburizing or by a liquid bath carburizing. A relatively deep case is often applied by nitriding or by similar procedures.

13. No. The base material must contain sufficient carbon to harden by itself without adding more for surface hardening.

14. Three methods of introducing carbon into heated steel are roll, pack, and liquid carburizing.

15. Nitriding.

section d unit 4 annealing, normalizing, and stress relieving

SELF-TEST ANSWERS

1. Medium carbon steels that are not uniform, have hardened areas from welding, or prior heat treating need to be normalized so they can be machined. Forgings, castings, and tool steel in the as-rolled condition are normalized before any further heat treatments or machining is done.

2. 1550°F (843°C). 50°F (10°C) above the upper critical limit.

3. The spheroidization temperature is quite close to the lower critical temperature line, about 1300°F (704°C).
4. The full anneal brings carbon steel to its softest condition as all the grains are reformed (recrystallized), and any hard carbide structures become soft pearlite as it slowly cools. Stress relieving will only recrystallize distorted ferrite grains and not the hard carbide structures or pearlite grains.
5. Stress relieving should be used on severely cold worked steels or for weldments.
6. High carbon steels (.8 to 1.7 percent C).
7. Process annealing is used by the sheet and wire industry and is essentially the same as stress relieving.
8. In still air.
9. Very slowly. Packed in insulating material or cooled in a furnace.
10. Low carbon steels tend to become gummy when spheroidized so the machinability is worse than in the as-rolled condition. Spheroidization sometimes is desirable when stress relieving weldments on low car-steels.

section d unit 5 rockwell and brinell hardness testers

SELF-TEST ANSWERS

1. Resistance to penetration is the one category that is utilized by the Rockwell and Brinell testers. The depth of penetration is measured when the major load is removed on the Rockwell tester and the diameter of the impression is measured to determine a Brinell hardness number.
2. As the hardness of a metal increases, the strength increases.
3. The "A" scale and a Brale marked "A" with a major load of 60 kgf should be used to test a tungsten carbide block.
4. It would become deformed or flattened and give an incorrect reading.
5. No. The Brale used with the Rockwell superficial tester is always marked or prefixed with the letter "N."
6. False. The ball penetrator is the same for all the scales using the same diameter ball.
7. The diamond spot anvil is used for superficial testing on the Rockwell tester. When used, it does not become indented, as is the case when using the spot anvil.
8. Roughness will give less accurate results than would a smooth surface.
9. The surface "skin" would be softer than the interior of the decarburized part.
10. A curved surface will give inaccurate readings.
11. A 3000 kg weight would be used to test steel specimens on the Brinell tester.
12. A 10 mm steel ball is usually used on the Brinell tester.

section e unit 1 basic semi-precision layout practice

SELF-TEST ANSWERS

1. The workpiece should have all sharp edges removed by grinding or filing. A thin, even coat of layout dye should be applied.
2. The towel will prevent spilling layout dye on the layout plate or layout table.
3. The punch should be tilted so that it is easier to see when the point is located on the scribe mark. It should then be moved to the upright position before it is tapped with the layout hammer.
4. The combination square can be positioned on the rule for measurements. The square head acts as a positive reference point for measurements. The rule may be removed and used as a straight edge for scribing.
5. The divider should be adjusted until you feel the tip drop into rule engraving.

section e unit 2 basic precision layout practice

SELF-TEST ANSWERS

1. (Figure 30) Inch—5.030 in. Metric—127.76mm
2. (Figure 31) Inch—8.694 in. Metric—220.82 mm

3. (Figure 32) Inch—5.917 in. Metric—150.30 mm
4. (Figure 33) Inch—4.086 in.
5. (Figure 34) Inch—1.456 in.
6. Zero reference is checked by bringing the scriber to rest on the reference surface and then checking the alignment of the beam and vernier zero lines.
7. The position of the vernier scale may be adjusted on height gages with this feature.
8. By turning the workpiece 90°.
9. 10 to 72 in.
10. The sine bar.

section f unit 1 using reciprocating and horizontal band cutoff machines

SELF-TEST ANSWERS

1. Raker, wave, and straight.
2. Workpiece material, cross section shape, and thickness.
3. On the back stroke.
4. The tooth offset on either side of the blade. Set provides clearance for the back of the blade.
5. Standard skip and hook.
6. Cutoff material can bind the blade and destroy the set.
7. The horizontal band saw.
8. Cooling, lubrication, and chip removal.
9. Scoring and possible blade breakage.
10. The workpiece must be turned over and a new cut started.

section f unit 2 abrasive and cold saws

SELF-TEST ANSWERS

1. Probably not.
2. Fast cutting and they can be used to cut nonmetallic materials.
3. Aluminum oxide, silicon carbide, and diamond.
4. Shellac, resinoid, and rubber.
5. Length tolerance of cutoff stock can be held very close.

    This unit has no post-test.

section f unit 3 preparing to use the vertical band machine

SELF-TEST ANSWERS

1. The ends of the blade should be ground with the teeth opposed. This will insure that the ends of the blade are square.
2. The blade ends are placed in the welder with the teeth pointed in. The ends must contact squarely in the gap between the jaws. The welder must be adjusted for the band width to be welded. You should wear eye protection and stand to one side during the welding operation. The weld will occur when the weld lever is depressed.
3. The weld is ground on the grinding wheel attached to the welder. Grind the weld on both sides of the band until the band fits the thickness gage. Be careful not to grind the saw teeth.
4. The guides support the band. This is essential to straight cutting.

5. Band guides must fully support the band except for the teeth. A wide guide used on a narrow band will destroy the saw set as soon as the machine is started.
6. The guide setting gage is used to adjust the band guides.
7. Annealing is the process of softening the band weld in order to improve strength qualities.
8. The band should be clamped in the annealing jaws with the teeth pointed out. A small amount of compression should be placed on the movable welder jaw prior to clamping the band. The correct annealing color is dull red. As soon as this color is reached, the anneal switch should be released and then operated briefly several times to slow the cooling rate of the weld.
9. Band tracking is the position of the band as it runs on the idler wheels.
10. Band tracking is adjusted by tilting the idler wheels until the band just touches the backup bearing.

section f unit 4 using the vertical band machine

SELF-TEST ANSWERS

1. The three sets are straight, wave, and raker. Straight set may be used for thin material, wave for material with a variable cross section, and raker for general purpose sawing.
2. Scalloped and wavy edged bands might be used on nonmetallic material where blade teeth would tear the material being cut.
3. Band velocity is measured in feet per minute.
4. The variable speed pulley is designed so that the pulley flanges may be moved toward and away from each other. This permits the belt position to be varied, resulting in speed changes.
5. The job selector provides information about recommended saw velocity, saw pitch, power feed, saw set, and temper. Band filing information is also indicated.
6. Speed range is selected by shifting the transmission.
7. Speed range shift must be done with the band speed set at the lowest setting.
8. The upper guidepost must be adjusted so that it is as close to the workpiece as possible.
9. Band pitch must be correct for the thickness of material to be cut. Generally, a fine pitch will be used on thin material. Cutting a thick workpiece with a fine pitch band will clog saw teeth and reduce cutting efficiency.
10. Band set must be adequate for the thickness of the blade used in a contour cut. If set is insufficient, the blade may not be able to cut the desired radius.

section g unit 1 the drill press

SELF-TEST ANSWERS

1. **a.** The sensitive, upright, and radial-arm drill presses are the three basic types. The sensitive drill press is made for light duty work and it provides the operator with a sense or "feel" of the feed on the drill. The upright is a similar, but heavy duty, drill press equipped with power feed. The radial-arm drill press allows the operator to position the drill over the work where he needs it, rather than to position the work under the drill as with other drill presses.
   b. The sensitive, upright, and radial-arm drill presses all perform much the same functions of drilling, reaming, counterboring, countersinking, spot facing, and tapping, but the upright and radial machines do heavier and larger jobs. The radial-arm drill can support large, heavy castings and work can be done on them without the workpiece being moved.
2. Sensitive drill press

a	Spindle
g	Quill lock handle
e	Column
l	Switch
b	Depth Stop

n	Head
o	Table
h	Table lock
f	Base
c	Power feed
j	Motor
d	Variable speed control
k	Table lift crank
i	Quill return spring
m	Guard

3. Radial drill press

b	Column
c	Radial arm
a	Spindle
d	Base
e	Drill head

section g unit 2 drilling tools

SELF-TEST ANSWERS

1.

t	Web
u	Margin
d	Drill point angle
p	Cutting lip
k	Flute
o	Body
e	Lip relief angle
g	Land
b	Chisel edge angle
j	Body clearance
i	Helix angle
y	Axis of drill
n	Shank length
c	Tang
x	Taper shank
w	Straight shank

2.

	Decimal Diameter	Fractional Size	Number Size	Letter Size	Metric Size
a	.0781	$\frac{5}{64}$			
b	.1495		25		
c	.272			I	
d	.159		21		
e	.1969				5
f	.323			P	
g	.3125	$\frac{5}{16}$			
h	.4375	$\frac{7}{16}$			
i	.201		7		
j	.1875	$\frac{3}{16}$			

section g unit 4 work locating and holding devices on drilling machines

SELF-TEST ANSWERS

1. The purpose for using workholding devices is to keep the workpiece rigid, from turning with the drill, and for operator safety.

2. Included in a list of workholding devices would be strap clamps, T-bolts and step blocks. Also used are C-clamps, V-blocks, vises, jigs and fixtures, and angle plates.
3. Parallels are mostly used to raise workpieces off the drill press table or to lift the workpiece higher in a vise, thus providing a space for the drill breakthrough. They are made of hardened steel so care should be exercised in their use.
4. Thin limber materials tend to be sprung downward from the force of drilling until drill breakthrough begins. The drill then "grabs" as the material springs upward and a broken drill is often the result. This can be avoided by placing the support or parallels as near the drill as possible.
5. Angle drilling is done by tilting a drill press table (not all types tilt), or by using an angular vise. If no means of setting the exact angle is provided, a protractor head with level may be used.
6. Vee blocks are suited to hold round stock for drilling. The most frequent use of vee blocks is for cross drilling holes in shafts, although many other setups are used.
7. The wiggler is used for locating a center punch mark under the center axis of a drill spindle.
8. Some odd shaped workpieces, such as gears with extending hubs that need holes drilled for set screws, might be difficult to set up without an angle plate.
9. One of the difficulties with hand tapping is the tendency for taps to start crooked or misaligned with the tap drilled hole. Starting a tap by hand in a drill press with the same set up as used for the tap drilling assures a perfect alignment.
10. Since jigs and fixtures are mostly used for production manufacturing, small machine shops rarely have a use for them.

section g unit 5 operating drilling machines

SELF-TEST ANSWERS

1. The three considerations would be speeds, feeds, and coolants.
2. The RPMs of the drills would be
   a. $\frac{1}{4}$ in. diameter is 1440 RPM.
   b. 2 in. diameter is 180 RPM.
   c. $\frac{3}{4}$ in. diameter is 480 RPM.
   d. $\frac{3}{8}$ in. diameter is 960 RPM.
   e. $1\frac{1}{2}$ in. diameter is 240 RPM.
3. Worn margins and outer corners broken down. The drill can be ground back to its full size and resharpened.
4. The operator will increase the feed in order to produce a chip. This increased feed is often greater than the drill can stand without breaking.
5. The feed is about right when the chip rolls into a close helix. Long, stringy chips can indicate too much feed.
6. Feeds are designated by a small measured advance movement of the drill for each revolution. A .001 in. feed for a $\frac{1}{8}$ in. diameter drill, for example, would move the drill .001 in. into the work for every turn of the drill.
7. The water soluble oil types and the cutting oils, both animal and mineral.
8. Besides having the correct cutting speed, a sulfurized oil based cutting fluid helps to reduce friction and cool the cutting edge.
9. Drill "jamming" can be avoided by a "pecking" procedure. The operator drills a small amount and pulls out the drill to remove the chips. This is repeated until the hole is finished.
10. The depth stop is used to limit the travel of the drill so it will not go on into the table or vise. The depth of blind holes is preset and drilled. Countersink and counterbore depths are set so that several can be easily made the same.

section g unit 6 countersinking and counterboring

SELF-TEST ANSWERS

1. Countersinks are used to chamfer holes and to provide tapered holes for flat head fasteners such as screws and rivets.

2. Countersink angles vary to match the angles of different flat head fasteners or different taper hole requirements.
3. A center drill is used to make a 60° countersunk hole in workpieces for lathes and grinders.
4. A counterbore makes a cylindrical recess concentric with a smaller hole so that a hex head bolt or socket head cap screw can be flush mounted with the surface of a workpiece.
5. The pilot diameter should always be a few thousands of an inch smaller than the hole, but not more than .005 in.
6. Lubrication of the pilot prevents metal to metal contact between it and the hole. It will also prevent the scoring of the hole surface.
7. A general rule is to use approximately one third of the cutting speed when counterboring as when using a twist drill with the same diameter.
8. Feeds and speeds when counterboring are controlled to a large extent by the condition of the equipment, the available power, and the material being counterbored.
9. Spotfacing is performed with a counterbore. It makes a flat bearing surface, square with a hole to seat a nut, washer, or bolt head.
10. Counterboring requires a rigid setup with the workpiece being securely fastened and provisions made to allow the pilot to protrude below the bottom surface of the workpiece.

section g unit 7 reaming in the drill press

SELF-TEST ANSWERS

1. Machine reamers are identified by the design of the shank, either a straight or tapered shank and usually a 45° chamfer on the cutting end.
2. A chucking reamer is a finishing reamer, the fluted part is cylindrical, and the lands are relieved. A rose reamer is a roughing reamer. It can remove a considerable amount of material. The body has a slight back taper and no relief on the lands. All cutting takes place on the chamfered end.
3. A jobbers reamer is a finishing reamer like a chucking reamer but it has a longer fluted body.
4. Shell reamers are more economical to produce than solid reamers, especially in larger sizes.
5. An accurate hole size cannot be obtained without a high quality surface finish.
6. As a general rule the cutting speed used to ream a hole is about one third to one half of the speed used to drill a hole of the same size in the same material.
7. The feed rate, when reaming as compared to drilling the same material, is approximately two to three times as great. As an example, for a 1 in. drill the feed rate is about .010 to .015 in. per revolution. A 1 in. reamer would have a feed rate of between .020 and .030 in.
8. The reaming allowance for a $\frac{1}{2}$ in. diameter hole would be $\frac{1}{64}$ in.
9. Coolants cool the tool and workpiece and act as lubricants.
10. Chatter may be eliminated by reducing the speed, increasing the feed, or using a piloted reamer.
11. Oversized holes may be caused by a bent reamer shank or buildup on the cutting edges. Check also if there is a sufficient amount and if the correct kind of coolant is being used.
12. Bell-mouthed holes are usually caused by a misaligned reamer and workpiece setup. Piloted reamers, bushings, or a floating holder may correct this problem.
13. Surface finish can be improved by decreasing the feed and checking the reaming allowance. Too much or not enough material will cause poor finish. Use a large volume of coolant.
14. Carbide tipped reamers are recommended for long production runs where highly abrasive materials are reamed.
15. Cemented carbides are very hard, but also very brittle. The slightest amount of chatter or vibration may chip the cutting edges.

section h unit 2 toolholders and tool holding for the lathe

SELF-TEST ANSWERS

1. A toolholder is needed to rigidly support and hold a cutting tool during the actual cutting operation. The cutting tool is often only a small piece of high speed steel or other cutting material that has to be clamped in a much larger toolholder in order to be usable.
2. On a left-hand toolholder, when viewed from above, the cutting tool end is bent to the right.
3. For high speed tools the square tool bit hole is angled upward in relation to the base of the toolholder, where it is parallel to the base for carbide tools.
4. Turning close to the chuck is usually best accomplished with a left-hand toolholder.
5. Tool height adjustments on a standard-type toolholder are made by swiveling the rocker in the tool post ring.
6. Quick-change toolholders are adjusted for height with a micrometer collar.
7. Tool height on turret toolholders is adjusted by placing shims under the tool.
8. Toolholder overhang affects the rigidity of a setup; too much overhang may cause chatter.
9. The difference between a standard type toolholder and a quick-change type toolholder is in the speed with which tools can be interchanged. The tools are usually fastened more securely in a quick-change toolholder and height adjustments on a quick-change toolholder do not change the effective back rake angle of the cutting tool.
10. Drilling machine tools are used in a lathe tailstock.
11. The lathe tailstock is bored with a Morse taper hole to hold Morse taper shank tools.
12. When a series of repeat tailstock operations are to be performed on several different workpieces, a tailstock turret should be used.

section h unit 3 cutting tools for the lathe

SELF-TEST ANSWERS

1. High speed steel is easily shaped into the desired shape of cutting tool. It produces better finishes on low speed machines and on soft metals.
2. Its geometrical form: the side and back rake, front and side relief angles, and chip breakers.
3. Unlike single point tools, form tools produce their shape by plunging directly into the work.
4. When a "chip trap" is formed by improper grinding on a tool, the chip is not able to clear the tool; this prevents a smooth flow across the face of the tool. The result is tearing of the surface on the workpiece and a possible broken tool.
5. Some tool holders provide a built-in back rake of about 16°; to this is added any back rake on the tool to make a total back rake that is excessive.
6. A zero rake should be used for threading tools. A zero to slightly negative back rake should be used for plastics and brass, since they tend to "dig in."
7. The side relief allows the tool to feed into the work material. The end relief angle keeps the tool end from rubbing on the work.
8. The side rake directs the chip flow away from the cut and it also provides for a keen cutting edge. The back rake promotes smooth chip flow and good finishes.
9. The angles can be checked with a tool grinding gage, a protractor, or an optical comparator.
10. Long, stringy chips or those that become snarled on workpieces, tool post, chuck or lathe dog are hazardous to the operator. Chip breakers and correct feeds can produce an ideal chip that does not fly off but will simply drop to the chip pan and is easily handled.
11. Chips can be broken up by using coarse feeds and maximum depth of cuts for roughing cuts and by using tools with chipbreakers on them.
12. Overheating a tool causes small cracks to form on the edge. When a stress is applied, as in a roughing cut, the tool end may break off.

section h unit 4 lathe spindle tooling

SELF-TEST ANSWERS

1. The lathe spindle is a hollow shaft that can have one of three mounting devices machined on the spindle nose. It has an internal Morse taper that will accommodate centers or collets.
2. The spindle nose types are the threaded, long taper key drive, and the camlock.
3. The independent chuck is a four-jaw chuck in which each jaw can move separately and independently of the others. It is used to hold odd-shaped workpieces.
4. The universal chuck is most often a three-jaw chuck although they are made with more or less jaws. Each jaw moves in or out by the same amount when the chuck wrench is turned. They are used to hold and quickly center round stock.
5. Combination chucks and Adjust-Tru three-jaw chucks.
6. A drive plate.
7. The live center is made of soft steel so it can be turned to true it up if necessary. It is made with a Morse taper to fit the spindle taper or special sleeve if needed.
8. A hardened drive center is serrated so it will turn the work between centers, but only light cuts can be taken. Face drives use a number of driving pins that dig into the work and are hydraulically compensated for irregularities.
9. Workpieces and fixtures are mounted on face plates. These are identified by their heavy construction and the T slots. Drive plates have only slots.
10. Collet chucks are very accurate workholding devices. Spring collets are limited to smaller material and to specific sizes.

section h unit 5 operating the machine controls

SELF-TEST ANSWERS

1. Very low speeds are made possible by disengaging the spindle by pulling out the lockpin and engaging the back gear.
2. The varidrive is changed while the motor is running, but the back gear lever is only shifted with the motor off.
3. Levers that are located on the headstock can be shifted in various arrangements to select speeds.
4. The feed reverse lever.
5. These levers are used for selecting feeds or threads per inch for threading.
6. The quick approach and return of the tool; this is used for delicate work and when approaching a shoulder or chuck jaw.
7. Since the cross feed is geared differently (about one third of the longitudinal feed), the outside diameter would have a coarser finish than the face.
8. The half-nut lever is used only for threading.
9. They are graduated in English units. Some metric conversion collars are being made and used that read in both English and metric units at the same time.
10. You can test with a rule and a given slide movement such as .125 or .250 in. If the slide moves one half that distance, the lathe is calibrated for double depth and reads the same amount as that taken off the diameter.

section h unit 6 facing and center drilling

SELF-TEST ANSWERS

1. A lathe board is placed on the ways under the chuck and the chuck is removed, since it is the wrong chuck to hold rectangular work. The mating parts of an independent chuck and the lathe spindle are cleaned and the chuck is mounted. The part is roughly centered in the jaws and adjusted to center by using the back of a toolholder or a dial indicator.

2. The tool should be on the center of the lathe axis.
3. No. The resultant facing feed would be approximately 0.3 to 0.5 × .012 in., which would be a finishing feed.
4. The compound must be swung to either 30° or 90° so that the tool can be fed into the face of the work by a measured amount. A depth micrometer or a micrometer caliper can be used to check the trial finish cut. A right-hand facing tool is used for shaft ends. It is different from a turning tool in that its point is only a 58° included angle to fit in the narrow space between the shaft face and the center.
6. RPM $= \dfrac{300 \times 4}{4} = 300$

7. Center drilling is done to prepare work for turning between centers and for spotting workpieces for drilling in the lathe.
8. Round stock is laid out with a center head and square or rectangular with diagonal lines. Stock can be held vertically in a vise or angle plates and vee blocks for center drilling in a drill press.
9. Center drills are broken as a result of feeding the drill too fast and having the lathe speed too slow. Other causes result from having the tailstock off center, the work off center in a steady rest, or lack of cutting oil.
10. The sharp edge provides a poor bearing surface and soon wears out of round, causing machining problems such as chatter.

section h unit 7 turning between centers

SELF-TEST ANSWERS

1. A shaft between centers can be turned end for end without loss of concentricity and it can be removed from the lathe and returned without loss of synchronization between thread and tool. Cutting off between centers is not done as it would break the parting tool and ruin the work. Steady rest work is not done with work mounted in a center in the headstock spindle.
2. The other method is where the workpiece is held in a chuck on one end and in the tailstock center on the other end.
3. Coarser feeds, deeper cuts, and smaller rake angles all tend to increase chip curl. Chip breakers also make the chip curl.
4. Dead centers are hardened 60° centers that do not rotate with the work but require high pressure lubricant. Ball bearing centers turn with work and do not require lubricant. Pipe centers turn with the work and are used to support tubular material.
5. With no end play in the workpiece and the bent tail of the lathe dog free to click against the sides of the slot.
6. Because of expansion of the workpiece from the heat of machining, it tightens on the center, thus causing more friction and more heat. This could ruin the center.
7. Excess overhang promotes lack of rigidity. This causes chatter and tool breakage.

8. RPM $= \dfrac{90 \times 4}{1\frac{1}{2}} = 360 \div \dfrac{3}{2} = 360 \times \dfrac{2}{3} = 240$ or $\dfrac{360}{1.5} = 240$

9. The spacing would be .010 in. as the tool moves that amount for each revolution of the spindle.
10. The feed rate for roughing should be $\frac{1}{5}$ to $\frac{1}{10}$ as much as the depth of cut. This should be limited to what the tool, workpiece, or machine can stand without undue stress.
11. For most purposes where liberal tolerances are allowed, .015 to .030 in. can be left for finishing. When closer tolerances are required, two finish cuts are taken with .005 to .010 in. left for the last finish cut.
12. After roughing is completed, .015 to .030 in. is left for finishing. The diameter of the workpiece is checked with a micrometer and the remaining amount is dialed on the cross feed micrometer collar. A short trial cut is taken and the lathe is stopped. This diameter is again checked. If the diameter is within tolerance, the finish cut is taken.

section h unit 8 alignment of the lathe centers

SELF-TEST

1. The workpiece becomes tapered.
2. The workpiece is tapered with the small end at the tailstock.

3. By the witness mark on the tailstock, by using a test bar, and by taking a light cut on a workpiece and measuring.
4. The dial indicator.
5. With a micrometer. The tailstock is set over with a dial indicator.

section h unit 9 drilling, boring, reaming, knurling, recessing, parting and tapping in the lathe

SELF-TEST ANSWERS

1. Drilled holes are not sufficiently accurate for bores in machine parts as they would be loose on the shaft and would not run true.
2. The workpiece is center spotted with a center drill at the correct RPM and, if the hole is to be more than a $\frac{3}{8}$ in. diameter, a pilot drill is put through. Cutting oil is used. The final size drill is put through at a slower speed.
3. The chief advantage of boring in the lathe is the bore runs true with the center line of the lathe and the outside of the workpiece, if the workpiece has been set up to run true (with no runout). This is not always possible when reaming bores that have been drilled, since the reamer follows the eccentricity or runout of the bore.
4. Ways to eliminate chatter in a boring bar are:
   a. Shorten the bar overhang, if possible.
   b. Reduce the spindle speed.
   c. Make sure the tool is on center.
   d. Use as large a diameter bar as possible without binding in the bore.
   e. Reduce the nose radius on the tool.
   f. Apply cutting oil to the bore.
   g. Use tuned or solid carbide boring bars.
5. Through boring is making a bore the same diameter all the way through the part. Counterboring is making two or more diameters in the same bore, usually with 90° or square internal shoulders. Blind holes are bores that do not go all the way through.
6. Grooves and thread relief are made in bores by means of specially shaped or ground tools in a boring bar.
7. A floating reamer holder will help to eliminate the bell mouth, but it does not remove the runout.
8. Hand reamers produce a better finish than machine reaming.
9. Cutting speeds for reaming are *one-half* that used for drilling; feeds used for reaming are *twice* that used for drilling.
10. Large internal threads are produced with a boring tool. Heavy forces are needed to turn large hand taps, so it is not advisable to use large taps in a lathe.
11. A tap drill can be used as a reamer by first drilling with a drill that is $\frac{1}{32}$ to $\frac{1}{16}$ in. undersized. This procedure assures a more accurate hole size by drilling.
12. A spiral point tap works best for power tapping.
13. The variations in pitch of the hand cut threads would cause the micrometer collar to give erroneous readings. A screw used for this purpose and for most machine parts must be threaded with a tool guided by the leadscrew on the lathe.
14. Thread relief and external grooves are produced by specially ground tools that are similar to internal grooving tools except that they have less end relief. Parting tools are often used for making external grooves.
15. Parting tools tend to seize in the work, especially with deep cuts or heavy feeds. Without cutting oil seizing is almost sure to follow with the possibility of a broken parting tool and misaligned or damaged work.
16. You can avoid chatter when cutting off with a parting tool by maintaining a rigid setup and keeping enough feed to produce a continuous chip, if possible.
17. Knurling is used to improve the appearance of a part, to provide a good gripping surface, and to increase the diameter of a part for press fits.
18. Ordinary knurls make a straight or diamond pattern impression by displacing the metal with high pressures.
19. When knurls produce a double impression, they can be readjusted up or down and moved to a new position. Angling the toolholder 5° may help.
20. You can avoid producing a flaking knurled surface by stopping the knurling process when the diamond points are almost sharp. Also use a lubricant while knurling.

section h unit 10 sixty degree thread information and calculations

SELF-TEST ANSWERS

1. The sharp V thread can be easily damaged while handling if it is dropped or allowed to strike against a hard surface.
2. The pitch is the distance between a point on a screw thread to a corresponding point on the next thread measured parallel to the axis. "Threads per inch" is the number of threads in one inch.
3. American National Standard and Unified Standard threads both have the 60° included angle and are both based on the inch measure with similar pitch series. The depth of the thread and the classes of thread fits are different in the two systems.
4. To allow for tolerancing of external and internal threads to promote standardization and interchangeability of parts.
5. This describes a diameter of $\frac{1}{2}$ in. 20 threads per inch, and Unified coarse series external thread with a class 2 thread tolerance.
6. The flat on the end of the tool for 20 threads per inch should be P = .050 × .144 = .007 in. for Unified threads and P = .050 × .125 = .006 in. for American National threads.
7. The compound at 30° will move in $\frac{.708}{20}$ = .0354 in. for Unified threads.

   The compound will move in $\frac{.75}{20}$ = .0375 in. for American National threads.

8. The fit of the thread refers to classes of fits and tolerances, while percent of thread refers to the actual minor diameter of an internal thread, a 100 percent thread being full depth internal threads.
9. The Systéme International (SI) thread and the British Standard ISO Metric Screw Threads are two metric thread systems in use.
10. In metric tolerance symbols, smaller numbers refer to smaller tolerances.

section h unit 11 cutting unified external threads

SELF-TEST ANSWERS

1. A series of cuts are made in the same groove with a single point tool by keeping the same ratio and relative position of the tool on each pass. The quick-change gearbox allows choices of various pitches or leads.
2. The chips are less likely to bind and tear off when feeding in with the compound set at 29° and the tool is less likely to break.
3. The 60° angle on the tool is checked with a center gage or optical comparator.
4. The number of threads per inch can be checked with a screw pitch gage or by using a rule and counting the threads in one inch.
5. A center gage is used to align the tool to the work.
6. No. The carriage is moved by the thread on the leadscrew when the half-nuts are engaging it.
7. Even numbered threads may be engaged on the half-nuts at any line and odd numbered threads at any numbered line. It would be best to use the same line every time for fractional numbered threads.
8. The spindle should be turning slow enough for the operator to maintain control of the threading operation, about one-fourth turning speeds usually.
9. The leadscrew rotation is reversed, which causes the cut to be made from the left to the right. The compound is set at 29° to the left. The threading tool and lathe settings are set up in the same way as for cutting right-hand threads.
10. Picking up the thread or resetting the tool is a procedure that is used to position a tool to existing threads.

section h unit 12 cutting unified internal threads

SELF-TEST ANSWERS

1. The minor diameter of the thread.
2. By varying the bore size, usually larger than the minor diameter. This is done to make tapping easier.
3. 75 percent.
4. A drill just under the tap drill size should first be used; thus the tap drill acts as a reamer.
5. $.500 - \left(\dfrac{.65}{13} \times 2 \times .6\right) = .440$

   Since the nearest fractional drill size is $\frac{7}{16}$ in. $= .4375$, the tap drill for 60% threads is $\frac{7}{16}$ in.
6. Large internal threads of various forms can be made, and the threads are concentric to the axis of the work.
7. To the left of the operator.
8. A screw pitch gage should be used.
9. Boring bar and tool deflection cause the threads to be under size from the calculations and settings on the micrometer collars.
10. The minor diameter equals $D - (P \times .541 \times 2)$.
   $\frac{1}{8}$ in. $= .125$ in.
   $d = 1$ in. $- (.125 \times .541 \times 2) = .8648$ or $.865$ in.

section h unit 13 taper turning, taper boring and forming

SELF-TEST ANSWERS

1. Steep tapers are quick release tapers and slight tapers are self-holding tapers.
2. Tapers are expressed in taper per foot, taper per inch, and by angles.
3. Tapers are turned by hand feeding the compound slide, by offsetting the tailstock and turning between centers, or by using a taper attachment. A fourth method is to use a tool that is set to the desired angle and form cut the taper.
4. No. The angle on the workpiece would be the included angle, which is twice that on the compound setting. The angle on the compound swivel base is the angle with the work centerline.
5. The reading at the lathe centerline index would be 55°, which is the complementary angle.

6. Offset $= \dfrac{10 \times (1.125 - .75)}{2 \times 3} = \dfrac{3.75}{6} = .625$ in.

7. Four methods of measuring the offset on the tailstock are centers and a rule, witness marks and a rule, the dial indicator, and toolholder–micrometer dial.
8. The two types of taper attachments are the plain and the telescopic. Internal and external and slight to fairly steep tapers can be made. Centers remain in line, and power feed is used for good finishes.
9. The taper plug gage and the taper ring gage are the simplest and most practical means to check a taper. Four methods of measuring tapers are the plug and ring gages; using a micrometer on layout lines; using a micrometer with precision parallels and drill rod on a surface plate; and using a sine bar, gage block, and a dial indicator.
10. Chamfers, V-grooves, and very short tapers may be made by the form tool method.

section h unit 14 using steady and follower rests in the lathe

SELF-TEST ANSWERS

1. When workpieces extend from the chuck more than four or five workpiece diameters and are unsupported by a dead center; when workpieces are long and slender.

2. Since they are useful for supporting long workpieces, heavier cuts can be taken or operations such as turning, threading, and grooving may be performed without chattering. Internal operations such as boring may be done on long workpieces.

3. The steady rest is placed near the tailstock end of the shaft, which is supported in a dead center. The steady rest is clamped to the lathe bed and the lower jaws are adjusted to the shaft finger tight. The upper half of the frame is closed and the top jaw is adjusted with some clearance. The jaws are locked and lubricant is applied.

4. The jaws should be readjusted when the shaft heats up from friction in order to avoid scoring. Also soft materials are sometimes used on the jaws to protect finishes.

5. A center punch mark is placed in the center of the end of the shaft. The lower two jaws on the steady rest are adjusted until the center punch mark aligns with the point of the dead center.

6. No.

7. No. When the surface is rough, a bearing spot must be turned for the steady rest jaws.

8. By using a cat head.

9. A follower rest.

10. The shaft is purposely made one or two inches longer and an undercut is machined on the end to clear the follower rest jaws.

section h unit 15 additional thread forms

SELF-TEST ANSWERS

1. Translating-type screws are mostly used for imparting motion or power and to position mechanical parts.

2. Square, modified square, Acme, stub Acme, and Buttress are five basic translating thread forms.

3. Since the pitch for 4 TPI would be .250 in., the depth of thread would be $\frac{P}{2} = \frac{.250}{2} = .125$ in.

4. Since the pitch for 4 TPI would be .250 in., the depth of the Acme thread would be .5P + .010 in. or .5 $\times$ .250 in. + .010 in. = .135 in.

5. General purpose Acme threads bear on the flanks and centralizing Acme threads bear at the major diameter.

6. 29°.

7. The general use for stub Acme threads is where a coarse pitch thread with a shallow depth is required.

8. The modified square thread.

9. The Acme thread form.

10. Buttress threads are used where great forces or pressures are exerted in one direction.

section h unit 16 cutting acme threads on the lathe

SELF-TEST ANSWERS

1. The thread angle.

2. The included angle, the relief angle, and the flat on the end of the tool.

3. Coarse threading on a lathe imposes heavy loads on these parts. Lubrication prior to threading helps to reduce excess wear.

4. The compound is usually set at $14\frac{1}{2}$°. Some machinists prefer to set the compound at 90°.

5. $P = \frac{1}{6} = .1666$ in.; depth − .5P + .010 in. = .93 in.

6. The tool is aligned by using the Acme tool gage. Refer to Figure 6.

7. $P = .1666$ in.; minor diameter = .750 in. − .1666 in. = .583 in.

8. With an Acme tap set.

9. An Acme thread plug gage.

10. Light finishing cuts with a honed tool and a good grade of sulfurized cutting oil will help make good thread finishes. A rigid setup and low speeds will also help.

section h unit 17 using carbides and other tool materials on the lathe

SELF-TEST ANSWERS

1. Tungsten carbide and cobalt.
2. Decreased hardness and increased toughness.
3. Edge or flank wear is considered normal wear.
4. Yes, if it is set at 90° to the axis of the work.
5. No. The cutting edge engagement length is greater than the depth of cut.
6. Antiweld, anticratering for machining steels.
7. Higher red hardness.
8. Increasing the nose radius will give good finishes even with an increased feed rate.
9. Tools are stronger.
10. Chatter may develop between tool and work.
11. Relief is ground just below the cutting edges of the carbide and clearance is ground primarily on the shank of the tool.
12. Aluminum oxide.
13. When considerable wear is evident on carbide tools. Equipment and setups must be very rigid for ceramics.
14. C-2
15. C-5
16. 44-A
17. C-7
18. The stock removal rate is low but very high finishes are obtained with these tools.
19. Very high tool life on some abrasive, difficult to machine materials.
20. Only silicon carbide (or diamond) wheels may be used for grinding carbide tools. The shank is made of steel, however, and clearance may be ground first on an aluminum oxide wheel.

section i unit 1 the vertical spindle milling machine

SELF-TEST ANSWERS

1. The column, knee, saddle, table, ram, and toolhead.
2. The table traverse handwheel and the table power feed.
3. The cross traverse handwheel.
4. The quill feed hand lever and handwheel.
5. The table clamp locks the table rigidly and keeps it from moving while other table axes are in movement.
6. The spindle brake locks the spindle while tool changes are being made.
7. The spindle has to stop before speed changes from high to low are made.
8. The ram movement increases the working capacity of the toolhead.
9. Loose machine movements are adjusted with the big adjustment screws.
10. The quill clamp is tightened to lock the quill rigidly while milling.

section i unit 2 vertical milling machine operations

SELF-TEST ANSWERS

1. Accurate centering of a cutter over a shaft is done with the machine dials.
2. The feed direction against the cutter rotation assures positive dimensional movement. It also prevents the workpiece from being pulled into the cutter because of any backlash in the machine.
3. End mills can work themselves out of a split collet if the cut is too heavy or when the cutter gets dull.

4. Angular cuts can be made by tilting the workpiece or by tilting the workhead.

5. Circular slots can be milled by using a rotary table or an index head.

6. Squares, hexagons, or other shapes that require surfaces at precise angles to each other are made by using a dividing head.

7. Square holes or other internal hole shapes can be made by using a vertical shaping attachment.

8. With a right angle milling attachment, milling cuts can be made in very inaccessible places.

9. Layout lines are used as guides to indicate where machining should take place. Layouts should be made prior to machining any reference surfaces away.

10. A T-slot cutter only enlarges the bottom part of a groove. The groove has to be made before a T-slot cutter can be used.

section i unit 3 cutting tools for the vertical milling machine

SELF-TEST ANSWERS

1. When viewed from the cutting end, a right-hand cut end mill will rotate counterclockwise.

2. An end mill has to have center cutting teeth to be used for plunge cutting.

3. End mills for aluminum usually have a fast helix angle and also highly polished flutes and cutting edges.

4. Carbide end mills are very effective when milling abrasive or hard materials.

5. Roughing mills are used to remove large amounts of material.

6. Tapered end mills are mostly used in mold or die making to obtain precisely tapered sides on workpieces.

7. Carbide insert tools are used because new cutting edges are easily exposed. They are available in grades to cut most materials and they are very efficient cutting tools.

8. Straight shank mills are held in collets or adapters.

9. Shell end mills are mounted on shell mill arbors.

10. Quick change tool holders make presetting of a number of tools possible and tools can be changed with a minimum loss of time.

section i unit 4 setups on the vertical milling machine

SELF-TEST ANSWERS

1. Workpieces can be aligned on a machine table by measuring their distance from the edge of the table, by locating against stops in the T-slots, or by indicating the workpiece side.

2. To align a vise on a machine table, the solid vise jaw needs to be indicated.

3. Tool head alignment is checked when it is important that machining takes place square to the machine table.

4. When the knee clamping bolts are loose, the weight of the knee makes it sag. But when the knee clamps are tightened, the knee is pulled into its normal position in relation to the column.

5. When the toolhead clamping bolts are tightened, it usually produces a small change in the toolhead position.

6. A machine spindle can be located over the edge of a workpiece with an edge finder or with the aid of dial indicator.

7. The spindle axis is one-half of the tip diameter away from the workpiece edge when the tip walks off sideways.

8. An offset edge finder works best at 600 to 800 RPM.

9. To eliminate the effect of backlash, always position from the same direction.

10. The center of a hole is located with a dial indicator mounted in the machine spindle.

section i unit 5 using end mills

SELF-TEST ANSWERS

1. Lower cutting speeds are used to machine hard materials, tough materials, abrasive material, on heavy cuts, and to get maximum tool life.
2. Higher cutting speeds are used to machine softer materials, to obtain good surface finishes, with small diameter cutters, for light cuts, on frail workpieces and on frail setups.
3. The calculated RPM is a starting point and may change depending on conditions illustrated in the answers to Problems 1 and 2.
4. Cutting fluids are used with HSS cutters except on materials such as cast iron, brass, and many nonmetallic materials.
5. Cutting with carbide cutters is performed without a cutting fluid, unless a steady stream of fluid can be maintained at the cutting edge of the tool.
6. The thickness of the chips affects the tool life of the cutter. Very thin chips dull a cutting edge quickly. Too thick chips cause tool breakage or the chipping of the cutting edge.
7. The depth of cut for an end mill should not exceed one-half of the diameter of the cutter.
8. Limitations on the depth of cut for an end mill are the amount of material to be removed, the power available, and the rigidity of the tool and setup.
9. The RPM for a $\frac{3}{4}$ in. diameter end mill to cut brass is:

$$\text{RPM} = \frac{\text{CS } 4}{\text{D}} = \frac{200 \times 4}{\frac{3}{4}} = 1067 \text{ RPM}$$

10. The feedrate for a two flute, $\frac{1}{4}$ in. diameter carbide end mill in medium alloy steel is:

$$\text{feedrate} = f \times \text{RPM} \times n$$
$$f = .0005$$
$$\text{RPM} = \frac{\text{CS} \times 4}{\text{D}} = \frac{150 \times 4}{\frac{1}{4}} = 2400$$
$$n = 2$$
$$\text{Feedrate} = .0005 \times 2400 \times 2 = 2.4 \text{ IPM}$$

section i unit 6 using the offset boring head

SELF-TEST ANSWERS

1. An offset boring head is used to produce standard and nonstandard size holes at precisely controllable hole locations.
2. Parallels raise the workpiece off the table or other workholding device to allow through holes to be bored.
3. Unless the locking screw is tightened after toolslide adjustments are made, the toolslide will move during the cutting operation, resulting in a tapered or odd-sized hole.
4. The toolslide has a number of holes so that the boring tool can be held in different positions in relation to the spindle axis for different size bores.
5. It is important that you know if one graduation is one-thousandths of an inch or two-thousandths of an inch in hole size change.
6. The best boring tool to use is the one with the largest diameter that can be used and the one with the shortest shank.
7. It is very important that the cutting edge of the boring tool is on the centerline of the axis of the toolslide. Only in this position are the rake and clearance angles correct as ground on the tool.
8. The hole size obtained for given amount of depth of cut can change depending on the sharpness of the tool, the amount of tool overhang (boring bar length), and the amount of feed per revolution.
9. Boring tool deflection changes when the tool gets dull, the depth of cut increases or decreases, or the feed is changed.
10. The cutting speed is determined by the kind of tool material and the kind of work material, but boring vibrations set up through an unbalanced cutting tool or a very long boring bar may require a smaller than calculated RPM.

section j unit 2 types of spindles, arbors, and adaptors

SELF-TEST ANSWERS

1. Face mills over 6 in. in diameter.
2. The two classes of taper are self-holding, with a small included angle, and self-releasing, with a steep taper.
3. 3½ IPF.
4. Any small nick or chip between shank and socket or between spacers will cause the cutter to run out and will mar these contact surfaces.
5. Where small diameter cutters are used, on light cuts, and where little clearance is available.
6. As close to the cutter as the workpiece and workholding device permit.
7. A Style C arbor is a shell end mill arbor.
8. Tightening or loosening an arbor nut without the arbor support in place will bend or spring the arbor.
9. To increase the range of cutters that can be used on a milling machine with a given size spindle socket.
10. Arbor extensions beyond the outer arbor support are often the cause of vibration and chatter.

section j unit 3 arbor-driven milling cutters

SELF-TEST ANSWERS

1. Profile sharpened cutters and form relieved cutters.
2. Light duty plain milling cutters have many teeth. They are used for finishing operations. Heavy duty plain mills have few but coarse teeth, designed for heavy cuts.
3. Plain milling cutters do not have side cutting teeth. This would cause extreme rubbing if used to mill steps or grooves. Plain milling cutters should be wider than the flat surface they are machining.
4. Side milling cutters, having side cutting teeth, are used when grooves are machined.
5. Straight tooth side mills are used only to mill shallow grooves because of their limited chip space between the teeth and their tendency to chatter. Stagger tooth mills have a smoother cutting action because of the alternate helical teeth; more chip clearance allows deeper cuts.
6. Half side milling cutters are efficiently used when straddle milling.
7. Metal slitting saws are used in slotting or cut off operations.
8. Gear tooth cutters and corner rounding cutters.
9. To mill V-notches, dovetails, or chamfers.
10. A right-hand cutter rotates counterclockwise when viewed from the outside end.

section j unit 4 setting speeds and feeds on the horizontal milling machine

SELF-TEST ANSWERS

1. Cutting speed is the distance a cutting edge of a tool travels in one minute. It is expressed in feet per minute (FPM).
2. Starting a cut at the low end of the speed range will save the cutter from overheating.
3. Carbide tools are operated at two to six times the speed of HSS tools. If a HSS tool is 100 FPM, the carbide tool would be 200 to 600 FPM.
4. Too low a cutting speed is inefficient because a cutter can do more work in a given time period.
5. The feedrate on a milling machine is given in inches per minute (IPM).
6. The feedrate is the product of RPM times the number of teeth of the cutter times the feed per tooth.
7. Feed per revolution does not consider the number of teeth on different cutters.
8. Too low a feedrate causes the cutter to rub and scrape the surface of the work instead of cut. Because of the high friction, the tool will dull quickly.

9. $RPM = \dfrac{CS \times 4}{D} = \dfrac{60 \times 4}{3} = 80$ RPM.

10. $RPM = \dfrac{CS \times 4}{D} = \dfrac{150 \times 4}{4} = 150$ RPM.

Feed per tooth = .003 in.

Number of teeth = 5

Feedrate = RPM × feed per tooth × number of teeth = 150 × .003 × 5 = 2.25 = 2¼ IPM.

section j unit 5 workholding and locating devices on the milling machine

SELF-TEST ANSWERS

1. A clamping bolt should be close to the workpiece and a greater distance away from the support block.
2. Finished surfaces should be protected with shims from being marked by clamps or rough vise jaws.
3. Screwjacks are used to support workpieces or to support the end of a clamp.
4. A stop block prevents a workpiece from being moved by cutting pressure.
5. Quick action jaws are two independent jaws mounted anywhere on a machine table to form a custom vise.
6. Swivel vise has one movement in a horizontal plane, where a universal vise swivels both horizontally and vertically.
7. All-steel vises are used to hold rough workpieces such as castings or forgings.
8. A rotary table is used to mill gears, circular grooves, or for angular indexing.
9. A dividing head is used to divide accurately the circumference of a workpiece into any number of equal divisions.
10. A fixture is used when a great number of pieces have to be machined in exactly the same way, and when the cost of making the fixture can be justified in savings resulting from its use.

section j unit 6 plain milling on the horizontal milling machine

SELF-TEST ANSWERS

1. Because the movable jaw is not solidly held. It can move and swivel slightly to align itself with the work to some extent.
2. The dial indicator is more accurate. It will show you the amount of misalignment when you make adjustments, you can see when alignment is achieved.
3. Keys are designed to align a vise or other attachments with the T-slots on a machine table.
4. No, indicating the table from the column only measures the table sliding in its ways, but it does not show if it travels parallel to the column.
5. Yes, if possible the cutting tool pressure should be against the solid jaw. This makes the most rigid setup.
6. The smallest diameter cutter that will do the job will be the most efficient, because it requires a shorter movement to have the workpiece move clear of the cutter.
7. In conventional milling, the cutting pressure is against the feed direction and also up, where in climb milling the cutting pressure is down and the cutter tends to pull the workpiece under itself.
8. A good depth for a finish cut is .015 to .030 in. Less than .010 in. makes the cutter rub, which causes rapid wear.
9. Cutting vibrations and cutting pressures may make the table move when it should be rigidly clamped.
10. When a revolving cutter is moved over a just machined surface, it will leave tool marks.

section j unit 7 using side milling cutters on the horizontal milling machine

SELF-TEST ANSWERS

1. Full side milling cutters are used to cut slots and grooves and where contact on both sides of the cutter is made.
2. Half side milling cutters make contact on one side only, as in straddle milling where a left-hand and a right-hand cutter are combined to cut a workpiece to length.
3. The best cutter is the one with the smallest diameter that will work, considering the clearance needed under the arbor support.
4. Usually a groove is wider than the cutter.
5. A layout shows the machinist where the machining is to take place. It helps in preventing errors.
6. Accurate positioning is done with the help of a paper feeler strip. Often adequate accuracy is achieved by using a steel rule or by aligning by sight with the layout lines.
7. If a workpiece is measured while it is clamped in the machine, additional cuts can be made without additional setups being made.
8. Shims and spacers control workpiece width in straddle milling operations.
9. The diameters of the individual cutters determine the relationship of the depth of the steps in gang milling.
10. Interlocking side mills are used to cut slots over 1 in. wide and also when precise slot width is to be produced.

section k unit 1 dividing head and rotary table

SELF-TEST ANSWERS

1. When accurate spacings are made as with gears, splines, keyways, or precise angular spacings.
2. The use of a worm and worm wheel unit.
3. To make the spindle freewheeling and when direct indexing.
4. For direct indexing.
5. The most common ratio is 40:1.
6. Different hole circles are used to make precise partial revolutions with the index crank.
7. The sector arms are set to the number of holes that the index crank is to move. They eliminate the counting of these holes for each spacing.
8. The spindle lock is tightened after each indexing operation and before a cut is made to prevent any rotary movement of the spindle.
9. With the use of high number index plates or with a wide range divider.
10. When the rotation of the index crank is reversed, the backlash between the worm and worm wheel affects the accuracy of the spacing.

section k unit 2 direct and simple indexing

SELF-TEST ANSWERS

1. Direct indexing is performed from index holes on the spindle nose. Simple indexing uses the worm and worm wheel drive and the side index plate.
2. Use a marking pen or layout dye to mark the holes to be used in direct indexing.
3. Twenty-four index holes let you make equal divisions of 2, 3, 4, 6, 8, 12, and 24 parts.
4. The sector arms can be adjusted so that the number of holes to be indexed plus one are between the beveled edges.
5. For the highest degree of accuracy use the largest possible hole circle.
6. $\frac{40}{6} = \frac{20}{3} = 6\frac{2}{3}$ turns. Use the 57 hole circle for the $\frac{2}{3}$ turn. $\frac{2}{3} \times \frac{19}{19} = \frac{38}{57}$. The fraction is 38 holes in the 57 hole circle.
7. $\frac{40}{15} = \frac{8}{3} = 2\frac{2}{3}$ turns. Use the same hole circle as for Problem 6 above. Two turns and $\frac{38}{57}$ turn.

8. $\frac{40}{25} = \frac{8}{5} = 1\frac{3}{5}$ turns. $\frac{3}{5} \times \frac{6}{6} = \frac{18}{30}$. One turn and 18 holes in the 30 hole circle.

9. $\frac{40}{47}$. There is a 47 hole circle, so use 40 holes in the 47 hole circle.

10. $\frac{40}{64} = \frac{5}{8}$ turns. The only hole circle divisible by 8 is 24. So, $\frac{5}{8} \times \frac{3}{3} = \frac{15}{24}$. = 15 holes in the 24 hole circle.

section k unit 3 angular indexing

SELF-TEST ANSWERS

1. 15 degrees.
2. Three holes.
3. Nine degrees.
4. All those hole circles are divisible by 9.
5. $\frac{17}{9} = 1\frac{8}{9}$ turn of the index crank.
6. 30 min or $\frac{1}{2}$ degrees.
7. 15 min or $\frac{1}{4}$ degrees.
8. 10 min or $\frac{1}{6}$ degrees.
9. 540 min.
10. Converting 54°30′ into minutes = 3270 min.

$$\frac{\text{Required minutes}}{540} = \frac{3270}{540} = 6\frac{30}{540} = 6\frac{3}{54} = 6 \text{ turns}$$

and 3 holes in the 54 hole circle.

section l unit 1 introduction to gears

SELF-TEST ANSWERS

1. Spur gears and helical gears.
2. Helical gears run more smoothly than spur gears because more than one tooth is in mesh at all times. Helical gears, however, generate axial thrust that has to be offset with thrust bearings.
3. The pinion in mesh with an internal gear rotates in the same direction as the gear.
4. The shafts will be at 90 degrees to each other.
5. 50:1.
6. No, a single start worm can only be replaced with a single start worm. To change the gear ratio, the number of teeth in the gear need to be changed.
7. Hardened steel.
8. Nonmetallic materials.
9. By making the pinion harder than the gear.
10. Nonferrous materials.

section l unit 2 spur gear terms and calculations

SELF-TEST ANSWERS

1. Pressure angles on gears vary from 14½ to 20 to 25 degrees.
2. Larger pressure angles make stronger teeth. They also allow gears to be made with fewer teeth.
3. $C = \dfrac{N_1 + N_2}{2P} = \dfrac{20 + 30}{2 \times 10} = \dfrac{50}{20} = 2.500$ in.

4. $C = \dfrac{D_1 + D_2}{2} = \dfrac{3.500 + 2.500}{2} = \dfrac{6.000}{2} = 3.000$ in.

5. The whole depth of a tooth is how deep a tooth is cut. The working depth gives the distance the teeth from one gear enter the opposing gear in meshing.

6. The addendum is above the pitch diameter and the dedendum is below the pitch diameter of a gear.

7. $D_o = \dfrac{N + 2}{P} = \dfrac{50 + 2}{5} = \dfrac{52}{5} = 10.400$ in.

$t = \dfrac{1.5708}{P} = \dfrac{1.5708}{5} = .314$ in.

8. $P = \dfrac{N}{D} = \dfrac{36}{3} = 12$

9. $D_o = \dfrac{N + 2}{P} = \dfrac{40 + 2}{8} = \dfrac{42}{8} = 5.250$ in.

$h_t = \dfrac{2.250}{P} = \dfrac{2.250}{8} = .2812$ in.

$D = \dfrac{N}{P} = \dfrac{40}{8} = 5.000$ in.

$b = \dfrac{1.250}{P} = \dfrac{1.250}{8} = .1562$ in.

10. $D_o = \dfrac{N + 2}{P} = \dfrac{48 + 2}{6} = \dfrac{50}{6} = 8.3333$ in.

$c = \dfrac{.157}{P} = \dfrac{.157}{6} = .0261$ in.

$h_t = \dfrac{2.157}{P} = \dfrac{2.157}{6} = .3595$ in.

$t = \dfrac{1.5708}{P} = \dfrac{1.5708}{8} = .2618$ in.

$D = \dfrac{N}{P} = \dfrac{48}{6} = 12.000$ in.

section 1 unit 3 cutting a spur gear

SELF-TEST ANSWERS

1. Gears cut on a milling machine lack a high degree of accuracy and are expensive to make.
2. Eight.
3. No. 3.
4. No, the tooth profile differs with different pressure angles.
5. For a 17 tooth gear.
6. The number of the cutter, the diametral pitch, the number of teeth the cutter can cut, the pressure angle, and the whole depth of tooth.
7. To get a setup as rigid as possible.
8. The number of holes to be indexed plus one.
9. To check the correctness of the number of spaces required.
10. A center rest is used under gear blanks, mandrels, or shafts to help prevent chatter and deflection.

section l unit 4 gear inspection and measurement

SELF-TEST ANSWERS

1. Gear measurements with a gear tooth vernier caliper and with a micrometer over two wires or pins.
2. The chordal tooth thickness.
3. The chordal addendum.
4. The chordal addendum.
5. The circular thickness.
6. By calculation or from *Machinery's Handbook* tables.
7. No, the pin sizes are specifically calculated for each differing diametral pitch.
8. By consulting a *Machinery's Handbook* table.
9. The dimension given is divided by the diametral pitch used; in this question, the divisor is 12.
10. The optical comparator magnifies the gear tooth profile and projects it on a screen. On the screen there is also a transparent drawing with the tooth profile; by aligning the shadow with the drawn profile, any variation between the two can be seen and measured.

section m unit 1 features and tooling on the horizontal shaper

SELF-TEST ANSWERS

1. A. Tilt table
   B. Apron
   C. Cross rail
   D. Crossfeed engagement lever
   E. Rail elevating crank
   F. Stroke adjusting shaft
   G. Ram
   H. Ram adjusting shaft
   I. Tool (swivel) head
   J. Tool lifter
2. It permits the cutting tool to tilt up on the return (feeding) stroke of the ram without damaging the cutting edge.
3. The crank shaper uses a crank gear with a movable pivot that drives a rocker arm connected to the ram. The return stroke is faster than the cutting stroke by about 3:2. The hydraulic shaper uses high pressure oil to drive a piston connected to the ram. The cutting and return rates are independently set by the operator.
4. The feeding of the table is done by a ratchet mechanism that is driven by a cam or adjustable eccentric. On many larger shapers, vertical feed can be selected as an alternative to table crossfeed.
5. The size of a shaper is the maximum tool movement that can be obtained. Shapers up to about 16 in. can be expected to hold a cubic workpiece of the shaper's specified size. For larger sizes, vertical height is essentially limited to 16 in.
6. The single screw vise, usually with a swivel base and the two or multiple screw vises for out-of-parallel parts, with or without a swivel base.
7. The use of a stop to arrest end motion and poppets or bunters and toe dogs to secure the part down to the table.
8. The fixture permits more rapid location and holding of the workpiece than direct attachment to the machine table.
9. Shaper and planer tools prepared for cast iron usually have less side and back rake angles than those prepared for steel.
10. Placing the tool on the ram side is better for roughing cuts because it permits the tool to deflect without increasing the load. Where visibility is more important and cuts are light, it is acceptable to place the tool on the front side.

section m unit 2 cutting factors on shapers

SELF-TEST ANSWERS

1. Negative back and side rake give better support beneath the cutting edge, so that at the instant of tool contact with the workpiece there is a lessened tendency for tool breakage.
2. M-2 and M-3 high speed steels are typically recommended. For carbides, the C-2 grade is recommended for cast iron and nonferrous materials and the C-6 grade for steel cutting.
3. The large amount of tool contact on planer finishing tools can lead to chatter at higher speeds.
4. *Machining Data Handbook* is the chief source of useful data for speeds and feeds on planer (and shaper) applications.
5. The depth of cut and the feed rate should be as much as can be tolerated by the weakest link of cutting tool, workholding, part strength, or machine tool rigidity and power, consistent with the required workpiece surface characteristics.

7. $\dfrac{CS \times 7}{L} = \dfrac{12 \times 7}{15 + 1} = \dfrac{84}{16} = 5.25$ strokes per minute.

8. $\dfrac{CS \times 7}{L} = \dfrac{60 \times 7}{4\frac{1}{4} + \frac{3}{4}} = \dfrac{420}{5} = 84$ strokes per minute.

9. Deep roughing = depth × 10 percent = .450 × .10 = .045 in.; shallow roughing = depth × $\frac{1}{3}$ to $\frac{1}{2}$ = .100 × .33 to .5 = .033 to .050 in.
10. Finishing feed with square nose finishing tool = approximately three quarter tool width. $\frac{1}{2} \times \frac{3}{4}$ = .50 × .75 = .375 in. desired feedrate.

section m unit 3 using the shaper

SELF-TEST ANSWERS

1. Positive pressure lubrication is usually supplied by a lubricating pump, or from the main hydraulic system.
2. Have your instructor inspect it.
3. At least $\frac{1}{2}$ in. before the cut and $\frac{1}{4}$ in. after (unless you are working to a shoulder).
4. The table should be indicated in the crossfeed direction. The tilt plate should be indicated (if that side is up).

5. $\dfrac{CS \times 7}{L} = \dfrac{35 \times 7}{L + 1} = \dfrac{245}{6.5} = 37.6$ strokes per minute.

6. The line contact provided by the round bar forces the reference surface into full contact with the solid jaw of the vise.
7. With vertical squaring the vise could be out of square considerably without affecting part squareness. When squaring from the end, the squareness of the part is completely dependent on the squareness of the vise.
8. Reversing the work while maintaining the table position assures that both angles will be exactly the same and will intersect at the center of the workpiece.

section n unit 1 selection and identification of grinding wheels

SELF-TEST ANSWERS

1. Use silicon carbide for the bronze, aluminum oxide for the steel, either manufactured or natural diamond for the tungsten carbide inserts, and cubic boron nitride or aluminum oxide for the high speed steel. Cubic boron nitride could be eliminated.

2. For a straight wheel the third dimension is the hole size. On a cylinder wheel the third dimension is the wall thickness.
3. Side grinding: cylindrical (Type 2), straight cup (Type 6), flaring cup (Type 11), and dish (Type 12). Peripheral grinding: straight (Type 1) and dish (Type 12). The saucer or dish wheel is the only shape on which both peripheral and side grinding are rated safe.
4. These cutoff wheels would be rubber-bonded, because they are the only ones that can be made so thin.
5. Ball grinding requires the hardest wheels, then cylindrical grinding (cylinder ground with a peripheral wheel), then flat surface grinding with a peripheral wheel, then internal grinding and, finally, the softest, flat surfacing with a side grinding wheel.
6. Wheel 1 is a resinoid-bonded (B) aluminum oxide (A) wheel, very coarse grit (14), very hard (Z) and dense (3) structure. It is a typical specification for grinding castings. Wheel 2 is a vitrified-bonded (V) silicon carbide (C) wheel, same grit size but with medium (J) grade and structure (6), for grinding some soft metal like copper.
7. Wheel 1 is the peripheral grinding wheel, because it is a harder (K) and denser (8) wheel. Both wheels have the same abrasive, silicon carbide, and the same bond, vitrified. The abrasive in wheel 2 is coarser, 24 grit as against 36, and may be used for side grinding.
8. This is a vitrified aluminum oxide wheel, H grade, 8 structure, in a 46 grit size. The 32 indicates a particular kind of aluminum oxide, and the "BE" a particular vitrified bond.
9. In a straight wheel for flat grinding, you would use a softer (J or I) and more open structure (7 or 8) wheel. With a cup or segmental wheel, you would use a still softer grade, like H. For flat grinding, probably a size coarser grit, 46, would be used. The important thing is to go softer.
10. The material to be ground usually determines the abrasive to be used, although it is not practical to stock cubic boron nitride or diamond wheels unless you have a good deal of work to be done with them.

   Hard materials require fine grit sizes and soft grades. With soft, ductile materials, coarser grit sizes and harder, longer-wearing wheels are needed.

   Except for very hard materials, the rule is that grit sizes are coarser as the amount of stock to be removed becomes greater.

   Generally, the finer the grit, the better the finish.

section n unit 2 care of abrasive wheels: trueing, dressing, and balancing

SELF-TEST ANSWERS

1. Trueing a grinding wheel means making the OD of the wheel concentric with the center of the spindle after the wheel is mounted. This means that the wheel will be a little out-of-round with its own center. On aluminum oxide wheels and silicon carbide wheels this usually means dressing abrasive grain off the periphery of the wheel. On diamond wheels the adjustment is made as close as possible by tapping the wheel into place. Trueing is needed to insure a good finish on the workpiece.
2. Wheels, like automobile tires, must be balanced when they are mounted, particularly on mounting flanges, as is true for many large wheels. In fact, the reasons and the procedure are very similar to the ones for tires. Balancing is required on all wheels over 12 in. in diameter; it is useful on smaller wheels when finish is critical.
3. Trueing is a dimensional job, making the wheel's OD concentric with the machine's spindle. Dressing is resharpening the wheel by removing dull grain and bits of work material. Wheels should be dressed when the grain becomes dulled or loaded (filled with bits of metal), and also when you change from roughing to finishing or back to roughing.
4. Form dressing is simply shaping the grinding surface of a wheel to produce something other than a flat surface on the workpiece. It can be done by crush roll, diamond-plated form block or roll, or single point diamond tools mounted so that it can accurately generate a shape in the wheel face by following a pattern.
5. Diamond and cubic boron nitride wheels have to be mounted to much closer tolerances than other types of wheels. With a common abrasive wheel, after the first mounting and trueing, it is often enough to remount the wheel and dress it for whatever type of work you want to do. With diamond wheels, even though the core is machined to a much better fit than is possible with any abrasive, the maximum runout is about .0005 in., and that takes time to achieve. It is just a fussier job.

6. The essential point in placing the diamond for wheel dressing is that it contact the wheel just a little past the vertical or horizontal center line of the wheel, depending on the placement of the dresser, and that the diamond be pointing in the direction of wheel travel at an angle of about 15 degrees. On most dressers, the angle is taken care of. If in doubt, however, move it about ⅛ in. ahead. If the wheel is rotating clockwise, then the diamond should be at about 12:05 p.m., 3:20 p.m., or 6:35 p.m.

7. The steps in trueing a wheel are:
   a. Assuming the diamond is properly placed, adjust the wheel so that the diamond is touching the high point of the wheel.
   b. Turn on the coolant (if possible) and start the wheel, allowing it to run for about a minute before starting to true the wheel.
   c. Use light infeed, .001 in. or less per pass across the wheel. For trueing, the speed of traverse does not matter.
   d. If dressing is done dry, stop traversing after every three or four passes to allow the diamond to cool.
   e. Continue trueing only until the diamond is contacting the complete grinding surface of the wheel. Going further wastes abrasive, and stopping short of that point means there is still a low point on the grinding surface of the wheel.

8. For roughing, you want the grinding surface to be as sharp and open as possible, so you take quick passes across the wheel. For finish grinding, the passes are slower, not so deep, and you probably will finish with two or three passes without any downfeed or infeed at all.

9. Assuming that bearings and other machine components are running true, if diamond wheel runout is still too much, the wheel must be trued. This can be done with a brake trueing device or, in the case of a resinoid-bonded diamond wheel, by grinding a piece of low carbon steel.

10. Generally, any wheel 12 in. or over in diameter should be balanced the first time it is mounted, particularly when it is mounted on flanges, and any other time it is remounted. If there is a pattern of recurring marks on the finish of the workpiece, then rebalancing may be in order also.

section n unit 3 grinding fluids

SELF-TEST ANSWERS

1. The direct answer to this question is that most grinding is done wet, and the grinder operator is in the best position to see whether the workpieces from the grinder are coming from the grinder in the condition they should. Fluids are necessary to reduce heat and he can observe the grinding operation; with a little experience he can begin to "sense" whether or not the coolant is doing its job. Finally, there are many brands of grinding fluids, and the operator who knows something of their uses can understand why some of the instructions are given. He will not be following them blindly.

2. The three major functions of the grinding fluid are to cool, lubricate, and clean the surface being ground. Cooling is essential to keep the heat from distorting and perhaps warping the workpiece, particularly thin work. It also prevents "burning" or heat discoloration of the ground surface. Lubrication affects the grinding action of the wheel. The quality of the finished surface is directly affected by the lubricating qualities of the fluid. Furthermore, lubrication probably tends to make the wheel act a little softer, so it can have some effect on wheel selection. The third action is cleaning, and the effectiveness here depends on both the volume and the force of the coolant stream. Water based fluids are probably superior in this function.

3. The design and placement of the flood coolant nozzle are important because together they ensure that the coolant penetrates the curtain of air that surrounds the revolving wheel and actually gets to the grinding or cutting point. It makes no difference how much fluid is splashing about. It is the amount that actually gets to the grinding line that counts.

4. Water soluble chemical fluids are the fluids used most with excellent cooling ability, additives that make them satisfactory in lubrication on most grinding jobs, and quite satisfactory in removing grinding swarf. An added benefit is that they are transparent, probably giving the best view of the work of all the fluids. Swarf settles out easily.

5. Water soluble oil grinding fluids are probably most often used on light to medium stock removal jobs. They are on a par with chemical fluids in cooling and cleaning, are slightly better in lubrication, and the swarf settles out easily. The addition of emulsifying agents causes the closest thing to an actual oil-water mixture. The result is a milky solution that cuts down on visibility.

6. Straight oils rate well in lubrication, but low in practically every other comparison. They are used only on the heaviest stock removal jobs, where lubrication is the top requirement that makes up for below average cooling ability, tendency to retain chips instead of allowing them to settle out, and other shortcomings. Oil does a pretty good job of cleaning but, as mentioned above, it is used mainly where its lubricating ability is a must.

7. A mist coolant system is simply one with a small container, a hose with a nozzle equipped so that a fine mist of the coolant is directed against the grinding area. The heat evaporates the mist, so there is nothing to recirculate; and consumption is closer to pints per day than gallons per hour. These systems are used mainly on dry grinders that need an occasional job done with a grinding fluid. These are usually portable and need only an air connection after they are mounted on the machine to be ready to go to work.

8. **a.** Settling

   Advantages. Easy to maintain. Needs only occasional cleaning out of accumulated swarf in the bottom, but usually comes equipped with a "drag-out" cleaner that makes this easy. Virtually no maintenance cost. With water based coolants, it does a good job if the tank is large enough and some recirculation of fines is acceptable.

   Disadvantages. Needs a larger total volume of coolant to allow swarf time to settle and more floor space than almost any other method. Does not work well if coolant volume is down. Needs time when fluid is quite low for the swarf to settle out.

   b. Filtering

   Advantages. Effective if mesh size of filter provides removal of enough swarf to satisfy finish requirements. Floor space and coolant volume requirements moderate. Simple piece of equipment that rarely breaks down, and easy to fix if it does.

   Disadvantages. Filter cloth with small enough mesh to remove very fine particles would also slow down fluid flow too much. Filter cloth can not be reused. On large volumes of swarf, the cost of cloth can become prohibitive.

   c. Magnetic separator

   Advantages. Probably the most positive protection against the recirculation of "tramp iron" in the fluid. Properly installed, it should remove all iron and steel from the fluid.

   Disadvantages. Almost requires use in combination with another separator to remove nonmagnetic materials such as brass or abrasive particles. Not efficient in shops where general grinding, including a large percentage of nonmagnetic metals, are ground.

   d. Centrifugal

   Advantages. Both the cyclonic and the bowl-type centrifugal do a good, fast job of separating all the swarf from the fluid. Removes both ferrous and nonferrous metal and abrasive. Floor space requirement is least for the bowl type, and perhaps moderate for the cyclonic type.

   Disadvantages. For the bowl type, the cleaning of bowls may present a probelm. There must be periodic shutdown, either to change bowls (if there is a spare) or, otherwise, to clean bowls. These are probably the most complicated and expensive of the cleaners. Furthermore, maintenance cost of the bowl types, because of the high speed or rotation, is likely to be high.

9. The principal argument for having the operator involved in the selection of the coolant is probably that he is in the best position to judge how it is working. In a small shop with perhaps six or less operators, the owner is likely to feel that there is not that much difference between water based coolants or oil coolants. Thus, he may not be inclined to buy something that he thinks his operators do not like or feel is effective. In a large shop, it is not usually practical to take a vote on what the operators feel is best. Usually in such operations, which are most often production rather than toolroom-type jobs, there is a person who has specialized in coolants and the plant's products and, in such cases, would make the decision. A good engineer always asks the opinions of operators, however. If the shop has a central coolant system, then someone in management must make the decision.

10. **a.** Make sure that there is always enough fluid in the system.

    b. Add water or concentrate as directed to keep the solution in proper balance.

    c. Check frequently for lubricating oil in the coolant and report it to the proper person.

    d. Check frequently to see that the nozzle is always in proper adjustment and accurately aimed.

section n unit 4 horizontal spindle, reciprocating table surface grinders

SELF-TEST ANSWERS

1. There are several reasons for studying this grinder:
   a. It is a very popular machine in machine shops.
   b. Ability to operate one is expected of most machinists.
   c. The same principles of abrasive and coolant selection and wheel dressing apply to it as much as they do to other machines.
   d. Workholding for many parts is usually simple.
2. These are characteristics of the electromagnetic chuck:
   a. It virtually eliminates clamping of parts.
   b. Ordinarily the entire top surface of the workpiece is available to be ground.
   c. It is probably the fastest way to set up a workpiece for grinding.
   d. It holds iron and steel readily and, with simple blocking, will hold nonferrous work as well.
3. The wheelhead moves up and down in the column and always grinds in a plane parallel to the top of the magnetic chuck and parallel to the front edge of the chuck. The magnetic chuck moves from right to left and left to right (reciprocates), and both toward and away from the operator (crossfeed).
4. It means that the plane of abrasive action or grinding action of the wheel is always parallel to the top of the chuck. Even when the wheel is formed in some way with several hills and valleys, each point across the wheel is always grinding parallel to the top of the chuck.
5. The downfeed is the most critical motion of all, because the amount of material taken off is often measured in "tenths" (ten-thousandths of an inch), sometimes even less. The crossfeed is much less critical; a couple of thousandths one way or the other are not usually critical. The traverse is not critical at all. The wheel must definitely clear the edge of the workpiece, but whether by half an inch or an inch is not critical. More than an inch, however, and you are wasting too much time.
6. Zeroing slip rings make calculating downfeed and crossfeed much easier and less likely to be in error. It eliminates all calculation in determining the total amount of stock to be removed and the amount of crossfeed per pass.
7. Vertical spindle machines that grind on the flat side of the wheel have tiny scratches caused by the abrasive which are usually seen as overlapping circles. The form on a horizontal spindle must run parallel to the path of the wheel as the workpiece passes under it. Of course, if you want to make a form with grooves crossing at right angles, all you have to do is to grind one set, turn the workpiece 90 degrees, and grind the other set.
8. a. Rotary chuck. Turns the grinder into a horizontal spindle, rotating table type.
   b. Centerless grinding attachment. Makes possible centerless grinding of relatively small parts.
   c. Center-type grinding attachment. Similar to centerless grinder.
   d. High speed attachment. Increases spindle speed to a point where internal grinding or any other type requiring a small, very high speed wheel can be used.
9. a. Vacuum chuck. Holds all kinds of work, both magnetic and nonmagnetic. Particularly good on thin work.
   b. Magnetic sine chucks. Provides workholding for surfaces of many different angles that are not parallel.
10. These grinders are very economical of floor space, probably requiring less than any other type. They are probably the most inexpensive grinders although, of course, the price varies with the quality of the grinder. Within the size capacity of the grinder, they can handle, with attachments, a wide variety of both flat and cylindrical work. Finally, they are probably the most suited of all grinders for teaching the principles of grinding.

section n unit 5 workholding on the surface grinder

SELF-TEST ANSWERS

1. An electromagnetic chuck is made up of alternating strips (rectangular) or rings (rotary) of steel and some nonmagnetic metal like brass or lead. Chucks operate on direct current, 24, 110, or 220 V, depending on the size of the machine. This type of construction has proved itself adequate for the various sizes of grinders in

use today. There is a special control that demagnetizes the workpiece quickly; otherwise magnetism remaining in large parts could hold them on the machine, possibly for a long time.

2. While it is said that about three quarters of all work done on surface grinders could be done on a 6 × 12 in. grinder, there is a size limit that is imposed by the dimensions of the chuck. The shape of the workpiece determines how much extra fixturing must be done. If there is a side for chucking parallel to the side to be ground, there is no problem. However, if there is no such side for chucking, then fixtures must be used to hold the surface to be ground parallel to the top face of the chuck. As to weight, the heavier the workpiece, the easier it is to hold. Nonmagnetic materials cannot be held on the chuck without other blocking or fixturing.

3. **a.** Wipe the top of the chuck clean before each use, preferably with a squeegee. Cloth can be used, but there is a possibility of picking up chips of metal that could later cut your hand.
   b. Run your hand lightly and carefully over the chuck to detect burrs or scratches. If any of these are noticed, remove them with either a black granite deburring stone or a fine grit abrasive stone.
   c. The deburring operation should probably be done at the beginning of each shift, also.
   d. Always be careful in placing workpieces or accessories on the chuck. It is a precision instrument, and its top surface should always be treated with care.

4. **a.** Use plenty of coolant.
   b. Use a white (friable) wheel, aluminum oxide, medium grit, medium structure, vitrified bond. (Friable A46-HV was the specification listed in the text.) Dress the wheel "open."
   c. Thinly cover the top of the chuck with layout dye, Prussian blue, or something similar for quick detection of low spots. Continue grinding until all the marking is gone.
   d. Downfeed about two "tenths" per pass. A few passes should ordinarily make the chuck flat again.
   e. Test flatness with a dial indicator mounted on the wheelguard.
   f. Deburr the chuck.

5. The electromagnetic chuck is one of the easiest workholding devices to use in the machine tool field. It is fast; most of the work, that with parallel sides for chucking and grinding, can simply be laid on the chuck top. After the current is turned on, you are often ready to grind. It is much faster than clamping or holding work between centers, for example. It will handle the bulk of the work (any iron or steel) without any added fixturing, if the geometry is right. It cuts setup time considerably.

6. Metals like aluminum and brass are nonmagnetic; for that reason they need to be held in place with either
   a. Steel blocking.
   b. Magnetic vises backed up by steel blocking.
   c. Clamping, preferably to a steel plate.
   d. If thin pieces, can sometimes be held by double-faced tape stuck to the chuck. No power is needed for this.

7. The basic precautions for grinding thin pieces are to use only the power that will hold the work in place and to take light cuts. Excessive power may flatten the work on the chuck, only to let it spring out of flat if rolling stresses are relieved. Too heavy cuts could knock the work off the chuck.

8. **a.** Magnetic sine chuck.
   b. Magnetic parallels or steel blocking of equal height, probably placed crosswise of the chuck.
   c. Requires at least an angle plate and a clamp. Depending on the squareness tolerance and the parallelism of the opposite end, might need a second angle plate and a steel ball under the part and on the chuck.

9. The surface to be ground must be parallel with the top of the chuck and the saddle ways; hence, parallel with the cutting or grinding line of the wheel as the work is traversed beneath it. The surface grinder is essentially a parallel grinder.

   There must be a parallel opposite face for chucking and, if not, the workpiece must be fixtured to make the surface to be ground parallel with the chuck top.

   The fixturing should be kept as simple as it can be to do the job to specifications.

10. Use an auxiliary holding device such as a precision vise or an angle plate. These devices present sufficient area to the magnetic chuck to provide a basic surface for good workholding.

section n unit 7 problems and solutions in surface grinding

SELF-TEST ANSWERS

1. Careful dressing of the wheel, keeping coolant tank full, checking coolant filters, thorough cleaning of chuck before each loading, checking tightness of wheel.
2. A surface defect is anything that is out of pattern; any kind of unwanted scratches or discoloration.

3. Vibration, heat, and dirt, plus poor quality wheel dressing.
4. Vibration is the principal cause. Off-grade wheels and wheels principally too hard.
5. Fishtails are random scratches caused most often by dirt and grit in the coolant. Sliding the work off the chuck might also be considered a cause; and less often, a wheel that is very much too soft. Usually, the starting point in getting rid of them is to check the coolant level and filters.
6. Burning, checking, or discoloration is one. Low spots (out of flatness) is the other. Probably you would check first to see that there was enough coolant and that it was actually getting to the cutting area. Speeding up the table speed, if this is possible, could also help.
7. A burnished surface is one in which the hills of the scratches have been rubbed over into the valleys. It is bad because the surface will not resist wear. Probably the wheel is too hard and you should change it.
8. Poor chuck alignment is probably the first cause. You might also look for dirt or swarf on the chuck or insufficient coolant.
9. Overheating, particularly of thin work, is one. Dressing a wheel too fine could also be a cause. A wheel that is too hard is another cause. In fact, although work can be parallel and not flat, or flat and not parallel, the two conditions usually go together.
10. Mostly because it can not absorb the heat, but also because the grinding may release stresses rolled into the workpieces.

section n unit 8 cylindrical grinders

SELF-TEST ANSWERS

1. The basic difference between the universal grinder and other center-type grinders is that on the universal grinder all major components swivel; certainly the wheelhead and the swivel table do. However, the swivel table is a feature of most grinders of this type.
2. The center-type cylindrical grinder was probably the first developed because it was essentially a matter of replacing the cutting tool on a lathe with a grinding wheel. In fact, the first cylindrical grinders were called grinding lathes.
3. Aside from the fact that surface grinding is flat grinding and cylindrical grinding is the grinding of shafts and similar parts, there are these differences, at least.
   a. Work is more solidly supported on the surface grinder because it is backed up by the magnetic chuck.
   b. On a surface grinder, the reduction in thickness is about the same as the downfeed (infeed) of the wheel; in cylindrical grinding, the reduction in diameter of a part is roughly twice the infeed.
   c. The wheel and the work always move in the same direction with relation to each other in cylindrical grinding. In surface grinding, the work moves alternately with and against the direction of wheel travel.
4. The first step in grinding a taper (or, for that matter, in correcting one you do not want) is to swivel the table to the desired taper. This is assuming that a square or A face is on the wheel. With a universal grinder, when you reach the limit of table swivel, you can also swivel the wheel without redressing the face of the wheel. Otherwise, you would have to dress the wheel to something other than a right angle or A face. Of course, even with a universal grinder you might reach a taper steep enough to require angle dressing of the wheel.
5. Obviously, as indicated in number 4, the steepness or degree of taper must be considered. Most tapers are usually ground by just swiveling the table.
6. In addition to supporting one end of the workpiece, the headstock also supplies the power to rotate it. The headstock can also be modified with accessories to hold work that cannot be held by centers. This might be a part that could be held in a chuck or on a faceplate.
7. Basically, on a center-type cylindrical grinder, the work is supported on two conical work centers that project into two matching conical holes in the ends of the workpiece.
8. A steadyrest is a support to back up a long, thin workpiece against the outward and downward pressure of the grinding wheel. They keep the workpiece from bending. With proper use of steadyrests in odd numbers (one, three, five, etc.) it is possible to support a very thin workpiece so that it will be ground straight. Steadyrests are always used in odd numbers, so that as each pair is added, after the first one, they are placed in the center of the remaining length to be supported.
9. Generally, the greater the variety of work that can be done on a cylindrical grinder, the less its rigidity. The universal grinder, with a swiveling wheelhead, is the least rigid, and the straight plunge grinder, with nothing swiveling, is the most rigid. Also, most workpieces that are plunge-ground are shorter than the thickness of the wheel, which provides a great deal of stiffness.

10. If the center holes are not very accurate and round, it is going to be impossible to make the workpiece they help support accurate and round. In other words, any lack of precision in the center holes will be reflected in the quality of the finished workpiece.

section o unit 1 numerical control dimensioning

SELF-TEST ANSWERS

1. See Figure 14.
2. Rotational axes might be used to define the motion of a rotary table or part indexer that operates rotationally around one of the basic axes.
3. -Z axis.
4. -X axis, with respect to the spindle. As the worktable moves right, the apparent spindle motion is left or in a -X direction.
5. Z.

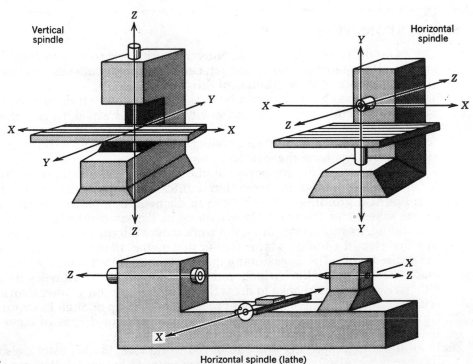

Figure 14.

Horizontal spindle (lathe)

section o unit 2 numerical control tape, tape preparation, tape code systems, and tape readers

SELF-TEST ANSWERS

1. Paper and plastic laminates.
2. More durable and less subject to damage and wear in the tape reader.

3. Preparation of new tapes, tape duplication, error correction, and typed records of tape information.
4. EIA and ASCII.
5. Electromechanical, photoelectric, and pneumatic.

section o unit 3 operating the N/C machine tool

SELF-TEST ANSWERS

1. After the tape has been punched, a printout should be obtained and checked against the program manuscript.
2. The dry tape run will verify the accuracy of the tape and prevent possible damage to the machine tool.
3. Be sure that the spindle and cutter are clear of all obstructions before positioning the worktable.
4. The tool length gage is used to adjust tool lengths as well as diameters on boring bars and insert cutters.
5. Micrometers and dial indicator types, electronic digital types.

index